AF410867

CITIES: Energetic Efficiency, Sustainability; Infrastructures, Energy and the Environment; Mobility and IoT; Governance and Citizenship

CITIES: Energetic Efficiency, Sustainability; Infrastructures, Energy and the Environment; Mobility and IoT; Governance and Citizenship

Editors

Luis Hernández-Callejo
Víctor Alonso Gómez
Sergio Nesmachnow
Vicente Leite
Javier Prieto
Ângela Ferreira

MDPI • Basel • Beijing • Wuhan • Barcelona • Belgrade • Manchester • Tokyo • Cluj • Tianjin

Editors

Luis Hernández-Callejo
Department of Agricultural
Engineering and Forestry
University of Valladolid
Soria
Spain

Víctor Alonso Gómez
Applied Physics Department
University of Valladolid
Soria
Spain

Sergio Nesmachnow
Computer Science Institute
Universidad de la República
Montevideo
Uruguay

Vicente Leite
School of Technology and
Management
Polytechnic Institute of Braganza
Braganza
Portugal

Javier Prieto
BISITE Research Group
University of Salamanca
Salamanca
Spain

Ângela Ferreira
Electrical Engineering
Department
Polytechnic Institute of Braganza
Braganza
Portugal

Editorial Office
MDPI
St. Alban-Anlage 66
4052 Basel, Switzerland

This is a reprint of articles from the Special Issue published online in the open access journal *Applied Sciences* (ISSN 2076-3417) (available at: www.mdpi.com/journal/applsci/special_issues/cities).

For citation purposes, cite each article independently as indicated on the article page online and as indicated below:

LastName, A.A.; LastName, B.B.; LastName, C.C. Article Title. *Journal Name* **Year**, *Volume Number*, Page Range.

ISBN 978-3-0365-2745-1 (Hbk)
ISBN 978-3-0365-2744-4 (PDF)

Contents

About the Editors

Luis Hernández-Callejo

Professor Dr. Luis Hernández Callejo is a professor and researcher at the University of Valladolid. His areas of interest are: renewable energy, microgrids, photovoltaic, wind, smart cities, and artificial intelligence.

Víctor Alonso Gómez

Professor Dr. Víctor Alonso Gómez is a professor and researcher at the University of Valladolid. His areas of interest are: renewable energy, microgrids, photovoltaic, wind, smart cities, and artificial intelligence.

Sergio Nesmachnow

Sergio Nesmachnow, is a full professor and researcher at Universidad de la República, Uruguay. His main research areas: scientific computing, high-performance computing, metaheuristics, and smart cities. He has had more than 300 articles published in scientific journals and conferences.

Vicente Leite

Vicente Leite received his diploma, MSc and PhD degrees in Electrical Engineering, from the Faculty of Engineering of University of Porto, in 1993, 1997 and 2004, respectively. He became an Assistant Professor at the School of Technology and Management (ESTiG) of Polytechnic Institute of Bragança (IPB) in 1994 where he is a professor in the field of Electronics, Power Electronics and Converters, Electrical Drives and Photovoltaic Systems. Currently, he is a sub-director of the ESTiG-IPB. Previously, he was President of the Scientific Council, President of the Assembly of Representatives, and Coordinator of the Electrical Engineering Department. He is the author of more than three dozen publications in his research fields, namely: power electronic converters; system identification, control and condition monitoring of electric machines and drives; renewable energies systems. He has participated in four projects funded by the European Community.

Javier Prieto

Dr. Javier Prieto earned his PhD in Information and Communication Technologies and the Extraordinary Performance Award for Doctorate Studies from the University of Valladolid in 2012. At this university, he also received a Telecommunication Engineer degree (2008) and the Marketing Research and Techniques degree (2010). Since 2007, he has worked with different public and private research centers, such as the Massachusetts Institute of Technology. Currently, he is a distinguished researcher of the University of Salamanca. He has published more than 85 research papers, participated in more than 50 research projects, and co-authored 2 Spanish patents. He serves as the Editor-in-Chief for the *Smart Cities* journal for its *Internet of Things* section, as an associate editor for the IEEE Communications Letters and for the *Wireless Communications and Mobile Computing* journal. His research interests include artificial intelligence, signal processing and blockchain.

Ângela Ferreira

Ângela Paula Barbosa da Silva Ferreira received a PhD degree in Electrical and Computer Engineering from the Faculty of Engineering of the Oporto University, Portugal, in 2012. Since 1997, she has been with the Polytechnic Institute of Bragança, where she is currently an Assistant Professor at the Department of Electrical Engineering. She is a member of the Research Centre in Digitalization and Intelligent Robotics (CeDRI). Her research interests are focused on the design and modelling of novel permanent magnet machines and the optimization of energy conversion systems using renewable energy sources.

Preface to "CITIES: Energetic Efficiency, Sustainability; Infrastructures, Energy and the Environment; Mobility and IoT; Governance and Citizenship"

Cities are changing; this is a reality. As this change takes place, advances in research must propose solutions to the numerous problems that arise with it. Administration, universities, and companies must cooperate to establish the foundations of the Smart City.

In this sense, the main pillars of a city are:

- Energetic efficiency and sustainability;

- Infrastructures, energy and the environment;

- Mobility and IoT;

- Governance and citizenship.

Therefore, this SI will be focused on thse topics. Submissions of both original research and review articles are invited. Additionally, invited papers based on excellent contributions to the 2020 CIUDADES INTELIGENTES TOTALMENTE INTEGRALES, EFICIENTES Y SOSTENIBLES (CITIES) will be included. We hope that this collection of papers will serve as an inspiration for all those interested in the prosperity of smart cities. This book integrates scientific articles on energy, mobility, artificial intelligence and sensors. All of them are within the scope of smart cities.

Luis Hernández-Callejo, Víctor Alonso Gómez, Sergio Nesmachnow, Vicente Leite, Javier Prieto, Ângela Ferreira

Editors

Article

A Thermal Discomfort Index for Demand Response Control in Residential Water Heaters

Rodrigo Porteiro [1,*,†], Juan Chavat [2,†] and Sergio Nesmachnow [2,*,†]

1   Administración Nacional de Usinas y Transmisiones Eléctricas, Montevideo 11300, Uruguay
2   Computer Science Institute, Engineering Faculty, Universidad de la República, Montevideo 11300, Uruguay; juan.pablo.chavat@fing.edu.uy
*   Correspondence: rporteiro@ute.com.uy (R.P.); sergion@fing.edu.uy (S.N.)
†   These authors contributed equally to this work.

Featured Application: The methodology described in this article is applicable to design proper management strategies for demand response in smart electricity grids to fairly select water heaters to intervene while guaranteeing the lower discomfort of users.

Abstract: Demand-response techniques are crucial for providing a proper quality of service under the paradigm of smart electricity grids. However, control strategies may perturb and cause discomfort to clients. This article proposes a methodology for defining an index to estimate the discomfort associated with an active demand management consisting of the interruption of domestic electric water heaters. Methods are applied to build the index include pattern detection for estimating the water utilization using an Extra Trees ensemble learning method and a linear model for water temperature, both based on analysis of real data. In turn, Monte Carlo simulations are applied to calculate the defined index. The proposed approach is evaluated over one real scenario and two simulated scenarios to validate that the thermal discomfort index correctly models the impact on temperature. The simulated scenarios consider a number of households using water heaters to analyze and compare the thermal discomfort index for different interruptions and the effect of using different penalty terms for deviations of the comfort temperature. The obtained results allow designing a proper management strategy to fairly decide which water heaters should be interrupted to guarantee the lower discomfort of users.

Keywords: demand response; smart grid; discomfort index; water heaters; thermal model; computational intelligence

Citation: Porteiro, R.; Chavat, J.; Nesmachnow, S. A Thermal Discomfort Index for Demand Response Control in Residential Water Heaters. *Appl. Sci.* **2021**, *11*, 10048. https://doi.org/10.3390/app112110048

Academic Editor: Amjad Anvari-Moghaddam

Received: 30 June 2021
Accepted: 6 September 2021
Published: 27 October 2021

1. Introduction

Energy demand management is a crucial idea for the modern paradigm of smart cities. The concept of smart electricity networks, or *smart grids*, refers to electrical grids enhanced by including operation and management features to improve the controlling of production and distribution of energy [1]. Smart grids are mainly oriented to maintain a reliable and secure infrastructure to allow properly satisfying the demand growth, the integration of distributed energy resources, smart storage, and other features related to smart devices and real-time information provided to clients [2]. Information and Communication Technologies (ICT) are very closely connected to smart grids, as they provide the basis for communicating and processing information that is very useful at different levels to implement the aforementioned services [3].

The daily consumption pattern of electricity involves periods with higher-than-average electricity consumption (*peaks*) and other periods with lower-than-average electricity consumption (*valleys*). A common situation in the daily operation of electric grids is that electricity generation and transmission systems may not always meet peak demand requirements. In these situations, power demand management strategies are helpful tools for operation of electric grids.

Power demand management refers to the proper administration of power consumption for end consumers in a smart grid in order to promote better energy utilization. Two of the most widely applied actions for power demand management are load management, whose main goal is modifying, reducing, or shifting the demand, and energy conservation, which is mainly focused on reducing the demand, e.g., via technological improvements. In turn, several other actions have been applied for demand management, including fuel substitution and load building [4]. Among load management techniques, the most used are peak reduction, oriented to reduce power consumption in periods of maximum demand; valley filling, whose main goal is to promote energy utilization in off-peak periods,; and load shifting from peak to off-peak periods.

Several demand response and demand management tools can be applied to mitigate overloads in the electrical system. One of the simplest yet most effective methods for direct load control is allowing the electric company to remotely control user devices. This method is properly applied to control those devices with a thermostat, especially those that have important thermal inertia. On the one hand, remote control is a very effective technique to achieve peak reduction and load shifting at critical periods. On the other hand, the benefits of the achieved reduction of energy consumption in the overall operating cost of the electrical system must be weighted against the loss of comfort that the users of the controlled devices may have. In order to define an economic value to the loss of comfort associated with an intervention, to be taken into account in the business model of the electric company, the discomfort of users must be properly evaluated with quantifiable metrics in advance.

In this line of work, this article proposes and evaluates a methodology to calculate a Thermal Discomfort Index (TDI) associated with a remote intervention for load management from the electric company. The computed index evaluates the discomfort for users, generated by the intervention of electric water heater appliances. The main goal of the index is to be a valuable tool for defining a method to manage a set of water heaters to defer electrical demand by properly identifying those appliances that have the lower impact on user comfort to be interrupted first. This way, the values of the computed TDI make it possible to decide in which order water heaters should be interrupted to minimize the overall total discomfort of a set of users. The key aspect of the proposed strategy is to know in advance the value of TDI in order to decide if it is economically profitable to carry out an intervention.

The proposed methodology for computing TDI is based on real data, a linear temperature model, and a forecasting model for water utilization applying ensemble learning and Monte Carlo simulation. The proposed TDI was developed and evaluated with data from the electrical system in Montevideo, Uruguay. This is a relevant case study since electricity is the main resource used for heating and other domestic activities in Uruguay, far superior to natural gas and other sources. In Montevideo, as in other main Uruguayan cities, more than 90% of households have a thermostat-controlled electric water heater (according to the 2019 household survey by National Statistics Institute [5]). Since the electric water heater is one of the most energy-intensive household appliance (accounting for 34% of residential energy consumption, on average), it is an ideal candidate for remote load control as a demand management technique.

The experimental analysis was performed both on real and simulated scenarios built using data from real electric water heaters in Uruguay, gathered in the ECD-UY dataset [6]. ECD-UY includes utilization of time series data about power consumption of several appliances, including water heaters, and aggregated consumption for representative households in the main Uruguayan cities.

Results reported for the considered case studies demonstrate that the proposed index managed to capture the impact of thermal discomfort, fulfilling the goal of sorting electric water heaters to be properly managed by applying a direct control strategy, and allows special cases to be defined where a particular water heater is required to never be interrupted for security reasons.

This article extends our previous conference article "Demand response control in electric waterheaters: evaluation of impact on thermal comfort" [7], presented at III Iberoamerican Congress on Smart Cities. The main contributions beyond those of the previous conference article include: (i) an extension of the developed temperature model for electric water heaters, only measuring the electrical state of the device; (ii) an improved procedure to predict water utilization by applying data analysis to real electricity consumption data from the ECD-UY dataset; (iii) the definition of an index to approximate the discomfort associated with an active demand management interruption of the water heater; and (iv) an extended evaluation of the proposed methodology for several cases of active demand management interruption of water heaters over different realistic scenarios.

The article is organized as follows. Section 2 presents the formulation of the demand management problem through direct control of electric appliances. Section 3 reviews related works. Section 4 describes the proposed approach to define a TDI for direct control of electric water heaters. Section 5 reports the experimental validation of the water utilization forecasting, the temperature model, and the proposed TDI for realistic case studies. Finally, Section 7 presents the main conclusions and lines for future work.

2. Demand Management and Direct Control of Electric Water Heaters

This section presents the main concepts of demand management strategies. In particular, direct load control strategies applied to load shifting are described, and the problem of affecting comfort of the end-user is discussed.

2.1. Demand Management and Direct Load Control

The traditional model of an electrical system supplies electricity to end consumers through a unidirectional flow of energy, which is delivered by centrally controlled generators. However, in the last thirty years, power grids all over the world became decentralized systems, thus resulting in distributed energy resources that have fostered new business models and specific transformations of energy markets. The concept of energy demand management emerged within these new business models.

Energy demand management involves a set of techniques oriented to modify the energy demand of consumers of an electric grid to fulfill specific goals [8]. The subset of techniques oriented to reduce the energy demand of consumers in the short term are known as demand response methods. A specific technique within demand response methods is *direct load control*. The main idea of direct load control is to provide the energy company the permission to control (i.e., switch off) the devices of end-users, which is usually obtained via specific agreements that grant users a monetary incentive. Load control is considered an effective technique to achieve immediate power reduction in a very short time and is very useful to deal with peak reduction needs [9] in order to obtain a more stable grid operation. Another situation where load control methods are very useful is to provide frequency regulation services [10], with the main goal of maintaining the system frequency close to the utility frequency (i.e., the nominal oscillation frequency of alternating current, 60 Hz in the Americas and Asia, and 50 Hz in other sites), preventing deviations that affect generators and also make the grid unstable. This article addresses some aspects related to the direct load control of electric water heaters, mainly focusing on the impact of using direct load control in the thermal comfort of end-users.

2.2. Load Shifting by Direct Control of Electric Water Heaters

In most countries, the profile of total electricity consumption shows two pronounced peaks: early in the morning, before starting the working day, and near two hours after people return to their homes from work. Usually, the power consumption of electric water heaters has a high correlation with the total consumption: both present coincident peaks, as shown in Figure 1, for a representative analysis using data from the *electric water heater consumption* subset, part of the ECD-UY dataset. This correlation is explained by the high impact of water heaters compared with other domestic appliances. For example, in the case study analyzed in this article (Uruguay), water heaters account for 27% of the total residential electricity consumption, on average, peaking at 35%, according to the subset data. A specific feature of water heaters is that they have the ability to accumulate energy in the form of heat inside the water tank. This thermal inertia allows proper planning of switch on/switch off periods to help the grid operation while trying to affect the thermal comfort of users as little as possible.

According to the correlation between total energy consumption and water heaters' energy consumption, the amount of energy associated with electric water heaters within a demand peak can be deferred by switching off the devices in a proper moment and switching them on in the future. This strategy allows implementing a load shifting on the demand curve, since the total amount of energy consumed in the period remains equal but the load profile is modified. Several studies have addressed the load shifting problem using direct control of devices [11,12], but few articles have focused on quantifying the thermal discomfort generated by the real application of this strategy.

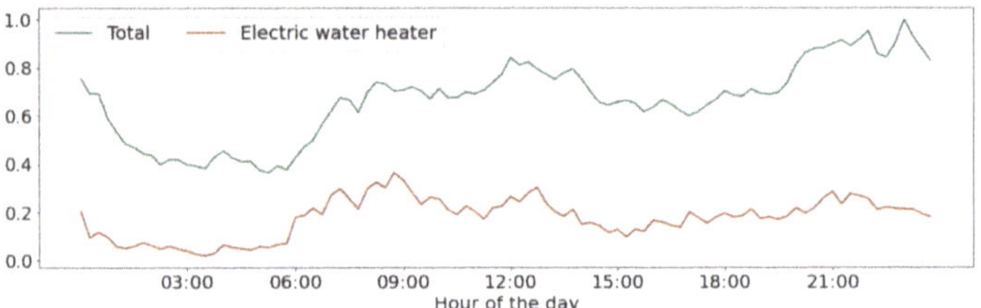

Figure 1. (Normalized) total power and electric water heater demand in a representative workweek (11 November to 15 November 2019) in Uruguay. Data from the ECD-UY dataset [6].

2.3. Problem Formulation

The problem proposes determining an index for the quantitative evaluation of the discomfort generated by the application of load shifting using a direct control of electric water heaters. Two main challenges must be addressed to solve the problem: determining proper variables for the analysis and, since no-deterministic behavior is involved, proposing an effective methodology to estimate their values.

The first step is evaluating relevant variables for the problem. In the case of electric water heaters, the main variables affecting the comfort of the users is the temperature of the water in the tank and variables that determine the utilization pattern of water heaters. Regarding user comfort, installing a remote device to the water heater that allows measuring its power and switching it off or on is reasonable, since it can be achieved via specific Internet of Things controllers without modifying the water heater structure [13,14]. However, installing a remote thermometer to measure the temperature of the tank requires modifying the structure of the water heater, which implies a significant monetary investment. Thus, non-intrusive methods [15] are preferred for the analysis.

The main goal of defining a proper direct control strategy is to determine the discomfort introduced by the same controlling action among several electric water heaters to decide which of them intervene and minimize the probability of generating discomfort. Therefore, it is not crucial to compute the temperature exactly, but a good approximation is enough for analyzing differences. Another relevant aspect for the analysis is that the user perceives thermal discomfort only when using water and the temperature is below a given threshold. At any other time, or if the water temperature is still above the threshold despite the intervention, users are not aware of the switching-off actions.

Two variables must be estimated to define a discomfort index in the event of an intervention: the water utilization and the temperature of the water in the tank. By estimating these two variables, the proposed TDI takes into account the period of time that the water temperature is below the comfort temperature for a person using the water. This way, the proposed TDI properly models the human discomfort associated with an intervention.

3. Related Works

This section reviews relevant related works about load reduction techniques and thermal discomfort evaluation.

Several works in the literature have described the main concepts of peak load reduction strategies and specific implementations in smart grids [9,16]. In particular, direct load control allows utilities to remotely manage electricity demand by modifying the operation of end-use devices to perform load shifting [11,17]. A special type of devices to perform load shifting are thermostat-controlled appliances (TCA). The main feature of TCA is that they provide flexibility to select the desired temperature, i.e., the *thermostat set point* [18,19]. In addition, some TCAs can also store energy in the form of heat, providing a great advantage over non-storage appliances when performing load shifting strategies. This is the case of domestic electric water heaters, which are the focus of this article.

Nehrir et al. [20] proposed and analyzed an interactive demand-side management strategy for electric water heaters, but the research did not elaborate on specific aspects of thermal discomfort of users. A recent article by Xiang et al. [21] introduced a complex strategy to minimize thermal discomfort related with electric water heater control. The proposed strategy requires information to build a time-varied weight matrix based on detected utilization patterns of domestic electric water heaters. In turn, the weight matrix and other information are used to generate a customer satisfaction prediction index. Due to the large amount of information required to build the model, the approach is difficult to apply in practice.

Demand response control strategies for TCA can be effectively implemented provided that thermal comfort is not compromised. Thus, quantifying the impact on user comfort is a crucial aspect, and it has been the focus of several works. Kampelis et al. [22] evaluated the thermal discomfort in demand response control of several appliances used for heating, ventilation, and air conditioning in a university building. A *daily discomfort score* was proposed for demand response events to reduce the cost of energy, but the proposed strategy requires knowing the real temperature. Regarding electric water heaters, the study of the hot water utilization profile by end-users is a key aspect for estimating discomfort. Tabatabaei and Klein [23] studied whether a smart heating system can benefit from good predictions of the user behavior. Pirow et al. [24] proposed an algorithm for estimating domestic hot water utilization, but the technique requires the installation of temperature and vibration sensors.

The main factor that defines comfort is the water temperature. Thus, a model to estimate water temperature from measured information is crucial for the effectiveness of the comfort evaluation. Paull et al. [25] proposed a water heater model to estimate the temperature of the water in the tank as a function of time and the related variables, including the thermal losses and the water utilization. Data from smart meters, recorded at 15 min intervals, were used for validation. The model was proposed to be applied in a multiobjective demand-side management program. The study by Lutz et al. [26] provided a comprehensive empirical analysis of a simplified energy consumption model for water heaters considering the variation of the temperature of the water in the tank. Results of the proposed Water Heater Analysis Model (WHAM) model were compared with data from water heaters simulation programs (TANK, WATSIM, and WATSMPL). WHAM obtained very accurate prediction results, while being significantly faster. In addition, WHAM requires less detailed engineering information about the water heaters. Finally, comfort evaluation must be considered in the problem of controlling a subset of electric water heaters, e.g., by building a ranking to sort all interruptible devices according to appropriate criteria. Yin et al. [27] proposed a scheduling strategy based on a temperature state priority list. In turn, Al-Jabery et al. [28] analyzed a scheduling strategy for electric water heaters based on approximate dynamic programming techniques and q-learning.

The analysis of related works allows concluding that few articles have studied the thermal comfort effect when applying direct load control of electric water heaters. Existing approaches are based on installing specific sensors or devices to monitor water temperature or on building sophisticated algebraic formulations for modeling utilization patterns. This article contributes in this line of works by proposing an approach to evaluate the thermal discomfort of an intervention on electric water heaters, without requiring installing additional devices (e.g., a thermometer to measure the water temperature). Instead, the proposed model follows a non-intrusive approach, applying data analysis and computational intelligence and demonstrates its effectiveness on realistic problem instances.

4. The Proposed Approach for Defining a Discomfort Index

This section describes the proposed approach for defining a discomfort index for demand response via direct control of water heaters. The approach applies ideas described in the previous sections and follows a data analysis approach [29,30] considering inforation from a group of remotely controlled electric water heaters located in Uruguay.

4.1. Data Preparation

The data used in this article were provided by the Uruguayan National Electricity Company (UTE). Data are available in the "Electric water heater consumption" dataset, one of the three subsets included in EDC-UY [6], an effort to build a national database of energy consumption by gathering data from several households located in the main Uruguayan cities.

In this article, only the electric water heater consumption records were used. These records have a sample period of one minute and cover a date range from 2 July 2019 to 26 October 2020. Customer records were filtered by the recording length, keeping only those with more than 5 months of recording (i.e., at least 216,000 records).

The disaggregated electric water heater data have several gaps caused by different problems that arose during the data collection process (e.g., malfunctioning of the data transmission network, power failures, etc.). The gaps were filled using two techniques: resampling and refilling. The resampling technique normalizes the sample period to an exact minute. First, the records are grouped by customers to build one-minute record containers, starting from the date and time of the first record. Then, records whose date and time match with the date range of the container are assigned to it. In case one or more records match the same container, the minimum consumption value is set; otherwise, a null value is set (i.e., it corresponds to a missing record). The resulting data are taken as the input of the refilling technique. First, refilling detects the data gaps (i.e., consecutive missed records) and refills each one of them according to the following criteria. Starting from both extremes of the gap up to seven minutes forward/backwards, the missing data is recreated by a linear interpolation method. Finally, if missing values are still present at the gap (i.e., the gap is larger than 14 min), the null value is assigned to all missing values. The described process results in normalized time series of consumption values without gaps. For data preparation purposes, a Jupyter notebook was implemented, and the basis of the scripts was provided by the ECD-UY dataset. The Jupyter notebook uses Python (version 3) programming language and libraries Pandas and Numpy. The resulting notebook is available to download from https://bit.ly/3h133qu (accessed on 23 October 2021).

An example of the effects of the refilling procedure on electric water heater activations is presented in Figure 2 (missing records) and Figure 3 (after processing). Both figures show the same water heater activations in the same date and time range. The first graphic shows the dataset before refilling the gaps and include the missing records. The second graphic was captured after the data were processed.

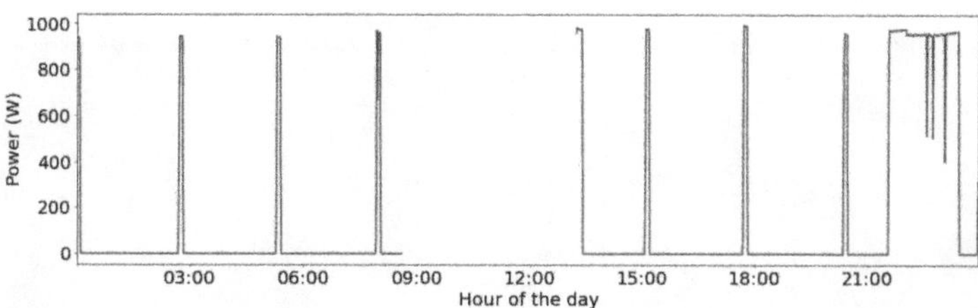

Figure 2. A day (14 November) of the electricity consumption of an electric water heater (customer id. 115747 from dataset ECD-UY) before refilling the data gaps.

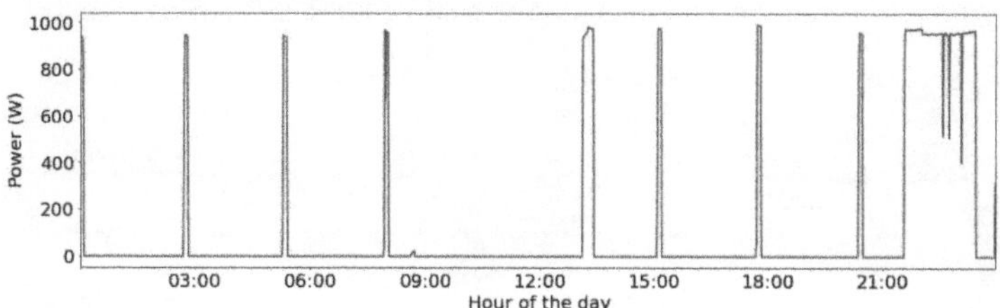

Figure 3. A day (14 November) of the electricity consumption of an electric water heater (customer id. 115747 from dataset ECD-UY) after refilling the data gaps.

4.2. Water Utilization Forecasting Model

4.2.1. Overall Description

The proposed model for water utilization forecasting is based on a specific characteristic of electric water heaters: when this appliance is on, its power consumption can be considered as constant, since it just has slight variations. Therefore, the load curve of an electric water heater can be represented in a binary format (0 when the appliance is off and a given value C when the appliance is on). A pattern similarity approach [15] is proposed to estimate water utilization. To properly characterize the utilization patterns, let us define an *on block* as the time interval in which the electric water heater is switched on continuously. From the analysis of the power consumption time series, some of these on blocks are associated with water utilization by the user and other (shorter) on blocks correspond to thermal recoveries to maintain the target temperature of the water.

4.2.2. Methodology

The proposed forecasting model is based on identifying the on blocks associated with water utilization and discarding those corresponding to thermal recoveries. In this regard, a threshold duration is defined, and any on block shorter than the threshold duration is considered to be a thermal-recovery on block and not considered for the analysis. The proposed approach is robust, since discarding short blocks is not relevant for the main goal of identifying long-term utilization blocks, which are generally associated with showers. The analysis considers as a baseline an electric water heater with a capacity of 60 L and an average water outlet flow rate, which is representative of a highly efficient appliance. In fact, these electric water heaters are the main target of a campaign to promote energy efficiency in residential buildings in Uruguay. For an electric water heater with a capacity of 60 L, the empirical duration of an on block is on average eight times the duration of the utilization period. The proposed approximation is applied to convert the information about on blocks into an estimation of the water utilization by users. An example of the identified on blocks and the inferred water utilization patterns obtained with the aforementioned procedure is presented in Figure 4. In the analysis, the proposed model is applied to forecast the water utilization each minute in a period of two hours. Thus, the output of the model is a vector of 120 Boolean values, indicating the water utilization (or lack thereof) in each minute.

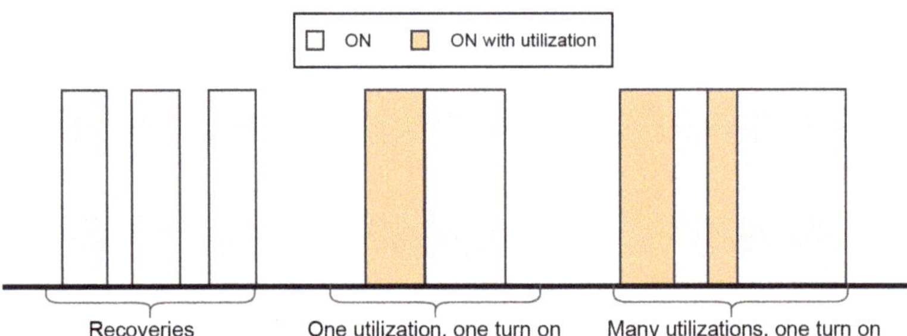

Figure 4. On blocks and water utilization patterns in the electric water heater consumption.

4.2.3. Formulation

The forecasting model for water utilization is based on an extremely randomized trees (ExtraTrees) regression model. ExtraTrees is an ensemble learning method for solving classification and regression problems that operates as a Random Forest (RF) technique. ExtraTrees defines a set of decision trees that is trained with training data, and the resulting output class (for classification problems) or prediction (for regression problems) is defined as the mode (for classification) or mean/average prediction (for regression) of the considered decision trees [31]. Using a large number of trees allows properly dealing with overfitting problems that arise when using few trees. ExtraTrees has two main differences with the standard procedure defined by RF: (i) each tree is trained using all the training data (and not using just a bootstrap sample, as the standard RF method), and (ii) a further step of randomization is included in the top-down splitting in the tree learner by selecting a random cut point for each considered feature in the problem (according to a uniform empirical distribution to select between the values for each feature in the training set). The split that computes the best result is then used to split the considered node of the tree [32]. ExtraTrees have proven to be an accurate predictor for electricity-related problems (e.g., for demand forecasting in industrial and residential facilities [7]).

Input features considered for the proposed ExtraTrees regression method include:

- *Use* ($\vec{u}$, 120 Boolean values): indicating whether a water utilization occurs in the past 120 min.
- *Month* (m, integer): indicating the month of the horizon to forecast.
- *Day* (d, integer): indicating the day of the horizon to forecast.
- *Hour* (h, integer): indicating the hour of the horizon to forecast.
- *Dayofweek* (dw, integer): indicating the day of the horizon to forecast.
- *Workingday* (wd, Boolean): indicating if the horizon to forecast is a working day or not.

Fine-tuning of the proposed ExtraTrees regression method was performed using grid search techniques for hyperparameter settings. Hyperparameters are external parameters, inherent to the learning model, whose values cannot be set or estimated from training data. Hyperparameter values affect the quality of the resulting model, and appropriate values must be set before launching the learning process. Grid search techniques define a search space as a grid of different combinations of candidate hyperparameter values and proceed to evaluate every combination in the grid. In this article, the GridSearchCV method from `scikit-learn` was used. GridSearchCV (the *CV* stands for *cross validation*) was applied with varying parameters: *number of estimators* and *max tree depth*, considering 10-folds cross-validation and the predetermined evaluation metrics over the model. After the model is trained, a vector with 120 Boolean values ($\vec{F}$), representing the water utilization forecast for the next two hours, is obtained according to the Equation: $Pred(\vec{u}, m, d, h, dw, wd) = \vec{F}$.

4.3. Water Temperature Model

4.3.1. Overall Description

The general formulation of equations for heating and cooling water in thermal tank devices, such as water heaters for domestic use, include a large number of variables. Existing models, even when they simplify the formulations, depend on several variables, including the insulation factor, the ambient temperature, the flow of water used, the time of use, and the tank volume, among other factors [26]. To overcome the difficulties of knowing beforehand all these variables, which are often difficult to determine in practice, the pattern similarity approach proposed for estimating water utilization is applied to define a linear temperature model, which provides a good approximation in order to estimate the TDI. The linear model provides a reasonable approximation, since the cooling and heating curves in this model are straight lines.

Five parameters are defined to build the temperature model from gathered data about on blocks and water utilization:

1. T_{min} is the temperature at which the electric water heater is turned on by the action of the thermostat when the water is cooling.
2. T_{max} is the temperature at which the electric water heater is turned off by the action of the thermostat when the water is heating.
3. c_{heat} is the slope of the line when the electric water heater is turned on.
4. c_{cool} is the slope of the line when the electric water heater is turned off and no water is being used.
5. c_{use} is the slope of the line when using water, whether or not water is being used.

The considered parameters depend on several factors. In this article, a specific approach is proposed to compute a robust approximation for each parameter value, i.e., to guarantee that all possible temperature approximation errors always produce underestimated temperature values. Thus, the proposed model is conservative about comfort estimation, in order to not affect the quality of service provided to the user.

The model assumes that the user sets the thermostat at a temperature value of 60°, which is the suggested temperature not only to achieve energy efficiency in households, but also due to health concerns, such as avoiding the proliferation of Legionella bacteria [33,34]. Without loss of generality, the variation range defined by $T_{min} = 55°$ and $T_{max} = 65°$ is considered. In any case, the proposed model is fully extensible to work with other values of T_{min} and T_{max}. The use of this model in an industrial context should consider: (i) variations in the set point of the thermostat (e.g., values extracted from statistics or provided by the users via survey or web/mobile application), and (ii) variations in the temperature of the room, in the average water utilization, and in the temperature of supplied water. Variations are captured by recomputing the coefficients after variations occur. Since the model is estimated from on-and-off data of the water heater in real time, the variations of T_{min} and T_{max} should be updated frequently. After that, coefficients c_{heat}, c_{cool}, and c_{use} should be recomputed using the updated values of T_{min} and T_{max}. These dynamics allow properly modeling different climate conditions and user preferences.

Figure 5 presents a schema of the proposed model definition from on blocks and water utilization. The black line on the upper graphic is the water temperature. On the bottom, grey on blocks represents thermal recoveries, and on blocks caused by water utilization are marked in orange. The analysis of the water temperature curve indicates that the water cools with a slope c_{cool} until it reaches the value T_{min}; at that moment, the electric water heater turns on. The heating phase starts with a slope c_{heat} until the temperature reaches the value T_{max}. Then, another cooling phase occurs until a water utilization causes a much faster cooling with a slope c_{use}. During the water utilization, the electric water heater turns on almost immediately after opening the water stream. When the water utilization ends, the electric water heater remains on because the water temperature is below T_{min}, so it heats the water with a slope c_{heat} until temperature T_{max} is reached, where the gray *on block* ends. Assuming the described behavior, an algebraic approach can be applied to compute the three slopes.

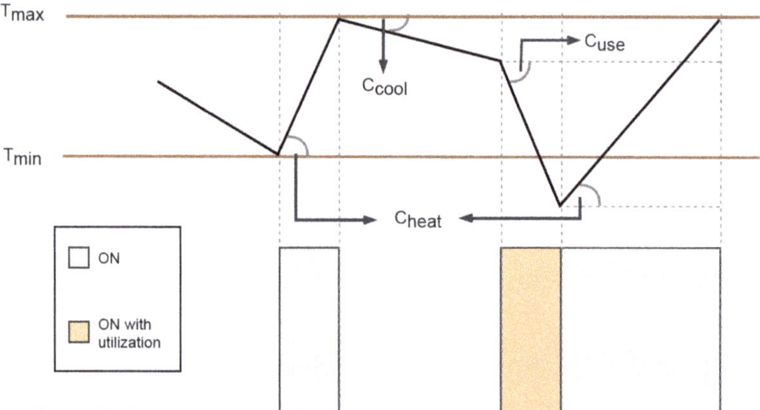

Figure 5. Linear temperature model.

The described procedure allows computing an approximation of the water temperature in a given interval from a set of on blocks and water utilization data in that interval. Therefore, a temperature forecast can be obtained from a set of on blocks predicted for a future time interval. The forecasting method can be applied to other situations, e.g., to simulate a remote switch off of the electric water heater, associated with a demand response action, and obtain the water temperature forecast for this event.

4.3.2. Formulation

The linear temperature model is applied to determine the values of coefficients c_{cool}, c_{heat}, and c_{use}. The input data for the temperature model are the values of T_{min} and T_{max} and the information about on blocks and water utilization.

The coefficients of the model are calculated as follows:

- *Coefficient c_{cool}.* First, two consecutive temperature recoveries are identified. Then, Δ_{rec} is calculated as the time between the the end of the first recovery and the beginning of the next recovery. Finally, c_{cool} is calculated by Equation (1). The procedure is described in the left box of Figure 6. From the graphic, $c_{cool} < 0$.

$$c_{cool} = (T_{min} - T_{max})/\Delta_{rec} \tag{1}$$

- *Coefficient c_{heat}.* First, a recovery is identified and dur_{rec} is defined as its duration. Then, c_{heat} is computed by Equation (2). The procedure is described in the middle box of Figure 6. From the graphic, it is clear that $c_{cheat} > 0$.

$$c_{heat} = (T_{max} - T_{min})/dur_{rec} \tag{2}$$

- *Coefficient c_{use}.* The first step is identifying a temperature recovery followed by a water utilization event. Then, T_{ini}^{use} and T_{end}^{use} are defined as the temperature at the beginning and end of the utilization, respectively. To compute T_{ini}^{use}, the value of c_{cool} (already computed) is used as the slope to draw the line that passes through T_{max} at the end of the recovery, and intersects with the start of the utilization. Similarly, to compute T_{end}^{use}, the value of c_{heat} (already computed) is used as the slope to draw the line that passes through the end of the on block associated with the utilization and intersects with the end of the utilization. dur_{use} is defined as the duration of the utilization. Finally, c_{use} is calculated by Equation (3). The procedure is described in the right box of Figure 6.

$$c_{use} = (T_{end}^{use} - T_{ini}^{use})/dur_{use} \tag{3}$$

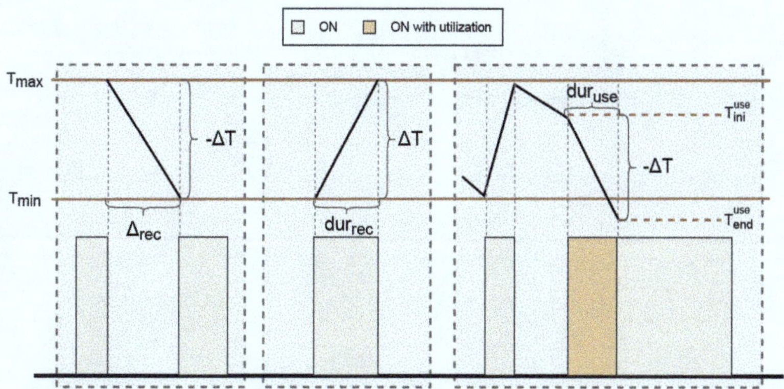

Figure 6. Graphic representation of c_{cool} (**left box**), c_{heat} (**middle box**), and c_{use} (**right box**).

4.4. Defining the Thermal Discomfort Index

The proposed index is conceived to capture the thermal impact that a user suffers due to an intervention by the electrical company in the electric water heater. The TDI is defined using the defined water use forecasting model and the temperature model. Since every forecast has uncertainty, the index is defined in terms of the expected value of the difference of the aforementioned temperatures, as expressed by Equations (4) and (5).

$$TDI(I, u, w) = \int_{t_{ini}(u)}^{t_{end}(u)} (T_n(t, w) - T_{int}(t, w))dt + \rho \int_{t \in \tau} (T_{comf} - T_{int}(t, w))dt \qquad (4)$$

In Equation (4), $TDI(I, u, w)$ represents the discomfort index of an interruption I, a water utilization u, and a realization w that defines a single scenario of water use and temperature evolution. The values $t_{ini}(u)$ and $t_{end}(u)$ are the starting and finishing time of utilization u. For every realization w, $T_n(t, w)$ is the temperature curve without interruption and $T_{int}(t, w)$ is the temperature curve with interruption. Both curves are obtained using the temperature model described in Section 4.3 for the realization w.

Finally, T_{comf} is the lowest water temperature that does not produce discomfort to the users, and τ represents the time interval in which $T_{int}(t, w) \leq T_{comf}$. The value ρ is a parameter that acts as a penalty over the area below the comfort temperature.

$$TDI(I) = E_w \left[\sum_{u \in U(w)} TDI(I, u, w) \right] \qquad (5)$$

In Equation (5), I denotes the interruption of the electric water heater by the electric company. In turn, $E_w[\cdot]$ represents the expected value of the expression inside parenthesis with respect to the random variable w. Each realization of the random variable w generates a different forecast of water use. Therefore, a forecast of the temperature obtained using the model described in Section 4.3 is also generated. $U(w)$ is the set of water utilization intervals in the analyzed time horizon associated with the forecast generated by w.

Figure 7 presents a visual representation of $TDI(I, u, w)$. Two temperature curves are represented on the upper graphic. The full black line represents the water temperature when no interruption occurs. In turn, the dotted black line represents the water temperature when an interruption occurs at time t^*. The green area between the curve of water temperature without interruption and the curve of water temperature with interruption (defined by the polygon PQRTU) is computed by the integral $\int_{t_{ini}(u)}^{t_{end}(u)} (T_n(t, w) - T_{int}(t, w))dt$. This area represents the heat loss due to the interruption. In turn, the red area below the comfort temperature (defined by the polygon RST) is computed by the integral $\int_{t \in \tau} (T_{comf} - T_{int}(t, w))dt$

and weighted by the penalty ρ. Introducing the penalty term ρ is interesting to provide the model the flexibility for adjusting the weight of the red area below the comfort temperature with respect to the whole area of temperature reduction (PQSU).

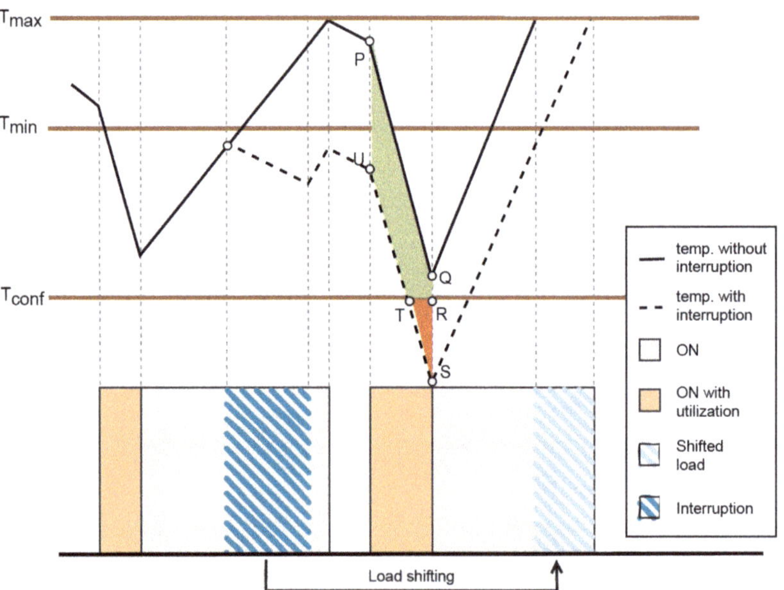

Figure 7. Graphical representation of TDI.

The expected value of TDI, as defined in Equation (4), is computed by applying a Monte Carlo (MC) simulation method. First, the MC method samples 100 realizations of the value w, with a normal distribution $N(0,1)$. Then, for each value of w, the following procedure is applied:

- The next 12 h are forecasted using the model described in Section 4.2. Six iterations are applied, considering that the Extra Trees regressor is defined for a period of two hours (120 observations).
- Using the temperature model described in Section 4.3 and the water utilization forecast, the water temperature is obtained for the next 12 h.
- An interruption of k minutes is simulated and the temperature for the next 12 h is obtained using the proposed temperature model.
- Since T_{max}, T_{min}, T_{comf} are known, Equation (5) is applied to compute $TDI(I, u, w)$ for the current realization w and all uses.
- An auxiliary variable $S_{uses}(w)$ is defined by Equation (6).

$$S_{uses}(w) = \sum_{u \in U(w)} TDI(I, u, w) \tag{6}$$

Finally, after computing $S_{uses}(w)$ for each realization, TDI is computed as the empirical expected value defined in Equation (7).

$$TDI(I) = \sum_{w=1}^{w=100} S_{uses}(w)/100 \tag{7}$$

5. Experimental Validation

This section presents the experimental validation of the proposed approach for defining a TDI.

5.1. Methodology

The methodology for the experimental evaluation includes two steps: validation of forecasting models and validation of the proposed index.

5.1.1. Validation of Forecasting Models

The first step of the experimental evaluation consists of validating the two models required to calculate the TDI (water utilization forecasting and water temperature). For the validation of the aforementioned models, the standard mean absolute percentage error (*MAPE*) metric is used to evaluate the forecasting capabilities. *MAPE* is defined in Equation (8), where $actual_i$ represents the measured value for $t = i$, $pred_i$ represents the predicted value, and n is the predicted horizon length.

$$MAPE = 100 \times \frac{1}{n} \sum_{i=1}^{n} \left| \frac{actual_i - pred_i}{actual_i} \right| \tag{8}$$

5.1.2. Validation of the Proposed TDI

After determining the forecasting accuracy of the proposed models, the second step of the experimental evaluation consists of validating the TDI calculation and utilization in realistic scenarios in order to properly evaluate the thermal discomfort of users. For this purpose, three experiments were designed based on scenarios that adequately represent the real operation of water heaters. A water heater dynamic simulator was developed and used to generate real scenarios. The description of the simulator and the experiments performed are presented in Section 6.

5.2. Development and Execution Platforms

Data processing algorithms and the proposed models to build the TDI were implemented using Python and well-known open source libraries such as Pandas, Numpy and Tensorflow. Data processing and the experimental analysis were performed on the high performance platform of National Supercomputing Center (Cluster-UY), Uruguay [35].

5.3. Evaluation of the Water Utilization Forecasting

Metrics defined in Section 4.2 were applied to evaluate the implementation of the water utilization forecasting model. A subset of ECD-UY was used, consisting of ten electric water heaters with more than five months of measurements. The grid search was performed on a two-dimensional grid to determine the best values for the number of trees in the forest and the maximum depth of the tree. The best parameter setting found applying the grid search configuration was $n\_estimators = 50$, $max\_depth = 200$.

Using the best parameter configuration, the ExtraTrees regressor achieved a *MAPE* value of 11.79 in just 4.09 s of execution time. This accuracy is adequate for estimation purposes to compute TDI, considering the high variance of individual water utilization. The method is useful for generating scenarios to apply the Monte Carlo simulation approach in order to estimate the empirical probability distribution of water utilization.

5.4. Evaluation of the Water Temperature Model

The linear model described in Section 4.3 was evaluated for a real case study corresponding to an electric water heater with a thermometer to measure the temperature of the water in the tank. The defined setting of the thermostat allowed knowing in advance the values of parameters T_{min} and T_{max}. The corresponding values are $T_{min} = 55\,°\text{C}$ and $T_{max} = 65\,°\text{C}$. Then, the other parameters of the model were calculated as described in Section 5.3. Finally, data of twelve hours on blocks of the electric water heater were used to estimate the temperature and compared with the real temperature measured. Table 1 reports the comparison of the real and the estimated temperature, and the largest difference in the three long utilization periods in the twelve hours analyzed.

Table 1. Accuracy of the water temperature model.

	1st Utilization	2st Utilization	3st Utilization
measured temperature	59.09 °C	53.03 °C	58.34 °C
linear temperature	59.88 °C	55.12 °C	59.41 °C
difference	0.79 °C	2.09 °C	1.07 °C

The second utilization had the largest temperature difference (2.7 °C, marked in light blue in Table 1), which represents a percentage error of 4.5% in the worst case. The other utilizations had a significantly lower error. The accuracy of the temperature model is adequate for the purpose of estimating TDI.

6. Application of the Proposed Model: TDI Calculation

This section describes the application of the proposed model for TDI calculation over relevant sample scenarios.

6.1. Overall Description

The main challenge when designing the TDI was to properly capture the differences between demand response strategies in order to fairly select water heaters to interrupt while minimizing the discomfort of users. Therefore, evaluating the TDI calculation is important to have real scenarios that capture the utilization profile of the users of electric water heaters. Evaluating the proposed methodology over a real scenario is not an easy task. The scenario must include a set of real water heaters large enough to carry out experiments on real water uses, and real interventions must be set up. To overcome these difficulties, a common approach in the related literature [22,26] consists of using simulations of thermal appliances.

A simulator was implemented to perform the experimental evaluation in those scenarios where real data is not available. The consists of two modules: the *individual water heater module* simulates the energy dynamics of a water heater, based on the work by Lutz [26] and the *household utilization* module generates scenarios of water utilization for a group of households, using a vector of hourly probabilities of water usage as input.

The proposed TDI is evaluated in three different scenarios accounting for different number of water heaters, households, and priorities. Scenario #1 considers two water heaters and real data for both temperature and the electrical state (ON/OFF) of the water heaters. Scenarios #2 and #3 considers a large number of water heaters, for which real data about the electrical state are available. In turn, the developed temperature and water utilization models, implemented in the simulator, were applied to validate the proposed TDI. The main details of the evaluation are reported in the following subsections.

6.2. Scenario #1: Evaluation of a Simple Case Study with Two Real Water Heaters

One of the main challenges related to the definition of TDI is modeling the differences of temperature (ΔT, a quantitative factor) between performing an interruption in different moments. A relevant case is analyzing the ΔT values situations in the interruption affects the most to comfort.

As a relevant sample study, the comparison of the TDI for two different values of ρ and two particular electric water heaters (EWH_1 and EWH_2) is presented. The considered electric water heaters model two different utilization patterns from two different users. On weekdays, EWH_1 has two consecutive utilizations, and EWH_2 is only used once. The TDI associated with a 20 min interruption between 20:10 and 20:30 was considered for this case. Figure 8 presents the empirical distribution of uses ($P(u)$) from 19:00 to 22:00 for EWH_1

(left) and EWH_2 (right). For each case, the interruption period is represented by a green band.

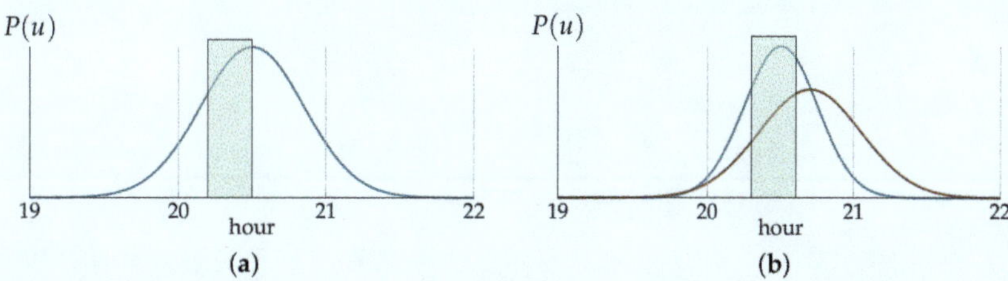

Figure 8. Probability distribution for the water utilization of two electric water heaters. (**a**) EWH_1; (**b**) EWH_2.

For the presented example, it is expected for the TDI value to be higher for EWH_2 than for EWH_1, because in the hours immediately after the interruption analyzed, the average historical utilization is higher for EWH_2. On the other hand, as the value of ρ increases, it is expected that the gap between the TDI of both electric water heaters becomes larger. Table 2 reports the TDI values computed for each electric water heater, considering two different values of ρ ($\rho = 1$, and $\rho = 2$).

Table 2. TDI applied for the interruption for EWH_1 and EWH_2.

Appliance	$\rho = 1$	$\rho = 2$
EWH_1	3362.3 °C s	4108.2 °C s
EWH_2	8109.6 °C s	11,041.7 °C s

The results in Table 2 confirm that the proposed index correctly models discomfort. The TDI value is higher for EWH_2, and the difference widens when considering larger penalty values (ρ). Results show that TDI properly models the differences of temperature between a scenario with an interruption and a scenario without interruption, as expected.

6.3. Scenario #2: Evaluation on a Group of Households with Water Heaters

Another challenge related to the definition of TDI is to prioritize which group of water heaters should be interrupted to reduce electricity demand while minimizing discomfort (e.g., for demand response management in a peak situation). In this experiment, a group of twenty households H with electric water heaters was studied using the developed simulator. Ten households ($H_1, \ldots, H_{10}$) have a water use profile concentrated in the morning hours and the other ten households ($H_{11}, \ldots, H_{20}$) in the evening hours. Figure 9 presents the average water use profiles for each type of household.

A simulation of a complete day for the twenty defined households was performed. In turn, the TDI associated with a 20 min interruption between 7:10 and 7:30 (TDI_m) and the TDI associated with a 20 min interruption between 20:10 and 20:30 (TDI_a) were computed for the twenty households using the method presented in Section 4.4.

Table 3 reports the computed TDI values for each household. Table 4 presents the two rankings defined: $Ranking_m$ sorts households according to the values of TDI_m and $Ranking_a$ sorts households according to the values of TDI_a, both from lowest to higher values. $H_1, \ldots, H_{10}$ households are highlighted in green and $H_{11}, \ldots, H_{20}$ in blue.

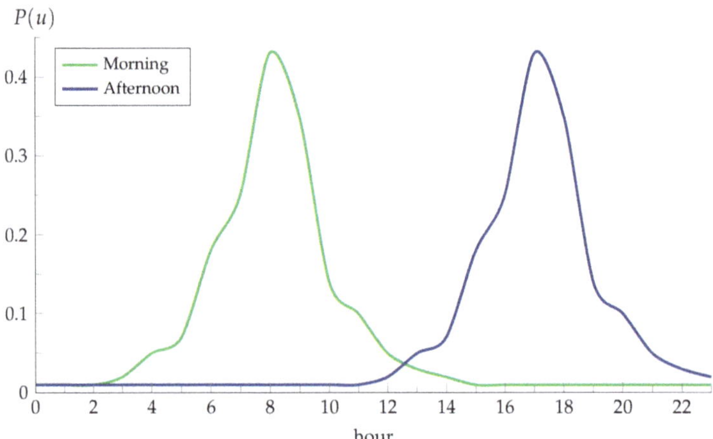

Figure 9. Water use profiles for a day.

Table 3. TDI_m and TDI_a for each water heater.

	TDI_m (°C s)	TDI_a (°C s)
H_1	3140.2	64.1
H_2	3170.4	55.9
H_3	3131.0	67.2
H_4	3158.6	77.1
H_5	3171.4	79.8
H_6	3176.7	77.4
H_7	3161.5	31.2
H_8	3110.8	74.7
H_9	3100.5	66.8
H_{10}	3146.9	77.2
H_{11}	71.6	1962.8
H_{12}	87.3	1996.9
H_{13}	97.8	2017.7
H_{14}	15.4	2019.5
H_{15}	55.8	1993.2
H_{16}	96.6	1986.6
H_{17}	29.4	2018.9
H_{18}	38.1	1954.8
H_{19}	82.6	1977.3
H_{20}	72.6	1956.2

Table 4. Rankings of water heaters sorted by TDI in ascending order.

	Order																			
	1	2	3	4	5	6	7	8	9	10	11	12	13	14	15	16	17	18	19	20
$Ranking_m$	H_{14}	H_{17}	H_{18}	H_{15}	H_{11}	H_{20}	H_{19}	H_{12}	H_{16}	H_{13}	H_9	H_8	H_3	H_1	H_{10}	H_4	H_7	H_2	H_5	H_6
$Ranking_a$	H_7	H_2	H_1	H_9	H_3	H_8	H_4	H_{10}	H_6	H_5	H_{18}	H_{20}	H_{11}	H_{19}	H_{16}	H_{15}	H_{12}	H_{13}	H_{17}	H_{14}

For the presented case study, and considering that the probability of affecting comfort when interrupting water heaters in the morning is higher for households $H_1, \ldots, H_{10}$, these ten households are expected to be in the last ten positions of $Ranking_m$. Analogously, $H_{11}, \ldots, H_{20}$ are expected to be in the last ten positions of $Ranking_a$. Results in Table 4 confirm that the TDI properly sorts the water heaters according to the real discomfort produced for users, as expected. These results allow for very precise information to decide about specific demand-management interventions.

6.4. Scenario #3: Setting ρ to Tune the Interruption Priority of a Special Water Heater

In several circumstances, an electric water heater is installed with specific objectives and should not be considered to be interrupted in a demand-management event. An example of this situation is a nursing home, where hot water is used for sanitary purposes. A simple approach to deal with this scenario is excluding the electric water heater from the list of appliances to be interrupted by the energy utility. This approach is not flexible because it only allows inserting or removing a device from the interruptible list and does not allow prioritizing special appliances. However, using the proposed TDI definition, the parameter ρ is used to independently weight those *special* electric water heaters. In this experiment, the case presented in Section 6.3 is extended by adding a household (H_E) with an special water heater. The hypothesis of the experiment is that the special water heater in household H_E has a water use profile almost constant throughout the day, and it cannot be interrupted in demand-response events. Figure 10 shows in orange the hourly probability vector of this special water heater. Green and blue dotted lines correspond to the profiles of households with water use concentrated in the morning and in the evening, respectively. The experiment consists of adding the device with the same value of the parameter $\rho = 1$ used in the base experiment and calculating $Ranking_m$ and $Ranking_a$. Then, the value of the parameter ρ for the special device is set to a very large value ($\rho = 1000$), and differences are analyzed.

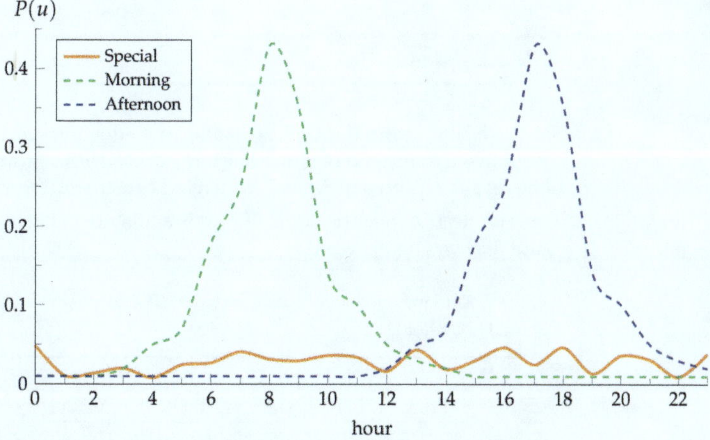

Figure 10. Water use profiles for a day in the special household.

The proposed scenario was studied via simulations. Two special household values were included, $H_E^{(1)}$ using $\rho = 1$ and $H_E^{(1000)}$ using $\rho = 1000$. Table 5 reports the TDI values obtained in the simulations, considering the two special households $H_E^{(1)}$ (highlighted in red) and $H_E^{(1000)}$ (highlighted in orange). Table 6 reports the order defined by $Ranking_m$ and $Ranking_a$ considering $H_E^{(1)}$, and Table 7 reports the order defined by $Ranking_m$ and $Ranking_a$ considering $H_E^{(1)}$ $H_E^{(1000)}$.

Table 5. TDI_m and TDI_a for each water heater with $\rho = 1$, adding a special household using $\rho = 1$ (red), and another special household $\rho = 1000$ (orange).

	TDI_m(°C s)	TDI_a(°C s)
H_1	3140.2	64.1
H_2	3170.4	55.9
H_3	3131.0	67.2
H_4	3158.6	77.1
H_5	3171.4	79.8
H_6	3176.7	77.4
H_7	3161.5	31.2
H_8	3110.8	74.7
H_9	3100.5	66.8
H_{10}	3146.9	77.2
H_{11}	71.6	1962.8
H_{12}	87.3	1996.9
H_{13}	97.8	2017.7
H_{14}	15.4	2019.5
H_{15}	55.8	1993.2
H_{16}	96.6	1986.6
H_{17}	29.4	2018.9
H_{18}	38.1	1954.8
H_{19}	82.6	1977.3
H_{20}	72.6	1956.2
$H_E^{(1)}$	354.4	309.5
$H_E^{(1000)}$	72,305.4	71,588.1

Results reported in Table 6 demonstrate that the special electric water heater $H_E^{(1)}$ is not ranked at the end of either *Ranking$_m$* or *Ranking$_a$*. A different situation is observed in the results reported in Table 7, which indicate that using the value $\rho = 1000$ only for the special electric water heater $H_E^{(1000)}$, this appliance appears last in both rankings *Ranking$_m$* and *Ranking$_a$*, as expected.

Table 6. Rankings of water heaters sorted by TDI using $\rho = 1$ in ascending order, considering the special household.

	Order																				
	1	2	3	4	5	6	7	8	9	10	11	12	13	14	15	16	17	18	19	20	21
Ranking$_m$	H_{14}	H_{17}	H_{18}	H_{15}	H_{11}	H_{20}	H_{19}	H_{12}	H_{16}	H_{13}	$H_E^{(1)}$	H_9	H_8	H_3	H_1	H_{10}	H_4	H_7	H_2	H_5	H_6
Ranking$_a$	H_7	H_2	H_1	H_9	H_3	H_8	H_4	H_{10}	H_6	H_5	$H_E^{(1)}$	H_{18}	H_{20}	H_{11}	H_{19}	H_{16}	H_{15}	H_{12}	H_{13}	H_{17}	H_{14}

Table 7. Rankings of water heaters sorted by TDI in ascending order using $\rho = 1000$ for the special household and $\rho = 1$ for the rest of the households.

	Order																				
	1	2	3	4	5	6	7	8	9	10	11	12	13	14	15	16	17	18	19	20	21
Ranking$_m$	H_{14}	H_{17}	H_{18}	H_{15}	H_{11}	H_{20}	H_{19}	H_{12}	H_{16}	H_{13}	H_9	H_8	H_3	H_1	H_{10}	H_4	H_7	H_2	H_5	H_6	$H_E^{(1000)}$
Ranking$_a$	H_7	H_2	H_1	H_9	H_3	H_8	H_4	H_{10}	H_6	H_5	H_{18}	H_{20}	H_{11}	H_{19}	H_{16}	H_{15}	H_{12}	H_{13}	H_{17}	H_{14}	$H_E^{(1000)}$

Overall, results obtained for scenario #3 validate the proposed approach and confirm the effect of modifying the ρ parameter for some water heaters to allow flexible demand-response strategies to be designed. For instance, a possible strategy is to define a classification of the complete set of households in three classes and associate each class with a different ρ value to define water heaters that are *not interruptible*, *interruptible if necessary*, and *no restrictions for interruption*. The flexibility of the proposed TDI using different values of ρ facilitates, in a real scenario that is continually changing in structure, performing active demand management to quickly adapt to the needs of the electrical system.

7. Conclusions and Future Work

This article presented an approach to evaluate the impact on the thermal comfort of direct demand response control using electric water heaters. An index associated with the thermal discomfort was defined according to the following procedure. First, a water utilization forecasting model was built using real power data from a set of electric water heaters and applying an ensemble learning technique. Then, a linear model was developed to estimate the temperature of the water in the electric water heater tank. Finally, the TDI associated with an intervention on the electric water heater was defined stochastically, and calculated via Monte Carlo simulation. A specific water heater simulator was developed for the evaluation of the proposed TDI.

The computational models and the reliability of the proposed index were evaluated in three real case studies. The first case considered two electric water heaters from different users with different average historical utilization. The TDI values were analyzed for both electric water heaters for different penalization factors ρ. Results confirmed that the proposed index correctly models discomfort since higher TDI values were computed for the electric water heater with the higher average historical utilization. The difference in TDI values increased when considering larger penalty values. The second case analyzed the use of TDI for sorting water heaters according to the discomfort caused by their intervention. Two sets of ten households were generated and simulated, considering two different utilization patterns and two interruptions (in the morning and the evening). Results confirmed that in the ranking generated with the TDI computed for the morning interruption, households with high probability of water utilization in the morning were in the last ten positions of the ranking. A similar result was obtained for the afternoon interruption, as expected. The third case study explored the use of parameter ρ to tune the interruption priority, considering an additional special household and two values of ρ. Results demonstrated that the new special household was in eleventh position in both rankings when $\rho = 1$ and changed to the last position in both rankings when $\rho = 1000$, properly modeling a non-interruptible appliance (e.g., for sanitary reasons). The main lines of future work are related to estimating an economic value of the TDI index (in USD/MWh), useful to characterize the profit of reducing the energy demanded by a set of electric water heaters by applying the interruption action in order to compare this strategy with other demand response techniques (e.g., using fuel generators or batteries). Expanding the developed simulator to consider all generators in the system is another line for future work to fairly compare demand response strategies by using TDI values and the economic impact of an interruption.

Author Contributions: R.P.: research conceptualization, data analysis, code development, manuscript writing, manuscript revision and correction. J.C.: data analysis, data cleansing, manuscript writing, manuscript revision and correction. S.N.: research conceptualization, data analysis, manuscript writing, manuscript revision and correction. All authors have read and agreed to the published version of the manuscript.

Funding: This research received no external funding.

Institutional Review Board Statement: Not applicable.

Informed Consent Statement: Not applicable.

Acknowledgments: The work of S. Nesmachnow was partly supported by ANII and PEDECIBA, Uruguay.

Conflicts of Interest: The authors declare no conflicts of interest.

References

1. Momoh, J. *Smart Grid: Fundamentals of Design and Analysis*; Wiley-IEEE Press: Hoboken, NJ, USA, 2012.
2. *Assessment of Demand Response & Advanced Metering*; Technical Report AD-06-2-00; Federal Energy Regulatory Commission: Washington, DC, USA, 2006.
3. European Automotive Research Partner Association. Smart Grids European Technology Platform. Available online: https://www.etip-snet.eu/ (accessed on 21 January 2021).
4. Bhattacharyya, S. Energy Demand Management. In *Energy Economics*; Springer: London, UK, 2011; pp. 135–160.
5. Instituto Nacional de Estadística, Uruguay. Microdatos de la Encuesta Continua de Hogares. 2019. Available online: http://www.ine.gub.uy/microdatos (accessed on 17 August 2020).
6. Chavat, J.; Nesmachnow, S.; Graneri, J.; Alvez, G. ECD-UY: Detailed household electricity consumption dataset of Uruguay. *Sci. Data* **2020**, submitted.
7. Porteiro, R.; Nesmachnow, S.; Hernández-Callejo, L. Electricity demand forecasting in industrial and residential facilities using ensemble machine learning. *Rev. Fac. Ing. Univ. Antioq.* **2020**, *102*, 9–25.
8. Deng, R.; Yang, Z.; Chow, M.; Chen, J. A Survey on Demand Response in Smart Grids: Mathematical Models and Approaches. *IEEE Trans. Ind. Inform.* **2015**, *11*, 570–582. [CrossRef]
9. Hassan, N.U.; Khalid, Y.I.; Yuen, C.; Tushar, W. Customer engagement plans for peak load reduction in residential smart grids. *IEEE Trans. Smart Grid* **2015**, *6*, 3029–3041. [CrossRef]
10. Tang, R.; Wang, S.; Yan, C. A direct load control strategy of centralized air-conditioning systems for building fast demand response to urgent requests of smart grids. *Autom. Constr.* **2018**, *87*, 74–83. [CrossRef]
11. Ozoh, P.; Apperley, M.; Olayiwola, M. Modelling of household power consumption and its effects on load shifting strategies. *J. Sci. Arts* **2018**, *18*, 1067–1072.
12. Witherden, M.; Rayudu, R.; Tyler, C.; Seah, W. Managing peak demand using direct load monitoring and control. In Proceedings of the Australasian Universities Power Engineering Conference, Hobart, TAS, Australia, 29 September–3 October 2013; pp. 1–6.
13. Orsi, E.; Nesmachnow, S. Smart home energy planning using IoT and the cloud. In Proceedings of the IEEE URUCON, Montevideo, Uruguay, 23–25 October 2017.
14. Orsi, E.; Nesmachnow, S. IoT for smart home energy planning. In Proceedings of the XXIII Congreso Argentino de Ciencias de la Computación, La Plata, Argentina, 9–13 October 2017; pp. 1091–1100.
15. Chavat, J.P.; Nesmachnow, S.; Graneri, J. Non-intrusive energy disaggregation by detecting similarities in consumption patterns. *Rev. Fac. Ing. Univ. Antioq.* **2020**, *98*, 27–46. [CrossRef]
16. Yoon, A.; Kang, H.; Moon, S. Optimal Price Based Demand Response of HVAC Systems in Commercial Buildings Considering Peak Load Reduction. *Energies* **2020**, *13*, 862. [CrossRef]
17. Huang, D.; Billinton, R. Effects of load sector demand side management applications in generating capacity adequacy assessment. *IEEE Trans. Power Syst.* **2011**, *27*, 335–343. [CrossRef]
18. Lu, N.; Katipamula, S. Control strategies of thermostatically controlled appliances in a competitive electricity market. In Proceedings of the IEEE Power Engineering Society General Meeting, San Francisco, CA, USA, 16 June 2005; pp. 202–207.
19. Perfumo, C.; Braslavsky, J.H.; Ward, J. Model-based estimation of energy savings in load control events for thermostatically controlled loads. *IEEE Trans. Smart Grid* **2014**, *5*, 1410–1420. [CrossRef]
20. Nehrir, M.; LaMeres, B.; Gerez, V. A customer-interactive electric water heater demand-side management strategy using fuzzy logic. In Proceedings of the Winter Meeting IEEE Power Engineering Society, New York, NY, USA, 31 January–4 February 1999; Volume 1, pp. 433–436.
21. Xiang, S.; Chang, L.; Cao, B.; He, Y.; Zhang, C. A Novel Domestic Electric Water Heater Control Method. *IEEE Trans. Smart Grid* **2019**, *11*, 3246–3256. [CrossRef]
22. Kampelis, N.; Ferrante, A.; Kolokotsa, D.; Gobakis, K.; Standardi, L.; Cristalli, C. Thermal comfort evaluation in HVAC Demand Response control. *Energy Procedia* **2017**, *134*, 675–682. [CrossRef]
23. Tabatabaei, S.; Klein, M. The role of knowledge about user behaviour in demand response management of domestic hot water usage. *Energy Effic.* **2018**, *11*, 1797–1809. [CrossRef]
24. Pirow, N.; Louw, T.; Booysen, M. Non-invasive estimation of domestic hot water usage with temperature and vibration sensors. *Flow Meas. Instrum.* **2018**, *63*, 1–7. [CrossRef]
25. Paull, L.; MacKay, D.; Li, H.; Chang, L. Awater heater model for increased power system efficiency. In Proceedings of the Canadian Conference on Electrical and Computer Engineering, St. John's, NL, Canada, 3–6 May 2009.
26. Lutz, J.; Whitehead, C.; Lekov, A.; Winiarski, D.; Rosenquist, G. *WHAM: A Simplified Energy Consumption Equation for Water Heaters*; American Council for and Energy-Efficient Economy Summer Study on Energy Efficiency in Buildings: Washington, DC, USA, 1998
27. Yin, Z.; Che, Y.; Li, D.; Liu, H.; Yu, D. Optimal scheduling strategy for domestic electric water heaters based on the temperature state priority list. *Energies* **2017**, *10*, 1425. [CrossRef]
28. Al-Jabery, K.; Xu, Z.; Yu, W.; Wunsch, D.C.; Xiong, J.; Shi, Y. Demand-side management of domestic electric water heaters using approximate dynamic programming. *IEEE Trans. Comput.-Aided Des. Integr. Circuits Syst.* **2016**, *36*, 775–788. [CrossRef]
29. Massobrio, R.; Nesmachnow, S.; Tchernykh, A.; Avetisyan, A.; Radchenko, G. Towards a cloud computing paradigm for big data analysis in smart cities. *Program. Comput. Softw.* **2018**, *44*, 181–189. [CrossRef]

30. Luján, E.; Otero, A.; Valenzuela, S.; Mocskos, E.; Steffenel, L.; Nesmachnow, S. An integrated platform for smart energy management: the CC-SEM project. *Rev. Fac. Ing. Univ. Antioq.* **2019**, *97*, 41–55. [CrossRef]
31. De Ville, B. Decision trees. *Wiley Interdiscip. Rev. Comput. Stat.* **2013**, *5*, 448–455. [CrossRef]
32. Zhang, C.; Ma, Y. (Eds.) *Ensemble Machine Learning*; Springer: New York, NY, USA, 2012.
33. Lévesque, B.; Lavoie, M.; Joly, J. Residential Water Heater Temperature: 49 or 60 Degrees Celsius? *Can. J. Infect. Dis.* **2004**, *15*, 11–12. [CrossRef]
34. Sangoi, J.; Ghisi, E. Energy Efficiency of Water Heating Systems in Single-Family Dwellings in Brazil. *Water* **2019**, *11*, 1068. [CrossRef]
35. Nesmachnow, S.; Iturriaga, S. Cluster-UY: Collaborative Scientific High Performance Computing in Uruguay. In *High Performance Computing*; Springer International Publishing: Berlin/Heidelberg, Germany, 2019; pp. 188–202.

Article

The Influence of Energy Certification on Housing Sales Prices in the Province of Alicante (Spain)

Maria-Francisca Cespedes-Lopez, Raul-Tomas Mora-Garcia *, V. Raul Perez-Sanchez and Pablo Marti-Ciriquian

Building Sciences and Urbanism Department, University of Alicante, 03690 San Vicente del Raspeig, Alicante, Spain; paqui.cespedes@ua.es (M.-F.C.-L.); raul.perez@ua.es (V.R.P.-S.); pablo.marti@ua.es (P.M.-C.)
* Correspondence: rtmg@ua.es

Received: 9 September 2020; Accepted: 10 October 2020; Published: 13 October 2020

Abstract: This work examines the implementation of energy labelling by the residential real estate sector. First, it considers the interest by real estate sellers in not publishing energy certification information, and then, it quantifies the impact of the housing's energy certification on the asking price. The results are compared with those obtained from other studies conducted in distinct European countries. The study's final sample was collected, including information from 52,939 multi-family homes placed on the real estate market in the province of Alicante (Spain). One-way analysis of variance (ANOVA) was used, as well as an ordinary least squares regression model. This study highlights the fact that, in the current market, owners and sellers have no incentive to reveal the energy certification, since this permits them to sell homes with low energy ratings at prices similar to those of more energy efficient homes. In addition, it was found that homes with better energy ratings (letters A and B) are not sold at higher prices than homes with other rating letters, unlike the case of other European countries that were examined.

Keywords: energy performance certificate (EPC); hedonic pricing method; real estate market; property prices; Alicante

---

1. Introduction

High energy consumption by buildings is a global issue, and the implementation of policies promoting energy performance improvement measures is challenging. According to Hirsch, et al. [1], the European Union (EU) has taken a leadership role in this area, implementing directives [2–5] and financing various energy efficiency research and innovation projects (Project SP-0165/CEENEREPC [6], EPLABEL [7], ENPER EXIST [8], SHARE [9], ASIEPI [10], RentalCal [11], RESPOND [12], CRREM [13], among others).

The European directives [2,3] establish a mandatory certification system, the so-called "ABCDEFG qualification", which rates buildings based on their energy efficiency, similar to the classification used for household appliances. In addition, these directives require the publication of an energy performance certificate (EPC), to be included in the documentation supplied by owners to purchasers or renters. These policies are an attempt to offer increased transparency and information to consumers, to assist in decision-making related to property purchase or rental.

In Spain, these directives result from the enforcement of a series of decrees [14,15] requiring energy certification and the presentation of an energy efficiency label in properties placed on the rental or sales market. On the other hand, the level of compliance with the directive varies depending on the specific case. In the case of real estate sales, compliance is significant, given that certification documentation is required by the notary public when formalizing a deed of sale. In the case of publishing energy certification in advertisements for sales or rentals, there is a high level of non-compliance with the directive [16].

With regard to the financed projects, studies [6,17] have analyzed the relationship between the certification system and the property's value, in an attempt to determine whether or not properties with higher sales or rental prices correspond to those that are more energy efficient. Mudgal, et al. [6] confirmed a positive relationship between the information published on energy performance and the property's value. Taltavull de La Paz, et al. [17] obtained contrary results, that is, homes that were more energy efficient did not have higher sales prices than other homes with poorer energy qualifications. However, these are not the only studies that have been conducted with this objective and that have obtained contradictory results. In studies carried out in Denmark [18], France [19,20], the Netherlands [21,22], Ireland [23], Italy [24], the United Kingdom [25–27] or Spain [16,28], it has been found that homes having better ratings (letters A, B or C) have higher market prices than less efficient homes (letters D, E, F or G). On the other hand, studies conducted in Ireland [29], Norway [30] and Spain [16,31] have obtained unexpected results. A thorough review of the literature on EPC and housing prices was carried out by Cespedes-Lopez, et al. [32].

This study has two main objectives. The first is to consider the interest by real estate sellers in not publishing energy qualifications. The second is to determine the economic impact of energy qualifications on the asking prices of homes sold on the real estate market in the province of Alicante (Spain), so that it is possible to compare the results of this study with past works.

The first hypothesis proposed in the study (H_1), is that housing that do not publish the energy qualification are sold at prices that are similar to housing having high energy qualifications. The second hypothesis (H_2) states that housing that published their energy qualifications have higher sales prices than those that do not publish them. The third hypothesis (H_3) is that housing with high energy qualifications will be offered at higher prices than less efficient homes.

The housing market serves major economic and social functions. Firstly, this sector is very significant in the economy of developed countries. Secondly, it determines the access of households to housing. The real estate crisis in the Spanish residential market was severe, resulting in a dramatic decline in housing prices. However, following the crisis, from 2014 until late 2018, real estate sales prices increased by 31%, mainly in the second-hand segment. During this period, foreigners who were not Spanish residents represented approximately 20% of all purchases, with a greater majority centering in the Mediterranean and the islands [33]. On a national level, 71.7% of all housing is intended for primary use, whereas in Alicante, it is 58.0%, with 25.6% of all housing being used as second residences (as compared to the 14.6% national rate). The main tenancy regime is ownership (85.4%) as opposed to rental (14.6%). The building stock consists mainly of collective housing (buildings), 68–77% of the total, as compared to single-family homes, which only represent 23–32%. See Table A1 of the Appendix A for more details on these statistics.

If the data on real estate transactions from 2019 is analyzed [34], on a national level, 41.7% of the housing transactions were carried out by Spanish residents, as compared to 26.1% in the Alicante province. The Alicante market has a significant international exposure, being a destination for foreigners, participating in 23.2% of all of the province's transactions. In the Valencian community, in a breakdown of the number of real estate transactions per province, it is seen that Alicante has a more dynamic market than Valencia and Castellón (Figure 1a). Analyzing the real estate transactions in terms of purchaser place of residence (Figure 1b), it may be observed that the volume of transactions by foreign residents in the province of Alicante, except during the second quarter of 2020 where data is provisional (*), exceeds 45%. In the other two provinces, the values are approximately 20%. This difference highlights a distinct behavior of the real estate market in the Alicante province as compared to the other two Valencia provinces, which serves as a motivation for its consideration in this work.

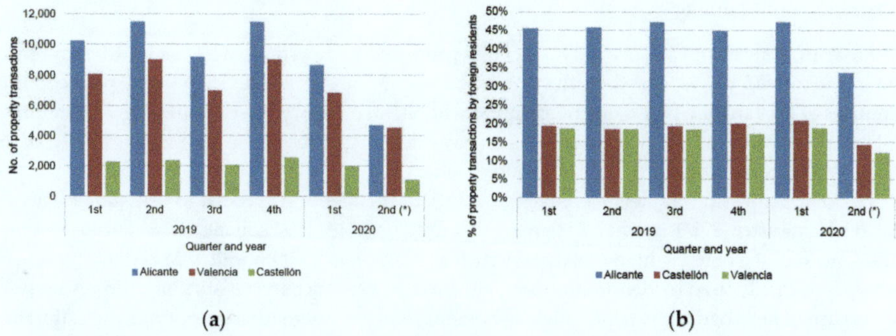

(a)   (b)

Figure 1. (**a**) Number of real estate transactions by province; (**b**) number of real estate transactions by foreign residents in the Valencian community.

In terms of energy and climate (Figure 2), Spain's climatic diversity should be considered (CTE-HE 2013) [35], since it has always, and continues to, significantly condition the energy characteristics of the building stock. In Spain, within the same region, it is possible to find mild climates in Mediterranean coastal areas and continental climates with more extreme temperatures, as occurring in the Alicante province. This characteristic may even be found in cases in which the separation between climate zones is barely 70 km, as occurs in the coast and the interior of the province, where mountain chains exceeding 1000 m in height are found. This climatic diversity leads to a distinct energy allocation of the buildings. Therefore, electricity is frequently used to heat and cool buildings on the Mediterranean coast, via heating pumps, whereas in the interior, it is more common to find buildings that are heated with community or individual heating systems using natural gas. As for the heating system used in the Valencian community [36], centralized or collective systems represent only 4.6% (10.56% on a national level), while heating in the entire housing represents only 27.4% (as compared to 46.30% across Spain). In this province, it is common to use small heating devices for single rooms (54.2%) or even, to have no heating devices at all (13.8%). (see Table A1 of the Appendix A).

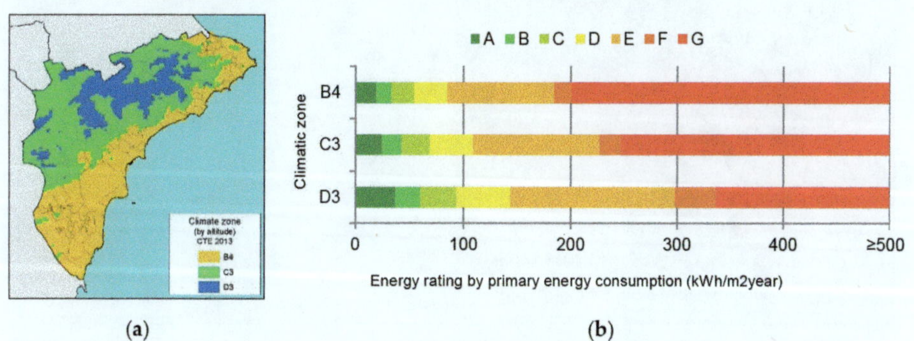

(a)   (b)

Figure 2. (**a**) Climatic areas in the Alicante province; (**b**) energy rating by primary energy consumption (kWh/m^2 year) for homes and buildings, according to climate zone, obtained from the CE3 programme and the CTE 2013 regulations.

The document is organized as follows: In the second section, the materials and methods are described, detailing the sources used and the database generated. The third section offers the results. The fourth section contains the discussion and finally, a summary of the conclusions obtained is presented.

2. Materials and Methods

First, an analysis of variance (ANOVA) was proposed to determine whether or not differences exist in the offered prices, based on the published energy qualification. To examine the economic premiums of the housing based on energy qualification, an ordinary least squares regression model has been proposed. For this, various estimates have been made, based on the reference used in the housing's energy qualification (letter or group of letters).

Hedonic regression models have been used since the "New Approach to Consumer Theory" created by Lancaster [37]. Ridker and Henning [38] used ordinary least squares (OLS) for the first time in the context of the housing market. Authors such as Zietz, et al. [39] indicate that hedonic regression analysis is normally used to identify the marginal effect of a set of characteristics on the housing price. For the case of heterogeneous goods such as housing, hedonic methodology permits the estimation for the contribution of each characteristic on the price [40]. Currently, this methodology is the most commonly used to determine the economic premium generated by distinct characteristics. In Table A2 of the Appendix A, the variables that are the most commonly used by other authors to determine housing price are shown.

2.1. Population and Sample

The database consists of multi-family housing placed on the market in the province of Alicante (Valencian Community, Spain) see Figure 3. The interest and selection criteria were based on the significant activity of the construction sector in this area, which is the third province in the country in terms of having the largest number of property transactions (purchases), after Barcelona and Madrid. On the other hand, it is fourth in terms of number of unsubsidized housing sales initiated in 2017 [41].

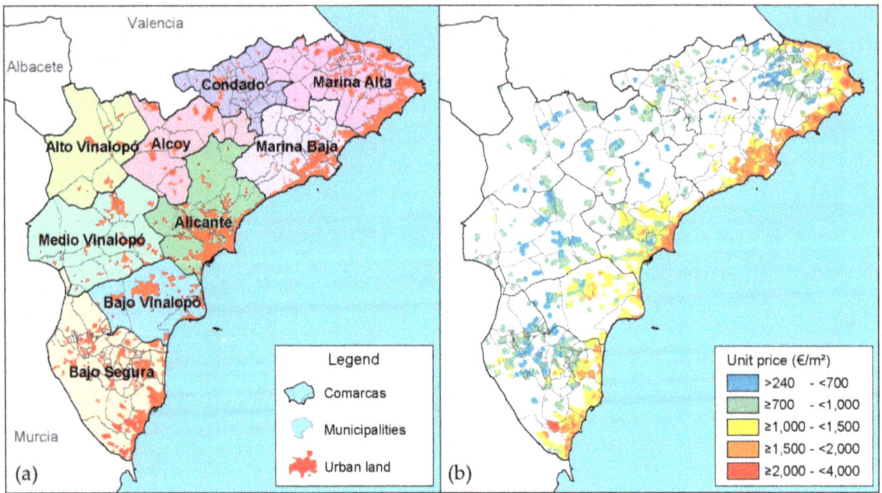

Figure 3. (**a**) Map of the province of Alicante with the delimitation of the *comarcas*, municipal boundary delimitation and the stain of continuous urban land; (**b**) kernel density map with unit prices of multi-family houses (euros/m^2).

The study sample consists of housing properties that were placed on the market via the *idealista.com* real estate portal between June 2017 and May 2018. During this period, information was collected on 97,279 properties placed on the market, extracting data on the characteristics of the housing and buildings. Subsequently, via GIS, information was provided on the location, neighborhood and market, obtained from other information sources. The final database was subject to a univariate analysis of outliers, discarding properties that differed by more or less than three standard deviations in their respective variables (Z scores). This process was performed on the following variables: natural log of the property price, age, height in stories, constructed surface area, number of bedrooms and bathrooms. To identify the multivariate atypical cases, the regression model was calibrated by calculating the Mahalanobis distance (MD) and its statistical significance, discarding any files in which the significance was less than 0.001, in accordance with Hair, et al. [42]. Finally, those properties having missing data on any of the variables that were subject of the analysis were discarded, obtaining a final sample of 52,939 observations, of which, 9194 included information on energy qualification.

The sample's representativeness was verified via the Equation (1), designed for large or infinite populations when the exact size of the units making it up is unknown [43].

$$n = \frac{(z_{\alpha/2})^2 * p * (1-p)}{E^2} \tag{1}$$

where: $z_{\alpha/2}$ is the Z score corresponding to the selected level of confidence, p is the probability that event p takes place (when not having sufficient information the least favorable value is assigned, $p = 0.50$), and E is the maximum admissible error or the maximum error that is committed in the sample.

Using a 95% confidence level ($z_{\alpha/2} = 1.96$), a probability of $p = 0.50$ and samples sizes of $n = 52{,}939$ and $n = 9194$, clearing E, a maximum estimated error of 0.4% (0.004) and 1.0% (0.010) were obtained, respectively, ensuring the high statistical precision of the sample.

2.2. Sources of Information

The main source of information is the real estate portal *idealista.com*, which publishes the asking prices along with the characteristics of the housing and the building in which it is located. Other studies have also considered real estate portals with the same objective, given the lack of official information available [44–48], with the real estate asking prices being a suitable substitute for the transaction prices [49]. In Figure 4, the distribution of energy certificates in the Alicante province is shown, as well as details on the province's two largest cities (Alicante and Elche).

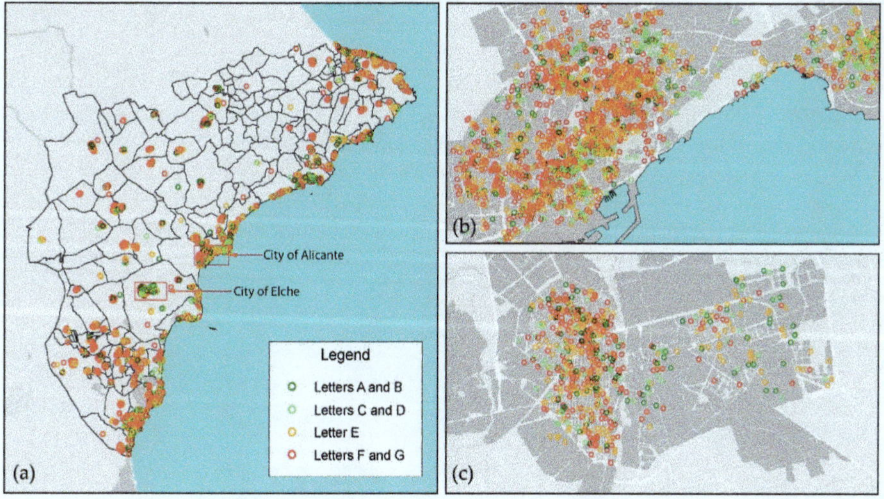

Figure 4. (**a**) Energy certificate distribution in the Alicante province; (**b**) detail of Alicante city; (**c**) detail of Elche city.

Other sources used have included the General Land Registry Directorate (DGC), the National Statistics Institute (INE), the National Geographic Institute (IGN), the Department of Education, Culture and Sports (CECD), the Directorate General for Planning, Assessment and Patient Care of the Health

Department (DGOEAPCS) and the Basic Document on Energy Saving from the Technical Code for Buildings (CTE-DB-HE).

Based on the alphanumeric and vectorial information from the DGC [50,51], a raster map was created to estimate the age of the housing (Figure 5a) and the ratio of the constructed surface area in the proximity of each building (150 m around the same). With population census and INE housing data [36] and IGN mapping, the type of occupation was collected for each census tract (vacant, main and secondary), as well as the type of tenancy regime (rented, mortgaged and owned) and the population's sociodemographic characteristics (dependency, ageing, foreign population and education level). With the information from the CECD and the DGOEAPCS, distances between the housing and the public services or the following points of interest were calculated: hospitals, health centers, pharmacies, schools (Figure 5b), as well as proximity to the coast. Distances have been calculated by network, that is, based on the length of origin and destination using a layout of pre-established streets and intersections, simulating the reality of the urban network. The CTE-DB-HE is used to determine the climatic zone of the town where the property is located (climatic severity of summer and winter seasons) [35,52].

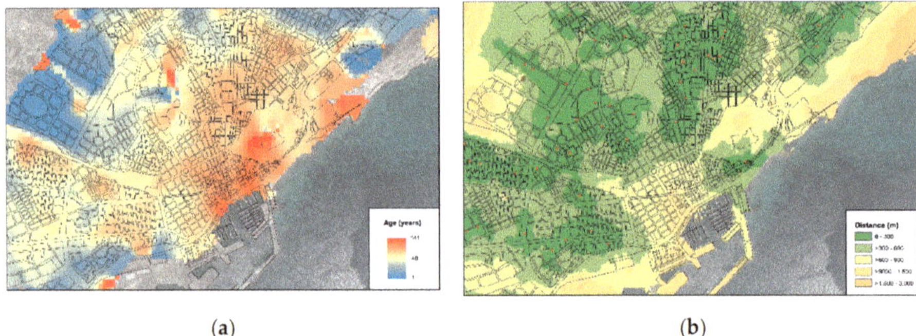

(a)   (b)

Figure 5. Maps of the city of Alicante: (**a**) age of the buildings and (**b**) distance from primary and secondary schools.

2.3. Data

Variables were selected based on a literature review (see Table A2 of the Appendix A). Based on the information received, 63 variables were obtained, as summarized in Table 1. The variables are ordered based on five categories: *Housing characteristics (A)*, *Building characteristics (B)*, *Location characteristics (C)*, *Neighborhood characteristics (D)*, and *Market characteristics (E)*. The unit with which each variable has been measured is also indicated, as well as a brief description of the same and verification as to whether or not it was used in model estimation.

Table 1. Set of variables that make up the study, with their units and description.

Category	Characteristics	Variable	Unit	Description of the Variable	Used	Expected Sign
	Age	*A_age*	numerical	Age of the building (years), number of years that have passed since it was built.	Yes	−
Dwelling characteristics (A)	Size	*A_area_m2*	numerical	Built dwelling surface (sqm), gross square meters of the dwelling.	Yes	+
		A_bedrooms	numerical	Number of bedrooms in the dwelling.	Yes	−
		A_bathrooms	numerical	Number of bathrooms.	Yes	+
	Extras	*A_wardrobe*	dummy	Availability of built-in wardrobes (=1).	Yes	+

Table 1. *Cont.*

Category	Characteristics	Variable	Unit	Description of the Variable	Used	Expected Sign
		A_air_cond	dummy	Availability of air conditioning (=1).	Yes	+
		A_terrace	dummy	Availability of balcony or terrace (=1).	Yes	+
	Floor	*A_floor*	numerical	Floor the dwelling was located on within the building.	Yes	+
	Status	*A_new_construction*	dummy	Newly build housing that can be: a project, under construction, or less than 3 years old.	Yes	+
		A_state_to_reform	dummy	Requires refurbishment.	Yes	−
		A_good_condition	dummy	Classification that the seller assigns to the state of the dwelling, such as "good".	Reference	
	Typology	*A_flat*	dummy	Indicates whether the property has this typology: Flat or apartment, studio flat, penthouse, duplex	Reference	
		A_studio_flat	dummy		Yes	−
		A_penthouse	dummy		Yes	+
		A_duplex	dummy		Yes	+
	Energy Rating	A	dummy	Indicates if the dwelling has an energy rating: Letters A, B, C, D, E, F or G, or has no label (NT).	Yes	+
		B	dummy		Yes	+
		C	dummy		Yes	+
		D	dummy		Yes	(Ref.)
		E	dummy		Yes	−
		F	dummy		Yes	−
		G	dummy		Yes	−
		NT	dummy		Yes	−
Building characteristics (B)	Equipment	*B_elevator*	dummy	Availability of elevator (=1).	Yes	+
		B_parking	dummy	Availability of garage slot (=1).	Yes	+
		B_storeroom	dummy	Availability of storage room (=1).	Yes	+
		B_pool	dummy	Availability of swimming pool (=1).	Yes	+
		B_garden	dummy	Availability of garden (=1).	Yes	+
Location characteristics (C)	Comarca	*C_Alicante*	dummy	Identifier of the comarca: Alicante, Marina Alta, Marina Baja, Bajo Vinalopó, Bajo Segura, El Condado, Alcoy, Alto Vinalopó and Medio Vinalopó. (*Comarcas* are administrative units equivalent to the districts in England or the Kreise in Germany).	Reference	
		C_Marina_Alta	dummy		Yes	+
		C_Marina_Baja	dummy		Yes	+
		C_Bajo_Vinalopo	dummy		Yes	−
		C_Bajo_Segura	dummy		Yes	−
		C_Condado	dummy		Yes	−
		C_Alcoy	dummy		Yes	−
		C_Alto_Vinalopo	dummy		Yes	−
		C_Medio_Vinalopo	dummy		Yes	−
	Climatic zone	*Zone_B4*	dummy	Identifier of the climatic zone according to the municipality in accordance with the CTE-DB-HE of 2019.	No	
		Zone_C3	dummy		No	
		Zone_D3	dummy		No	
	Location	*C_dist_pharmacy*	numerical	Distance from the dwelling to the nearest pharmacy, in km.	Yes	−
		C_dist_health	numerical	Distance from the dwelling to the health centre, in km.	Yes	−

Table 1. *Cont.*

Category	Characteristics	Variable	Unit	Description of the Variable	Used	Expected Sign
		C_dist_hospital	numerical	Distance from the dwelling to the hospital, in km.	Yes	−
		C_dist_educ1	numerical	Distance from the dwelling to level 1 educational centres (infant and primary), in km.	Yes	−
		C_dist_educ2	numerical	Distance from the dwelling to level 2 educational centres (secondary and high school), in km.	Yes	−
		C_coastalregion	dummy	Identification of property location within a coastal region.	Yes	+
		C_FAR	numerical	Floor Area Ratio (total building floor area/gross sector area), 150 m alrededor del edificio, in m^2 floor area/m^2 sector area.	Yes	−
Neighborhood characteristics (D)	Neighborhood	*D_dependency*	numerical	Dependency ratio (sum of the population aged >64 and <16/population aged 16–64).	Yes	+
		D_elderly	numerical	Aging Index (population aged >64/population aged 0–15).	Yes	+
		D_foreigners	numerical	Percentage of foreign population.	Yes	+
		D_no_studies	numerical	Percentage of population without education.	Yes	−
		D_students	numerical	Percentage of the population with primary, secondary studies and high school.	No	
		D_university	numerical	Percentage of the population with university studies.	Yes	+
Market characteristics (E)	Price	*Ln_price*	numerical	Dependent variable. The natural log of the property price offered by the seller (in Euro).	Yes	
	Seller	*E_professional*	dummy	Identifier of the seller: professional, private or bank.	Yes	+
		E_private	dummy		Reference	
		E_bank	dummy		Yes	−
	Occupancy	*E_vacant_dw*	numerical	Percentage of vacant dwellings, main and secondary.	No	
		E_main_dw	numerical		No	
		E_secondary_dw	numerical		Yes	+
	Housing tenure	*E_rented_dw*	numerical	Percentage of dwellings for rent, mortgaged or owned.	Yes	+
		E_mortgaged_dw	numerical		No	
		E_homeownership	numerical		No	

2.4. Descriptive Statistics

The descriptive statistics of the variables are shown in Table 2.

Table 2. Descriptive statistics for the variables.

Category	Variable	Continuous Variables				Dummies Variables		
		Mean	SD	Min.	Max.	Coding.	Freq.	%
Dwelling (A)	A_age	31.460	11.169	3	68			
	A_area_m2	93.760	28.314	31	192			
	A_bedrooms	2.570	0.865	0	5			
	A_bathrooms	1.550	0.532	1	3			
	A_wardrobe					(0) Without	20,899	39.5
						(1) With	32,040	60.5
	A_air_cond					(0) Without	29,807	56.3
						(1) With	23,132	43.7
	A_terrace					(0) Without	22,392	42.3
						(1) With	30,547	57.7
	A_floor	2.880	2.396	0	12			
	A_new_construction						539	1.0
	A_state_to_reform						2730	5.2
	A_good_condition						49,670	93.8
	A_flat						47,610	89.9
	A_studio_flat						549	1.0
	A_penthouse						3146	5.9
	A_duplex						1634	3.1
	A						807	1.5
	B						325	0.6
	C						488	0.9
	D						587	1.1
	E						3083	5.8
	F						864	1.7
	G						3040	5.8
	NT						43,745	82.6
Building (B)	B_elevator					(0) Without	13,033	24.6
						(1) With	39,906	75.4
	B_parking					(0) Without	32,526	61.4
						(1) With	20,413	38.6
	B_storeroom					(0) Without	40,133	75.8
						(1) With	12,806	24.2
	B_pool					(0) Without	32,207	60.8
						(1) With	20,732	39.2
	B_garden					(0) Without	37,472	70.8
						(1) With	15,467	29.2
Location (C)	C_Alicante						20,601	38.9
	C_Marina_Alta						6244	11.8
	C_Marina_Baja						5980	11.3
	C_Bajo_Vinalopo						6368	12.0
	C_Bajo_Segura						10,956	20.7
	C_Condado						189	0.4
	C_Alcoy						1021	1.9
	C_Alto_Vinalopo						402	0.8
	C_Medio_Vinalopo						1178	2.2
	Zone_B4						50,265	94.9
	Zone_C3						2475	4.7

Table 2. *Cont.*

Category	Variable	Continuous Variables				Dummies Variables		
		Mean	SD	Min.	Max.	Coding.	Freq.	%
	Zone_D3						199	0.4
	C_dist_pharmacy	0.517	0.739	0	9.51			
	C_dist_health	1.211	1.396	0	18.86			
	C_dist_hospital	6.244	5.707	0.02	30.33			
	C_dist_educ1	0.888	1.036	0	13.32			
	C_dist_educ2	1.402	1.555	0.01	18.18			
	C_coastalregion					(0) Non-coastal	19,349	36.5
						(1) Coastal	33,590	63.5
	C_FAR	1.155	0.861	0	7.62			
Neighborhood (D)	D_dependency	0.528	0.187	0	1.81			
	D_elderly	1.841	1.815	0	11.56			
	D_foreigners	24.065	20.998	0	92.52			
	D_no_studies	7.232	5.202	0	43.78			
	D_students	60.601	9.736	0	85.51			
	D_university	17.218	9.650	0	54.01			
	Ln_price	11.628	0.537	10.32	12.94			
	E_professional						41,533	78.5
	E_private						10,272	19.4
Market (E)	E_bank						1134	2.1
	E_vacant_dw	16.189	12.802	0	67.73			
	E_main_dw	57.053	26.879	9.49	100.00			
	E_secondary_dw	26.690	24.782	0	84.18			
	E_rented_dw	13.616	10.590	0	84.62			
	E_mortgaged_dw	38.996	16.930	3.70	96.15			
	E_homeownership	41.616	15.536	0	82.76			

2.5. Methodology

The analysis of variance allows for the contrasting of the null hypothesis that the means of K populations ($K > 2$) are equal, with the alternative hypothesis that at least one of the populations differs from the others in terms of its expected value (Equation (2)). The one-way analysis of variance consists of three parts. The first part of the analysis permits contrasting of the null hypothesis of equality of means in the groups through the F statistic. The second contrasts the equality of the variances of the dependent variable in the groups using Levene's test. The third and final part of the analysis determines which of the distinct levels of the factor differ from the others, based on different post hoc tests.

$$H_0 : \mu_1 = \mu_2 = \ldots = \mu_K = \mu$$
$$H_1 : \exists \mu_j \neq \mu; j = 1, 2, \ldots, K \tag{2}$$

The regression model is estimated using ordinary least squares (OLS), and its specification is semilogarithmic, based on the following expression:

$$ln(P_i) = \alpha + \sum_{j=1}^{n} \beta_j X_{ij} + \sum_{k=1}^{m} \gamma_k D_{ik} + \varepsilon_i \tag{3}$$

where $ln\,(P_i)$ is the natural logarithm of the advertised asking price for housing "i"; α is the fixed component, it does not depend on the market; β_j is the parameter to estimate related to the characteristic "j"; X_{ij} is the continuous variable that collects the characteristic "j" of the observation "i"; γ_k is the parameter to estimate related to the characteristic "k"; D_{ik} is the dummy variable that collects the characteristic "k" of the observation "i"; and ε_i is the error term in the observation "i".

The semilogarithmic functional form was selected, since according to [53,54], this form offers certain advantages. First, it facilitates the interpretation of the coefficients. That is, for each increase in unit of the explanatory variable (X_j and D_k), the dependent variable (P)—in this case, the asking price—varies on average ($100 \cdot \beta$). And second, it minimizes the problem of heteroscedasticity, improving the goodness of fit of the estimates.

The model is estimated on distinct occasions, based on the energy qualification characteristic (Table 3), such that the results obtained may be compared with other studies. For this analysis, the SPSS statistics package for Windows, version 24 was used [55], based on the method of "excluding cases listwise". This leads to the elimination of observations with missing data.

Table 3. Summary of estimates made based on the reference used, Ref.(qualification letter), and sample size.

Estimates	Energy Rating Variables	Final Sample
1	ABCDEFG/**Ref. NT**	52,939
2	A/**Ref. NT**	44,552
3	A/B/C/D/E/F/G/**Ref. NT**	52,939
4	A/B/C/**Ref. D**/E/F/G	9194
5	ABC/**Ref. D**/EFG	9194
6	AB/C/**Ref. D**/E/F/G	9194
7	A/B/C/D/E/F/**Ref. G**	9194
8	ABC/D/E/F/**Ref. G**	9194
9	AB/C/D/E/F/**Ref. G**	9194

3. Results

3.1. One-Way Analysis of Variance (ANOVA)

In the database created, from a sample of 52,939 homes, only 9194 published their energy qualifications (17.4%), despite the fact that Royal Decree 235/2013 [15] requires the publication of energy rating of homes that are being sold or rented. This low percentage leads us to believe that the failure to publish an energy qualification may have some sort of advantage for real estate sellers. In order to examine this supposition, a statistical test was created for a one-way analysis of variance (ANOVA), graphically revealing the data in Figure 6. By evaluating the homogeneity of the variance of each group using Levene's test ($F_{(7, 52931)} = 68.8$, $p = 0.000$), the variance of the groups is found to differ. This result supports the use of robust tests of equality of means, specifically, those by Welch [56] ($F_{(7, 2185)} = 314.2$, $p = 0.000$) and Brown-Forsythe [57] ($F_{(7, 3835)} = 237.6$; $p = 0.000$), which confirm that the mean asking prices differ between energy qualification letters.

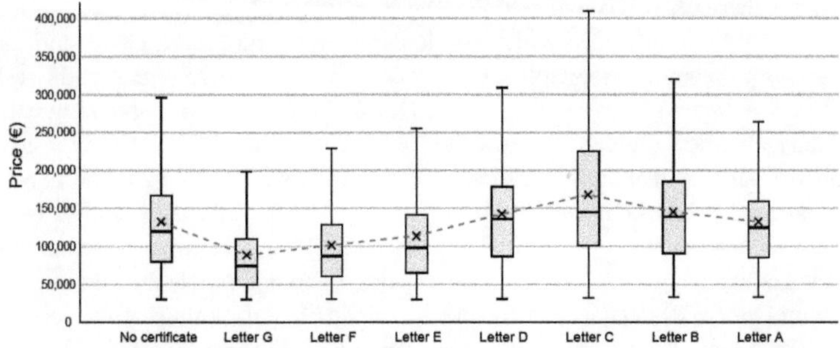

Figure 6. Graph representing the mean asking price and box plot, based on energy qualification.

To identify the relationships between the groups, a "post hoc" test was performed with Scheffé's method and a classification of the groupings (or homogenous subgroupings) based on the value of the means. For each subset, a test was carried out on the equality of means hypothesis, with significances of over 0.05 (no difference in means), in accordance with that observed in Table 4. Subset 1 is formed by housing with letters G and F, whose means do not differ significantly ($p = 0.055$). Subset 2 includes housing with letters F and E, whose means do not differ significantly ($p = 0.123$). Subset 3 is made up of homes with letters NT, A, D and B, whose means do not differ significantly ($p = 0.053$), and subset 4 consists of only those homes with letter C, which, obviously, do not differ from themselves ($p = 1.0$).

Table 4. Homogenous subsets of the mean asking prices by qualification letters.

Energy Rating	N	Subset for Alpha = 0.05			
		1	**2**	**3**	**4**
Letter G	3040	89,216.5			
Letter F	864	102,037.9	102,037.9		
Letter E	3083		113,680.5		
NT (no label)	43,745			132,699.2	
Letter A	807			133,694.0	
Letter D	587			142,756.2	
Letter B	325			145,554.3	
Letter C	488				168,049.7
Sig.		0.055	0.123	0.053	1.000

3.2. Regression Analysis

Upon introducing the variables in the regression model, problems of self-correction were observed between some of the same. Therefore, a total of eight variables have been discarded. Three correspond to the climate area (*Zone_B4*, *Zone_C3*, and *Zone_D3*), one is referred to the percentage of the population with primary and secondary school educations (*D_students*), two are referred to the percentages of vacant and main homes (*E_vacant_dw* and *E_main_dw*), and two more are referred to the percentages of mortgaged homes and properties with homeowners (*E_mortgaged_dw* and *E_homeownership*). Figure A1 in the Appendix A shows a graph with the most relevant correlations.

In order to determine if the estimates achieved suitable quality criteria, the following were examined: the normality of the population, the lack of problems of specification in the estimates (no multicollinearity, heteroscedasticity or autocorrelation), the statistical significance of the estimates, and finally, that the proportion of the estimated variance was high (R^2). The normality of the population is verified through a histogram (Figure 7a,d) and a graph of normality of the residuals (Figure 7b,e), revealing that the sample has a normal distribution. The multicollinearity was verified via the VIF statistic (Variance Inflation Factor), with various authors suggesting that there are collinearity problems if any VIF exceeds 10 [58,59]. In the new estimations made, the majority of the VIF values are between 1 and 4.6, therefore it is considered that there are no problems arising from multicollinearity. The heteroscedasticity was analyzed with a residual dispersion plot (Figure 7c,f) and there was no evidence of serious problems of heteroscedasticity, given the random distribution of the residuals. The existence of autocorrelation was verified using the Durbin-Watson statistic, obtaining values close to two in all of the estimations, which suggests the absence of autocorrelation in the residuals [60,61]. The significance of each estimation is measured with Snedecor's F-test, being found to be statistically significant. The coefficient of determination (adjusted R^2) of the estimates is indicated in Table 5 and all of these have an explanatory power approaching 71%. In summary, the estimations have a sufficient level of robustness and significance, making them acceptable for purposes of inference making. To control the fixed effects due to the spatial location of the data, the *comarcas* location variables are used. A positive spatial autocorrelation is detected with the Moran's I test [62,63] (residuals from estimation 3, $I = 0.274$, $z = 99.89$, $p < 0.001$; inverse distance squared, bandwidth 500 m), a common result in global regression models [64].

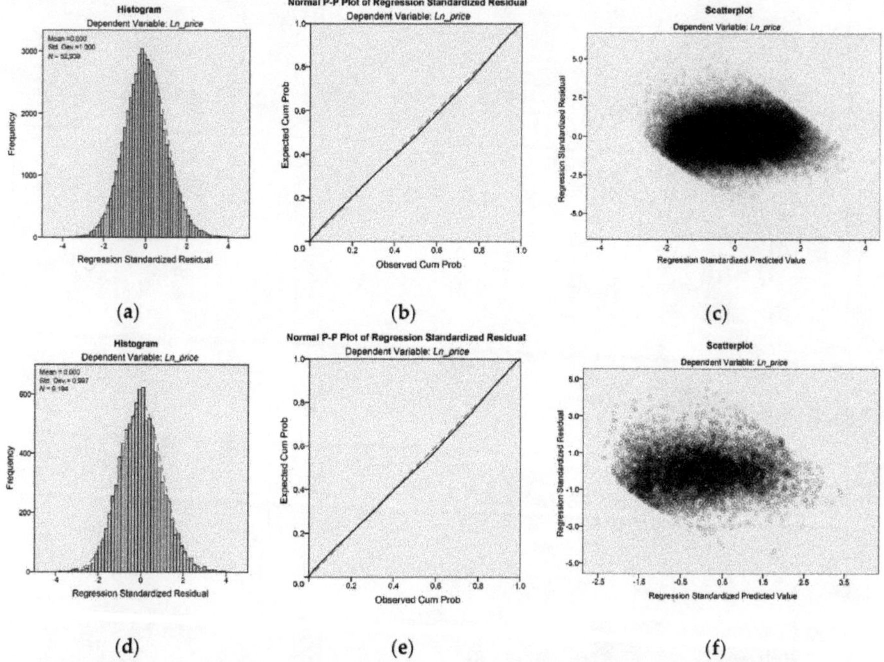

Figure 7. Graphs of the estimation 3: (**a**) histogram and normal curve of the standardized residual error; (**b**) P-P plot of residual normality; (**c**) scatter plot of the predicted values and standardized errors. Graphs of the estimation 4: (**d**) histogram and normal curve of the standardized residual error; (**e**) P-P plot of residual normality; (**f**) scatter plot of the predicted values and standardized errors.

Table 5. Results of the model estimations according to the energy characteristic.

Variable	Estimates								
	1	2	3	4	5	6	7	8	9
	ABCDEFG/Ref.NT	A/Ref.NT	A/B/C/D/E/F/G/Ref.NT	A/B/C/Ref.D/E/F/G	ABC/Ref.D/EFG	AB/C/Ref.D/E/F/G	A/B/C/D/E/F/Ref.G	ABC/D/E/F/Ref.G	AB/C/D/E/F/Ref.G
Intercept	10.099 *** (0.011)	10.082 *** (0.011)	10.102 *** (0.011)	10.225 *** (0.029)	10.216 *** (0.029)	10.228 *** (0.029)	10.137 *** (0.026)	10.131 *** (0.026)	10.141 *** (0.026)
	Dwelling characteristics (A)								
A_age	−0.0005 *** (0.0001)	0.0001 (0.0002)	−0.0005 *** (0.0001)	−0.0029 *** (0.0003)	−0.0029 *** (0.0003)	−0.0029 *** (0.0003)	−0.0029 *** (0.0003)	−0.0029 *** (0.0003)	−0.0029 *** (0.0003)
A_area_m2	0.0058 *** (0.0001)	0.0058 *** (0.0001)	0.0058 *** (0.0001)	0.0058 *** (0.0002)	0.0058 *** (0.0002)	0.0058 *** (0.0002)	0.0058 *** (0.0002)	0.0058 *** (0.0002)	0.0058 *** (0.0002)
A_bedrooms	−0.002 (0.002)	0.004 (0.003)	−0.002 (0.002)	−0.021 *** (0.006)	−0.021 *** (0.006)	−0.021 *** (0.006)	−0.021 *** (0.006)	−0.021 *** (0.006)	−0.021 *** (0.006)
A_bathrooms	0.217 *** (0.003)	0.214 *** (0.004)	0.216 *** (0.003)	0.224 *** (0.008)	0.225 *** (0.008)	0.225 *** (0.008)	0.224 *** (0.008)	0.225 *** (0.008)	0.225 *** (0.008)
A_wardrobe	0.013 *** (0.003)	0.015 *** (0.003)	0.012 *** (0.003)	0.002 (0.009)	0.003 (0.009)	0.002 (0.009)	0.002 (0.009)	0.002 (0.009)	0.002 (0.009)
A_air_cond	0.077 *** (0.003)	0.074 *** (0.003)	0.076 *** (0.003)	0.084 *** (0.008)	0.086 *** (0.008)	0.084 *** (0.008)	0.084 *** (0.008)	0.086 *** (0.008)	0.084 *** (0.008)
A_terrace	0.009 ** (0.003)	0.006 * (0.003)	0.008 ** (0.003)	0.028 *** (0.008)	0.029 *** (0.008)	0.028 *** (0.008)	0.028 *** (0.008)	0.029 *** (0.008)	0.028 *** (0.008)
A_floor	0.004 *** (0.001)	0.004 *** (0.001)	0.004 *** (0.001)	0.0002 (0.002)	0.0002 (0.002)	0.0001 (0.002)	0.0002 (0.002)	0.0002 (0.002)	0.0001 (0.002)
A_new_construction	0.224 *** (0.013)	0.224 *** (0.014)	0.222 *** (0.013)	0.208 *** (0.038)	0.191 *** (0.038)	0.189 *** (0.038)	0.208 *** (0.038)	0.190 *** (0.038)	0.189 *** (0.038)
A_state_to_reform	−0.151 *** (0.006)	−0.168 *** (0.007)	−0.152 *** (0.006)	−0.095 *** (0.013)	−0.094 *** (0.013)	−0.095 *** (0.013)	−0.095 *** (0.013)	−0.095 *** (0.013)	−0.095 *** (0.013)
A_good_condition (Ref.)									
A_flat (Ref.)									
A_studio_flat	−0.223 *** (0.013)	−0.214 *** (0.014)	−0.223 *** (0.013)	−0.236 *** (0.033)	−0.235 *** (0.033)	−0.233 *** (0.033)	−0.236 *** (0.033)	−0.236 *** (0.033)	−0.233 *** (0.033)
A_penthouse	0.108 *** (0.006)	0.112 *** (0.006)	0.108 *** (0.006)	0.095 *** (0.016)	0.091 *** (0.016)	0.094 *** (0.016)	0.095 *** (0.016)	0.091 *** (0.016)	0.094 *** (0.016)
A_duplex	0.044 *** (0.008)	0.049 *** (0.008)	0.044 *** (0.008)	0.013 (0.021)	0.010 (0.021)	0.012 (0.021)	0.013 (0.021)	0.010 (0.021)	0.012 (0.021)
ABCDEFG	−0.032 *** (0.004)								
ABC					−0.025 (0.015)			0.062 *** (0.011)	
AB						−0.046 ** (0.016)			0.041 ** (0.013)
A		−0.003 (0.010)	−0.002 (0.010)	−0.026 (0.017)			0.061 *** (0.014)		
B			−0.080 *** (0.016)	−0.091 *** (0.021)			−0.003 (0.019)		
C			0.049 *** (0.013)	0.014 (0.019)		0.015 (0.019)	0.101 *** (0.016)		0.102 *** (0.016)
D			0.033 ** (0.012)	Ref.	Ref.	Ref.	0.087 *** (0.014)	0.088 *** (0.014)	0.087 *** (0.014)
E			−0.043 *** (0.006)	−0.081 *** (0.014)		−0.081 *** (0.014)	0.006 (0.008)	0.006 (0.008)	0.006 (0.008)
F			−0.038 *** (0.010)	−0.080 *** (0.016)		−0.079 *** (0.016)	0.008 (0.012)	0.007 (0.012)	0.008 (0.012)
G			−0.053 *** (0.006)	−0.087 *** (0.014)		−0.087 *** (0.014)	Ref.	Ref.	Ref.
EFG					−0.083 *** (0.013)				

Table 5. *Cont.*

Variable	Estimates								
	1	2	3	4	5	6	7	8	9
	ABCDEFG/Ref.NT	A/Ref.NT	A/B/C/D/E/F/G/Ref.NT	A/B/C/Ref.D/E/F/G	ABC/Ref.D/EFG	AB/C/Ref.D/E/F/G	A/B/C/D/E/F/Ref.G	ABC/D/E/F/Ref.G	AB/C/D/E/F/Ref.G
NT (no label)	Ref.	Ref.	Ref.						
	Building characteristics (B)								
B_elevator	0.185 *** (0.003)	0.179 *** (0.004)	0.183 *** (0.003)	0.195 *** (0.008)	0.196 *** (0.008)	0.195 *** (0.008)	0.195 *** (0.008)	0.195 *** (0.008)	0.195 *** (0.008)
B_parking	0.111 *** (0.003)	0.111 *** (0.003)	0.110 *** (0.003)	0.102 *** (0.009)	0.102 *** (0.009)	0.101 *** (0.009)	0.102 *** (0.009)	0.102 *** (0.009)	0.101 *** (0.009)
B_storeroom	0.048 *** (0.003)	0.048 *** (0.003)	0.047 *** (0.003)	0.049 *** (0.009)	0.048 *** (0.010)	0.049 *** (0.010)	0.049 *** (0.009)	0.048 *** (0.010)	0.049 *** (0.010)
B_pool	0.092 *** (0.004)	0.093 *** (0.004)	0.092 *** (0.004)	0.094 *** (0.011)	0.093 *** (0.011)	0.093 *** (0.011)	0.094 *** (0.011)	0.092 *** (0.011)	0.093 *** (0.011)
B_garden	0.034 *** (0.004)	0.037 *** (0.004)	0.034 *** (0.004)	−0.002 (0.011)	−0.002 (0.011)	−0.002 (0.011)	−0.002 (0.011)	−0.002 (0.011)	−0.002 (0.011)
	Location characteristics (C)								
C_Alicante (Ref.)									
C_Marina_Alta	0.004 (0.005)	0.031 *** (0.006)	0.006 (0.005)	−0.097 *** (0.013)	−0.099 *** (0.013)	−0.097 *** (0.013)	−0.097 *** (0.013)	−0.099 *** (0.013)	−0.097 *** (0.013)
C_Marina_Baja	0.098 *** (0.005)	0.115 *** (0.006)	0.098 *** (0.005)	0.018 (0.014)	0.014 (0.014)	0.016 (0.014)	0.018 (0.014)	0.014 (0.014)	0.016 (0.014)
C_Bajo_Vinalopo	0.016 ** (0.005)	0.020 *** (0.005)	0.015 ** (0.005)	−0.004 (0.013)	−0.005 (0.013)	−0.003 (0.013)	−0.004 (0.013)	−0.004 (0.013)	−0.003 (0.013)
C_Bajo_Segura	−0.207 *** (0.005)	−0.211 *** (0.006)	−0.206 *** (0.005)	−0.194 *** (0.012)	−0.194 *** (0.012)	−0.194 *** (0.012)	−0.194 *** (0.012)	−0.194 *** (0.012)	−0.194 *** (0.012)
C_Condado	−0.164 *** (0.021)	−0.164 *** (0.023)	−0.164 *** (0.021)	−0.186 *** (0.056)	−0.192 *** (0.056)	−0.188 *** (0.056)	−0.186 *** (0.056)	−0.192 *** (0.056)	−0.188 *** (0.056)
C_Alcoy	−0.210 *** (0.010)	−0.214 *** (0.011)	−0.211 *** (0.010)	−0.193 *** (0.022)	−0.192 *** (0.022)	−0.191 *** (0.022)	−0.193 *** (0.022)	−0.192 *** (0.022)	−0.191 *** (0.022)
C_Alto_Vinalopo	−0.135 *** (0.016)	−0.116 *** (0.019)	−0.134 *** (0.016)	−0.155 *** (0.028)	−0.156 *** (0.028)	−0.156 *** (0.028)	−0.155 *** (0.028)	−0.156 *** (0.028)	−0.156 *** (0.028)
C_Medio_Vinalopo	−0.185 *** (0.009)	−0.186 *** (0.010)	−0.186 *** (0.009)	−0.196 *** (0.020)	−0.196 *** (0.020)	−0.196 *** (0.020)	−0.196 *** (0.020)	−0.196 *** (0.020)	−0.196 *** (0.020)
C_dist_pharmacy	−0.017 *** (0.003)	−0.021 *** (0.003)	−0.017 *** (0.003)	−0.005 (0.007)	−0.004 (0.007)	−0.005 (0.007)	−0.005 (0.007)	−0.004 (0.007)	−0.005 (0.007)
C_dist_health	0.008 *** (0.001)	0.007 *** (0.001)	0.008 *** (0.001)	0.017 *** (0.003)	0.016 *** (0.003)	0.017 *** (0.003)	0.017 *** (0.003)	0.016 *** (0.003)	0.017 *** (0.003)
C_dist_hospital	0.002 *** (0.000)	0.002 *** (0.000)	0.002 *** (0.000)	0.003 *** (0.001)	0.003 *** (0.001)	0.003 *** (0.001)	0.003 *** (0.001)	0.003 *** (0.001)	0.003 *** (0.001)
C_dist_educ1	0.027 *** (0.002)	0.026 *** (0.003)	0.026 *** (0.002)	0.021 *** (0.006)	0.021 *** (0.006)	0.021 *** (0.006)	0.021 *** (0.006)	0.021 *** (0.006)	0.021 *** (0.006)
C_dist_educ2	−0.021 *** (0.001)	−0.022 *** (0.001)	−0.021 *** (0.001)	−0.012 *** (0.003)	−0.012 *** (0.003)	−0.012 *** (0.003)	−0.012 *** (0.003)	−0.012 *** (0.003)	−0.012 *** (0.003)
C_coastalregion	0.146 *** (0.004)	0.140 *** (0.004)	0.145 *** (0.004)	0.166 *** (0.009)	0.165 *** (0.009)	0.165 *** (0.009)	0.166 *** (0.009)	0.165 *** (0.009)	0.165 *** (0.009)
C_FAR	−0.026 *** (0.002)	−0.027 *** (0.002)	−0.027 *** (0.002)	−0.018 *** (0.005)	−0.017 ** (0.005)	−0.018 *** (0.005)	−0.018 *** (0.005)	−0.017 *** (0.005)	−0.018 *** (0.005)
	Neighborhood characteristics (D)								
D_dependency	0.221 *** (0.008)	0.237 *** (0.009)	0.220 *** (0.008)	0.157 *** (0.021)	0.161 *** (0.021)	0.159 *** (0.021)	0.157 *** (0.021)	0.161 *** (0.021)	0.159 *** (0.021)
D_elderly	0.009 *** (0.001)	0.008 *** (0.001)	0.009 *** (0.001)	0.011 *** (0.002)	0.010 *** (0.002)	0.011 *** (0.002)	0.011 *** (0.002)	0.011 *** (0.002)	0.011 *** (0.002)
D_foreigners	0.001 *** (0.000)	0.001 *** (0.000)	0.001 *** (0.000)	0.001 *** (0.000)	0.001 *** (0.000)	0.001 *** (0.000)	0.001 *** (0.000)	0.001 *** (0.000)	0.001 *** (0.000)
D_no_studies	−0.008 *** (0.000)	−0.008 *** (0.000)	−0.008 *** (0.000)	−0.005 *** (0.001)	−0.005 *** (0.001)	−0.005 *** (0.001)	−0.005 *** (0.001)	−0.005 *** (0.001)	−0.005 *** (0.001)
D_university	0.009 *** (0.000)	0.009 *** (0.000)	0.009 *** (0.000)	0.008 *** (0.000)	0.009 *** (0.000)	0.009 *** (0.000)	0.008 *** (0.000)	0.009 *** (0.000)	0.009 *** (0.000)

Table 5. *Cont.*

Variable	Estimates								
	1	2	3	4	5	6	7	8	9
	ABCDEFG/Ref.NT	A/Ref.NT	A/B/C/D/E/F/G/Ref.NT	A/B/C/Ref.D/E/F/G	ABC/Ref.D/EFG	AB/C/Ref.D/E/F/G	A/B/C/D/E/F/Ref.G	ABC/D/E/F/Ref.G	AB/C/D/E/F/Ref.G
	Market characteristics (E)								
$E\_professional$	−0.028 *** (0.003)	−0.028 *** (0.004)	−0.025 *** (0.003)	−0.006 (0.010)	−0.002 (0.009)	−0.010 (0.010)	−0.006 (0.010)	−0.002 (0.009)	−0.010 (0.010)
$E\_private(Ref.)$									
E_bank	−0.003 (0.010)	−0.103 *** (0.029)	0.005 (0.010)	−0.006 (0.015)	−0.0004 (0.014)	−0.010 (0.015)	−0.006 (0.015)	−0.002 (0.015)	−0.010 (0.015)
$E\_secondary\_dw$	0.003 *** (0.000)	0.003 *** (0.000)	0,003 *** (0.000)	0.002 *** (0.000)	0.002 *** (0.000)	0.002 *** (0.000)	0.002 *** (0.000)	0.002 *** (0.000)	0.002 *** (0.000)
E_rented_dw	0.003 *** (0.000)	0.002 *** (0.000)	0.003 *** (0.000)	0.003 *** (0.000)	0.003 *** (0.000)	0.003 *** (0.000)	0.003 *** (0.000)	0.003 *** (0.000)	0.003 *** (0.000)
N	52,939	44,552	52,939	9194	9194	9194	9194	9194	9194
R^2	0.708	0.702	0.708	0.716	0.716	0.716	0.716	0.716	0.716
adj. R^2	0.708	0.702	0.708	0.715	0.714	0.715	0.715	0.714	0.715
Std. Error	0.29	0.29	0.29	0.30	0.30	0.30	0.30	0.30	0.30
F (sig.)	2979 ***	2442 ***	2621 ***	481 ***	523 ***	490 ***	481 ***	500 ***	490 ***
Durbin-Watson	1.89	1.91	1.89	1.79	1.79	1.79	1.79	1.79	1.79

Notes: dependent variable *Ln_price*; signification: *** $p < 0.001$, ** $p < 0.01$, * $p < 0.05$; standard errors in parentheses.

Currently, studies carried out have been based on distinct scenarios and the literature has revealed a certain diversity in terms of determining the letter (or set of letters) of reference for measuring and comparing the impact of energy qualifications on housing prices (Figure 8). This circumstance hinders comparisons between the premium resulting from going from one value to another within the ABCDEFG qualification scale. To facilitate comparison between studies, some authors have recommended that letters not be grouped and that D be considered the letter of reference [32,65], since it is in the middle of the scale.

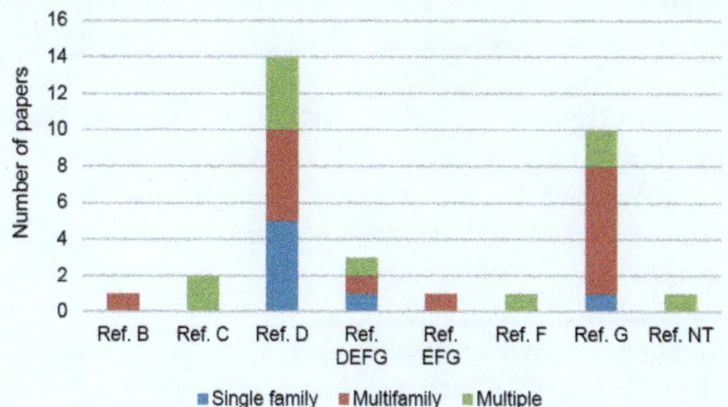

Figure 8. Letters of reference used to estimate the price premium (graph elaborated from 20 papers).

Based on this background, the results of the model obtained for the estimate 4 are presented, since it is the one that complies with the recommendations of using letter D as a reference. As for the characteristics of the housing, the model estimates that for each additional year (in terms of age of the housing), the asking price will be reduced a mean of 0.29%. As for size, the estimated impact implies that an increase of one square meter in surface area results in a 0.58% price increase, whereas the addition of another bedroom leads to a reduction of 2.09%. However, an additional bathroom represents a mean price increase of 22.44%. If the property has extras, such as built-in wardrobe, air conditioning or a terrace, the mean impact on prices estimated by the model is 0.25%, 8.42% and 2.81%, respectively. The results reveal that a home situated on an additional story has a price increase of 0.02%. Using as a reference a second-hand home in good state, the model estimates a mean discount in the asking price of 9.49% for a second-hand home that needs renovation. On the other hand, if the home is a new construction, the results reveal a price increase of 20.78%. Within the typology of homes and using apartments as the reference, a duplex or attic apartment has a price increase of 1.30% and 9.45%, respectively, whereas studio apartments have a discount of 23.56%.

The values obtained in the estimation of the parameters related to the building characteristics, such as having an elevator, garage, storage space or swimming pool, imply a mean price increase of 19.48%, 10.18%, 4.88% and 9.38%, respectively. On the other hand, having a garden has a contrary effect, leading to a mean price reduction of 0.16% (not significant).

As for characteristics related to location, for properties situated in neighborhoods with a higher gross development (those in which there are more homes per sector surface area), the model estimates a mean reduction in price of approximately 1.80%. As for the geographic distances, they are all statistically significant, except for the distance to pharmacies. The results reveal that for each kilometer that the housing is distanced from pharmacies or level 2 schools (secondary and high schools), the price decreases by 0.50% and 1.18%. The opposite occurs when the housing is distanced from health centers, hospitals and level 1 schools (infant and primary schools). Homes that are in coastal towns have a price increase of 16.57%. Finally, the estimated impact on prices of homes situated in the Marina Baja district have an increase of 1.81% with regard to the reference district (Alicante). As for the rest of the districts, the effect that is estimated by the model implies a reduction in asking prices, reaching discounts of between 15% and 20% in interior and southern districts of the province [66,67].

As for the neighborhood characteristics, an increase of 1% was found for dependency and ageing, implying an increase in sales price of 0.16% and 0.01%, respectively. As for the percentage of foreigners or the percentage of individuals with university studies, an increase of 1% for these variables implies a 0.12% and 0.85% price increase, respectively. On the other hand, with a 1% increase in the percentage of the population without an education, there is a price reduction of 0.54%.

As for market characteristics, the model estimations reveal increases in prices in areas having a higher percentage of homes in rent and secondary homes, at 0.32% and 0.21%, respectively. The sale of homes indicates that when the properties are sold by professionals or banks, the price is reduced by a mean of 0.63% and 0.62%, respectively, with these values not being statistically significant.

As for the characteristics having a greater impact on asking prices, the five variables from the estimates having the greatest explanatory power, according to the standardized beta coefficients (not included due to problems of extension) are: (A) housing characteristics—constructed surface area and number of bathrooms; (B) building characteristics—having an elevator; and (C) location characteristics—percentage of individuals with university studies and being situated in a coastal town.

As for the energy qualification, the results for the entire sample are summarized in Figure 9, where it is observed that the housing with any qualification type (ABCDEFG grouping) and homes with high energy qualifications (letter A) had lower prices, respectively, 3.22% and 0.30% lower. Estimate 3 reveals that housing with high qualifications (letters A and B) do not have better economic premiums than other homes with lower qualifications or those that have not published their qualifications. This suggests that by not publishing the energy qualification, sellers may ask for higher prices than those asked for other homes with lower qualifications (E, F or G).

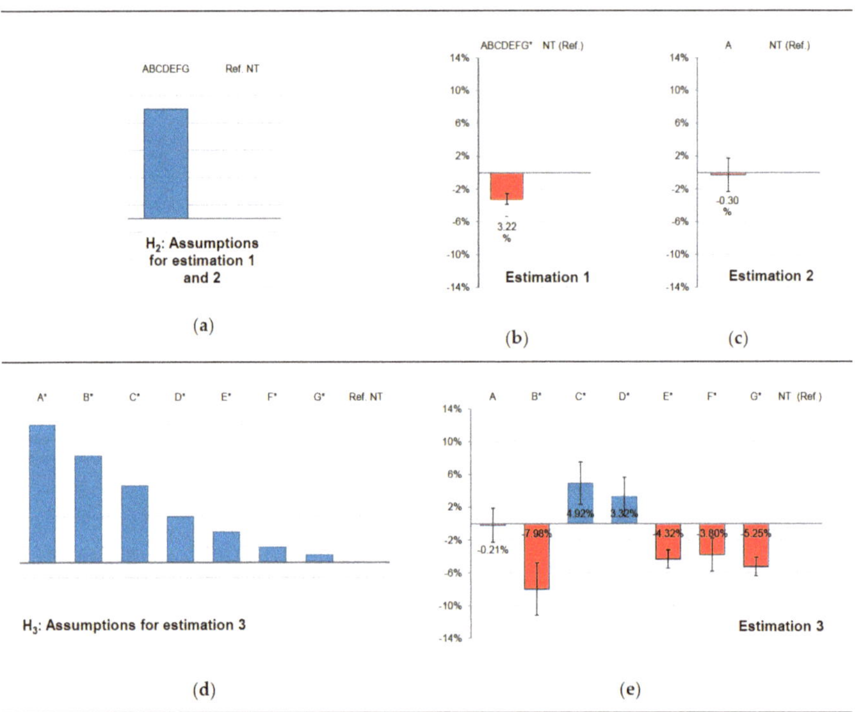

Figure 9. Bar graph with the asking price premiums (%) and CI (95%) for estimations 1 to 3 of the model. (**a**) Assumptions for estimations 1 and 2; (**b**) estimation 1; (**c**) estimation 2; (**d**) Assumptions for estimation 3; (**e**) estimation 3. Note: * statistically significant result.

Estimations 3 to 9 are carried out with the sample of homes that published their energy qualifications (Figures 9 and 10). If letter D is used as the reference (estimations 4, 5 and 6), it is observed that letters A, B (and the AB grouping) do not have better premiums than those of letters C and D. In the case of homes qualified as E, F or G, they have very similar negative premiums, a decrease of approximately 8%. If adopting the letter G as a reference (estimations 7, 8 and 9), the positive price premiums for letters C and D are of special note, as well as the similarity of prices for the lower qualifications (E, F and G).

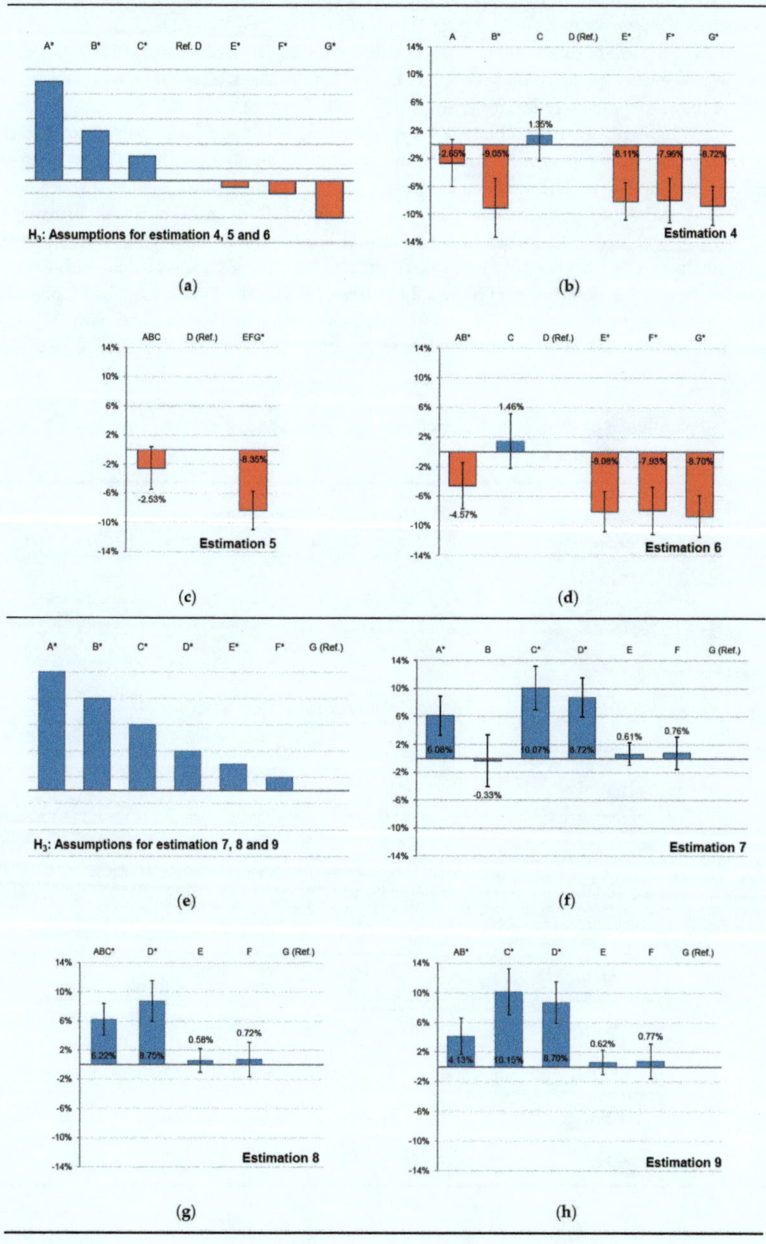

Figure 10. Bar graphs with asking price premiums (%) and CI (95%) for estimations 4 to 9 of the

model. (**a**) Assumptions for estimations 4, 5 and 6; (**b**) estimation 4; (**c**) estimation 5; (**d**) estimation 6; (**e**) Assumptions for estimates 7, 8 and 9; (**f**) estimation 7; (**g**) estimation 8; (**h**) estimation 9. Note: * statistically significant result.

4. Discussion

The results of the one-way analysis of variance (ANOVA) support the first hypothesis (H$_1$), since it reveals that the mean of the asking prices for the properties that do not publish their energy qualification (NT) are similar to those of homes with higher ratings, letters A, B or D (subset 3). Therefore, the sellers of these properties have no interest in publishing the qualification letter. These results are in line with other works [16,31]. It is very likely that homes hiding the energy qualification have letters E, F or G, since this segment represents 86% of the labelled building stock of the autonomous community [68].

The results of the regression model obtained from estimations 1, 2 and 3 are contradictory to the second hypothesis (H$_2$) proposed in this document. Homes with an energy qualification—ABCDEFG grouping (estimation 1)—or a high qualification—letter A (estimation 2)—have a discount in price of 3.22% or 0.30%, respectively, with respect to the homes that do not publish their energy qualification (NT). In addition, if comparing estimations 1 and 2 with the results obtained by Cespedes-Lopez, et al. [32], it is observed that having an energy qualification, as compared to not having one, does not have a positive effect on the asking price, as it does in Europe, in general (2.32%). In estimation 3, it is observed that homes that do not publish their energy qualifications have higher prices than those qualified with letters B, E, F and G. This estimation may be compared with [22] (see Table 6), where a positive impact was found on prices for the high qualifications (A and B), and a decrease of between 0.8 and 1.6% for the poorer qualified (D, E, F and G).

Table 6. Comparative of premiums in % from this study (estimations 4 to 9) with studies published in Europe on energy qualifications.

Study Data			Economic Price Premium in % According to the Reference-Ref.-(Certificate Letter or Group of Letters)												Compared with
Paper	Country	Estim.	ABC	AB	A	B	C	D	E	F	G	FG	EFG	NT	
[21]	Netherlands	2 (Table 3)			10.2*	5.6*	2.2*	Ref.	−0.5	−2.5*	−5.1*				Estim. 4
[69]	Portugal	2 (Table 5)	5.94*					Ref.					−4.03*		Estim. 5
[25]	United Kingdom	4 (Table 4)		1.6*			0.8*	Ref.	−1.4*	−2.9*	−7.2*				Estim. 6
[26]	United Kingdom	7 (Table 2)		3.6			3.9	Ref.	−8.2	−10.5	−15.0				Estim. 6
[18]	Denmark	3 (Table 1)		6.6*			0.2	Ref.	−1.5*	−3.5*	−9.3*				Estim. 6
		4 (Table 1)		6.2*			5.1*	Ref.	−5.4*	−12.9*	−24.3*				Estim. 6
[23]	Ireland	1 (Table 4)			9.3*	5.2*	1.7*	Ref.	−0.4			−10.6*			-
[19]	France	Occitainie			14.0*		3.0*	Ref.	−4.0*			−6.0*			-
[20]	France	Provence			6.0*		2.0*	Ref.	−3.0*			−10.0*			-
[70]	Germany	8 (Table II)			-	0.76*	0.65*	0.75*	0.87*	0.30*	Ref.				-
[24]	Italy	1 (Table 2)			21.9*	20.2*	17.4*	17.1*	9.5*	2.3	Ref.				Estim. 7
[31]	Spain	5BCN			10.0*	-	6.0*	7.0*	2.0	1.0*	Ref.				Estim. 7
		5VIC			29.0*	-	18.0	16.0*	4.0*	−2.0	Ref.				Estim. 7
		5ALC			8.0*	-	−23.0*	2.0	−5.0*	−5.0*	Ref.				Estim. 7

Table 6. *Cont.*

| Study Data | | | Economic Price Premium in % According to the Reference-Ref.-(Certificate Letter or Group of Letters) | | | | | | | | | | | | Compared with |
Paper	Country	Estim.	ABC	AB	A	B	C	D	E	F	G	FG	EFG	NT	
[16]	Spain	3B (Table 4)			9.62 *	-	−3.0	3.87 *	2.0	1.0	Ref.				Estim. 7
[17]	Spain	3 (Table 3)	−6.3 *					1.9	1.1 *	1.8 *	Ref.				Estim. 8
[27]	United Kingdom	5 (Table 5)		11.6 *			10.4 *	9.3 *	8.0 *	5.6 *	Ref.				Estim. 9
[22]	Netherlands	4 (Table 2)			5.6 *	1.1 *	−0.2	−0.8 *	−1.4 *	−1.6 *	−0.8			Ref.	Estim. 3
This research	Spain	3			−0.2	−8.0 *	4.9 *	3.3 *	−4.3 *	−3.8 *	−5.3 *			Ref.	[22]
		4			−2.7	−9.1 *	1.4	Ref.	−8.1 *	−8.0 *	−8.7 *				[21]
		5	−2.5					Ref.					−8.4 *		[69]
		6		−4.6 *			1.5	Ref.	−8.1 *	−7.9 *	−8.7 *				[18,25,26]
		7			6.1 *	−0.3	10.1 *	8.7 *	0.6	0.8	Ref.				[16,24,31]
		8	6.2 *					8.8 *	0.6	0.7	Ref.				[17]
		9		4.1 *			10.2 *	8.7 *	0.6	0.8	Ref.				[27]

Note: * indicates that the coefficient is statistically significant.

Results of the regression model for estimations 4 to 9 (Figure 10) are contrary to the third hypothesis (H_3). For estimations 4, 5 and 6 (ref. D) a negative sign was anticipated for letters below the reference letter, and a positive impact for letters above the reference letters (Figure 10a). Estimations 7, 8 and 9 (ref. G) anticipated that the premium of letter A would be positive and the sign of the subsequent letters would also be positive and with a decreasing impact on the prices until reaching the reference letter (Figure 10e). However, upon comparing the expected and obtained results, it can be seen that they do not comply with the initially proposed hypothesis (H_3), since homes with better qualifications do not have better premiums than those with poorer qualifications. For housing with high qualifications—letters A, B and the AB and ABC groupings—a discount was obtained with respect to the reference housing –letter D– (estimations 4, 5 and 6). For estimations 7, 8 and 9, it is seen that housing qualified as C and D are the best valued of the market segment, with the highest premiums as compared to the reference G.

If comparing these results with those obtained from other studies (Table 6), it may be observed that, in general, housing with higher qualifications have higher sales prices. For example, estimation 6 from this study may be compared to the results of [18,25,26], where high qualifications (the AB grouping) obtained a positive price premium, and for the housing with low qualifications, the premiums are negative and decrease as the qualification decreases.

5. Conclusions

This work seeks to examine the effect of energy qualifications on the asking price of housing located in the Alicante (Spain) real estate market. To do so, a database was constructed based on 52,939 observations, of which 9194 offered information on energy qualifications (17.4%). The information contained in the database has permitted the creation of 63 variables that are used to estimate the regression model. In order to compare the results of this work with those of other studies, the model has been estimated 9 times.

The first objective attempts to determine if an interest exists in not publishing energy qualification information for homes being sold. Two initial hypotheses are proposed—H_1 and H_2. The one-way analysis of variance (ANOVA) reveals that the H_1 hypothesis is supported, since hiding the qualification may lead to higher asking prices. The second proposed hypothesis—H_2—attempts to contrast whether or not energy qualification is a determining characteristic of the asking price. Estimations 1 and 2 reveal that this hypothesis is rejected, since homes with an energy qualification—ABCDEFG grouping—or those having a high qualification—letter A—as compared to those that did not publish their rating —NT—, have a negative premium of 3.2% (significant) and of 0.3% (not sig.) respectively.

The second objective proposes quantifying the economic impact of energy qualification on the asking price, offering the hypothesis—H_3. Estimations 4 to 9 suggest the contrary, finding that housing qualified with letters C and D have higher premiums than housing with higher qualifications (A or B).

This study reveals that, in multi-family housing sold in the Alicante (Spain) province, a positive relationship does not exist between the energy certification system and the housing's asking price. This is due to a variety of reasons.

- First, real estate sellers and owners who do not publish energy qualifications offer their homes at prices that are similar to those having high qualifications.
- Second, there is the lack of sanctions placed by the public administration on companies, owners and real estate portals that do not publish the energy qualifications of the housing that is for sale or rent, motivating owners to not publish the letter and generating distorted asking prices for the housing. Therefore, it is important for the administration to closely supervise compliance with regulations and assign the necessary resources to local authorities to ensure said compliance, and if needed, to impose sanctions.
- Third, owners are not interested in improving energy qualifications, since, according to [71,72] there is no compensation for the additional investment needed to improve this qualification. And fourth and finally, the current regulations for housing only require that these homes obtain energy qualification if they are going to sell, rent or publish. However, there is no obligation to obtain a minimum qualification, so the improved energy performance of the homes is not encouraged [73,74].
- Fifth and finally, society's perception of EPC is negative, as revealed by several studies relying on surveys completed by professional real estate agents [74,75] or energy certifiers [73]. Regardless, these studies suggest that the main criteria used to select a home is price and location [74–76].

Currently, both nationally and regionally, economic incentives exist in order to offer value to housing with higher qualifications and to promote renovation. On a national level, the PAREER II (2014–2020) [77] program was financed with 204 million euros. Regionally speaking (Valencian community), there are distinct plans such as RENHATA [78], which intends to offer 4.95 million euros between 2020–2021 to improve the preservation of housing, accessibility and energy rehabilitation. Given that in Spain, there are over 9.5 million buildings, it is unlikely that a country's building stock will be renewed thanks to public budgets. Therefore, it will be necessary to rely on private initiatives, based on market incentives (higher sales prices, higher rents) that encourage investments in energy renewal of buildings. In this way, not only would property owners directly benefit from these renewals, the entire population would also receive benefits. This would ensure a more sustainable and environmentally respectful building stock, helping to create cleaner cities and an improved quality of life.

Author Contributions: Conceptualization, M.-F.C.-L. and R.-T.M.-G.; methodology, M.-F.C.-L., R.-T.M.-G., V.R.P.-S. and P.M.-C.; formal analysis, investigation, data curation, and writing—original draft preparation, M.-F.C.-L., R.-T.M.-G. and V.R.P.-S.; writing—review and editing, M.-F.C.-L., R.-T.M.-G., V.R.P.-S. and P.M.-C. All authors have read and agreed to the published version of the manuscript.

Funding: This research and the ACP were partially funded by Universidad de Alicante (http://dx.doi.org/10.13039/100009092).

Acknowledgments: To the Spanish General Council of Technical Architecture (http://www.arquitectura-tecnica.com) for the granting of a subsidy to promote the doctorate training in the course year 2018/19, granted to the author, M.-F.C.-L. The contents of this publication are the exclusive responsibility of the authors and do not necessarily reflect the opinion of the Universidad de Alicante or the Spanish General Council of Technical Architecture.

Conflicts of Interest: The authors declare no conflict of interest.

Appendix A

Table A1. Principales indicadores del sector inmobiliario para España y la provincia de Alicante.

Block	Indicator	España	Alicante	Source
Use of the housing (2011)	Main housing	18,083,692 (71.7%)	738,367 (58.0%)	[36]
	Secondary housing	3,681,565 (14.6%)	326,705 (25.6%)	
	Empty housing	3,443,365 (13.7%)	209,024 (16.4%)	
Tenancy regime (2011)	Ownership	14,274,987 (85.4%)	613,626 (88.6%)	[36]
	Rental	2,438,574 (14.6%)	79,165 (11.4%)	
Housing real estate transactions (2019)	Total number of transactions	569,993	42,418	[34]
	Residents in Spain (Spaniards)	474,102 (83.2%)	21,815 (51.4%)	
	Residents in Spain (foreigners)	91,668 (16.1%)	19,447 (45.8%)	
Type of building (2019)	Single-family home	5,945,000 (32.1%)	299,487 (22.9%)	[36]
	Collective housing	12,591,200 (67.9%)	1,006,882 (77.1%)	
Availability of heating (2011)	Collective or central	10.56%	5.26%	[36]
	Individual	46.30%	24.87%	
	Without installation, but with some device permitting heating of a room	29.48%	54.83%	
	Without heating	13.66%	15.04%	

Table A2. Variables used by other authors for the determination of the price of housing. Own elaboration from [66].

Category	Characteristics	References
Dwelling characteristics (A)	Dwelling typology	[17,21,23,26,44,79–88]
	Age of the dwelling	[17,26,45,46,79,82–112]
	Dwelling surface area	[16,17,21,28,44–46,80,82–88,90–93,95–101, 103–108,110,112–124]
	Number of bedrooms	[17,21,23,26,28,45,80,81,84,85,87,88,93,96,103, 113,116,125,126]
	Number of bathrooms	[16,17,24,81,84,87,93,100,113,115,121]
	Floor of the dwelling	[16,17,28,44,85,91,99–101,104,110,113,121,124]
	Terrace	[45,87,93,113,118]
	Wardrobe	[105,123]
	State of conservation	[21,45,46,82,86–88,96,123]
Features of the building (B)	Garage slot	[24,45,80,82,83,87,93,100,106–108,113,115, 120–123]
	Elevator	[16,80,82,86,87,92,93,105,118,123]
	Swimming pool in the building	[16,80,83,86,93,112,113,120,122–124]
Characteristics of the location (C)	Location within the territory or the city	[17,24,28,80,82,93,97,98,100,101,103,106–108, 111–114,120,122,125,127]

Table A2. *Cont.*

Category	Characteristics	References
Characteristics of the neighbourhood (D)	Age of the population	[44,45]
	Number of Foreigners	[44,45,47,86,87,104,117]
	Level of studies	[16,44,47,85,89,116,128]
Market, occupation and sale characteristics (E)	Price	In all studies this is the dependent variable
	Use of the dwelling	[17,28,96,118]
	Housing tenure	[26]

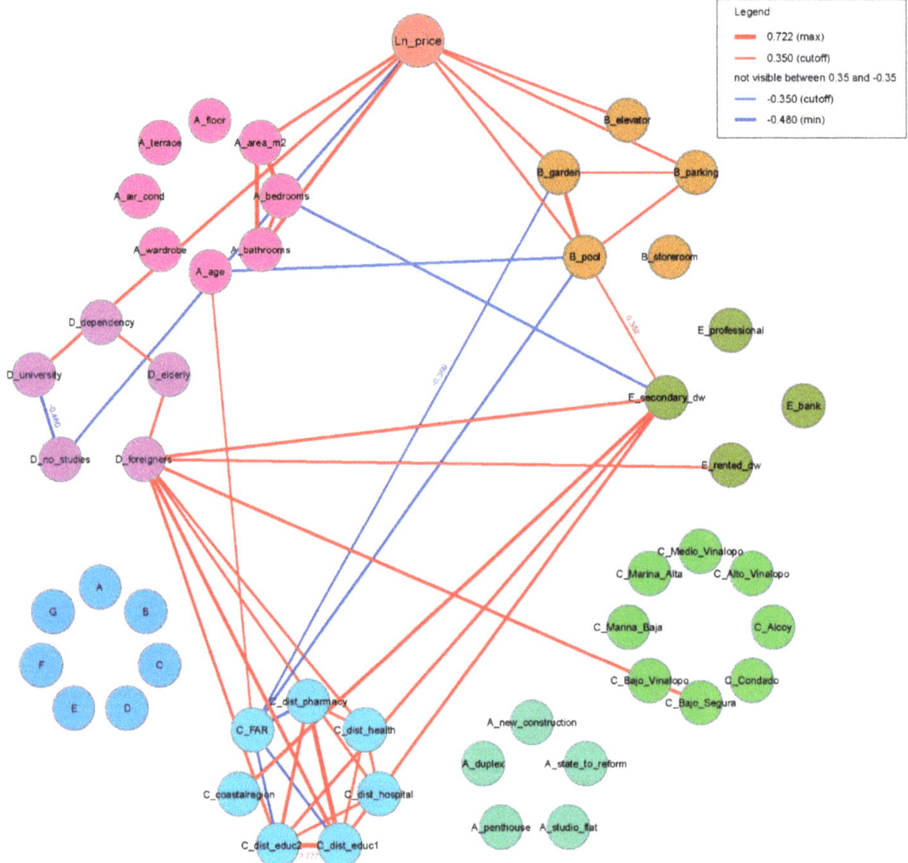

Figure A1. Correlation between the characteristics of the properties—independent variables—and the asking prices. Only correlations greater than 0.35 (in absolute value) are shown.

References

1. Hirsch, J.; Lafuente, J.J.; Spanner, M.; Geiger, P.; Haran, M.; McGreal, S.; Davis, P.T.; Recourt, R.; de la Paz, P.T.; Perez-Sanchez, V.R.; et al. *Stranding Risk & Carbon. Science-Based Decarbonising of the EU Commercial Real Estate Sector*; CRREM Report No.1; IIÖ Institut für Immobilienökonomie GmbH: Wörgl, Austria, 2019; p. 143. Available online: https://www.crrem.eu/wp-content/uploads/2019/09/CRREM-Stranding-Risk-Carbon-Science-based-decarbonising-of-the-EU-commercial-real-estate-sector.pdf (accessed on 3 September 2019).

2. The European Parliament and the Council of the European Union. Directive 2002/91/EC of the European Parliament and of the Council of 16 December 2002 on the energy performance of buildings. *Off. J. Eur. Communities* **2003**, *46*, 7.

3. The European Parliament and the Council of the European Union. Directive 2010/31/EU of the European Parliament and of the Council of 19 May 2010 on the energy performance of buildings. *Off. J. Eur. Communities* **2010**, *53*, 23.

4. The European Parliament and the Council of the European Union. Directive 2012/27/EU of the European Parliament and of the Council of 25 October 2012 on energy efficiency, amending Directives 2009/125/EC and 2010/30/EU and repealing Directives 2004/8/EC and 2006/32/EC. *Off. J. Eur. Union* **2012**, *55*, 56.

5. The European Parliament and the Council of the European Union. Directive (EU) 2018/844 of the European Parliament and of the Council of 30 May 2018 amending Directive 2010/31/EU on the energy performance of buildings and Directive 2012/27/EU on energy efficiency. *Off. J. Eur. Union* **2018**, *61*, 17.

6. Mudgal, S.; Lyons, L.; Cohen, F.; Lyons, R.C.; Fedrigo-Fazio, D. *Energy Performance Certificates in Buildings and Their Impact on Transaction Prices and Rents in Selected EU Countries*; Bio Intelligence Service: Paris, France, 2013; p. 158. Available online: https://ec.europa.eu/energy/sites/ener/files/documents/20130619-energy_performance_certificates_in_buildings.pdf (accessed on 5 January 2018).

7. European Commission. A Programme to Deliver Energy Certificates for Display in Public Buildings across Europe within a Harmonising Framework (EPLABEL). Available online: https://ec.europa.eu/energy/intelligent/projects/en/projects/eplabel (accessed on 9 October 2019).

8. European Commission. Applying the EPBD to Improve the ENergy PErformance Requirements to EXISTing Buildings (ENPER EXIST). Available online: https://ec.europa.eu/energy/intelligent/projects/en/projects/enper-exist (accessed on 9 October 2019).

9. European Commission. Social Housing Action to Reduce Energy Consumption (SHARE). Available online: https://ec.europa.eu/energy/intelligent/projects/en/projects/share (accessed on 9 October 2019).

10. European Commission. Assessment and Improvement of the EPBD Impact (for New Buildings and Building Renovation) (ASIEPI). Available online: https://ec.europa.eu/energy/intelligent/projects/en/projects/asiepi (accessed on 9 October 2019).

11. European Commission. Incentives through Transparency: European Rental Housing Framework for Profitability Calculation of Energetic Retrofitting Investments (RentalCal). Available online: https://cordis.europa.eu/project/id/649656 (accessed on 9 October 2019).

12. European Commission. RESPOND: Integrated Demand REsponse Solution towards Energy POsitive NeighbourhooDs. Available online: https://cordis.europa.eu/project/id/768619 (accessed on 9 October 2019).

13. European Commission. Carbon Risk Real Estate Monitor-Framework for Science Based Decarbonisation Pathways, Toolkit to Identify Stranded Assets and Push Sustainable Investments (CRREM). Available online: https://cordis.europa.eu/project/id/785058 (accessed on 9 October 2019).

14. Ministerio de la Presidencia. Real Decreto 47/2007, de 19 de enero, por el que se aprueba el Procedimiento básico para la certificación de eficiencia energética de edificios de nueva construcción. *Bol. Estado Madr.* **2007**, *27*, 4499–4507.

15. Ministerio de la Presidencia. Real Decreto 235/2013, de 5 de abril, por el que se aprueba el procedimiento básico para la certificación de la eficiencia energética de los edificios. *Bol. Estado Madr.* **2013**, *89*, 27548–27562.

16. Marmolejo-Duarte, C. The incidence of the energy rating on residential values: An analysis for the multifamily market in Barcelona. *Inf. Constr.* **2016**, *68*, e156. [CrossRef]

17. Taltavull-de-la-Paz, P.; Perez-Sanchez, V.R.; Mora-Garcia, R.T.; Perez-Sanchez, J.C. Green Premium Evidence from Climatic Areas: A Case in Southern Europe, Alicante (Spain). *Sustainability* **2019**, *11*, 686. [CrossRef]

18. Jensen, O.M.; Hansen, A.R.; Kragh, J. Market response to the public display of energy performance rating at property sales. *Energy Policy* **2016**, *93*, 229–235. [CrossRef]

19. Notaries-France. *La Valeur Verte des Logements en 2016*; Étude Statistiques Immobilières: France Métropolitaine, France, 2017; p. 4. Available online: https://immobilier.statistiques.notaires.fr/sites/default/contrib/valeur%20_verte.pdf (accessed on 14 March 2019).

20. Notaries-France. *La Valeur Verte des Logements en 2017*; Étude Statistiques Immobilières: France Métropolitaine, France, 2018; p. 11. Available online: https://www.notaires.fr/sites/default/files/Valeur%20verte%20-%20Octobre%202018.pdf (accessed on 14 March 2019).

21. Brounen, D.; Kok, N. On the economics of energy labels in the housing market. *J. Environ. Econ. Manag.* **2011**, *62*, 166–179. [CrossRef]
22. Chegut, A.; Eichholtz, P.; Holtermans, R. Energy efficiency and economic value in affordable housing. *Energy Policy* **2016**, *97*, 39–49. [CrossRef]
23. Hyland, M.; Lyons, R.C.; Lyons, S. The value of domestic building energy efficiency: Evidence from Ireland. *Energy Econ.* **2013**, *40*, 943–952. [CrossRef]
24. Bonifaci, P.; Copiello, S. Price premium for buildings energy efficiency: Empirical findings from a hedonic model. *Valori Valutazioni* **2015**, *14*, 5–15.
25. Fuerst, F.; McAllister, P.; Nanda, A.; Wyatt, P. Does energy efficiency matter to home-buyers? An investigation of EPC ratings and transaction prices in England. *Energy Econ.* **2015**, *48*, 145–156. [CrossRef]
26. Fuerst, F.; McAllister, P.; Nanda, A.; Wyatt, P. Energy performance ratings and house prices in Wales: An empirical study. *Energy Policy* **2016**, *92*, 20–33. [CrossRef]
27. Fuerst, F.; McAllister, P.; Nanda, A.; Wyatt, P. *An Investigation of the Effect of EPC Ratings on House Prices*; 13D/148; Department of Energy & Climate Change: London, UK, 2013; p. 41. Available online: https://www.gov.uk/government/publications/an-investigation-of-the-effect-of-epc-ratings-on-house-prices (accessed on 30 November 2019).
28. De Ayala, A.; Galarraga, I.; Spadaro, J.V. The price of energy efficiency in the Spanish housing market. *Energy Policy* **2016**, *94*, 16–24. [CrossRef]
29. Stanley, S.; Lyons, R.C.; Lyons, S. The price effect of building energy ratings in the Dublin residential market. *Energy Effic.* **2016**, *9*, 875–885. [CrossRef]
30. Olaussen, J.O.; Oust, A.; Solstad, J.T. Energy performance certificates–Informing the informed or the indifferent? *Energy Policy* **2017**, *111*, 246–254. [CrossRef]
31. Marmolejo-Duarte, C.; Chen, A. The impact of EPC rankings on the Spanish residential market: An analysis for Barcelona, Valence and Alicante. *Ciudad Territ. Estud. Territ.* **2019**, *51*, 101–118.
32. Cespedes-Lopez, M.F.; Mora-Garcia, R.T.; Perez-Sanchez, V.R.; Perez-Sanchez, J.C. Meta-Analysis of Price Premiums in Housing with Energy Performance Certificates (EPC). *Sustainability* **2019**, *11*, 6303. [CrossRef]
33. Banco de España. El mercado de la vivienda en España entre 2014 y 2019. *Doc. Ocas.* **2020**, *2013*, 55.
34. MITMA, Ministerio de Transportes, Movilidad y Agenda Urbana. Transacciones Inmobiliarias (Compraventa). Available online: https://apps.fomento.gob.es/BoletinOnline2/?nivel=2&orden=34000000 (accessed on 20 February 2020).
35. Ministerio de Vivienda. Real Decreto 314/2006, de 17 de marzo, por el que se aprueba el Código Técnico de la Edificación. *Bol. Estado Madr.* **2006**, *74*, 11816–11831.
36. INE, Instituto Nacional de Estadística. Censo de Población y Vivienda de. 2011. Available online: https://www.ine.es/censos2011_datos/cen11_datos_resultados.htm (accessed on 10 October 2019).
37. Lancaster, K.J. A new approach to consumer theory. *J. Polit. Econ.* **1966**, *74*, 132–157. [CrossRef]
38. Ridker, R.G.; Henning, J.A. The determinants of residential property values with special reference to air pollution. *Rev. Econ. Stat.* **1967**, *49*, 246–257. [CrossRef]
39. Zietz, J.; Zietz, E.N.; Sirmans, G.S. Determinants of house prices: A quantile regression approach. *J. Real Estate Financ. Econ.* **2008**, *37*, 317–333. [CrossRef]
40. Sirmans, G.S.; Macpherson, D.A.; Zietz, E.N. The composition of hedonic pricing models. *J. Real Estate Lit.* **2005**, *13*, 3–43.
41. MITMA, Ministerio de Transportes, Movilidad y Agenda Urbana. Vivienda Libre. Series Anuales. 3.1. Número de Viviendas Libres Iniciadas. Available online: https://apps.fomento.gob.es/BoletinOnline2/?nivel=2&orden=32000000 (accessed on 21 October 2019).
42. Hair, J.F.; Black, W.C.; Babin, B.J.; Anderson, R.E. *Multivariate Data Analysis*, 7th ed.; Pearson Education Limited: Harlow, UK, 2014.
43. Johnson, R.R.; Kuby, P.J. *Elementary Statistics*, 11th ed.; Cengage Learning: Boston, MA, USA, 2011.
44. Chasco-Yrigoyen, C.; Sánchez-Reyes, B. Externalidades ambientales y precio de la vivienda en Madrid: Un análisis con regresión cuantílica espacial. *Rev. Galega Econ.* **2012**, *21*, 1–21.
45. Brandt, S.; Maennig, W. The impact of rail access on condominium prices in Hamburg. *Transportation* **2012**, *39*, 997–1017. [CrossRef]

46. Bauer, T.K.; Feuerschütte, S.; Kiefer, M.; an de Meulen, P.; Micheli, M.; Schmidt, T.; Wilke, L.-H. Ein hedonischer Immobilienpreisindex auf Basis von Internetdaten: 2007–2011. *AStA Wirtsch. Soz. Arch.* **2013**, *7*, 5–30. [CrossRef]

47. Agnew, K.; Lyons, R.C. The impact of employment on housing prices: Detailed evidence from FDI in Ireland. *Reg. Sci. Urban Econ.* **2018**, *70*, 174–189. [CrossRef]

48. Limsombunchai, V.; Gan, C.; Lee, M. House price prediction: Hedonic price model vs. Artificial neural network. *Am. J. Appl. Sci.* **2004**, *1*, 193–201. [CrossRef]

49. Lyons, R.C. Can list prices accurately capture housing price trends? Insights from extreme markets conditions. *Financ. Res. Lett.* **2019**, *30*, 228–232. [CrossRef]

50. SEC, Sede Electrónica del Catastro Inmobiliario. Información Alfanumérica y Cartografía Vectorial. Available online: https://www.sedecatastro.gob.es/ (accessed on 1 October 2019).

51. Mora-Garcia, R.T. Modelo Explicativo de las Variables Intervinientes en la Calidad del Entorno Construido de las Ciudades. Ph.D. Thesis, Universidad de Alicante, San Vicente del Raspeig, Spain, 2016. Available online: http://hdl.handle.net/10045/65829 (accessed on 15 February 2018).

52. Ministerio de Fomento. Real Decreto 732/2019, de 20 de diciembre, por el que se modifica el Código Técnico de la Edificación, aprobado por el Real Decreto 314/2006, de 17 de marzo. *Bol. Estado Madr.* **2019**, *311*, 140488–140674.

53. Kain, J.F.; Quigley, J.M. *Housing Markets and Racial Discrimination: A Microeconomic Analysis*; National Bureau of Economic Research: New York, NY, USA, 1975; p. 393.

54. Malpezzi, S. Hedonic Pricing Models: A Selective and Applied Review. In *Housing Economics and Public Policy*; O'Sullivan, T., Gibb, K., Eds.; Blackwell Science: Great Britain, UK, 2003. [CrossRef]

55. IBM Corporation. *IBM SPSS Statistics for Windows*; IBM Corporation: Armonk, NY, USA, 2016.

56. Welch, B.L. On the Comparison of Several Mean Values: An Alternative Approach. *Biometrika* **1951**, *38*, 330–336. [CrossRef]

57. Brown, M.B.; Forsythe, A.B. The Small Sample Behavior of Some Statistics Which Test the Equality of Several Means. *Technometrics* **1974**, *16*, 129–132. [CrossRef]

58. Kleinbaum, D.; Kupper, L.; Nizam, A.; Rosenberg, E. *Applied Regression Analysis and Other Multivariable Methods*, 5th ed.; Cengage Learning: Boston, MA, USA, 2013; p. 1072.

59. Chatterjee, S.; Simonoff, J.S. *Handbook of Regression Analysis*; John Wiley & Sons Inc.: Hoboken, NJ, USA, 2013; p. 240.

60. Montgomery, D.C.; Peck, E.A.; Vining, G.G. *Introduction to Linear Regression Analysis*, 5th ed.; John Wiley & Sons Inc.: Hoboken, NJ, USA, 2012; p. 645.

61. Yan, X.; Gang-Su, X. *Linear Regression Analysis: Theory and Computing*; World Scientific Publishing Company Pte. Limited: Singapore, 2009; p. 349.

62. Moran, P.A.P. The Interpretation of Statistical Maps. *J. R. Stat. Soc. Ser. B Methodol.* **1948**, *10*, 243–251. [CrossRef]

63. Moran, P.A.P. Notes on continuous stochastic phenomena. *Biometrika* **1950**, *37*, 17–23. [CrossRef]

64. Fotheringham, A.S. "The Problem of Spatial Autocorrelation" and Local Spatial Statistics. *Geogr. Anal.* **2009**, *41*, 398–403. [CrossRef]

65. Fizaine, F.; Voye, P.; Baumont, C. Does the Literature Support a High Willingness to Pay for Green Label Buildings? An Answer with Treatment of Publication Bias. *Rev. D'écon. Polit.* **2018**, *128*, 1013–1046. [CrossRef]

66. Mora-Garcia, R.T.; Cespedes-Lopez, M.F.; Perez-Sanchez, V.R.; Marti-Ciriquian, P.; Perez-Sanchez, J.C. Determinants of the Price of Housing in the Province of Alicante (Spain): Analysis Using Quantile Regression. *Sustainability* **2019**, *11*, 437. [CrossRef]

67. Perez-Sanchez, V.R.; Mora-Garcia, R.T.; Perez-Sanchez, J.C.; Cespedes-Lopez, M.F. The influence of the characteristics of second-hand properties on their asking prices: Evidence in the Alicante market. *Inf. Constr.* **2020**, *72*, 12. [CrossRef]

68. MITECO, Ministerio para la Transición Ecológica y el Reto Demográfico. *Estado de la Certificación Energética de los Edificios (7° Informe)*; Instituto para la Diversificación y Ahorro de la Energía: Madrid, Spain, 2018; p. 11. Available online: https://energia.gob.es/desarrollo/EficienciaEnergetica/CertificacionEnergetica/Documentos/Documentos%20informativos/informe-seguimiento-certificacion-energetica.pdf (accessed on 24 September 2019).

69. Ramos, A.; Pérez-Alonso, A.; Silva, S. *Valuing Energy Performance Certificates in the Portuguese Residential*; Economics for Energy: Vigo, Spain, 2015; pp. 1028–3625. Available online: https://ideas.repec.org/p/efe/wpaper/02-2015.html (accessed on 14 March 2019).

70. Cajias, M.; Piazolo, D. Green performs better: Energy efficiency and financial return on buildings. *J. Corp. Real Estate* **2013**, *15*, 53–72. [CrossRef]

71. García-Navarro, J.; González-Díaz, M.J.; Valdivieso, M. Assessment of construction costs and energy consumption resulting from house energy ratings in a residential building placed in Madrid: "Precost&e Study". *Inf. Constr.* **2014**, *66*, e026. [CrossRef]

72. Kholodilin, K.A.; Mense, A.; Michelsen, C. The market value of energy efficiency in buildings and the mode of tenure. *Urban Stud.* **2017**, *54*, 3218–3238. [CrossRef]

73. Checa-Noguera, C.; Biere-Arenas, R. Approach to the influence of energy certifications on Real Estate values. *ACE Archit. City Environ.* **2017**, *12*, 165–190. [CrossRef]

74. Pascuas, R.P.; Paoletti, G.; Lollini, R. Impact and reliability of EPCs in the real estate market. *Energy Procedia* **2017**, *140*, 102–114. [CrossRef]

75. Marmolejo-Duarte, C.; Spairani-Berrio, S.; Del Moral-Ávila, C.; Delgado-Méndez, L. The Relevance of EPC Labels in the Spanish Residential Market: The Perspective of Real Estate Agents. *Buildings* **2020**, *10*, 27. [CrossRef]

76. Amecke, H. The impact of energy performance certificates: A survey of German home owners. *Energy Policy* **2012**, *46*, 4–14. [CrossRef]

77. Ministerio para la Transición Ecológica y el Reto Demográfico; Instituto para la Diversificación y Ahorro de la Energía (IDAE). *Programa de Ayudas Para la Rehabilitación Energética de Edificios Existentes (Programa PAREER II)*, Fondo Europeo de Desarrollo Regional (FEDER): 2014–2020; Instituto para la Diversificación y Ahorro de la Energía: Madrid, Spain.

78. Vicepresidencia Segunda y Conselleria de Vivienda y Arquitectura Bioclimática. *Ayudas Plan RENHATA. Generalidad Valenciana*; Diari Oficial de la Generalitat Valenciana: Valencia, Spain, 2020.

79. Igbinosa, S.O. Determinants of Residential Property Value in Nigeria—A Neural Network Approach. *Int. Multidiscip. J. Ethiop.* **2011**, *5*, 152–168. [CrossRef]

80. Selím, S. Determinants of House Prices in Turkey: A Hedonic Regression Model. *Doğuş Üniv. Derg.* **2008**, *9*, 65–76.

81. Galvis, L.A.; Carrillo, B. Un índice de precios espacial para la vivienda urbana en Colombia: Una aplicación con métodos de emparejamiento. *Rev. Econ. Rosario* **2012**, *16*, 25–29.

82. Ferreira-Vaz, A.J. La Dimensión de la Subjetividad en la Formación del Valor Inmobiliario: Aplicación del Método de Análisis de Ecuaciones Estructurales al Mercado Residencial de Lisboa. Ph.D. Thesis, Universidad Politécnica de Madrid, Madrid, Spain, 2013. Available online: http://oa.upm.es/15577/ (accessed on 11 April 2018).

83. Yayar, R.; Demir, D. Hedonic estimation of housing market prices in Turkey. *Erciyes Univ. J. Fac. Econ. Adm. Sci.* **2014**, *43*, 67–82. [CrossRef]

84. Zahirovich-Herbert, V.; Gibler, K.M. The effect of new residential construction on housing prices. *J. Hous. Econ.* **2014**, *26*, 1–18. [CrossRef]

85. Chasco-Yrigoyen, C.; Gallo, J.L. The impact of objective and subjective measures of air quality and noise on house prices: A multilevel approach for downtown Madrid. *Econ. Geogr.* **2013**, *89*, 127–148. [CrossRef]

86. Fernández-Durán, L. Análisis del Impacto de los Aspectos Relativos a la Localización en el Precio de la Vivienda a Través de Técnicas de Soft Computing. Una Aplicación a la Ciudad de Valencia. Ph.D. Thesis, Universidad Politécnica de Valencia, Valencia, Spain, 2016. Available online: http://hdl.handle.net/10251/63253 (accessed on 15 February 2018).

87. Baudry, M.; Guengant, A.; Larribeau, S.; Leprince, M. Formation des prix immobiliers et consentements à payer pour une amélioration de l'environnement urbain: L'exemple rennais. *Rev. D'Écon. Rég. Urb.* **2009**, *2*, 369–411. [CrossRef]

88. Evangelista, R.; Ramalho, E.A.; Andrade-e-Silva, J. On the use of hedonic regression models to measure the effect of energy efficiency on residential property transaction prices: Evidence for Portugal and selected data issues. *Energy Econ.* **2020**, *86*, 104699. [CrossRef]

89. Humaran-Nahed, I.; Roca-Cladera, J. Towards a integrated measure of the location factor in the Real Estate valuation: The case of Mazatlan. *ACE Arquit. Ciudad Entorno* **2010**, *13*, 185–218. [CrossRef]

90. Sagner, A. Determinantes del precio de viviendas en la región metropolitana de Chile. *Trimest. Econ.* **2011**, *78*, 813–839. [CrossRef]

91. Bohl, M.T.; Michels, W.; Oelgemöller, J. Determinanten von Wohnimmobilienpreisen: Das Beispiel der Stadt Münster. *Jahrbuch Reg.* **2012**, *32*, 193–208. [CrossRef]

92. Nicodemo, C.; Raya, J.M. Change in the distribution of house prices across spanish cities. *Reg. Sci. Urban Econ.* **2012**, *42*, 739–748. [CrossRef]

93. Kaya, A.; Atan, M. Determination of the factors that affect house prices in Turkey by using Hedonic Pricing Model. *J. Bus. Econ. Financ.* **2014**, *3*, 313–327.

94. Wen, H.; Bu, X.; Qin, Z. Spatial effect of lake landscape on housing price: A case study of the West Lake in Hangzhou, China. *Habitat Int.* **2014**, *44*, 31–40. [CrossRef]

95. Alkan, L. Housing market differentiation: The cases of Yenimahalle and Çankaya in Ankara. *Int. J. Strateg. Prop. Manag.* **2015**, *19*, 13–26. [CrossRef]

96. Fitch-Osuna, J.M. Sistema de valuación masiva de inmuebles para tasaciones. *Contexto. Rev. Fac. Arquit. Univ. Autón. Nuevo León* **2016**, *10*, 51–63.

97. Van-Dijk, D.; Siber, R.; Brouwer, R.; Logar, I.; Sanadgol, D. Valuing water resources in Switzerland using a hedonic price model. *Water Resour. Res.* **2016**, *52*, 3510–3526. [CrossRef]

98. Zambrano-Monserrate, M.A. Formation of housing rental prices in Machala, Ecuador: Hedonic prices analysis. *Cuad. Econ.* **2016**, *39*, 12–22. [CrossRef]

99. Keskin, B.; Watkins, C. Defining spatial housing submarkets: Exploring the case for expert delineated boundaries. *Urban Stud.* **2017**, *54*, 1446–1462. [CrossRef]

100. Lara-Pulido, J.A.; Estrada-Díaz, G.; Zentella-Gómez, J.C.; Guevara-Sanginés, A. The costs of urban expansion: An approach based on a hedonic price model in the Metropolitan Area of the Valley of Mexico. *Estud. Demogr. Urbanos* **2017**, *32*, 37–63. [CrossRef]

101. Park, J.; Lee, D.; Park, C.; Kim, H.; Jung, T.; Kim, S. Park accessibility impacts housing prices in Seoul. *Sustainability* **2017**, *9*, 185. [CrossRef]

102. Wen, H.; Xiao, Y.; Zhang, L. School district, education quality, and housing price: Evidence from a natural experiment in Hangzhou, China. *Cities* **2017**, *66*, 72–80. [CrossRef]

103. Xiao, Y.; Chen, X.; Li, Q.; Yu, X.; Chen, J.; Guo, J. Exploring Determinants of Housing Prices in Beijing: An Enhanced Hedonic Regression with Open Access POI Data. *ISPRS Int. J. Geo-Inf.* **2017**, *6*, 358. [CrossRef]

104. Li, R.; Li, H. Have housing prices gone with the smelly wind? Big data analysis on landfill in Hong Kong. *Sustainability* **2018**, *10*, 341. [CrossRef]

105. Lama-Santos, F.A.D. Determinación de las Cualidades de Valor en la Valoración de Bienes Inmuebles. La Influencia del nivel Socioeconómico en la Valoración de la Vivienda. Ph.D. Thesis, Universidad Politécnica de Valencia, Valencia, Spain, 2017. Available online: http://hdl.handle.net/10251/90526 (accessed on 6 March 2018).

106. Landajo, M.; Bilbao, C.; Bilbao, A. Nonparametric neural network modeling of hedonic prices in the housing market. *Empir. Econ.* **2012**, *42*, 987–1009. [CrossRef]

107. Núñez-Tabales, J.M.; Caridad-y-Ocerín, J.M.; Ceular-Villamandos, N.; Rey-Carmona, F.J. Obtención de precios implícitos para atributos determinantes en la valoración de una vivienda. *Rev. Int. Adm. Finanz.* **2012**, *5*, 41–54.

108. Núñez-Tabales, J.M.; Caridad-y-Ocerín, J.M.; Rey-Carmona, F.J. Artificial Neural Networks for predicting real estate prices. *Rev. Métodos Cuantitativos Para Econ. Empresa* **2013**, *15*, 29–44.

109. Wen, H.; Xiao, Y.; Zhang, L. Spatial effect of river landscape on housing price: An empirical study on the Grand Canal in Hangzhou, China. *Habitat Int.* **2017**, *63*, 34–44. [CrossRef]

110. Yu, C.-M.; Chen, P.-F. House Prices, Mortgage Rate, and Policy: Megadata Analysis in Taipei. *Sustainability* **2018**, *10*, 926. [CrossRef]

111. Wu, H.; Jiao, H.; Yu, Y.; Li, Z.; Peng, Z.; Liu, L.; Zeng, Z. Influence Factors and Regression Model of Urban Housing Prices Based on Internet Open Access Data. *Sustainability* **2018**, *10*, 1676. [CrossRef]

112. Seo, D.; Chung, Y.; Kwon, Y. Price Determinants of Affordable Apartments in Vietnam: Toward the Public–Private Partnerships for Sustainable Housing Development. *Sustainability* **2018**, *10*, 197. [CrossRef]

113. Cebula, R.J. The hedonic pricing model applied to the housing market of the city of Savannah and its Savannah historic Landmark district. *Rev. Reg. Stud.* **2009**, *39*, 9–22.

114. Stetler, K.M.; Venn, T.J.; Calkin, D.E. The effects of wildfire and environmental amenities on property values in northwest Montana, USA. *Ecol. Econ.* **2010**, *69*, 2233–2243. [CrossRef]
115. Duque, J.C.; Velásquez, H.; Agudelo, J. Public infrastructure and housing prices: An application of geographically weighted regression within the context of hedonic prices. *Ecos Econ.* **2011**, *15*, 95–122.
116. Moreno-Murrieta, R.E.; Alvarado-Lagunas, E. El entorno social y su impacto en el precio de la vivienda: Un análisis de precios hedónicos en el Área Metropolitana de Monterrey. *Trayectorias Rev. Cienc. Soc.* **2011**, *14*, 131–147.
117. Fernández-Durán, L.; Llorca-Ponce, A.; Valero-Cubas, S.; Botti-Navarro, V.J. Incidencia de la localización en el precio de la vivienda a través de un modelo de red neuronal artificial. Una aplicación a la ciudad de Valencia. *Catastro* **2012**, *74*, 7–25.
118. McGreal, W.S.; Taltavull-de-la-Paz, P. Implicit house prices: Variation over time and space in Spain. *Urban Stud.* **2013**, *50*, 2024–2043. [CrossRef]
119. Wen, H.; Goodman, A.C. Relationship between urban land price and housing price: Evidence from 21 provincial capitals in China. *Habitat Int.* **2013**, *40*, 9–17. [CrossRef]
120. Rey-Carmona, F.J. Alternativas Determinantes en Valoración de Inmuebles Urbanos. Ph.D. Thesis, Universidad de Córdoba, Córdoba, Spain, 2014. Available online: http://hdl.handle.net/10396/12473 (accessed on 11 April 2018).
121. Quispe-Villafuerte, A. Una aplicación del modelo de precios hedónicos al mercado de viviendas de Lima Metropolitana. *Rev. Econ. Derecho* **2012**, *9*, 85–121.
122. Núñez-Tabales, J.M.; Rey-Carmona, F.J.; Caridad-y-Ocerín, J.M. Artificial Intelligence (AI) techniques to analyze the determinants attributes in housing prices. *Intell. Artif.* **2016**, *19*, 23–38. [CrossRef]
123. Casas-del-Rosal, J.C. Métodos de Valoración Urbana. Ph.D. Thesis, Universidad de Córdoba, Córdoba, Spain, 2017. Available online: http://hdl.handle.net/10396/15417 (accessed on 11 April 2018).
124. Zhang, L.; Yi, Y. Quantile house price indices in Beijing. *Reg. Sci. Urban Econ.* **2017**, *63*, 85–96. [CrossRef]
125. Ezebilo, E. Evaluation of House Rent Prices and Their Affordability in Port Moresby, Papua New Guinea. *Buildings* **2017**, *7*, 114. [CrossRef]
126. Liu, J.-G.; Zhang, X.-L.; Wu, W.-P. Application of Fuzzy Neural Network for Real Estate Prediction. In *Advances in Neural Networks–ISNN 2006*; Springer: Berlin/Heidelberg, Germany, 2006; pp. 1187–1191. [CrossRef]
127. Keskin, B.; Dunning, R.; Watkins, C. Modelling the impact of earthquake activity on real estate values: A multi-level approach. *J. Eur. Real Estate Res.* **2017**, *10*, 73–90. [CrossRef]
128. Fitch-Osuna, J.M.; Soto-Canales, K.; Garza-Mendiola, R. Assessment of Urban-Environmental Quality. A Hedonic Modeling: San Nicolás de los Garza, Mexico. *Estud. Demogr. Urbanos* **2013**, *28*, 383–428. [CrossRef]

Publisher's Note: MDPI stays neutral with regard to jurisdictional claims in published maps and institutional affiliations.

applied sciences

Article

Battery Energy Storage System Dimensioning for Reducing the Fixed Term of the Electricity Access Rate in Industrial Consumptions

Jorge Nájera *, Miguel Santos-Herran, Marcos Blanco, Gustavo Navarro, Jorge Torres and Marcos Lafoz

Centro de Investigaciones Energéticas, Medioambientales y Tecnológicas (CIEMAT), Government of Spain, 28040 Madrid, Spain; miguel.santos@ciemat.es (M.S.-H.); marcos.blanco@ciemat.es (M.B.); gustavo.navarro@ciemat.es (G.N.); jorgejesus.torres@ciemat.es (J.T.); marcos.lafoz@ciemat.es (M.L.)
* Correspondence: jorge.najera@ciemat.es; Tel.: +34-91-335-7194

Abstract: Industrial buildings account for very few high peaks of power demand. This situation forces them to contract a high fixed electricity term to cover it. A more intelligent use of the energy in industrial buildings, together with an improved efficiency of the transmission and distribution of the energy along the electric power grid, can be achieved by reducing the peak consumption of industrial buildings. Energy storage systems, and lithium ion (Li-ion) batteries in particular, are one of the most promising technologies for reducing this peak consumption. However, selecting a proper Li-ion battery requires a dimensioning process in terms of energy and power which is not straightforward. This paper proposes a dimensioning methodology that takes into consideration both technical and economic implications, and applies it to a case example with real industrial consumption data and a commercial battery. Results show that implementing batteries for reducing this peak consumption can lead to a cost–benefit improvement.

Keywords: battery energy storage system; dimensioning methodology; industrial consumption; electricity fixed term

Citation: Nájera, J.; Santos-Herran, M.; Blanco, M.; Navarro, G.; Torres, J.; Lafoz, M. Battery Energy Storage System Dimensioning for Reducing the Fixed Term of the Electricity Access Rate in Industrial Consumptions. *Appl. Sci.* **2021**, *11*, 7395. https://doi.org/10.3390/app11167395

Academic Editors:Luis Hernández-Callejo, Víctor Alonso Gómez, Sergio Nesmachnow, Vicente Leite, Javier Prieto and Ângela Ferreira

Received: 30 June 2021
Accepted: 6 August 2021
Published: 11 August 2021

Publisher's Note: MDPI stays neutral with regard to jurisdictional claims in published maps and institutional affiliations.

1. Introduction

The fee that electric energy consumers pay for being able to make use of it, commonly known as electricity access rate, is split into two terms in the majority of European countries: a variable term and a fixed term [1]. The variable term is associated with the energy consumed by the user (kWh), while the fixed term corresponds to the contracted power (kW), defined as the maximum power that the user can consume before the power-limit switch is triggered.

Commonly, the fixed term is over-dimensioned, meaning that the usual consumption along a day is considerably lower than the power consumption limit, the latter being reached only during a few instants per day. This behaviour is specially representative of industrial consumption, forcing them to contract a high fixed electricity term due to very few high power consumption peaks [2].

Modifying this electrical consumption pattern could lead to a more intelligent use of the energy in industrial buildings, to the fostering of the "prosumer" behaviour among industrial consumers, and to the improvement of the efficiency of transmission and distribution of the electricity across the power grid. Among the different solutions proposed in the literature for achieving this goal, the inclusion of photovoltaic panels is presented as a feasible solution in [3,4], although the vast majority of studies opt for implementing an energy storage system (ESS) so that the joint consumption profile of the ESS and the industrial building is modified, as in [5–7] among other studies.

Battery energy storage systems (BESS) arise as the most promising ESS candidates for their connection in buildings and infrastructures, both industrial and residential, considering that this is one of their most prominent and leading uses [8]. The benefits of including BESS encompass the majority of the electricity value chain, including peak shaving and continuity of supply for residential and industrial customers, ancillary and power quality services for transmission and distribution system operators, and flexibility and curtailment minimization for renewable generation companies [9].

Nevertheless, selecting a BESS for its connection to an industrial consumption for reducing the peak power consumption is not straightforward. A wide variety of solutions is offered by battery manufacturers, so a dimensioning process based on a techno-economic assessment is required in order to help decision makers. This techno-economic assessment will incorporate the load characteristics, the BESS technical features and its cost.

In this regard, BESS dimensioning for reducing the electricity access fee in industrial buildings has been studied together with photovoltaic in [10,11], where a dimensioning process is proposed for a BESS and photovoltaic installation by using a brute force methodology, and applied to peak shaving and prosumer purposes, and for working isolated from the grid. In addition, the authors in [12] analyse the economic viability of BESS integration under different tariff structures and systems configurations, although a generalized and simplified ageing for the Li-ion BESS has been assumed for the yearly cost calculations. Thus, the importance of the BESS charge/discharge profile has not been taken into account properly. A similar approach is proposed in [13,14] for residential consumption. Both papers assess the economic viability of integrating BESS at a residential level with and without subsidies. However, these studies lack a detailed analysis of the BESS lifespan, since they neglect its influence in the overall yearly costs. Further studies, such as [15], analyse the peak shaving of decentralised residential BESS for load peak shaving. The proposed BESS model includes the battery efficiency and SoC estimation, but an over-simplified battery ageing assumption is made, where the payback period of the battery varies linearly with the battery size.

Regarding alternative dimensioning methodologies, stochastic models for improving the decision making are developed in [16], where the methodologies are applied to reach the optimal BESS investment for industrial buildings. As identified in the previously mentioned papers, there is no detailed battery ageing analysis, so the influence of this variable is underestimated.

The BESS dimensioning studies mentioned above employ different selection criteria. Those criteria are based only on technical issues, or on techno-economic issues but using generic ageing data to calculate the BESS lifespan. Moreover, they analyse the complete compensation of technical issues under study, e.g., feed an isolated load with a combination of a BESS together with photovoltaic or eliminate the consumption peaks via smart charging of electric vehicles, which may not be the most feasible scenario faced by residential or industrial consumers. As a main contribution and further development, this paper proposes a BESS dimensioning methodology for diminishing the fixed term of the electricity access rate in real industrial consumption, employing a techno-economic criterion for selecting the optimum solution. The proposed methodology includes a detailed evaluation of the BESS ageing, which is included in the cost–benefit analysis as a key component. This improved BESS ageing assessment highly influences the time for a BESS replacement and, hence, the yearly cost (BESS cost and fixed electricity access fee).

Hence, the applicability of this methodology encompasses industrial consumers with an irregular consumption accounting for several peaks of consumption, as well as BESS manufacturers aiming for specific developments for industrial consumption. The applicability of the developed methodology is exemplified in this paper for an existing industrial consumption and on a typical day in summer and in winter, considering a commercial battery and standard prices for the economic analysis.

The paper is organized as follows: the BESS dimensioning methodology is described in Section 2, including a series of steps to be followed. The different simulation models used along the different steps determined by methodology are detailed in Section 3, including a power grid model, an industrial consumption model, a power converter model, and a battery model. At last, a case example with real industrial consumption data, commercial BESS and standard prices is presented in Section 4, while conclusions are drawn in Section 5.

2. BESS Dimensioning Methodology

The BESS dimensioning methodology described in this section aims at reducing the fixed term of the electricity access rate by introducing a BESS that deals with the peaks of the industrial consumption. A few assumptions made by the authors in the development of the BESS dimensioning methodology are presented in the following.

Firstly, both the BESS (P_{BESS}) and the electrical grid (P_{GRID}) must cover the customers' demand (P_{LOAD}) and the losses (P_{LOSS}):

$$P_{BESS} + P_{GRID} = P_{LOAD} + P_{LOSS} \tag{1}$$

Moreover, the BESS must account for enough energy to cover the energy to fulfill the extra energy demand not supplied/covered by the grid. Besides, a BESS must be able to fully recharge the same amount of energy as dispatched in a specific scenario. If not, the mean SoC will be decaying as the scenario repeats over time, being eventually unable to fulfil the load power demand.

The methodology comprises a set of steps, as seen in the flow chart of Figure 1.

1. Step 1: Models parameterizations. The different models, which are mathematical representations of physical systems or devices, account for variables (inputs and outputs) and parameters. In order to particularize the models (grid, BESS, and industrial consumption) to a specific system or device, those parameters must be adjusted to a fixed value.
2. Step 2: Define test cases. The variable that defines the test cases is the maximum fixed power that the grid can deliver (P_{LIMIT}). Reducing this power limit implies adding a greater contribution from the BESS, which yields to the different scenarios, which consequently will result in different BESS configurations and different BESS ageing.
3. Step 3: Simulate test case, calculate BESS requirements, and BESS configuration. Once the scenarios have been defined, a simulation is performed for every value of P_{LIMIT}. Each scenario implies different necessities from the BESS, regarding power and energy. This requirements give different BESS configurations, in terms of number of cells in series and in parallel. The charge/discharge profile for the BESS is also obtained from the simulation. If the energy available for charging is unable to match the amount of energy dispatched by a BESS in a specific scenario, that scenario will not be considered for the dimensioning process. As previously noted, this situation will force the BESS to, eventually, not being able to feed the load when the scenario repeats over time.
4. Step 4: Calculate BESS ageing. Once the minimum BESS configuration that satisfies the power and energy requirements is selected for each scenario, a BESS ageing analysis is performed, based on the BESS model and configuration, and on the charge/discharge profile for each specific scenario. As a result of this step, the number of days until the BESS loses a 20% of capacity retention is obtained.
5. Step 5: Cost–benefit analysis. Prices and fees for both a BESS (as a function of the kW and kWh of the system) and the fixed access power are needed for their incorporation into a cost function. BESS standard prices as well as Spanish fixed access rates are provided in Section 4, with a case example. The cost can be computed yearly, taking into account the net present cost, and the capital recovery factor. The base scenario, i.e., when there is no BESS, is then compared to the rest of scenarios in terms of cost, and the final solution can be chosen.

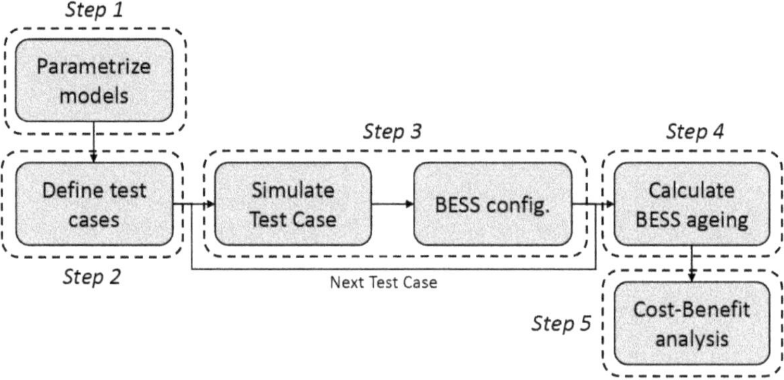

Figure 1. Flow chart of the proposed BESS dimensioning methodology.

3. Industrial Consumption, BESS, and Grid Modelling

The mathematical and simulation models used in Section 4 for performing a case example of the aforementioned methodology are described in this section. Three main models are considered: an industrial consumption, a BESS, and a power grid. The BESS is connected to the same point as the industrial consumption, the point of common coupling (PCC), and both of them are connected to the grid, as it can be seen in the general model diagram in Figure 2. The complete model outputs several electrical-variables time series (P, Q, V, and I) for the three main components. Additional electrical variables such as frequency or harmonic distortion are also evaluated and calculated by the model, but the authors decided not to output them since they are not relevant for the purpose of this paper.

The models are developed with MATLAB Simulink, under the Simscape Electrical environment. When developing simulation models, a higher complexity and completeness usually implies a higher accuracy [17,18]. However, complex models tend to require a substantial computational effort to run them, and a huge amount of information to parametrize them. Moreover, they are commonly valid only for a specific device, making them non-generalizable [19,20].

On the other hand, simplified models, usually composed by combination of resistors, inductors, and capacitors, are easily parametrized, they require low computational effort, and are valid for several devices with similar characteristics. Furthermore, even though they are not as accurate as complex models, they can mimic the behaviour of the device with sufficient precision, allowing for the extraction of general conclusions [21,22].

Hence, given the objective of this paper, the approach taken by the authors was to develop simplified models, that are generalizable and easily parametrized, which is sufficient for obtaining accurate general conclusions when dimensioning BESS.

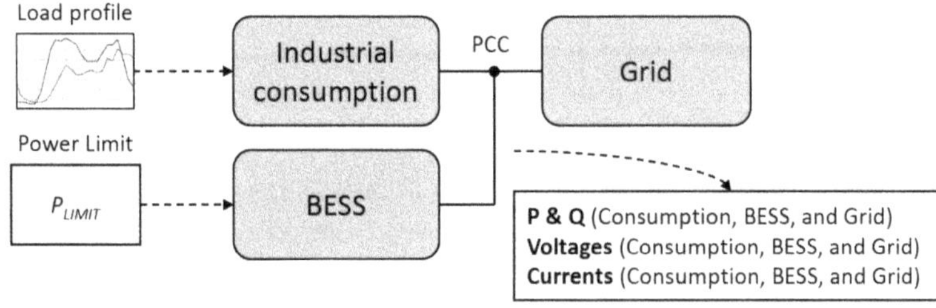

Figure 2. General model diagram.

3.1. Industrial Consumption Model

The industrial consumption is modelled as a PQ load, i.e., it consumes the desired active and reactive power independently of the voltage at the PCC. Hence, it needs to be modelled as a controlled current source, as depicted in Figure 3. The inputs of the model, named as P and Q in Figure 3, correspond to the 24-hour active and reactive power profile on a 1-second basis.

The Simulink diagram in Figure 3 shows, from left to right, the inputs of the model P and Q together with the measured voltage (Vabc). From those inputs, the complex values of the phase currents are calculated, and then redirected to the corresponding current controlled source. A resistor with a very high value is also connected in parallel, which acts as a snubber. At last, the industrial consumption model is linked to the rest of the model via connectors A, B, C, and N.

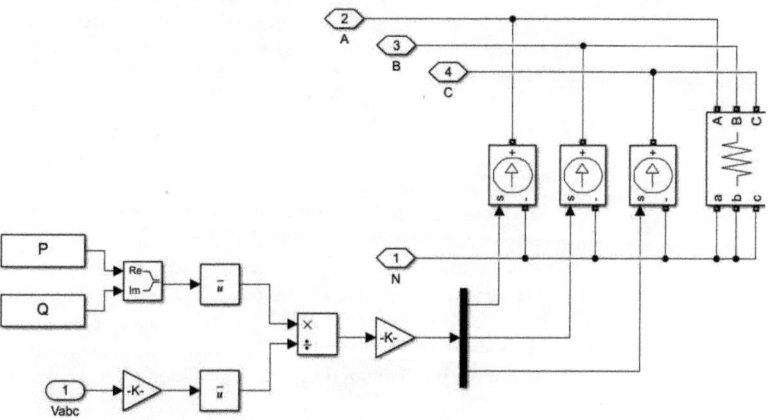

Figure 3. Industrial consumption model in MATLAB Simulink.

3.2. BESS Model

The BESS model includes both a battery model and a power electronic converter model, the latter being necessary to manage the battery behaviour and to connect the battery to the AC grid.

3.2.1. Electronic Converter Model

For the purpose of this work, there is no need to model the power electronic converter in detail, since neither efficiency nor power converter ageing are incorporated into the economic evaluation of the methodology. Thus, an ideal voltage source converter (VSC) has been selected, with no conduction or switching losses and no power limitations, so the battery can dispatch as much power as required. As a result, the VSC is modelled as a controlled three-phase voltage source whose reference voltage is given by the converter control, which manages the power that flows in and out of the battery.

The control of a VSC can be modelled as explained in [23], consisting of two levels: a high-level control and a low-level control. The high-level control is in charge of selecting, based on a specific strategy, the power dispatched or consumed by the battery. Several strategies have been proposed in the literature, from those based on rules [24] to more complex strategies [25,26]. For this paper, a simple strategy based on rules has been selected, which sets the reference power for the battery (P_{BATT}) based on the maximum power that the industrial consumption can absorb from the grid (P_{LIMIT}), the actual power that the industrial load is demanding (P_{LOAD}), and the battery state of charge (SoC_{BATT}). The flow diagram for the high-level control is shown in Figure 4.

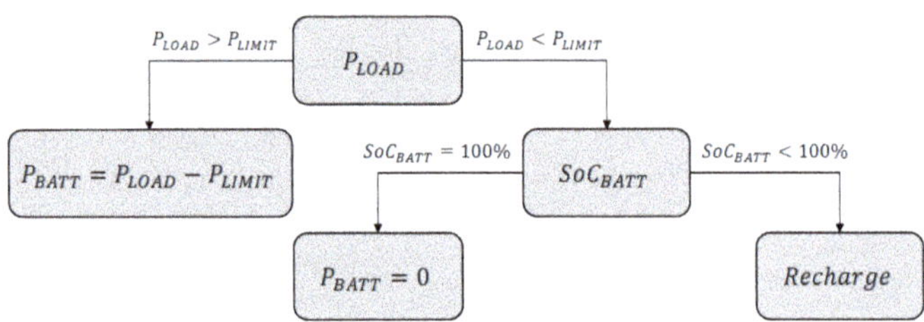

Figure 4. Flow diagram for the VSC high-level control.

The BESS recharge profile has been defined as the minimum current that allows the BESS to recharge the same amount of energy that it has provided to the load along the day. Moreover, reactive power reference for the battery has been set to zero, so no contribution is expected from it.

On the other hand, the low-level control is in charge of setting the voltage at the VSC terminals so that the power management commanded by the high-level control can be performed. The low-level control is a dq axis control, shown in Figure 5, which has been developed building on [27]. The control is duplicated for P and Q, with minor variations. Each branch accounts for an outer PI loop control, which sets the battery power equal to the reference power coming from the high-level control, outputting a reference current. Then, there is an inner PI loop control that adjusts the current flow through the VSC with the reference current, so a reference voltage is outputted. This voltage is finally regulated so that it corresponds to the VSC voltage at its terminals. In order to properly control the VSC, the inner control must be considerably faster than the outer control. Detailed explanation about the implemented dq axis control for a VSC can be found in [23,27].

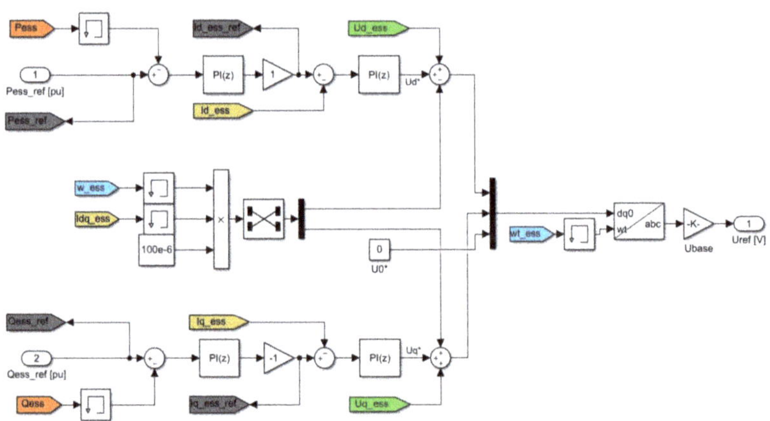

Figure 5. VSC low-level control model in MATLAB Simulink.

3.2.2. Battery Model

The battery model controlled by the VSC is represented as an equivalent circuit model which, as mentioned above, sacrifices accuracy in favour of generality. Nevertheless, equivalent circuit models account for errors lower than 5% [28], sufficient for obtaining general conclusions. The battery equivalent circuit is depicted in Figure 6, where u is the instantaneous battery voltage [V], i is the battery current [A] ($i > 0$ discharging; $i < 0$ charging), I_{SELF} is the self-discharge current, R_{OHM} is the ohmic internal resistance [Ω], R_{POL} is the polarization internal resistance [Ω], and C_{POL} is the polarization capacitor [F].

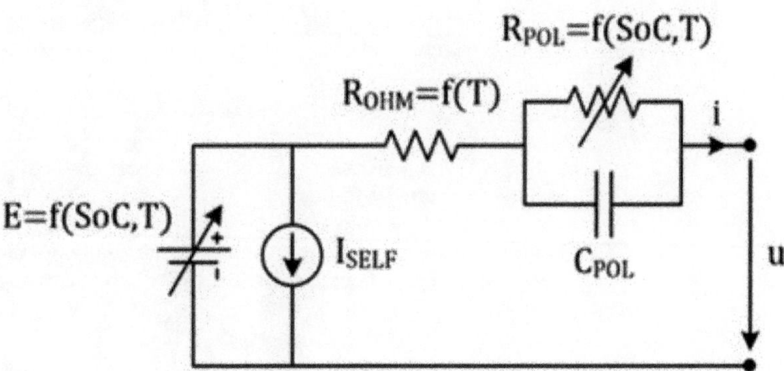

Figure 6. Battery equivalent circuit model.

The selected battery model has been developed and validated by the authors in [18,29]. The model comprises three parts: a voltage/runtime model, a thermal model, and an ageing model.

Battery Voltage/Runtime and Thermal Model

The lithium-ion battery model used in this paper is a modification of the MATLAB Simulink model in [30]. The battery equivalent circuit is based on the Shepherd equation [31], which was experimentally validated later in [32]. The Shepherd model keeps the error between 1% and 5% and it can be parametrizable without testing the battery in the laboratory: it suffices with the manufacturer datasheet information. For the work developed in this paper, both voltage/runtime and thermal models have been parametrized via a least-square method. Data can be obtained from typical manufacturer's curves: voltage-capacity curve as a function of different C rates (with constant temperature), and voltage-capacity curve at different temperatures (with constant C rate). The detailed parametrization process is described in [33]. The voltage/runtime equations that govern the model, considering thermal dependencies, can be written as follows:

$$E(SoC, T) = E_0(T) - K(T) \cdot Q_{MAX}(T) \cdot \left(\frac{100}{SoC} - 1\right) + A \cdot e^{\left(-B \cdot Q_{MAX}(T) \cdot \left(1 - \frac{SoC}{100}\right)\right)} \tag{2}$$

$$R_{POL-DISCHARGE}(SoC, T) = K(T) \cdot \frac{100}{SoC} \tag{3}$$

$$R_{POL-CHARGE}(SoC, T) = K(T) \cdot \frac{1}{1.1 - \frac{SoC}{100}} \tag{4}$$

$$E_0(T) = E_0(T_0) - \frac{dE}{dT} \cdot (T - T_0) \tag{5}$$

$$Q_{MAX}(T) = Q_{MAX}(T) - \frac{dQ}{dT} \cdot (T - T_0) \tag{6}$$

$$K(T) = K(T_0) - e^{\alpha \cdot \left(\frac{1}{T} - \frac{1}{T_0}\right)} \tag{7}$$

$$R_{OHM}(T) = R_{OHM}(T_0) - e^{\beta \cdot \left(\frac{1}{T} - \frac{1}{T_0}\right)} \tag{8}$$

where E is the open-circuit nonlinear voltage (V), E_0 is the open-circuit constant voltage (V), K is the polarization parameter (Ah^{-1}), Q_{MAX} is the maximum capacity (Ah), A is the exponential voltage constant (V), B is the exponential capacity constant (Ah^{-1}), T is the temperature (K), T_0 is the parameter reference temperature (K), and SoC is the state of charge (%).

The thermal model comprises two different parts: a heat generation model and a heat evacuation model. The heat generation model can be expressed as follows [34]:

$$H = (E_0 - E) \cdot i + T \cdot \frac{dE}{dT} \cdot i + (R_{OHM} + R_{POL}) \cdot i^2 \tag{9}$$

where H is the generated heat (W), and dE/dT the change of the equilibrium potential with temperature (V·K^{-1}).

The heat evacuation model occurs due to convection and radiation mechanisms, the latter being negligible [35]. Assuming that the temperature in the battery is uniform (reasonable approximation according to [36]), the following expression is obtained [35]:

$$H = m \cdot c_p \cdot \frac{dT}{dt} + \frac{1}{R_{OUT}} \cdot (T - T_0) \tag{10}$$

$$R_{OUT} = \frac{1}{h \cdot Ar} \tag{11}$$

where m is the mass of the battery (kg), c_p is the specific heat capacity (J·kg^{-1}·K^{-1}), h is the convective heat transfer coefficient (W·m^{-2}·K^{-1}), Ar is the external surface area (m^2), and R_{OUT} is the equivalent thermal resistance with the ambient (K·W^{-1}).

Hence, the temperature variation can be calculated as:

$$T(s) = \frac{H \cdot R_{OUT} + T_0}{1 + m \cdot c_p \cdot R_{OUT} \cdot s} = \frac{H \cdot R_{OUT} + T_0}{1 + t_{th} \cdot s} \tag{12}$$

where t_{th} is the thermal time constant (s), and s is the Laplace complex frequency parameter.

Battery Ageing Model

The ageing estimator implemented in the battery model comprises two parts: a cycling ageing model and a calendar ageing model. In order to parametrize both models, a least-square method has been applied, so the error between the model output and the real battery data is minimized. The data for parametrizing the models need to be obtained from experimental tests. The complete parametrization process is detailed in [33]. A minimum set of 4 tests at different SoC and temperature are needed for the calendar ageing, and a minimum of 5 tests with different combinations of C rate, DoD and temperature are needed for the cycling model. Cycling ageing is related to the capacity fade when the battery is subject to charging and discharging cycles, i.e., when the current is flowing through the battery. It depends on several factors, among which C_{rate}, temperature, and DoD are the most relevant [37]. The cycling ageing model published in [38] for lithium-ion batteries has been implemented:

$$Q_{loss-cyc} = (a \cdot T^2 + b \cdot T + c) \cdot e^{(d \cdot T + e) \cdot C_{rate}} \cdot Ah \tag{13}$$

where $Q_{loss-cyc}$ is the cycling loss capacity (Ah), Ah the Ah that have flown through the battery (in or out), and $a, b, c, d,$ and e are constants depending on the battery chemistry.

On the other hand, calendar ageing is related to the capacity fade in the battery because of the time passed from the moment it was manufactured. It is dependent on the SoC of the battery, the temperature, and the time [37]. The calendar ageing model published in [39] has been implemented:

$$Q_{loss-cal} = f \cdot e^{g \cdot SoC} \cdot e^{\frac{h}{T}} \cdot t^z \tag{14}$$

where $a, b,$ and c are constants that depend on the battery chemistry, and z can take values between 0.5 and 1 and depends on the chemistry as well.

Apart from the ageing factors previously mentioned, additional ageing factors such as DoD or mean SoC along a cycle are indirectly considered by the model. The Ah dependency included in the cycling ageing is indirectly considering the DoD, since the higher the DoD the higher the Ah that flow through the battery. Moreover, the mean SoC along a cycle is considered in the calendar ageing equation, since the latter is constantly computing the battery SoC contribution.

3.3. Grid Model

The grid is modelled as an infinite power grid, so both the frequency and voltage at the PCC are kept constant. A Three-Phase Source model has been selected for this purpose, with a swing generator type, high short-circuit power, and low X/R ratio.

4. Case Example

In order to illustrate how the described methodology could be implemented, a case example is analysed. The models developed in Section 3 are used for performing the simulations. The same structure of steps defined in Section 2 is applied:

4.1. Step 1. Parametrize Models

At first, the simulations models need to be parametrized. For the grid model, the following parameters have been selected, so an almost infinite power grid is implemented (Table 1).

Table 1. Grid model parameters.

Parameter	Value
Reference voltage	400 V
Short-circuit ratio	25
X/R ratio	20

The battery model can be parametrized following the methodology defined in [29,32,33]. A commercial lithium polymer battery cell from Kokam, model SLPB100255255HR2, has been selected for this case study, which is a standard lithium-ion pouch cell [40]. The experimental data for calculating the ageing model parameters have been obtained from [40] and [33], where the complete theoretical and experimental validation for the selected battery cell has been performed. The battery ageing model performs with an error below 5%, as detailed in [33]. The selected battery cell main parameters are shown in Table 2.

Table 2. Battery model data and parameters [33,40].

Data	Value	Remarks
Reference capacity	55 Ah	0.2 C, 23 °C
Impedance	Max. 0.60 mΩ	AC 1 kHz
Average voltage	3.7 V	
Lower limited voltage	2.7 V	
Upper limited voltage	4.2 V	

Parameter	Value	Parameter	Value
a	1.3961×10^{-7}	b	-7.7116×10^{-5}
c	0.0107	d	-1.2959×10^{-3}
e	0.4404	f	164.7155
g	0.6898	h	-4.2126×10^{3}
z	0.5		

The load needs a power consumption profile as input. For this case example, two 24-h profile on a 1-second basis have been selected, corresponding to a typical day in summer and a typical day in winter in an industrial building. The selected building is the CEDEX facility in Madrid, located at Julián Camarillo 30, 28037 Madrid (Spain). For the analysis performed in this paper, the load corresponds to a three-phase load with no unbalance between phases, whose active power profiles are shown in Figure 7 for both summer and winter. The cumulative energy consumed by the load for each typical day is also shown in Figure 8.

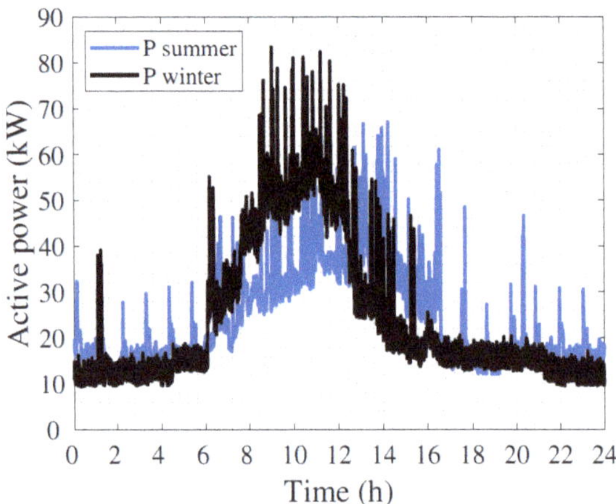

Figure 7. Industrial consumption profiles for summer and winter scenarios.

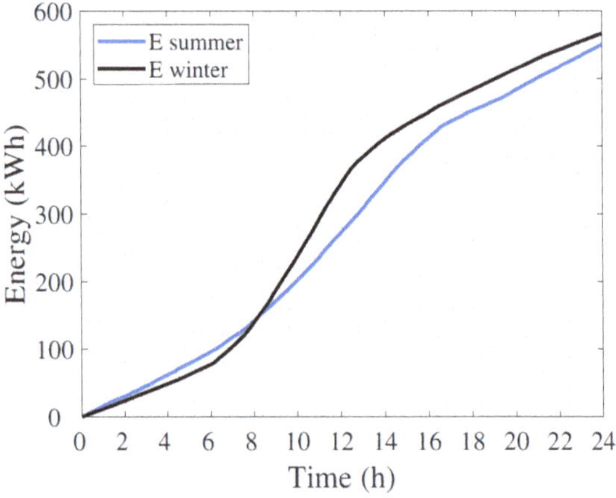

Figure 8. Cumulative energy consumed by the load in summer and winter scenarios.

4.2. Step 2. Define Test Cases

Test cases are defined by the BESS control parameter, which is the power limit that the load is allowed to consume from the grid (P_{LIMT}). Hence, as P_{LIMIT} decreases, the battery will increase its contribution. Scenarios are divided between summer scenarios (S) and winter scenarios (W). SB and WB correspond to the base case scenario for each season, which is the current scenario at the real world facility, with a contracted power of 90 kW. Steps of 10 kW have been selected, so the different scenarios are defined in Table 3.

Table 3. Scenarios defined for the case example.

Scenario	P_{LIMIT}
SB	70 kW
S6	60 kW
...	...
S2	20 kW
S1	10 kW
WB	90 kW
W8	80 kW
...	...
W2	20 kW
W1	10 kW

4.3. Step 3. Simulate Scenarios and Obtain BESS Requirements and Configuration

As previously explained in Sections 2 and 3, BESSs that are not able to recharge as much energy as they have delivered along the day, are considered as not valid. Once the simulations are performed, BESS power profiles for the valid configurations are obtained. For the case example presented in this paper, valid BESS power profiles are shown in Figures 9 and 10.

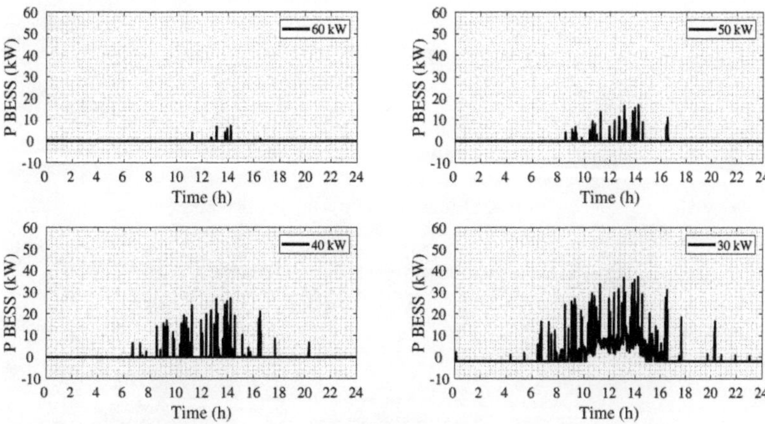

Figure 9. BESS power profile in summer for different values of P_{LIM}.

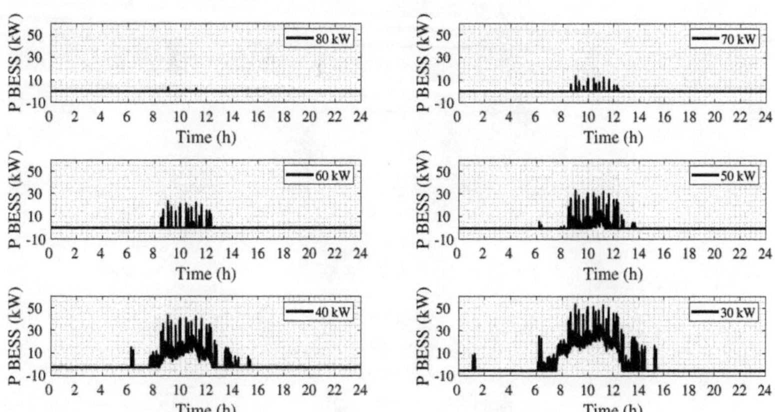

Figure 10. BESS power profile in winter for different values of P_{LIM}.

As expected, the lower the P_{LIMIT}, the higher the contribution from the BESS. Since the BESS deals with high demand peaks, it follows a similar distribution as the top part of the industrial load profile for the corresponding scenario. Scenarios SB and WB, i.e., when the grid delivers as much power as demanded by the load, implies no contribution from the BESS.

For P_{LIMIT} values below 30 kW, BESSs are not able to recharge the same amount of energy delivered during the high-demand period, both for summer and winter scenarios. It must be noticed that P_{LIMIT} also affects the maximum power available for recharging, i.e., for a P_{LIMIT} value of 20 kW, the maximum power available for recharging is 20 kW since that recharging power is coming from the grid, which is not sufficient for the case with P_{LIMIT} values below 30 kW.

Based on the BESS power profiles, the power and energy requirements ($P_{BESS-req}$ and $E_{BESS-req}$) for the valid scenarios are shown in Table 4.

Once the requirements are calculated, the most suitable BESS configuration must be selected for each scenario (number of cells connected in series and in parallel). Commercial three-phase inverters for 400 V grids demand 48 V in the DC side. Hence, based on the standard cell parameters, the number of cells in series so the DC side voltage reaches 48 V equals:

$$\frac{48V_{DC-side-voltage}}{3.7V_{Nominal-cell-voltage}} \approx 13 \tag{15}$$

The number of cells in parallel depends on the BESS energy requirements. Given that one branch of 13 cells accounts for:

$$13 \text{ cells} \cdot 3.7 \text{ V/cell} \cdot 55 \text{ Ah} = 2.64 \text{ kWh} \tag{16}$$

The following table can be filled:

Table 4. BESS requirements for the different scenarios.

Scenario	P_{LIMIT}	$P_{BESS-req}$	$E_{BESS-req}$	Min. BESS Config.	E_{BESS}
-	[kW]	[kW]	[kWh]	-	[kWh]
SB	70	-	-	-	
S6	60	7.13	0.015	13s1p	2.64
S5	50	17.13	0.12	13s1p	2.64
S4	40	27.13	2.07	13s1p	2.64
S3	30	37.13	34.79	13s14p	36.96
WB	90	-	-	-	
W8	80	3.45	0.01	13s1p	2.64
W7	70	13.45	0.05	13s1p	2.64
W6	60	23.45	0.43	13s1p	2.64
W5	50	33.45	11.51	13s5p	13.2
W4	40	43.45	49.78	13s19p	50.16
W3	30	53.45	101.14	13s39p	102.96

Scenarios S6 to S4 and W8 to W6 can satisfy their energy requirements with just one BESS branch in parallel. It must be noticed that, given the BESS configuration and the power demand shown in Table 4, the BESS in scenario S4 will be more demanded than in scenarios S5 and S6, and the BESS in scenario W6 will also be more demanded than scenarios W7 and W8, i.e., it will work with higher C rates.

4.4. Step 4. Calculate BESS Ageing

BESS ageing is closely related to its C rate and temperature (among other factors), as previously explained in Section 3. The higher the C rate, the higher the temperature augmentation in the battery. Hence, the BESS configuration has a huge influence on the BESS ageing, since batteries with a higher number of cells in parallel work with lower C rates, i.e., with less stress. This implies that "bigger" BESSs may last longer than "smaller" BESSs, which may be beneficial from a cost–benefit point of view.

Four configurations have been tested for each BESS: the minimum number of cells so that the energy requirements are satisfied, and three more configurations increasing one extra branch each time, so that the C rate is decreased.

Commonly, a BESS is considered as aged when it loses 20% of its capacity retention.

As seen in Figure 11, Scenario S4 presents a very different behaviour from the minimum configuration to the minimum plus one. This happens because the C rate is diminished to a half when the branches in parallel are increased by one (from 1 to 2), which is a big difference according to the ageing model. A higher number of branches in parallel do not improve that much the behaviour, although the ageing is further reduced, as expected.

Scenario S3, which is the summer scenario with more energetic demand from the BESS, does not present a high ageing. This happens because the highest C rates are close to 1C, and there are few moments where those C rates are reached, even though the BESS is dispatching power during longer periods along a day. However, ageing for Scenario S3 is higher than Scenario S6, the BESS is working with a less intermittent profile, implying that temperature is higher during a cycle. Scenario S5 presents a higher ageing that Scenario S3 for the minimum BESS configuration, given that the C rate is higher (close to 3C), although it works for shorter times with that C rate. In any case, a low ageing is expected for every scenario except for Scenario S4.

For winter scenarios, as seen in Figure 12, Scenarios W8 and W7 present a low ageing due to the few periods of time under which the BESS is dispatching power, even though the C rate for W7 and the minimum BESS configuration reaches 5C. Besides, a low ageing is expected for Scenarios W3 and W4 which, being the most energetic form the point of view of the BESS, are hugely over-dimensioned in terms of power (very low C rates).

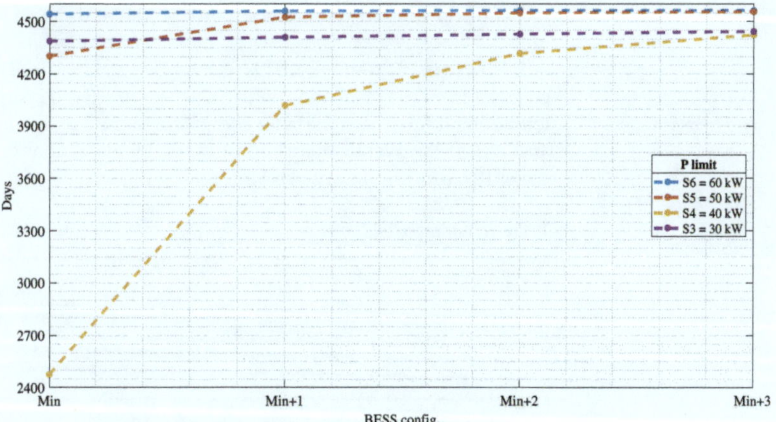

Figure 11. Days until $\nabla Q = 20\%$ as a function of the BESS configuration in summer.

Scenario W6 presents a similar behaviour as S4, where increasing the number of branches in parallel from 1 to 2 prevents from an excessive ageing, which is produced as a consequence of working with C rates higher than 8C for a few moments along a day. Scenario W5 steadily improves its ageing as the branches in parallel increase, although it suffers from the highest ageing except for the minimum BESS configuration.

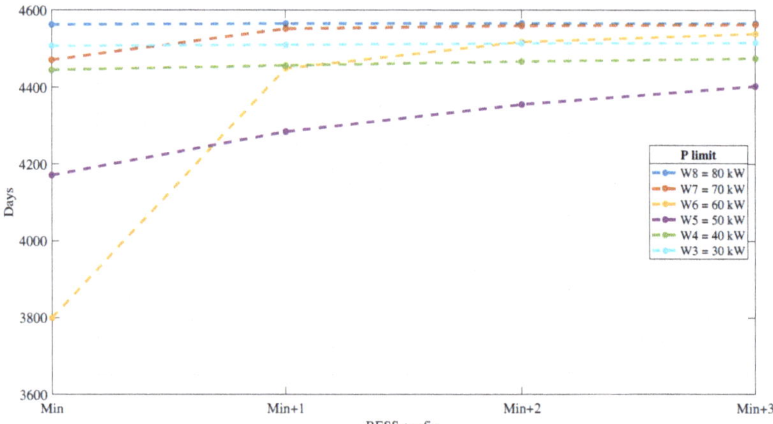

Figure 12. Days until $\nabla Q = 20\%$ as a function of the BESS configuration in winter.

4.5. Step 5. Cost–Benefit Analysis

Once the BESS ageing for different configurations and scenarios has been calculated, a cost–benefit analysis must be performed, considering both the BESS cost and fixed term energy cost. Among the different reports found in the literature that analyse energy storage costs, the one published in [41] has been selected as the most reliable study.

Moreover, electricity access fees for industrial consumption for the regulated market in Spain can be found in [42]. The fixed term of the access fee considers three time periods of time-discrimination (P1, P2, and P3). For further information about the Spanish regulated market, and time discrimination periods, the authors recommend the work published in [42].

The variable term of energy is kept constant for every scenario, i.e., the BESS is recharged from the grid, and the BESS energy recharged form the grid plus the industrial load demand matches the total energy consumed by the industrial load in scenarios SB and WB, so the overall energy consumption remains constant. Hence, it is not included in the cost. Costs are summarized in Table 5 [42].

Table 5. Summarized costs.

Item	Cost
-	[€/kW and Year]
P1	59.1734
P2	39.4906
P3	8.3677

Considering this, the cost equation can be written as:

$$Cost(\frac{euro}{year}) = (P1 + P2 + P3) \cdot P_{LIMIT} + BESS_{project-cost} \tag{17}$$

The contracted power (P_{LIMIT}) is considered constant for the three time-discrimination periods (as is requested to industrial consumption).

The yearly BESS project cost has been calculated based on [43–45]. The annualized cost of the BESS project has been defined as:

$$C_{ann} = CRF(i, R_{proj}) \cdot C_{NPC} \tag{18}$$

where C_{ann} is the annualized cost, C_{NPC} is the net present cost, and CRF is the capital recovery factor, which depends on the annual real discount rate (i) and on the project lifetime (R_{proj}). C_{NPC} is calculated as the present value of all the costs of installing and operating the battery over the project lifetime, minus the present value of all the revenues earned over the project lifetime.

Hence, the total yearly cost is calculated as follows:

$$Cost(\frac{euro}{year}) = (P1 + P2 + P3) \cdot P_{LIMIT} + C_{ann} = (P1 + P2 + P3) \cdot P_{LIMIT} + CRF(i, R_{proj}) \cdot C_{NPC} \tag{19}$$

The constants for calculating the BESS project cost are listed in Table 6 [41,43],

Table 6. Summarized costs for the BESS project.

Item	Value [Unit]
BESS cost (€/kW)	1876 €/kW
BESS cost (€/kWh)	469 €/kWh
C_{OM}	10 €/kW and year
i	3.41%
N_{proj}	25 years

where C_{OM} is the operation and maintenance cost, and N_{proj} is the number of years over which the cost is annualized. The total cost for each scenario is shown in Tables 7 and 8, where:

$$Total(\frac{euro}{year}) = Fixed - access - fee + BESS - project \tag{20}$$

Table 7. Costs for the summer scenarios.

Scenario	P_{LIMIT}	BESS Config.	Fixed Access Fee	BESS Project	Total Cost
-	[kW]	-	[€/Year]	[€/Year]	[€/Year]
SB	90	-	7492.2	-	7492.2
S6	60	13s1p	6421.9	940.3	7362.2
		+1p		989.8	7411.7
		+2p		1039.7	7461.6
		+3p		1089.6	7511.5
S5	50	13s1p	5351.6	2074	7425.6
		+1p		2118.3	7469.9
		+2p		2167.5	7519.1
		+3p		2217.1	7568.7
S4	40	13s1p	4281.3	3291	7572.3
		+1p		3265.5	7546.8
		+2p		3305.5	7586.8
		+3p		3351.6	7632.9
S3	30	13s14p	3211	6098.5	9309.5
		+1p		6141.6	9352.6
		+2p		6185.9	9396.9
		+3p		6231.1	9442.1

Table 8. Costs for the winter scenarios.

Scenario	P_{LIMIT}	BESS Config.	Fixed Access Fee	BESS Project	Total
-	[kW]	-	[€/Year]	[€/Year]	[€/Year]
WB	90	-	9632.9	-	9632.9
W8	80	13s1p	8562.5	525.1	9087.6
		+1p		574.9	9137.4
		+2p		624.8	9187.3
		+3p		674.7	9234.2
W7	70	13s1p	7492.2	1654.6	9146.8
		+1p		1702.6	9194.8
		+2p		1752.2	9244.4
		+3p		1802	9294.2
W6	60	13s1p	6421.9	2802.5	9224.4
		+1p		2833.4	9255.3
		+2p		2881.3	9303.2
		+3p		2930.5	9352.4
W5	50	13s5p	5351.6	4487.7	9839.3
		+1p		4523.9	9875.5
		+2p		4565	9916.6
		+3p		4609.1	9960.7
W4	40	13s19p	4281.3	7460	11,741.3
		+1p		7509	11,790.3
		+2p		7554.7	11,836
		+3p		7601.4	11,882.7
W3	30	13s39p	3211	11,224	14,435
		+1p		11,271	14,482
		+2p		11,319	14,530
		+3p		11,367	14,578

As it can be seen in Tables 7 and 8, results indicate that introducing a BESS into an industrial consumption for diminishing the fixed term of the electricity fee, can reduce the overall yearly cost. Moreover, this reduction can enhance an intelligent use of the energy in industrial buildings, as well as the improvement of the energy efficiency along the electric power grid, since consumption peaks are eliminated.

Tables 7 and 8 show that the minimum reduction in the fixed power limit is the best solution for this particular case example, both for summer and winter, from the point of view of the industrial consumer. Every configuration in scenario S6, and two of them in Scenario S5, reduce the yearly price of having a higher contracted power in summer. A similar behaviour is obtained in Table 8 for Scenarios W8, W7 and W6. The rest of the scenarios do not improve the yearly cost for this case example, although some interesting conclusions can be drawn.

Scenario S4 shows that the minimum configuration might not be the optimal solution, since configuration 13s2p provides a higher cost reduction than the minimum BESS configuration. A higher number of branches in parallel allows for a reduced C-rate, DoD, and temperature in each scenario. Hence, the BESS lifespan is extended and a higher cost reduction is achieved. A similar result would be expected for Scenario W6, given the ageing behaviour shown in Figure 12. However, in this case, the ageing is sufficiently low with the minimum BESS configuration, so the extra benefit of the further reduction in ageing does not improve that much the yearly cost.

Costs are not reduced in scenarios S4, S3, W5, W4, and W3 with respect to the base case scenario, since the BESS is dedicated not only to reducing the peak consumption, but to dealing with an important part of the load demand. This forces an over-dimensioning in the BESS's number of cells in parallel (given that the number of cells in series is fixed), and indicates that those BESS configurations may not be the best option for feeding an industrial infrastructure by themselves since, even though technically viable, it is costly.

5. Conclusions

This paper presents a BESS dimensioning methodology for reducing the fixed term of the electricity access fee for industrial consumption, considering battery ageing and the cost of the BESS and electricity access fees. By connecting a BESS together with a consumption, the power demand peaks can be reduced, so there is no need for contracting a high fixed term power to cover the high demand peaks.

A five-step dimensioning methodology is described, finishing with a cost–benefit analysis to facilitate the final decision making. The methodology is easily implemented and can be adapted to different electricity market conditions, as well as particularized to different BESS or load conditions. The parameter that defines the different scenarios of the methodology is the power contracted from the grid, being the decision variable the combined cost of the BESS and the fixed electricity fee.

Results drawn from the two case examples show that implementing a small battery to limit the consumption peaks can lead to considerable cost reduction for the industrial consumer. In fact, the best techno-economical solution corresponds to a small BESS that limits the peak consumption. The low C rate, DoD, and steady temperature experienced by a BESS that only limits the top part of the peak consumption, guarantees a long lifespan that highly reduces the overall yearly costs. Given that BESS are over-dimensioned in energy in several scenarios (S6, S5, W8, W7, and W6), a higher contribution from the BESS could be imposed in order to reduce the fixed power contracted from the grid, achieving a higher peak reduction. However, this leads to an increased ageing (higher C rate, DoD, and temperature profile) that has an impact on the techno-economical decision, turning those solutions into non-optimum.

Apart from the previous conclusion, depending on the case example and scenario under study, it could be beneficial from the economical point of view to increase the size of the BESS with more cells in parallel. This situation occurs when the BESS energy is suited to the BESS energy requirements (scenario S4). This forces the BESS to work with a high DoD and high C rates, deriving in a premature ageing. Thus, increasing the BESS size will allow it to work under less stress, extending the time needed to a BESS replacement and reducing the overall yearly costs.

One important conclusion drawn form the case examples results is that the optimum BESS is over-dimensioned in both power and energy. This may indicate that other energy storage technologies different than Li-ion batteries could be more beneficial from the techno-economic point of view. Supercapacitors, which can last millions of cycles until they need a replacement, or flow batteries, which can dissociate energy and power, may be suitable for their inclusion in a further study. In addition, if the industrial consumption is facing a scenario with low consumption peaks, the optimum solution could be discarding the usage of an energy storage system at all. As previously stated, the presence of high consumption peaks is the reason that justifies the inclusion of a BESS together with the industrial consumption.

Consequently, the proposed methodology is of potential interest for industrial consumers, since the cost-benefit analysis may reduce the yearly cost of their infrastructure, and also for TSOs and DSOs, since the peak reduction improves the efficiency of the transmission and distribution of the energy along the electric power grid. Moreover, BESS manufacturers could also find benefits in this methodology for the development of specific products related to this industrial demand peak and fixed term fee reduction.

Suggested future works include the parametrizations of the grid with frequency and voltage variations, so that further analysis considering BESS contributions to the grid in terms of frequency and voltage regulation can be performed. Moreover, time discrimination pricing can be added to electricity energy prices, so that the BESS can recharge during low price periods, providing an even higher reduction than the one shown in this work. At last, a techno-economic evaluation considering other energy storage technologies such as supercapacitors or flow batteries may result in interesting conclusions.

Author Contributions: Conceptualization, J.N. and M.L.; methodology, J.N., M.B. and G.N.; software, J.N. and M.B.; validation, J.T. and M.B.; formal analysis, J.N.; investigation, J.N.; data curation, M.S.-H., J.T. and M.B.; writing—original draft preparation, J.N.; writing—review and editing, M.S.-H., M.L. and G.N.; visualization, J.N., J.T. and M.S.-H.; supervision, M.L. All authors have read and agreed to the published version of the manuscript.

Funding: This research received no external funding.

Institutional Review Board Statement: Not applicable.

Informed Consent Statement: Not applicable.

Conflicts of Interest: The authors declare no conflicts of interest.

Abbreviations

The following abbreviations are used in this manuscript:

BESS	Battery Energy Storage System
DoD	Depth of Discharge
ESS	Energy Storage System
PCC	Point of common coupling
SoC	State of Charge
VSC	Voltage Source Converter

References

1. Eurostat, S.E. Electricity Price Statistics. Available online: https://ec.europa.eu/eurostat/statistics-explained/index.php?title=Electricity_price_statistics#Electricity_prices_for_non-household_consumers (accessed on 11 August 2021).
2. Jardini, J.A.; Tahan, C.M.; Gouvea, M.R.; Ahn, S.U.; Figueiredo, F. Daily load profiles for residential, commercial and industrial low voltage consumers. *IEEE Trans. Power Deliv.* **2000**, *15*, 375–380. [CrossRef]
3. Svetozarevic, B.; Begle, M.; Jayathissa, P.; Caranovic, S.; Shepherd, R.F.; Nagy, Z.; Hischier, I.; Hofer, J.; Schlueter, A. Dynamic photovoltaic building envelopes for adaptive energy and comfort management. *Nat. Energy* **2019**, *4*, 671–682. [CrossRef]
4. Biyik, E.; Araz, M.; Hepbasli, A.; Shahrestani, M.; Yao, R.; Shao, L.; Essah, E.; Oliveira, A.C.; Del Cano, T.; Rico, E.; et al. A key review of building integrated photovoltaic (BIPV) systems. *Eng. Sci. Technol. Int. J.* **2017**, *20*, 833–858. [CrossRef]
5. Niu, J.; Tian, Z.; Lu, Y.; Zhao, H. Flexible dispatch of a building energy system using building thermal storage and battery energy storage. *Appl. Energy* **2019**, *243*, 274–287. [CrossRef]
6. Lizana, J.; Chacartegui, R.; Barrios-Padura, A.; Ortiz, C. Advanced low-carbon energy measures based on thermal energy storage in buildings: A review. *Renew. Sustain. Energy Rev.* **2018**, *82*, 3705–3749. [CrossRef]
7. e Silva, G.d.O.; Hendrick, P. Pumped hydro energy storage in buildings. *Appl. Energy* **2016**, *179*, 1242–1250. [CrossRef]
8. Durand, J.M.; Duarte, M.J.; Clerens, P. European energy storage technology development roadmap towards 2030. *Int. Energy Storage Policy Regul Work* **2017**, *108*, 1–128.
9. EASE—Storage Applications. Available online: http://ease-storage.eu/energy-storage/applications/ (accessed on 15 April 2021).
10. Yassin, M.A.; Kolhe, M.; Sharma, A.; Garud, S. Battery capacity estimation for building integrated photovoltaic system: Design study for different geographical location (s). *Energy Procedia* **2017**, *142*, 3433–3439. [CrossRef]
11. Salpakari, J.; Lund, P. Optimal and rule-based control strategies for energy flexibility in buildings with PV. *Appl. Energy* **2016**, *161*, 425–436. [CrossRef]
12. Milis, K.; Peremans, H.; Van Passel, S. Steering the adoption of battery storage through electricity tariff design. *Renew. Sustain. Energy Rev.* **2018**, *98*, 125–139. [CrossRef]
13. Davis, M.; Hiralal, P. Batteries as a service: A new look at electricity peak demand management for houses in the UK. *Procedia Eng.* **2016**, *145*, 1448–1455. [CrossRef]
14. Mulder, G.; Six, D.; Claessens, B.; Broes, T.; Omar, N.; Van Mierlo, J. The dimensioning of PV-battery systems depending on the incentive and selling price conditions. *Appl. Energy* **2013**, *111*, 1126–1135. [CrossRef]
15. Jankowiak, C.; Zacharopoulos, A.; Brandoni, C.; Keatley, P.; MacArtain, P.; Hewitt, N. Assessing the benefits of decentralised residential batteries for load peak shaving. *J. Energy Storage* **2020**, *32*, 101779. [CrossRef]
16. Pandžić, H. Optimal battery energy storage investment in buildings. *Energy Build.* **2018**, *175*, 189–198. [CrossRef]
17. Fotouhi, A.; Auger, D.J.; Propp, K.; Longo, S.; Wild, M. A review on electric vehicle battery modelling: From Lithium-ion toward Lithium–Sulphur. *Renew. Sustain. Energy Rev.* **2016**, *56*, 1008–1021. [CrossRef]
18. Zhang, L.; Hu, X.; Wang, Z.; Sun, F.; Dorrell, D.G. A review of supercapacitor modeling, estimation, and applications: A control/management perspective. *Renew. Sustain. Energy Rev.* **2018**, *81*, 1868–1878. [CrossRef]

19. Amiribavandpour, P.; Shen, W.; Kapoor, A. Development of thermal-electrochemical model for lithium ion 18650 battery packs in electric vehicles. In Proceedings of the 2013 IEEE Vehicle Power and Propulsion Conference (VPPC), Beijing, China, 15–18 October 2013; pp. 1–5.
20. Rao, V.; Singhal, G.; Kumar, A.; Navet, N. Battery model for embedded systems. In Proceedings of the 18th International Conference on VLSI Design held jointly with 4th International Conference on Embedded Systems Design, Kolkata, India, 3–7 January 2005; pp. 105–110.
21. Lam, L.; Bauer, P.; Kelder, E. A practical circuit-based model for Li-ion battery cells in electric vehicle applications. In Proceedings of the 2011 IEEE 33rd International Telecommunications Energy Conference (INTELEC), Amsterdam, The Netherlands, 9–13 October 2011; pp. 1–9.
22. Tremblay, O.; Dessaint, L.A.; Dekkiche, A.I. A generic battery model for the dynamic simulation of hybrid electric vehicles. In Proceedings of the 2007 IEEE Vehicle Power and Propulsion Conference, Arlington, TX, USA, 9–12 September 2007; pp. 284–289.
23. Fan, L. *Control and Dynamics in Power Systems and Microgrids*; CRC Press: Boca Raton, FL, USA, 2017.
24. Allègre, A.L.; Trigui, R.; Bouscayrol, A. Different energy management strategies of Hybrid Energy Storage System (HESS) using batteries and supercapacitors for vehicular applications. In Proceedings of the 2010 IEEE Vehicle Power and Propulsion Conference, Lille, France, 1–3 September 2010; pp. 1–6.
25. Laldin, O.; Moshirvaziri, M.; Trescases, O. Predictive algorithm for optimizing power flow in hybrid ultracapacitor/battery storage systems for light electric vehicles. *IEEE Trans. Power Electron.* **2012**, *28*, 3882–3895. [CrossRef]
26. Zhou, F.; Xiao, F.; Chang, C.; Shao, Y.; Song, C. Adaptive model predictive control-based energy management for semi-active hybrid energy storage systems on electric vehicles. *Energies* **2017**, *10*, 1063. [CrossRef]
27. Yazdani, A.; Iravani, R. *Voltage-Sourced Converters in Power Systems*; Wiley Online Library: Hoboken, NJ, USA, 2010; Volume 39.
28. Concha, P.M.T. Analysis and Design Considerations of an Electric Vehicle Powertrain regarding Energy Efficiency and Magnetic Field Exposure. Ph.D. Thesis, ETSI Industriales (UPM), Madrid, Spain, 2016.
29. Nájera, J.; Moreno-Torres, P.; Lafoz, M.; De Castro, R.M.; R Arribas, J. Approach to hybrid energy storage systems dimensioning for urban electric buses regarding efficiency and battery aging. *Energies* **2017**, *10*, 1708. [CrossRef]
30. Omar, N.; Monem, M.A.; Firouz, Y.; Salminen, J.; Smekens, J.; Hegazy, O.; Gaulous, H.; Mulder, G.; Van den Bossche, P.; Coosemans, T.; et al. Lithium iron phosphate based battery–Assessment of the aging parameters and development of cycle life model. *Appl. Energy* **2014**, *113*, 1575–1585. [CrossRef]
31. Shepherd, C.M. Design of primary and secondary cells: II. An equation describing battery discharge. *J. Electrochem. Soc.* **1965**, *112*, 657. [CrossRef]
32. Tremblay, O.; Dessaint, L.A. Experimental validation of a battery dynamic model for EV applications. *World Electr. Veh. J.* **2009**, *3*, 289–298. [CrossRef]
33. Nájera, J. Study and Analysis of the Behavior of LFP and NMC Electric Vehicle Batteries Concerning Their Ageing and Their Integration into the Power Grid. Ph.D. Thesis, Universidad Politécnica de Madrid, Madrid, Spain, 2021.
34. Saw, L.; Somasundaram, K.; Ye, Y.; Tay, A. Electro-thermal analysis of Lithium Iron Phosphate battery for electric vehicles. *J. Power Sources* **2014**, *249*, 231–238. [CrossRef]
35. Chen, Y.; Evans, J.W. Three-dimensional thermal modeling of lithium-polymer batteries under galvanostatic discharge and dynamic power profile. *J. Electrochem. Soc.* **1994**, *141*, 2947. [CrossRef]
36. Rao, L.; Newman, J. Heat-generation rate and general energy balance for insertion battery systems. *J. Electrochem. Soc.* **1997**, *144*, 2697. [CrossRef]
37. Redondo-Iglesias, E.; Venet, P.; Pelissier, S. Calendar and cycling ageing combination of batteries in electric vehicles. *Microelectron. Reliab.* **2018**, *88*, 1212–1215. [CrossRef]
38. Wang, J.; Purewal, J.; Liu, P.; Hicks-Garner, J.; Soukazian, S.; Sherman, E.; Sorenson, A.; Vu, L.; Tataria, H.; Verbrugge, M.W. Degradation of lithium ion batteries employing graphite negatives and nickel–cobalt–manganese oxide+ spinel manganese oxide positives: Part 1, aging mechanisms and life estimation. *J. Power Sources* **2014**, *269*, 937–948. [CrossRef]
39. Iglesias, E.R. Étude du Vieillissement des Batteries Lithium-ion Dans Les Applications "Véhicule Électrique": Combinaison des Effets de Vieillissement Calendaire et de Cyclage. Ph.D. Thesis, Université de Lyon, Lyon, France, 2017.
40. Kokam. *SLBP 55Ah High Power Superior Lithium Polymer Battery*; Technical Report; Kokam: Siheung, Korea, 2015.
41. Mongird, K.; Viswanathan, V.V.; Balducci, P.J.; Alam, M.J.E.; Fotedar, V.; Koritarov, V.S.; Hadjerioua, B. *Energy Storage Technology and Cost Characterization Report*; Technical Report; Pacific Northwest National Lab. (PNNL): Richland, WA, USA, 2019.
42. Gobierno de España. *Informe de Precios Energéticos Regulados*; Technical Report; Instituto para la Diversificación y Ahorro de la Energía (IDAE): Madrid, Spain, 2019.
43. Mongird, K.; Viswanathan, V.; Balducci, P.; Alam, J.; Fotedar, V.; Koritarov, V.; Hadjerioua, B. An evaluation of energy storage cost and performance characteristics. *Energies* **2020**, *13*, 3307. [CrossRef]
44. Budes, F.B.; Ochoa, G.V.; Escorcia, Y.C. An Economic Evaluation of Renewable and Conventional Electricity Generation Systems in a Shopping Center Using HOMER Pro®. *Contemp. Eng. Sci.* **2017**, *10*, 1287–1295. [CrossRef]
45. Mehta, S.; Basak, P. A case study on pv assisted microgrid using homer pro for variation of solar irradiance affecting cost of energy. In Proceedings of the 2020 IEEE 9th Power India International Conference (PIICON), Sonepat, India, 28 February–1 March 2020; pp. 1–6.

Article

Conversion of a Network Section with Loads, Storage Systems and Renewable Generation Sources into a Smart Microgrid

Oscar Izquierdo-Monge [1], Paula Peña-Carro [1], Roberto Villafafila-Robles [2], Oscar Duque-Perez [3,*], Angel Zorita-Lamadrid [3] and Luis Hernandez-Callejo [4,*]

[1] CEDER-CIEMAT, Autovía de Navarra A15 Salida 56, 422290 Lubia (Soria), Spain; oscar.izquierdo@ciemat.es (O.I.-M.); paula.pena@ciemat.es (P.P.-C.)

[2] Centre d'Innovació Tecnològica en Convertidors Estàtics i Accionaments (CITCEA-UPC), Departament d'Enginyeria Elèctrica, Universitat Politècnica de Catalunya, ETS d'Enginyeria Industrial de Barcelona, Av. Diagonal 647, 2nd Floor, 08028 Barcelona, Spain; roberto.villafafila@citcea.upc.edu

[3] ADIRE-ITAP, University of Valladolid, Paseo del Cauce 59, 47011 Valladolid, Spain; zorita@eii.uva.es

[4] ADIRE-ITAP, University of Valladolid, Campus Universitario Duques de Soria, 42004 Soria, Spain

[*] Correspondence: oscar.duque@eii.uva.es (O.D.-P.); luis.hernandez.callejo@uva.es (L.H.-C.); Tel.: +34-975129418 (L.H.-C.)

Citation: Izquierdo-Monge, O.; Peña-Carro, P.; Villafafila-Robles, R.; Duque-Perez, O.; Zorita-Lamadrid, A.; Hernandez-Callejo, L. Conversion of a Network Section with Loads, Storage Systems and Renewable Generation Sources into a Smart Microgrid. *Appl. Sci.* **2021**, *11*, 5012. https://doi.org/10.3390/app11115012

Academic Editors: Víctor Alonso Gómez, Sergio Nesmachnow, Vicente Leite, Javier Prieto, Ângela Ferreira and Hannu Laaksonen

Received: 1 May 2021
Accepted: 26 May 2021
Published: 28 May 2021

Publisher's Note: MDPI stays neutral with regard to jurisdictional claims in published maps and institutional affiliations.

Abstract: This paper shows an experimental application case to convert a part of the grid formed by renewable generation sources, storage systems, and loads into a smart microgrid. This transformation will achieve greater efficiency and autonomy in its management. If we add to this the analysis of all the data that has been recorded and the correct management of the energy produced and stored, we can achieve a reduction in the electricity consumption of the distribution grid and, with this, a reduction in the associated bill. To achieve this transformation in the grid, we must provide it with intelligence. To achieve this, a four steps procedure are proposed: identification and description of the elements, integration of the elements in the same data network, establishing communication between the elements and the control system, creating an interface that allows control of the entire network. The microgrid of CEDER-CIEMAT (Renewable Energy Centre in Soria, Spain) is presented as a real case study. This centre is made up of various sources of generation, storage, and consumption. All the elements that make up the microgrid are incorporated into free software, Home Assistant, allowing real-time control and monitoring of all of them thanks to the intelligence that has been provided to the grid. The novelty of this paper is that it describes a procedure that is not reported in the current literature and that, being developed with Home Assistant, is free and allows the control and management of a microgrid from any device (mobile, PC) and from any place, even though not on the same data network as the microgrid.

Keywords: smart microgrid; Home Assistant; monitoring and control system

1. Introduction

Energy has become a key factor in the development of our lives, leading to an increase in energy needs at a global level, presenting a challenge for the supply of the traditional electricity system. For years, different solutions have been implemented to guarantee supply to all citizens, trying to ensure that the electricity supply comes from different renewable energy sources, making the system as distributed as possible. With the implementation of this diversification, the aim is to improve efficiency, reduce transport losses, and make better use of renewable energy sources, giving rise to microgenerators and with them options such as microgrids and smartgrids.

Microgrid is a term that can be defined as the U.S. Department of Energy [1] proposes: "a group of interconnected loads and distributed energy resources within clearly defined electrical boundaries that act as a single controllable entity respect to the grid. A microgrid can be connected and disconnected from the distribution grid to allow it to operate in grid or island mode. A remote microgrid is a variation of a microgrid operating in island conditions".

Another option is the smartgrid, a newer concept than microgrid but increasingly implemented due to the advantages it offers. It can be defined as a microgrid that has been equipped with an intelligent communications infrastructure, allowing it to operate and act semi-autonomously and/or autonomously on all the elements that make up the microgrid. This management method improves reliability, safety, and efficiency. It optimizes energy flows and improves the detection of grid supply shortages, making quick decisions to provide the necessary supply to the microgrid.

In either case, whether it is a microgrid, or a smartgrid, the installation of storage systems (such as batteries, super-capacitors, or flywheels) is highly recommended. They can counteract the energy imbalances produced [2] and curb energy fluctuations, managing to adjust the generation and overall load curve of the microgrid [3,4], which is affected by the uncontrollable generation associated with such systems [5–8].

The current trend is either the creation of smartgrid or the conversion of an already created microgrid into a smartgrid due to the operational improvements that can be obtained from the whole.

Nowadays, there are many installations or buildings that have micro-generators and storage systems but do not have them integrated in an intelligent way, which means that these installations are not efficient. The problem that this work aims to solve is to show how these networks can be made intelligent in a straightforward way and using free software.

The procedure described in this paper has already been implemented in practice, being the main objective of this work to present the procedure followed at CEDER to convert a grid section with renewable generation sources and different storage systems and loads into a smart microgrid. To achieve this, the Home Assistant tool has been used to create a control software (CMEMS-CEDER Microgrid Energy Management System).

All the elements that make up the microgrid are integrated in CMEMS, which allows real-time monitoring and management to achieve greater efficiency and autonomy.

This procedure consists of four steps: identification and description of the elements, integration of the elements in the same data network, establishment of communication between the elements and the control system, creation of an interface that allows control of the entire network.

Although the work describes a procedure for an "experimental application case", it could be applied generically to other similar network sections with generation, storage, and consumption elements, since the steps described are valid for any grid.

The development of the article continues in Section 2, which presents a revision of literature. Section 3 present the four steps procedure to transform a grid section into a microgrid. Section 4 presents the experimental application case. Finally, conclusions are drawn.

2. Revision of Literature

To manage the entire microgrid more efficiently, it is necessary to install a metering, communication, and data processing system to allow semi-autonomous or autonomous operation. In this way, possible supply problems can be solved in a shorter period [9,10]. Within this field, we can find various communication technologies, such as Zigbee, WLAN/Wi-Fi, serial communication (Ethernet/RS-232), WiMAX, power line communication, GSM/GPRS, and DASH 7 smart [11]. Among them, the most used are the first mentioned, due to the security, reliability, and scope they offer [12–15]. The implementation of one or another technology is influenced by the communication distance and the budget available for this component [16].

Independently of the communication technology chosen, every microgrid must have the following communication elements: a defined local communication structure, a hierarchical system of supervision, control and management of the different elements, and intelligent controllers for loads, consumption, and storage systems.

As discussed above, a microgrid can operate either connected to the main distribution grid via the point of common coupling (PCC) or disconnected from it, in island mode. A stable and economically efficient mode of operation is required [16].

The microgrid is governed by a central controller (MGCC). It provides set point signals to the equipment controllers, such as generation systems, storage systems, and loads. The control system is responsible for regulating the operating voltage and frequency. It is also responsible for redistributing the load between the different distributed generation (DG) and storage elements, managing the flow with the main grid, and optimising operating costs.

In order to control all the generation systems and consumption systems that make up the microgrid and to ensure stable operation and economic efficiency [17], it is necessary to have the installation of smart meters in each piece of the equipment (in case the device does not have an integrated measurement and communication system). The smart meters will monitor in real time different parameters such as voltage, current, power, consumption, etc. to know instantly the operating status, and the existence of any problem that may arise if communication with the generation and/or storage device that records all these parameters is impossible, or to monitor more parameters than those offered by the device. To transmit the data collected by the smart meters to an HMI (Human Machine Interface) in which users can read these data and send operating instructions, it is necessary to install another device.

Arduino, which consists of a board with a microcontroller and open software, can be used for this purpose. This device can read the data collected by the Smart Meters, perform calculations if necessary, and send the results where the user wishes. Arduino can communicate with a Raspberry Pi [14,18], a computer with a reduced board with which greater versatility and calculation power is achieved. Raspberry has Wi-Fi and Ethernet connectivity integrated in the board. These devices are responsible for collecting the data registered by Arduino and communicating all these values with other devices, platforms the user has established, and applications and/or local or online databases to store and visualize all those values. Users who want to collect historical data of all the parameters that the installed devices record, have two different options: the manual download when the device memory is full, or the automated sending to a database (normally in the cloud) [13,19–21].

For everything to work, communication protocols (mentioned above) must be established between the different installed devices. The protocol selected differs for each case, device, and location.

In order to manage all the equipment, it is necessary to display all the registers in an interface and allow the user to interact with each one of them. For this purpose, the software is used to act semi-autonomously and/or autonomously on the different equipment that integrates the microgrid. This operation mode will give the user the possibility to optimize the generation and demand curve and therefore reduce the cost of the electricity bill, one of the final targets of this proposal.

We can find different tools, like Matlab/Simulink, Python or another programming language (R, C++, etc.), Labview, etc. that enable the development of software to monitor and control the elements (generation sources, storage systems, and load) that make up a microgrid and turn it into a smart microgrid.

Among all the tools proposed [8,22–24], there are characteristics common to all of them that can present themselves as an advantage: custom code development. Thanks to the ability to adapt the code to each of the particular cases, they are widely used software for the configuration of all the elements of the microgrid. For the specific tool, Home Assistant, or programming languages (R, C++, Python, etc.), we can add another advantage, related to the costs derived from the control and monitoring of the microgrid [25,26]. To use the full license, it is not necessary to pay for it; it is a free program available to any user.

Concerning the disadvantages of its use in any of the cited software, it is the high amount of time dedicated to its preparation, due to the complexity of writing the code.

Home Assistant stands out from this aspect, as it does not require long programming codes for its configuration, nor high programming knowledge to implement all the elements of the microgrid in the software.

For correct management of the microgrid, as mentioned in the previous point, and to optimize, both globally and individually, all the elements that make it up, it is necessary to have software that allows the user to interact with each of them within the same interface.

Different software allows these actions. Examples of some of them are Advance EMS-Platform, ETAP Energy Management System, Monarch TM-Open Systems International, Wattics, etc. In our case, we have developed our own software (CMEMS) using Home Assistant due to its advantages and the total coverage of the required needs. Table 1 shows the advantages and disadvantages of each of the applications mentioned above.

Table 1. Advantages and disadvantages of different software. Source: prepared by the authors from various documents [22–26]. Own elaboration.

Software	Advantages	Disadvantages
Advance EMS–Platform [27]	Alarm notification Real-time readings and management Access control Historical data Desktop version	Paid license Only for Wind, PV, and Storage No smartphone version Development and configuration by the company
ETAP Energy Management System [28]	Alarm notification Real-time readings and management Access control Historical data Desktop version	Paid license No smartphone version Development and configuration by the company
Monarch TM–Open Systems International [29]	Alarm notification Real-time readings and management Access control Historical data Desktop and smartphone version	Paid license No readings and management for storage systems Development and configuration by the company
Wattics [30]	Read data from meters and sensors Alarm notification Real-time readings and management Access control Historical data Desktop and smartphone version	Paid license Development and configuration by the company
CMEMS (developed with Home Assistant [31]	Open-source No limitation on the type and number of equipment to be monitored and managed Alarm notification Real-time readings and management Access control Historical data Desktop and smartphone version Development and configuration by yourself	

The software mentioned above have several features in common, such as real-time monitoring, alarm notification, secure access, or access from a computer. Home Assistant has been the software of choice mainly for three reasons that make it stand out from the other options: it does not require a license fee to use any of the levels offered by the program, which reduces the overall cost of installation and control of the microgrid [25,26]; it can be configured and edited at any time, without dependence on the company that developed it; and any generation, storage, or load system can be registered without any limitation, making it very suitable for this case study.

3. Procedure for the Conversion of a Network Section into a Smart Microgrid

As mentioned in Section 1, this document aims at presenting an experimental application case to provide intelligence to an electricity grid with different generation sources, storage systems, and loads and transform it into a smart microgrid that can operate with high efficiency.

Although the described procedure is applied to an "experimental application case", it could be applied in a generic way in similar grids, since the described steps are valid for any grid of these characteristics.

The procedure to convert an electrical grid with renewable generation sources, storage systems, and consumption that do not communicate with each other, into a smart microgrid is shown in Figure 1.

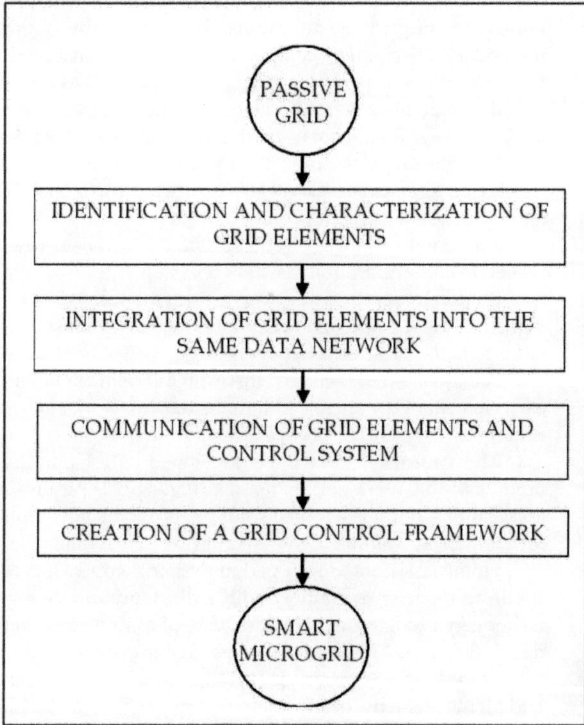

Figure 1. Stages of the procedure to transform passive grid into a smart microgrid. Source: prepared by authors.

Step 1. Identification and characterization of grid elements: the first step to provide intelligence to an electricity grid is to identify and characterize all the elements connected: generation sources, storage systems, and loads.

There are different generation sources (renewable and non-renewable) that can be integrated in an electrical microgrid. The most common are photovoltaic, small wind turbines, generator sets, and mini/micro-hydraulic turbines. For each one, it is necessary to know the nominal installed power and the range of power in which it can operate. It is also necessary to know if it is a dispatchable or non-dispatchable generation source. As shown later, if they are non-dispatchable sources, it will be enough to monitor the parameters to be integrated into the control system that is developed (mainly instantaneous power). In the case of dispatchable generation sources, it will not only be necessary to monitor the desired parameters, but it will also be necessary to develop a software (SCADA-

Supervisory Control and Data Acquisition) to operate them and integrate it into the control system of the microgrid.

It is also convenient to know the type of generation source, renewable or non-renewable, to monitor (if possible) the resource in case of renewable energies (wind speed for wind energy, solar radiation in case of photovoltaic, water level for hydraulic turbines, etc.) or fuel in case of non-renewable energies (diesel for generator sets, biomass, etc.).

In case of storage systems, the most used are batteries, or pumping associated to a mini-hydraulic turbine. This procedure can be applied to them or to any storage system such as flywheels, compressed air, etc. All of them are dispatchable, so in addition to monitoring their capacity, having energy that can store, power, performance, etc., it will be necessary, as in the case of dispatchable energies, to develop a software to operate them that must also be integrated into the microgrid control system.

In addition to the generation and storage systems, it is necessary to have a deep knowledge of grid consumptions. It is important to know the power contracted with the energy distribution company, the energy consumed monthly (if possible, at least twelve months to avoid seasonal effects), as well as the most significant loads. As with non-dispatchable generation elements, the microgrid control system cannot control the loads—they will either start or stop depending on the microgrid users. Therefore, it will only be necessary to monitor the instantaneous power and some other parameter that is considered appropriate. It will also be interesting to establish a ranking of loads by priority and influence of the load profile.

Along with all these electrical issues, it will be necessary to know and detail aspects related to communications as shown in step 3.

Step 2. Integrate grid elements into the same data network: once the elements that will be part of the microgrid are well known, they must be connected (including the control system) to the same data network.

As explained in Section 1, there are different communication technologies to integrate each element of the microgrid into the data network of the control system, such as: ZigBee, Ethernet, Wi-Fi, GRPS, GSM, WLAN.

The control system will be developed with Home Assistant, and here lies the main novelty of this work and its main advantage. Typically, it is used for home automation applications, but it is a robust solution, economically affordable, and with great potential for monitoring and managing microgrids in real time.

Home Assistant, developed in Python, is open-source software which reduces costs. It allows the user to connect with almost any device regardless of the communication technology used, through a wide range of communication protocols. In addition, it allows developing a SCADA for each dispatchable element and its integration in an interface.

Once Home Assistant is installed in a Raspberry Pi, a computer or even a server, and all the elements of the microgrid defined, it can be managed from every device and everywhere due to its web server, including a computer connected or not to the same data network or from a mobile phone due to the Home Assistant app and the web server.

To integrate each element of the grid to the data network, it is necessary to know some issues related to communications. First, it is necessary to know if each element has a communication card, which allows access to the data network, either locally or remotely (for many years, all devices have had it).

If the element does not allow communication, which is not very common except in loads and in non-dispatchable outdated generation sources, it will be necessary to install a grid analyzer with a communication card (TCP/IP, Wi-Fi, etc.) to measure parameters seen in step 1 (instantaneous power, etc.). In the case of loads, it may be interesting to monitor the most significant, but it would be enough to monitor the total consumption of the microgrid.

If the element only allows connection in local mode (usually through RS232 or RS485), it will be necessary to install some intermediate device (Arduino, data acquisition system,

Raspberry Pi, etc.) to connect it to the data network or a converter to switch local mode to remote mode (for instance RS232 to TCP/IP).

If the element allows remote connection through a network card, it can be connected directly to the data network.

Step 3. Establish communication between the elements of the grid and the control system: once the components of the grid have been identified and described, and connected to the microgrid control system data network, communication must be established between all of them, so that they do not work independently, but as a whole, and there can be an interaction between them.

To connect each device with the Home Assistant's control system, it is also necessary to know their communication protocols. There are multiple communication protocols that allow the transmission of information, provided that all the devices are connected to the same data network or to equipment in local mode connected to the network. All communication protocols are developed under the framework known as OSI Model (Open System Interconnection).

Among all the protocols, Modbus (RTU, TCP, or RTU over TCP) stands out, which is widely extended and is the one used by most equipment (Programmable Logic Controllers-PLC, photovoltaic inverters, wind inverters, battery chargers, network analyzers for measuring loads, etc.) since it is robust, easy, open-source, and therefore free and above all reliable.

Finally, once all the equipment is connected to the same data network and the protocol used by each one is known (mainly Modbus), communication must be established between them and Home Assistant in the following way:

1. The element (generation source, storage system, or load) must be in Home Assistant configuration file, in a different way depending on the protocol used, as follows:

- Modbus TCP/IP (Transmission Control Protocol/Internet Protocol): it is based on a client/server architecture and allows communication over an Ethernet grid, no CRC required. It is the most common protocol. A name must be assigned to the element, and the IP address defined, the type of communication, and the communication port (502), as follows:

  - name: Photovoltaic1   # name assigned to the element.
  type: tcp   # protocol type
  host: 192.168.15.75   # IP address
  port: 502   # communication port

- Modbus RTU: this is a master/slave architecture for linking a control system to a Remote Terminal Unit (RTU) via a serial port. Commonly, the master is an HMI or a SCADA system that sends a request and the slave is a sensor or PLC that returns a response. A CRC (Cyclic Redundancy Checksum) is used as an error-checking mechanism and as a procedure to ensure data reliability. A name must be assigned to the element, the type defined, and method of communication, the communication port, baudrate, stopbits, bytesize and parity, all given, as follows:

  - name: Battery1   # name assigned to the element.
  type: serial   # type of communication.
  method: rtu   # method of communication.
  port:/dev/ttyUSB0   # port (USB in this case).
  baudrate: 19200
  stopbits: 1
  bytesize: 8
  parity: E

- Modbus RTU over TCP: it is a combination of Modbus TCP and Modbus RTU. It is based on client/server architecture for communications over Ethernet as Modbus TCP/IP but uses a CRC as Modbus RTU. The same issue as in TCP/IP must be defined.

  - name: SmallWindTurbine1
  type: rtuovertcp

host: 192.168.15.87
port: 7128

- HTTP protocol: The element must be defined, a name assigned, the http address for the information required, and the scan interval provided: sensor:
- platform: command_line
name: "Pump1"
command: "curl -s 'http://admin:password@192.168.15.107/io.cgi?' (accessed on 1 January 2021) | awk '/^relays/{print substr($2, 1, 1)}'"
scan_interval: 1

Despite that there are other many communication protocols, most of the elements that can be connected to a microgrid use one of those described above.

Once all elements are defined, the registers with the required information must be read from the addresses of each one. This procedure is also carried out in the configuration file of Home Assistant.

In Modbus protocol, it is necessary to have the Modbus frame (it should be in the manufacturer's manual, but many times it is not and it is necessary to contact the manufacturer to provide it) to know the addresses where the variables to be read are.

For instance, to read instantaneous power in the photovoltaic defined above, the following definition must be included in the configuration:

```
- platform: Modbus                          #Modbus
scan_interval: 1                            #scan interval
registers:
- name: Photovoltaic1_Instant_Power         #monitored parameter
hub: Photovoltaic 1                         #element defined previously
register_type: input                        #type of register (input, holding, etc.)
unit_of_measurement: W                      #unit of the parameter
slave: 1                                    #Modbus Id
register: 44                                #Modbus address
count: 1                                    #number of address affected
scale: 1                                    #scale (x1, x10, etc.)
```

In http protocol, we do not need to include all this information; in the http address defined before is included the position of the parameter to monitor (and password if required):

http://admin:password@192.168.15.107/io.cgi?' | awk '/^relays/{print substr($2, 1, 1)}' (accessed on 1 January 2021)

This will have to be done for each parameter to monitor.

In dispatchable generation elements, such as the photovoltaic, wind turbines, and loads, it is just necessary to read power (instantaneous values) since action orders cannot be sent to them. By contrast, in dispatchable generation elements and storage systems, it is necessary to read all the registers. With them, a SCADA is developed that allows for control of their operation through the control panel, integrated with all the elements of the grid.

The steps to integrate grid elements into the same data network and establish communication with the Home Assistant control system are shown in Figure 2.

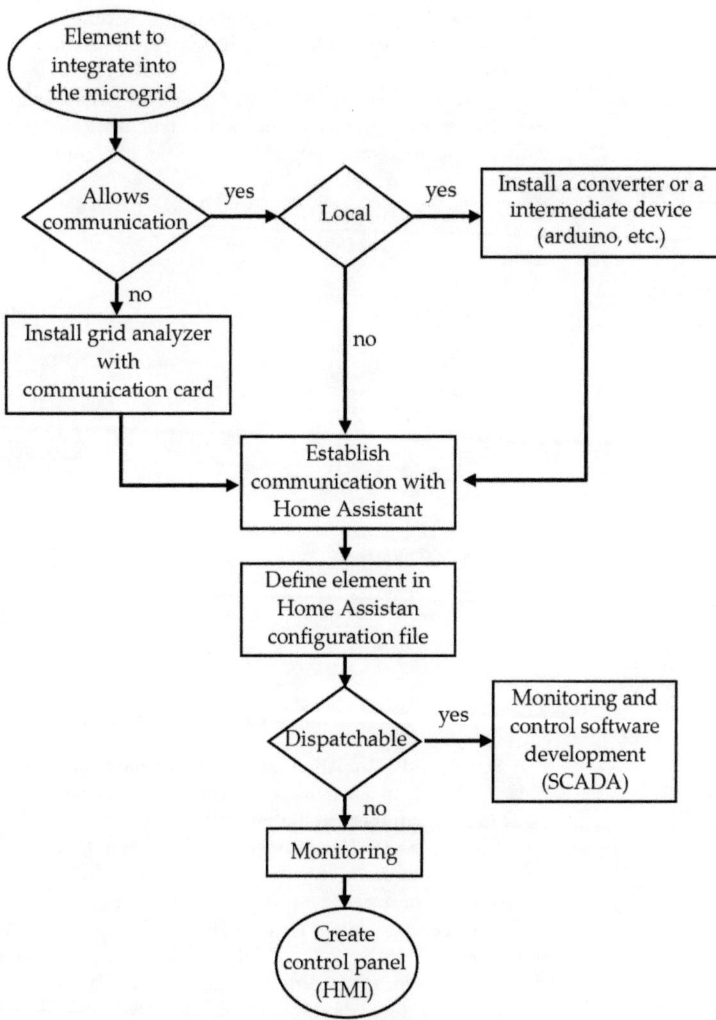

Figure 2. Stages of the procedure to set up communication with each element of the grid. Source: prepared by authors.

Step 4. Create a control framework: once the communication has been established between all the elements of the grid, a control interface (HMI) must be created to visualise, simply and intuitively, the working mode of each element of the grid, and to send operating instructions to achieve optimisation in the performance of the microgrid, maximising its efficiency.

With Home Assistant, it is easy to create an interface to control the grid after establishing communication with all the grid elements and defining them in the configuration file. The starting point is a graphical interface that gives the possibility to insert cards with different functionalities. It is possible to add cards with maps or weather forecasts that facilitate the estimation of renewable generation systems' power production. There is also the option to include cards with all the registered values (in the configuration file, as seen above) numerically (entities), or to graphically represent these values (historical graph), or to jointly represent (graphically and numerically) a value using the sensor card.

Home Assistant also allows including scripts to send orders to each dispatchable element of the microgrid, start/stop generation elements, charging/discharging storage systems, etc.

In addition, automations can be defined (we can turn them on or off with a switch) to schedule the performance of each dispatchable element of the microgrid based on the values of the monitored parameters, as seen in Figure 3.

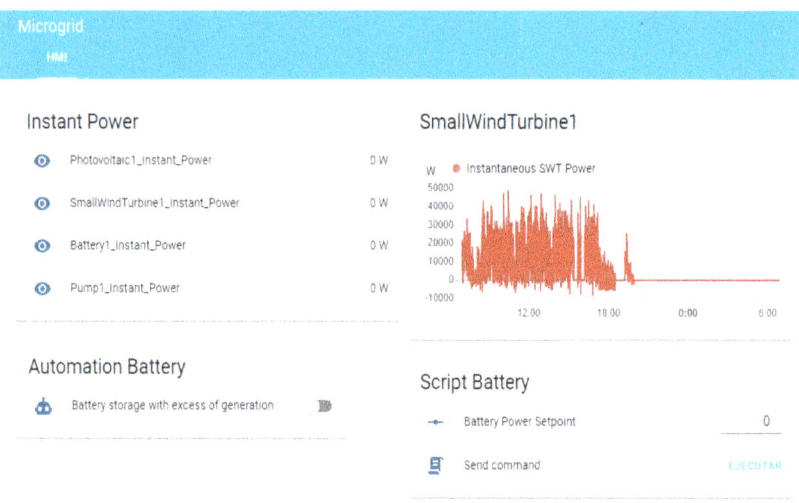

Figure 3. Home Assistant control panel. Source: prepared by authors.

Scripts and automation can be programmed via code in Home Assistant configuration files, or directly in a frame special for that purpose which makes it easier since there is no need to program them in Python.

Home Assistant has a Telegram integration (also free software) that allows sending text messages to the cell phone of the microgrid supervisor to inform him of relevant events to take the appropriate actions or perform the necessary maintenance tasks.

It is very convenient for later study, although it is not necessary for the operation of the microgrid, to store all the data monitored by the system. Home Assistant stores the data, but for a more efficient management, it also allows to store them (thus establishing them in its configuration file) in one of the multiple existing databases, such as MySQL, which makes easier their later treatment.

Once the control software is fully developed in Home Assistant, ensuring its functioning and analyzing the data registered, microgrid management strategies must be defined to improve its efficiency. In this case, the main strategy for the batteries is to charge them when there is surplus of generation, that is, when the generation systems produce more energy than the demand for the consumption elements of the microgrid.

To discharge the battery, the best strategy based on the analysis of the data recorded for months is to provide power to the microgrid for two hours from 7 am to 9 am (if possible), because is when the start-up of the main loads takes place, and therefore the moment of greatest consumption of CEDER.

Once these strategies are implemented, new data must be analyzed to validate their effectiveness.

4. Case Study

CEDER (Centro para el Desarrollo de las Energías Renovables) is the experimental application case. This centre belongs to CIEMAT (Centro de Investigaciones Energéticas, Medioambientales y Tecnológicas), which is a Spanish public research body, assigned to the Ministry of Science and Innovation, focusing on energy and environment and the technologies related to them. It is located in Lubia, in the province of Soria (Spain), with a built-up area of 13,000 m^2 in three distinct areas, out of a total of 640 ha (see Figure 4).

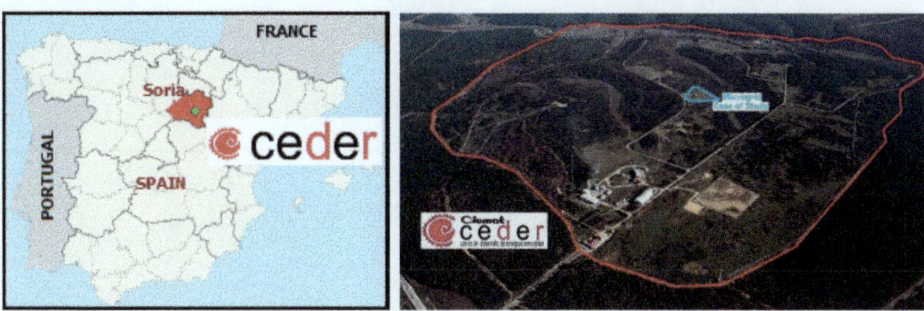

Figure 4. CEDER location and distribution. Source: prepared by authors.

CEDER grid is connected to a 45 kV distribution network and transforms at its input to 15 kV. Eight transformation centres make up the grid, reducing the voltage to 400 V. The centre has various non-controllable renewable (photovoltaic and wind), controllable renewable (hydraulic turbine), and non-renewable (diesel generator) generation systems, several storage systems, mechanical (pumping system with tanks at different levels) and electrochemical (lithium-ion and Pb-acid batteries), as well as various consumption elements connected to each of the transformation centres monitored with grid analyzers (PQube) installed in the low voltage part of each of the transformation centres.

To simplify the control system, the case study focuses on a single transformer substation, so there will be no redundancies as much of the installed equipment is similar in the different cases.

4.1. Identification and Specification of the Grid Components

The first step to develop as indicated in the previous section is to identify and describe all the elements that are connected to the network (see Figure 5). In this case, these are detailed below:

- Photovoltaic: four series of six photovoltaic panels (210 W each one) for a total of 5 kW. Panels are polycrystalline silicon and are assembled on a floor structure with a variable inclination angle and are connected to a 5 kW inverter (Ingeteam-Ingecon Sun Lite). It is necessary to connect with the inverter to be able to read instantaneous power values. This model of inverter has a network card included which allows to connect it directly to the CEDER data network (Ethernet) and enables communication via Modbus TCP/IP.
- Wind power: there is a three-blade horizontal axis small wind turbine. Diameter 15 m and 50 kW power. It is operating leeward. A National Instruments FieldPoint is installed to measure variables such as wind speed, power, etc. and can be connected to the CEDER data network via Modbus TCP/IP protocol. With the installation of this element, it is possible to have a data register as this wind turbine does not have a previous connection.
- Battery energy storage system (BESS): formed by 120 Pb-acid cells. Total voltage: 240 V, capacity: 1080 Ah at 120 h (C_{120}). BESS is connected to a 50 kW charger/regulator/inverter. The inverter is connected with a SCADA for battery control via RS485 to the serial port of a computer. The control system of the microgrid to be developed will communicate locally with the charger using Modbus RTU protocol.
- Consumption: loads of the microgrid come from the buildings located inside the part of the CEDER where the laboratories and workshops are located. Consumption is measured by means of power quality/grid analyzers (PQube) that allow connection to the CEDER's data network via Ethernet and Modbus TCP/IP communication. These

are installed in the low voltage part of the transformation centre. For this case study, there are no critical loads that need to be supplied in the event of failure or serious disturbance of the microgrid.

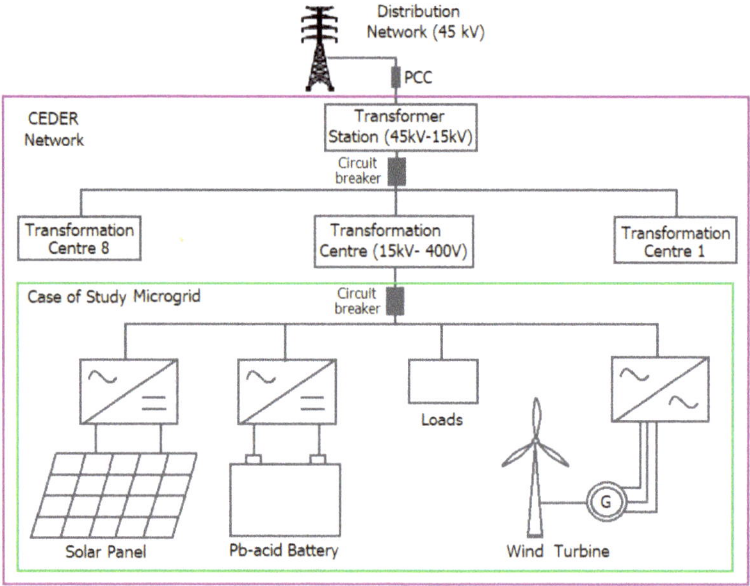

Figure 5. Case of study microgrid. Source: prepared by authors.

4.2. Integration of Grid Elements into the Same Data Network

After identifying, characterizing, and establishing the form of communication of all the elements of the microgrid, the next step is the incorporation of each of the elements into the same data network.

In these cases, we have to connect the following elements to the data network: photovoltaic, wind turbine, and battery inverters, and a grid analyzer (PQube) to measure the loads.

Photovoltaic inverter and grid analyzer (PQube) have a TCP/IP Ethernet communication card. Battery control system allows local communication (RS232), so, as seen previously, we have to install a converter RS232 to TCP/IP with Modbus. Finally, the wind turbine inverter is quite old and does not allow any communication, so we have installed a data acquisition system (National Instrument Compact FieldPoint, which allows TCP/IP communication with Modbus) to measure different parameters and communicate with Home Assistant control system.

4.3. Establish Communication between Grid Elements

Subsequently, the communication configuration of each of the elements of the microgrid must be carried out with Home Assistant. There are different communication protocols that allow the transmission of information when all the devices are connected to the same data grid, or to other devices in local mode, and these in turn are connected to the same grid.

Among the most common communication protocols, Modbus stands out from the rest. With it, we can control a network of devices (in our case, generation, storage, and consumption systems) and communicate between them with a control system (Raspberry Pi 4 with Home Assistant). It is the standard communication protocol in the sector as it is robust, easy to use, free and, above all, reliable.

The microgrid scheme communications are shown in Figure 6.

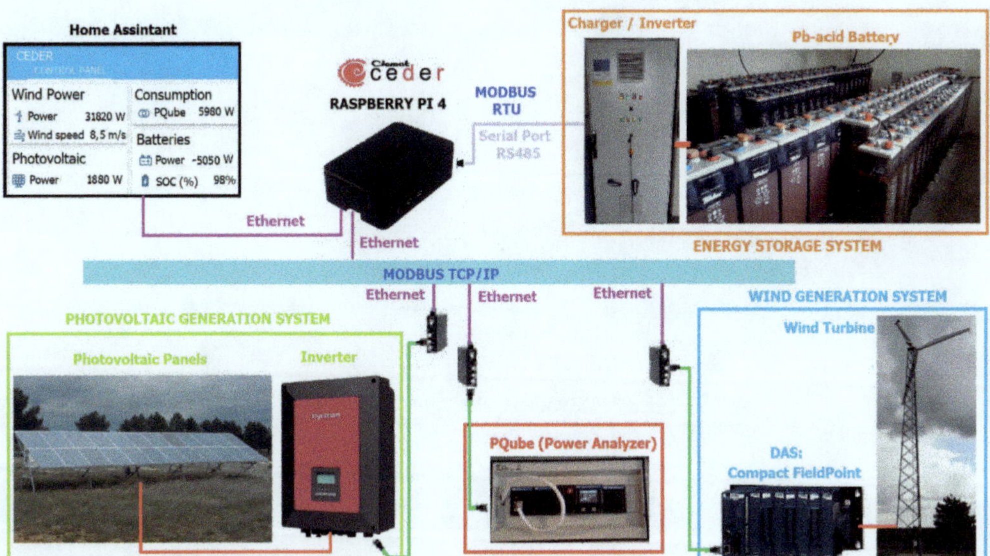

Figure 6. Elements of the CEDER's microgrid. Source: prepared by authors.

To communicate between each element (photovoltaic panels, wind turbine, Pb-acid batteries, and loads) with Home Assistant, and monitoring and recording required information, we have to proceed as seen in step 3, Section 2.

For the non-dispatchable generation elements, such as the wind turbine, photovoltaic inverter, and loads (PQube grid analyzer), only the instantaneous power will be read, as it is not possible to send them any setpoint values. On the other hand, for the controllable generation elements and storage systems (Pb-acid batteries), all the records will be read, with which a SCADA will be developed to control their operation via the control panel in which they will be integrated with all the elements of the microgrid.

4.4. Control Grid Framework Creation

After configuring the communications with all the equipment, we have to design an interface that provides the user with the ability to see all the registers in real time and to have the possibility to execute commands on all of them, such as start/stop actions of generation equipment, load/unload of storage systems, etc.

Figure 7 shows instantaneous values monitored in real-time for the microgrid. As shown, there are 1320 W of photovoltaic generation, 8020 W of wind generation, with a wind speed of 5.1 m/s, compared to a demand of 9880 W. In this case, the batteries are at a standstill as there is no high demand and no surplus energy production to charge them. Otherwise, the storage system is almost fully charged, with a State of Charge (SOC) of 98%.

The SCADA for the control of the batteries also has to be created, as can be seen in Figure 8. It shows how the batteries are in a charging process. At that moment, the microgrid is injecting 1780 W to the batteries, which have a SOC of 98%, and 266 V.

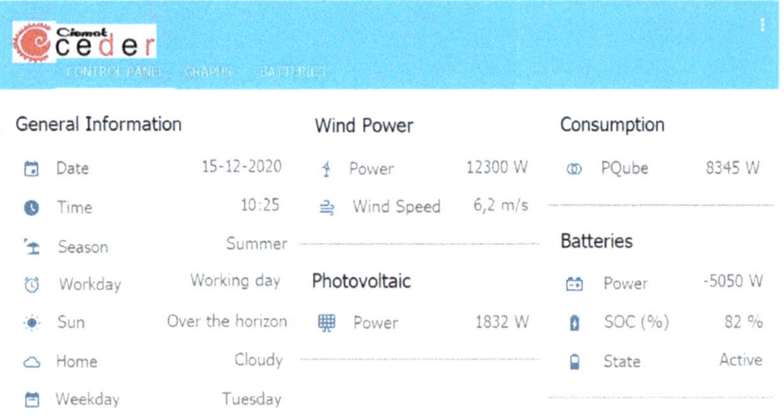

Figure 7. CEDER's microgrid control framework developed in Home Assistant. Source: prepared by authors.

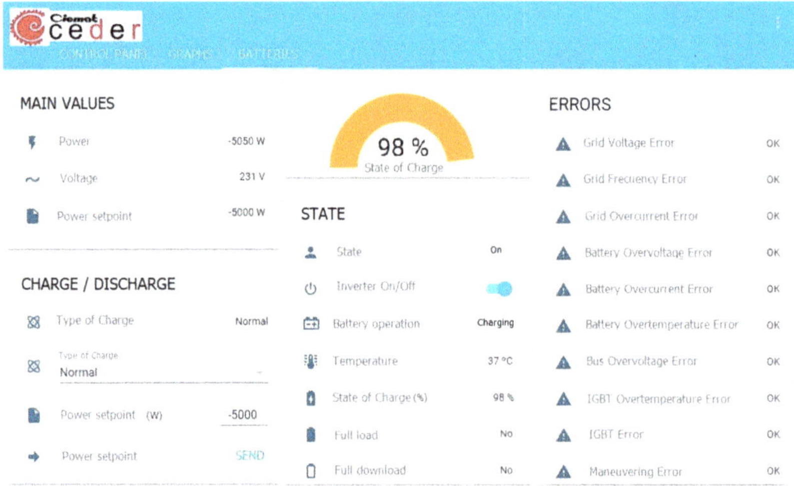

Figure 8. CEDER's microgrid battery SCADA developed in Home Assistant. Source: prepared by authors.

Home Assistant allows visualising in real time the value of the variables collected from each of the elements, as shown in Figure 9.

The initial situation was that of an electrical grid with two renewable generation sources (wind turbine and photovoltaic panels), a Pb-acid batteries storage system, and loads. All these elements worked independently. After applying the procedure proposed, all the elements are integrated into a single control system, and they can work in a coordinated way, which allows establishing management strategies. For example, we can set a strategy to charge the batteries when there is excess generation (See Figures 10–12).

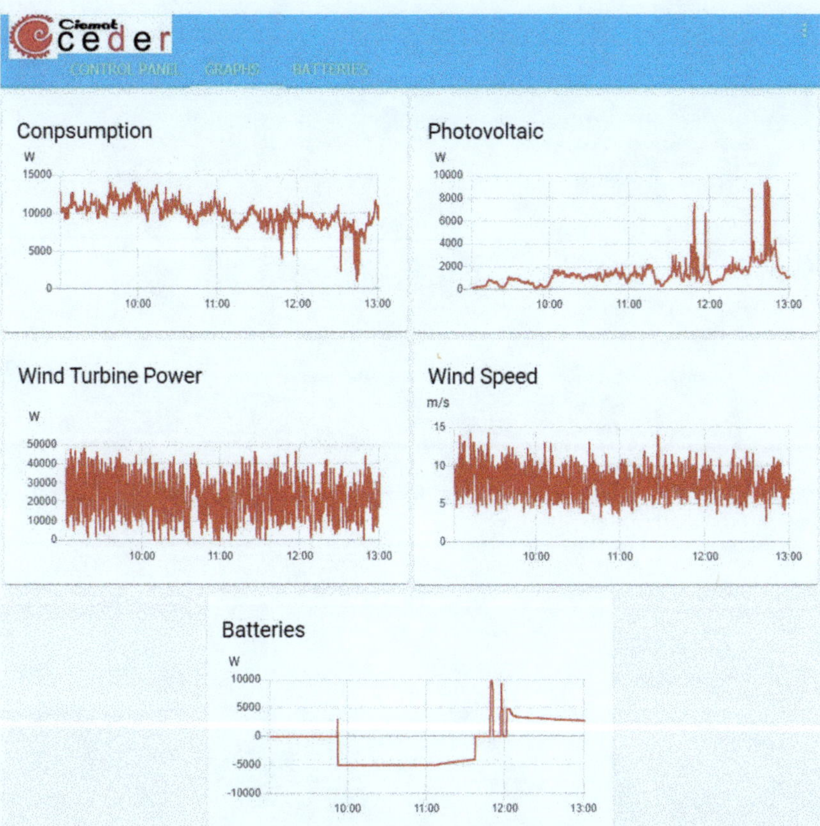

Figure 9. CEDER's microgrid elements' instantaneous values graphically represented. Source: prepared by authors.

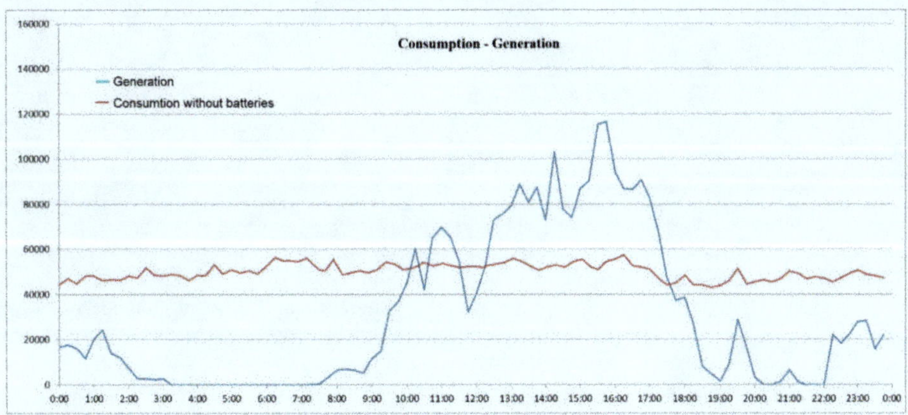

Figure 10. Daily grid generation–consumption graphic (before implementation). Source: prepared by authors based on the results of this study.

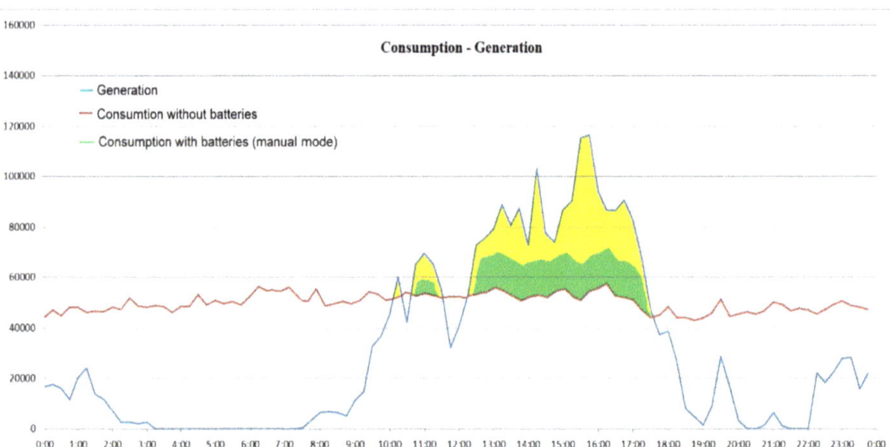

Figure 11. Daily grid generation–consumption graphic with batteries in manual mode (before implementation). Source: prepared by authors based on the results of this study.

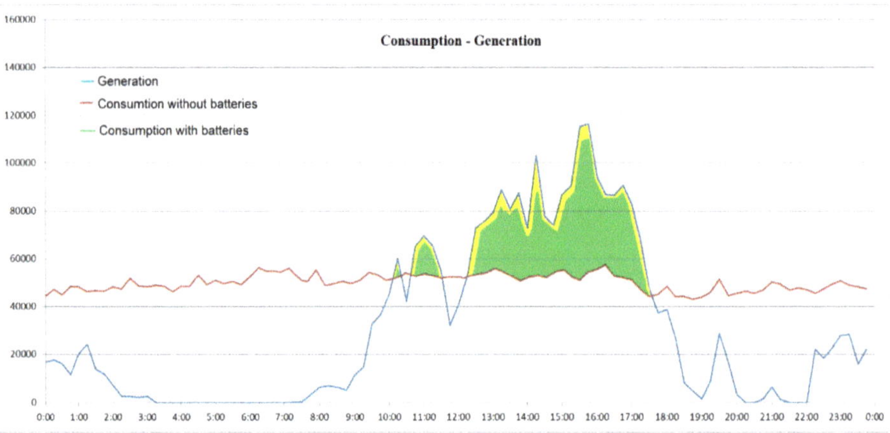

Figure 12. Daily microgrid generation–consumption after implementation. Source: prepared by authors based on the results of this study.

Figure 10 shows grid one-day generation and consumption curves before the conversion into a smart microgrid. The yellow area represents the power surplus for that day. As there is no communication between the elements of the grid, batteries cannot see there is a power surplus and they do not do anything.

Figure 11 shows grid one-day generation and consumption curves (before the conversion into a smart microgrid), including batteries' manual charge when there is a power surplus, that is, an operator sees the power surplus and active batteries charge with a fix power value. The green area represents energy stored in batteries. The yellow area represents the power surplus after the batteries' manual charge.

Figure 12 shows microgrid one-day generation and consumption curves after the implementation of the proposed procedure to convert the network section into a smart microgrid, including automatic batteries' charges when there is a power surplus. All the elements of the grid are integrated in the same control system developed with Home Assistant; therefore, the control system can detect the power surplus and send to the batteries the best charging instructions at all times. The green area represents energy stored in batteries (optimized). The yellow area represents the power surplus after the batteries' automatic charge.

5. Conclusions

This work presents an experimental application case for the conversion of a grid section with the generation, storage, and load systems into a smart microgrid that can be replicated in any grid section. It is based on a four-stage procedure: (1) identification and characterization of grid elements, (2) integration of grid elements into the same data network, (3) establishment of communication between grid elements, and (4) creation of a control grid framework that allows the management of the microgrid.

The most significant advantage is the use of the free tool Home Assistant to develop a software (CMEMS) to manage the microgrid. Although Home Assistant is designed for home automation, it offers all the necessary features to monitor, manage, and integrate in real time, and in a single HMI (CMEMS), all the renewable generation sources, storage systems, and consumption elements are connected to a grid to turn it into a smart microgrid. It has to be taken into account that Home Assistant is an open software, and therefore free; behind it, there is a large community of users who continually develop new capabilities, solve problems from previous versions, and answer questions in forums for free and without the need to hire any technical service. In addition, it can be managed from every device and everywhere due to its web server and its app for mobile phones.

The software developed is open, so new elements (generation sources, storage systems, or loads) can be included easily.

All this is an important advantage compared to other software shown in Section 2. All of them require the payment of a license fee, and have some limitations to connect some kind of devices. Moreover, the development and configuration are done by the company, so there is no option to modify it to include new systems on our own.

Finally, the developed software allows to implement strategies, after the analysis of the recorded data, to optimize the performance of the microgrid by maximising the use of renewable generation sources and reducing the consumption of the microgrids as much as possible with the help of storage systems. This will allow a reduction in the cost of the electricity bill.

Author Contributions: Conceptualization, O.I.-M.; methodology, O.I.-M., P.P.-C. and L.H.-C.; software, O.I.-M.; writing—original draft preparation O.I.-M. and P.P.-C.; writing—review and editing, L.H.-C., A.Z.-L. and O.D.-P.; supervision, R.V.-R. All authors have read and agreed to the published version of the manuscript.

Funding: This research received no external funding.

Institutional Review Board Statement: Not applicable.

Informed Consent Statement: Not applicable.

Data Availability Statement: Confidential data.

Acknowledgments: The authors are grateful and would like to acknowledge the support granted by Carlos Barrera and Gonzalo Martín for valuable discussions.

Conflicts of Interest: The authors declare no conflict of interest.

References

1. Ton, D.T.; Smith, M.A. The U.S. Department of Energy's Microgrid Initiative. *Electr. J.* **2012**, *25*, 84–94. [CrossRef]
2. Divya, K.C.; Østergaard, J. Battery energy storage technology for power systems-An overview. *Electr. Power Syst. Res.* **2009**, *79*, 511–520. [CrossRef]
3. Sachs, T.; Gründler, A.; Rusic, M.; Fridgen, G. Framing Microgrid Design from a Business and Information Systems Engineering Perspective. *Bus. Inf. Syst. Eng.* **2019**, *61*, 729–744. [CrossRef]
4. Jirdehi, M.A.; Tabar, V.S.; Ghassemzadeh, S.; Tohidi, S. Different aspects of microgrid management: A comprehensive review. *J. Energy Storage* **2020**, *30*, 101457. [CrossRef]

5. Li, Q.; Xu, Z.; Yang, L. Recent advancements on the development of microgrids. *J. Mod. Power Syst. Clean Energy* **2014**, *2*, 206–211. [CrossRef]

6. Jia, Y.; Lyu, X.; Lai, C.S.; Xu, Z.; Chen, M. A retroactive approach to microgrid real-time scheduling in quest of perfect dispatch solution. *J. Mod. Power Syst. Clean Energy* **2019**, *7*, 1608–1618. [CrossRef]

7. Kaur, A.; Kaushal, J.; Basak, P. A review on microgrid central controller. *Renew. Sustain. Energy Rev.* **2016**, *55*, 338–345. [CrossRef]

8. Ahmad Khan, A.; Naeem, M.; Iqbal, M.; Qaisar, S.; Anpalagan, A. A compendium of optimization objectives, constraints, tools and algorithms for energy management in microgrids. *Renew. Sustain. Energy Rev.* **2016**, *58*, 1664–1683. [CrossRef]

9. Gharavi, H.; Ghafurian, R. *Smart Grid: The Electric Energy System of the Future*; Gharavi, H., Ghafurian, R., Eds.; IEEE: New York, NY, USA, 2011; Volume 99, pp. 917–921.

10. Teufel, S.; Teufel, B. The Crowd Energy Concept. *J. Electron. Sci. Technol.* **2014**, *12*, 263–269. [CrossRef]

11. Aslam, W.; Soban, M.; Akhtar, F.; Zaffar, N.A. Smart meters for industrial energy conservation and efficiency optimization in Pakistan: Scope, technology and applications. *Renew. Sustain. Energy Rev.* **2015**, *44*, 933–943. [CrossRef]

12. Khan, K.R.; Siddiqui, M.S.; Saawy, Y.A.; Islam, N.; Rahman, A. Condition Monitoring of a Campus Microgrid Elements using Smart Sensors. *Procedia Comput. Sci.* **2019**, *163*, 109–116. [CrossRef]

13. Vargas-Salgado, C.; Aguila-Leon, J.; Chiñas-Palacios, C.; Hurtado-Perez, E. Low-cost web-based Supervisory Control and Data Acquisition system for a microgrid testbed: A case study in design and implementation for academic and research applications. *Heliyon* **2019**, *5*, e02474. [CrossRef]

14. Kunicki, M.; Borucki, S.; Zmarzły, D.; Frymus, J. Data acquisition system for on-line temperature monitoring in power transformers. *Measurement* **2020**, *161*, 107909. [CrossRef]

15. Nikolic, D.; Negnevitsky, M. Smart Grid in Isolated Power Systems–Practical Operational Experiences. *Energy Procedia* **2019**, *159*, 466–471. [CrossRef]

16. Eissa, M.M.; Awadalla, M.H.A. Centralized protection scheme for smart grid integrated with multiple renewable resources using Internet of Energy. *Glob. Transit.* **2019**, *1*, 50–60. [CrossRef]

17. Shuai, Z.; Sun, Y.; Shen, Z.J.; Tian, W.; Tu, C.; Li, Y.; Yin, X. Microgrid stability: Classification and a review. *Renew. Sustain. Energy Rev.* **2016**, *58*, 167–179. [CrossRef]

18. Hanah, A.; Farook, R.; Elias, S.J.; Rejab, M.R.A.; Fadzil, M.F.M.; Husin, Z. IoT Room Control And Monitoring System Using Rasberry Pi. In Proceedings of the 2019 4th International Conference and Workshops on Recent Advances and Innovations in Engineering (ICRAIE), Kedah, Malaysia, 27–29 November 2019; pp. 1–4. [CrossRef]

19. Singh, S.; Weeber, M.; Birke, K.P.; Sauer, A. Development and Utilization of a Framework for Data-Driven Life Cycle Management of Battery Cells. *Procedia Manuf.* **2020**, *43*, 431–438. [CrossRef]

20. Li, W.; Rentemeister, M.; Badeda, J.; Jöst, D.; Schulte, D.; Sauer, D.U. Digital twin for battery systems: Cloud battery management system with online state-of-charge and state-of-health estimation. *J. Energy Storage* **2020**, *30*, 101557. [CrossRef]

21. Kim, T.; Makwana, D.; Adhikaree, A.; Vagdoda, J.; Lee, Y. Cloud-Based Battery Condition Monitoring and Fault Diagnosis Platform for Large-Scale Lithium-Ion Battery Energy Storage Systems. *Energies* **2018**, *11*, 125. [CrossRef]

22. Anandan, N.; Sivanesan, S.; Rama, S.; Bhuvaneswari, T. Wide area monitoring system for an electrical grid. *Energy Procedia* **2019**, *160*, 381–388. [CrossRef]

23. Petrollese, M.; Valverde, L.; Cocco, D.; Cau, G.; Guerra, J. Real-time integration of optimal generation scheduling with MPC for the energy management of a renewable hydrogen-based microgrid. *Appl. Energy* **2016**, *166*, 96–106. [CrossRef]

24. Kennedy, J.; Ciufo, P.; Agalgaonkar, A. Intelligent load management in microgrids. *IEEE Power Energy Soc. Gen. Meet.* **2012**, 1–8. [CrossRef]

25. Kornas, T.; Wittmann, D.; Daub, R.; Meyer, O.; Weihs, C.; Thiede, S.; Herrmann, C. Multi-Criteria Optimization in the Production of Lithium-Ion Batteries. *Procedia Manuf.* **2020**, *43*, 720–727. [CrossRef]

26. Augustins, E.; Jaunzems, D.; Rochas, C.; Kamenders, A. Managing energy efficiency of buildings: Analysis of ESCO experience in Latvia. *Energy Procedia* **2018**, *147*, 614–623. [CrossRef]

27. GE Digital. Advanced Energy Management System (AEMS). Available online: https://www.ge.com/digital/applications/transmission/advanced-energy-management-system-aems (accessed on 1 March 2021).

28. Operation Technology, Inc. Etap Powering Success. Available online: https://etap.com/es/home (accessed on 3 March 2021).

29. Open Systems International (OSI). Available online: https://www.osii.com/index.asp (accessed on 2 January 2021).

30. Wattics. Available online: https://www.wattics.com/dashboard/ (accessed on 1 January 2021).

31. Home Assistant. Available online: https://www.home-assistant.io/ (accessed on 3 March 2021).

applied sciences

Article

Substitutability and Complementarity of Municipal Electric Bike Sharing Systems against Other Forms of Urban Transport

Michał Suchanek [1,*], Aleksander Jagiełło [1] and Justyna Suchanek [2]

[1] Faculty of Economics, University of Gdansk, Ul. Armii Krajowej 119/121, 81-824 Sopot, Poland; aleksander.jagiello@ug.edu.pl

[2] Independent Department of EU Projects and Mobility Management, Ul. 10 Lutego 33, 81-364 Gdynia, Poland; j.suchanek@gdynia.pl

* Correspondence: michal.suchanek@ug.edu.pl

Abstract: The current quantitative and qualitative development of bike-sharing systems worldwide involves particular implications regarding the level of sustainability of urban development and city residents' quality of life. To make these implications as large as possible as well as the most positive, it is essential that the people who use municipal bikes on a regular basis to the largest extent possible abandon car travel at the same time. Thanks to their operational characteristics, electric bikes should enable meeting the transport needs of a wider group of city residents compared with traditional bicycles. The main aim of this study was therefore to check whether the municipal electric bike system (MEVO) in Gdańsk-Gdynia-Sopot metropolitan area of Poland lived up to the hopes placed upon it by policymakers. Therefore, the article tests the hypothesis indicating that the municipal electric bike systems constitute a substitutable form of transportation against passenger cars to a larger extent than against collective urban transport and walking trips. The analysis was performed based on the results of primary studies conducted among the users of MEVO. The data show that the MEVO was a substitutable form of transportation against collective transport and walking trips to a larger extent than against passenger cars. Through logistic regression analysis, the variables concerning the probability of replacing car trips by MEVO bicycles were determined. Among the analyzed variables, the following turned out to be statistically significant: age, the number of people in the household, the number of cars in the household, the distance from work, and gender. The results therefore indicate that substituting in favor of electro bikes was more probable for younger people with fewer people in the household and a distance to travel below 3 km, whereas it was less probable for people with more cars in the household or traveling a distance longer than 10 km. Additionally, females were more likely to choose the bike system.

Keywords: municipal electric bike system; electric bike-sharing; sustainable urban transport; travel behavior

Citation: Suchanek, M.; Jagiełło, A.; Suchanek, J. Substitutability and Complementarity of Municipal Electric Bike Sharing Systems against Other Forms of Urban Transport. *Appl. Sci.* **2021**, *11*, 6702. https://doi.org/10.3390/app11156702

Academic Editor: Luis Hernández-Callejo

Received: 24 June 2021
Accepted: 20 July 2021
Published: 21 July 2021

1. Introduction

The current urban transport policies focus mostly on the external costs of transport, such as the ones related to air pollution, noise, road congestion, and road accidents. These costs are, to a large extent, a consequence of the dominant share of individual transport, mostly car transport, in the city residents' municipal travels. The political decisions that are made indicate the intention to reduce the share of passenger cars in daily travels. However, it requires providing city residents with an attractive alternative. Fulfilling the city residents' transport needs with the use of bicycles matches the concept of sustainable urban development, since travel by bicycle, apart from walking trips, constitutes the most environmentally friendly mode of transportation of the city residents. To increase the share of bike travels in the modal split of particular cities worldwide, bike-sharing systems have been introduced.

In the Gdańsk-Gdynia-Sopot metropolitan area, the bike-sharing system (called MEVO) was introduced in 2019. In 2019, there were 96 shared bike systems in operation in Poland. Access to bike-sharing services operating in 108 cities had 2.6 million registered users [1]. MEVO was the first bike-sharing system in Poland using only electrical bicycles (with electric power-assisted steering). The main reason for the use of electric bicycles in the MEVO system rather than traditional bicycles was the large differences in the ground levels in the Gdańsk-Gdynia-Sopot metropolitan area. Electric bicycles were therefore intended to enable almost everyone to use the bike-sharing system.

The introduction of the MEVO system into operation was accompanied by the hope of reduced car trips and an increased share of public transport [2]. The increase in the share of public transport was to result from the use of city bikes by residents for the first and last mile of their journeys. Furthermore, the reference literature indicates numerous advantages to fulfilling transport needs with the use of bikes, and consequently, bike-sharing systems, the advantages of which include the following [3–11]:

- Reduced consumption of fuel necessary for meeting the city residents' transport needs;
- Reduced congestion;
- Reduced level of noise and pollution emitted by fuel-powered vehicles;
- Complementing the offer of an urban transport system;
- Supporting park-and-ride systems;
- Offering the "last mile" solution while traveling to places where transportation is limited or prohibited;
- Improved transport availability;
- Increased tourist attractiveness of the city;
- Improved physical and psychological health of the city residents;
- Increased possibilities to meet the city residents' transport needs;
- Increased social acceptance for the introduction of rigorous transport policy toward passenger cars (e.g., reduced number of parking places).

To make it possible for bike-sharing systems to have a positive effect to the largest extent possible through their advantages on the city residents' quality of life, it is preferable for the users of bike-sharing systems to abandon using passenger cars for urban travels. Therefore, it is preferable for municipal bikes to become the substitute for passenger cars to the largest extent possible (particularly owned cars). To make this possible, it is necessary to facilitate not only the bike renting process itself but also traveling by bike. One of the feasible solutions involves using electric bicycles in the bike-sharing systems. Thanks to such a solution, bike travel requires less physical effort, and consequently, it is possible for a larger group of potential users. Definitely less favorable is a situation where the municipal bike users abandon walking trips or traveling by public transport in favor of municipal bikes.

The main objective of this study is to verify whether municipal electric bike systems constitute a substitutable form of transportation against passenger cars. If that is not the case and they substitute public transport instead, then they do not contribute to sustainable urban development to a large extent. Given the foregoing, this article verifies the following hypothesis: municipal electric bike systems constitute a substitutable form of transportation against passenger cars to a larger extent than against collective public transport and walking trips. This study thus shows the real effects of implementing an electric bike-sharing system in the largest metropolitan area in northern Poland and analyzes whether the effects of the implementation of the MEVO system are consistent with the expectations of policymakers. Based on the obtained data, the factors affecting the residents' willingness to substitute car travel with municipal bike travel were analyzed as well.

2. Literature Review

2.1. History of Bike Sharing—From "White" to Electric Bikes

Bike-sharing systems are based on the idea of sharing economy. Since the concept of sharing economy is only a recent concept in the economic sciences (one of the first persons who used this term in 2008 was Prof. L. Lessig of Harvard Law School [12]), we have not yet observed one commonly applied definition of this term. The sharing economy is most frequently defined from the angle of its usability, as providing consumers with temporary access to unused physical resources is usually performed in return for financial means [13].

The concept of municipal bikes was introduced for the first time in the 1960s [14]. Over the years, the concept evolved from the so-called "white bicycles" into the currently introduced fourth generation municipal bikes. The first attempts to introduce bike sharing should be considered a failure [15]. Even though the bike's gratuitousness and availability was supposed to be its best advantage, quickly (i.e., only within several days), the bikes operating within these systems were damaged by vandals, stolen by thieves, or confiscated by the police [8,16]. In response to the observed problems with the first-generation municipal bike systems, attempts were made to introduce second-generation systems. These attempts were first made in the first half of the 1990s, mainly in Denmark [16]. These bikes were specially adjusted for intensive use, and as a result, they were less prone to accidental damage resulting from normal use. The greatest breakthrough involved applying docking stations, which enabled renting a bike against a deposit [8]. The problem affecting the second-generation systems included the still too significant anonymity of the bike users. This anonymity led to acts of vandalism and bike theft incidents. The third-generation municipal bike systems were successful mainly thanks to the development of technology, which was observed in the second half of the 1990s and later. The drawback of the third-generation municipal bike systems compared with subsequent generations is their dependence on docking stations. This means that the user of a third-generation municipal bike is forced to rent it from and return it to one of the docking stations located in a particular city. The possibility to start and finish the ride outside the docking station in any location (within the area of system operation) is provided by fourth-generation bike-sharing systems. Moreover, they still feature all the advantages of the previous generations' systems. The drawback of the fourth-generation systems refers to the need to reposition the bikes, which was more broadly presented in [17–20].

Each subsequent generation contributed to changing the operational scheme of the municipal bike system. In the case of bike-sharing systems with stations, their success depends on the solution of three optimization problems: the number and locations of stations, the capacity of the stations, and the bicycle distribution [21]. The operational schemes of subsequent generations of municipal bikes from the user perspective are presented in Figure 1. This shows that the subsequent generations of bike-sharing systems not only reduced the bike renting process but also added more elements into the system. Nevertheless, thanks to the application of IT solutions (that allow, for example, to find the accurate location of a vacant bike, pay automatically, and register in the system online), the service of fourth-generation municipal bikes should create no barriers to their use, whereas people who do not have a smartphone are partially excluded from using fourth-generation municipal bikes. In Poland in 2018, 63% of people had a smartphone (the median for the developed countries totaled 76%) [22]. Fourth-generation municipal bike systems are the first to use not only conventional but also electric bikes.

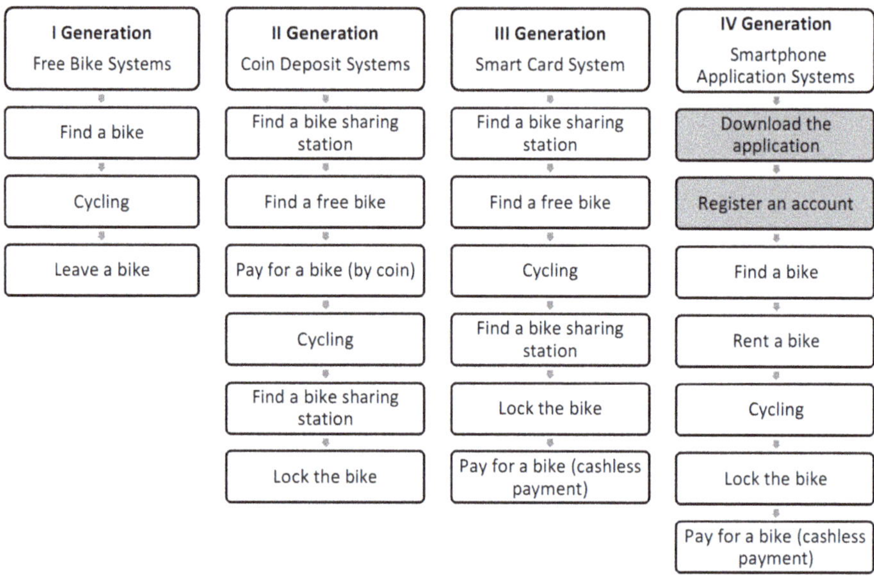

Figure 1. Operational schemes of municipal bikes of different generations from the user perspective. Source: own elaboration. Explanatory note: gray elements refer to stages the user is required to perform only once.

2.2. Place of Bike Sharing in the Urban Transport System

Between the systems of municipal bikes and other forms of urban transport, we can observe one of the three relations. Municipal bikes may be complementary, substitutable, or unrelated to the other forms of transportation [23]. Some researchers studying this issue analyzed the relations within the municipal bike systems from the angle of their complementarity and substitutability against passenger cars and public transport [24,25]. The relations between municipal bike systems and public transport are particularly complicated since, in such case, we may observe both substitutability and complementarity against public transport.

The complementarity between municipal bike systems and public transport results from the possibility to access the public transport station (first mile) by bike and travel the "last mile" by bike from the station to the destination [26]. Since public transport terminals constitute significant traffic generators, we can observe located in their vicinity the stations of bike-sharing systems [8,27]. Studies show that variables related to public transport (e.g., the distance between the municipal bike stations and public transport terminals or the number of passengers of a particular public transport terminal) affect, in a statistically significant manner, the level of demand for bike-sharing system services [28,29]. Therefore, they constitute a good predictor of the volume of demand for the services of particular stations of municipal bike systems.

On the other hand, municipal bike systems constitute the substitutable means of transportation against public transport, particularly for people who do not own passenger cars. Studies have proven that the substitutability of bicycles against public transport is particularly high in the case of short travels with distances of up to 5 km [30–33]. The use of electric bikes in municipal bike systems extends the average travel distance, which involves the substitution of public transport [34]. Numerous studies have proven that the average distance of electric bike travel is longer by roughly 50% than travel by traditional bikes [35,36].

The situation where the municipal bike system substitutes public transport, namely the situation where the city residents abandon using public transport services in favor of bike-sharing services, involves two consequences. First, the decrease in the number of public transport users will result in a decrease in the public transport fee rate. The public transport service fee rate shows to what extent the total costs of services are covered by the revenue from ticket sales. Therefore, a decrease in the public transport service fee rate means that public transport must be subsidized to a larger extent from the local authority resources. At present, within the area where the MEVO municipal bike system was operational, there are two main public transport operators: ZKM in Gdynia and ZTM Gdańsk. In the case of these two operators, the value of the fee rate does not exceed 50% (ZKM in Gdynia: 43.8%; ZTM Gdańsk: 48%). This means that more than half of the public transport operational costs within the network managed by the above-mentioned operators is subsidized from the local authority budgets. Further outflow of public transport passengers and a decrease in the fee rate may lead to a situation where it is necessary to reduce the public transport offered. Such a situation may lead to a decrease in the attractiveness of the offered public transport and consequently to a further decrease in the number of public transport users. Therefore, fulfilling transport needs with the use of municipal bikes and abandoning the fulfilment of those needs through public transport may lead to increasing the phenomenon defined in the reference literature as "the public transport vicious circle". Despite the fact that "the public transport vicious circle" most frequently occurs during the debate on the place and importance of passenger cars in the municipal transport system, substituting public transport with municipal bikes may lead to increasing this phenomenon, since a decrease in the number of public transport passengers resulting in a decrease in the quality of public transport offered or an increase in ticket prices may lead to the situation where people using public transport services up to this point to meet their transport needs abandon these services and swap for passenger cars (referring in particular to people disinterested in meeting their transport needs with the use of municipal bike systems). Such a situation is currently observed, inter alia, in Latin American cities [37]. It is worth noting that due to the car being a generally less sustainable mode of transport than public transport, if a bike-sharing system substitutes public transport rather than the car, then the overall positive effect in terms of sustainability might be far less significant than if it substitutes the car.

2.3. Differences between Conventional and Electric Bikes in the Context of Bike-Sharing Services

Studies have proven that there are certain factors affecting the intention to make use of bicycles by the city residents. The frequency of using municipal bike systems is affected, for example, by age, BMI (body mass index, one of the main health-related indicators), and the users' physical fitness [38–44]. In this context, it should be noted that the analyzed MEVO municipal bike system included only bicycles equipped with electric power-assisted steering (pedelec models, or pedal electric cycles). In these bikes, in order to activate the power-assisting system, the user must push on the pedals. Since the bicycles were equipped with electric power-assisted steering, it is likely that the users of the system also included people who would not decide to use the traditional city bikes (those without the electric power-assisted steering). The people who considered electric power-assisted steering as particularly important included the obese, the elderly, and people with poor physical fitness or health problems (e.g., knee problems, arthritis, asthma, and back pain) [45–47]. These are people who prefer using passenger cars to meet their transport needs. Taking into account the specificity of the MEVO municipal bike system, it can be assumed that the system substitutes, to a larger extent, travel by passenger car compared with the traditional municipal bike systems (with no electric power-assisted steering). It is important in so far as under the review of a conducted survey, it can be concluded that the municipal bike systems using traditional bikes (with no electric power-assisted steering) substitute the sustainable forms of transportation (e.g., public transport, walking trips, and travel by private bicycle) to a larger extent than travel by passenger car [15,23,48–50]. However, it must be emphasized that some analyses (in particular these related to mid-sized cities) indicate the reverse dependency [51]. The fact that electric bikes can replace passenger cars in urban travels to a larger extent than public transport is confirmed by the results of studies conducted among the bike owners [35,52,53]. It should be noted that these studies included people who owned electric bikes and therefore consciously made the decision to change their transport behavior. In his work, J. Arendsen emphasizes that there are particular (often significant) barriers to changing city residents' behavior and transport preferences [54]. These barriers are usually overcome as a result of serious life changes (e.g., change of work, place of residence, or health condition). Regarding the city residents' strong habits of fulfilling their transport needs, and in particular strong attachment to owning passenger cars, it is justifiable to conclude that the shift from passenger cars to bicycles in daily urban travel should be easier in the case of electric bikes. Thanks to the electric power-assisted steering, such travel is easier, more convenient, and faster, which reduces the discrepancy in the perceived difficulty between traveling by bike and passenger car [55,56].

3. Methodology and Data Source

In order to verify the above hypothesis and achieve the main goal of the article, the transport behavior of MEVO municipal bike system users was analyzed. The analysis was conducted based on the method of individual, direct interviews with the use of an original questionnaire. The interviews were conducted at the turn of August and September 2019 (i.e., in the months when the weather in Poland does not hinder the use of bicycles to meet transport needs and also pre-COVID-19). The survey sample included 500 respondents. Before the survey was taken, there had been a pilotage of the questionnaire, which included 12 local experts discussing the questionnaire. The experts included 6 researchers, 2 bike activists, and 4 employees of the town halls specializing in the public transport domain. The questionnaire consisted of 18 questions, including one screening question. The given screening question allowed for the omission of respondents who did not live in the area served by the MEVO system (e.g., tourists). The pool of the final respondents included the municipal bike system's actual users who left or rented municipal bikes at one of six stations located in the three main cities (Gdynia, Gdańsk, and Sopot) where the MEVO municipal bike system subject to analysis was operational. The stations where the survey was conducted included six stations which were all relatively close to public station stops (bus, trolley, or municipal rail):

– Galeria Bałtycka (station No. 11365);
– Olivia Business Centre (station No. 11358);
– Gdynia Główna (station No. 12000);
– Gdynia City Museum (station No. 12053);
– Skwer Kuracyjny Sopot (station No. 10100);
– Bohaterów Monte Cassino (station No. 10124).

It should be emphasized that MEVO functioned as a point-area system (i.e., the return of the bicycle can take place at the station but also by leaving the bike anywhere in the designated area after paying an additional fee). The stations surveyed were selected so that their operational scope covered some of the main traffic generators located within the area covered by the MEVO system of various specificity (e.g., traffic generators related to work, recreation, or shopping). Thereby, the survey covered people who used the municipal bike system to differentiate the goals (meeting the transport needs related to diversified destinations). According to the data of the system operator, these stations were the most popular stations of the MEVO system in individual cities [57]. It should also be noted that an equal number of interviews were conducted at each station.

Upon conducting the survey, the MEVO system comprised 1224 bicycles equipped with electric power-assisted steering. Ultimately, the system was supposed to include 4080 electric bicycles, which would make it the largest bike-sharing system of this type in Poland and Europe [58]. The lack of experience in the exploitation of electric bikes by the system operator (Nextbike, the European bike-sharing market leader) resulted in the suspension of the whole system after seven months of operation. A particularly big problem for the system operator was the charging of the batteries in the bicycles and the redistribution of the charged vehicles. Despite the attempts made, the system was not brought back into operation due to an ongoing legal dispute.

Table 1 presents the most important characteristics of the survey sample. The respondents comprised a similar number of women and men. The gender parity with slightly more women among the MEVO municipal bike users complied with the gender structure of the inhabitants of Poland and was similar to the parity among cyclists in the European conditions [59]. The majority of respondents had a driver's license and a passenger car in their household. Therefore, they could make a decision to fulfil their transport needs by municipal bike, urban transport, the passenger car they owned, or a car rented based on the car-sharing system. The data show that the majority of MEVO municipal bike users subjected to the survey included regular users who used municipal bikes to meet their transport needs at least several times a week.

Table 1. Characteristics of the survey sample ($n = 500$).

Respondent Characteristics	Mean	SD	Min	Max
Age (in years)	28	5.7	16	59
Household size (persons)	2.9	1.5	1	7
Household bicycle ownership	1.2	1.4	0	7

Respondent Characteristics	Characteristic Value (%)
Gender	(Male—48.21%; female—51.56%)
Education	(Higher (bachelor's or master's degree)—64.73%; middle school—18.08%; in the course of bachelor or master studies—13.61%; primary education only—3.23%)
Social and professional status *	(Working on a contract—78.79%; student—13.83%; pensioner—2.45%; freelancing—6%; unemployed—4%)
Driver's license	(Yes—84.38%; No—15.62%)
Car ownership	(Yes—65.4%; No—34.6%)
Disposable income per month per capita in household	(Below 250 EUR—7.14%; 251–500 EUR—6.02%; 501–750 EUR—26.79%; 751–1000 EUR—25.45%; 1001–1500 EUR—15.63%; above 1500 EUR—18.75%)
Main purpose of using MEVO	(Traveling to work—63.84%; traveling to the place of study—9.60%; sport or recreation—6.03%; shopping—14.29%; visiting friends or family—3.79%; other—2.23%)
Distance between place of residence and place of work or study	(Below 2 km—11.38%; 2–3 km—30.13%; 4–6 km—22.32%; 7–9 km—16.52%; 10–13 km—6.47%; 14–20 km—8.04%; above 20 km—4.91%)
How often do you travel by MEVO bike?	(Every day—10%; several times a week—51%; once a week—6%; several times a month—21%; once a month—3%; less than once a month—9%; first time—0.4%)

* Multiple answers possible. Source: own elaboration based on the data obtained in the study.

4. Results

4.1. Substitutability of MEVO Municipal Electric Bike Systems against the Other Forms of Urban Transport

To verify the hypothesis formulated in this article, the respondents' answers to the three following questions were absolutely essential:

–What form of transportation do you replace most frequently with the MEVO municipal bike system?
–If for some reason you would not be able to travel by MEVO bike today (e.g., no bike at the start station or system failure), how would you travel to your destination?
–How did your previous transport behavior change after the launch of the MEVO municipal bike system?

The data presented in Table 2 shows that based on the respondents' answers, travel with the use of municipal electric bikes replaced, to a large extent, the sustainable forms of transportation. In the case of over 50% of the respondents, the municipal bike replaced their travels by public transport. Over 30% of the following respondents abandoned non-motorized forms of transportation (walking and travel by private bicycle) in favor of the municipal bikes. Only 14% of the respondents declared that, for them, the municipal electric bike most frequently replaced the passenger car (including private passenger car, taxi services, car-sharing, and carpooling services). It should also be noted that every fifth surveyed MEVO system user declared that the municipal bike was used instead of walking. In their case, the change in transport behavior, namely walking being partially replaced by travel with the use of municipal electric bikes, should be considered unfavorable for sustainable urban development.

Table 2. What does the metropolitan bike replace for you most often? (n = 500).

Mode of Transportation	Percentage of Cases	Mode of Transportation	Percentage of Cases	Mode of Transportation	Percentage of Cases
passenger car (incl. taxi, car-sharing, and carpooling)	14.1	passenger car (incl. taxi, car-sharing, and carpooling)	14.1	passenger car (incl. taxi, car-sharing, and carpooling)	14.1
public railway transport	18.1	total public transport	52.4	sustainable forms of transportation	85.9
public transport (bus, trolleybus, and tram)	34.3				
private bicycle	12.5	non-motorized forms of transportation	33.5		
walking trips	21.0				

Source: own elaboration based on the data obtained in the study.

Moreover, Table 3 presents data on the respondents' declarations regarding the form of transportation most frequently replaced by municipal electric bikes in their urban travels. However, in this case, the survey covered only the respondents who had a driver's license and a passenger car in their household at the time of the survey. Therefore, the data presented in the table include only those respondents who may truly be in a dilemma over traveling by private passenger car or using other forms of transportation. This is very specific, because people who carry out the significant investment of purchasing a car are then naturally more inclined to use it as the vast part of the costs had already been spent. The results related to people with a driver's license and access to a passenger car were similar to the results obtained for the entire population. Therefore, only for every fifth surveyed respondent who had a real possibility to travel by passenger car was travel by a municipal bike involved in the reduction in car travel.

Table 3. What does the metropolitan bike replace for you most often? Answers are from people who have a private passenger car in the household and a driver's license (n = 275).

Mode of Transportation	Percentage of Cases	Mode of Transportation	Percentage of Cases	Mode of Transportation	Percentage of Cases
passenger car (incl. taxi, car-sharing, and carpooling)	21.8	passenger car (incl. taxi, car-sharing, and carpooling)	21.8	passenger car (incl. taxi, car-sharing, and carpooling)	21.8
public railway transport	17.1	total public transport	40.7	sustainable forms of transportation	78.2
public transport (bus, trolleybus, and tram)	23.6				
private bicycle	14.2	non-motorized forms of transportation	37.5		
walking trips	23.3				

Source: own elaboration based on the data obtained in the study.

The fact that MEVO municipal electric bikes substituted the sustainable forms of transportation to the largest extent is confirmed by the data presented in Table 4. The data prove that over 80% of the respondents declared that if for some reason they were not able to make their planned travel by the municipal electric bike, they would travel by one of the sustainable forms of transportation. If for some reason travel by municipal bike was impossible, nearly 50% of the respondents would travel by public transport. Less than 20% of the respondents declared that if travel by MEVO municipal bike was impossible, they would travel by passenger car.

Table 4. If for some reason you would not be able to travel by a MEVO municipal bike today (e.g., no bike at the start station or system failure), how would you travel to your destination? (*n* = 500).

Mode of Transportation	Percentage of Cases	Mode of Transportation	Percentage of Cases
by passenger car	17.4	by passenger car	17.4
by public transport	48.5		
by private bicycle	10.4	sustainable forms of transportation	81.0
on foot	21.4		
in another way	0.7		
I would abandon traveling	1.6	I would abandon traveling	1.6

Source: own elaboration based on the data obtained in the study.

As concluded above, with regard to the transport policies of cities and their sustainable development and maximization of positive effects resulting from the implementation of municipal bike systems, the people who should travel by municipal bikes most often should be those who substitute the passenger car with the municipal bike most often in their urban travels. Table 5 shows that in the analyzed municipal electric bike system, such a relation was not observed. The largest number of regular users was observed in a group of people for whom the municipal bikes substituted public transport to the largest extent.

Table 5. What does the metropolitan bike replace for you most often? This regards the frequency of travel by municipal bikes.

How Often Do You Travel by MEVO Bike?		Total Public Transport (%)		Passenger Car (Incl. Taxi, Car-Sharing, and Carpooling) (%)		Non-Motorized Forms of Transportation (%)	
every day	regular users	12.8	75.4	9.5	66.7	6.7	53.0
several times a week		58.3		42.9		42.3	
once a week		4.3		14.3		4.0	
once a month	occasional users	4.3	24.7	0.0	33.3	3.4	47.0
several times a month		14.9		23.8		29.5	
less than once a month		5.1		9.5		14.1	
first time		0.4		0.0		0.0	

Source: own elaboration based on the data obtained in the study.

One of the factors that should positively affect the city residents' willingness to change their mode of urban transportation, swapping a passenger car for a municipal bike, was a low fee for using a bike-sharing system. The data presented in Table 6 show that the monthly cost incurred by the system user who traveled using the system for no longer than 90 min per day totalled EUR 2.3. The running costs of a municipal electric bike system are presented in Table 7.

Table 6. Prices of particular tariffs of the MEVO system.

Tariff	Monthly	Annual	Annual Plus	Minute Fee	2-Day	2-Day Plus	5-Day	5-Day Plus
Subscription (EUR)	2.3	23.3	35	0.023 per 1 min	4.7	9.3	9.3	18.7
Initial fee (one-off) (EUR)	2.3	2.3	2.3	2.3	2.3	2.3	2.3	2.3
Tariff time (days)	30	365	365	unlimited	2	2	5	5
Daily time in subscription (min)	90	90	120	-	300	700	300	700
Rate after exceeding the daily time of rental (EUR per 1 min)	0.012	0.012	0.012	0.023	0.012	0.012	0.012	0.012

Source: own elaboration based on [60].

Table 7. Monthly costs of urban trips made by particular forms of transportation within the analyzed area (EUR per month).

Form of Transportation	30 min/Day	60 min/Day	90 min/Day	120 min/Day
MEVO (bike left at the station)	2.3	2.3	2.3	9.9
MEVO (bike left outside the station)	13.8	13.8	13.8	21.4
Public transport (ticket of one organizer)	35.3	35.3	35.3	35.3
Public transport (rail and municipal metropolitan ticket of all organizers)	54.1	54.1	54.1	54.1
Passenger car	43.2	86.4	129.6	172.8
Car sharing	126.5	253.0	379.5	506.0

Source: own elaboration based on [60–62].

The data presented in Table 7 show the monthly costs of urban trips within the operational area of the MEVO municipal bike scheme by various forms of transport. The analysis covered the costs of travel by MEVO bikes, public transport (within the area managed by one public transport operator and within the entire metropolitan area), owned passenger cars, and passenger cars within a car-sharing scheme. During the analysis, the following assumptions were applied: a city resident travels 21 days per month (this is the monthly average number of working days in Poland) and makes 2 trips per day. In the case of individual motorized forms of transportation, it was assumed that the trip would be characterized by an average speed of 20 km/h. The data presented in the table show that fulfilling the transport needs with the use of municipal bikes was definitely the least expensive alternative. The fact that municipal bikes, as proven above, substituted passenger cars only to a small extent, even though fulfilling urban transport needs with the use of municipal bikes even involved dozens of times lower costs compared with passenger cars, proved that the cost of travel was not the main factor affecting the change of city residents' transport behavior. This conclusion is confirmed by the data presented in Table 8, presenting the mode of using municipal bikes by the respondents. The data show that most of the respondents used municipal bikes for direct trips to their destinations. Therefore, they left the bikes outside the stations of the MEVO system, which involved the need to incur an additional fee to the amount of EUR 0.5. Furthermore, the article presented analyses related to non-cost factors affecting the substitutability of municipal electric bikes against passenger cars.

Table 8. In what circumstances do you use MEVO municipal bikes most often? ($n = 500$).

Manner of Using Municipal Electric Bike	Percentage of Cases
"I travel directly to my destination, leaving the bike outside a docking station"	66.2
"I travel to a docking station closest to my destination"	21.8
"I travel to public transport terminals"	11.9

Source: own elaboration based on the data obtained in the study.

4.2. Complementarity of MEVO Municipal Electric Bike Systems against Other Forms of Urban Transportation

As indicated above in the literature review, municipal bike schemes aim, for example, to make it easier for passengers to travel to and from public transport terminals or stations (the so-called first and last mile). Therefore, municipal bike schemes should stimulate the demand for public transport services through increasing their time-based availability, since the total urban travel time in regard to door-to-door systems, thanks to the municipal bikes, should be reduced (referring to people who walked the first and the last mile before introducing the municipal bike scheme). The reduced total travel time in regard to the door-to-door system resulted from the higher average speed of electric bike travel compared with walking trips or trips by traditional bicycles. The average speed of walking trips totaled roughly 4.5 km/h [63], whereas the trips by traditional bicycles amounted to 18.8 km/h. Meanwhile, bikes with electric power-assisted steering (Pedelec) reached 21.9 km/h, and electric bikes (S-pedelec) reached 27.9 km/h [64]. The data show that traveling by a bike equipped with electric power-assisted steering was nearly, on average, five times faster than walking. In order to increase the time-based availability of public transport through offering intermodal travels combined with municipal bikes, it is necessary to ensure proper quality of the bike-sharing systems. In this context, the particularly important features of bike-sharing systems include the number of municipal bikes per city resident or system user (which affects their availability) and the mode of bike relocation in systems with no docking stations (which affects the distance between the user and the closest vacant bike).

The data presented in Table 8 indicate that only every ninth surveyed respondent used a MEVO municipal bike most frequently for travel combined with public transport. The majority of the surveyed respondents used the municipal bike for direct trips to their destinations. Therefore, a MEVO municipal bike was used only to a small extent as a means of transportation complementary to public transport.

4.3. Variables Determining the Probability of Replacing Car Journeys with the MEVO Bicycle

The logistic regression models were then constructed to verify the effect of independent variables on the distribution of users substituting commuting by car with the MEVO bicycle. A sequence of two hundred models were assessed in Statistica 13.1. software with all the possible different combinations of independent variables. The Wald Chi-square test was used to assess the significance of these models. The pseudo R-squared values were calculated to determine the explanatory power of the models, and the Akaike information criterion (AIC) was used to compare the relative quality of the models and to help us investigate if any variables which were omitted could have been included in the model, providing further added value. Logistic regression has been widely used to analyze the determinants of transport behavior and is the generally accepted method in similar research (e.g., [65–68]).

A binary variable showing whether someone substituted a car or sustainable means of transport was chosen as the dependent variable. The respondents declaring that they substituted public transport (road or rail), their private bike, or commuting on foot with the MEVO bike system were aggregated together. Thus, a value of one for the dependent variable indicated that the trip made substituted a car trip. A value of zero meant that it did not or that it substituted another type of trip, including public transport trips. The independent variables were divided into two groups: factors and covariates. The factors included the following:

– Reason behind traveling;
– Distance between the place of residence and the place of work;
– Most common use of the MEVO system;
– Gender;
– Education;
– Socioeconomic status;
– Owning a driver's license;

– Disposable income.

The covariates included the following:

– Year of birth;
– Number of people in the household;
– Number of cars in the household;
– Number of bikes in the household.

The results of logistic regression are presented in Table 9. Only the best model with all the significant variables is presented.

Table 9. Logistic regression model ($n = 500$).

Variable	Odds Ratios
Year of birth	-0.071 (0.034) **
Number of people in the household	-0.451 (0.143) ***
Number of cars in the household	0.344 (0.146) ***
Distance from home to work (less than 3 km)	-0.891 (0.338) ***
Distance from home to work (more than 10 km)	1.430 (0.595) **
Gender (female)	-0.278 (0.126) *
Wald Chi-square test (p-value)	0.028
Pseudo R-squared	0.323
AIC	289.492

Note: logistic regression coefficients with standard errors are in parentheses (*** $p < 0.01$; ** $p < 0.05$; * $p < 0.1$). The dependent variable is the transport choice flag (1 = car, 0 = sustainable means of transport). Source: own computation in Statistica.

Only a few of all the variables turned out to be statistically significant. These included the age, number of people in the household, number of cars in the household, distance from work, and gender. The results prove that for the analyzed group, the younger participants substituted their cars with MEVO rather than substituting public transport. With each year of age, people were relatively 7% more likely to opt for the substitution of transport modes. Moreover, a higher number of people in a household were correlated with a 45% chance (c.p.) of someone substituting a car with MEVO, possibly due to the lower relative availability of cars within the household as perceived by the respondent. On the other hand, a higher number of cars was correlated with a statistical increase of substituting public transport with MEVO, indicating the well-known fact that car owners tend to use their cars, and the bike-sharing system played a mostly recreational role for them. Each additional car in the household was correlated with a decrease in likelihood by a further 34%. Moreover, people living far away from their places of work more often substituted public transport with MEVO rather than people living within a 3-km radius of their place of work who, in comparison with the average respondent, commonly substituted a car. In this case, they probably were not that dependent on cars in their commutes and thus could allow themselves to choose MEVO despite its initial problems with availability. Finally, women substituted cars with MEVO 28% more often than men.

Interestingly enough, the reason behind the choice of MEVO systems as a mode of transport was not correlated with the willingness to swap the car for a bike. This might result from the fact that at the time of the survey, the MEVO system was in its early stages, and it was very unlikely to have become someone's main mode of transport in everyday commuting. The willingness to substitute cars was not correlated with the respondent's education, socioeconomic status, or disposable income. This stands in contradiction with the international consensus but goes in line with the fact that the passenger car in Poland is not a luxury in any way. It also plays a key role as some social symbol, crossing the borders of socioeconomic divisions. Further studies should probably include the ages of the owned cars so as to mediate for this effect. Interestingly enough, holding a driver's license was not correlated with the choice, perhaps due to a very small number of people being without a driver's license within the analyzed group.

5. Discussion and Conclusions

The analysis presented in the article provides several important implications and conclusions that can be used when planning the implementation of similar electric bike-sharing systems in other metropolitan areas. The survey presented in the article showed that the MEVO municipal electric bike scheme constituted the substitutable form of transportation against collective urban transport to a larger extent than against passenger cars. This means that the operation of a MEVO scheme only limited the use of passenger cars in the city residents' daily trips within the scheme's operational area to a small extent. Consequently, a small degree of car travel was substituted with municipal bike travel, meaning that the MEVO system contributed to a small extent to reducing the external costs generated as a result of fulfilling the city residents' transport needs, therefore contributing only to a small extent to improving the sustainability of cities where the system operated. Moreover, the survey presented in the article proves that the MEVO municipal electric bike system constituted a substitutable form of transportation against walking to a larger extent than against passenger cars. Therefore, as a result of using electric bikes, the system contributed to the increase in external costs of the transport system. Additional costs arose mainly from the need to charge batteries used in the bikes and the redistribution of bikes between stations (with the use of conventional vehicles). The above results mean that the effects of the implementation of the MEVO system met the expectations of policymakers to a very limited extent. This shows that despite the significant advantages that electric bikes have over conventional bikes, this study did not prove that electric bikes could replace more private car trips than conventional bikes. This is particularly important given the significant differences between the cost of implementing and maintaining traditional and electric bike-sharing systems. These differences were so significant that they contributed to the temporary suspension of the MEVO system's operation. The findings of the study partially confirm research being carried out in the already existing body of literature on the subject. More specifically, the results are in line with the results obtained in previous similar studies regarding the strength and direction of substitution between bike-sharing systems and public transport as well as personal cars [2,69,70]. However, there are also studies that show the opposite result, specifically that there is a high or even total rate of substitution of car trips by bike-sharing systems. These studies have been mostly carried out within countries with an already strong culture of active commuting (e.g., Sweden), and the results of our study might be partially contradictory due to general social and cultural factors [71,72].

Moreover, the factors affecting the residents' willingness to substitute car travel with municipal bike travel were analyzed, too. It turned out that the factors increasing the probability to swap the car for a municipal electric bike included gender (women were more willing to abandon car travel), young age, living in a multi-person household (which potentially may adversely affect the possibility to use a passenger car), and living within a small distance (less than 3 km) from the place of work or study.

Taking into account all the results mentioned above, it should be noted that 6 months passed since the launch of the MEVO system when the survey was conducted. Due to the organizational difficulties on the part of the system operator, slightly more than 1200 bikes were used on a regular basis out of the scheduled 4080 municipal electric bikes. This means that the system did not reach its full capacity, which would contribute to the increased availability of bikes and consequently to the increased reliability of this mode of transport perceived by the city residents in urban travel within the system's operational area. It can be assumed that if the number of available municipal bikes increased, the willingness to abandon car travel in favor of bike travel would also increase. It is also possible that the long-term effects of introducing an electric bike-sharing system are different from the short-term effects presented in this article. Another limitation of this study results from its nature as a case study (the sample of participants was not representative and should therefore not be generalized to the wider population). Therefore, it cannot be ruled out that a city's electric bike system operating in different conditions will be characterized by a different degree of substitutability and complementarity against other forms of urban transport. An additional limitation of this study is the fact that only the declarations of the MEVO system users were examined and not their actual behavior. Furthermore, the very nature of the topic indicates that the results might change if the study is carried out in different weather conditions, during different seasons, and also during different days of the week, because mobility patterns, especially regarding various types of active commuting including cycling, change depending on these factors. In the long run, the results might also change based on different transport policies as well as sociocultural aspects, such as the social perception of a personal car.

Recent studies show that most respondents stated that COVID-19 would not affect their intention to use bike-sharing systems [73–75]. Moreover, during the pandemic, many countries promoted the use of bicycles as a safe way to meet transportation needs [76]. This means that despite the collapse in demand for bike-sharing services, there is a good chance that these systems will not lose their relevance [77,78]. It also means that as electric vehicles become more popular, more bike-sharing systems using only electric bikes will be introduced in Poland and Europe. Thus, new opportunities will arise to study the role of electric bike-sharing in urban mobility. Future research may use big data analysis and real behavioral data instead of declarative data. There is also still a need for research into the elements of the urban transport system, the change or improvement of which may lead to a greater shift from car trips to electric city bikes. This is confirmed by the contradictory results of studies of individual conventional and electric bike-sharing systems. We are currently unable to determine to what extent the COVID-19 pandemic will delay the restart of the MEVO system.

Author Contributions: M.S.: conceptualization, data curation, methodology, formal analysis, and writing—original draft; A.J.: conceptualization, data curation, formal analysis, project administration, writing—original draft, visualization, and funding acquisition; J.S.: conceptualization, data curation, formal analysis, writing—original draft, and funding acquisition. All authors have read and agreed to the published version of the manuscript.

Funding: This research was funded by the Faculty of Economics at the University of Gdansk, project No. 539-3210-B362-19.

Institutional Review Board Statement: The study was conducted according to the guidelines of the Declaration of Helsinki, and approved by the Ethics Committee of Faculty of Economics, University of Gdańsk on 1 February 2019.

Informed Consent Statement: Informed consent was obtained from all subjects involved in the study.

Data Availability Statement: Dataset is available at https://ekonom.ug.edu.pl/pp/download.php?OpenFile=34895.

Conflicts of Interest: The authors declare that they have no known competing financial interests or personal relationships that could have appeared to influence the work reported in this paper.

References

1. Jędrzejewski, A. *Ostre Hamowanie Roweru Miejskiego Raport*; Bikesharing 2019/2020; Mobilne Miasto: Warsaw, Poland, 2020.
2. Bieliński, T.; Dopierała, Ł.; Tarkowski, M.; Ważna, A. Lessons from Implementing a Metropolitan Electric Bike Sharing System. *Energies* **2020**, *13*, 6240. [CrossRef]
3. Midgley, P. Bicycle-Sharing Schemes: Enhancing Sustainable Mobility in Urban Areas. *Commun. Sustain. Dev.* **2011**, *8*, 24.
4. Andersen, L.B.; Lawlor, D.A.; Cooper, A.R.; Froberg, K.; Anderssen, S.A. Physical fitness in relation to transport to school in adolescents: The Danish Youth and Sports Study. *Scand. J. Med. Sci. Sports* **2008**, *19*, 406–411. [CrossRef]
5. Eriksson, M.; Uddén, J.; Hemmingsson, E.; Agewall, S. Impact of physical activity and body composition on heart function and morphology in middle-aged, abdominally obese women. *Clin. Physiol. Funct. Imaging* **2010**, *30*, 354–359. [CrossRef] [PubMed]
6. Xu, Y.; Chen, D.; Zhang, X.; Tu, W.; Chen, Y.; Shen, Y.; Ratti, C. Unravel the landscape and pulses of cycling activities from a dockless bike-sharing system. *Comput. Environ. Urban Syst.* **2019**, *75*, 184–203. [CrossRef]
7. Mátrai, T.; Tóth, J. Comparative Assessment of Public Bike Sharing Systems. *Transp. Res. Procedia* **2016**, *14*, 2344–2351. [CrossRef]
8. Shaheen, S.; Guzman, S.; Zhang, H. UC Davis Recent Work Title Bikesharing in Europe, the Americas, and Asia: Past, Present, and Future. *Transp. Res. Rec. J. Transp. Res. Board* **2010**, *2143*, 159–167. [CrossRef]

9. Shaheen, S.A.; Martin, E.W.; Chan, N.D.; Cohen, A.P.; Pogofzinski, M. *Public Bikesharing in North America: Understanding Impacts, Business Models, and Equity Effects of Bikesharing Systems During Rapid Industry Expansion | MTI Research in Progress*; Mineta Transportation Institute: San Jose, CA, USA, 2014; p. 218.

10. Ricci, M. Bike sharing: A review of evidence on impacts and processes of implementation and operation. *Res. Transp. Bus. Manag.* **2015**, *15*, 28–38. [CrossRef]

11. Qiu, L.Y.; He, L.Y. Bike sharing and the economy, the environment, and health-related externalities. *Sustainability* **2018**, *10*, 1145. [CrossRef]

12. Kim, J.; Yoon, Y.; Zo, H. Why people participate in the sharing economy: A social exchange perspective. In Proceedings of the 19th Pacific Asia Conference on Information Systems, PACIS, Singapore, 5–9 July 2015; Volume 76, p. 3.

13. Frenken, K.; Schor, J. Putting the sharing economy into perspective. *Environ. Innov. Soc. Transit.* **2017**, *23*, 3–10. [CrossRef]

14. Cerutti, P.S.; Martins, R.D.; Macke, J.; Sarate, J.A.R. "Green, but not as green as that": An analysis of a Brazilian bike-sharing system. *J. Clean. Prod.* **2019**, *217*, 185–193. [CrossRef]

15. Fishman, E. Bikeshare: A Review of Recent Literature. *Transp. Rev.* **2016**, *36*, 92–113. [CrossRef]

16. DeMaio, P. Bike-sharing: History, Impacts, Models of Provision, and Future. *J. Public Transp.* **2009**, *12*, 41–56. [CrossRef]

17. Berbeglia, G.; Cordeau, J.-F.; Gribkovskaia, I.; Laporte, G. Static pickup and delivery problems: A classification scheme and survey. *TOP* **2007**, *15*, 1–31. [CrossRef]

18. Ghosh, S.; Varakantham, P.; Adulyasak, Y.; Jaillet, P. Dynamic Repositioning to Reduce Lost Demand in Bike Sharing Systems. *J. Artif. Intell. Res.* **2017**, *58*, 387–430. [CrossRef]

19. Ho, S.C.; Szeto, W.Y. Solving a static repositioning problem in bike-sharing systems using iterated tabu search. *Transp. Res. Part E Logist. Transp. Rev.* **2014**, *69*, 180–198. [CrossRef]

20. Forma, I.A.; Raviv, T.; Tzur, M. A 3-step math heuristic for the static repositioning problem in bike-sharing systems. *Transp. Res. Part B Methodol.* **2015**, *71*, 230–247. [CrossRef]

21. Boufidis, N.; Nikiforiadis, A.; Chrysostomou, K.; Aifadopoulou, G. Development of a station-level demand prediction and visualization tool to support bike-sharing systems' operators. *Transp. Res. Procedia* **2020**, *47*, 51–58. [CrossRef]

22. Silver, L.; Cornibert, S. *Smartphone Ownership Is Growing Rapidly Around the World, But Not Always Equally*; Pew Research Center: Washington, DC, USA, 2019.

23. Campbell, K.B.; Brakewood, C. Sharing riders: How bikesharing impacts bus ridership in New York City. *Transp. Res. Part A Policy Pract.* **2017**, *100*, 264–282. [CrossRef]

24. Suchanek, M.; Wołek, M. The relations between the bikesharing systems systems and public expenditures. Cluster analysis of the Polish bikesharing. In *New Research Trends in Transport Sustainability and Innovation: TranSopot 2017 Conference*; Suchanek, M., Ed.; Springer Proceedings in Business and Economics; Springer International Publishing: Cham, Switzerland, 2018; pp. 37–45.

25. Sato, H.; Miwa, T.; Morikawa, T. A study on use and location of community cycle stations. *Res. Transp. Econ.* **2015**, *53*, 13–19. [CrossRef]

26. Poliaková, B.; Kubasáková, I. Public passenger transport and integration with cycling. *Tech. Transp. Szyn.* **2014**, *21*, 34–36.

27. Shaheen, S.; Guzman, S. Worldwide Bikesharing. *Access Mag.* **2011**, *39*, 22–27.

28. Fishman, E.; Washington, S.; Haworth, N.; Mazzei, A. Barriers to bikesharing: An analysis from Melbourne and Brisbane. *J. Transp. Geogr.* **2014**, *41*, 325–337. [CrossRef]

29. Zhang, Y.; Zhang, Y. Associations between public transit usage and bikesharing behaviors in the United States. *Sustainability* **2018**, *10*, 1868. [CrossRef]

30. Pucher, J.; Buehler, R. *City Cycling*; Pucher, J., Buehler, R., Eds.; The MIT Press: London, UK, 2012.

31. Dekoster, J.; Schollaert, U. Cycling: The way ahead for towns and cities. *Eur. Commun.* **2004**, *11*, 309–317.

32. Volinski, J. *Implementation and Outcomes of Fare-Free Transit Systems*; The National Academies Press: Washington, DC, USA, 2012.

33. ECMT. *Implementing Sustainable Urban Travel Policies: Moving Ahead: National Policies to Promote Cycling*; European Conference of Ministers of Transport: Paris, France, 2004; ISBN 9789282103296.

34. Campbell, A.A.; Cherry, C.R.; Ryerson, M.S.; Yang, X. Factors influencing the choice of shared bicycles and shared electric bikes in Beijing. *Transp. Res. Part C Emerg. Technol.* **2016**, *67*, 399–414. [CrossRef]

35. Helms, H.; Kämper, C.; Lienhop, M. *Pedelction Mobilitätsmuster, Nutzungsmotive und Verlagerungseffekte*; Institute for Energy and Environmental Research: Heidelberg, Germany, 2015.

36. Engelmoer, W. The E-Bike: Opportunities for Commuter Traffic. The Potentials of Using Electric Bicycles and Scooters in Commuting Traffic in Relation to the Accessibility and Quality of the Local Environment of a Compact Dutch City. Master's Thesis, University of Groningen, Groningen, The Netherlands, 2012.

37. Rivas, M.E.; Suarez-Aleman, A.; Serebrisky, T. *Stylized Urban Transportation Facts in Latin America and the Caribbean*; Inter-American Development Bank: Washington, DC, USA, 2019.

38. Barbour, N.; Zhang, Y.; Mannering, F. A statistical analysis of bike sharing usage and its potential as an auto-trip substitute. *J. Transp. Heath* **2019**, *12*, 253–262. [CrossRef]

39. Menghini, G.; Carrasco, N.; Schüssler, N.; Axhausen, K.W. Route choice of cyclists in Zurich. *Transp. Res. Part A Policy Pract.* **2010**, *44*, 754–765. [CrossRef]

40. Efthymiou, D.; Antoniou, C.; Waddell, P. Factors affecting the adoption of vehicle sharing systems by young drivers. *Transp. Policy* **2013**, *29*, 64–73. [CrossRef]

41. Bachand-Marleau, J.; Lee, B.; El-Geneidy, A. Better understanding of factors influencing likelihood of using shared bicycle systems and frequency of use. *Transp. Res. Rec.* **2012**, *2314*, 66–71. [CrossRef]

42. Fishman, E.; Washington, S.; Haworth, N.; Watson, A. Factors influencing bike share membership: An analysis of Melbourne and Brisbane. *Transp. Res. Part A Policy Pract.* **2015**, *71*, 17–30. [CrossRef]

43. Wang, K.; Akar, G.; Chen, Y.J. Bike sharing differences among Millennials, Gen Xers, and Baby Boomers: Lessons learnt from New York City's bike share. *Transp. Res. Part A Policy Pract.* **2018**, *116*, 1–14. [CrossRef]

44. Ji, Y.; Fan, Y.; Ermagun, A.; Cao, X.; Wang, W.; Das, K. Public bicycle as a feeder mode to rail transit in China: The role of gender, age, income, trip purpose, and bicycle theft experience. *Int. J. Sustain. Transp.* **2017**, *11*, 308–317. [CrossRef]

45. Dill, J.; Rose, G. Electric bikes and transportation policy: Insights from early adopters. *Transp. Res. Rec.* **2012**, *2314*, 1–6. [CrossRef]

46. MacArthur, J.; Dill, J.; Person, M. Electric bikes in North America: Results of an online survey. *Transp. Res. Rec.* **2014**, *2468*, 123–130. [CrossRef]

47. Johnson, M.; Rose, G. Extending life on the bike: Electric bike use by older Australians. *J. Transp. Health* **2015**, *2*, 276–283. [CrossRef]

48. Fishman, E. *Bike Share*; Routledge: London, UK, 2020; ISBN 978-1138682498.

49. Shaheen, S.; Chan, N. *Mobility and the Sharing Economy: Potential to Overcome First- and Last-Mile Public Transit Connections*; Transportation Sustainability Research Center: Berkeley, CA, USA, 2016; pp. 573–588.

50. Cherry, C.R.; Yang, H.; Jones, L.R.; He, M. Dynamics of electric bike ownership and use in Kunming, China. *Transp. Policy* **2016**, *45*, 127–135. [CrossRef]

51. Veeneman, W. Public transport in a sharing environment. In *The Sharing Economy and the Relevance for Transport*, 1st ed.; Elsevier: Amsterdam, The Netherlands, 2019; Volume 4, p. 50. ISBN 9780128162118.

52. Wolf, A.; Seebauer, S. Technology adoption of electric bicycles: A survey among early adopters. *Transp. Res. Part A Policy Pract.* **2014**, *69*, 196–211. [CrossRef]

53. Hiselius, L.W.; Svenssona, Å. Could the increased use of e-bikes (pedelecs) in Sweden contribute to a more sustainable transport system? In Proceedings of the 9th International Conference on Environmental Engineering, Vilnius, Lithuania, 22–24 May 2014.

54. Arendsen, J. *Shared Mobility for the First and Last Mile: Exploring the Willingness to Share*; Delft University of Technology: Delft, The Netherlands, 2019.

55. Langford, B.C.; Cherry, C.R.; Bassett, D.R.; Fitzhugh, E.C.; Dhakal, N. Comparing physical activity of pedal-assist electric bikes with walking and conventional bicycles. *J. Transp. Health* **2017**, *6*, 463–473. [CrossRef]

56. Dozza, M.; Werneke, J.; Mackenzie, M. e-BikeSAFE: A Naturalistic Cycling Study to understand how electrical bicycles change cycling behaviour and influence safety. In Proceedings of the International Cycling Safety Conference, Helmond, The Netherlands, 20–21 November 2013; pp. 1–10.

57. Szymczewski, A. Najpopularniejsze Trasy Przejazdów i Stacje Mevo. Available online: https://www.trojmiasto.pl/wiadomosci/Mevo-najpopularniejsze-stacje-i-trasy-przejazdow-n135077.html (accessed on 1 June 2021).

58. Galatoulas, N.F.; Genikomsakis, K.N.; Ioakimidis, C.S. Spatio-temporal trends of e-bike sharing system deployment: A review in Europe, North America and Asia. *Sustainability* **2020**, *12*, 4611. [CrossRef]

59. Garrard, J.; Crawford, S.; Hakman, N. *Revolutions for Women: Increasing Women's Participation in Cycling*; School of Health and Social Development, Deakin University: Geelong, Australia, 2006.

60. Szymczewski, A. Rower Mevo Wystartował. Jak go Wypożyczyć i ile to Kosztuje? Available online: https://www.trojmiasto.pl/wiadomosci/Mevo-oficjalnie-wystartowalo-Test-aplikacji-n132966.html (accessed on 1 April 2020).

61. MZKZG Season Tickets. Available online: https://mzkzg.org/bilety-okresowe (accessed on 1 April 2020).

62. Jaspers. *Blue Book. Road Infrastructure*; European Commision: Brussels, Belgium, 2015.

63. Bohannon, R.W.; Williams Andrews, A. Normal walking speed: A descriptive meta-analysis. *Physiotherapy* **2011**, *97*, 182–189. [CrossRef]

64. Schleinitz, K.; Petzoldt, T.; Franke-Bartholdt, L.; Krems, J.; Gehlert, T. The German Naturalistic Cycling Study—Comparing cycling speed of riders of different e-bikes and conventional bicycles. *Saf. Sci.* **2017**, *92*, 290–297. [CrossRef]

65. Koohsari, M.J.; Cole, R.; Oka, K.; Shibata, A.; Yasunaga, A.; Hanibuchi, T.; Owen, N.; Sugiyama, T. Associations of built environment attributes with bicycle use for transport. *Environ. Plan. B Urban Anal. City Sci.* **2019**, *47*, 1–13. [CrossRef]

66. Boulange, C.; Gunn, L.; Giles-Corti, B.; Mavoa, S.; Pettit, C.; Badland, H. Examining associations between urban design attributes and transport mode choice for walking, cycling, public transport and private motor vehicle trips. *J. Transp. Health* **2017**, *6*, 155–166. [CrossRef]

67. Adams, E.J.; Goodman, A.; Sahlqvist, S.; Bull, F.C.; Ogilvie, D. Correlates of walking and cycling for transport and recreation: Factor structure, reliability and behavioural associations of the perceptions of the environment in the neighbourhood scale (PENS). *Int. J. Behav. Nutr. Phys. Act.* **2013**, *10*, 1–15. [CrossRef]

68. Ismail, A.; Elmloshi, A.E. Logistic Regression Models to Forecast Travelling Behaviour in Tripoli City. *Int. J. Adv. Sci. Eng. Inf. Technol.* **2011**, *1*, 618. [CrossRef]

69. Bigazzi, A.; Wong, K. Electric bicycle mode substitution for driving, public transit, conventional cycling, and walking. *Transp. Res. Part D Transp. Environ.* **2020**, *85*, 102412. [CrossRef]

70. Bieliński, T.; Kwapisz, A.; Ważna, A. Electric bike-sharing services mode substitution for driving, public transit, and cycling. *Transp. Res. Part D Transp. Environ.* **2021**, *96*, 102883. [CrossRef]

71. Söderberg, F.K.A.; Andersson, A.; Adell, E.; Winslott Hiselius, L. What is the substitution effect of e-bikes? A randomised controlled trial. *Transp. Res. Part D Transp. Environ.* **2021**, *90*, 102648. [CrossRef]
72. Winslott Hiselius, L.; Svensson, Å. E-bike use in Sweden—CO_2 effects due to modal change and municipal promotion strategies. *J. Clean. Prod.* **2017**, *141*, 818–824. [CrossRef]
73. Nikiforiadis, A.; Ayfantopoulou, G.; Afroditi, S. Assessing the impact of COVID-19 on bike-sharing usage: The case of thessaloniki, Greece. *Sustainability* **2020**, *12*, 8215. [CrossRef]
74. Jobe, J.; Griffin, G.P. Bike share responses to COVID-19. *Transp. Res. Interdiscip. Perspect.* **2021**, *10*, 100353.
75. Hua, M.; Chen, X.; Cheng, L.; Chen, J. Should bike sharing continue operating during the COVID-19 pandemic? Empirical findings from Nanjing, China. *arXiv* **2020**, arXiv:2012.02946.
76. Torrisi, V.; Ignaccolo, M.; Inturri, G.; Tesoriere, G.; Campisi, T. Exploring the factors affecting bike-sharing demand: Evidence from student perceptions, usage patterns and adoption barriers. *Transp. Res. Procedia* **2021**, *52*, 573–580. [CrossRef]
77. Shang, W.; Chen, J.; Bi, H.; Sui, Y.; Chen, Y.; Yu, H. *Since January 2020 Elsevier has Created a COVID-19 Resource Centre with Free Information in English and Mandarin on the Novel Coronavirus COVID-19. The COVID-19 Resource Centre Is Hosted on Elsevier Connect, the Company's Public News and Information*; Elsevier: Amsterdam, The Netherlands, 2020.
78. Bergantino, A.S.; Intini, M.; Tangari, L. Influencing factors for potential bike-sharing users: An empirical analysis during the COVID-19 pandemic. *Res. Transp. Econ.* **2021**, *86*, 101028. [CrossRef]

Article

Urban Mobility Data Analysis for Public Transportation Systems: A Case Study in Montevideo, Uruguay

Renzo Massobrio [1,2,*,†] and **Sergio Nesmachnow** [1,*,†]

[1] Facultad de Ingeniería, Universidad de la República, Montevideo11300 , Uruguay
[2] Escuela Superior de Ingeniería, Universidad de Cádiz, Puerto Real, 11519 Cádiz, Spain
* Correspondence: renzom@fing.edu.uy (R.M.); sergion@fing.edu.uy (S.N.)
† These authors contributed equally to this work.

Received: 25 June 2020; Accepted: 29 July 2020; Published: 5 August 2020

Abstract: Transportation systems play a major role in modern urban contexts, where citizens are expected to travel in order to engage in social and economic activities. Modern transportation systems incorporate technologies that generate huge volumes of data, which can be processed to extract valuable mobility information. This article describes a proposal for studying public transportation systems following an urban data analysis approach. A thorough analysis of the transportation system in Montevideo, Uruguay, and its usage is outlined, combining several sources of urban data. Furthermore, origin-destination matrices, which describe mobility patterns in the city, are generated using ticket sales data. The computed results are validated with a recent mobility survey. Finally, a visualization web application is presented, which allows conveying mobility information in an intuitive way.

Keywords: urban mobility; data analysis; origin-destination; ITS

---

1. Introduction

Mobility of citizens is a critical issue emerging from the urbanization process. The geographical organization of urban scenarios demands citizens to travel for engaging in social and economic activities. Public transportation systems are the cornerstone of urban mobility, as they represent the most efficient, sustainable, and socially fair mode of transportation [1]. Understanding the synergy between citizens and public transportation is a key factor to improve mobility in a city.

Modern smart cities use technology in order to improve urban services [2]. Related to smart cities are Intelligent Transportation Systems (ITS), using technology to improve mobility. ITS collect large volumes of urban data [3] that allow understanding the mobility of citizens. For this purpose, urban data analysis arises as a valuable tool to derive information from raw urban data sources.

Understanding the dynamics of mobility is crucial to improve transportation systems. Mobility is described through origin-destination (OD) matrices that indicate the number of passengers traveling between relevant locations. Traditionally, OD matrices are generated based on surveys or manual passenger counts. However, these methods are very expensive to be carried out regularly, so they offer a partial and outdated view of mobility patterns in a city [4]. ITS incorporate technology to locate vehicles and pay for tickets in public transportation. As a by-product these technologies generate valuable data that can be processed to estimate OD matrices.

This manuscript extends our previous conference article "Urban data analysis for the public transportation system of Montevideo, Uruguay" [5], presented at the II Ibero-American Congress on Smart Cities. The new scientific contributions in this article include a characterization of the public transportation system in Montevideo and its users derived from ITS data, extending the previous work by focusing on characterizing the mobility patterns of citizens. In addition, a destination estimation algorithm is proposed and applied to ticket sales data to generate OD matrices that describe mobility in the city. The computed results are compared against the findings of a recent mobility household survey. Finally, a visualization tool is also presented which allows presenting the computed OD matrices to stakeholders in an intuitive fashion.

The article is structured as follows. Section 2 presents the methodology for urban data analysis and reviews related works. Section 3 describes the studied scenario and results from the data analysis to describe the use of the system. The estimation of OD matrices is presented in Section 4, along with a comparison against the mobility survey and a description of the visualization tool. Finally, Section 5 presents the conclusions and the main lines of future work.

2. Methodology and Related Works

This section outlines the urban data analysis methodology applied to characterize mobility using ITS data as well as the main related works on the topic. The methodology used to analyze ITS data is presented in Section 2.1 and a review of related works is outlined in Section 2.2.

2.1. Methodology

This subsection presents the methodology applied for the analysis of ITS data.

2.1.1. Urban Data Analysis

Data analysis is the process of collecting and processing raw data to extract meaningful information to provide supporting evidence to help decision-making. Alternative workflow proposals exist to describe the data analysis process. This work applies a workflow proposed by Schutt and O'Neil [6] (Figure 1).

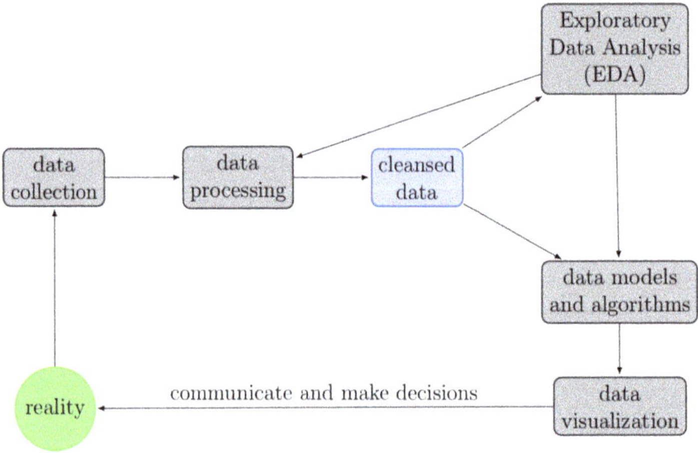

Figure 1. Data analysis workflow applied in this work.

The data analysis process starts and ends in the current reality. In urban contexts, this implies collecting raw data from a given city and, in the end, communicating findings to authorities and citizens. To this end, the data analysis process is comprised of several phases. Initially, raw data must be processed, including placing data into tables, inspecting datasets, and cleansing data to detect corrupt or inaccurate records. After that, Exploratory Data Analysis (EDA) [7] is performed. EDA aims at describing what data can tell, beyond a formal modeling and hypothesis testing phase. Urban data tends to come from a variety of diverse and dynamic sources (e.g., sensors, mobile phones, social media), thus EDA becomes mandatory for urban data analysis to detect inaccuracies. After EDA, statistical models and algorithms (e.g., inferential statistics, machine learning) are applied to identify relationships among the data. Finally, results are communicated, usually through visualization techniques.

EDA makes an intensive use of data visualization with the goal of efficiently displaying measured quantities through graphics. Traditionally, data visualization techniques were mainly dominated by charts and diagrams comprised of numerical data. However, areas such as urban data analysis, which demand combining quantitative and qualitative data, require more advanced means of visualizing results for effective communication. Since urban data usually has a prevalence of geographic components, urban data visualization combines classic statistical graphics with Geographic Information Systems (GIS).

2.1.2. Origin-Destination (OD) Matrices

Mobility is usually described using OD matrices, which indicate the number of trips between relevant locations in a city [8]. Each trip can have multiple legs, if a passenger makes intermediate stops and transfers between vehicles to get to their final destination. Thus, when building OD matrices, the destination of a trip is considered as the final destination of the sequence of legs, where a passenger is assumed to go to perform an activity. Different divisions can be used to analyze mobility at a finer (e.g., specific locations, bus stops) or coarser grain (e.g., municipalities, neighborhoods). OD matrices can be built for specific periods of time to characterize mobility in different days (e.g., working days vs. weekends) or times of the day (e.g., peak vs. non-peak hours).

Traditionally, OD matrices are generated using information from mobility surveys. Unless performed regularly, surveys offer a partial and outdated view of mobility patterns. Additionally, in large cities, where mobility analysis requires detailed zonification and time disaggregation, surveys demand very large sample sizes to compute results with statistical significance. As a consequence, surveys are usually a very expensive mean to characterize urban mobility. Thus, there is a growing interest in using data analysis to estimate OD matrices from available sources of urban data.

Automatic Fare Collection (AFC) systems automate ticketing of a public transportation network. Most AFC systems are comprised of fare media, read/write devices for these media, networks for communication, and back-office systems. Contactless smart cards are the de facto fare media in AFC systems. Pelletier et al. [9] provided a thorough review on the use of smart cards in public transportation systems. Different alternatives to estimate OD matrices using AFC data are reviewed next. The trip-chaining method used in this work is described in Section 4.

2.2. Related Works

A variety of sources have been used to estimate OD matrices in transportation systems. Li et al. [10] identified three models for destination estimation: (i) the probability model, which computes the alighting probability based on the traveled distance and the number of passengers on board, but without identifying pairs (board–alight) corresponding to the same passenger; (ii) the deep learning model, which requires boarding and alighting data for training, being more suitable to railway/subway systems where passengers are required to validate their cards both to enter and exit stations; and (iii) the trip chaining model. The most relevant works based on the trip chaining method, i.e., the one applied in this article, are reviewed next.

The trip chaining model [11] infers destinations by looking at the history of trips of each cardholder. Two hypotheses are considered: the origin of a new trip is the destination of the previous one, and at the end of the day, users return to the origin of their first trip of the day. The proposed model was applied to the subway system of New York, where nearly 80% of riders use smart cards. The computed OD matrix was validated using station exit counts at different times of the day and using peak load passenger volume data and a trip assignment model. The authors estimated that 90% of destinations can be accurately inferred for a 78% share of the total number of subway users.

Trépanier et al. [12] proposed using the trip chaining model for estimating the destination of passengers boarding buses with smart cards, following a database programming approach. Trips for which chaining is not possible were compared with all other trips of the month for the same user to find similar trips with known destination. The experimental evaluation used real datasets from Gatineau, Quebec. The proposed approach allowed estimating the destination of 66% of the trips. However, the real estimation accuracy could not be assessed due to the lack of a second source of data (e.g., surveys) for comparison.

Wang et al. [13] applied the trip chaining method to infer bus passenger origin-destination from smart card transactions from London, UK. Results were compared against the passenger intercept survey performed every five to seven years for each bus route and includes the number of people boarding and alighting at each bus stop. The analysis showed that destinations could be estimated for nearly 57% of all trips. When compared to the survey, the difference on the estimated destinations were below 4% on the worst case.

Munizaga and Palma [14] estimated OD matrices in the multimodal transportation system of Santiago, Chile, where passengers can use their smart cards to pay for tickets at metros, buses, and bus stations. The proposed approach was evaluated using smart card datasets corresponding to two different weeks, with over 35 million transactions each. The destination and time of alighting was estimated for over 80% of the transactions. Later [15], the authors validated the main assumptions of the model by comparing the estimated OD matrices with data from surveys and personal interviews to passengers. The authors concluded that the proposed model was highly reliable, accurately estimating 84.2% of the inferred destinations.

Alsger et al. [16] analyzed one week of smart card data (628,479 transactions) from bus, train, and ferry networks of South East Queensland, Australia. The dataset contained both origin and destination records, since passengers are required to validate their smart cards when boarding and alighting. Therefore, the authors were able to study different variants of the trip chaining method and compare the resulting OD matrices against the real data from AFC records. Results showed that for nearly 88% of the passengers the last destination of the day was within a walkable distance of their first origin, thus validating one of the key assumptions of the trip chaining model.

The analysis of related works allows identifying several proposals for using ITS data analysis to understand and improve urban mobility. This article expands the original trip chaining method by also considering transfers between bus lines. Thus, the OD matrix estimation considers trips that may include several bus trips involving transfers as well as walks between bus stops to do those transfers.

This article proposes applying existing knowledge about urban data analysis and OD matrix generation to understand mobility using ITS data. As a case study, data from the ITS in Montevideo, Uruguay, is analyzed in Section 3 and an OD matrix estimation procedure is outlined in Section 4. There are no previous works using ITS data to understand and improve urban mobility in Montevideo. Therefore, the research contributes with a novel proposal to assess transportation systems and understand mobility patterns, and applies it to real data from the ITS in Montevideo, Uruguay.

3. Characterizing Public Transportation System

This section presents the urban data analysis process aimed at characterizing public transportation systems using ITS data. An overview of the case study is presented in Section 3.1. Then, Section 3.2 presents the urban data analysis process and its results are discussed in Section 3.3. Finally, Section 3.4 presents two practical use cases to show the advantages of using data analysis for authorities and transport planners.

3.1. Overview of the Case Study

This section presents an overview of the case study: the public transportation system of Montevideo, Uruguay.

3.1.1. Montevideo, Uruguay

Montevideo extends to an area of only 530 km^2 and is comprised of eight municipalities and 1063 census tracts, which are defined by the National Institute of Statistics [17]. Census tracts are the administrative division used in census and surveys performed by the state and, consequently, most socioeconomic indicators available are aggregated using this zonification.

Montevideo has an estimated population of 1,319,108 unevenly distributed, with high population densities near the coastline bordering the Río de la Plata estuary. A socioeconomic description of the population can be obtained by studying Unsatisfied Basic Needs (UBNs), which identify the lack of goods or services that prevent citizens from exercising their social rights. The choropleth map in Figure 2 indicates the percentage of households with one or more UBNs. The most vulnerable citizens are located farther away from the coast and the city center, in sparsely populated areas.

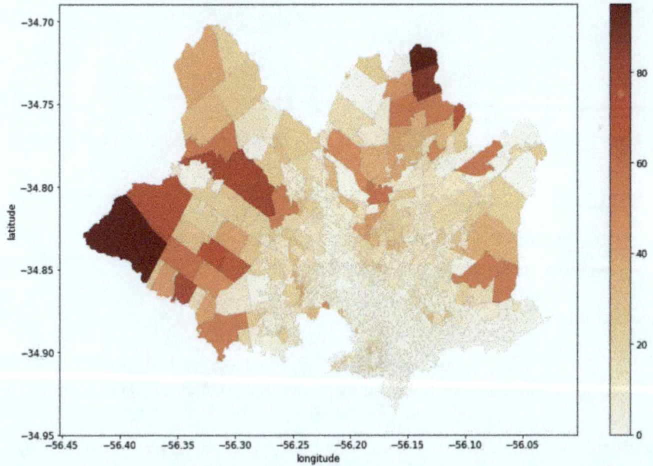

Figure 2. Percentage of households with one or more UBNs in Montevideo, Uruguay.

3.1.2. The Public Transportation System In Montevideo, Uruguay

The public transportation system in Montevideo is comprised of 1528 buses operating in 145 main bus lines with different variants, accounting for outward and return trips, as well as shorter versions of the main line. The total number of different bus lines is 1383 (Figure 3). The average bus line length is 16.7 km (median: 16.4 km, longest line: 39.6 km). Intuitively, these figures strike as remarkably large. The bus network is comprised of 4718 bus stops, most of them located in the city center. This fact remarks the important role of this area within the bus network.

Contact-less top-up smart cards are used to allow passengers to pay for tickets without using physical money. Smart cards are are linked to the identity of the owner (a valid ID is required to get one). Two different types of bus tickets exist: one-hour tickets allow boarding up to two buses within an hour, while two-hours tickets grant unlimited bus transfers within a period of two hours. Passengers may transfer between any bus line at any bus stop. In practice, this means that a passenger can even make an outward and return trip in the same line, as long as the boarding time of the second bus is within the validity period of the ticket. Passengers do not validate their smart cards when alighting a bus. This constitutes one of the main challenges for building OD matrices.

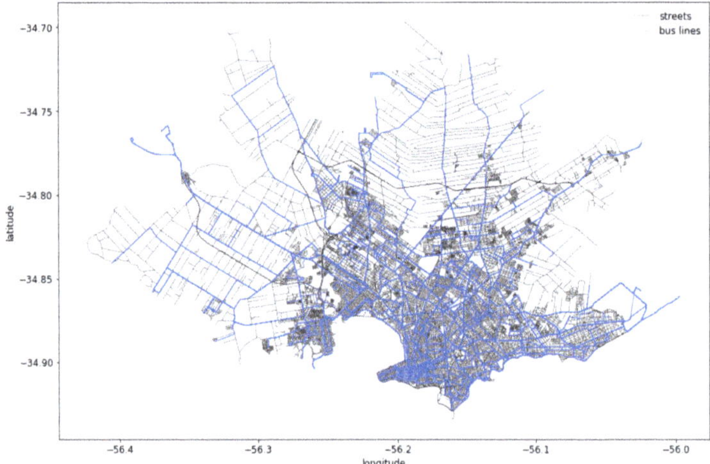

Figure 3. Bus lines of the public transportation system of Montevideo, Uruguay.

3.2. Urban Data Analysis Process

This section describes the urban data analysis process performed with the goal of characterizing how citizens of Montevideo use the public transportation system.

3.2.1. Data Collection and Processing

The data analysis process used national [18] and city [19] open data, and public transportation system data (GPS bus location and bus ticket sales payed with smart cards during 2015, over 150 GB of raw data).

The bus location dataset holds the position of each bus sampled every 10–30 s, including the following information: bus line identifier, trip identifier to set appart different trips of the same bus line, GPS coordinates, instant speed, and time stamp corresponding to the GPS measure. Ticket sales data contain smart card transaction records, including: trip identifier (which allows linking to the bus location dataset), GPS coordinates of the smart card validation, bus stop identifier, time stamp of the smart card validation, unique smart card identifier (hashed for privacy purposes), number of passengers traveling with the same smart card, and leg number (for trips involving transfers).

The data collection process was straightforward in the case of open datasets. The main efforts on this phase were related to data provided by Intendencia de Montevideo. Several meetings with authorities were celebrated, until an agreement was signed granting access and use to the data for research purposes.

Regarding the processing phase, the studied data was structured in Python pandas dataframes. Among the many transformations performed to the datasets, the most significant one was related to the Coordinate Reference System (CRS). In order to be able to combine different datasets, all geospatial data was transformed to WGS 84 (EPSG:4326), which is the standard CRS used by GPS.

To present clear visualizations, the reported results are from tickets sold during May 2015. Pre-hoc analysis of the full dataset suggest that this month is representative of the trends in the complete dataset. The source code for the analysis is configurable to process any subset of the full dataset.

3.2.2. Exploratory Data Analysis

An initial EDA was performed to characterize the dataset of sales with smart cards. Figure 4 shows an aggregated visualization of the geolocation of 20.4 million sales (interactive version available at www.fing.edu.uy/~renzom/msc). Considering the active population in Montevideo (between

15 and 64 years old, ~830 K people), this corresponds to 25 transactions per inhabitant per month. Dividing the total number of transactions by the total number of unique smartcards used at least once during May (~654 K), we get a ratio of nearly 30. In Figure 4, the location of each smart card transaction was projected on to a grid of bins of size equal to one pixel of the 900×750 image. Then, transactions on the same bin were aggregated and a color mapping was applied to generate the final image, where brighter areas indicate high concentration of ticket sales.

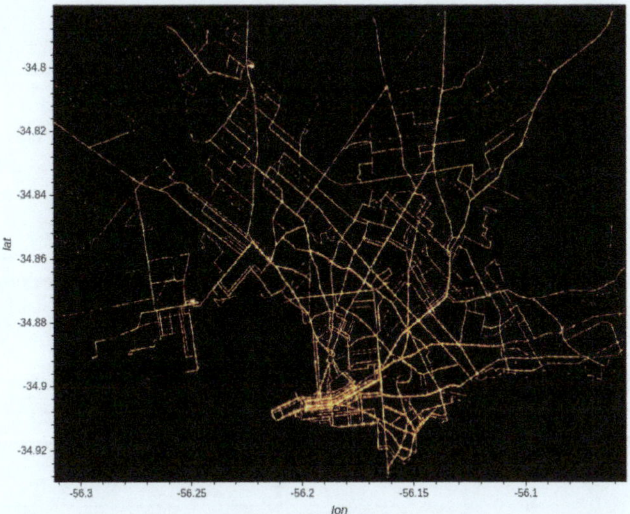

Figure 4. Aggregated sales with smart cards. Whiter pixels indicate more ticket sales, redder fewer ticket sales. Black pixels indicate no tickets were sold at that location.

The initial visualization of aggregated sales location data uncovers several interesting facts of the underlying dataset. Firstly, the city center is clearly different from other zones, with a significant higher number of smart card transactions. Additionally, the main avenues can be clearly identified due to the higher number of ticket sales. Furthermore, some sales activity is registered outside of the limits of Montevideo. This is an important insight that guided the data cleansing process described in the following section.

3.2.3. Data Cleansing

Data cleansing is mandatory to detect and correct corrupt or inaccurate records [20]. Since no backup source of information was available, the chosen strategy was to delete records that appeared to be corrupted. Filtered data included: records with no corresponding bus line, consecutive tap-ins in the same bus, transactions on May 1st (since they correspond to Labour Day, when the public transport system is mostly inoperative), and transactions occuring in bus lines or bus stops that no longer exist (since sales data correspond to 2015). During the complete data cleansing process 311,772 records were filtered, accounting for 1.53% of a total of 20,359,835 records.

3.3. Results and Discussion

This section outlines the main results of the urban data analysis process to characterize the use of the public transportation system in Montevideo, Uruguay. A description of the use patterns of smart cards is presented, as well as a spatial and temporal analysis of the use of the transportation system.

3.3.1. Cardholders

The sales dataset holds transactions made with 654,228 different smart cards. As explained in Section 3.2.1, several passengers may travel together using a single smart card. However, the vast majority of passengers use their own personal smart card: over 97% of transactions correspond to individual tickets. Therefore, smart cards can be confidently assumed to represent a single passenger. This is a key assumption used in the OD matrix estimation presented in Section 4, where all passengers under the same smart card are assumed to travel from origin to destination together. Thus, the fact that few group trips are performed using the same smart card provides a certain level of robustness to the OD matrix estimation model.

Data analysis can also give insight into the frequency of use of the transportation system. Table 1 reports descriptive statistics of daily and monthly transactions per smart card, including the minimum (min) and maximum (max) values, the 25th (Q1), 50th (Q2), and 75th (Q3) percentiles, and the Median Absolute Deviation (MAD). The 50th percentile corresponds to the median of the distribution of transactions per smart card. Monthly statistics consider all transactions done by each cardholder. Daily statistics only consider days for which at least one transaction was made. Values corresponding to the complete dataset are presented (all weekdays) and also considering only working days and only weekends. Additionally, daily and monthly sales distributions considering all weekdays are displayed in Figures 5 and 6, respectively. Plots are limited to the most occurring values for better visualization.

Table 1. Descriptive statistics of daily and monthly smart card transactions.

		min	Q1 (25%)	Q2 (50%)	Q3 (75%)	max	MAD
all weekdays	daily	1	2	2	4	54	1.2
	monthly	1	8	22	47	528	22.5
working days	daily	1	2	2	4	54	1.2
	monthly	1	7	19	40	492	19.0
weekends	daily	1	2	2	3	32	1.1
	monthly	1	3	5	11	151	5.3

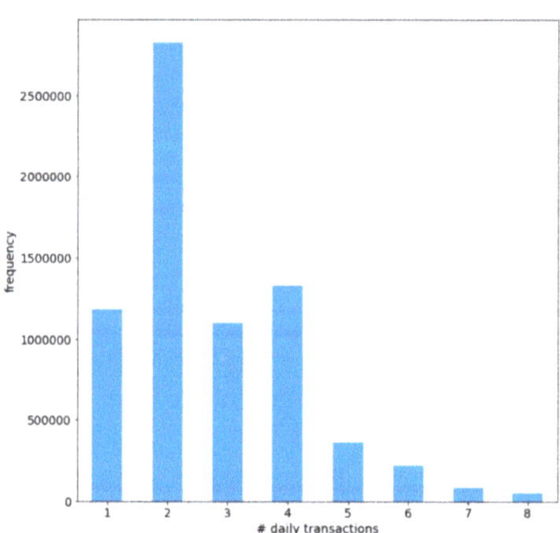

Figure 5. Distribution of daily smartcard transactions.

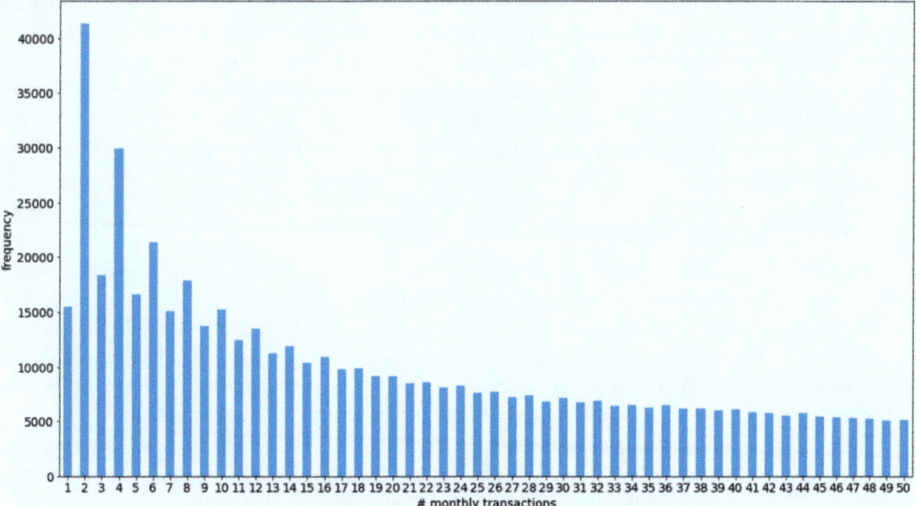

Figure 6. Distribution of monthly smartcard transactions.

Regarding monthly use, the median cardholder performs 22 transactions per month, nearly one transaction per working day in the month. However, the MAD is 22.5, suggesting a significant difference between regular and sporadic users of the public transportation system. Regarding daily use, the median cardholder performs two smart card transactions each active day (i.e., each day with at least one transaction). Additionally, more cardholders perform four rather than three transactions (see Figure 5), which could be explained by passengers using trips involving a transfer. Thus, two transactions correspond to the outward trip and the remaining two transactions to the return trip. Daily usage is higher on working days than weekends when looking at the top quartiles. Additionally, monthly usage is also sensibly higher on working days than on weekends. Taking into account that the studied month had nine days in weekends, users perform (in median) roughly one transaction in the whole weekend vs. one transaction per day on working days. This is consistent with the information from the 2016 mobility survey, which states that commutes to work are the main purpose of traveling, accounting for nearly 30.9% of all trips. An in-depth analysis of the effect of the public transportation system in employment in Montevideo is studied in [21].

Identifying outliers within the smart card use statistics can be a useful tool for authorities of the public transportation system. On the one hand, cardholders with very few monthly transactions can be identified by their card ID. In the studied dataset, 15,440 cardholders performed a single trip during the whole month of May 2015. Targeted marketing campaigns could be designed to encourage disengaged citizens to use public transportation more frequently. On the other hand, cardholders with a large number of transactions can also be identified. In the studied dataset a single card was found to perform 54 transactions in a single day. This information can help authorities to further investigate and identify possible abuses to the system.

3.3.2. Transfers

As introduced in Section 3.1.2, the fare scheme allows transfers between any bus line at any bus stop. Thus, a trip can be comprised of several legs, with bus transfers between each leg. Results show that 55.99% of all transactions involve a single direct trip. Similarly, 40.26% of smart card transactions correspond to a trip comprised of two legs and involving one transfer. The number of transactions involving more than two bus transfers are less than 4% of the total dataset. The average number of legs for the studied dataset is 1.37. According to the household mobility survey, the average number of legs when travelling by bus is 1.5. The slight difference between both estimations might be explained due to the fact that the mobility survey considers the walks to/from the bus stop as separate legs (if they are longer than 500 m). Since the cardholders identity is not included in the study dataset for privacy issues, personal information (e.g., home address) cannot be used to infer the walked distance to/from the bus stop. Thus, direct trips requiring the passenger to walk more than 500 m to reach the bus stop are counted as two-legged trips in the mobility survey and as one-legged trips in the urban data analysis approach.

3.3.3. Temporal Analysis of Transactions

The AFC in the public transportation system records the date and time of each transaction, which allow analyzing the distribution of transactions across time.

Firstly, the number of transactions occurring each day of the week was analyzed. As expected, working days show the largest concentration of transactions with an average of ~3.31 M of transactions and a median of ~3.44 M. In contrast, transactions during weekends drop significantly, with a clear difference between Saturdays (~2.19 M transactions) and Sundays (~1.28 M transactions).

Then, a finer-grain analysis was performed to study the distribution of transactions across time. Figure 7 shows an histogram with the number of smart card transactions at each hour of the day during May 2015. Two clear peaks of smart card transaction activity were detected during the morning (7:00–8:00) and the afternoon (16:00–18:00), probably due to commuting. The morning peak is preceded by an increasing trend of sales starting at 3:00 while the afternoon peak gradually decays as the night approaches. However, an interesting observation is that another peak occurs at midday (12:00–13:00) which might not be foreseen prior to the analysis. In fact, the overall largest amount of transactions occur at 13:00. Finally, it is worth noting that the lowest number of ticket sales happen at 3:00. This finding is used for the OD matrix estimation algorithm presented in Section 4, which considers each new day as starting at 3:00, when fewer sales are made. Results are in line with the 2016 household mobility survey.

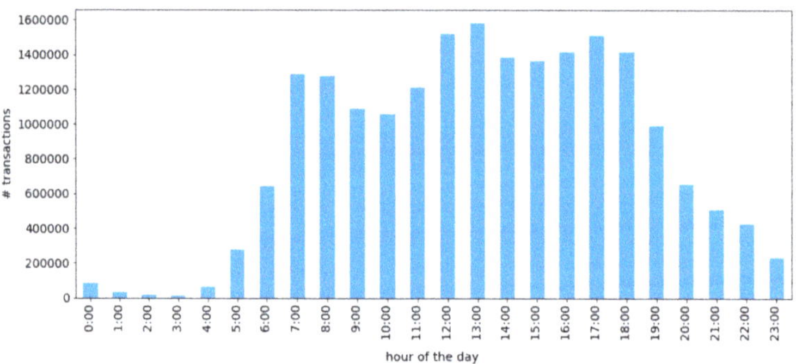

Figure 7. Histogram of sales with smart cards at different times of the day.

3.3.4. Spatiotemporal Analysis of Transactions

Spatial and temporal dimensions of smartcard transactions can be combined to gain insights that might not arise when studying each dimension independently. Figure 8 shows an aggregated visualization of the spatiotemporal distribution of sales in Montevideo. Each transaction occurring at a given pixel in the image is categorized according to its time stamp. Then, the color of the pixel is set considering the amount of transactions on each category. The color mapping, which is detailed in the visualization, corresponds to: red (0:00), yellow (4:00), green (8:00), cyan (12:00), blue (16:00), and purple (20:00). An interactive version of the visualization is available at www.fing.edu.uy/~renzom/msc.

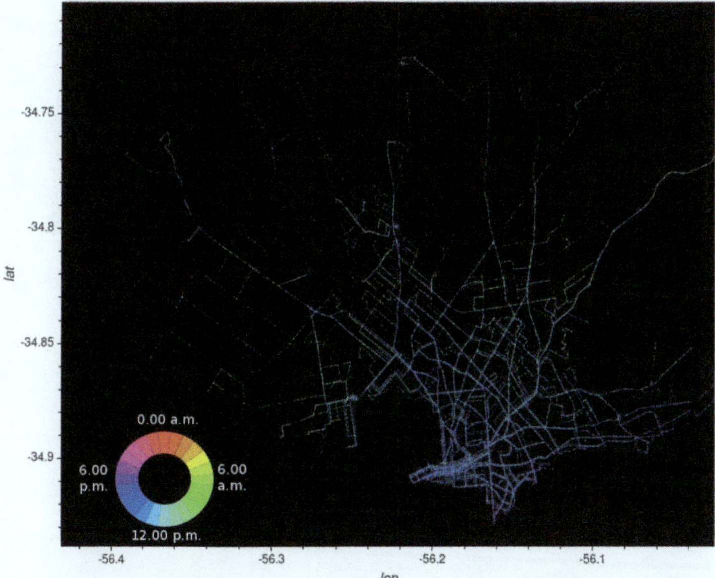

Figure 8. Spatiotemporal distribution of trips in Montevideo.

The city center has most transactions taking place between noon and the afternoon, which might be explained by the fact that many offices and public entities are located in this area. Consequently, most transactions correspond to people commuting back to their homes by the end of the office-hours. A clear difference can also be noticed between areas near the coast and areas farther away. The majority of transactions in areas farther away from the coast occur earlier in the day than those near the coast. This might be explained by people commuting early in the day from these areas to workplaces located closer to the city center. It is worth noting that, as outlined in Section 3.1, areas farther away from the coast are usually more vulnerable from a socioeconomic point of view.

Figures 9 and 10 show choropleth maps of the number of transactions occurring in each census tract in the morning and evening, respectively.

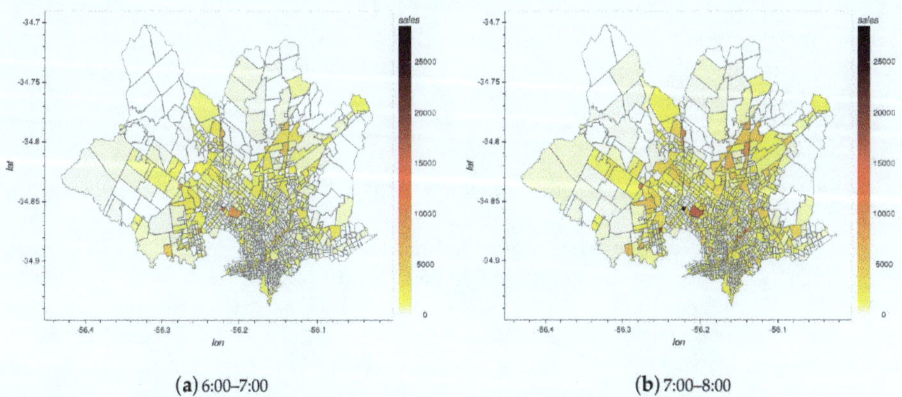

(**a**) 6:00–7:00　　　　　(**b**) 7:00–8:00

Figure 9. Choropleth map of smart card transactions in the morning.

In the morning, those areas farther away from the city center and the coastline have higher smart card transaction activity early (6:00–7:00) than those near the coast. Transaction activity in the city center and near the coastline intensifies an hour later. Between 7:00 and 8:00 large amounts of transactions occur in most areas of Montevideo. A few census tracts show a specially large number of transactions. These areas correspond to the location of bus terminals, where several bus lines converge and many transfers between bus lines occur.

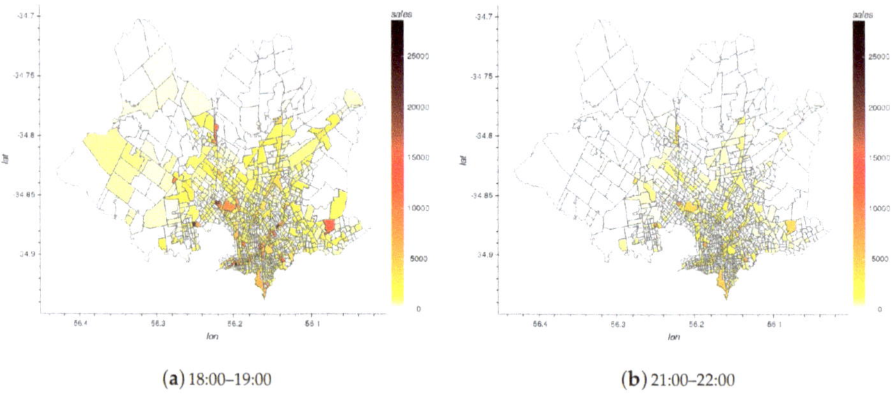

(**a**) 18:00–19:00   (**b**) 21:00–22:00

Figure 10. Choropleth map of smart card transactions in the evening.

In the evening, a large number of transactions occur in the city center. This is explained by people returning to their homes from workplaces in the city center at the end of office hours (18:00–19:00). Between 21:00 and 22:00 the amount of sales in the whole territory significantly drops. The areas with some remaining transaction activity are, once again, those located farther away from the city center and the coastline. This might be explained by people living in poorly connected areas taking longer to commute back to their homes by the end of the working day or also due to citizens working during night shifts and commuting to their workplace.

Spatiotemporal analysis can be combined with the population and socioeconomic description. Areas with transactions occurring early in the morning/late at night are also more vulnerable from a socioeconomic point of view, as outlined in Section 3.1. This study helps understanding the variation of mobility patterns for citizens with different socioeconomic levels [22].

3.4. Practical Use Cases

This subsection presents two relevant case studies to illustrate different ways in which the proposed methodology applying data analysis can contribute to help authorities with the task of operating public transportation service and improving the quality of service.

3.4.1. Event Detection

The analysis of anomalous registers in either time and/or location data of ticket sales can help authorities to identify special events taking place in the city. The image in Figure 11 presents an aggregated visualization combining spatial and temporal data of smart card transactions for a neighborhood of Montevideo. A small cluster of red pixels (highlighted with a white circle in the figure) is detected on the map. This cluster corresponds to bus ticket sales occurring at midnight, representing a clear outlier from the rest of ticket sales in the dataset. Taking in consideration the location of those ticket sale records (near an outdoor venue), it is reasonable to assume that the transactions correspond to a special social event (e.g., a concert) held at night in that venue. In case the situation repeats periodically, specific action can be taken to satisfy that mobility demand. This relevant case exemplifies how city authorities can take advantage of a methodology using urban data analysis to detect periodical special events in the city and plan the transportation services in response to those events.

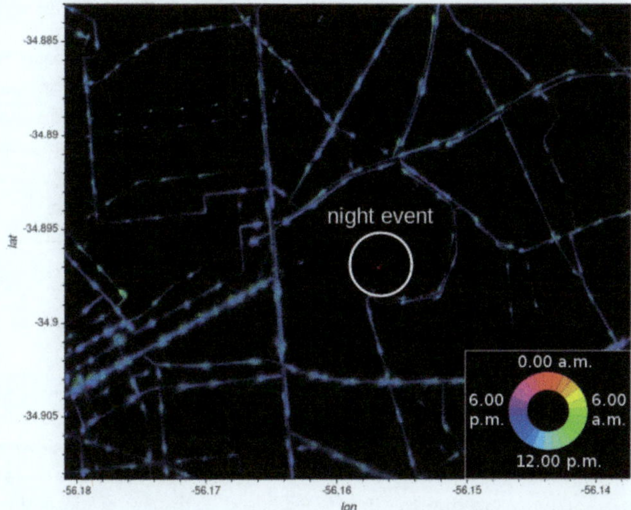

Figure 11. Event detection: smartcard transactions at midnight near an outdoor venue.

3.4.2. Driving Behavior and Safety

Another relevant application of urban data analysis is related to safety in the public transportation service. The heatmap on Figure 12 reports the location of smartcard transactions near a roundabout. The bus stops nearby are marked in blue. The figure shows that a large number of transactions are recorded when the bus is within the roundabout. This action might be related to a relevant safety issue, since passengers validate their cards standing in front of the ticket machine. Furthermore, this driving pattern is related to a more serious safety issue in those buses where drivers are also in charge of operating the smart card terminal and selling tickets, as in more than 60% of the bus fleet in Montevideo. Data analysis can be applied to analyze and audit driving behavior, to detect anomalous situations that can impact on safety. This way, the proposed methodology helps improving the safety of passengers, bus drivers, pedestrians, and drivers of other vehicles.

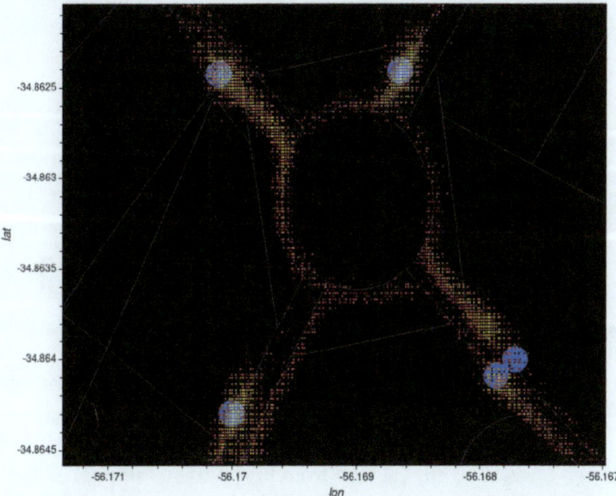

Figure 12. Driving behavior and safety: spatial distribution of smartcard transactions in a roundabout.

4. Origin-Destination Matrices Estimation

This section outlines the details of the generation of OD matrices using data from the ITS in Montevideo, Uruguay. Section 4.1 describes the destination estimation algorithm used to build OD matrices. Then, Section 4.2 presents the computed OD matrix and its validation against a mobility survey is presented in Section 4.3. Finally, an online visualization tool for the computed OD matrix is outlined in Section 4.4.

4.1. Implemented Solution

This subsection presents the destination estimation algorithm using trip chaining and its adaptation to the case study of the ITS in Montevideo.

4.1.1. Destination Estimation Algorithm

The origin of trips is identified by combining smart card and GPS location data, since the location of the bus is recorded whenever a passenger pays for a ticket using a smart card. However, since passengers are only required to validate their smart cards when boarding and not when alighting the bus, the destination of each trip is unknown and must be estimated in order to generate OD matrices.

A destination estimation algorithm was developed based on the assumptions of the trip chaining method: (i) the origin of a new trip is near the destination of the previous one; and (ii) at the end of the day, users return to the origin of their first trip of the day. Figure 13 shows an example of the proposed method: the passenger performs three smart card transactions throughout the day. The boarding bus stops associated to each transaction are marked in green, and the estimated destinations of trips and trip legs are marked in orange.

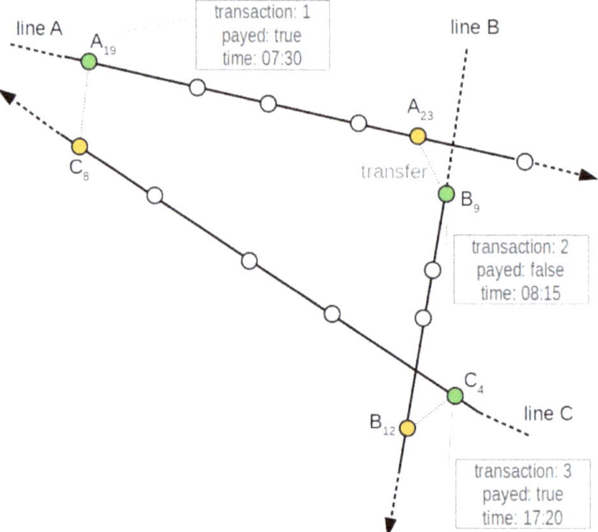

Figure 13. Example of the trip chaining algorithm to estimate destinations.

In the example, the first transaction of the day occurs at 07:30, when the passenger boards bus line A at bus stop A_{19}. Later, at 08:15, the passenger boards bus line B at bus stop B_9 without paying for a new ticket. Since the boarding occurred within the validity of the previous ticket, the trip is assumed to be a transfer between buses. The closest stop from line A to bus stop B_9 is A_{23}, which is assumed to be the destination of the leg trip starting at 07:30. The last transaction of the day occurs at 17:20, when the passenger boards line C at bus stop C_4 and pays for a new ticket. Bus stop B_{12} is identified as the destination of the leg trip starting at 08:15, since it is the closest stop from line B to bus stop C_4.

Since a new ticket was payed for, no further transfers are considered. Thus, an OD pair is identified between bus stops A_{19} and B_{12}. Finally, the destination of the last trip of the day is assumed to be bus stop C_8, since it is the closest bus stop of line C to the origin of the first transaction of the day (A_{19}). As a result, two OD pairs are identified, one consisting of two leg trips with a bus transfer and the other being a direct trip.

4.1.2. Configuration for the ITS in Montevideo

The destination estimation algorithm processes sales data grouped in chunks corresponding to 24 h periods. Records are split at the time of the day when the lowest sales activity is observed, as recommended by Munizaga et al. [15]. In the studied scenario, the lowest amount of sales occurs at 3:00.

The destination estimation algorithm limits the search of a possible destination bus stop to a configurable radius. The search is sensitive to this parameter: large values may incorrectly identify destinations when other transport modes are used within the chain of bus trips, while a small radius might miss to identify destinations for trips that involve large walks from the bus stop to the destination. In the reviewed works of the related literature, several values were found for this parameter: 800 m [16], 1000 m [13,14], and 2000 m [12]. In this work the maximum distance to search for a destination bus stop was set to 1000 m, which is the median of the values found in the related literature. Additionally, 1000 m is also the maximum distance used to classify a walk as "short" according to the urban mobility survey [23].

4.2. Numerical Results

After the cleansing process, 311,772 records were discarded from the dataset corresponding to May 2015, leading to a cleansed dataset comprised of 20,048,063 records. For the destination estimation process, this dataset was split into chunks, where each chunk held the information for an entire day starting and ending at 3:00. Additionally, since the destination estimation algorithm requires at least two transactions to perform trip-chaining, the records associated to cardholders that only performed one transaction within a given day were filtered from the dataset. As a result, the destination estimation algorithm was applied to a set of 18,885,711 records. Out of these records, the implemented algorithm was able to assign a destination to 15,414,230 trips, achieving a success rate of 81.62%. This is a highly competitive result, considering the success rates achieved by other works in the related literature, e.g., 57% [13], 66% [12], 80% [14]. Each identified trip holds the following information: boarding bus stop, time stamp at boarding, bus line identifier, and alighting bus stop.

Computed results allowed identifying 9,485,904 OD pairs. At the finest grain, OD matrices were generated considering each pair of bus stops (size 4718 × 4718). At a more coarse grain, OD matrices were built at the census tract level (size 1063 × 1063). Both OD matrices are available at www.fing.edu.uy/~renzom/msc in CSV files with their corresponding metadata. For the sake of visualization, OD matrices in this article are aggregated by municipality (size 8 × 8). Table 2 outlines the estimated OD matrix corresponding to the studied dataset (each municipality is represented by its identifying code).

Table 2. Estimated OD matrix by municipalities.

		Destination								
		A	**B**	**C**	**CH**	**D**	**E**	**F**	**G**	**total**
	A	626,388	199,196	184,905	98,087	30,108	40,370	21,875	73,390	1,274,319
	B	154,358	662,993	224,578	366,865	108,640	173,898	119,306	108,469	1,919,107
	C	174,040	260,526	320,368	111,113	102,244	64,691	62,188	101,337	1,196,507
	CH	100,348	334,040	131,089	362,377	101,433	156,685	115,310	66,461	1,367,743
origin	D	48,502	222,110	148,581	130,733	321,610	71,018	93,969	64,253	1,100,776
	E	27,463	138,400	46,288	110,868	86,344	287,243	133,179	28,827	858,612
	F	21,038	127,429	51,570	108,017	155,355	82,811	315,573	20,427	882,220
	G	74,482	141,380	120,539	57,388	41,670	29,779	21,068	379,724	866,030
	total	1,226,619	2,086,074	1,227,918	1,345,448	947,404	906,495	882,468	842,888	

The largest values are located in the diagonal of the computed OD matrix, which represent trips starting and ending within the same municipality. Municipality B stands out as both the largest generator and attractor of trips when considering the total number of OD pairs. This is consistent with the fact that the city center and other surrounding areas are within municipality B, where multiple workplaces, public offices, and services are located. Considering that most trips correspond to people commuting to their workplaces (30.9% according to the 2016 mobility survey), these observations could suggest that the majority of citizens work either within their own municipalities or travel to municipality B where most job opportunities are located. The lowest number of transactions occur in municipalities F and G, which have large rural areas with lower population density.

4.3. Comparison to the 2016 Mobility Survey

According to the best practices reviewed in the related literature, results of the OD matrix estimation must be compared with other sources of information. To this end, the results from the household urban mobility survey carried out in 2016 [23] were used. The comparison is done at the municipality level, since the survey data does not allow comparing at a finer-grain. The unavailability of more disaggregated sources of OD data somewhat limits the comparison with our proposed OD matrix.

The Spearman correlation coefficient was applied to quantify the similarities between the OD matrix estimated from ITS data and the OD matrix from the mobility survey. To compute this coefficient, matrices were vectorized in row-major order, without losing information since proximity in the matrix does not imply geographical proximity between municipalities. Results show that the estimated OD matrix and the mobility survey OD matrix have a Spearman correlation coefficient of 0.895 (p-value 2.026×10^{-23}), indicating a strong correlation between them, thus validating the proposed approach for OD matrix estimation based on ITS data.

Figure 14 presents a visual comparison between the OD matrices derived from ITS data and from the mobility survey. Each OD matrix is represented as a two-dimensional grid with colors mapped according to the number of transactions occurring in each OD pair.

The visual representation of OD matrices as heatmaps on two-dimensional grids in Figure 14 allows identifying similarities between the results computed with ITS data and those from the mobility survey. Trips within municipalities A and B are the most dominant according to both estimations, followed by trips within municipality G. Both figures show that trips from B to CH and vice versa are also highly dominant. The diagonal of the grid is mapped to more intense colors in Figure 14a than in Figure 14b. This might be a consequence of the larger number of trips considered in the OD matrix generated from ITS data. Despite this observation, an outstanding number of similar patterns are found when comparing the grids both row-wise and column-wise.

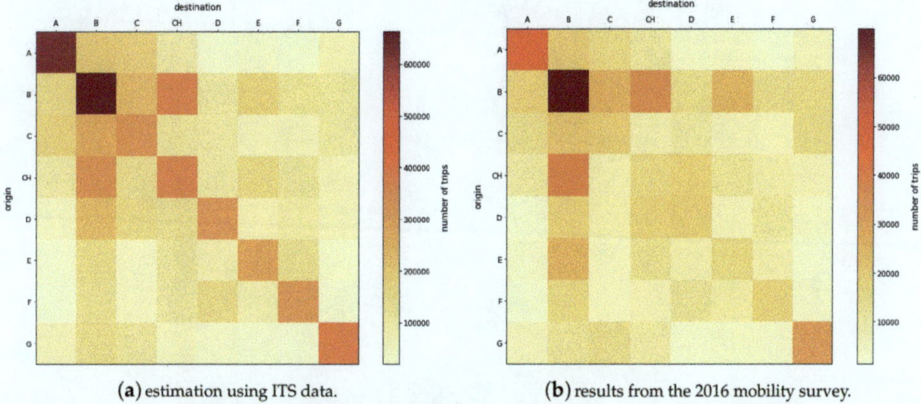

(**a**) estimation using ITS data.　　　　　　(**b**) results from the 2016 mobility survey.

Figure 14. Comparison of OD matrices (ITS data processing vs. mobility survey).

Results show that OD matrices generated from ITS data are a valid alternative to understand mobility in a city. The proposed approach for building OD matrices has several advantages: (i) due to the large volume of data generated by ITS compared to the number of individuals that participate in a survey, a finer-grain OD matrix is obtained using data analysis (e.g., bus stop and census tract levels), whereas the mobility survey results only apply to municipalities; and (ii) data analysis allows computing different OD matrices applying different criteria (e.g., days of the week, hours of the day, etc) and the mobility survey refers to working days only. Thus, in order to gain insight on the mobility of citizens under different conditions (e.g., during weekends) a new survey ought to be carried out, with the associated costs and delays.

Regarding costs, the proposed approach for OD matrix estimation provides an attractive alternative for public administrations to characterize mobility in a city. This alternative takes advantage of valuable data that arise from the infrastructure deployed in modern ITS. This is the case of Montevideo, where the ITS infrastructure has been deployed in the last decade. It is worth noting that the proposed approach can be easily applied whenever new data becomes available. This represents a clear advantage in comparison to surveys, which demand a long time to plan, carry out the survey, and process the results. As a consequence, the proposed approach allows easily obtaining an up-to-date view on the mobility of a city while surveys offer a partial and mostly outdated picture.

4.4. OD Matrix Visualization Tool

The last step of every urban data analysis workflow involves presenting results visually to communicate the main findings to help stakeholders make decisions that can shape the studied reality [6]. For this purpose, an interactive web application was developed to show the computed OD matrices in an intuitive and friendly manner.

The OD visualization tool allows selecting a geographical zone and creates a heatmap indicating the number of passengers traveling from the selected area to all other areas in the map. The tool was developed using open source software: Python, Pandas for data processing, Geopandas to display the map of the city and the administrative divisions, and the Bokeh library to provide interactivity to the visualization. The web application is freely available at www.fing.edu.uy/~renzom/msc. Figure 15 shows the user interface of the developed tool.

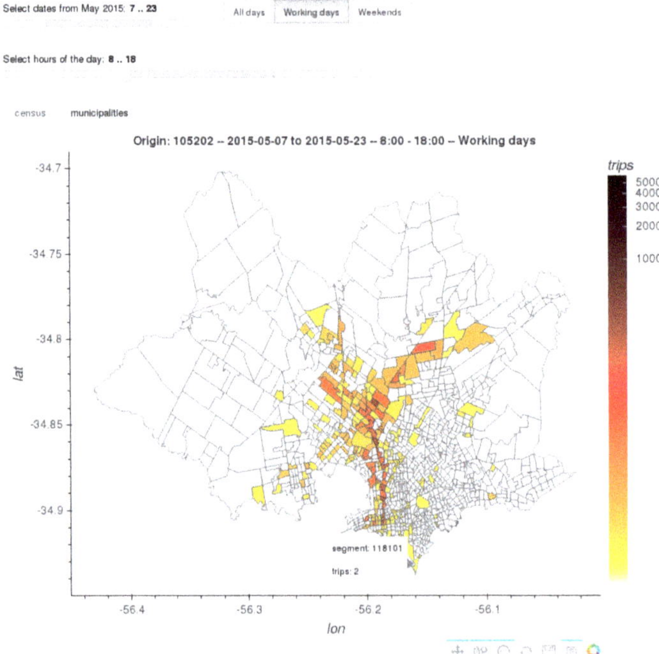

Figure 15. User interface of the OD matrix visualization tool.

5. Conclusions and Future Work

This article presented an urban data analysis approach to study mobility using ITS data. As a case study, the ITS in Montevideo, Uruguay was analyzed by studying a dataset of GPS bus location and smart card ticket sales. Several insights were obtained through data analysis, including: number of passengers traveling with the same smart card, frequency of use of the smart cards, and number of bus transfers. A temporal analysis of ticket sales was performed, identifying three peak hours during working days. Then, a spatiotemporal analysis revealed that citizens from areas farther away from the coastline start trips earlier than those near the coast. Additionally, two practical use cases were presented: event detection in the city and reckless driving behavior identification.

Besides a purely descriptive utilization of ITS data to characterize a public transportation system, a methodology for building OD matrices using trip chaining was applied to estimate destinations. The implemented algorithm was able to estimate the destination for 81.62% of trips in the studied dataset, a highly competitive result when compared to the ones reported in the related literature.

The OD matrix computed for Montevideo was compared against the one from the 2016 urban mobility survey. Results showed a Spearman correlation coefficient of 0.895, suggesting that the proposed approach is a valid alternative to understand mobility in the city. The proposed approach allows studying mobility at a finer grain, obtaining OD matrices between pairs of bus stops and census tracts, whereas the OD matrix from the mobility survey only applies to municipalities. The implemented solution is inexpensive if the ITS infrastructure is already deployed and it allows computing OD matrices considering different criteria and providing up-to-date mobility information.

An interactive web application was developed to visually display the computed OD matrices. The visualization tool allows selecting a geographical area and displays a heatmap indicating the number of passengers traveling from the selected area to all other in the city. The application supports working at different aggregation levels for OD matrices and offers several tools to filter data.

The main lines for future work include studying other interesting aspects of mobility in the city, e.g., the quality of service offered by the transportation system in terms of punctuality, frequency of lines, and load of passengers. Regarding OD matrices estimation, the proposed approach can be extended to tickets sold without smart cards, to account for all passengers of the transportation system. The destination estimation algorithm can be further refined by using historical passenger data and machine learning techniques to infer frequent destinations when trip chaining fails. Furthermore, activity detection could be used to discriminate between short individual trips using the same ticket from multi-leg trips involving bus transfers [24]. Parameters tuning of the destination estimation algorithm can be studied in case that fine-grain mobility data sources are available. Finally, the results of the analysis can be applied to solve optimization problems, e.g., synchronization of bus schedules [25], demand-based fleet size optimization, bus stops location, and bus line network redesign.

Author Contributions: Conceptualization, R.M. and S.N.; writing, review, and editing, R.M. and S.N. Data analysis, R.M. Both authors have read and agreed to the published version of the manuscript.

Funding: This research received no external funding.

Acknowledgments: The work of S. Nesmachnow has been partly supported by ANII and PEDECIBA, Uruguay. R. Massobrio would like to thank Fundación Carolina, Spain.

Conflicts of Interest: The authors declare no conflict of interest.

References

1. Grava, S. *Urban Transportation Systems*; McGraw-Hill: New York, NY, USA, 2000.
2. Deakin, M.; Waer, H.A. From intelligent to smart cities. *Intell. Build. Int.* **2011**, *3*, 140–152.
3. Figueiredo, L.; Jesus, I.; Tenreiro, J.; Ferreira, J.; Martins, J. Towards the development of intelligent transportation systems. In Proceedings of the 2001 IEEE Intelligent Transportation Systems, Oakland, CA, USA, 25–29 August 2001; pp. 1206–1211.
4. Ortúzar, J.; Armoogum, J.; Madre, J.L.; Potier, F. Continuous Mobility Surveys: The State of Practice. *Transp. Rev.* **2011**, *31*, 293–312.
5. Massobrio, R.; Nesmachnow, S. Urban Data Analysis for the Public Transportation System of Montevideo, Uruguay. In *Smart Cities*; Nesmachnow, S., Hernández Callejo, L., Eds.; Springer: Cham, Switzerland, 2020; pp. 199–214.
6. Schutt, R.; O'Neil, C. *Doing Data Science: Straight Talk from the Frontline*; O'Reilly Media, Inc.: Sebastopol, CA, USA, 2013.
7. Tukey, J. *Exploratory Data Analysis*; Addison-Wesley Publishing Company: Boston, MA, USA, 1977.
8. Ortúzar, J.; Willumsen, L. *Modelling Transport*; John Wiley & Sons: Hoboken, NJ, USA, 2011.
9. Pelletier, M.P.; Trépanier, M.; Morency, C. Smart card data use in public transit: A literature review. *Transp. Res. Part C Emerg. Technol.* **2011**, *19*, 557–568.
10. Li, T.; Sun, D.; Jing, P.; Yang, K. Smart Card Data Mining of Public Transport Destination: A Literature Review. *Information* **2018**, *9*, 18.
11. Barry, J.; Newhouser, R.; Rahbee, A.; Sayeda, S. Origin and destination estimation in New York City with automated fare system data. *Transp. Res. Rec. J. Transp. Res. Board* **2002**, *1817*, 183–187.
12. Trépanier, M.; Tranchant, N.; Chapleau, R. Individual Trip Destination Estimation in a Transit Smart Card Automated Fare Collection System. *J. Intell. Transp. Syst.* **2007**, *11*, 1–14.
13. Wang, W.; Attanucci, J.; Wilson, N. Bus Passenger Origin-Destination Estimation and Related Analyses Using Automated Data Collection Systems. *J. Public Transp.* **2011**, *14*, 131–150.
14. Munizaga, M.; Palma, C. Estimation of a disaggregate multimodal public transport Origin-Destination matrix from passive smartcard data from Santiago, Chile. *Transp. Res. Part C Emerg. Technol.* **2012**, *24*, 9–18.
15. Munizaga, M.; Devillaine, F.; Navarrete, C.; Silva, D. Validating travel behavior estimated from smartcard data. *Transp. Res. Part C Emerg. Technol.* **2014**, *44*, 70–79.
16. Alsger, A.; Mesbah, M.; Ferreira, L.; Safi, H. Use of Smart Card Fare Data to Estimate Public Transport Origin–Destination Matrix. *Transp. Res. Rec. J. Transp. Res. Board* **2015**, *2535*, 88–96.
17. Instituto Nacional de Estadística, Uruguay. Resultados del Censo de Población 2011: PoblacióN, Crecimiento y Estructura por Sexo y Edad. Available online: http://www.ine.gub.uy (accessed on 30 December 2018).

18. Catálogo de Datos Abiertos. Available online: https://catalogodatos.gub.uy/ (accessed on 16 August 2020).
19. Sistema de Información Geográfica. Available online: http://sig.montevideo.gub.uy/ (accessed on 16 August 2020).
20. Rahm, E.; Do, H.H. Data Cleaning: Problems and Current Approaches. *IEEE Data Eng. Bull.* **2000**, *23*, 3–13.
21. Hernandez, D.; Hansz, M.; Massobrio, R. Job accessibility through public transport and unemployment in Latin America: The case of Montevideo (Uruguay). *J. Transp. Geogr.* **2020**, *85*, 102742. doi:10.1016/j.jtrangeo.2020.102742.
22. Nesmachnow, S.; Baña, S.; Massobrio, R. A distributed platform for big data analysis in smart cities: combining Intelligent Transportation Systems and socioeconomic data for Montevideo, Uruguay. *EAI Endorsed Trans. Smart Cities* **2017**, *2*, 1–18.
23. Mauttone, A.; Hernández, D. Encuesta de Movilidad del área Metropolitana de Montevideo. Principales Resultados e Indicadores. 2017. Available online: http://scioteca.caf.com/handle/123456789/1078 (accessed on 30 December 2018).
24. Nassir, N.; Hickman, M.; Ma, Z.L. Activity detection and transfer identification for public transit fare card data. *Transportation* **2015**, *42*, 683–705.
25. Nesmachnow, S.; Muraña, J.; Goñi, G.; Massobrio, R.; Tchernykh, A. Evolutionary Approach for Bus Synchronization. In *High Performance Computing*; Crespo-Mariño, J.L., Meneses-Rojas, E., Eds.; Springer: Cham, Switzerland, 2020; pp. 320–336.

Article

Towards Energy-Efficient Mobile Ad Optimization: An App Developer Perspective

Ahmad Raza Hameed [1], Saif ul Islam [2,*], Ahmad Almogren [3,*], Hasan Ali Khattak [4], Ikram Ud Din [5] and Abdullah Bin Gani [6,7]

1   Department of Computer Science, National University of Computer and Emerging Sciences, FAST, Islamabad 44000, Pakistan; i171025@nu.edu.pk
2   Department of Computer Science, KICSIT, Institute of Space Technology, Islamabad 45000, Pakistan
3   Department of Computer Science, College of Computer and Information Sciences, King Saud University, Riyadh 11633, Saudi Arabia
4   School of Electrical Engineering and Computer Science, National University of Sciences and Technology (NUST), H12 Islamabad 45000, Pakistan; hasan.alikhattak@seecs.edu.pk
5   Department of Information Technology, The University of Haripur, Haripur 22620, Pakistan; ikramuddin205@yahoo.com
6   Faculty of Computer Science and Information Technology, University of Malaya, Kuala Lumpur 50603, Malaysia; abdullahgani@ums.edu.my
7   Faculty of Computing and Informatics, University Malaysia Sabah, Labuan 88400, Malaysia
*   Correspondence: saiflu2004@gmail.com (S.u.I.); ahalmogren@ksu.edu.sa (A.A.)

Received: 5 September 2020; Accepted: 28 September 2020; Published: 1 October 2020

Featured Application: The proposed approach can be efficiently used to enhance the overall user experience while using mobile phone applications for consuming content.

Abstract: Advertising over smart devices is one of the growing trends in the information technology domain. Most of the Android application (app) developers generate revenue through the use of ads, but on the other hand, the end users get the free app. However, the excessive number of ads infers hidden costs with respect to energy consumption, network utilization, and user comfort. These factors affect the app rating and reviews. Consequently, developers require a technique to balance app performance through optimized mobile ad usage. Therefore, in this paper, we extend an existing work and propose an energy-efficient method that uses gamma correction to reduce the hidden costs of a mobile app. In this approach, gamma correction efficiently balances the app performance by minimizing the size of ads. The size of the mobile ad is reduced by adjusting its pixels, reducing background color, and illuminating the content of the ad. After several experiments, it is deduced that our proposed approach efficiently saves mobile battery and developers can apply this approach to improve app rank and feedback.

Keywords: Android application; energy-efficient; mobile ads; gamma correction

---

1. Introduction

In the modern era, the smartphone has a considerable impact on communications, studying, infotainment, finance, and various other factors of human life. According to Gartner statistics, the number of mobile phone users has been increased from 12% to 116% from 2000 to 2018—and is projected to reach 453.19 billion in 2022 [1,2]. In order to make the usability of mobile gadgets better, various solutions have been proposed. However, due to its user-friendly interface and cost-effectiveness, Android is considered one of the most popular platforms in the smart world. Therefore, with the increase in Android usability, the range of Android applications is growing exponentially with the passage of time [3,4]. Therefore, Android developers focus on app development to get huge revenue. In this context, advertisements through smartphone apps have gained a great attention. A recent analysis shows that more than $30 billion were spent on marketing through smartphones in 2014. Furthermore, the analysis predicts the amount of mobile advertisements even exceed TV advertisement [5]. Based upon this work, it was felt that overall optimization is required for enhancing the overall experience.

Typically, the Android developer gets ads from Google Ads Services and uses the mobile app to generate revenue; on the other hand, end users get the free app. It is a "win–win" situation for developers and users. Presently, a lot of Android developers develop an app without considering energy efficiency. Therefore, smart devices with limited resources of battery expend energy rapidly. Because of the quick power dissipation of mobile, end users under-rate an app, meaning that the developer needs to redesign an app, which is costly in terms of time and money [6].

Recently, a lot of research works have been conducted on energy-efficient techniques within software, application, and hardware level. More specifically, researchers have focused on application level because the developer is unable to buy and operate specialized hardware while designing a mobile app. On the other hand, a recent survey of the mobile ecosystem shows that the perception of free apps is misleading. In fact, almost half of the apps on the Google Play Store contain an excessive amount of ads [7,8]. Consequently, mobile ads contain hidden costs both for an end user and developer. The hidden ad cost affects end users as they slow down the response (consume an excessive amount of CPU memory), increase network usage, and increase energy cost (an excessive amount of battery usage). For the developer, the hidden costs may result in bad feedback [9]. Hence, the developer wants an approach to estimate hidden ad energy cost on a mobile app. Therefore, Gui et al. [5] have proposed an approach for the estimation of hidden power cost of ads at the application level of a smartphone. The estimation approach works in two phases: the first one is before the app implementation phase, where static modeling is applied to static ad configuration to generate energy consumption of mobile ads. The second one is after app implementation, where a dynamic model is applied on an implemented app to compute the energy consumption of ads. However, this work only estimates the hidden ad energy costs. This motivates us to propose an approach that reduces the hidden energy cost of ads over mobile app.

In the proposed work, gamma correction is applied over mobile ads, which reduces the hidden energy cost in terms of power dissipation and network usage. Furthermore, the main contribution of the work is summarized below:

- Gamma correction reduces the size of the mobile ad by adjusting pixels and reducing background color
- After size reduction, there is illumination of the content of the mobile ad

The experiment validates that the proposed approach efficiently reduces power dissipation of the mobile app caused by mobile ads; therefore the developer applies this approach to improve app rating and feedback.

The remainder of this paper is structured in such a way that Section 2 presents the related work. In Section 3, we discuss the problem statement. Section 4 elaborates proposed work. The evaluation and experimentation is presented in Section 5. Finally, Section 6 concludes the paper.

2. Related Work

There are several research contributions to developing a technique that efficiently saves power consumption of smart devices. However, the hidden energy cost of mobile ads over apps is overlooked. Therefore, in this context, Gui et al. [5] use a statistical method to compute the energy utilization related to the ads. This technique operates in two ways: First, they built a statistical model which provides the assessment. This model assumes the static values of ads that are SIZE, TYPE, and RRATE before the implementation of an app, where the SIZE is the size of the ad, TYPE is the behavior of the app either text or video and RRATE is the refresh time of the ad content. Second, for more precise energy measurement, they introduce a run-time energy measurement technique. This technique works after app implementation. This technique captures key run-time metrics such as system energy model, network energy model, and display energy model. This technique provides information related to the power consumption of an app. Therefore, this proposed work predicts 31% energy consumption before the implementation and 14% after the app development.

Performance-based energy-efficient guidelines for Android mobiles are proposed in [8]. In this work, the authors provide the best set of practices to develop an energy-efficient app. This practice includes static analysis of the mobile app. The static analysis includes allocating an object upfront, efficient wake call, recycles, reduce over layout, useless parents, using fewer resources, reducing view call and overdrawing. Through this analysis, the developer reduces a considerable amount of mobile app energy consumption. Furthermore, energy modeling provides the source of energy dissipation; therefore, developers correct a particular part of the code to achieve energy savings. However, this static analysis is unable to reduce the hidden energy cost by mobile ads.

EcoDroid, an energy-based ranking approach, is proposed in [9]. In this work, Jabbarvand et al. efficiently calculate the energy consumption of mobile apps by dynamic and static analysis. In dynamic analysis, a test case is generated by interaction with the application, and converts it to path information. This information is combined with an analyzer for the estimation of consumed power of the dynamic path in the app. However, static analysis extracts the app call graph which contains the different possible invocation sequences of Android application. Through this energy consumption information, EcoDroid ranks the same category applications according to their energy consumption.

Hao et al. [10] propose a lightweight fine-grained power estimation *eLens* approach, which efficiently calculates the power consumed in smart devices at the software level. In order to estimate energy dissipation, *eLens* combines two energy estimation processes, program analysis and per-instruction energy modeling. In this approach, the workload is generated to find the path of the user action, and is incorporated with the energy model to calculate per-instruction energy consumption. Furthermore, *eLens* uses energy annotation to display energy consumption per-instruction graphically; through this, the developer efficiently finds energy consumption of the app and reduces it. However, this approach enables the reduction of the hidden cost of an app such as mobile ads. This hidden cost affects the mobile app in terms of energy dissipation and user rating.

In [11], Hao et al. propose a programmable user interface (UI) automation (PUMA) for mobile applications. PUMA is a programmable framework which efficiently separates the exploring logic of app pages from analyzing the logic of the app. The authors implemented PUMA to perform dynamic analysis of smartphone apps using event handler Monkey (UI automation)—to monitor security, energy consumption, performance, and the correctness. Through this, the developer modifies apps at the run-time and enables the validation of app activities from the security breach. However, the behavior of hidden mobile app activities is unseen, which affects mobile devices in terms of energy consumption and security.

Corral et al. [12] proposes a kernel customization approach for reducing the power consumption in a smartphone application. In this approach, authors customize the kernel by optimizing CPU frequency scale, input–output (I/O) scheduling, under-lacking/under-voting and timer coalescing to adjust the power and workings of an app. The authors perform several tests on the customize kernel. Therefore, the modified kernel efficiently reduces the power consumption of the app.

In [13], the authors studied the optimal service allocation for a group of mobile applications in mobile cloud computing. They proposed a novel framework named location–time workflows (LTW) that is used to model the mobile applications for handling the service allocation during mobility. Furthermore, the framework optimally partitioned the workflow over mobile applications in 2-tier architecture based on the utility metrics, energy consumption, cost of the services, and delay of the mobile applications. The proposed system is evaluated using varying mobility models include Random

Waypoint and Manhattan models. It achieves 20% less mobile energy consumption and a reduced (30%) network delay.

Rong et al. [14], studied wireless sensor deployment and monitoring problems. They also discussed infrastructures and technologies to support the use of sensors in the smart city. For efficient network deployment configuration, they investigate different aspects including coverage, lifetime, and connectivity. Similarly, in the case of monitoring (mobile and static sensors), they also analyze sensing time, location, devices, and power consumption. Finally, the authors identify some research opportunities and directions to further explore sensor deployment and monitoring.

The authors in [15] proposed a knowledge-aware proactive node selection (KPNS) framework for an IoT environment. In KPNS, the selection of proactive nodes is based on their predicted preceding position. Furthermore, the KPNS system monitors the quality and energy efficiency of nodes. It also reduces the hot spot regions by efficiently utilizing the power of nodes.

Considering energy wastage and ignorance of processing requests by the network, the authors proposed an EAR-ADS algorithm (including energy-aware routing and an adaptive delayed shutdown mechanism) in [16]. This algorithm deployed dynamic service function chains (SFC), which means offline and on–off nodes in the network. Furthermore, this technique saves the power consumption of servers. Due to the adaptive delay shutdown mechanism, the energy wastage is further reduced.

Table 1 presents a comparative analysis of related work. As we see, existing approaches focus either on estimating the hidden ad energy consumption on the mobile app or to reducing the energy consumption of app in different perspectives such as code optimization, kernel optimization, etc. However, the hidden cost of ads over mobile apps is not addressed. Therefore, we propose an energy-efficient mechanism to reduce the hidden cost of ads on the mobile app.

Table 1. Comparison of state-of-art work.

Protocol Name	Features	Achievements	Deficiencies
Static approach for measuring ad-related energy cost [5]	Static modeling and run-time dynamic modeling	Estimate energy consumption 31% before implementation and 14% after implementation.	Unable to reduce hidden energy cost
The hidden cost of mobile ads for software developers [7]	Static mobile ads model	Estimate the energy consumption of ads before app implementation phase.	Focus on identity of hidden energy cost of ads
Performance-based energy-efficient [8]	Static analysis include object upfront, efficient wake call, recycles, reduce over layout, useless parents, useless resources, reduce view call and overdrawing	Minimize the energy consumption of app.	Still ad hidden energy cost
EcoDroid: an energy-based ranking approach [9]	Dynamic and static analysis	Dynamic analysis estimates the energy consumption of ads by interaction path analyzer while static analysis uses history of mobile data.	Absence of mechanism to reduce hidden ad cost
eLens app energy estimation [10]	power modeling	power consumption estimation.	Unable to mitigate ads hidden cost

Table 1. *Cont.*

Protocol Name	Features	Achievements	Deficiencies
PUMA [11]	Separate the exploring logic of app pages from analyzing logic of app	It verifies security breach, energy consumption and correctness of activities in response.	Absence of hidden cost
Software-based kernel customization approach [12]	Customize the kernel and balance between energy and performance	This phenomena reduces the energy consumption of app running on it.	Hidden energy cost
An optimal service allocation approach for mobile applications. [13]	A location–time workflow (LTW) model for mobile apps	Services are offloaded during mobility and the workload is partitioned to minimize the energy utilization of apps.	Hidden ads energy cost
A survey on wireless sensors for smart city environment [14]	Deployment strategies and monitoring techniques	Analyze scheduling techniques to reduce energy consumption of network and mobile devices.	Illustrate ads energy consumption
KPNS [15]	The law of target movement for prediction	Maintain balanced workload to reduce the energy cost of mobile devices.	Performance degradation of mobile devices

3. Problem Statement

For better understanding of the problem, we illustrate problem statement in the design science approach (DSA) way.

3.1. Motivation

In recent years, mobile devices become an important part of our daily life. The emergence of mobile devices such as smartphones, tablets, etc. provides us with useful features. In addition, it helps us to acquire information in no time. Android is considered as a ruling platforms in the mobile industry. The number of Android applications has increased, which makes them versatile and necessary for us [3]. Moreover, ads are an integral part of mobile applications. Typically, ads are used a source of income, while the subscribers get the free app in return. Due to the use of inefficient mobile ads, there is an energy depletion problem which directly reduces the performance and battery life of the smartphone. Consequently, the end user underrates the developer app, which directly affects a lot of the information technology (IT) community. Therefore, recent research has focused on this problem to compute the energy utilization of ads in apps.

3.2. Problem

In [5], the authors calculate the energy utilization of mobile ads in two phases, as portrayed in Figure 1—pre- and post-implementation. In the pre-implementation phase, static modeling is applied on the static configuration of the ads which estimates its energy consumption on a mobile app. Due to this, the developer gets early feedback before application implementation. Thus, the developer designs an application according to the feedback solution. In the post-implementation phase, the developer provides a workload as an input to calculate the power consumed through ads. To find the precise information of ads, the authors make a duplicate copy of the app and remove an ad from it then apply the workload on both apps with or without ads. The run-time energy calculation model includes system, network and display model, which works parallel to the workload to find precise mobile ad energy consumption after application implementation. However, existing work only focuses on estimating mobile ads energy consumption. This motivates us to propose a solution for the app developer which reduces the size of ads by applying gamma correction on mobile ads.

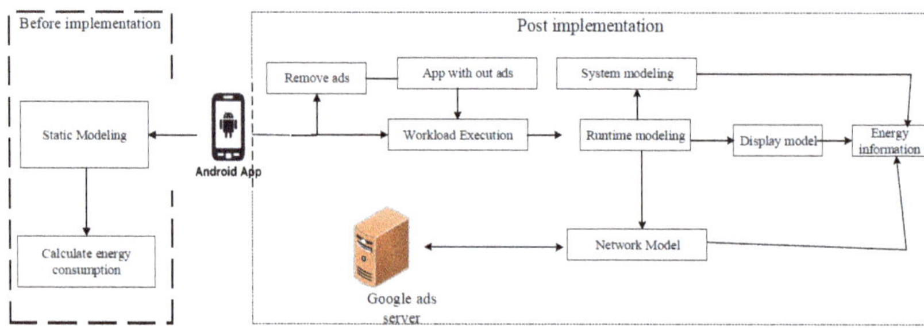

Figure 1. Existing work.

3.3. Evaluation

For fair comparison, we used state of the art tools and devices for experimentation [5]. Furthermore, cam scan, nature photo editor, blind traveler app, and karvan card applications are used for the proposed work evaluation.

- **RERAN tool**: Recording the manual generated workload. Through this we use the same workload at any time.
- **Android profiler**: Recording energy consumption of applications.
- **Trapen profiler**: Recording energy consumption of applications.
- **Android studio**: For Code.
- **Device for experiment**: Q mobile, Samsung core prime and Huawei Mate 10 lite.
- **Matlab tool**: Pictorial representation of findings.
- **Android applications for experimentation**: Cam Scan, Nature Photo Editor, Blind traveler app, and Karvan Card.

3.4. Hypothesis

The proposed mechanism is expected to reduce mobile energy consumption by 50% by optimizing ads over the mobile application. Furthermore, our work has an impact on a considerable part of the Android community, in particular app developers who optimize applications to improve rank and feedback.

4. Overview of the Proposed Work

The aim of the proposed work is to reduce energy use along with estimating the energy cost of ads all around the life cycle of app development. The proposed study uses the same techniques as presented in [5], which estimates the power cost of ads before and after application development as shown in Figure 1. In order to balance battery depletion, we applied gamma correction on mobile ads. Gamma correction reduces the size of mobile ads by adjusting the pixels and reduces the background color. For better visibility of mobile ads, we illuminate its content. The process is applied both before the implementation phase and after the app development phase.

The proposed work is shown in Figure 2, which operates in two ways:

- Before the implementation phase
- post-implementation phase

In the first phase, our approach takes the static model value as an input to gamma correction, which further reduces the quality and size of the ad over the mobile app. This process provides early feedback to the developer to optimize mobile ads before app development. Similarly, in the post-implementation phase, the proposed work takes the display model as an input to gamma correction, which customizes mobile ads to make it energy-efficient for an app. We validate our approach by comparing the modified ad application with other existing application.

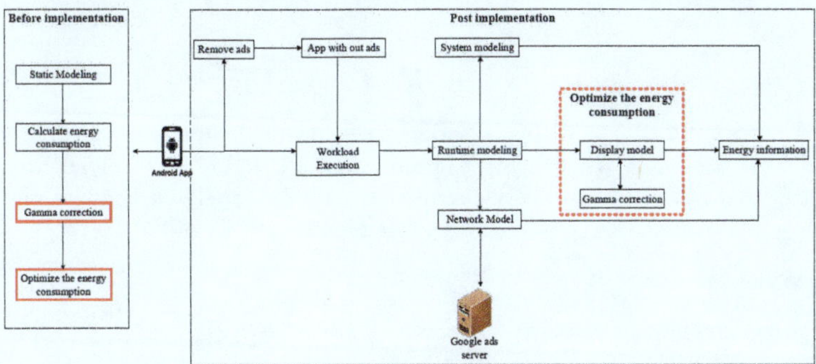

Figure 2. Proposed work.

4.1. Static Model

To acquire the energy consumption of mobile ads in app, existing work [5] uses static modeling. The model provides information regarding the ads energy utilization before the development phase. Before app implementation, the developer includes some static variables of ads such as size s, type t, and refresh rate r. Therefore, these values of a variable are taken as an input to the static model result in high-level energy estimation of ads in the mobile app. In order to build efficient estimation through the static model, we conduct several numerical experiments. In these experiments, we considered static ad configuration, varying one configuration and keeping the others the same, then calculating the average energy consumption of mobile ads over application. In order to find accurate average energy consumption of ads, we repeat this numerical experiment several times. Thus, finally we conclude that energy consumption of ads is linear when r when s and t are kept fixed. We formulate the relation as:

$$E = \beta - \alpha * (r - 30) \tag{1}$$

In the above Equation (1), α is a coefficient, β presents the power consumption (average) and the lowest refresh rate r is 30.

4.2. Dynamic Model

The approach proposed in [5] provides developers with energy consumption information about mobile ads. The proposed approach takes the implemented apps and generated workload by the developer as input and then computes the energy cost of a mobile ad at run-time. For manual workload generation, the RERAN tool is used [5]. This tool keeps a record of user activities which are repeated later on. Furthermore, for clear differences between the app and mobile ad energy consumption, the proposed approach compares both the with-and without-ads application and computes the utilization of energy by the ads. This model considers the utilization of CPU at different frequencies, network usage, and screenshots with the time-stamp. The energy model is formulated in Equation (2):

$$\bar{E}_{total} = \bar{E}_{system} + \bar{E}_{network} + \bar{E}_{display} \tag{2}$$

4.2.1. System Model

The system model computes the energy consumption of the CPU at a different frequencies. This model generates information regarding the power consumption of ads. This model finds a linear relationship between CPU time and frequency, which is calculated as:

$$\bar{E}_{system} = \sum_{f=1}^{n} \left(\bar{E}_f \times \bar{T}_f \right) + \left(\bar{E}_m \times \bar{M} \right) \tag{3}$$

In Equation (3), f presents the CPU frequency and n shows the total number of variations in CPU frequency. $\bar{E}_f$ is the energy usage while mobile ad running on CPU and $\bar{T}_f$ is the total time at which mobile ad running on the CPU. The mobile ad also use memory while running; therefore, this model computes memory energy usage and total memory usage respectively, $\bar{E}_m$ and $\bar{M}$.

4.2.2. Network Model

The network model computes the energy utilization of mobile ads by considering the total number of bytes sent over the network. The model formulates a linear relationship in Equation (4):

$$\bar{E}_{network} = \bar{C}_n \times \bar{B}_{total-bytes} \tag{4}$$

where C_n is the coefficient that shows the energy utilization per unit byte transferred over the network and $B_{total-bytes}$ is a total number of bytes that the ads have sent.

4.2.3. Display Model

The display model estimates the energy consumption of ads by taking screenshots of mobile ads as an input when the workload is manually generated then efficiently estimates the energy utilization of mobile ads. The energy of mobile ads is computed for a specific interval of time as per Equation (5):

$$\bar{E}_{display} = \sum_{s=0}^{n} \left(\bar{P}_{Screen-shot}(s) \times \bar{T}_{screen-shot}(s) \right) \tag{5}$$

where $\bar{P}_{Screen-shots}(s)$ is the total power consumed by the ads at specific interval $\bar{T}_{screen-shots}(s)$, herewith s is the number of screenshots.

$$\bar{P}_{Screen-shot}(s) = \sum_{Kes} \left(\bar{C}(R_k G_k B_k) \right) \tag{6}$$

Furthermore, the power consumption of each screenshot is the sum of the cost of pixel values as in Equation (6). The prior pixel values are found by RGB (red, green, blue) values of the screenshot.

$$\bar{C}(R_k G_k B_k) = rR + gG + bB + c \tag{7}$$

4.3. Gamma Correction

Gamma correction is a nonlinear operation used to encode and decode luminance in images and videos. The digital camera captures an image; its intensity is not balanced as human perception [17]; therefore, gamma correction is applied to it. The purpose of this correction is to balance the pixels of the image according to human perception (eye intensity) by applying the gamma correction formula:

$$\bar{V}_c = \bar{A} \times \bar{V}_{uc}^{\gamma} \tag{8}$$

here $\bar{A}$ is the arbitrary constant, $\bar{V}_{uc}$ is the uncompressed image obtained from the Google Play Store and γ is the gamma correction value. In order to compress and reduce the pixels of an image, γ value is considered to be less than 1 [18]. However, in mobile apps we apply gamma correction on mobile ads to reduce the hidden energy consumption of mobile devices. In order to apply gamma correction, first, the mobile ad converts the image into a bitmap image (it is a vector drawing in Android), then this image is passed to the gamma correction function. In this function, we set the threshold value of 250 and reduce the width and height of the image up to less than equal to the threshold value. After the image-reduction step, image decoding starts in which the compressed matrix applies to each image pixel. Through this process, illumination of pixels is adjusted, which reduces the size and quality of the image. The modified image is encoded and converted into a bitmap object which passes to the mobile app. Due to this process, a considerable amount of hidden energy is saved, which directly increases the battery lifetime of a mobile device.

5. Evaluation and Experiments

In this section, we describe and evaluate the experimentation of the proposed gamma correction algorithm as presented in Algorithm 1. In particular, we categorize the experiment to address the following research questions.

- **RQ1:** Can image-compression technique (gamma correction) reduce the energy consumption of the mobile app?
- **RQ2:** Does gamma correction efficiently increase battery lifetime and performance the of a mobile device?

Algorithm 1: Gamma correction on mobile ads.

Initially;
Image $\leftarrow$ get-image(URL);
Bit-image $\leftarrow$ Bitmap-covert(Image);
Gamma correction (Bit-image)
W $\leftarrow$ Bit-image.Width;
H $\leftarrow$ Bit-image.Height;
if $W \geq$ *threshold value* & $H \geq$ *threshold value* **then**
  SW $\leftarrow$ W/half;
  SH $\leftarrow$ H/half;
  Goto $IF(W \& H \geq threshold - value)$
D-image $\leftarrow$ decode(Bit-image);
foreach $Pixels \in Bit - image$ **do**
  C-Image $\leftarrow$ Compress-matrix(Bit-image);
E-image $\leftarrow$ encode(C-Image);
B-image $\leftarrow$ Bitmap-covert(E-image)
return Gamma correction$(B - image)$;

5.1. Experiment Setup

To examine the performance of the proposed approach, the experiments are performed on the Samsung Core Prime, Huawei Mate 10 lite and Q mobiles, as listed in Table 2. The Android application along with ad (considered as an image) builds on the Android studio. The gamma correction (image compression) technique is applied on the image to reduce the hidden energy cost of mobile ads in the application. For the sake of fare energy comparison, we use the same application with or without an image-compression technique. Furthermore, both applications run for 7 min on a mobile device and we compute the hidden energy cost of mobile ads using the Android profiler. Furthermore, this energy is also computed for verification by trapen energy profiler. For pictorial representation of result, Matlab is used. Therefore, calculation shows that the proposed mechanism is suitable for the app developer while developing an app in the presence of mobile ads.

Table 2. Tools and subjected apps for experiment.

Name	Purpose
Android profiler	Recording energy consumption
Trapen profiler	Recording energy consumption
Android studio	For Code
Device for experiment	Q mobile, Samsung core prime, Huawei mate 10 lite
Matlab	Pictorial representation of findings
RERAN tool	Recording the manual generated workload
Android applications for experimentation	Cam Scan, Nature Photo Editor, Blind traveler app, Karvan Card

5.2. Rq1: Gamma Correction Reduces the Energy Consumption of the Mobile App

Traditionally, gamma correction is used to manipulate the pixels of a digital camera image, due to which the human eye can view a fully illuminated image [18]. However, in this work, we use correction on mobile ads to reduce the size and adjust the pixels of the image to reduce the quality of mobile ads. Moreover, the quality is maintained up to the point where the content can read easily, hence, it is a "win–win" situation for the developer, end user, and third party Google Store. Figure 3 shows that the energy consumption of gamma-corrected applications is low compared to other non-corrected versions of the applications.

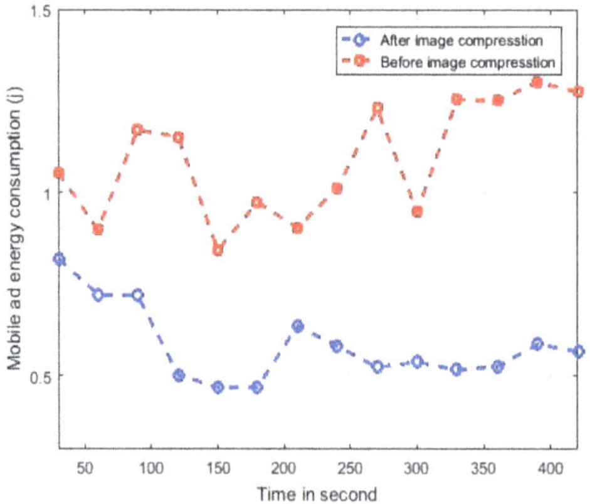

Figure 3. Mobile ad energy consumption.

5.3. Rq2: Gamma Correction Efficiently Increases the Battery Lifetime and Performance of the Mobile Device

According to existing research on the mobile ecosystem [3–7], we can say that the energy consumption of the mobile device has a nonlinear relationship with the lifetime of battery and performance of the mobile device. Similarly, Figure 4 illustrates, after gamma correction on mobile applications, that it consumed less mobile battery. Hence, the gamma correction increases the lifetime of mobile battery and the performance of the mobile device. Thus, our technique provides a way for a developer to reduce the hidden energy cost of the mobile application.

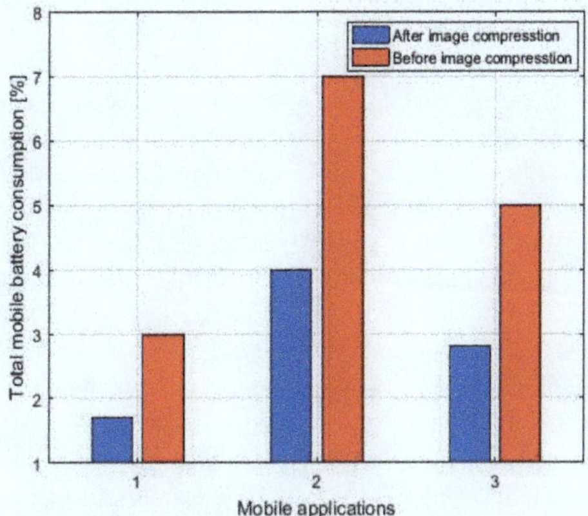

Figure 4. The impact of gamma correction on mobile applications.

6. Conclusions and Future Work

Developing an energy-efficient mobile application in the presence of mobile ads is a daunting task for the software developer. Unfortunately, much existing work focuses either on finding and estimation the energy consumption of a mobile app. Furthermore, other online resources focus on the performance of the mobile application. Thus, due to lack of guidance, the developer is unable to design an energy-efficient mobile application. Therefore, in this paper, we proposed a way to optimize mobile ads in applications. In this work, we apply gamma correction to a mobile application to reduce the size of mobile ads. Furthermore, in order to reduce the quality of mobile ads, gamma correction adjusts every pixel of the mobile ad. Due to this, our proposed work minimizes the hidden energy cost of mobile ads, which directly increases the lifetime and performance of the mobile app. The simulation result validates that our proposed work efficiently reduces the energy consumption of a mobile application. In the future, we plan to integrate other image-compression techniques with gamma to further optimize the energy depletion of mobile applications and extend our research for the iOS platform.

Author Contributions: Conceptualization, A.R.H., S.u.I. and H.A.K.; Formal analysis, I.U.D. and A.B.G.; Funding acquisition, A.A.; Investigation, A.R.H.; Methodology, A.R.H., S.u.I. and H.A.K.; Software, A.R.H., Project administration, S.u.I. and A.A.; Resources, S.u.I.; Supervision, S.u.I. and H.A.K.; Validation, I.U.D. and A.B.G.; Writing—Original draft, A.R.H., H.A.K. and S.u.I.; Writing—Review & editing, A.B.G., I.U.D. and A.A. All authors have read and agreed to the published version of the manuscript.

Funding: This work was supported by King Saud University, Riyadh, Saudi Arabia, through Researchers Supporting Project number RSP-2020/184.

Acknowledgments: Authors acknowledge the support by COMSATS University Islamabad and National University of Computer and Emerging Sciences, FAST, Islamabad for providing means to conduct the study.

Conflicts of Interest: The authors declare no conflict of interest.

References

1. Capponi, A.; Fiandrino, C.; Kantarci, B.; Foschini, L.; Kliazovich, D.; Bouvry, P. A Survey on Mobile Crowdsensing Systems: Challenges, Solutions, and Opportunities. *IEEE Commun. Surv. Tutor.* **2019**, *21*, 2419–2465. [CrossRef]
2. Wang, X.; Athanasios, V.; Vasilakos, M.C.; Yunhao, L.; Ted, T.K. A survey of green mobile networks: Opportunities and challenges. *Mob. Netw. Appl.* **2012**, *17*, 4–20. [CrossRef]
3. Lee, S.; Go, M.; Ha, R.; Cha, H. Provisioning of energy consumption information for mobile ads. *Pervasive Mob. Comput.* **2019**, *53*, 49–61. [CrossRef]

4.   Cai, H.; Gu, Y.; Vasilakos, A.V.; Xu, B.; Zhou, J. Model-driven development patterns for mobile services in cloud of things. *IEEE Trans. Cloud Comput.* **2016**, *6*, 771–784. [CrossRef]

5.   Gui, J.; Li, D.; Wan, M.; Halfond, W.G. Lightweight measurement and estimation of mobile ad energy consumption. In Proceedings of the 2016 IEEE/ACM 5th International Workshop on Green and Sustainable Software (GREENS), Austin, TX, USA, 16 May 2016; IEEE: Piscataway, NJ, USA, 2016; pp. 1–7.

6.   Corral, L.; Georgiev, A.B.; Sillitti, A.; Succi, G. Can execution time describe accurately the energy consumption of mobile apps? an experiment in Android. In Proceedings of the 3rd International Workshop on Green and Sustainable Software, Hyderabad, India, 1 June 2014; ACM: New York, NY, USA, 2014; pp. 31–37.

7.   Gui, J.; Mcilroy, S.; Nagappan, M.; Halfond, W.G. Truth in advertising: The hidden cost of mobile ads for software developers. In Proceedings of the 37th International Conference on Software Engineering, Florence, Italy, 16–24 May 2015; IEEE Press: Piscataway, NJ, USA, 2015; pp. 100–110.

8.   Cruz, L.; Abreu, R. Performance-based guidelines for energy-efficient mobile applications. In Proceedings of the 2017 IEEE/ACM 4th International Conference on Mobile Software Engineering and Systems (MOBILESoft), IEEE, Buenos Aires, Argentina, 22–23 May 2017; pp. 46–57.

9.   Behrouz, R.J.; Sadeghi, A.; Garcia, J.; Malek, S.; Ammann, P. Ecodroid: An approach for energy-based ranking of Android apps. In Proceedings of the Fourth International Workshop on Green and Sustainable Software, Florence, Italy, 18 May 2015; IEEE Press: Piscataway, NJ, USA, 2015; pp. 8–14.

10. Hao, S.; Li, D.; Halfond, W.G.; Govindan, R. Estimating mobile application energy consumption using program analysis. In Proceedings of the 2013 35th International Conference on Software Engineering (ICSE), San Francisco, CA, USA, 18–26 May 2013; IEEE: Piscataway, NJ, USA, 2013; pp. 92–101.

11. Hao, S.; Liu, B.; Nath, S.; Halfond, W.G.; Govindan, R. Puma: Programmable ui-automation for large-scale dynamic analysis of mobile apps. In Proceedings of the 12th Annual International Conference on Mobile Systems, Applications, and Services, Bretton Woods, NH, USA, 16–19 June 2014; ACM: New York, NY, USA, 2014; pp. 204–217.

12. Corral, L.; Georgiev, A.B.; Janes, A.; Kofler, S. Energy-aware performance evaluation of Android custom kernels. In Proceedings of the 2015 IEEE/ACM 4th International Workshop on Green and Sustainable Software (GREENS), Florence, Italy, 18 May 2015; IEEE: Piscataway, NJ, USA, 2015; pp. 1–7.

13. Rahimi, M.R.; Venkatasubramanian, N.; Mehrotra, S.; Vasilakos, A.V. On optimal and fair service allocation in mobile cloud computing. *IEEE Trans. Cloud Comput.* **2015**, *6*, 815–828. [CrossRef]

14. Du, R.; Santi, P.; Xiao, M.; Vasilakos, A.V.; Fischione, C. The sensable city: A survey on the deployment and management for smart city monitoring. *IEEE Commun. Surv. Tutor.* **2018**, *21*, 1533–1560. [CrossRef]

15. Liu, X.; Zhao, S.; Liu, A.; Xiong, N.; Vasilakos, A.V. Knowledge-aware proactive nodes selection approach for energy management in Internet of Things. *Future Gener. Comput. Syst.* **2019**, *92*, 1142–1156. [CrossRef]

16. Sun, G.; Zhou, R.; Sun, J.; Yu, H.; Vasilakos, A.V. Energy-efficient provisioning for service function chains to support delay-sensitive applications in network function virtualization. *IEEE Internet Things J.* **2020**, *7*, 6116–6131. [CrossRef]

17. Jobson, D.J.; Rahman, Z.U.; Woodell, G.A. A multiscale retinex for bridging the gap between color images and the human observation of scenes. *IEEE Trans. Image Process.* **1997**, *6*, 965–976. [CrossRef] [PubMed]

18. Gupta, B.; Tiwari, M. Minimum mean brightness error contrast enhancement of color images using adaptive gamma correction with color preserving framework. *Optik* **2016**, *127*, 1671–1676. [CrossRef]

Sample Availability: Data can be made available upon reasonable request to the corresponding author.

Review

How Can Smart Mobility Innovations Alleviate Transportation Disadvantage? Assembling a Conceptual Framework through a Systematic Review

Luke Butler [1], **Tan Yigitcanlar** [1,*] and **Alexander Paz** [2]

1   School of Built Environment, Queensland University of Technology, Brisbane, QLD 4000, Australia; luke.butler@hdr.qut.edu.au
2   School of Civil and Environmental Engineering, Queensland University of Technology, Brisbane, QLD 4000, Australia; alexander.paz@qut.edu.au
*   Correspondence: tan.yigitcanlar@qut.edu.au; Tel.: +61-7-3138-2418

Received: 26 August 2020; Accepted: 9 September 2020; Published: 10 September 2020

Abstract: Transportation disadvantage is about the difficulty accessing mobility services required to complete activities associated with employment, shopping, business, essential needs, and recreation. Technological innovations in the field of smart mobility have been identified as a potential solution to help individuals overcome issues associated with transportation disadvantage. This paper aims to provide a consolidated understanding on how smart mobility innovations can contribute to alleviate transportation disadvantage. A systematic literature review is completed, and a conceptual framework is developed to provide the required information to address transportation disadvantage. The results are categorized under the physical, economic, spatial, temporal, psychological, information, and institutional dimensions of transportation disadvantage. The study findings reveal that: (a) Primary smart mobility innovations identified in the literature are demand responsive transportation, shared transportation, intelligent transportation systems, electric mobility, autonomous vehicles, and Mobility-as-a-Services. (b) Smart mobility innovations could benefit urban areas by improving accessibility, efficiency, coverage, flexibility, safety, and the overall integration of the transportation system. (c) Smart mobility innovations have the potential to contribute to the alleviation of transportation disadvantage. (d) Mobility-as-a-Service has high potential to alleviate transportation disadvantage primarily due to its ability to integrate a wide-range of services.

Keywords: smart mobility; demand responsive transport; connected and autonomous vehicle; Mobility-as-a-Service (MaaS); electric mobility; shared transportation; intelligent transportation systems; smart city; transportation disadvantage; social exclusion

1. Introduction

In recent decades, rural-to-urban migration influenced by factors such as increased employment opportunities, access to services, education, and communication networks has led to a period of rapid urbanization [1]. Over 50% of the world's population live in cities with this number expected to increase to 68% by 2050 [2]. While the environmental impact of providing transportation infrastructure in growing cities remains a primary concern in research [3], another important challenge relates to the provision of an inclusive, accessible, and affordable transportation for all individuals [4]. This is important as having access to transportation is crucial to improve social inclusion and allow people to access essential services, employment, and recreational facilities. Access to transportation is a critical component in achieving quality of life—particularly among vulnerable groups such as the elderly and disabled [5].

Transportation disadvantage relates to an individual's ability to access transport and is particularly prevalent in areas without good access to public transportation. In these areas, individuals must rely on "private motor vehicles" (PMV), which typically come with higher costs than public transport due to purchasing, fuel, maintenance, insurance, and storage costs [6]. This combined with increased population growth has had a significant impact on property values with areas around public transport nodes experiencing higher property values [7]. Lower income earners are then forced into surrounding fringe areas, further increasing transportation costs and exacerbating issues surrounding transport disadvantage [8].

Smart mobility has been identified as a potential solution to alleviate many of the issues associated with transport disadvantage [9]. Smart mobility, a general term used to describe many of the transport-related technologies that have been implemented in urban areas, represents a new way of thinking about transportation including the creation of a more sustainable system that is able to overcome some of the issues associated with PMV [10,11]. While the number of research articles that focus on smart mobility is growing, little research to date has focused on how smart mobility can address transport disadvantage. Similarly, where specific smart mobility innovations, such as "autonomous vehicles" (AV), "flexible transportation services" (FTS), and "free-floating e-mobility" (FFM), or the integration of intelligent technologies have been investigated as a potential solution to transport disadvantage, they are often treated as separate entities with only a few comprehensive attempts to conceptualize how their integration can contribute to or alleviate the issue [12]. This requires explicit consideration as these changes do not happen in a silo, but are rather concurrent, or even dependent, on each other.

This paper attempts to contribute to existing research by analyzing the way that smart mobility innovations can address transport disadvantage in cities. Using a systematic literature review as the research methodology, this paper seeks to answer the research question: How can smart mobility contribute to alleviate transport disadvantage? To answer this question and ensure all technological advances are considered, we first reviewed the literature to determine the innovations relevant to the smart mobility field, how they relate to each other, and what the major benefits of these systems are to urban areas. Then, by looking through the lens of transport disadvantage, major contributions were identified and associated with our research aim and question. From the literature review, a conceptual framework representing the relationship between the benefits of smart mobility innovations and the various aspects of transport disadvantage was developed with the view that it could help researchers better understand the relationship between the two concepts. This paper also highlights future areas of research that can help other look to smart mobility innovations to alleviate issues regarding transport disadvantage.

2. Background to Smart Mobility

Smart mobility as a concept has its roots within the smart cities model: driven by policy, technology, and community, the primary goal of smart cities is to deliver productivity, innovation, livability, wellbeing, sustainability, accessibility, and good governance and planning [13]. The conceptual framework shown in Figure 1 demonstrates this concept through a simple input–output–impact model. In the context of smart cities, the transportation system could be considered an asset of the city which is implemented through various drivers, including technology, policy, and community. When successfully implemented, these drivers should lead to more desirable outputs (or outcomes), the result (or impact) being a smarter city—or in this case a smarter mobility system [13].

Built into this concept of smart cities is the notion of smart mobility [10,14]. Similar to the broader smart city concept, smart mobility is partially driven by community and policy; however, much of the focus is on using technology as a way to transform the transportation system while addressing the societal, economic, and environmental impacts associated with PMV, including issues regarding transport disadvantage [15,16]. Some of these innovations such as "demand responsive transportation" (DRT) have been implemented by local governments as a way of offering services to those most in need or replace underutilized public transport systems. They are often viewed more as an extension of the existing public transport network than a stand-alone system [17]. Similarly, ubiquitous infrastructure ("U-Infrastructure") harnesses technological advances in ICT, "intelligent transportation systems" (ITS), and digital networks to improve efficiency of urban infrastructure [18].

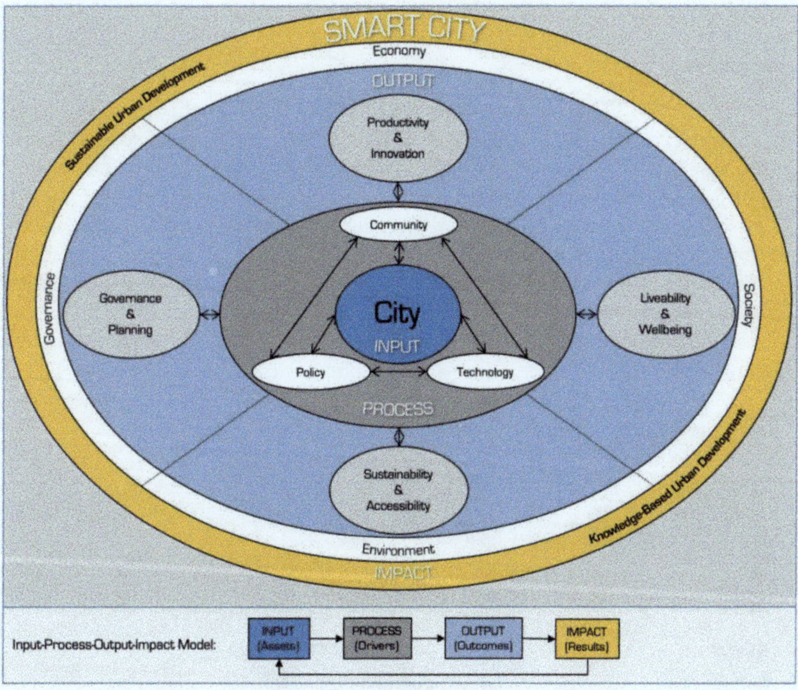

Figure 1. Smart city conceptual framework (derived from [13]).

Other systems including car sharing, ride sharing, FFM, AV, and alternative fuel vehicles are driven by private industry and with rapid advances in technology they are likely to disrupt the transport system, whether their benefits are harnessed by governments or they are left to evolve organically [19,20]. This was seen in 2015 in Australia following the introduction of Uber's ride sharing platform. Regulators were effectively left playing catch-up to a disruptive technology that was implemented and already in widespread use prior to the appropriate legislation being developed. The impact of this lack of foresight not only led to issues regarding overnight loss of value to taxi licenses [21] but has also led to concerns about the underpayment of workers [22], and an eventual oversaturation of the market [23].

There is an important distinction to be made here about how smart mobility innovation are introduced into the market and the importance of managing disruptive technology. While modern visions of smart mobility are generally optimistic and show a transportation system where everybody has equal access and PMV travel is replaced with services that users can access on-demand, the reality could be very different. In fact, as with the introduction of the automobile in the early 1900s, there is a risk that this new technology will create even greater issues that we were unable to see or predict due to a persistent cloud of optimism that shades our judgement. Thus, critical in the realm of urban governance is to develop an understanding of the potential contributions of smart mobility so that its impacts can be managed effectively and the societal, economic, and environmental objectives of the smart city are achieved [24].

The conceptual background for smart mobility outlined above underlines the importance of further investigating the contribution smart mobility can make to urban areas. This is particularly true with regards to the issue of transport disadvantage which could potentially risk further decline if smart mobility innovations are implemented into urban areas without any actions taken by decision-makers. Similarly, misunderstanding the potential benefits of smart mobility could lead to missing opportunities to improve the equity of the transportation system.

3. Materials and Methods

For this study, a systematic literature review was utilized as the methodology and based on the three-stage approach implemented by Yigitcanlar et al. [15]. The purpose of the review was to address the research question: How can smart mobility contribute to alleviate transport disadvantage?

The first stage in this process was planning. Our research objectives were defined as being to identify any relationship between smart mobility innovations and transport disadvantage. Based on this objective research aim and question, several keywords relevant to the subject area were developed. Primary inclusion criteria included articles that were peer-reviewed, published online, and in English. Secondary inclusion criteria were to only include articles that were relevant to the research aim. Exclusionary criteria were articles that did not meet the inclusion criteria. The keywords were then used to undertake an open-ended search to September 2020 using a university library search engine with access to 393 academic databases. Boolean search query was used with keywords, as shown in Figure 2. The initial search yielded 2136 articles.

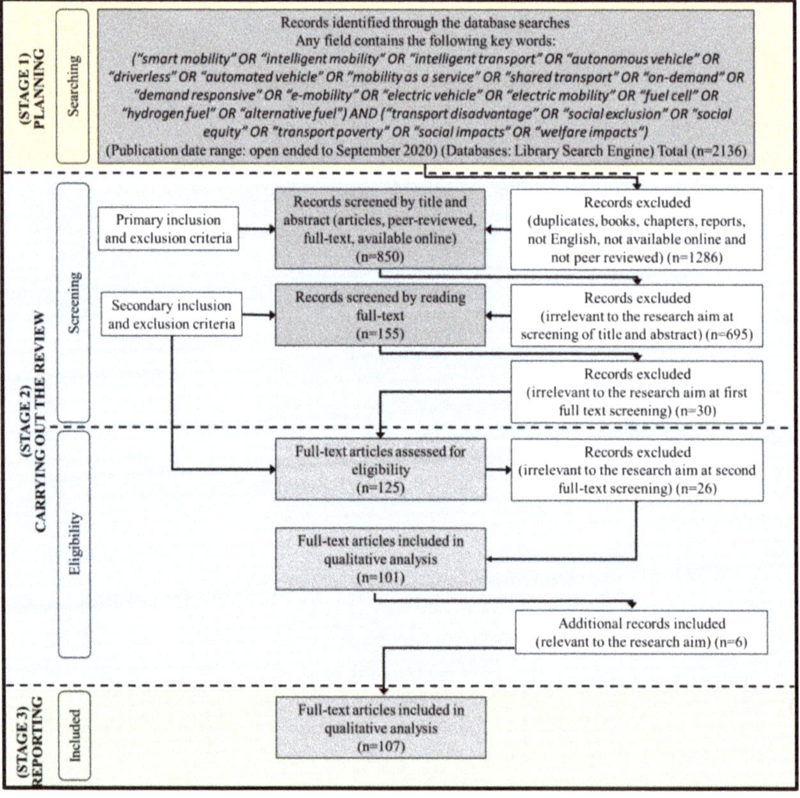

Figure 2. Literature selection procedure (source: authors).

The second stage of the process was performing the review. Abstracts of the proposed articles were scanned against the primary inclusion and exclusion criteria; duplicates and articles that did not comply with the criteria were removed. Following this, the full text of the remaining articles was read twice to ensure compliance with the secondary inclusion criteria. Articles irrelevant to the research aim were removed. In total, 99 articles were considered relevant to the research aim and included in the final qualitative review. Figure 1 provides a step by step outline of the literature selection and review.

The remaining articles ($n = 101$) were categorized using a directed content analysis method, whereby major themes were selected based on a theory or framework identified in the literature. An additional six publications relevant to research topic but not otherwise meeting the search criteria were also added to the study. As the purpose of the review was to determine the contribution smart mobility innovations could make to alleviate transport disadvantage, a framework was selected to ensure all relevant dimensions were considered. Based on a previous review of transport disadvantage by Yigitcanlar et al. [4], it was decided that Suhl and Carreno's [25] six dimensions for transport disadvantage—i.e., physical, economic, spatial, temporal, psychological, and information—would be used as it was the most comprehensive. The articles were then reviewed using a descriptive rather statistical technique. Pattern matching and other qualitative techniques, such as scanning for common subjects, were also used to group the articles into the pre-defined categories. As a result, relationships were identified between the transport disadvantage dimensions and the number of final categories reduced to three: (a) Physical and Economic ($n = 39$); (b) Spatial and Temporal ($n = 33$); and (c) Psychological and Information ($n = 28$). Following a review of the literature a seventh category, "institutional disadvantage", was also discussed ($n = 7$). A description of the relevant dimensions is shown in Table 1.

Table 1. Dimensions of transportation disadvantage (derived from [4,25]).

Dimension	Description
Physical	Relates to the physical barriers that may limit a person from accessing transportation. Limitations may include inability to operate a motor vehicle and inability to physically access a vehicle or public transport due to a disability.
Economic	Relates to the economic barriers that may limit a person from accessing transportation. Specifically concerns the personal cost of transportation and can include ticket price, fuel, insurance, storage, purchase, and travel time.
Spatial	Relates to the spatial barriers that may limit a person from accessing transportation. Often associated with geographic-related transport disadvantage in areas where public transport coverage is inadequate, and individuals are forced to own PMV to satisfy transportation needs.
Temporal	Relates to the temporal barriers that may limit a person from accessing transportation. Often associated with geographic-related transport disadvantage in areas where public transport frequency is inadequate, and individuals are forced to own PMV to satisfy transportation needs.
Psychological	Relates to the psychological barriers that may limit a person from accessing different transportation modes. This barrier can include issues associated with perception and safety.
Information	Relates to the information barriers that may limit a person from accessing different transportation modes. These barriers relate to an individual's ability to use and understand how to use transportation modes.
Institutional	Relates to the institutional and governance barriers that may limit a person from accessing different transport modes. These barriers include policy, regulations, registration requirements, and other local laws that may limit an individual's ability to use transportation.

The third and final stage of the process was reporting. In this stage, the analysis of the 107 articles completed during the screening stage was used to present the results by preparing and writing the final article. Finally, additional publications ($n = 31$) were used to support our findings, elaborate on our results, and provide a contextual background to this research.

4. Results

4.1. General Observations

Interest in the social issues surrounding smart mobility has grown over the past two decades. In fact, while only 2 of the selected articles were published before 2005, that number has continued to grow with 4 articles published during 2006–2008, 5 articles during 2009–2011, 13 articles during 2012–2014, 20 articles during 2015–2017, and 63 since 2018. Leading authors are affiliated with universities in Europe ($n = 58$), Oceania ($n = 19$), North America ($n = 18$), Asia ($n = 9$), South America ($n = 1$), and Middle East ($n = 2$). The articles were published in a wide-range of journal including Research in Transportation Economics ($n = 9$), Sustainability ($n = 9$), Transport Policy ($n = 7$), Journal of Transport Geography ($n = 6$), Transport Research Part A ($n = 5$), Transportation ($n = 5$), Travel, Behaviour & Society ($n = 4$), Transport Planning and Technology ($n = 4$), Transport Reviews ($n = 4$), Journal of Transport & Health ($n = 3$), Energies ($n = 2$), Energy Research & Social Science ($n = 2$), Land Use Policy ($n = 2$), Local Economy ($n = 2$), and Transportation Research Part D ($n = 2$). The remaining 41 articles were published in 36 different journals from a range of research areas including urban planning and policy, transportation, ethics, sociology, and health.

Articles were categorized into three groups based on the defined categories: Physical and Economic ($n = 33$), Temporal and Spatial ($n = 31$), and Psychological and Information ($n = 26$). With reference to the main smart mobility innovations, DRT were discussed in 40 articles, followed by AV ($n = 38$), ITS ($n = 25$), shared mobility ($n = 17$), "Mobility-as-a-Service" (MaaS) ($n = 12$), and "alternative fuel vehicles" ($n = 12$). Twelve articles discussed smart mobility generally but were not specific regarding technological innovations.

4.2. Smart Mobility Impacts

This section discusses the main innovations identified in the literature that are associated with smart mobility and what impacts these innovations will make to transportation. Understanding the broad impacts each of the innovations will have on the transportation system is important so that the flow on effects can be analyzed against each of the transport disadvantage dimensions.

The six major smart mobility innovations identified in the literature are: (a) DRT; (b) shared mobility; (c) ITS; (d) alternative fuel vehicles; (e) AV; and (f) MaaS. ITS, alternative fuel vehicles, and AV represent direct technological advances that will affect vehicles and infrastructure. On the other hand, while technology is critical to the development of DRT, shared mobility, and MaaS, they are more associated with innovations to the way transportation services are provided to the community rather than a direct impact to the vehicles and infrastructure in the transport system. A description of each of these innovations and relevant literature is shown in Table 2.

Table 2. Smart mobility innovations (source: authors).

Innovation	Description	Reference
DRT	DRT provides a transportation options distinct from traditional fixed-route services in that they utilize dynamic, semi-fixed, or fixed routes with users able to pre-book based on travel needs and services operated on-demand. The main impacts on the transportation system relate to ability to provide greater coverage and flexibility. Although existing DRT services have been criticized for their inability to manage high demand and provide the required service coverage at an appropriate cost unless supported by other innovations including ITS, AV, and shared mobility.	[26–33]

Table 2. *Cont.*

Innovation	Description	Reference
ST	Shared mobility refers to services where rides are shared with other users (e.g., ride-sharing) or vehicles are shared but used at different times (e.g., car sharing or bike sharing). Traditional types of shared mobility include car rentals, public transport, or taxis. Recent shared mobility innovations that make use of ITS, DRT, or battery-operated systems include FFM and peer-to-peer ride sharing apps such as Uber. The main advantage of shared mobility is that resources are shared among multiple users resulting in improved efficiencies in operation, storage, and cost. A move towards the shared mobility is a critical component for smart mobility innovations including DRT, AV, and MaaS to achieve sustainable city goals.	[34–42]
ITS	ITS refers to applications which utilize advances in ICT to effectively shared data between vehicles, infrastructure, and between users. ITS covers a broad range of applications including transport telematics, automatic information signs (ATIS), electronic ticketing, smart infrastructure, and in-car assistance, sensors, and other safety features. ITS is enhanced by advances in big data, Internet of things (IoT), cloud computing, and artificial intelligence (AI), which contribute to more efficient data collection, processing, and analysis. The primary advantage of ITS within the transportation field relates to its ability to optimize the performance of other technological innovations.	[43–46]
Alternative Fuel	Alternative fuel vehicles refer to those vehicles which do not rely on petroleum-based fuel sources. These vehicles use batteries for their primary fuel source, which in turn are fueled by non-petroleum sources such as the electrical grid, hydrogen, or solar power. Most literature focuses on the environmental benefits of this technology. However, with reference to transport disadvantage, the primary benefits of battery-operated vehicle relate to the convenience of FFM which provides users an assortment of conveniently placed powered transportation options which they can access on-demand.	[47–50]
AV	AV refer to vehicles that can be operated without input of a human driver. While there are various levels of autonomy for the purpose of this article any reference to AV assumes the vehicles can operate without human input in all conditions—unless otherwise stated. The primary advantage of AV relates to improved accessibility, operational efficiency, and safety.	[51–57]
MaaS	MaaS is a concept whereby a range of transportation options, including ride-sharing, car-sharing, public transport, and FFM, is offered to customers via a single online platform, or app. Users subscribe to the service which gives them access to various transport options that were traditionally offered separately. The advantage of MaaS is that as an integrated system it can provide the platform from which mobility providers are able to shared resources, and could contribute to better transport outcomes as any issues can be better considered by looking at the transport system as a whole rather than only concentrating on individual parts. MaaS can also provide the operational structure from which new transport innovations are released into the market.	[24,38,58–60]

The innovations often overlap to optimize the potential positive impacts. DRT systems and AV enabled by ITS technology are often referred to as real-time or dynamic FTS [61] and connected AV (CAV), respectively [55,62]. Shared AV (SAV) incorporates elements of shared mobility and AV [44], and FFM is essentially a combination of shared mobility, battery electric vehicles, and DRT [63]. MaaS forms an overarching platform in which each of these services can be bundled together [64]. A conceptual diagram is shown in Figure 3 to better understand the relationships between these innovations. This diagram is by no means exhaustive, for example free floating e-mobility incorporates elements of ITS, and, when offered as a bicycle service, it might not rely on electric powered engines. In addition, DRT are typically offered as shared mobility to improve efficiency and costs [65]. Nonetheless, the figure provides a conceptual outline to better understand the relationships between the various smart mobility innovations identified in the literature.

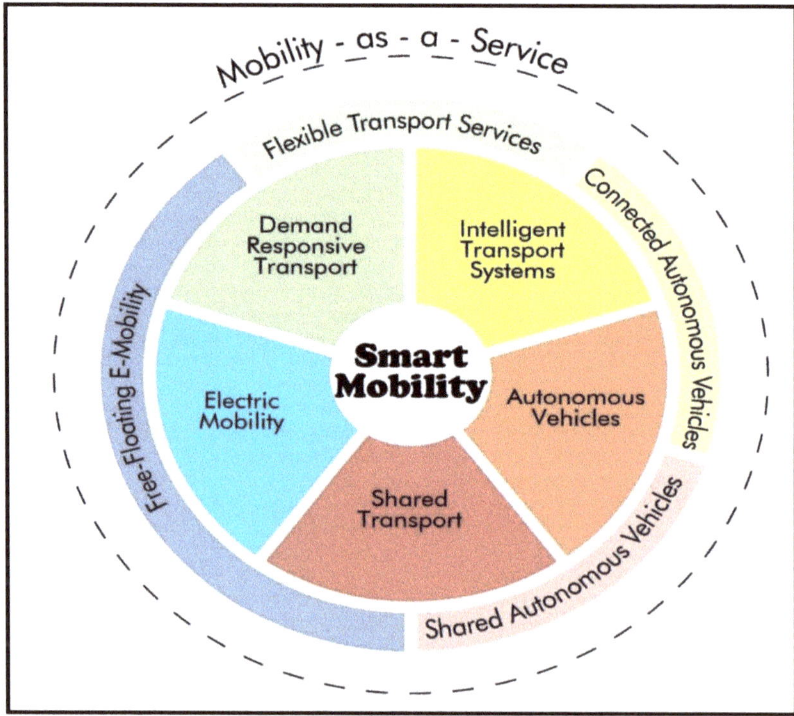

Figure 3. Relationship between smart mobility innovations (source: authors).

4.3. Physical and Economic Dimensions

This section discusses how smart mobility innovations can contribute to the alleviation of the physical and economic dimensions of transport disadvantage. Based on the reviewed literature smart mobility could alleviate the physical and economic dimensions of transport disadvantage by: (a) improving accessibility to transportation for those unable to access or operate a vehicle; (b) creating a transportation system in which services are more responsive to user needs; (c) reducing the cost for users by improve the efficiency of the transportation system and promoting a move towards shared mobility; and (d) improving the "value of time" (VOT) spent in transit. A list of all reviewed literature is shown in Appendix A.

Firstly, various smart mobility innovations have been shown to improve accessibility for those physically unable to access transportation or operate a vehicle. Access to a vehicle is an important factor in maintaining a good standard of living and providing security and freedom of movement to access social activities, employment, and other services, including healthcare [57,66], particularly in low-density areas [67]. DRT services that provide door-to-door transportation have been shown to improve user accessibility by reducing issues surrounding the first- and last-mile access of public transport [68]. In fact, when compared to traditional public transport services, one study showed high satisfaction with flexible DRT services resulting in a doubling of older users [69]. Furthermore, when operated as a shared service they have been shown to increase social interactions resulting in reduced feelings of isolation [68].

Despite the benefits of DRT, another commonly cited innovation to improve accessibility relates to AV [57,70]. As AV are able to drive without human input, the elderly, disabled, young, unlicensed, and those unfamiliar with local conditions may no longer be excluded from operating a PMV [51,71–73]. Even in a semi-AV setting, in-vehicle technologies such as crash avoidance; warnings for lane departure, collision, and blind spots; navigation systems; parking assistance; and adaptive cruise control may result in more elderly residents being able to hold onto their license for longer [52,74]. Some may also benefit from improved access to FFM including bike sharing, which would add additional accessibility options—particularly for short distance trips [75]. Due to these advantages, it is important that these services are implemented with regulations to ensure equal access [40].

Nevertheless, increased accessibility means that there is a risk of increasing accessibility to PMV. This could result in more demand for car ownership and increased per capita "vehicle kilometers traveled" (VKT) [10]. This premise is supported by research which predicts that AV will result in a mode shift away from public and active transport, increasing total VKT by 15–59% [57]. This could lead to increased externalities including congestion and urban sprawl and result in greater infrastructure and transit costs [53,57,76]. Due to this potential impact, researchers consistently highlight the benefits of shared mobility [53,56]. If fact, research shows that SAV would actually decrease VKT by 10–25% [57]. However, this shift is dependent on how shared mobility appeals to consumers and will require a significant cultural shift supported by policy and regulation, public awareness campaigns, and land use interventions—particularly in areas where PMV is the dominate mode choice [66,77].

Secondly, smart mobility innovations present an opportunity to provide services to the transport disadvantage populations that are more responsive to their specific needs. In fact, when enabled by other smart technology, DRT systems have been able to use advances in data collection, distribution, and analysis to improve decision-making, simplify ticketing, and enhance route planning, scheduling, and vehicle selection [29,78,79]. Data obtained from smart ticketing systems can be integrated and used to analyze the behavior of passengers and identifying service gaps [80]. AV are also important because without the need for a driver the internal layouts can be reconfigured to provide comfort and access based on special needs [72].

This is important as studies have found that, for shared mobility to be appealing, it needs to be flexible and able to satisfy individual needs—particularly with regards to having both on- and off-peak access to employment, healthcare, and recreational areas [81]. Integrated services such as MaaS can help by facilitating better multi-stakeholder collaboration and the sharing of information. The needs and trends of each individual user can then be used collaboratively to support the day-to-day operation of the entire system [82] and connect potential users with the most suitable providers [42].

Thirdly, there is potential for smart mobility to reduce transportation costs by improving the efficiency of the transportation system and promoting a move towards shared mobility. The integration of services through a MaaS-like system and the use of ITS has the potential to reduce administration and management costs. Cost savings related to the design of transportation systems could theoretically be passed onto the consumer or used to supply transportation services to the disadvantaged [24,79]. Using shared mobility as a replacement for PMV would also remove many of the economic barriers associated with ownership—e.g., purchasing, maintenance, insurance, and storage costs [39]. Parking costs would also be reduced under a shared system as vehicles would spend less time in idle [34,83]. However, the distribution of shared services is likely to favor areas with high demand and is unlikely to reduce issues associated with geographic-related transport disadvantage. Furthermore, disability related disadvantage is unlikely to benefit from car and bike schemes alone [34]. AV would be beneficial in this regard as removing the need for a driver would significantly reduce operational costs and help solve issues associated with accessibility. In fact, SAV have the potential to reduce total transportation cost by over 80% when compared to traditional PMV [39].

However, new technology also brings high upfront costs and low short-term return on investment. Thus, while research continues to show that there is a huge demand for more sustainable vehicles [84], residents are willing to pay extra for more environmentally friendly options [49], and shared mobility is cheaper than car ownership [85], it may be difficult to guarantee economic sustainability—particularly in the short term. For example, while the use of alternative environmentally friendly fuels often achieves better economic performance, the high cost of vehicles—particularly hydrogen fuel cell vehicles—can make the economic sustainability of such vehicles difficult [48]. Similarly, despite potential for lower maintenance costs, reduced accidents, and overall efficiency, the high upfront cost may limit potential for market penetration [85].

Finally, cost alone might be not be enough to sway users to a shared system. In fact, in many countries, the modes with the lowest cost of operation—e.g., public and active transportation—are often not the ones with the highest market share. Other factors, such as comfort and prestige, also play a part [39]. AV could transform interior of private vehicles into mobile offices, dwellings, or entertainment and communication hubs improving the VOT by facilitating the ability to work, eat, socialize, and rest while in transit [56]. This could improve work life balance and reduce stress—particularly among those who travel regularly. Conversely, increasing VOT may also lead to an increase in VKT, further exacerbating issues associated with infrastructure demand and urban sprawl [39,86].

From an economic perspective, the increase in demand for private AV may also lead to disadvantaged populations being priced out of the market, leaving them unable to benefit from the advantages of the technology [40,56,87]. Similarly, with alternative fuel vehicles, users unable to afford the new technology may be charged with a Pigouvian tax to discourage the use of fossil fuels [47,50,88,89]. Increased use in private AV and shared mobility may also lead to a reduction in public transport use, reducing revenue and resulting in higher costs and future degradation of services. This is likely to impact lower income and geographically disadvantaged residents the most [56,90].

There is also economic risk associated with an integrated transportation system such as MaaS. Where a single entity is responsible for the selection and distribution of mobility providers, the system itself may become a barrier to new transportation companies entering the market. This could result in monopolization, increasing the risk of uncompetitive markets, price gouging, and other unfair businesses practices [76,91]. Conversely, government control could create tension with the private sector, which is critical in the development and funding of new transportation innovations [24,76]. If we are to rely on private companies to provide most of the services, it is unlikely that off-peak and low demand services would be provided, and significant subsides, political engagement, and planning would be required to ensure that societal goals are being maintained [92].

4.4. Spatial and Temporal Dimensions

This section discusses how smart mobility innovations can contribute to the alleviation of the spatial and temporal dimensions of transport disadvantage. Due to the association with time and distance, this dimension is most closely related to issues surrounding geographic-related transport disadvantage. Based on the reviewed literature, smart mobility could alleviate the spatial and temporal dimensions of transport disadvantage by: (a) filling gaps in the public transport network by improving the coverage and frequency of services; (b) strengthening the connection with public services by designing services to act as a feeder system which connects to major public transport nodes and employment centers; (c) improving the flexibility of public transport by offering services on-demand; and (d) creating more transportation choices in areas where choice is traditionally limited. A list of all reviewed literature is shown in Appendix A.

Firstly, literature on smart mobility consistently identifies smart mobility innovations as a way to fill gaps in the public transport network. In doing so, smart mobility can contribute to improved coverage and frequency of services [43,93]. DRT services, in particular, have been highlighted as a way to provide door-to-door transportation by using fleets of smaller shared vehicles as opposed to fixed route services [94]. Other advantages of using smaller vehicles over traditional buses is that they have a lower operational cost per passenger and can access areas with smaller road widths [94]. However, these services often require significant government subsidies as they do not have the required number of users to support profitability over the required coverage [35,95]. While subsidizing these services may be more economical than providing fixed route public transport [96], ITS can also help better match supply and demand and develop locally specific strategies that also contribute to lower costs and better efficiency [95,97]. ITS has been shown to allow better real-time control over the networks and enhance the potential for DRT to provide increased flexibility and greater coverage while bringing costs closer to that of public transport [27,43]. SAV has also been identified as a way to improve coverage particularly by reducing the instance of dead runs [33,98].

Notwithstanding, there will also be issues associated with providing the necessary infrastructure to facilitate suitable network coverage [99,100]. Furthermore, when promoting alternative fuel vehicles that generate electricity from the grid, there may be issues associated with grid capacity. Infrastructure issues are intensified in low density and rural areas due to inadequate infrastructure and longer transmission distances [47,50,101]. As such, low-density areas would still attract higher transportation costs than high density areas, and significant investment is required for ensure geographic equity [101]. One solution relates to cross-subsidization where profits made in areas with high demand are used to subsidize and fund the required infrastructure in areas with lower demand [47,95]. By sharing information and resources across the transportation system MaaS can help facilitate this cross-subsidization to ensure maximum profitability and promote social equity [83]. Furthermore, since subsidies may make low density housing more attractive, planning interventions that promote walking, cycling, higher densities around employment and transit centers, and investment in high speed public transport remain important [59,102].

Secondly, smart mobility can be used to support investment in high-speed public transport by using innovative services to act as a feeder system, which acts as a first- and last-mile connection to major public transport nodes and employment centers [96]. Theoretically, improved access to public transport would reduce car dependency and therefore reduce transportation costs [103]. DRT systems could be timed to public transport hubs to ensure reductions in transfer times. Public transport would therefore form the backbone of these "pulse networks", which could also allow for integrated ticketing and services [103]. The overall coverage of these networks could be supported by shared mobility such as FFM that would provide connections for shorter distances and provide more transportation options [37,104]. By limiting long trips, directly into denser urban areas congestion will be reduced, which means individuals who are required to travel by PMV will likely see a reduction in fuel price and time spent in traffic [96].

Efficient trip chaining is also important as studies have shown that users are more sensitive to travel time than travel cost; thus, ensuring transfers are easy and free from unnecessary delays can contribute to improving the appeal of public transport [105]. ITS has a role in improving the efficiency of these transfer, by improving the ability to apply real-time alterations to routing [26,104,106]. In fact, studies have shown that DRT services that connect directly to major transportation hubs and are enabled by ITS contribute to increases in total public transport ridership [27,107]. Furthermore, significant modal shift away from PMV has been observed when "artificial intelligence" (AI) is used to configure routes to reduce travel time [108] or through the use of MaaS systems to create synergies between mobility providers [109].

Thirdly, by improving the flexibility of public transport and offering services on-demand, transportation systems can be designed to respond directly to the specific geographic and social characteristics of the local area [32,97]. For example, in some areas, such as those with large numbers of tourists, conventional public transport with fixed schedules and timetables may be more advantageous [28,110]. In addition, in areas with higher numbers of people unable to operate a vehicle, car sharing schemes should be limited in favor of more flexible routes and timetables [111]. Similarly, in very low-density areas, semi-fixed, as opposed to door-to-door, services may be more efficient [33]. In designing transportation systems, planners should consider how changes respond to local characteristics and ensure the optimal allocation of available resources [32]. Furthermore, any local transportation plan should be able to be scaled up if demand increases to ensure equal and equitable coverage [96].

Finally, using smart mobility to create more options for users in areas where transportation choice is limited can be beneficial [59]. Having more mobility options available to users has been identified as an important step to overcome the culture of PMV ownership. Similarly, supportive policies with awareness of shared services would be useful [112]. MaaS provides an opportunity to bundle services and offer a range of options to consumers through a single online platform [83,109]. Alternatively, transportation choice may also include the choice to not travel. Advances in the design of digital neighborhoods, smart homes, ICT, and home delivery has the potential to remove the need for physical trips—particularly those related to employment [94,112]. Similarly, with the view to reduce PMV, ICT and data obtained from ITS can be used to help residents make more informed decisions regarding residential or work location [113].

4.5. Psychological and Information Dimensions

This section discusses how smart mobility innovations can contribute to the alleviation of the psychological and information dimensions of transport disadvantage. Based on the reviewed literature, smart mobility could alleviate the spatial and temporal dimensions of transport disadvantage by: (a) improving the safety of travel; (b) improving the perception of existing transportation options; and (c) improving the ability to make informed decisions. A list of all reviewed literature is shown in Appendix A (Table A1).

Firstly, smart mobility innovations have been shown to contribute to improved safety in the transportation system. This is important as the perception of safety is critical to ensure individuals want to use smart mobility [54]. AV have the potential to significantly reduce the number of vehicular accidents caused by human error [54,55,114]. CAVs can use advances in ITS, ICT, and AI data processing to communicate with other vehicles, infrastructure, and sensors, identifying dangers early and further improving safety for drivers and pedestrians [55]. In addition, given that no driver is required in the internal configuration, it can be reconfigured to add to the safety of the vehicle [55]. Similarly, DRT that offers door-to-door transportation and shared mobility are perceived as a safer option than public transport—particularly at nighttime [30,31,115].

Nonetheless, from the perspective of the user, safety not only comes from feeling safe while engaged in journey, but also with regards to digital safety [36,116]. In fact, lack of trust in technology is consistently identified as a reason for not using new transport technologies, particularly among the elderly [54,117–119]. This is understandable as increased reliance on technology introduces additional risks including those related to data privacy, cyberterrorism, grounding of fleets due to grid failures, faulty data [55,120], unconscious bias [114], and questions of legal liability [121]. To build trust, significant investment is required in cyber and data safety. Information campaigns are also beneficial to garner support among late adopters [119].

Secondly, there is potential for smart mobility innovations to improve the perception of existing DRT and public transport systems. Many DRT have been implemented around the world; however, the perception of these services is often that they are for the old and disabled—even when they are offered to all in the community [122,123]. Furthermore, users who benefit the most from the services are often confused and unclear about how these new transportation services could serve them [123–125]. In fact, research has shown that attitudes towards smart mobility among those with disabilities was entirely dependent on having prior knowledge of the technology [125]. Those with more knowledge tended to be more positive [126].

More information about potential routes and scheduling could help users better navigate new transportation innovations [124]. MaaS can help with this by providing all services and relevant information through a single digital platform giving users unbiased choice of various modes [38,60]. In addition, as all services are effectively bundled together, any offerings that are targeted towards those with special needs may no longer be viewed as an entitlement but would instead be part of a city, regional, or nationwide system that is synergized to benefits all of society [58].

Finally, smart mobility could improve the ability for commuters to make informed decisions. Technological advances in ITS can facilitate the collection and analysis of large amounts of data from cameras, sensors, vehicle locations, smart ticketing systems, social media, credit cards, mobile phones, and many other sources [13,45,127]. Automating the analysis of this "big data" could help individuals with route planning and vehicle selection [44,46,128]. The ability to make informed decisions based on real-time data can help commuters reduce uncertainty, fear, discomfort, enhance user experience, and improve confidence [44,45,115].

However, given the reliance on smart technology, there are issues associated with technical literacy and the digital divide [36,38,116,117]. The digital divide refers to the gap between those who can access ICT and those who cannot. This issue is not only associated with the spatial distribution of network coverage or equality of access to physical smart devices but also the ability for particular socioeconomic groups to use and understand the technology [36,117,127]. Statistically, the elderly, lower income, female, and disabled are less familiar with new technology due to lower lifelong exposure to ICT. Therefore, they often struggle to quickly learn the required skills to access and pay for digital services [117,127,129]. This is where an integrated system such as MaaS can help. By integrating a range of mobility providers into a single platform, it could simplify the process for accessing transport by reducing complexity and the need to cycle through various mobility applications [127]. Stakeholder engagement and public participation is also important to understand existing challenges within the community [130].

4.6. Institutional Dimensions

Upon review of the literature, a seventh and final transport disadvantage dimensions has emerged. The "institutional" dimension includes institutional and governance related barriers including policy, regulation, and institutions that may limit an individual's ability to use a transport mode or service. Based on the reviewed literature, smart mobility innovations do not necessarily directly contribute to the alleviation of this barrier. However, given the fast pace nature of technological change within the transport sector—including widespread trials of smart mobility services including DRT, AV, and MaaS and the rapid emergence of new technologies associated with car-, bike-, and scooter-sharing—it is important that decision makers understand the strengths and weakness associated with them so that opportunities and risks can be identified. This is important because public sector does not necessarily function adequately in times of uncertainty [76] and a failure to address the short- and long-term issues associated with these transport services could exacerbate negative externalities associated with the transport system. It is therefore important that strategies remain flexible so that they can adapt to changing circumstances and community needs [30,131].

It is critical that institutional barriers do not inhibit the ability for users to access services which could have wider societal benefits including high cost and inconvenience of registering for new services [132], laws that explicitly ban the use or inhibits the ability to use a mode or services within a particular area [133], or lack of available infrastructure to support mode choice—e.g., lack of dedicated active and public transport infrastructure [134]. Of equal importance is the use of institutional measures to promote and support the development of smart mobility. These could include: (a) establishment of standards for data management and sharing, which should be established on a national or transnational level [135]; (b) institutional support structures to assist with community adaptation to new technology, particularly among disadvantaged groups including elderly, migrants, or disabled [136]; (c) development of parking restrictions to discourage private vehicle use [131], engaging the public in decision-making [130]; and (d) ensuring public value and societal goals are maintained [24,137].

5. Discussion

5.1. Key Findings

This review study investigated the impact of smart mobility innovations through the lens of transport disadvantage. Specifically, the review sought to answer the research question: How can smart mobility contribute to the alleviation of transport disadvantage? Firstly, some common smart mobility innovations were identified and the relationships between these innovations shown. These innovations include new vehicular and infrastructural innovations such AV, ITS, and alternative fuel vehicles, in addition to new and existing ways of offering services to the community including DRT, shared mobility, and MaaS. These innovations will likely benefit urban areas by improving accessibility, efficiency, coverage, flexibility, safety, and integration of the transportation system.

The study also showed how smart mobility innovations have the potential to contribute to the alleviation of all six dimensions of transport disadvantage: (a) physical; (b) economic; (c) spatial; (d) temporal; (e) psychological; and (f) information. We also discussed some implications associated with a seventh, "institutional", dimension. Potential risks have been identified, and there are a number of key actions that can be taken to alleviate these risks. Of these actions, the implementation of MaaS and shared mobility appears as a common thread to overcoming the risks associated with smart mobility.

Firstly, a move towards the shared mobility is critical to ensure resources are shared efficiency and services offered have the required accessibility, coverage, and flexibility to reach all users and do not result in excess consumer costs or reliance on government subsidies. This conclusion is reflected in studies on DRT [66], AV [39,40,52], and MaaS [40].

Secondly, the review showed that it is often a combination of innovations that will best benefit the transport disadvantage. For instance, DRT and AV are shown to work more efficiently, and safely, when enabled by ITS and other smart technology including big data and cloud computing. Furthermore, the negative externalities associated with AV use, including increased VKT, suburbanization, and infrastructure demand, are significantly reduced when operating within a shared economy. This highlights the specific advantages of MaaS, which as an integrated system can provide the operational structure from which new innovations are trialed and released into the market. It also can help connect users to shared mobility and provide a platform from which mobility providers share resources. Sharing data between mobility providers could help decision-makers achieve better outcomes as issues associated with transport disadvantage can be considered by looking at the transportation system as a whole rather than concentrating on individual parts. Similar conclusions regarding the importance of MaaS as an overarching operational structure is supported by a number of studies including Gonzalez-Feliu et al. [82], Mulley and Kronsell [58], Soares Machado et al. [38], and Beecroft et al. [116].

Lastly, a summary of smart mobility potential contribution and risks and their association with transport disadvantage dimensions is shown in Table 3.

5.2. Conceptual Framework

Within the realm of smart mobility, a key challenge to overcome transport disadvantage is to understand how the specific benefits of new transportation innovations can be harnessed to respond to each of the dimensions of transport disadvantage. The results of the literature review highlight important relationships between the benefits of smart mobility innovations and the different dimensions of transport disadvantages. Specifically, the review showed that the benefits of smart mobility can be specifically aligned with the corresponding transport disadvantage dimension. A conceptual framework showing the relationship between these factors is shown in Figure 4. For the purpose of providing a conceptual framework related to how smart mobility can alleviate transport disadvantage, the institutional barrier has been excluded from the framework as it is not a barrier that can be overcome by smart mobility innovations alone. Nevertheless, supportive policy, regulations, and other governance structures are critical to the implementation smart mobility in a way that strengthens its benefits while responding to issues of transport disadvantage.

Table 3. Summary of literature review findings (source: authors).

Dimension	Contribution	Risk	Potential Actions
Physical	Improved accessibility to vehicle Door-to-door transportation Connection to public transport Increased social interactions More transportation options More responsive to specific need	Unequal access to services Increased VKT per capita Increased suburbanization Unappealing to user	Integration of services (MaaS) Marketing and education Policy and Regulation Promote shared mobility
Economic	Improved efficiency of system Reduced consumer costs Increased VOT	Unequal access to services Increased VKT per capita Increased infrastructure demand Increased suburbanization Monopolization	Integration of services (MaaS) Promote shared mobility Land use planning interventions Subsidies Stakeholder engagement
Spatial	Improved coverage of services Fill gaps in public transport network Feeder system to public transport	Increased infrastructure demand Network coverage Grid capacity Unequal access to services	Active transportation infrastructure Cross-subsidization Digital neighborhoods Integration of services (MaaS) Marketing and education Promote shared mobility
Temporal	Improved flexibility of services Better real-time control of network Better match supply and demand Reduced transfer times Reduced congestion	Routing should be specific to needs	Analysis of local characteristics Digital neighborhoods Invest in intelligent technology (ITS) Integration of services (MaaS) Marketing and education
Psychological	Improved safety of vehicle Safety of door to door transportation Improved perception	Data safety Cyber safety Unconscious bias Legal liability	Invest in intelligent technology (ITS) Marketing and education Promote shared mobility
Information	Improved integration Improve decision-making	Digital divide Technology literacy	Integration of services (MaaS) Invest in intelligent technology (ITS) Stakeholder engagement
Institutional	Opportunity for change	Rapid technological change Increase negative externalities Lost opportunity	Adaptive policy and regulations Supportive governance structures

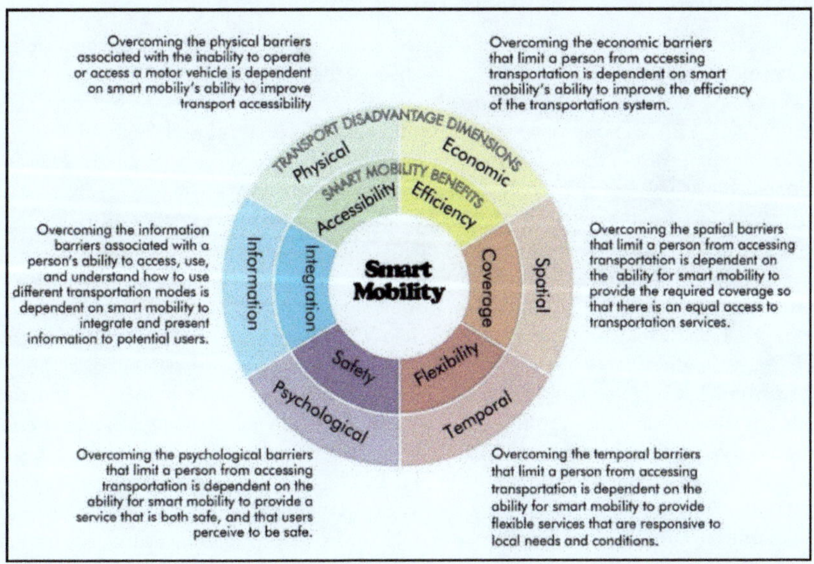

Figure 4. Conceptual framework of smart mobility and transportation disadvantage (source: authors).

Firstly, when looking through the *physical* dimension of transport disadvantage, the primary contribution of smart mobility is its ability to improve transportation *accessibility* through implementation of AV, flexible door-to-door transportation, strengthening connections with existing public transport networks, providing more mode options, or specifically targeting user needs. Similarly, when looking through the *economic* dimension of transport disadvantage, the primary contribution of smart mobility is its ability to improve transportation *efficiency*, which could contribute to reduced consumer costs—whether by reducing cost of actual travel or increasing the VOT spent in traffic.

Secondly, when looking through the *spatial* dimension of transport disadvantage, the primary contribution of smart mobility is its ability to improve transportation *coverage* by filling gaps in public transport or acting as a feeder system to major public transport nodes. Similarly, when looking through the *temporal* dimension of transport disadvantage, the primary contribution of smart mobility is its ability to improve *flexibility* by moving towards dynamic routing of transportation services, having more real-time control over the transportation network, better matching supply and demand, reducing transfer times and reducing congestion for those who are required to travel by PMV.

Finally, when looking through the *psychological* dimension of transport disadvantage, the primary contribution of smart mobility is its ability to improve transportation *safety*, whether through the use of AV which removes the need for a human driver, ITS that communicate with vehicles and drivers regarding potential hazards, or door-to-door transportation that removes safety concerns associated with accessing fixed-route public transport stops—particularly in low occupancy areas. Similarly, when looking through the *information* dimension of transport disadvantage, the primary contribution of smart mobility is its ability to integrate a wide range of data and services which can be used to improve decision-making whether those decisions are made autonomously or following analysis of available data regulators, mobility providers, and users.

Given the relationship between smart mobility and transport disadvantage, the challenge for decision-makers and mobility providers is to analyze specific case study areas to determine the issues associated with transport disadvantage that are most relevant. From there, the smart mobility benefits that most closely represent each of these dimensions can be used to identify which innovation is best suited for the local area.

5.3. Research Directions

Few studies identified in this review considered the six smart mobility innovations together as a broad driver for change in the transportation system. Given that alternative fuel vehicles, such as battery electric and hybrid electric, and ITS have already started to be introduced into urban areas, and trials of AV are prevalent throughout the world, it is problematic to analyze each of these technological drivers as individual entities that will not interact and influence the success, or failure, of each other. The management of these technological innovations is therefore necessary to harness their benefits in response to transport disadvantage. That is why new operational structures and ways of looking at the transportation system including DRT, shared mobility, and MaaS remain important.

Nonetheless, while DRT systems are not new and have been implemented throughout the world—as an alternative to public transport and targeted toward those experiencing disadvantage—it has developed a stigma whereby it is often viewed as an option for only the aged and disabled. Similarly, shared mobility offered by private industry including ride-sharing, car-sharing, and FFM are typically targeted towards users in centralized, denser areas where the highest demand is available to ensure maximum profit. These services, therefore, rarely benefit those experiencing transport disadvantage, and often only exacerbate existing issues with unequal accessibility. As an integrated service, MaaS represents a new way of branding DRT, while enhancing public transport, shared mobility, and other elements of the transportation system. Furthermore, MaaS presents a unique opportunity to provide the platform from which new innovations are introduced into market, the data analyzed, shared, and used to assess its suitability for alleviating transport disadvantage, and other related issues.

Prospective research should, hence, look at ways to use MaaS to harness the benefits of smart mobility innovations and attract users to shared mobility and public transport. MaaS is a relatively new topic so further research could focus on the barriers, and risks associated with implementing MaaS within urban areas. Analysis throughout a range of case study areas using transportation modeling, consumer surveys, expert opinion, and trials could also identify issues specific to the varying characteristics of different regions, including those associated with regulatory systems, policy frameworks, cultural differences, and geographic conditions.

Secondly, research could also focus on other innovative ways to integrate transportation modes, attract users to shared mobility, or develop alternatives systems. Research could explore the role of other technological advances outside the field of transportation including 5G, AI, digital twins, virtual reality, blockchain, IoT, big data, and cloud computing. For example, the use of virtual reality and augmented reality could be used to educate, market, and promote new transportation innovations towards individuals and business. Similarly, it could be used to let users experience new transportation technology prior to analyzing their attitudes.

Finally, given the recent events associated with the COVID-19 pandemic and its potential implication on consumer attitudes towards shared mobility, there is also a need to analyze whether the experience has changed user perspectives and willingness-to-ride shared, and public transport. This is important as attitudes may be changing due to increased awareness of vulnerabilities associated with virus transmission from passengers sharing close quarters in vehicles that often rely on centralized air-conditioning and little ventilation [138]. Furthermore, given these unprecedented events and the pressure on individuals and business to quickly adopt remote working and social environments transportation researchers may be more inclined to ask whether no mobility is smarter than smart mobility. From a transport disadvantage perspective research could be undertaken to compare individual transportation needs before, during, and after the lockdown experiences. Representatives from the commercial sector could be interviewed to discuss experiences with remote working, and how the experience will shape business models into the future, as one of the advantages of remote working is that for many jobs individuals may no longer be limited to employment opportunities due to location or issues with being able to afford or access transportation that is responsive to their needs.

Author Contributions: L.B., data collection, processing, investigation, analysis, writing—original draft preparation, and writing—review and editing; and T.Y. and A.P., supervision, conceptualization, writing—review and editing. All authors have read and agreed to the final version of the manuscript.

Funding: This research received no external funding.

Acknowledgments: This research did not receive any specific grant from funding agencies in the public, commercial or not-for-profit sectors. The authors thank the editor and three anonymous referees for their invaluable comments on an earlier version of the manuscript.

Conflicts of Interest: The authors declare no conflict of interest.

Appendix A

Table A1. Reviewed literature pieces.

Author	Year	Title	Journal	Country	Category	Findings	Relevance
Mageean, J., and Nelson, J. D.	2003	The evaluation of demand responsive transport services in Europe.	Journal of Transport Geography	UK	Temporal and Spatial	Finding presents results from an evaluation of DRT in sites across Europe	Identifies transportation telematics as a way to improve the efficiency DRT systems.
Brake, J., Nelson, J. D., and Wright, S.	2004	Demand responsive transport: towards the emergence of a new market segment.	Journal of Transport Geography	UK	Temporal and Spatial	Findings highlight issues relating to the development of DRT.	Describes how DRT services can be enhanced by ITS to better deal with high demand, route planning, and integration.
Brake, J., and Nelson, J. D.	2007	A case study of flexible solutions to transport demand in a deregulated environment.	Journal of Transport Geography	UK	Temporal and Spatial	Findings demonstrate the evolution of public transport in the case study area and highlights the potential for better integration if deregulated	Provides insights into the use of the DRT to fill gaps in public transport networks particularly in dispersed areas.
Ferreira, L., Charles, P., and Tether, C.	2007	Evaluating Flexible Transport Solutions.	Transportation Planning and Technology	Australia	Temporal and Spatial	Findings report on a recent study of the use of FTS in Brisbane, QLD.	Describes how FTS have the potential to increase public transport use by providing a more customer centric and adaptive solution to the first-last-mile problem.
Hensher, D. A.	2007	Some Insights into the Key Influences on Trip-Chaining Activity and Public Transport Use of Seniors and the Elderly.	International Journal of Sustainable Transportation	Australia	Psychological and Information	Findings show that "the loss of a driver's license and a partner have the potential to be major contributors to social isolation in the absence of inadequate flexible public transport and or support mechanisms that enable access to the car as a passenger."	Identifies the potential for ATIS and ITS signs to provide dynamic information targeted directly to elderly drivers, such as avoiding challenging routes.
Zografos, K., Androutsopoulos, K., and Sihvola, T.	2008	A methodological approach for developing and assessing business models for flexible transport systems.	Transportation	Greece	Physical and Economic	Develops a methodology for the development of flexible transport systems.	Describes how FTS allow flexibility in assigning routes, schedule, vehicles, and ticketing systems making them more responsive to local needs.
Battellino, H.	2009	Transport for the transport disadvantaged.	Transport Policy	Australia	Temporal and Spatial	Findings brings attention to the potential for scheduled transport services to fulfil transport needs in rural communities	Describes how DRT could be used to enhance the availability and scope of community transport to better service residents who need it most.

Table A1. *Cont.*

Author	Year	Title	Journal	Country	Category	Findings	Relevance
Mulley, C., and Nelson, J. D.	2009	Flexible transport services.	Research in Transportation Economics	Australia	Temporal and Spatial	Findings that if implemented correctly flexible transport systems have the potential to improve bus services	Describes how the use of ITS to better match supply and demand can improve the efficiency of DRT and lead to lower costs more reflective of PT.
Nelson, J. D., Wright, S., Masson, B., Ambrosino, G., and Naniopoulos, A.	2010	Recent developments in Flexible Transport Services.	Research in Transportation Economics	UK	Physical and Economic	Proposes the introduction of an organizational structure (FAMS) to help with the introduction of flexible transport services	Discusses the use of FTS to compliment conventional public transport by responding directly to user-demand.
Santos, G., Behrendt, H., and Teytelboym, A.	2010	Part II: Policy instruments for sustainable road transport.	Research in Transportation Economics	UK	Temporal and Spatial	An analysis of policies related to sustainable road transport which fall into three categories: physical, soft, and knowledge policies.	Discusses how subsidized demand responsive taxis could replace conventional public transport in rural and low-density areas.
O'Shaughnessy, M., Casey, E., and Enright, P.	2011	Rural transport in peripheral rural areas.	Social Enterprise Journal	Republic of Ireland	Physical and Economic	Findings show that users of DRT are typically long-term residents, female, elderly, and those who live alone in isolated areas.	Describes how DRT services have helped increase independence, reduced feelings of social isolation and improve access to services for residents in rural areas without access to a PMV.
Broome, K., Worrall, L., Fleming, J., and Boldy, D.	2012	Evaluation of flexible route bus transport for older people.	Transport Policy	Australia	Physical and Economic	Findings show that when replacing fixed route with flexible service in Australia the use by older people almost doubled.	Provides insights into how DRT has improved accessibility and social inclusion for elderly.
Lucas, K., and Currie, G.	2012	Developing socially inclusive transportation policy.	Transportation	UK	Temporal and Spatial	Findings identify that there are important differences between transport disadvantage in low income populations in UK and Australia.	Identifies that more flexible routes and timetabling is required to meet the needs of TDA.
Nelson, J. D., and Phonphitakchai, T.	2012	An evaluation of the user characteristics of an open access DRT service.	Research in Transportation Economics	UK	Psychological and Information	Results reveal that DRT system can improve accessibility, particularly for older residents.	Describes how users are more satisfied with DRT over conventional PT, particularly regarding safety of door-to-door services at night.
Shergold, I., and Parkhurst, G.	2012	Transport-related social exclusion amongst older people in rural Southwest England and Wales.	Journal of Rural Studies	UK	Psychological and Information	Findings reveal that the availability of private vehicles "is not a strong indicator of overall location, although non-availability was important in limiting access to particular types of location."	Describes issues relating to DRT systems reliability and its perception as being for old people.

Table A1. *Cont.*

Author	Year	Title	Journal	Country	Category	Findings	Relevance
Velaga, N. R., Beecroft, M., Nelson, J. D., Corsar, D., and Edwards, P.	2012	Transport poverty meets the digital divide: accessibility and connectivity in rural communities.	Journal of Transport Geography	UK	Psychological and Information	Concludes that the provision of adequate transportation services to rural communities presents significant challenges because of issues relating to transport poverty and the digital divide.	Describes the use of GPS, ICT and other technology to enhance DRT.
Wells, P	2012	Converging transport policy, industrial policy and environmental policy.	Local Economy	UK	Temporal and Spatial	The "article identifies inter- and intra-regional dimensions of inequality that are emerging around the convergence of transport policy, industrial policy and environmental policy."	Provides insights into social equity issues surrounding future EM introduction.
Newman, D.	2013	Cars and consumption	Capital and Class	UK	Physical and Economic	Concludes that electric vehicle will never be the ideal solution to promoting sustainable transport systems if they are used to promote increased consumption.	Discusses how electric vehicles are significantly more expensive than traditional vehicles.
Ward, M. R. M., Somerville, P., and Bosworth, G.	2013	Now without my car I don't know what I'd do'	Local Economy	UK	Psychological and Information	Findings show that, "while community transport services play a vital role in rural communities, many older people are confused or unclear about what these services do, how they can be used, and how to access them"	Describes how older people often confused and unclear about how DRT could serve them and that more information could help.
Akgöl, K., and Günay, B.	2014	Prediction of Modal Shift Using Artificial Neural Networks.	TEM Journal	Turkey	Temporal and Spatial	Findings reveal the potential of applying machine learning to calculate modal shift.	Provides insights into modal shift away from PMV when AI is used to optimize route planning.
Harrison, G., and Shepherd, S	2014	An interdisciplinary study to explore impacts from policies for the introduction of low carbon vehicles.	Transportation Planning and Technology	UK	Physical and Economic	Establishes an ethical framework to "balance obligations to reduced greenhouse gases (GHG) emissions and rights to car ownership"	Provides insights into the importance of having access to a PMV.
Martinez, L. M., Viegas, J. M., and Eiro, T.	2014	Formulating a New Express Minibus Service Design Problem as a Clustering Problem.	Transportation Science	Portugal	Temporal and Spatial	"Presents a novel and simple modelling approach to design innovative transportation services, such as the express minibus service."	Describes how future DRT models can be developed with real-time booking systems allow real-time routing changes to ensure the most direct routes for customers.

Table A1. *Cont.*

Author	Year	Title	Journal	Country	Category	Findings	Relevance
Stanley, J., and Lucas, K.	2014	Workshop 6 Report: Delivering sustainable public transport.	Research in Transportation Economics	Australia	Physical and Economic	Develops "a set of general principles intended to further promote sustainable public transport."	Provides insights into how technology can improve DRT systems.
Wang, C., Quddus, M., Enoch, M., Ryley, T., and Davison, L.	2014	Multilevel modelling of Demand Responsive Transport (DRT) trips in Greater Manchester based on area-wide socio-economic data.	Transportation	UK	Psychological and Information	Findings show that "the demand for DRT services was higher in areas with low car ownership, low population density, high proportion of white people, and high levels of social deprivation, measured in terms of income, employment, education, housing and services, health and disability, and living environment."	Describes how DRT appeals to areas of low population density, low car ownership and high levels of social deprivation. Explains that the perception of DRT is that it is safer than other forms of road transport.
Beecroft, M., and Pangbourne, K.	2015	Future prospects for personal security in travel by public transport.	Transportation Planning and Technology	UK	Psychological and Information	Develops "a set of policy recommendations, operator, and business opportunities, knowledge gaps and research priorities to support and enhance provision for personal security in travel by public transport."	Provides insights into the use of technology to improve safety and security on shared mobility services.
Bigerna and Polinori	2015	Willingness to Pay and Public Acceptance for Hydrogen Buses.	Sustainability	Italy	Physical and Economic	"The results confirm that residents in Perugia are willing to pay extra to support the introduction of H2B."	Provides evidence that users are willing to pay more for sustainable public transport (Hydrogen Buses)
Evans, G., Guo, A. W., Blythe, P., and Burden, M.	2015	Integrated smartcard solutions.	Transportation Planning and Technology	UK	Physical and Economic	"Findings suggest there is potential for an integrated TranCit card, facilitating easier access to services and travel options across boundaries, even at the international level."	Describes how data obtained from smart card ticketing systems can be used to better understand the behavior of passenger and improve services.
Gomes, R., Pinho de Sousa, J., and Galvão Dias, T.	2015	Sustainable Demand Responsive Transportation systems in a context of austerity.	Research in Transportation Economics	Portugal	Temporal and Spatial	Finding show that "service design is critical" to ensure DRT services "answer sustainability and social inclusion challenges" while keeping costs low.	Discusses how DRT services can reduce costs and improve efficiency of transport network by using fewer, smaller vehicles, incorporating dynamic route planning and passenger allocation, and reducing instances of dead-runs.

Appl. Sci. **2020**, *10*, 6306

Table A1. *Cont.*

Author	Year	Title	Journal	Country	Category	Findings	Relevance
Grieco, M.	2015	Social sustainability and urban mobility.	Social Responsibility Journal	UK	Temporal and Spatial	Findings show that "databases and methodologies around social sustainability have not been sufficiently developed to permit ready operationalisation" of advances in urban mobility.	Describes how intelligent data collection can help local authorities identify where transport services are required in order to reduce social inequalities resulting from physical or geographic conditions.
Haustein, S., and Siren, A.	2015	Older People's Mobility.	Transport Reviews	Denmark	Physical and Economic	Propose a hypothetical model based on the findings from a systematic comparison study. The modal "integrates the most relevant determinants of older people's mobility patterns and their interrelations.	Discusses how e-mobility, in-car assistance technology, and AVs will offer good opportunities for the elderly to remain mobile for longer.
Kammerlander, M., Schanes, K., Hartwig, F., Jäger, J., Omann, I., and O'Keeffe, M.	2015	A resource-efficient and sufficient future mobility system for improved well-being in Europe.	Journal of Futures Research	Austria	Temporal and Spatial	Findings show that to achieve the vision of resource efficiency in the transport section a new way of thinking about mobility is required. "It is not about travelling fastest and frequently, but unhurried, infrequently, and sustainably."	Describes how DRT, and door-to-door transport, can reduce demand for PMV travel by providing a viable option in low populated regions.
Mackett, R.	2015	Improving accessibility for older people—Investing in a valuable asset.	Journal of Transport and Health	UK	Physical and Economic	Findings show that the travel patterns of older people are consistent with the assumption that they contribute to society economically, by frequenting local shops and through volunteer work and childcare. The barriers older people face with regards to transportation may hinder an even greater contribution.	Describes how the potential ability for AVs to drive without human input means degenerative disabilities will no longer inhibit elderly and disabled individuals.
Thomopoulos, N., and Givoni, M.	2015	The autonomous car—a blessing or a curse for the future of low carbon mobility?	European Journal of Futures Research	UK	Physical and Economic	Concludes that the introduction of AV is only likely to create a more desirable transport system if it is accompanied by social change.	Provides insights into how AV could eliminate transport related exclusion.
Cheyne, C., and Imran, M.	2016	Shared transport.	Energy Research and Social Science	New Zealand	Physical and Economic	Findings from a survey, focus group, and analysis of census data in New Zealand highlight "a growing need for alternatives to private transport for residents of small towns."	Provides insights into the important of shared mobility being flexible in order to accommodate individual needs.

Table A1. *Cont.*

Author	Year	Title	Journal	Country	Category	Findings	Relevance
Clark, J., and Curl, A.	2016	Bicycle and Car Share Schemes as Inclusive Modes of Travel?	Social Inclusion	UK	Physical and Economic	"Argues that there is a need to consider the social inclusivity of sharing schemes and to develop appropriate evaluation frameworks accordingly."	Describes how shared bicycle and car schemes can remove economic barriers associated with owning your own car and how those with disabilities are unlikely to benefit from car and bike schemes alone.
Davidsson, P., Hajinasab, B., Holmgren, J., Jevinger, Å., and Persson, J.	2016	The Fourth Wave of Digitalization and Public Transport.	Sustainability	Sweden	Psychological and Information	Concludes that for transport operators, planners, and users to take advantage of the opportunities related to IoT and its impact on public transport a number of technical and non-technical challenges need to be addressed.	Discusses how smart mobility enabled by IoT could improve data collection and contribute to providing accurate, real-time information about vehicles, users, traffic, and air quality.
Leibert, T., and Golinski, S.	2016	Peripheralisation.	Comparative Population Studies	Germany	Temporal and Spatial	Findings argue "that the peripheralisation approach is a helpful tool to better understand how interaction of out-migration, dependence, disconnection, and stigmatisation shape the future of rural regions."	Describes there is a need for local specific strategies to address how DRT can be used more efficiently and equitably.
Petersen, T.	2016	Watching the Swiss.	Transport Policy	Australia	Temporal and Spatial	Analyses the characteristics of public timetable networks in the contest of rural transportation in Switzerland. Findings identify lessons for their potential application in other locations.	Discusses how first-mile, last-mile transport networks may provide better coverage and be more efficient they are timed to connect to public transport hubs to ensure reductions in transfer times.
Chen, Y., Ardila-Gomez, A., and Frame, G.	2017	Achieving energy savings by intelligent transportation systems investments in the context of smart cities.	Transportation Research Part D	USA	Psychological and Information	Findings from literature review, case studies and interviews has "found that the smart cities context has transformed traditional ITS into smart mobility with three major characteristics: people-centre, data-driven, and powered by bottom-up innovations."	Describes how technological improvements in ITS and ICT can facilitate the collection and analysis of data which can be used to improve the efficiency of the transportation system.
McLeod, S., Scheurer, J., and Curtis, C.	2017	Urban Public Transport.	Journal of Planning Literature	Australia	Temporal and Spatial	The "findings of this systematic review support the paradigm of PT oriented urban mobility and provide an optimistic insight into the future of sustainable travel in cities."	Discusses how DRT enabled by autonomous technology and shared mobility has the potential to increase the catchment of traditional public transport systems.

Table A1. *Cont.*

Author	Year	Title	Journal	Country	Category	Findings	Relevance
Melo, S., Macedo, J., and Baptista, P.	2017	Guiding cities to pursue a smart mobility paradigm.	Research in Transportation Economics	Portugal	Psychological and Information	Results show that traffic management systems that re-route can reduce travel times and enhance the efficiency of roads.	Describes how traffic management systems enabled by advances in ICT and data collection provide guidance information to drivers to assist route planning.
Milakis, D., van Arem, B., and van Wee, B.	2017	Policy and society related implications of automated driving.	Journal of Intelligent Transportation Systems	The Netherlands	Physical and Economic	Findings show that "first-order impacts (of autonomous vehicles) on road capacity, fuel efficiency, emissions, and accidents risk are expected to be beneficial. The magnitude of these benefits will likely increase with the level of automation and cooperation and with the penetration rate of these systems."	Describes how AV provide another transportation option for those unable to drive a PMV.
Newman, D.	2017	Automobiles and socioeconomic sustainability	Transfers	UK	Physical and Economic	Proposes a mobility bill of rights that states: (1) everybody should have access to affordable mobility which meets basic needs; (2) transport should not harm us or the environment; (3) transport should not threaten health, safety or the environment; (4) transport pricing should not penalize those who use it less; (5) transport should be accessibly so we are not excluded from society; (6) we should not have to rely on private vehicles for our travel; (7) everyone should have access to a public transport system; and (8) transport should not contribute to depletion of natural resource	Discusses how electric vehicles offer very little to overcome transport related social exclusion.
Sun, Y., Olaru, D., Smith, B., Greaves, S., and Collins, A.	2017	Road to autonomous vehicles in Australia.	Road and Transport Research	Australia	Physical and Economic	Findings identify a number of key issues associated with the introduction of AV in Australia.	Discusses how increased VOT in AV could have a positive impact on those impacted by geographic-related TDA and there is potential for SAV to significantly reduce the costs of DRT.

Table A1. *Cont.*

Author	Year	Title	Journal	Country	Category	Findings	Relevance
Adnan, N., Md Nordin, S., Bin Bahruddin, M. A., and Ali, M.	2018	How trust can drive forward the user acceptance to the technology?	Transportation Research Part A	Malaysia	Physical and Economic	Findings show "that the level of trust, which may vary on the sociodemographic profile of the users, has been studied as one of the factors for user acceptance."	Describes AV could improve accessibility for elderly and disabled including the potential for increased social interactions, greater connection to employment and health services, improved comfort, and increased VOT.
Docherty, I., Marsden, G., and Anable, J.	2018	The governance of smart mobility.	Transportation Research Part A	UK	Physical and Economic	Identifies public value as the key governance aim that should be implemented for the transition to smart mobility.	Describes how MaaS can facilitate the integration of a wide range of mobility providers and help strengthen the efficiency of public transport and DRT.
Gonzalez-Feliu, J., Pronello, C., and Salanova Grau, J.	2018	Multi-stakeholder collaboration in urban transport.	Transport	France	Physical and Economic	Provides an analysis and overview of a set of papers which focus on "the field of multi-stakeholder and collaboration in urban transport"	Discusses how an integrated system such as MaaS can facilitate multi-stakeholder collaboration, and the sharing of information and resources.
Graham, H., de Bell, S., Flemming, K., Sowden, A., White, P., and Wright, K.	2018	The experiences of everyday travel for older people in rural areas.	Journal of Transport and Health	UK	Psychological and Information	Identifies three themes related to older people and their experiences of everyday travel: (a) experience with inadequate transport system; (b) importance of everyday travel to maintain lives; and (c) the symbolic importance of travel.	Describes how DRT and community transport is often stigmatized within the community and there is confusion about how to, and who can access it.
Guo et al.	2018	Impacts of internal migration, household registration system, and family planning policy on travel mode choice in China.	Travel, Behavior and Society	USA	Institutional	Findings "suggest that—among other factors—continuing internal migration, relaxation of household registration system, and changes in family planning policy, are likely to affect travel mode choices."	Provides insights into the impact of policy and laws on travel mode choice.
Hopkins and Schwanen	2018	Automated Mobility Transitions.	Sustainability	UK	Institutional	Results "suggest that the UK has adopted a reasonably comprehensive approach to the governing of automated vehicle innovation but that this approach cannot be characterized as sufficiently inclusive, democratic, diverse and open."	Discusses importance of including general public in decision making

Table A1. *Cont.*

Author	Year	Title	Journal	Country	Category	Findings	Relevance
Howard, A., and Borenstein, J.	2018	The ugly truth about ourselves and our robot creations.	Science and Engineering Ethics	USA	Psychological and Information	Concludes that a range of measures should be taken to ensure bias is removed or mitigated from robotic technology - including self-driving vehicles.	Discusses how AVs will have to make decisions based on a range of alternative options and are therefore at risk of bias.
Illgen, S., and Höck, M.	2018	Establishing car sharing services in rural areas.	Transportation	Germany	Temporal and Spatial	"Findings indicate a certain feasibility of rural car sharing development, while highlighting the positive effect it could have on car sharing demand in urban areas."	Provides insights into how ride sharing can contribute to further TDA.
Jin, S. T., Kong, H., Wu, R., and Sui, D. Z.	2018	Ridesourcing, the sharing economy, and the future of cities.	Cities	USA	Psychological and Information	Findings describe how it is unlikely that ride sharing will reduce car ownership.	Describes how shared mobility and AVs can help smaller communities that do not have access to public transport by providing more options, more frequently.
Lam, D., and Givens, J.W.	2018	Small and smart.	New Global Studies	USA	Temporal and Spatial	Using South Bend, Indiana as an example the study looks at the potential for smart cities in smaller communities.	Discusses the use of free-floating bike sharing for first- and last-mile connection to PT.
Li, X., Zhang, Y., Sun, L., and Liu, Q.	2018	Free-floating bike sharing in Jiangsu.	Energies	Singapore	Temporal and Spatial	Findings show that: (a) bike sharing was mainly used for travelling short distances; (b) lower costs, more education, and promotion of health benefits could be used to promote bike sharing; and (c) bike sharing is more attractive to higher income residents.	Describes how important the perception of safety is to ensure successful operation and use of AV.
Lim, H. S. M., and Taeihagh, A.	2018	Autonomous vehicles for smart and sustainable cities.	Energies	Singapore	Psychological and Information	Findings describe how addressing privacy and cybersecurity related to AV is crucial to the development of smart and sustainable cities.	Describes how AV may lead to increased suburbanization or density.
Milakis, D., Kroesen, M., and van Wee, B.	2018	Implications of automated vehicles for accessibility and location choices.	Journal of Transport Geography	The Netherlands	Physical and Economic	Findings from Q-method study showed that experts expect AV to influence accessibility through all four level (land use, transport, temporal and individual)	Discusses advantage of a MaaS system would be that unlike existing DRT services the subsidized provision of transport may not be seen as an entitlement but instead be part of a larger system that benefits all.

Table A1. *Cont.*

Author	Year	Title	Journal	Country	Category	Findings	Relevance
Mulley, C., and Kronsell, A.	2018	The "uberisation" of public transport and mobility as a service (MaaS).	Research in Transportation Economics	Australia	Psychological and Information	Findings of workshop discussion show a difference between policy and mobility provider views, a need for flexibility, the importance of collaboration, and a need to address user safety.	Discusses how MaaS provides an opportunity to cross subsidize which could improve the transportation for disadvantaged groups (e.g., aged, disabled, and rural areas).
Mulley, C., Nelson, J. D., and Wright, S.	2018	Community transport meets mobility as a service.	Research in Transportation Economics	Australia	Temporal and Spatial	Findings show that CT operators in Australia are very enthusiastic about the potential for MaaS to offer mobility packages to services their users.	Discusses how AVs with the absence of strict policy measures could result in more demand for car ownership and miles travelled.
Noy, K., and Givoni, M.	2018	Is "Smart Mobility" Sustainable?	Sustainability	Israel	Physical and Economic	Findings from a survey of 117 entrepreneurs "shows that there is a mismatch between interpretation and understanding of what is 'smart' and what is 'sustainable'."	Discusses that for shared mobility to help achieve sustainable mobility objectives it is important to identify how existing public transport and shared mobility can be synergized to make them complementary and benefit the transport system as a whole.
Soares Machado, C., de Salles Hue, N. P. M., Berssaneti, F. T., and Quintanilha, J. A.	2018	An Overview of Shared Mobility.	Sustainability	Brazil	Psychological and Information	Findings determine that based on literature review the introduction of shared modes alone "will not solve transportation problems in large cities."	Discusses how people with visual impairment would greatly benefit if existing door-to-door transportation services were improved.
Wong, S.	2018	Traveling with blindness.	Health and Place	USA	Physical and Economic	Findings show "space-time constraints of people with visual impairments are closed linked to their access to transportation, assistive technologies, and mobile devices."	Describes how providing transportation for low density dispersed neighborhoods is challenging due to dispersal of individuals and destinations.
Allen, J., and Farber, S.	2019	Sizing up transport poverty.	Transport Policy	Canada	Temporal and Spatial	Recommends that future investments in major transportation infrastructure should be focused in areas with high density of low-income households and low levels of accessibility. In areas of low density, subsidized ride sharing and DRT should be considered.	Describes how SAV could reduce total cost of ownership by over 80% per km travelled compared to a conventional car.

Table A1. *Cont.*

Author	Year	Title	Journal	Country	Category	Findings	Relevance
Axsen, J., and Sovacool, B. K.	2019	The roles of users in electric, shared and automated mobility transitions.	Transportation Research Part D	Canada	Physical and Economic	Findings summarize "characteristics of early users, as well as practical insights for strategies and policies seeking societally-beneficial outcomes from mass deployment of" transport innovations.	Describes how FTS can contribute to rural connectivity by providing door-to-door transport that does not rely on fixed routes and how MaaS is promising because it provides integrated customer experiencing linking users with a range of transport options on demand.
Beecroft, M., Cottrill, C. D., Farrington, J. H., Nelson, J. D., and Niewiadomski, P.	2019	From infrastructure to digital networks.	Scottish Geographical Journal	UK	Psychological and Information	Identifies connectivity as a central theme when looking at the development and evolution of transport geography research at the University of Aberdeen.	Provides insights into attitudes towards AV among those with intellectual disability.
Bennett, R., Vijaygopal, R., and Kottasz, R.	2019	Willingness of people with mental health disabilities to travel in driverless vehicles.	Journal of Transport and Health	UK	Psychological and Information	Findings show "three categories of attitude towards AVs arose from the STM; respectively involving freedom, fear and curiosity."	Discusses how public transport providers should look to integrated systems such as MaaS which can help with sharing of data, identification of demand, and connect potential users with the most suitable providers.
Bennett, R., Vijaygopal, R., and Kottasz, R.	2019	Attitudes towards autonomous vehicles among people with physical disabilities.	Transportation Research Part A	UK	Psychological and Information	Findings show that "attitudes towards AVs among people with disabilities were significantly influenced by their levels of interest in new technology, generalized anxiety, intensity of a person's disability, prior knowledge of AVs, locus of control and action orientation."	Provides insights into how AV could perpetuate or create new social inequalities.
Canitez, F.	2019	Pathways to sustainable urban mobility in developing megacities.	Technological Forecasting and Social Change	Turkey	Physical and Economic	Findings "proposes a socio-technical transition perspective to examine and analyze the urban mobility systems in developing megacities. In addition, a multi-level perspective is offered to understand the dynamics of sustainable urban mobility transitions"	Provides insights into attitudes towards AV among those with intellectual disability and those without.
Creitzig et al.	2019	Leveraging digitalization for sustainability in urban transport.	Global Sustainability	Germany	Institutional	Concludes that "only strong public policies can steer digitalization towards fostering sustainability in urban transport."	Provides insights into the importance of policy in smart technology development.

Table A1. *Cont.*

Author	Year	Title	Journal	Country	Category	Findings	Relevance
Curtis, C., Stone, J., Legacy, C., and Ashmore, D.	2019	Governance of future urban mobility.	Urban Policy and Research	Australia	Physical and Economic	"Findings from industry engagement workshop highlight the complexity of issues and questions surrounding MaaS implementation."	Discusses how the change from a PMV to more sustainable system is reliant on the cost of transport, regulations, planning, land use, technology, public awareness and culture.
Dean, J., Wray, A., Braun, L., Casello, J., McCallum, L., and Gower, S.	2019	Holding the keys to health?	BMC Public Health	Canada	Psychological and Information	Findings show "there is general agreement that AVs will improve road safety overall, thus reducing injuries and fatalities from human errors in operating motorized vehicles. However, the relationships with air quality, physical activity, and stress, among other health factors may be more complex."	Discusses how when MaaS is implemented with AV there is a risk that those currently able to operate a PMV will have access to private AV.
Freemark, Y., Hudson, A., and Zhao, J.	2019	Are cities prepared for autonomous vehicles?	Journal of the American Planning Association	USA	Physical and Economic	Findings show that: (1) planning for AV is not widespread; (2) bigger cities are more likely to have started planning for AV; (3) there is optimism among local officials regarding the potential increase in safety, and decrease in costs and pollution associated with AV; and (4) over one-third of local officials are concerned about the impact AV will have on VKT and public transport ridership.	Discuss how AVs can contribute to the improved safety but increased reliance on technology opens up additional risks.
Goggin et al.	2019	Disability at the centre of digital inclusion.	Communication Research and Practice,	Australia	Psychological and Information	Concludes "that 'disability and digital inclusion' should be specifically also placed at the heart of digital economy policy and plans"	Provides insights into digital divide and disability
Groth, S.	2019	Multimodal divide: Reproduction of transport poverty in smart mobility trends.	Transportation Research Part A	Germany	Psychological and Information	Findings show that smart mobility can potentially contribute to transport poverty by: (a) providing an unequal distribution of mode options; (b) excluding those who are unable to use technology; and (c) excluding those who are unwilling to us technology due to privacy concerns.	Discusses how improved costs and accessibility associated with AV may reduce public transport use.

Table A1. *Cont.*

Author	Year	Title	Journal	Country	Category	Findings	Relevance
Hawkins, J., and Habib, K. N.	2019	Heterogeneity in marginal value of urban mobility.	Transportation	Canada	Temporal and Spatial	"Findings reveal the potential for social exclusion follow the adoption of MaaS."	Discuss how smart mobility may contribute to social exclusion.
Jokinen, J.-P., Sihvola, T., and Mladenovic, M. N.	2019	Policy lessons from the flexible transport service pilot Kutsuplus in the Helsinki Capital Region.	Transport Policy	Finland	Temporal and Spatial	Findings "provide a range of guidelines and lessons for future urban FMTS"	Discusses how MaaS would likely result in higher transport costs for those in low density areas and governments will likely need to provide subsidies and planning interventions in order to ensure equitable access to transport.
Kandt, J., and Leak, A.	2019	Examining inclusive mobility through smartcard data.	Journal of Transport Geography	UK	Temporal and Spatial	Findings show "first, the decline in patronage occurs in three waves across the study period according to distinct activity patterns; second, formerly frequent (daily) passengers tend to abandon the bus and thus show the largest impact on the overall trend; third, the neighbourhood context of withdrawing passengers indicates social disadvantage, higher instance of ethnic minorities and lower car ownership rates, in other words higher risk of social exclusion."	Provides insights into the importance of DRT to direct users to major transport hubs, how users are more sensitive to travel time than cost, and how technology can improve efficiency of DRT.
Kuzio, J.	2019	Planning for Social Equity and Emerging Technologies.	Transportation Research Record	USA	Physical and Economic	Findings show that 80% of metropolitan planning organizations have plans that included a response to social equity, however, only 20% of plans considered how new technologies would impact on social equity.	Describes how MaaS can contribute to improved public transport use in low density, less accessibility areas by providing a connection to major public transport hubs, and creating synergies between a range of transport options.
Le Boennec, R., Nicolaï, I., and Da Costa, P.	2019	Assessing 50 innovative mobility offers in low-density areas.	Transport Policy	France	Temporal and Spatial	Develops a two-step decision-making tool to assist local governments with planning and implementing transportation policies.	Describes how the benefits of AV including the ability to work, eat, and rest while in transit could be highly desirable and result in increased cost, VKT, and demand on infrastructure.

Table A1. *Cont.*

Author	Year	Title	Journal	Country	Category	Findings	Relevance
Legacy, C., Ashmore, D., Scheurer, J., Stone, J., and Curtis, C.,	2019	Planning the driverless city.	Transport Reviews	Australia	Physical and Economic	Findings reveal "the conceptual gaps in the framing of AV technology—the prospects and limits—and how these are conceived"	Discusses how despite new technology encouraging walking is still key to reducing social exclusion as walking is costless, and contributes to improved health.
Liu, C., Yu, B., Zhu, Y., Liu, L., and Li, P.	2019	Measurement of rural residents' mobility in western China.	Sustainability	China	Temporal and Spatial	Findings "show that Qingyang's rural mobility is at a low level, but differences in the types of rural residents, districts and counties, and dimensions of mobility are observed."	Discusses how for AV to reduce some of the social equity issues associated with the PMV it should be introduced as part of a system integrated with existing transport providers.
Martin, G.	2019	An Ecosocial Frame for Autonomous Vehicles.	Capitalism Nature Socialism	USA	Physical and Economic	"As is usually the case with a new technological consumer product, discourse centers on its promises, not its perils. Largely ignored are potential impacts on social justice and environmental sustainability."	Provides insights into how information availability in lower density areas is critical for the success of many smart mobility applications.
Martínez-Díaz, M., Soriguera, F., and Pérez, I.	2019	Autonomous driving: a bird's eye view.	IET Intelligent Transport Systems	Spain	Psychological and Information	Provides an overall state-of-the-art of the development of AV and identifies the issues critical for its success.	Discusses risk that AV will be expensive and exclude low income residents and how subsidized SAV may provide a viable alternative.
Meelen, T., Frenken, K., and Hobrink, S.	2019	Weak spots for car-sharing in The Netherlands?	Energy Research and Social Science	The Netherlands	Physical and Economic	Findings "demonstrate how the relation between niche innovation and the socio-technical regime of private car ownership affects adoption patterns."	Discusses how while AV are expected to bring a much safer driving environment, acceptability among people over 50 is still quite low.

Table A1. *Cont.*

Author	Year	Title	Journal	Country	Category	Findings	Relevance
Nordhoff, S., Kyriakidis, M., van Arem, B., and Happee, R.	2019	A multi-level model on automated vehicle acceptance (MAVA).	Theoretical Issues in Ergonomics Science	Switzerland	Psychological and Information	Findings reveal "that 6% of the studies investigated the exposure of individuals to AVs (i.e., knowledge and experience). 22% of the studies investigated domain-specific factors (i.e., performance and effort expectancy, safety, facilitating conditions, and service and vehicle characteristics), 4% symbolic-affective factors (i.e., hedonic motivation and social influence), and 12% moral-normative factors (i.e., perceived benefits and risks). Factors related to a person's socio-demographic profile, travel behavior and personality were investigated by 28%, 15% and 14% of the studies, respectively. "	Discusses how car sharing contributes to improved accessibility for those without access to a vehicle—satisfying basic needs relating to transportation.
Ortar, N., and Ryghaug, M.	2019	Should all cars be electric by 2025?	Sustainability	UK	Temporal and Spatial	Findings show that there is much uncertainty regarding how the transition between fuel-based and electric vehicles occurs including issues of efficiency, affordability and sustainability	Discusses how the general public perceives AV and identifies risks they associate with the technology.
Ruan et al.	2019	Social adaptation and adaptation pressure among the "drifting elderly" in China.	International Journal of Health Planning and Management,	China	Institutional	"The drifting elderly had poor adaptation regarding self-identity, daily activities, and social context."	Discusses issues of digital divide among elderly migrants
Sener, I. N., Zmud, J., and Williams, T.	2019	Measures of baseline intent to use automated vehicles	Transportation Research Part F	USA	Psychological and Information	Findings show "the strongest associations with intent to use (AVs) were observed for attitudes toward self-driving vehicles, performance expectation, perceived safety, and social influence."	Describes social equity issues surrounding the introduction of EM.
Sovacool, B., Martiskainen, M., Hook, A., and Baker, L.	2019	Decarbonization and its discontents.	Climatic Change	UK	Temporal and Spatial	Develops a framework for "energy justice" with four distinct dimensions: (a) distributive justice; (b) procedural justice; (c) cosmopolitan justice; and (d) recognition justice.	Discusses how AV bring a significant, and uncertain impact on the transport system.

Table A1. *Cont.*

Author	Year	Title	Journal	Country	Category	Findings	Relevance
Tang, C. S., and Veelenturf, L. P.	2019	The strategic role of logistics in the industry 4.0 era	Transportation Research Part E	USA	Psychological and Information	Concludes that companies must take measures to ensure underlying risks associated with technological advancements including: (a) cyber-attacks; (b) faulty data; (c) safety regulations; and (d) privacy.	Discusses how those in rural, or low density, sparsely populated areas may not be able to fully benefit from smart technology due to network coverage, and electricity prices.
Viergutz, K., and Schmidt, C.	2019	Demand responsive - vs. conventional public transportation.	Procedia Computer Science	Germany	Temporal and Spatial	Findings show that DRT services may not be the solution to public transport in rural areas and further research is need to balance access, financial, service, and pollution issues associated with DRT.	Discusses risks associated with AV and ITS.
Waseem et al.	2019	Integration of solar energy in electrical, hybrid, autonomous vehicle	SN Applied Science	India	Physical and Economic	"Overview of electric and hybrid vehicles suggests that in a developing country such as India, there is a huge demand for green-powered electric vehicles for the transportation sector."	Discusses demand for green-powered vehicles.
Yigitcanlar, T., Han, H., Kamruzzaman, M., Ioppolo, G., and Sabatini-Marques, J.	2019	The making of smart cities.	Land Use Policy	Australia	Psychological and Information	Findings "disclose the need for a comprehensive smart city conceptualization to inform policymaking and consequently the practice."	Discusses how AV could assist the development of DRT in low demand areas, reducing costs, and making the system operate more efficiently.
Zhou, J.	2019	Ride-sharing service planning based on smartcard data	Transport Policy	China	Physical and Economic	Findings show "that some low-demand transit routes can probably be replaced by Uber at a lower level of overall costs."	Provides insights into cities that use technology enabled smart traffic systems.
Becker et al.	2020	Assessing the welfare impacts of Shared Mobility and Mobility as a Service (MaaS).	Transportation Research Part A	Switzerland	Psychological and Information	"Results show that in Zurich, through less biased mode choice decisions alone, transport-related energy consumption can be reduced by 25%"	Discusses role of MaaS in providing users an unbiased choice of modes.
Bissell, D., Birtchnell, T., Elliott, A., and Hsu, E. L.	2020	Autonomous automobilities.	Current Sociology	Australia	Physical and Economic	Shows "how a mobilities approach provides an ideal conceptual lens through which the broader social impacts of autonomous vehicles might be identified and evaluated."	Discusses social issues surrounding introduction of AV

Table A1. *Cont.*

Author	Year	Title	Journal	Country	Category	Findings	Relevance
Ferdman, A.	2020	Corporate ownership of automated vehicles.	Transport Reviews	Germany	Physical and Economic	"Proposes a new angle on the relationship between ownership models of automated vehicles and implications for travel."	Describe how for shared mobility to work in rural areas a broad base of users is required, large investments from the start, and strong connection with the broader region - particularly cities.
Guo and Peeta	2020	Impacts of personalized accessibility information on residential location choice and travel behavior.	Travel, Behavior and Society	USA	Temporal and Spatial	Results "show that personalized accessibility information can potentially make relocators more informed about travel-related information, and assists them in selecting a residence that better addresses their travel needs based on higher accessibility to potential destinations."	Discusses the use of ICT to help residents make more informed choice.
Guo et al.	2020	Personal and societal impacts of motorcycle ban policy on motorcyclists' home-to-work morning commute in China.	Travel, Behavior and Society	USA	Institutional	"These results suggest that policy and infrastructural support for using public transit, walk, and bike modes, household mobility, and plan to purchase a car were likely to affect the personal and societal impacts of the motorcycle ban policy on travel mode shifts"	Provides insights into the impact of policy and laws on travel mode choice.
Hoque et al.	2020	Life Cycle Sustainability Assessment of Alternative Energy Sources for the Western Australian Transport Sector	Sustainability	Australia	Physical and Economic	"The results show that the environment-friendly and socially sustainable energy options, namely, ethanol-gasoline blend E55, electricity, electricity-E10 hybrid, and hydrogen, would need around 0.02, 0.14, 0.10, and 0.71 AUD/VKT of financial support, respectively, to be comparable to gasoline."	Discusses economic sustainability of alternative fuel vehicles
Liu et al.	2020	A tale of two social groups in Xiamen, China: Trip frequency of migrants and locals and its determinants.	Travel, Behavior and Society	Hong Kong	Institutional	"Highlights the importance of context and population differentiation and calls for more in-depth research on migrants' travel behaviors as well as their determinants."	Provides insights into barriers related to infrastructure provision.

Table A1. *Cont.*

Author	Year	Title	Journal	Country	Category	Findings	Relevance
Meng et al.	2020	Policy implementation of multi-modal (shared) mobility.	Transport Reviews	Australia	Institutional	"Suggests that policy entrepreneurship in collaboration with other partners, policy innovation, and the notions of merit goods and second-best policymaking can enable policy initiatives towards multi-modal shared mobility and provide supporting arguments if policies encounter failures."	Discusses the importance of policy in development of shared mobility
Rojas-Rueda et al.	2020	Autonomous Vehicles and Public Health.	Annual Review of Public Health	USA	Physical and Economic	Provides recommendations for the use of AV to improve public health.	Provides insights into some benefits and issues with AV from a public health perspective.
Soares Machado et al.	2020	Placement of Infrastructure for Urban Electromobility	Sustainability	Brazil	Temporal and Spatial	Results "shows that districts with the largest demand for charging stations are located in the central area, where the population also exhibits the highest purchasing power"	Discusses issues associated with electric mobility in low density and rural areas.
Tao et al.	2020	Investigating the impacts of public transport on job accessibility in Shenzhen, China.	Land Use Policy	China	Physical and Economic	"Highlights land use and transport policy countermeasures to improve job accessibility by public transport."	Discusses how job accessibility is greatly improved if one has access to a PMV
Tomej, K., and Liburd, J. J.	2020	Sustainable accessibility in rural destinations.	Journal of Sustainable Tourism	Austria	Temporal and Spatial	Findings "demonstrates the use of sustainable transport accessibility as a measure for transport evaluation that considers both environmental aspects and social justice framed as sustainable tourism participation for all."	Provides insights into the use of DRT in rural areas with high levels of tourism.
Turoń and Kubik	2020	Economic Aspects of Driving Various Types of Vehicles in Intelligent Urban Transport Systems, Including Car-Sharing Services and Autonomous Vehicles.	Applied Science	Poland	Physical and Economic	"Results indicate the relation of travel parameters (including vehicle type) to the total cost of travel in urban transport systems."	Discusses the economic sustainability of AV and shared transport

References

1. Ingrao, C.; Messineo, A.; Beltramo, R.; Yigitcanlar, T.; Ioppolo, G. How can life cycle thinking support sustainability of buildings? *J. Clean. Prod.* **2018**, *201*, 556–569. [CrossRef]
2. Arbolino, R.; De Simone, L.; Carlucci, F.; Yigitcanlar, T.; Ioppolo, G. Towards a sustainable industrial ecology. *J. Clean. Prod.* **2018**, *178*, 220–236. [CrossRef]
3. Lee, S.H.; Yigitcanlar, T.; Han, J.H.; Leem, Y.T. Ubiquitous urban infrastructure: Infrastructure planning and development in Korea. *Innovation* **2008**, *10*, 282–292. [CrossRef]
4. Yigitcanlar, T.; Mohamed, A.; Kamruzzaman, M.; Piracha, A. Understanding transport-related social exclusion. *Urban Pol. Res.* **2019**, *37*, 97–110. [CrossRef]
5. Currie, G.; Delbosc, A. Modelling the social and psychological impacts of transport disadvantage. *Transportation* **2010**, *37*, 953–966. [CrossRef]
6. Duvarci, Y.; Yigitcanlar, T.; Mizokami, S. Transportation disadvantage impedance indexing. *J. Transp. Geogr.* **2015**, *48*, 61–75. [CrossRef]
7. Yigitcanlar, T.; Dodson, J.; Gleeson, B.; Sipe, N. Travel self-containment in master planned estates. *Urban Pol. Res.* **2007**, *25*, 129–149. [CrossRef]
8. Delbosc, A.; Currie, G. The spatial context of transport disadvantage, social exclusion and well-being. *J. Transp. Geogr.* **2011**, *19*, 1130–1137. [CrossRef]
9. Yigitcanlar, T.; Wilson, M.; Kamruzzaman, M. Disruptive impacts of automated driving systems on the built environment and land use. *JOItmC* **2019**, *5*, 24. [CrossRef]
10. Noy, K.; Givoni, M. Is 'Smart Mobility' Sustainable? *Sustainability* **2018**, *10*, 422. [CrossRef]
11. Yigitcanlar, T.; Kamruzzaman, M. Smart cities and mobility. *J. Urban Technol.* **2019**, *26*, 21–46. [CrossRef]
12. Golbabaei, F.; Yigitcanlar, T.; Bunker, J. The role of shared autonomous vehicle systems in delivering smart urban mobility. *Int. J. Sustain. Transp.* **2020**. [CrossRef]
13. Yigitcanlar, T.; Han, H.; Kamruzzaman, M.; Ioppolo, G.; Sabatini-Marques, J. The making of smart cities. *Land Use Policy* **2019**, *88*, 104187. [CrossRef]
14. Yigitcanlar, T.; Kamruzzaman, M.; Foth, M.; Sabatini-Marques, J.; Da Costa, E.; Ioppolo, G. Can cities become smart without being sustainable? *Sustain. Cities Soc.* **2019**, *45*, 348–365. [CrossRef]
15. Yigitcanlar, T.; Desouza, K.; Butler, L.; Roozkhosh, F. Contributions and risks of artificial intelligence (AI) in building smarter cities. *Energies* **2020**, *13*, 1473. [CrossRef]
16. Yigitcanlar, T.; Butler, L.; Windle, E.; Desouza, K.; Mehmood, R.; Corchado, J. Can building 'artificially intelligent cities' protect humanity from natural disasters, pandemics and other catastrophes? *Sensors* **2020**, *20*, 2988. [CrossRef]
17. Kaufman, B. 1 Million Rides and Counting. The Conversation 2020. Available online: https://theconversation.com/1-million-rides-and-counting-on-demand-services-bring-public-transport-to-the-suburbs-132355 (accessed on 1 May 2020).
18. Mahbub, P.; Goonetilleke, A.; Ayoko, G.A.; Egodawatta, P.; Yigitcanlar, T. Analysis of build-up of heavy metals and volatile organics on urban roads in Gold Coast, Australia. *Water Sci. Technol.* **2011**, *63*, 2077–2085. [CrossRef]
19. Dowling, R. Smart Cities: Does This Mean More Transport Disruption. The Conversation 2016. Available online: https://theconversation.com/smart-cities-does-this-mean-more-transport-disruptions-63638 (accessed on 1 May 2020).
20. Kane, M.; Whitehead, J. How to ride transport disruption. *Aust. Plan.* **2017**, *54*, 177–185. [CrossRef]
21. De Percy, M. Taxi Driver Compensation for Uber Is Unfair and Poorly Implemented. The Conversation 2016. Available online: https://theconversation.com/taxi-driver-compensation-for-uber-is-unfair-and-poorly-implemented-64354 (accessed on 1 May 2020).
22. Munton, J.R. Explainer: What Rights Do Workers Have to Getting Paid in the Gig Economy? The Conversation 2017. Available online: https://theconversation.com/explainer-what-rights-do-workers-have-to-getting-paid-in-the-gig-economy-70281 (accessed on 1 May 2020).
23. Barratt, T.; Veen, A.; Goods, C.; Josserand, E.; Kaine, S. As Yet Another Ridesharing Platform Launches in Australia, How Does This All End? The Conversation 2018. Available online: https://theconversation.com/as-yet-another-ridesharing-platform-launches-in-australia-how-does-this-all-end-98389 (accessed on 1 May 2020).

24. Docherty, I.; Marsden, G.; Anable, J. The governance of smart mobility. *Trans. Res. Part A Policy Pract.* **2018**, *115*, 114–125. [CrossRef]

25. Suhl, K.; Carreno, M. Can transport-related social exclusion be measured? In Proceedings of the 8th International Conference, Vilnius, Lithuania, 19–20 May 2011; pp. 1001–1008.

26. Mageean, J.; Nelson, J.D. The evaluation of demand responsive transport services in Europe. *J. Transp. Geogr.* **2003**, *11*, 255–270. [CrossRef]

27. Brake, J.; Nelson, J.D.; Wright, S. Demand responsive transport. *J. Transp. Geogr.* **2004**, *12*, 323–337. [CrossRef]

28. Battellino, H. Transport for the transport disadvantaged. *Trans. Policy* **2009**, *16*, 123–129. [CrossRef]

29. Nelson, J.D.; Wright, S.; Masson, B.; Ambrosino, G.; Naniopoulos, A. Recent developments in Flexible Transport Services. *Res. Transp. Econ.* **2010**, *29*, 243–248. [CrossRef]

30. Nelson, J.D.; Phonphitakchai, T. An evaluation of the user characteristics of an open access DRT service. *Res. Transp. Econ.* **2012**, *34*, 54–65. [CrossRef]

31. Wang, C.; Quddus, M.; Enoch, M.; Ryley, T.; Davison, L. Multilevel modelling of Demand Responsive Transport (DRT) trips in Greater Manchester based on area-wide socio-economic data. *Transportation* **2014**, *41*, 589–610. [CrossRef]

32. Gomes, R.; Pinho de Sousa, J.; Galvão Dias, T. Sustainable Demand Responsive Transportation systems in a context of austerity. *Res. Transp. Econ.* **2015**, *51*, 94–103. [CrossRef]

33. Viergutz, K.; Schmidt, C. Demand responsive - vs. conventional public transportation. *Procedia Comput. Sci.* **2019**, *151*, 69–76. [CrossRef]

34. Clark, J.; Curl, A. Bicycle and car share schemes as inclusive modes of travel? *Soc. Incl.* **2016**, *4*, 83–99. [CrossRef]

35. Illgen, S.; Höck, M. Establishing car sharing services in rural areas. *Transportation* **2020**, *47*, 811–826. [CrossRef]

36. Jin, S.T.; Kong, H.; Wu, R.; Sui, D.Z. Ridesourcing, the sharing economy, and the future of cities. *Cities* **2018**, *76*, 96–104. [CrossRef]

37. Li, X.; Zhang, Y.; Sun, L.; Liu, Q. Free-floating bike sharing in Jiangsu. *Energies* **2018**, *11*, 1664. [CrossRef]

38. Soares Machado, C.; de Salles Hue, N.P.M.; Berssaneti, F.T.; Quintanilha, J.A. An overview of shared mobility. *Sustainability* **2018**, *10*, 4342. [CrossRef]

39. Axsen, J.; Sovacool, B.K. The roles of users in electric, shared and automated mobility transitions. *Transp. Res. D Transp. Environ.* **2019**, *71*, 1–21. [CrossRef]

40. Martin, G. An ecosocial frame for autonomous vehicles. *Capital. Nat. Social.* **2019**, *30*, 55–70. [CrossRef]

41. Meelen, T.; Frenken, K.; Hobrink, S. Weak spots for car-sharing in The Netherlands? *Energy Res. Soc. Sci.* **2019**, *52*, 132–143. [CrossRef]

42. Zhou, J. Ride-sharing service planning based on smartcard data. *Trans. Policy* **2019**, *79*, 1–10. [CrossRef]

43. Brake, J.; Nelson, J.D. A case study of flexible solutions to transport demand in a deregulated environment. *J. Transp. Geogr.* **2007**, *15*, 262–273. [CrossRef]

44. Davidsson, P.; Hajinasab, B.; Holmgren, J.; Jevinger, Å.; Persson, J. The fourth wave of digitalization and public transport. *Sustainability* **2016**, *8*, 1248. [CrossRef]

45. Chen, Y.; Ardila-Gomez, A.; Frame, G. Achieving energy savings by intelligent transportation systems investments in the context of smart cities. *Transp. Res. D Transp. Environ.* **2017**, *54*, 381–396. [CrossRef]

46. Melo, S.; Macedo, J.; Baptista, P. Guiding cities to pursue a smart mobility paradigm. *Res. Transp. Econ.* **2017**, *65*, 24–33. [CrossRef]

47. Wells, P. Converging transport policy, industrial policy and environmental policy. *Local Econ.* **2012**, *27*, 749–763. [CrossRef]

48. Hoque, N.; Biswas, W.; Howard, I. Life cycle sustainability assessment of alternative energy sources for the western australian transport sector. *Sustainability* **2020**, *12*, 5565. [CrossRef]

49. Bigerna, S.; Polinori, P. Willingness to pay and public acceptance for hydrogen buses. *Sustainability* **2015**, *7*, 13270–13289. [CrossRef]

50. Ortar, N.; Ryghaug, M. Should all cars be electric by 2025? *Sustainability* **2019**, *11*, 1868. [CrossRef]

51. Thomopoulos, N.; Givoni, M. The autonomous car—A blessing or a curse for the future of low carbon mobility? *Eur. J. Futures Res.* **2015**, *3*, 1–14. [CrossRef]

52. Milakis, D.; van Arem, B.; van Wee, B. Policy and society related implications of automated driving. *J. Intell. Transp. Syst.* **2017**, *21*, 324–348. [CrossRef]

53. Milakis, D.; Kroesen, M.; van Wee, B. Implications of automated vehicles for accessibility and location choices. *J. Transp. Geogr.* **2018**, *68*, 142–148. [CrossRef]

54. Lim, H.S.; Taeihagh, A. Autonomous vehicles for smart and sustainable cities. *Energies* **2018**, *11*, 1062. [CrossRef]
55. Dean, J.; Wray, A.; Braun, L.; Casello, J.; McCallum, L.; Gower, S. Holding the keys to health? *BMC Public Health* **2019**, *19*, 1258.
56. Bissell, D.; Birtchnell, T.; Elliott, A.; Hsu, E.L. Autonomous automobilities. *Curr. Sociol.* **2020**, *68*, 116–134. [CrossRef]
57. Rojas-Rueda, D.; Nieuwenhuijsen, M.J.; Khreis, H.; Frumkin, H. Autonomous vehicles and public health. *Annu. Rev. Public Health* **2020**, *41*, 329–345. [CrossRef]
58. Mulley, C.; Kronsell, A. The "uberisation" of public transport and mobility as a service (MaaS). *Res. Transp. Econ.* **2018**, *69*, 568–572. [CrossRef]
59. Hawkins, J.; Habib, K.N. Heterogeneity in marginal value of urban mobility. *Transportation* **2019**. [CrossRef]
60. Becker, H.; Balac, M.; Ciari, F.; Axhausen, K.W. Assessing the welfare impacts of shared mobility and mobility as a service (MaaS). *Trans. Res. Part A Policy Pract.* **2020**, *131*, 228–243. [CrossRef]
61. Van Engelen, M.; Cats, O.; Post, H.; Aardal, K. Enhancing flexible transport services with demand-anticipatory insertion heuristics. *Transp. Res. E Logist. Transp. Rev.* **2018**, *110*, 110–121. [CrossRef]
62. Yigitcanlar, T.; Kankanamge, N.; Vella, K. How are smart city concepts and technologies perceived and utilized? a systematic geo-twitter analysis of smart cities in Australia. *J. Urban Technol.* **2020**. [CrossRef]
63. Winter, K.; Cats, O.; Correia, G.; van Arem, B. Performance analysis and fleet requirements of automated demand-responsive transport systems as an urban public transport service. *IJTST* **2018**, *7*, 151–167. [CrossRef]
64. Matyas, M.; Kamargianni, M. The potential of mobility as a service bundles as a mobility management tool. *Transportation* **2019**, *46*, 1951–1968. [CrossRef]
65. Czioska, P.; Kutadinata, R.; Trifunović, A.; Winter, S.; Sester, M.; Friedrich, B. Real-world meeting points for shared demand-responsive transportation systems. *Public Transp.* **2019**, *11*, 341–377. [CrossRef]
66. Harrison, G.; Shepherd, S. An interdisciplinary study to explore impacts from policies for the introduction of low carbon vehicles. *Transp. Plan. Technol.* **2014**, *37*, 98–117. [CrossRef]
67. Tao, Z.; Zhou, J.; Lin, X.; Chao, H.; Li, G. Investigating the impacts of public transport on job accessibility in Shenzhen, China. *Land Use Policy* **2020**, *99*, 105025. [CrossRef]
68. O'Shaughnessy, M.; Casey, E.; Enright, P. Rural transport in peripheral rural areas. *Soc. Enterp. J.* **2011**, *7*, 183–190. [CrossRef]
69. Broome, K.; Worrall, L.; Fleming, J.; Boldy, D. Evaluation of flexible route bus transport for older people. *Trans. Policy* **2012**, *21*, 85–91. [CrossRef]
70. Faisal, A.; Yigitcanlar, T.; Kamruzzaman, M.; Currie, G. Understanding autonomous vehicles. *J. Transp. Land Use* **2019**, *12*, 45–72. [CrossRef]
71. Mackett, R. Improving accessibility for older people. *J. Transp. Health* **2015**, *2*, 5–13. [CrossRef]
72. Adnan, N.; Md Nordin, S.; Bin Bahruddin, M.A.; Ali, M. How trust can drive forward the user acceptance to the technology? *Trans. Res. Part A Policy Pract.* **2018**, *118*, 819–836. [CrossRef]
73. Wong, S. Traveling with blindness. *Health Place* **2018**, *49*, 85–92. [CrossRef]
74. Faisal, A.; Yigitcanlar, T.; Kamruzzaman, M.; Paz, A. Mapping two decades of autonomous vehicle research. *J. Urban Technol.* **2020**. [CrossRef]
75. Haustein, S.; Siren, A. Older people's mobility. *Transp. Rev.* **2015**, *35*, 466–487. [CrossRef]
76. Curtis, C.; Stone, J.; Legacy, C.; Ashmore, D. Governance of future urban mobility. *Urban Pol. Res.* **2019**, *37*, 393–404. [CrossRef]
77. Canitez, F. Pathways to sustainable urban mobility in developing megacities. *Technol. Forecast. Soc. Chang.* **2019**, *141*, 319–329. [CrossRef]
78. Zografos, K.; Androutsopoulos, K.; Sihvola, T. A methodological approach for developing and assessing business models for flexible transport systems. *Transportation* **2008**, *35*, 777–795. [CrossRef]
79. Stanley, J.; Lucas, K. Workshop 6 Report: Delivering sustainable public transport. *Res. Transp. Econ.* **2014**, *48*, 315–322. [CrossRef]
80. Evans, G.; Guo, A.W.; Blythe, P.; Burden, M. Integrated smartcard solutions. *Transp. Plan. Technol.* **2015**, *38*, 534–551. [CrossRef]
81. Cheyne, C.; Imran, M. Shared transport. *Energy Res. Soc. Sci.* **2016**, *18*, 139–150. [CrossRef]
82. Gonzalez-Feliu, J.; Pronello, C.; Salanova Grau, J. Multi-stakeholder collaboration in urban transport. *Transport* **2018**, *33*, 1079–1094. [CrossRef]

83. Mulley, C.; Nelson, J.D.; Wright, S. Community transport meets mobility as a service. *Res. Transp. Econ.* **2018**, *69*, 583–591. [CrossRef]

84. Waseem, M.; Sherwani, A.; Suhaib, M. Integration of solar energy in electrical, hybrid, autonomous vehicles. *SN Appl. Sci.* **2019**, *1*, 1459. [CrossRef]

85. Turoń, K.; Kubik, A. Economic aspects of driving various types of vehicles in intelligent urban transport systems, including car-sharing services and autonomous vehicles. *Appl. Sci.* **2020**, *10*, 5580. [CrossRef]

86. Sun, Y.; Olaru, D.; Smith, B.; Greaves, S.; Collins, A. Road to autonomous vehicles in Australia. *Road Trans. Res.* **2017**, *26*, 34–47.

87. Kuzio, J. Planning for social equity and emerging technologies. *Transp. Res. Rec.* **2019**, *2673*, 693–703. [CrossRef]

88. Newman, D. Cars and consumption. *Cap. Cl.* **2013**, *37*, 457–476. [CrossRef]

89. Newman, D. Automobiles and socioeconomic sustainability. *Transfers* **2017**, *7*, 100–106. [CrossRef]

90. Freemark, Y.; Hudson, A.; Zhao, J. Are cities prepared for autonomous vehicles? *JAPA* **2019**, *85*, 133–151. [CrossRef]

91. Ferdman, A. Corporate ownership of automated vehicles. *Transp. Rev.* **2020**, *40*, 95–113. [CrossRef]

92. Legacy, C.; Ashmore, D.; Scheurer, J.; Stone, J.; Curtis, C. Planning the driverless city. *Transp. Rev.* **2019**, *39*, 84–102. [CrossRef]

93. Santos, G.; Behrendt, H.; Teytelboym, A. Part II: Policy instruments for sustainable road transport. *Res. Transp. Econ.* **2010**, *28*, 46–91. [CrossRef]

94. Grieco, M. Social sustainability and urban mobility. *Soc. Responsib. J.* **2015**, *11*, 82–97. [CrossRef]

95. Mulley, C.; Nelson, J.D. Flexible transport services. *Res. Transp. Econ.* **2009**, *25*, 39–45. [CrossRef]

96. Allen, J.; Farber, S. Sizing up transport poverty. *Trans. Policy* **2019**, *74*, 214–223. [CrossRef]

97. Leibert, T.; Golinski, S. Peripheralisation. *Comp. Popul. Stud.* **2016**, *41*, 255–284.

98. Lam, D.; Givens, J.W. Small and smart. *New Glob. Stud.* **2018**, *12*, 21–36. [CrossRef]

99. Liu, C.; Yu, B.; Zhu, Y.; Liu, L.; Li, P. Measurement of rural residents' mobility in western China. *Sustainability* **2019**, *11*, 2492. [CrossRef]

100. Sovacool, B.; Martiskainen, M.; Hook, A.; Baker, L. Decarbonization and its discontents. *Clim. Chang.* **2019**, *155*, 581–619. [CrossRef]

101. Soares Machado, C.; Takiya, H.; Quintanilha, J. Placement of Infrastructure for Urban Electromobility. *Sustainability* **2020**, *12*, 6324. [CrossRef]

102. Le Boennec, R.; Nicolaï, I.; Da Costa, P. Assessing 50 innovative mobility offers in low-density areas. *Trans. Policy* **2019**, *83*, 13–25. [CrossRef]

103. Petersen, T. Watching the Swiss. *Trans. Policy* **2016**, *52*, 175–185. [CrossRef]

104. McLeod, S.; Scheurer, J.; Curtis, C. Urban Public Transport. *J. Plan. Lit.* **2017**, *32*, 223–239. [CrossRef]

105. Jokinen, J.P.; Sihvola, T.; Mladenovic, M.N. Policy lessons from the flexible transport service pilot Kutsuplus in the Helsinki Capital Region. *Trans. Policy* **2019**, *76*, 123–133. [CrossRef]

106. Martinez, L.M.; Viegas, J.M.; Eiro, T. Formulating a new express minibus service design problem as a clustering problem. *Transp. Sci.* **2014**, *49*, 85–98. [CrossRef]

107. Ferreira, L.; Charles, P.; Tether, C. Evaluating flexible transport solutions. *Transp. Plan. Techn.* **2017**, *30*, 249–269. [CrossRef]

108. Akgöl, K.; Günay, B. Prediction of Modal Shift Using Artificial Neural Networks. *TEM J.* **2014**, *3*, 223–229.

109. Kandt, J.; Leak, A. Examining inclusive mobility through smartcard data. *J. Transp. Geogr.* **2019**, *79*, 102474. [CrossRef]

110. Tomej, K.; Liburd, J.J. Sustainable accessibility in rural destinations. *J. Sustain. Tour.* **2020**, *28*, 222–239. [CrossRef]

111. Lucas, K.; Currie, G. Developing socially inclusive transportation policy. *Transportation* **2012**, *39*, 151–173. [CrossRef]

112. Kammerlander, M.; Schanes, K.; Hartwig, F.; Jäger, J.; Omann, I.; O'Keeffe, M. A resource-efficient and sufficient future mobility system for improved well-being in Europe. *Eur. J. Futures Res.* **2015**, *3*, 1–11. [CrossRef]

113. Guo, Y.; Peeta, S. Impacts of personalized accessibility information on residential location choice and travel behavior. *Travel Behav. Soc.* **2020**, *19*, 99–111. [CrossRef]

114. Howard, A.; Borenstein, J. The ugly truth about ourselves and our robot creations. *Sci. Eng. Ethics.* **2018**, *24*, 1521–1536. [CrossRef]
115. Beecroft, M.; Pangbourne, K. Future prospects for personal security in travel by public transport. *Transp. Plan. Techn.* **2015**, *38*, 131–148. [CrossRef]
116. Beecroft, M.; Cottrill, C.D.; Farrington, J.H.; Nelson, J.D.; Niewiadomski, P. From infrastructure to digital networks. *Scott. Geogr. J.* **2015**, *135*, 343–355. [CrossRef]
117. Groth, S. Multimodal divide. *Trans. Res. Part A Policy Pract.* **2019**, *125*, 56–71. [CrossRef]
118. Sener, I.N.; Zmud, J.; Williams, T. Measures of baseline intent to use automated vehicles. *Transp. Res. Part F Traffic Psychol. Behav.* **2019**, *62*, 66–77. [CrossRef]
119. Martínez-Díaz, M.; Soriguera, F.; Pérez, I. Autonomous driving: A bird's eye view. *IET Intell. Transp. Syst.* **2019**, *13*, 563–579. [CrossRef]
120. Tang, C.S.; Veelenturf, L.P. The strategic role of logistics in the industry 4.0 era. *Transp. Res. E Logist. Transp. Rev.* **2019**, *129*, 1–11. [CrossRef]
121. Nordhoff, S.; Kyriakidis, M.; van Arem, B.; Happee, R. A multi-level model on automated vehicle acceptance (MAVA). *Theor. Issues Ergon. Sci.* **2019**, *20*, 682–710. [CrossRef]
122. Shergold, I.; Parkhurst, G. Transport-related social exclusion amongst older people in rural Southwest England and Wales. *J. Rural Stud.* **2012**, *28*, 412–421. [CrossRef]
123. Graham, H.; de Bell, S.; Flemming, K.; Sowden, A.; White, P.; Wright, K. The experiences of everyday travel for older people in rural areas. *J. Transp. Health* **2018**, *11*, 141–152. [CrossRef]
124. Ward, M.R.; Somerville, P.; Bosworth, G. Now without my car I don't know what I'd do. *Local Econ.* **2013**, *28*, 553–566. [CrossRef]
125. Bennett, R.; Vijaygopal, R.; Kottasz, R. Willingness of people with mental health disabilities to travel in driverless vehicles. *J. Transp. Health* **2019**, *12*, 1–12. [CrossRef]
126. Bennett, R.; Vijaygopal, R.; Kottasz, R. Attitudes towards autonomous vehicles among people with physical disabilities. *Trans. Res. Part A Policy Pract.* **2019**, *127*, 1–17. [CrossRef]
127. Velaga, N.R.; Beecroft, M.; Nelson, J.D.; Corsar, D.; Edwards, P. Transport poverty meets the digital divide: Accessibility and connectivity in rural communities. *J. Transp. Geogr.* **2012**, *21*, 102–112. [CrossRef]
128. Hensher, D.A. Some insights into the key influences on trip-chaining activity and public transport use of seniors and the elderly. *Int. J. Sustain. Transp.* **2007**, *1*, 53–68. [CrossRef]
129. Goggin, G.; Ellis, K.; Hawkins, W. Disability at the centre of digital inclusion. *Commun. Res. Pract.* **2019**, *5*, 290–303. [CrossRef]
130. Yigitcanlar, T. Australian local governments' practice and prospects with online planning. *URISA J.* **2006**, *18*, 7–17.
131. Meng, L.; Somenahalli, S.; Berry, S. Policy implementation of multi-modal (shared) mobility. *Transp. Rev.* **2020**, *40*, 670–684. [CrossRef]
132. Guo, Y.; Wang, J.; Peeta, S.; Anastasopoulos, P.C. Impacts of internal migration, household registration system, and family planning policy on travel mode choice in China. *Travel Behav. Soc.* **2018**, *13*, 128–143. [CrossRef]
133. Guo, Y.; Wang, J.; Peeta, S.; Ch. Anastasopoulos, P. Personal and societal impacts of motorcycle ban policy on motorcyclists' home-to-work morning commute in China. *Travel Behav. Soc.* **2020**, *19*, 137–150. [CrossRef]
134. Liu, J.; Xiao, L.; Yang, L.; Zhou, J. A tale of two social groups in Xiamen, China: Trip frequency of migrants and locals and its determinants. *Travel Behav. Soc.* **2020**, *20*, 213–224. [CrossRef]
135. Creutzig, F.; Franzen, M.; Moeckel, R.; Heinrichs, D.; Nagel, K.; Nieland, S.; Weisz, H. Leveraging digitalization for sustainability in urban transport. *Glob. Sustain.* **2019**, *2*, e14. [CrossRef]
136. Ruan, Y.; Zhu, D.; Lu, J. Social adaptation and adaptation pressure among the "drifting elderly" in China. *Int. J. Health Plan. Manag.* **2019**, *34*, e1149–e1165. [CrossRef] [PubMed]
137. Hopkins, D.; Schwanen, T. Automated mobility transitions. *Sustainability* **2018**, *10*, 956. [CrossRef]
138. Wong, Y.Z. To Limit Coronavirus Risks on Public Transport, Here's What We Can Learn from Efforts Overseas. The Conversation 2020. Available online: https://theconversation.com/to-limit-coronavirus-risks-on-public-transport-heres-what-we-can-learn-from-efforts-overseas-133764 (accessed on 20 March 2020).

applied sciences

Article

Behavior of Traffic Congestion and Public Transport in Eight Large Cities in Latin America during the COVID-19 Pandemic

Renato Andara [1], Jesús Ortego-Osa [2], Melva Inés Gómez-Caicedo [3], Rodrigo Ramírez-Pisco [4,5], Luis Manuel Navas-Gracia [2,*], Carmen Luisa Vásquez [1] and Mercedes Gaitán-Angulo [6]

1 Directorate of Research and Graduate Studies, Universidad Nacional Experimental Politécnica Antonio José de Sucre, Barquisimeto 3001, Venezuela; randara@unexpo.edu.ve (R.A.); vasquez@unexpo.edu.ve (C.L.V.)
2 Universidad de Valladolid, 47002 Valladolid, Spain; jesus.ortego@uva.es
3 Fundación Universitaria Los Libertadores, Bogotá 16-63A, Colombia; migomezc@libertadores.edu.co
4 Universitat Carlemany, 4000 Sant Julià de Lòria, Andorra; rramirez@universitatcarlemany.com or rodrigo.ramirez.pisco@upc.edu
5 Universitat Politècnica de Catalunya, 08034 Barcelona, Spain
6 Fundación Universitaria Konrad Lorenz, Bogotá 62-43, Colombia; mercedes.gaitana@konradlorenz.edu.co
* Correspondence: luismanuel.navas@uva.es

Citation: Andara, R.; Ortego-Osa, J.; Gómez-Caicedo, M.I.; Ramírez-Pisco, R.; Navas-Gracia, L.M.; Vásquez, C.L.; Gaitán-Angulo, M. Behavior of Traffic Congestion and Public Transport in Eight Large Cities in Latin America during the COVID-19 Pandemic. *Appl. Sci.* **2021**, *11*, 4703. https://doi.org/10.3390/app11104703

Academic Editors: Luis Hernández-Callejo, Víctor Alonso Gómez, Sergio Nesmachnow, Vicente Leite, Javier Prieto and Ângela Ferreira

Received: 10 April 2021
Accepted: 14 May 2021
Published: 20 May 2021

Publisher's Note: MDPI stays neutral with regard to jurisdictional claims in published maps and institutional affiliations.

Abstract: This comparative study analyzes the impact of the COVID-19 pandemic on motorized mobility in eight large cities of five Latin American countries. Public institutions and private organizations have made public data available for a better understanding of the contagion process of the pandemic, its impact, and the effectiveness of the implemented health control measures. In this research, data from the IDB Invest Dashboard were used for traffic congestion as well as data from the Moovit© public transport platform. For the daily cases of COVID-19 contagion, those published by Johns Hopkins Hospital University were used. The analysis period corresponds from 9 March to 30 September 2020, approximately seven months. For each city, a descriptive statistical analysis of the loss and subsequent recovery of motorized mobility was carried out, evaluated in terms of traffic congestion and urban transport through the corresponding regression models. The recovery of traffic congestion occurs earlier and faster than that of urban transport since the latter depends on the control measures imposed in each city. Public transportation does not appear to have been a determining factor in the spread of the pandemic in Latin American cities.

Keywords: traffic congestion; public urban transport; daily infections by COVID-19

1. Introduction

At the end of 2019, a new coronavirus, identified as SARS-CoV-2, was detected in Wuhan, China, which causes a disease called COVID-19 [1]. On 9 March 2020 in Italy, the second outbreak of the virus was detected, leading to mobility restrictions, school closures and measures such as social distancing [2]. The virus quickly spread to other cities and regions in Asia, Europe, Africa, North America and Latin America.

On 11 March 2020, the World Health Organization declared this virus as a pandemic, leading governments to take measures to promote changes in mobility, the way people work and social relations [3]. However, as a result of these actions, there have been closures of businesses and offices, the prohibition of travel that is not strictly necessary, and even a compulsory quarantine at home has been imposed [4]. Its rapid spread led to congestion in health systems and the introduction of restrictions on people to slow down the rate of infection and death. This brought limitations to economic activities and a contraction of the gross domestic product (GDP) of the different economies of the countries [5].

The use of public transport around the world was significantly reduced due to the need to safeguard health and avoid an increase in the number of infections, leading to growth in the use of private vehicles as a means of transport [6]. Indeed, the pandemic has harmed mass transport, such as buses and subways, as users decided to use private means of transport, such as bicycles and other soft means of transport, adapting routes for these means of transport and walking [7–9] in order to reduce the spread of the virus. It is, therefore, relevant to analyze the impact of the measures adopted by governments and the consequences for the actors involved in the sector [10]. In addition, this pandemic significantly affected the development of higher education and included changes in the way of teaching (the rise of e-learning), especially in the area of student mobility (national and international) due to travel restrictions, campus closures and the need to maintain health and safety [11,12]. Additionally, the restrictions imposed by the COVID-19 pandemic [13] had a greater impact on mobility in low-income groups, compared to high-income groups [14].

The characteristic of COVID-19 has been analyzed by the international community based on reports of new cases as the pandemic has progressed. It is emphasized that the route of infection is by air and by contact with infected people or surfaces that have been in contact with them. Its mortality rate is not high (approximately 2–3%); however, its rapid spread encourages the implementation of protocols to stop its spread [15]. Within this framework, among the generalized measures to prevent the spread are mobility restrictions, such as the closure of borders and airports and, among others, suspension of flights and access to public and private urban transport systems [16]. However, the uncertainty of the forms of contagion and effectiveness of these measures persists [17,18].

The appearance of this disease is a serious alert for society and a challenge that transcends borders, with an effort that requires understanding and rationalizing of the reach and potential of such a threat. However, just as there are doubts about the pharmacological therapeutic measures required, there are also doubts regarding the adequacy and effectiveness of the control measures [18]. Due to the initial COVID-19 outbreak, mass transportation systems in China were closed, resulting in at least 40 million people stopping their travel plans, the economic and social costs of which escape the epidemiological estimates that have been done [19,20]. Similar to this situation, the economic consequences of the pandemic have proven to be very negative, as different levels of mobility restrictions have been implemented in response to the spread of the virus, with the transport of people and goods being among the most affected sectors [5,16].

At the international level, initiatives for infection treatment and public health measures have focused on strengthening non-pharmaceutical interventions, including intensive contact follow-up, the quarantine of individuals potentially exposed to infection, and the isolation of those infected or suspiciously asymptomatic [15]. The potential effects deserve coordinated, timely, and effective actions to prevent additional cases or worse health outcomes [21].

One of the regions that are later to suffer from COVID-19 is Latin America, where the first cases of infection and death from this cause occurred on 26 February [22] and 27 March 2020 [23], in Brazil and Argentina respectively. The control measures to curb infections in the region are not homogeneous [24], but all of them have directly impacted the mobility of people, among other effects [25–28].

The mobility of people is a social expression, giving rise to patterns of behavior due to different social, cultural, and economic experiences. This guarantees communication at different distances, affecting the regional and world economy. It is constantly changing, is part of urban life, and has an important cultural and political, and economic development component [29].

In the last decades, there has been an accelerated and uncontrolled urban growth in Latin America, producing a great increase in the need for services, among which include mobility and transport. Traffic congestion problems are evident, which has accelerated the need for higher levels of investment in urban infrastructure and transport modalities [29–31]. Currently, due to COVID-19, there is a decrease in the levels of congestion and use of urban transport in the region.

The purpose of this article is to analyze traffic congestion and the use of public urban transport concerning daily COVID-19 infections in eight of the main cities in Latin

America, belonging to five countries, for the period from 9 March to 30 September 2020, corresponding, therefore, to approximately seven months. Descriptive statistical techniques have been used to analyze the relationship between pandemic-associated infections and people's mobility from the point of view of individual mobility, which affects traffic congestion, as well as the relationship with the use of urban transport. The article is restricted to Ibero-America where the cities were chosen because they are the capital cities of their respective countries, because they account for a high proportion of the population of their countries, as shown in Table 1, and because they have consistently high levels of congestion. Additionally, except for Santiago de Chile, all of the selected cities are among the largest cities in the world in terms of the number of inhabitants in 2015 [32]. Figure 1 shows the geographic location of the cities and countries selected in this article.

Table 1. Population of the cities and countries under study.

City (Country)	Population (Number of Inhabitants) of City (Country)	Percentage of City's Population in Relation to Country (%)
Bogotá (Colombia)	8,770,058 (50,882,884)	17.24
Buenos Aires (Argentina)	14,338,718 (45,195,777)	31.73
Mexico City (Mexico)	22,381,714 (128,932,753)	17.36
Santiago (Chile)	5,688,218 (19,116,209)	29.76
Lima (Perú)	9,609,692 (32,971,846)	29.15
Brasilia (Brazil)	2,043,811 (212,559,409)	0.96
São Paulo (Brazil)	21,001,688 (212,559,409)	9.88
Rio de Janeiro (Brazil)	10,523,151 (212,559,409)	4.95

Figure 1. Cities and countries (in green) that are the subject of this article.

For the data, the information available on specific official web pages and public social networks was used. Different institutions and organizations have made the data available to the health authorities, the scientific community, and interested parties to understand better the contagion process of the pandemic, its impacts, and the effectiveness of the health control measures implemented. All of this was done in order to disseminate information on the progress of the infection and the basic measures of care, control of travelers, and local plans to deal with possible cases. For the case of this article, the traffic congestion data published by the Inter-American Development Bank (IDB) [24] were used and for urban transport, the data from the Moovit© application were used [33]. In addition, the information from Johns Hopkins Hospital University [34] was used for the number of daily infections by COVID-19.

The cities analyzed were Bogotá (Colombia), Buenos Aires (Argentina), Mexico City (Mexico), Santiago (Chile), Lima (Perú) and Brasilia, São Paulo and Rio de Janeiro, with the latter three in Brazil. These cities are distinguished by having the largest population in the region [35], levels of congestion [36] and number of COVID-19 infections [37]. Among the public urban transport systems, the metro, bus rapid transit (BRT), surface trains, trams, and trolleybuses [38,39] stand out. Table 2 shows for each of these cities the population density in the metropolitan area, the percentage of modal participation of public urban transport, and the approximate number of daily passengers served in the BRT and subway systems, which are the ones with the highest participation [40,41].

Table 2. Data on population density in the metropolitan area, the modal share of urban public transport and the number of daily passengers served in the BRT and subway systems of the cities under study.

City	Population Density (Inhab/km^2)	Modal Share of Urban Transport (%)	Daily Passengers in the BRT	Daily Passengers in the Subway
Bogotá	3347.4	59.0	2,192,009	-
Buenos Aires	50.7	44.7	1,419,000	1,500,000
Mexico City	2450.7	77.9	1,240,000	1,600,000
Santiago	462.0	29.1	340,800	1,975,000
Lima	3008.8	62.0	704,803	550,000
Brasilia	491.6	36.2	51,000	43,800
São Paulo	2714.4	36.8	3,300,000	4,600,000
Rio de Janeiro	2208.8	48.7	3,535,466	780,000

The metro of the city of Buenos Aires (SUBTE) is one of the first implemented in the region and the fleet of electric BRT of the city of Santiago is the second largest in the world in 2020, which shows that accelerated and giant steps have been taken for the development of public urban transport, in this way, guaranteeing sustainable urban mobility.

Additionally, [42], derived from ICSC-CITIES 2020 Congress, shows the impact of the COVID-19 pandemic on traffic congestion in 13 Latin American cities during the first five months of the pandemic declaration. The results are analyzed by grouping the countries into four conglomerates, showing that as social distancing measures are relaxed, there is a recovery of traffic congestion due to the use of private vehicles. In contrast to [42], the present work includes mobility with private vehicles and urban public transport in its different forms, in eight cities. This allows for a broader view by including all means of motorized mobility for a longer period (seven months), allowing us to analyze people's behavior based on their motorized mobility in a time window where a recovery in the number of infections is beginning to be observed.

2. Methodology

The data used in this article were obtained from the IDB Coronavirus Impact Dashboard [24], according to the methodological note [22], in accordance with the agreement of this institution with the Waze© platform [43]. These data come from the information of the mobile phones of its users, which are aggregated and geocoded in real-time every 2 min. The Coronavirus Impact Dashboard uses the week of 1 to 7 March 2020, as a reference for comparison, because traffic patterns were not affected by regional holidays and there were still very few reported cases of infections in the region. Additionally, governments have not yet issued restrictions or recommendations for social distancing.

The IDB Coronavirus Impact Dashboard methodology for creating urban centers as adjacent grid groups limits publication to urban centers with more than 750,000 inhabitants. In other words, it restricts the analysis to centers with sufficient Waze© activity and historical information, leaving only 64 metropolitan areas in 19 countries, using the Traffic Congestion Intensity indicator (TCI) as an indicator for the analysis, which is the sum of total times and lengths of congestion for periods of 24 h, compared with those carried out on the days of 1 to 7 March 2020 (ΔTCI). This indicator allows the percentage comparison between the same days of the current week, called ratio-20.

The IDB Coronavirus Impact Dashboard [24] publishes information from 64 metropolitan areas; however, for the present study, only the cities of Bogotá (Colombia), Buenos Aires (Argentina), Mexico City (Mexico), Santiago (Chile), Lima (Perú), and Brasilia, São Paulo and Rio de Janeiro, (Brazil) were considered because they are large cities where the impact of public transport is very important. The analysis period ranges from 9 March to 30 September of 2020, approximately 7 months.

In this work, the graphs of variation of the TCI (ΔTCI) and the percentage of reduction of public urban transport are analyzed as a function of time and, in turn, of the cases of infections detected daily for the same days. The contagion data are taken directly from those published by the Coronavirus Research Center of the Johns Hopkins Hospital University [34], corresponding to all countries.

For each area, graphs were made, and descriptive statistics were estimated, based on observing the recovery of traffic congestion and public urban transport as a function of the decrease in infections. These graphs determine the minimum vehicle mobility and, from this moment, the mobility recovery rate (MRR), as an indicator of the recovery rate of traffic congestion in the area analyzed through regression analysis of the data.

To perform the analysis of urban mobility in public transport during the COVID-19 pandemic, we used data collected by the Inter-American Development Bank (IDB) in its Coronavirus Impact Dashboard [24]. This dashboard is based on data from different sources on public transportation. For example, for the city of Bogotá, the information from the BRT Transmilenio open data website was used. For Lima, the information from the PROTRANSPORTE Metropolitan Institute of Lima of the Municipality of Lima was used. In both cases, the daily data were compared with the validations in the week of 2 to 8 March 2020. For São Paulo, the data of the Municipal Secretariado de Mobilidade e Transportes of the Municipality of São Paulo were used and compared with the validations from the week of 15 January 2020. For the rest, the data from the Moovit© public transport application were used and compared with the week of 2 to 8 March 2020.

3. Results

Figures 2–9 show the percentages of reduction of traffic congestion and of the use of urban transport, along with daily infections by COVID-19 for the cities of Bogotá, Buenos Aires, Mexico City, Santiago, Lima, Brasilia, São Paulo and Rio de Janeiro, respectively. The analysis period corresponds from 9 March to 30 September of 2020, approximately 7 months.

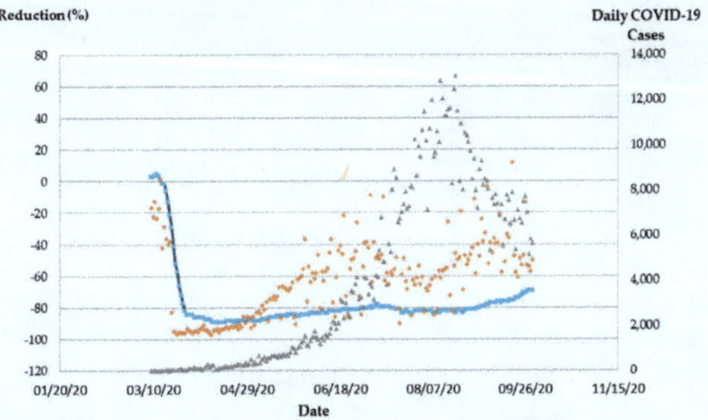

Figure 2. Percentage reduction (%) in traffic congestion (red) and urban transport use (blue), compared to the number of daily COVID-19 infections (gray). Bogota city, Colombia.

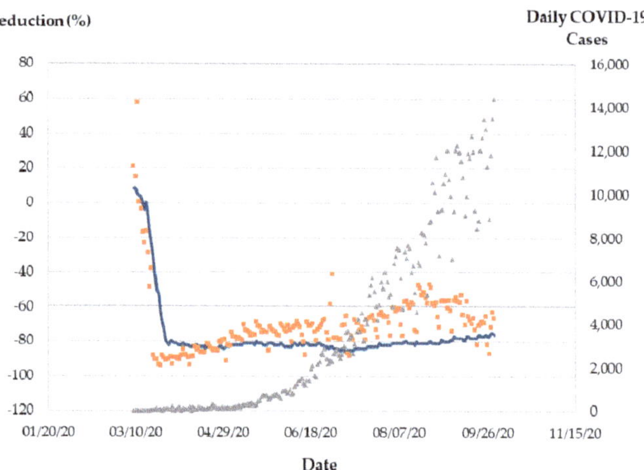

Figure 3. Percentage reduction (%) in traffic congestion (red) and urban transport use (blue), compared to the number of daily COVID-19 infections (gray). Buenos Aires, Argentina.

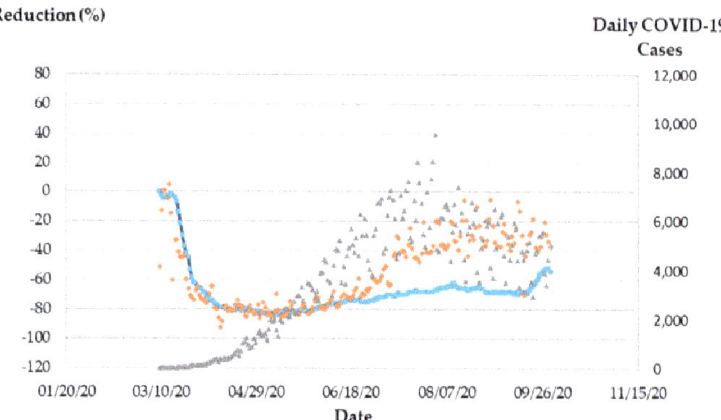

Figure 4. Percentage reduction (%) in traffic congestion (red) and urban transport use (blue), compared to the number of daily COVID-19 infections (gray). Mexico City, Mexico.

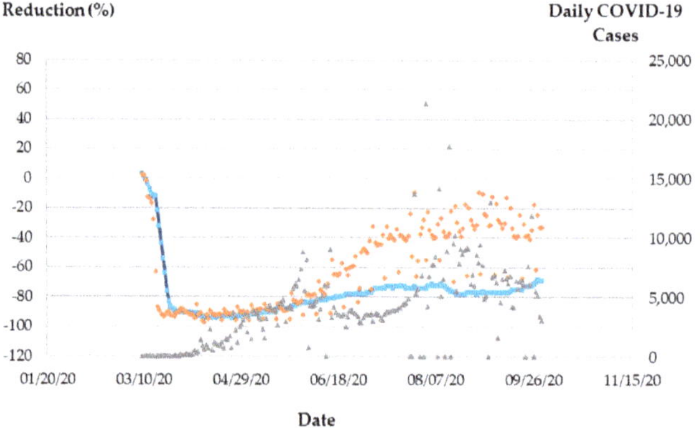

Figure 5. Percentage reduction (%) in traffic congestion (red) and urban transport use (blue), compared to the number of daily COVID-19 infections (gray). Santiago, Chile.

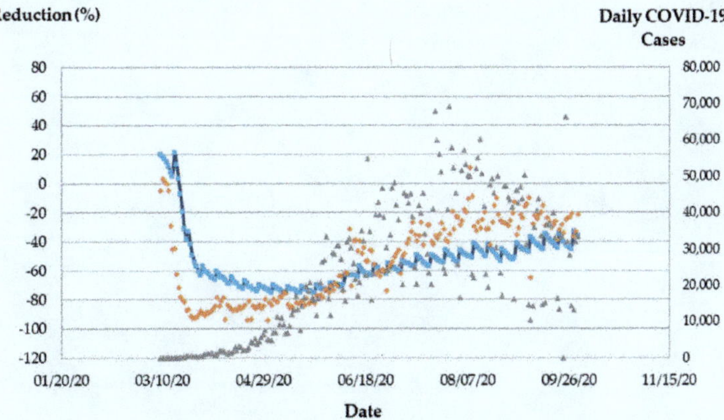

Figure 6. Percentage reduction (%) in traffic congestion (red) and urban transport use (blue), compared to the number of daily COVID-19 infections (gray). Lima, Perú.

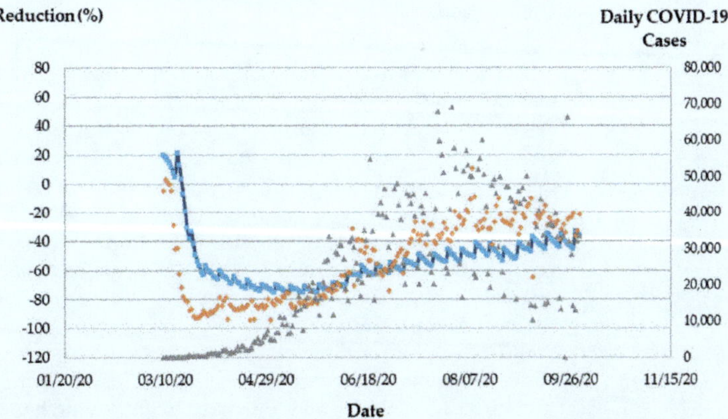

Figure 7. Percentage reduction (%) in traffic congestion (red) and urban transport use (blue), compared to the number of daily COVID-19 infections (gray). Brasilia, Brazil.

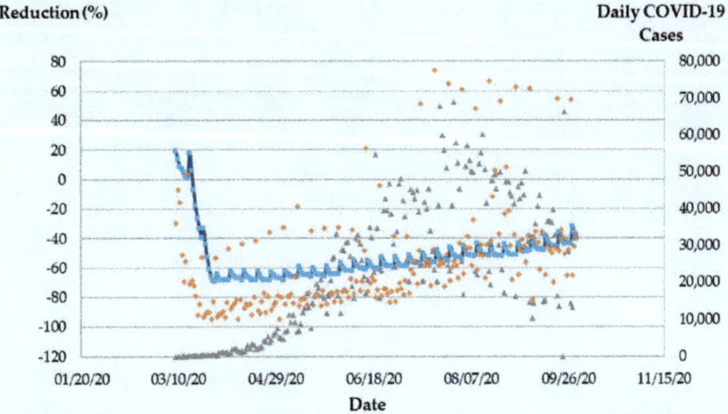

Figure 8. Percentage reduction (%) in traffic congestion (red) and urban transport use (blue), compared to the number of daily COVID-19 infections (gray). Sao Paulo, Brazil.

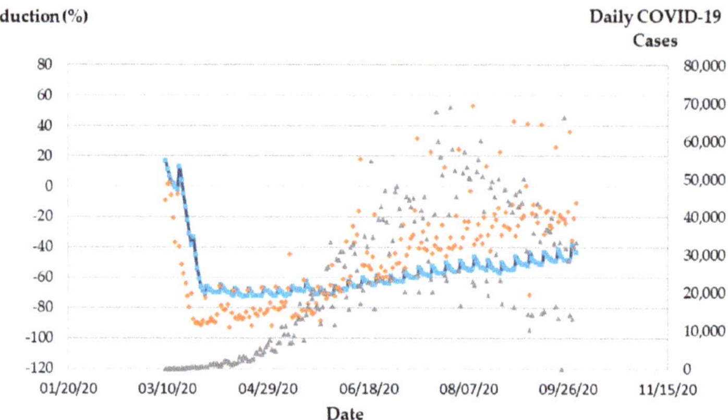

Figure 9. Percentage reduction (%) in traffic congestion (red) and urban transport use (blue), compared to the number of daily COVID-19 infections (gray). Rio de Janeiro, Brazil.

4. Discussion

Urban mobility is a key element for the economic and social development of cities, linked to the means of transport used (motorized or not) and to a set of externalities, including traffic congestion, correlated with the use of private vehicles and urban transport systems. In this context, urban transport is essential to ensure the social and economic well-being of cities, and the model that articulates order and fluidity and guarantees the comfort, saved time , economy and mobility of its inhabitants is, therefore, of very relevant importance.

Prior to the COVID-19 pandemic, Latin American cities were among the most congested in the world. According to the INRIX© ranking published in 2019 [44], cities such as Bogotá, Rio de Janeiro, Mexico City and São Paulo were ranked 1, 2, 3 and 5, respectively, as the most congested in a universe of 945 cities worldwide; in these cities, a driver loses more than 145 h a year sitting in his or her vehicle and the average speed is less than 21 km/h. However, in the publication of this ranking for the year 2020 [36], there is evidence of a decline in the rankings and hours lost, with driving hours decreasing by 31%, 69% and 66% in the cities of Bogotá, Mexico City and São Paulo, respectively, as a result of the COVID-19 pandemic.

Similarly, in [45] and for the year 2020, traffic congestion drops were reported, comparatively, with respect to 2019, in the cities of Bogotá, Mexico City, Rio de Janeiro, Santiago, São Paulo, Brasilia and Buenos Aires with 19%, 16%, 14%, 13%, 15%, 15% and 11%, respectively. As an example, government restrictions in Colombia reduced demand on transport systems [46] and eventually changed the modes of transport used by the population, in efforts to try to maintain social distance. Many people decided not to use public transport and replaced it with private or non-motorized vehicles, such as walking or cycling [13]. Additionally, the total number of taxi trips in the areas studied in China declined sharply during the pandemic. This decline was most significant for taxi use during night-time hours, with no variation in average trip distances.

The challenges of urban mobility in Latin American cities prior to the COVID-19 pan-demic were as follows [47]:

- Public transport is the most used mode of transport (buses and micro-buses, with 102 million trips per day, followed by metros and trains, with 19 million trips per day), accounting for 42% of trips in metropolitan areas. However, public transport is of poor quality and travel time and cost for users are high;
- Road safety affects the most vulnerable (pedestrians) who account for more than half of all traffic fatalities (10,000 deaths per year);

- The level of pollutant emissions due to transport is very high in cities (260,000 tonnes of CO_2 and 3600 tonnes of other pollutants, such as CO and NOx), damaging public health;
- Traffic management is very limited, which prevents the optimization of the existing road infrastructure. Effective priority for buses, pedestrians and cyclists is very low.

The figures previously introduced (Figures 2–9) show that the mobility restriction measures in the first months of contagion achieved a significant reduction in traffic congestion, being the lowest in the city of Lima with 87%. In turn, urban transport falls close to 90% for this same city. Table 3 shows a summary of the variables analyzed for all the cities considered.

Table 3. Percentages of maximum decline and mobility recovery rate applied to traffic congestion and urban public transport in the Latin American cities under study.

City	Traffic Congestion		Public Urban Transport	
	Maximum Decay (%)	Mobility Recovery Rate (%)	Maximum Decay (%)	Mobility Recovery Rate (%)
Bogotá	−97.00	27.9	−89.00	7.31
Buenos Aires	−94.08	14.4	−86.00	9.79
México City	−92.05	41.8	−84.00	12.2
Santiago	−92.52	44.2	−88.00	17.3
Lima	−97.24	45.2	−94.00	11.9
Brasilia	−95.04	35.9	−69.00	14.9
São Paulo	−92.74	50.4	−73.00	17.4
Rio de Janeiro	−94.22	45.1	−75.00	25.6

It can be recognized that as the number of daily infections has decreased there has been a recovery in the use of urban transport and private vehicles. Since urban transport is managed by the government, it is observed that its recovery happens slower than that of private vehicles, which is identified as the recovery of congestion. TCI has a higher correlation with infections than the use of urban transport.

Figure 10 shows for the months of February, April, and June the number of people or passengers using the BRT Transmilenio urban transport system in the city of Bogotá for the years 2017 to 2020, which, in this city, is the urban transport system of reference consisting of exclusive transit buses. The figure shows that the number of people using the system fell below 20,000,000 per day in April 2020, which is down 18%, compared to the same month in previous years. Additionally, there was a slight recovery in the number of passengers for the month of June 2020, but they did not exceed 25% of the same month in previous years [48]. According to "it can be argued that these findings do not support the effectiveness of suspending mass urban transport systems as a pandemic countermeasure aimed at reducing or slowing population spread because, whatever the relevance of public transport is to individual-level risk, household exposure most likely poses a greater threat" [48].

Interestingly, the point of greatest decline in the use of private vehicles was greater than that of public transport. However, in each city, the speed of recovery of the mobility of particular vehicles was greater than that of public transport. This can be due to two factors: public transport follows the guidelines dictated by the regional or national government; it is also possible that due to fear of infection, people use private vehicles instead of public transport. Likewise, the rebound in use was earlier for private vehicles than for public transport.

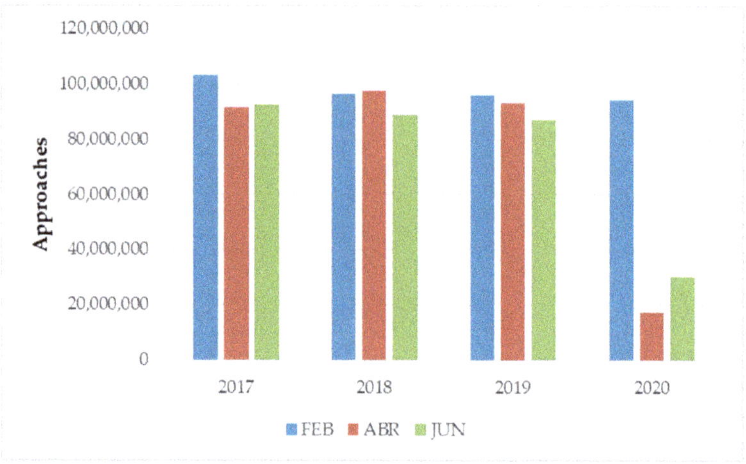

Figure 10. Validations in the BRT Transmilenio (Bogotá, Colombia) in the months of February, April and June, from the years 2017 to 2020.

An indicator of the change in the behavior model associated with mobility restrictions due to the pandemic is the increase in internet data traffic [49], affecting some cities, such as Bogotá, up to 40%. Another indicator is the behavior of people, as shown in Table 4, depending on the destination of their movements during the months of the COVID-19 pandemic. In this study, it is observed that trips to second homes only increased in the five countries considered in the study. In turn, there was a significant reduction in visits to parks, this being the main destination affected by the reduction in mobility in Argentina, Brazil and Chile, while in Peru and Colombia, respectively, the destination to shops and leisure, and to transport stations were reduced. The greatest average decrease in trips, regardless of destination, occurred in Chile, with an average decrease of −59%, followed by Perú and Argentina, both with decreases of over −50%. On the contrary, the lowest average decrease in trips was observed in Brazil, with a value of −28%, highlighting in this country the destinations of supermarkets and pharmacies, which practically reached the same values as those before the start of the pandemic. Table 4 shows in bold the maximum drop in the mobility percentage according to the destination for each country. The cities of Rio de Janeiro, Mexico City and São Paulo are ranked 2, 4 and 5, respectively, among the most populated cities in the world [32], which demonstrates the representativeness of the reduction of destinations.

Table 4. Percentage of variation (increase +, and reduction −) of mobility, according to destinations, during the COVID-19 pandemic in Argentina, Brazil, Chile, Colombia and Perú.

Percentage of Variation of Mobility According to Destinations (%)	Country				
	Argentina	Brazil	Chile	Colombia	Perú
Shops and Leisure	−61	−40	−66	−47	**−67**
Supermarkets and Drugstores	−21	−2	−46	−24	−35
Parks	**−83**	**−39**	**−67**	−38	−46
Transport stations	−54	−38	−66	**−48**	−58
Work	−28	−22	−50	−35	−54

5. Conclusions

This study analyzes the impact of the COVID-19 pandemic on motorized mobility in large megacities belonging to five of the most populated countries in Latin America, including Bogotá, Buenos Aires, Mexico City, Santiago, Lima, Brasilia, São Paulo, and Rio de Janeiro. For this, the public data made available to society by different public and private institutions, such as the Inter-American Development Bank, Johns Hopkins Hospital University, and the Moovit© platform, were used. These institutions have made the data public in order to better understand the contagion process of the pandemic, its impacts, and the effectiveness of the health control measures implemented.

For each city, a descriptive analysis of the recovery of mobility from traffic congestion and urban public transport is made, depending on the level of contagion. As expected, the recovery of traffic congestion occurs earlier and faster than that of urban transport, since the latter depends on the control measures imposed in each locality, highlighting that the decrease in the number of infections brings the flexibility of social distancing measures and the recovery of vehicular mobility, which is observed as a function of the increase in MRR. Only the city of Buenos Aires does not show a clear recovery in cases of contagion by COVID-19, unlike the other cities analyzed. However, with adequate sanitation measures, public transport is more efficient for reducing the risks of contagion.

In the work, a decoupling is observed between the evolution of the number of COVID-19 cases and the recovery rate of urban public transport, which is explained by the perception of the risk of the population being infected in this type of transport. Certainly, according to [50], the following factors affect the risk of contagion in public transport systems and must be considered to guarantee the mobility of citizens with the highest occupancy levels: (a) user behavior related to masks, eye protection, and quiet travel to reduce airborne transmission, (b) the type of ventilation system of the vehicle and frequency of air renewal, (c) proximity of drivers, (d) duration of the trip and, finally, (e) cleaning and disinfection of high-contact surfaces.

The main recommendations of the World Health Organization (WHO) [51] for the transformation of transport in cities establish lines of concrete actions for its future in cities, among which are the adoption of clean methods of electricity generation; the prioritization of rapid urban transport, pedestrian and bicycle paths in cities, interurban transport of cargo and passengers by rail; and the use of cleaner diesel-engine heavy-duty vehicles and low emission vehicles and fuels, especially with low sulfur fuels.

The pandemic and the different mobility restrictions have highlighted the need for more agile and less collective forms of transport services, representing an opportunity for the development of new public–private transport services that would contribute to sustainability, resilience, mitigation of climate change, and the health of the population. In the case of Latin America and the studied cities, several challenges must be addressed, such as the quality of public transport, road safety for the most vulnerable (pedestrians), emissions and their impact on health, as well as the high traffic congestion and optimization of road infrastructure and finally, the effective prioritization of buses, cyclists and pedestrians.

The COVID-19 pandemic has revealed less collective and more agile forms of mobility, which is an important opportunity for the region to develop new forms of transport, such as soft and personal forms of mobility which were observed as the link between the evolution of the number of COVID-19 cases and the recovery rate of urban public transport, which is explained by the perception of the risk of the population being infected in this type of transport.

Author Contributions: Conceptualization, R.A., J.O.-O., R.R.-P., L.M.N.-G. and C.L.V.; methodology, R.A., J.O.-O., C.L.V., L.M.N.-G., R.R.-P. and M.I.G.-C.; validation, R.A., R.R.-P. and M.G.-A.; formal analysis, J.O.-O., C.L.V., L.M.N.-G., R.R.-P., M.I.G.-C. and M.G.-A.; investigation, J.O.-O., C.L.V. and M.G.-A.; data curation, R.A., L.M.N.-G. and R.R.-P.; writing—original draft preparation, C.L.V., L.M.N.-G., R.R.-P., M.I.G.-C. and M.G.-A.; writing—review and editing, R.A., J.O.-O., M.I.G.-C. and C.L.V.; visualization, R.A., J.O.-O. and L.M.N.-G.; supervision, R.R.-P., C.L.V., M.I.G.-C. and M.G.-A.; project administration, L.M.N.-G., R.R.-P. and C.L.V. All authors have read and agreed to the published version of the manuscript.

Funding: This research was funded by the Sustainable Urban Transport and Mobility Network (RITMUS, 718RT0566) of the Science and Technology for the Development of Ibero-America Program (CYTED), Carlemany University and University Foundation "Los Libertadores".

Institutional Review Board Statement: This study did not involve humans or animals.

Informed Consent Statement: This study did not involve humans.

Data Availability Statement: This study did not report any data.

Acknowledgments: The authors of this article would like to thank the Science and Technology for the Development of Ibero-America Program (CYTED), since it was developed within the framework of the Sustainable Urban Transport and Mobility Network (RITMUS, 718RT0566). Additionally, the authors would also like to thank University Foundation "Los Libertadores" and Carlemany University for supporting the article.

Conflicts of Interest: The authors declare no conflict of interest.

References

1. Aragón-Nogales, R.; Vargas-Almanza, I.; Miranda-Novales, M.G. COVID-19 por SARS-CoV-2: La nueva emergencia de salud. *Rev. Mex. Pediatría* **2020**, *86*, 213–218. (In Spanish)
2. Charoenwong, B.; Kwan, A.; Pursiainen, V. CORONAVIRUS. Social connections with COVID-19–affected areas increase compliance with mobility restrictions. *Sci. Adv.* **2020**, *6*, eabc3054. [CrossRef]
3. Scorrano, M.; Danielis, R. Active mobility in an Italian city: Mode choice determinants and attitudes before and during the Covid-19 emergency. *Res. Transp. Econ.* **2020**, *86*, 101031. [CrossRef]
4. Baldwin, R.; Weder di Mauro, B. Economics in the Time of COVID-19: A New eBook. 2020. Available online: https://fondazionecerm.it/wp-content/uploads/2020/03/CEPR-Economics-in-the-time-of-COVID-19_-A-new-eBook.pdf (accessed on 10 October 2020).
5. Banco de España. *Escenarios Macroeconómicos de Referencia para la Economía Española Tras el COVID-19*; Banco de España: Madrid, Spain, 2020. (In Spanish)
6. Campisi, T.; Basbas, S.; Skoufas, A.; Akgün, N.; Ticali, D.; Tesoriere, G. The impact of COVID-19 pandemic on the resilience of sustainable mobility in Sicily. *Sustainability* **2020**, *12*, 8829. [CrossRef]
7. Aloi, A.; Alonso, B.; Benavente, J.; Cordera, R.; Echániz, E.; González, F.; Sañudo, R. Effects of the COVID-19 lockdown on urban mobility: Empirical evidence from the city of Santander (Spain). *Sustainability* **2020**, *12*, 3870. [CrossRef]
8. Badii, C.; Bellini, P.; Difino, A.; Nesi, P. Smart city IOT platform respecting gdpr privacy and security aspects. *IEEE Access* **2020**, *8*, 23601–23623. [CrossRef]
9. Moslem, S.; Campisi, T.; Szmelter-Jarosz, A.; Duleba, S.; Nahiduzzaman, K.M.; Tesoriere, G. Best–worst method for modelling mobility choice after COVID-19: Evidence from Italy. *Sustainability* **2020**, *12*, 6824. [CrossRef]
10. Gkiotsalitis, K.; Cats, O. Public transport planning adaption under the COVID-19 pandemic crisis: Literature review of research needs and directions. *Transp. Rev.* **2020**, 1–19. [CrossRef]
11. Tesar, M. Towards a post-COVID-19 'New Normality?': Physical and social distancing, the move to online and higher education. *Policy Futures Educ.* **2020**, *18*, 556–559. [CrossRef]
12. Mok, K.H.; Xiong, W.; Ke, G.; Cheung, J.O.W. Impact of COVID-19 pandemic on international higher education and student mobility: Student perspectives from mainland China and Hong Kong. *Int. J. Educ. Res.* **2020**, *105*, 101718. [CrossRef]
13. Nian, G.; Peng, B.; Sun, D.J.; Ma, W.; Peng, B.; Huang, T. Impact of COVID-19 on urban mobility during post-epidemic period in megacities: From the perspectives of taxi travel and social vitality. *Sustainability* **2020**, *12*, 7954. [CrossRef]
14. Wilder-Smith, A.; Freedman, D.O. Isolation, quarantine, social distancing and community containment: Pivotal role for old-style public health measures in the novel coronavirus (2019-nCoV) outbreak. *J. Travel Med.* **2020**, *27*, taaa020. [CrossRef] [PubMed]
15. Palacios, M.; Santos, E.; Velázquez, M.; León, M. COVID-19, a Worldwide Public Health Emergency. *Revis. Clínica Española* **2020**, *221*, 55–61. (In Spanish) [CrossRef]
16. Rodríguez, E. Colombia: Impacto económico, social y político del COVID-19. *Fund. Carol.* **2020**, *24*, 1–14. (In Spanish)
17. Moreno-Montoya, J. El desafío de comunicar y controlar la epidemia por coronavirus. *Biomédica* **2020**, *1*, 11–13. (In Spanish) [CrossRef] [PubMed]
18. Li, Q.; Guan, X.; Wu, P.; Wang, X.; Zhou, L.; Tong, Y.; Ren, R.; Med, M.; Leung, K.S.M.; Lau, E.H.Y.; et al. Early transmission dynamics in Wuhan, China, of novel coronavirus–infected pneumonia. *N. Engl. J. Med.* **2020**, *382*, 1199–1206. [CrossRef]
19. Tang, B.; Wang, X.; Li, Q.; Bragazzi, N.L.; Tang, S.; Xiao, Y.; Wu, J. Estimation of the transmission risk of 2019-nCov and its implication for public health interventions. *J. Clin. Med.* **2020**, *9*, 462. [CrossRef]
20. Jin, L.; Zhao, Y.; Zhou, J.; Tao, M.; Yang, Y.; Wang, X.; Ye, P.; Shan, P.; Yuan, H. Distribuciones de hora, sitio, y población de novel coronavirus enfermedad 2019 (COVID-19) desde enero 20 a febrero 2020, en China. *Rev. Clínica Española* **2020**, *20*, 495–500. (In Spanish) [CrossRef]
21. Ferrer, R. COVID-19 pandemic: The greats challenge in the history of critical care. *Med. Intensiva* **2020**, *44*, 323–324. [CrossRef] [PubMed]
22. Inter-American Development Bank (IDB). Coronavirus Impact Dashboard Methodological Note. 2020. Available online: https://www.iadb.org/en/topics-effectiveness-improving-lives/coronavirus-impact-dashboard-about (accessed on 3 October 2020).

23. Pierre, R.; Harris, P. COVID-19 en América Latina: Retos y oportunidades. *Rev. Chil. Pediatría* **2020**, *91*, 179–182. (In Spanish) [CrossRef]
24. Inter-American Development Bank (IDB). Tablero de Impacto del Coronavirus. 2020. Available online: https://www.iadb.org/es/topics-effectiveness-improving-lives/coronavirus-impact-dashboard (accessed on 3 October 2020). (In Spanish).
25. Iranzo, A. COVID-19: ¿(in)seguridad sin (in)movilidad? Acercando la política de la movilidad a los estudios críticos de seguridad. *Geopolítica(s) Rev. Estud. Sobre Espac. Poder* **2020**, *11*, 61–68. (In Spanish) [CrossRef]
26. Kraeme, M.; Yang, C.; Gutierrez, B.; Wu, C.; Klein, B.; Pigott, D.; Plesiss, L.; Faria, N.; Li, R.; Hanage, W.; et al. Coronavirus: The effect of human mobility and control measures on the COVID-19 epidemic in China. *Science* **2020**, *368*, 493–497. [CrossRef]
27. Sirkeci, I.; Murat, M. Coronavirus and migration: Analysis of human mobility and the spread of COVID-19. *Migr. Lett.* **2020**, *17*, 379–398. [CrossRef]
28. Peñafiel-Chang, L.; Camelli, G.; Peñafiel-Chang, P. Pandemia COVID-19: Situación política-económica y consecuencias sanitarias en América Latina. *Rev. Cienc. UNEMI* **2020**, *13*, 120–128. (In Spanish) [CrossRef]
29. Alonso, R.; Lugo-Morìn, D. El estado del arte de la movilidad del transporte en la vida urbana en ciudades latinoamericanas. *Rev. Transp. Territ.* **2018**, *19*, 133–157. (In Spanish) [CrossRef]
30. Rivasplata, C. Congestion pricing for Latin America: Prospects and constraints. *Res. Transp. Econ.* **2013**, *40*, 56–65. [CrossRef]
31. Gandelman, N.; Serebrisky, T.; Suárez-Alemán, A. Household spending on transport in Latin America and the Caribbean: A dimension of transport affordability in the region. *J. Transp. Geogr.* **2019**, *79*, 1–14. [CrossRef]
32. Molloy, J.; Tchervenkov, C.; Schatzmann, T.; Schoeman, B.; Hitermann, B.; Axhausen, K.W. MOBIS-COVID-19/13: Results as for 06/07/2020 (Post Lockdown). *Arb. Verk. Raumplan.* **2020**, 1531. Available online: https://www.research-collection.ethz.ch/bitstream/handle/20.500.11850/425344/ab1531.pdf?sequence=2&isAllowed=y (accessed on 1 October 2020).
33. Moovit©. El Coronavirus y su Viaje Diario: Cómo el COVID-19 Está Afectando el Transporte Público en Todo el Mundo. 2020. Available online: https://moovit.com/blog/coronavirus-effect-public-transportation-worldwide (accessed on 30 September 2020). (In Spanish)
34. Johns Hopkins Hospital University, Coronavirus Research Center. 2020. Available online: https://coronavirus.jhu.edu/map.html (accessed on 3 October 2020).
35. Vargas-Mendoza, F. The city global region for the economic and social development of Latin America. *Aibi Rev. Investig. Adm. Ing.* **2020**, *8*, 58–68. [CrossRef]
36. INRIX©. INRIX Global Traffic Scorecard. 2020. Available online: https://inrix.com/scorecard/ (accessed on 3 October 2020).
37. Moreno, A. Behavior adopted in Latin America due to COVID-19. *BILO* **2020**, *2*, 3–9.
38. Vásquez, C.; Ramírez-Pisco, R.; Viloria, A.; Martínez-Sierra, D.; Ruíz-Barrios, E.; Hernández, H. Conglomerates of bus rapid transit in Latin American countries. In *Proceedings of the International Conference on Intelligent Computing, Information and Control Systems*; 2019; pp. 220–228. Available online: https://link.springer.com/chapter/10.1007/978-3-030-30465-2_25 (accessed on 10 October 2020).
39. Vásquez, C.; Pérez, R.; Ramírez-Pisco, R.; Osal, W. Sistemas de transporte urbano en Latinoamérica. *TRIM Tordesillas Rev. Investig. Multidiscip.* **2019**, *17*, 31–44. (In Spanish)
40. BRTData. Global BRT Data. 2020. Available online: https://brtdata.org/ (accessed on 3 October 2020).
41. Mapa-metro. 2020. Available online: https://mapa-metro.com/es/ (accessed on 3 October 2020). (In Spanish).
42. Ortego, J.; Andara, R.; Navas, L.M.; Vásquez, C.; Ramírez-Pisco, R. Impact of the COVID-19 pandemic on traffic congestion in Latin American cities: An updated five-month study. In *ICSC-CITIES 2020*; Nesmachnow, S., Hernández Callejo, L., Eds.; Communications in Computer and Information Science Book Series (CCIS); Springer: Cham, Swizerland, 2021; Volume 1359, Available online: https://link.springer.com/chapter/10.1007/978-3-030-69136-3_15#citeas (accessed on 1 February 2021).
43. Waze©. WAZE for Cities. 2020. Available online: https://www.waze.com/es/ccp (accessed on 3 October 2020).
44. INRIX©. INRIX Global Traffic Scorecard. 2019. Available online: https://static.poder360.com.br/2019/02/INRIX_2018_Global_Traffic_Scorecard_Report__final_.pdf (accessed on 3 November 2020).
45. TomTom Traffic Index. The World Has Changed, Traffic Has Changed. 2020. Available online: https://www.tomtom.com/en_gb/traffic-index/ranking/ (accessed on 3 October 2020).
46. Arellana, J.; Márquez, L.; Cantillo, V. Outbreak in Colombia: An analysis of its impacts on transport systems. *J. Adv. Transp.* **2020**, *2020*, 8867316. [CrossRef]
47. Vasconcellos, A. Las condiciones de movilidad urbana en América Latina. In *Transporte y Desarrollo en América Latina*; CAF—Banco de Desarrollo de América Latina: Caracas, Venezuela, 2018; Volume 1, pp. 111–119.
48. Musselwhite, C.; Avineri, E.; Susilo, Y. The Coronavirus disease COVID-19 and implications for transport and health. *J. Transp. Health* **2020**, *16*, 100853. Available online: https://www.ncbi.nlm.nih.gov/pmc/articles/PMC7174824/ (accessed on 12 October 2020). [CrossRef] [PubMed]
49. Inter-American Development Bank (IDB). Coronavirus: Generando Nuevo Tráfico en América Latina. 2020. Available online: https://blogs.iadb.org/transporte/es/coronavirus-generando-nuevo-trafico-en-america-latina/ (accessed on 6 May 2020). (In Spanish)
50. NUMO. El Distanciamiento de los Pasajeros no es el Único Factor para Prevenir la Propagación del COVID-19 en el Transporte Público. 2020. Available online: https://www.numo.global/news/release-passenger-distancing-not-sole-factor-preventing-spread-covid-19-public-transportation (accessed on 20 August 2020). (In Spanish)
51. World Health Organization (WHO). Calidad del Aire y Salud. Available online: https://www.who.int/es/news-room/fact-sheets/detail/ambient-(outdoor)-air-quality-and-health (accessed on 3 April 2020). (In Spanish)

applied sciences

Article

Exact and Evolutionary Algorithms for Synchronization of Public Transportation Timetables Considering Extended Transfer Zones [†]

Sergio Nesmachnow *[ORCID] and Claudio Risso

Computer Science Institute, Engineering Faculty, University of the Republic, Montevideo 11200, Uruguay; crisso@fing.edu.uy
* Correspondence: sergion@fing.edu.uy
† This paper is an extended version of our paper published in III Ibero-American Congress on Smart Cities.

Featured Application: The planning methods presented in this article are specifically applicable to help decision makers in the processes of the configuration and operation of public intelligent transportation systems under the novel paradigm of smart cities.

Abstract: This article addresses timetable synchronization in public transportation, an important problem in modern smart cities, in order to guarantee a proper quality of service to citizens. Two variants of the bus timetabling synchronization problem considering extended transfer zones are studied: optimizing offsets and optimizing offsets and headways for each line. An exact mixed integer programming and an evolutionary algorithm are developed to solve both problem variants. The algorithms are evaluated on 45 instances of a real case study, the intelligent transportation system of Montevideo, Uruguay. Experimental results reported significant improvements over the current timetable implemented by the city administration. The number of successful synchronizations improved up to 66.6% and 179.9% for the first and second problem variant, respectively. The average waiting times for transfers improved, especially in tight problem instances (up to 57.8% and 158.3% for the first and second problem variant, respectively). The proposed planning methods are useful to help decision makers to configure public transportation systems.

Keywords: timetable synchronization; public transportation planning; mixed integer programming; evolutionary algorithms; real case study; smart cities

Citation: Nesmachnow, S.; Risso, C. Exact and Evolutionary Algorithms for Synchronization of Public Transportation Timetables Considering Extended Transfer Zones. *Appl. Sci.* **2021**, *11*, 7138. https://doi.org/10.3390/app11157138

Academic Editor: Jason K. Levy

Received: 3 July 2021
Accepted: 29 July 2021
Published: 2 August 2021

Publisher's Note: MDPI stays neutral with regard to jurisdictional claims in published maps and institutional affiliations.

1. Introduction

Public transportation is a key service in smart cities, allowing for efficient and low pollution mobility for citizens and reducing the dependency on cars and other motorized transportation modes [1,2]. Designing and operating an efficient public transportation system requires solving several relevant problems, including the proper design and management of routes, timetabling, and planning of buses and drivers, to provide a good quality of service to citizens [3] and also to promote sustainability [4].

The timetable synchronization problem consists of crafting a timetable that optimizes transfers between lines in a public transportation system. Timetable synchronization has been recognized as one of the most difficult problems for public transportation planning and optimization [5]. The problem has been often addressed intuitively, assuming that experienced operators are able to define *headways*, i.e., the time between consecutive departures of buses in a line to provide a proper quality of service. Conversely, this work proposes combinatorial optimization approaches to state the problem and the use of optimization techniques to solve real-world instances.

A flexible approach is proposed, entirely applicable to contexts where information about operating lines, traveling times, and transfer zones is available and when transfers between media are fairly regular along some reference time-window. The considered public transportation system consists of a mesh of routes operated by vehicles with similar capacities. Given origin-destination matrices for trips, expectations of passengers, technical features of buses, and operational costs, a historical design problem defines the lines to be deployed and their routes throughout each city in order to obtain a network with a proper trade-off between cost and quality of service [3]. Regarding the scope of this work, lines and routes are assumed fixed and known in advance, and timetable modifications are explored to allow for better overall efficiency [6].

Two variants of the bus timetabling synchronization problem considering extended transfer zones for every pair of bus stops in a city are formulated: optimizing the offset for each line and optimizing the offset and headways for each line. A realistic formulation is included to model the transfer demands of passengers. Two computational methods, exact mixed integer programming and the evolutionary algorithm (EA), are developed to efficiently solve both problem variants. The algorithms are evaluated on 45 instances of a real-world application case in Montevideo, Uruguay. Problem instances are defined using real data about lines, headways, and transfer demands, gathered from the intelligent transportation system of Montevideo. Results are compared with the real timetable currently implemented by the city administration.

The public transportation system of Montevideo is flexible, providing several options for passengers to reach a destination. Many passengers complete end-to-end trips by using a sole line, but either due to feasibility or convenience, many users transfer between lines to fulfill end-to-end trips. Transfers are implemented in the automatic Metropolitan Transportation System (STM). Users are charged through radio frequency identification cards, which allow them to pay and easily transfer between lines, even between different (geographically separated) bus stops.The STM keeps historical ticket sales and travel times via GPS-integrated controller units on buses. Records include user trip details along any day and time-stamped and geo-referenced transfers, etc. This dataset is a primordial asset for authorities, and it is used to adjust the configuration of lines. A relevant adjustment is the timetable, i.e., the schedule of departures each line must conform to. The STM dataset indicates that there are time-windows where the load of the system is regular, that is, with low deviation from mean values of traveling times, number of passengers in each bus, and the number of transfers between them [7]. By assuming that current headways are dimensioned to manage the load of the system along a time-window with regularity and that relatively slow deviations from those values linearly affect surveyed figures, a formulation is devised to allow tuning the system to increase the number of successful transfers. Timetable synchronization is a very relevant problem in this case study, since the bus system in Montevideo is extremely complex. The bus network consists of 145 main bus lines, and each one of them has multiple variants (e.g., outward and return trips, different origins/destinations, and shorter versions), so the total number of bus lines considering all variants is 1383, a remarkably large number when compared to bus networks in other similar cities [7].

In any case, the proposed models and algorithms are fully portable to other systems, as long as transfers are allowed and the main goal is optimizing the number of synchronized trips. The formulation is independent of data particulars, and it can be easily ported to another system. For instance, transfer zones can be arbitrarily set. They are not bound to infrastructure deployment, so an instance can, in fact, be determined from existing transfer stations or a mix of closed and on-street stations. In addition, there is no particular reason to prevent these models from being used with other means of transportation, such as light rail transit, metro networks, or arrangements thereof.

This article extends our previous conference article "Exact and metaheuristic approach for bus timetable synchronization to maximize transfers" [8]. New contents include: (i) an extended transfer zones model; (ii) the formulation of two problem variants, optimizing offset and headways (variable within a range); (iii) a new and more realistic objective function, properly modeling the transfer demands considering variable headways; (iv) the proposed exact and evolutionary approaches, adapted to solve both addressed problem variants; (v) an exhaustive experimental evaluation of the proposed optimization methods over 45 problem instances for each problem variant.

Regarding previous studies in the literature, this article contributes by considering an extended transfer zones proposal to model transfers between different (geographically separated) bus stops of different lines. This formulation is more useful than standard models to capture the reality of modern intelligent transportation systems that do not limit transfers to specific locations, instead allowing them to be performed at every pair of bus stops. The new problem model provides flexibility for passengers who have many realistic transfer options available and also for the bus system administration to design proper timetables. The new formulation also considers the explicit demand of transfer trips for each pair of bus stops that defines a transfer zone, unlike previous formulations that only considered the number of bus trips synchronized [9–11] or proposals focused on minimizing waiting times between bus trips [12]. Thus, the proposed formulation allows for focusing on the quality of service offered to passengers better than existing approaches. The proposed evolutionary approach is also a contribution, as no previous application of EAs to solve the problem was found in the review of related works. The EA is useful for solving large-dimension problem instances, e.g., by considering a large number of transfer zones in a city scenario. Besides the formulations proposed to model both problem variants, a relevant contribution of the reported research is related to the obtained results for the case study in Montevideo: both the exact and EA approaches are able to compute timetables that are significantly improved over the real timetable applied by the city administration, considered as a reference baseline for the comparison.

The article is organized as follows. Section 2 introduces the problem, the proposed model, and the two variants addressed. Section 3 presents a review of related articles. The exact and evolutionary approaches for bus synchronization are described in Section 5. Section 6 reports the evaluation of the proposed methods over realistic problem instances in Montevideo, and the main conclusions are outlined in Section 7.

2. The Bus Synchronization Problem

This section introduces the bus synchronization problem (BSP) model and the formulation of two specific variants of the problem.

Overall Description of the Problem Model

The BSP considers two of the most important purposes of a mass transportation system: offering an efficient means for the movement of citizens, while maintaining low costs and fares. The considered model focuses on citizens, offering them an efficient travel experience and short waiting times for passengers that use two or more buses to perform consecutive trips.

The concept of a *synchronization* event is defined as the action of providing passengers transfers whose waiting times are lower than the maximum time passengers are willing to wait. The research proposes addressing the BSP on real scenarios, built using real data about lines, bus stops, traveling times, and passengers performing transfers between lines. The problem model divides a day into several planning periods, considering the regularity of travel demands, travel times, and citizens' behavior. On each planning period, a data analysis approach can be applied to extract common characteristics and steady information to build BSP instances.

In the considered problem model, the bus network is represented by a set of bus lines and a set of relevant locations (the *synchronization nodes* or *transfer zones*), where passengers can transfer from one line to another. Unlike previous formulations of the problem [9,10], in the proposed model, nodes are not just bus stops but wider transfer zones, formed by separated bus stops for lines i and j. This way, the model can include several bus stops for several lines (see a graphical description for two generic lines i and j in Figure 1). The proposed model explicitly includes the distance between the bus stops for lines i and j,

$d_b^{i,j}$, drawn in red in Figure 1. For the instances solved in this article, it is assumed that the walking speed of pedestrians is constant $ws = 6$ km/h, and the walking time for a distance d between any pair of bus stops is given by $wt = d/ws$.

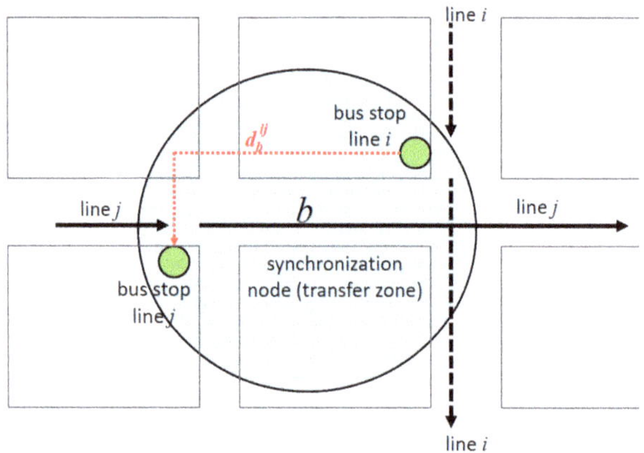

Figure 1. The considered model for the BSP with expanded transfer zones [blue circle].

The proposed problem model is useful for capturing the reality of contemporary intelligent transportation systems, which do not impose a limitation on the number of transfers for passengers and in which transfers are not limited to specific locations. Instead, transfers can be performed at every pair of bus stops, providing flexibility to passengers when planning a trip. This is the case of the intelligent transportation system in Montevideo, Uruguay, which is the case study in this article [7]. In this scenario, the corresponding synchronization problem is more complex, as many realistic transfers options are available to passengers, which in turn may require them to walk between bus stops.

Unlike traditional formulations, the proposed problem model is not focused on maximizing the number of passengers headed from origin to destination, mainly because the main focus is on the user experience when performing transfers, by considering the maximum time passengers are willing to wait for a transfer. This way, the objective function of the optimization model considers the demand of transfers in each transfer zone (pair of bus stops) for all synchronized trips of two bus lines. Previous articles addressing different variants of the timetable synchronization problem have worked under stronger assumptions, i.e., only considering trips for lines to synchronize.

Two cases are distinguished in the proposed model. The case with *uniform departing times* and the case with *non-uniform departing times*. These two cases are described next.

In the case with uniform departing times, there are no differences between departing times in the planning period; thus, r and s (trips of line i and line j, respectively) are not relevant for defining the time between consecutive departures (which in fact is constant, i.e., F_i^1). This is a realistic assumption when planning for short and medium periods (e.g., in the morning, in the afternoon, in rush hours, etc.). For example, in the case study solved in this article, i.e., the bus network of Montevideo, Uruguay, this assumption holds: in the time period 12:00–14:00 on working days, the time between consecutive buses is almost constant for all lines (the standard deviation of their values is between 0.80% and 1.37%). However, trips of line i and line j are relevant for defining the number of synchronizations and the number of synchronized passengers on each trip.

In the case with non-uniform departing times, the trip number r is relevant to determine synchronizations. The time between consecutive buses is given by $X_r^i - X_{r-1}^i$ for trip r of line i, and the problem formulation must be solved considering the number of trips that are synchronized. When computing the objective function in this model, the demand is split uniformly among the f_j trips of line j. This assumption is also realistic when passengers' demand has slight variations between consecutive trips. For modeling purposes, we are simply assuming that the planning period is determined because of its regularity. Whenever buses' interarrival times are uniformly distributed along a planning period, a fairly precise reference weight (i.e., average number of passengers) for each transfer zone is the ratio of the demand of transfers for the considered lines i and j over the number of trips of line i in the planning period T. However, due to the fact that the passengers transferring onto a bus previously alighted from another, they arrive in bursts. There are more accurate formulations for weights in transfer zones [13]. Defining and studying an accurate model to handle variable demands are proposed as future work.

3. Related Work

Timetable synchronization was stated as an important problem for public transportation systems in pioneering research articles by Ceder [3]. Daduna and Voß [14] presented one of the first proposals for synchronizing schedules on public transportation. The authors studied several objective functions, including weighted sum approaches with transfers and maximum waiting time, and proposed metaheuristics for simple problem versions with uniform frequencies. The approach was evaluated over a case study on Berlin and several cities in Germany. In the experiments, tabu search outperformed simulated annealing in randomized instances, and the authors highlighted the trade-off between operation costs and efficiency.

The transit network timetabling problem was studied by Ceder et al. [15] to optimize synchronization events at stops shared by several bus lines, by maximizing the number of simultaneous arrivals. A greedy approach was introduced to define custom timetables, properly selecting nodes from the bus network. The article focused on maximizing simultaneous bus arrivals over small problem instances involving few nodes and lines. A subjective metric was proposed by Fleurent et al. [16] to evaluate synchronizations, including expert knowledge. The proposed synchronization metric was applied to design a heuristic to minimize vehicle operation costs in a case study consisting of small problem instances from Montréal, Canada. Different timetables were found by weighting the components of the cost function.

Shafahi and Khani [17] presented a mixed integer programming model for optimizing transfers in a public transportation network with fixed transfer stations. The problem considered the offset optimization, i.e., setting the departure times of buses, which was solved by an exact method using CPLEX. A second problem variant was formulated, accounting for the extra stopping time of buses at transfer stations. A genetic algorithm was proposed to solve this problem variant over small problem instances involving 14 lines and just three transfer stations. A real case study was presented: the public transportation network in the city of Mashhad, Iran. The proposed methods were able to improve up to 14.5% with a business-as-usual (i.e., no explicit optimization) strategy.

A variant of the the synchronization problem including time-windows between travel times was addressed by Ibarra and Ríos [10]. A multi-start iterated local search (MILS) was applied to solve eight scenarios from the public transportation of Monterrey, Mexico, considering from 4 to 40 nodes. In less than 60 seconds, MILS computed accurate timetables for medium-size instances, regarding both upper bounds and an exact Branch & Bound algorithm. MILS was also applied by Ibarra et al. [11] to solve the multiperiod BSP, to optimize multiple trips of a given set of lines. MILS was able to compute results similar to a variable neighborhood search and a simple population-based algorithm on synthetic instances with few transfer zones. Results for a sample case study using data for a single line of Monterrey demonstrated that maximizing synchronizations for a specific node usually reduces the number of synchronizations for other nodes.

A recent article by Abdolmaleki et al. [18] proposed a model for timetable synchronization assuming a fixed headway for each line. The authors identified specific cases of the problem that are solvable in polynomial time. An approximation algorithm was proposed, based on the maximum directed cut problem. The proposed method was evaluated in a simple network and a large case study in Mashhad, Iran. In turn, a recursive algorithm that executes in quasi-linear time was proposed to minimize the total transfer waiting time in a problem variant that relaxes the fixed headway assumption.

An EA was proposed in a previous article [19] for a specific variant of the BSP, which outperformed real timetables and heuristics. This article extends our previous approach, solving a different BSP variant to determine optimal values for offset and headways, maintaining the number of trips of the real timetable, thus not impacting the provided quality of service.

Regarding the related literature, the approach proposed in this article contributes by considering the extended transfer zones model accounting for transfers between different (geographically separated) bus stops of different lines; by considering the explicit demand of transfer trips for each pair of bus stops that defines a transfer zone to better focus on the quality of service offered to passengers; and by simultaneously adjusting offsets and headways for each line.

4. The Proposed Problem Formulations

This section describes the proposed problem formulations for the considered problem variants.

4.1. Problem Data

The formulation of the studied problem considers:

- The planning period $[0, T]$, expressed in minutes.
- A set of bus lines $I = \{i_1, i_2, \ldots, i_n\}$, whose routes are fixed and known beforehand.
- The total trips needed to perform in order to fulfill the demand for each line i within the planning period $[0, T]$ is f_i. The demand considers both direct trips and transfers.
- A set of *synchronization nodes*, or *transfer zones*, $B = \{b_1, b_2, \ldots, b_m\}$. Each transfer zone $b \in B$ has three elements $<i, j, d_b^{ij}>$: i and j are the lines that may synchronize, and d_b^{ij} is the distance that separates the bus stops for lines i and j in b. Each b considers two bus stops with transfer demand between lines i and j, as described in Figure 1. The distance d_b^{ij} defines the time that a passenger must walk to transfer from the bus stop of line i to the bus stop of line j in the considered transfer zone.
- A *traveling time function* $TT : I \times B \to \mathbf{Z}$. $TT_b^i = TT(i, b)$ defining the time that buses in line i need to travel to reach the transfer zone b. The time is measured from the departure of the line. The traveling time depends of the studied scenario and is affected by several factors such as the maximum allowed speed, the traffic in the city, the travel demands, etc.
- A *demand function* $P : I \times I \times B \to \mathbf{Z}$. $P_b^{ij} = P(i, j, b)$ defines how many passengers perform a transfer from line i to line j in transfer zone b in $[0, T]$. As described in the previous subsection, a hypothesis of uniform demand is assumed. Thus, an effective number transfers from a trip of line i to a trip of line j is properly defined, taking into account the time between two consecutive trips of buses in line i. The uniform demand hypothesis is realistic for short periods, such as in the problem instances defined and solved for the addressed case study.
- The maximum time W_b^{ij} that passengers are willing to wait for line j, after alighting from line i and walking to the corresponding stop of line j in a transfer zone b. Two trips of line i and j are synchronized for transfers if and only if the waiting time of passengers that transfer from line i to line j is lower than or equal to W_b^{ij}.
- A set of departing times of each trip r of line i, X_r^i, which define the *headways* of the line as the time between two consecutive trips $F_r^i = (X_r^i - X_{r-1}^i)$. Headway values must be within a range of minimum (h_i) and maximum (H_i) headways for that line. Both extreme values are defined by the city administration or the bus system operator. The *offset* of each line is the departing time of the first trip of the line (X_1^i). Without

losing generality, the model assumes $X_0^i = 0$. All trips of each line must start within the planning period $[0, T]$ (i.e., $X_{f_i}^i \le T$).

Considering the previously defined elements, two variants of the BSP are formulated, accounting for the optimization of just offsets and both offsets and headways for each line, respectively. The next subsections describe these two problem variants.

4.2. Problem Variant #1: Offset Optimization

The first problem variant focuses on optimizing the offsets (i.e., the departing time of the first trip of each line). Subsequent departing times are fixed by the reference value defined by the city administration (F^i). Thus, the control variables of the problem are the offset of each line (X_1^i), which defines the whole set of departing times for all trips of each line. Auxiliary variables are needed to capture the synchronization events in each transfer zone. Binary variables Z_{rsb}^{ij} take a value 1 when trip r of line i and trip s of line j are synchronized in node b (i.e., trip r of line i arrives before trip s of line j and allows passengers to complete the transfer, i.e., walk between the corresponding bus stops and wait less than the waiting threshold for that transfer, W_b^{ij}). To guarantee that all lines perform the required number of trips in the planning period, possible values for the offset are limited to the interval $[0, T \,(\mathrm{mod}\ F^i)]$ (see a graphical description in Figure 2).

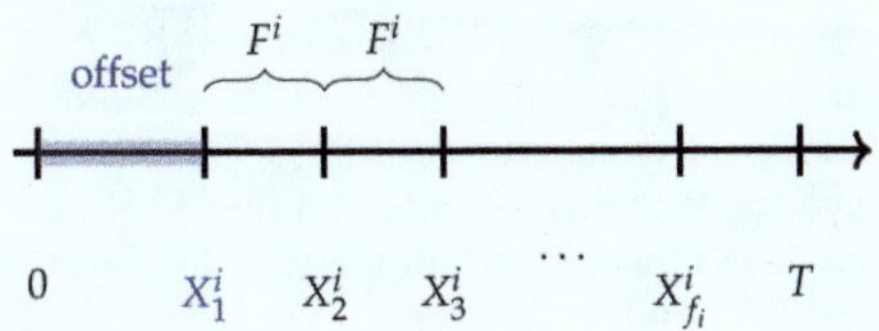

Figure 2. Graphical representation of BSP variant #1. The offset (in blue) is the control variable, and subsequent departures are separated by a fixed time F^i.

The mathematical model of BSP variant #1 as a mixed integer programming (MIP) problem is formulated in Equations (1)–(5).

$$\text{maximize} \quad \sum_{b \in B} \left(\sum_{r=1}^{f_i} \sum_{s=1}^{f_j} Z_{rsb}^{ij} \right) \cdot \frac{P_b^{ij} \times F^i}{T} \tag{1}$$

$$\text{subject to} \quad Z_{rsb}^{ij} \le 1 + \frac{(A_{rb}^i + d_b^{ij} + W_b^{ij}) - A_{sb}^j}{M}, \quad \begin{aligned} &\forall b = < i, j, d_b^{ij} > \in B, \\ &1 \le r \le f_i, 1 \le s \le f_j, \end{aligned} \tag{2}$$

$$Z_{rsb}^{ij} \le 1 + \frac{A_{sb}^j - (A_{rb}^i + d_b^{ij})}{M}, \quad \begin{aligned} &\forall b = < i, j, d_b^{ij} > \in B, \\ &1 \le r \le f_i, 1 \le s \le f_j, \end{aligned} \tag{3}$$

with $A_{rb}^i = X_1^i + (r-1)F^i + TT_b^i$ and $A_{sb}^j = X_1^j + (s-1)F^j + TT_b^j$

$$0 \le X_1^i \le \min(H_i, T \,(\mathrm{mod}\ F^i)), \quad \forall i \in I \tag{4}$$

$$Z_{rsb}^{ij} \in \{0, 1\}, \ \forall i \in I \tag{5}$$

The optimization problem formulates the maximization of the number of successful transfers completed in the planning period in every transfer zone (the objective function in Equation (1)). The expression $\sum_{r=1}^{f_i} \sum_{s=1}^{f_j} Z_{rsb}^{ij}$ is the total number of successful connections between trips of each pair of lines i and j in each transfer zone b. Assuming a uniform demand hypothesis, the number of transfers from a trip of line i to a trip of line j in the planning period is $P_b^{ij} \times F^i / T$. Equations (2)–(5) define the problem constraints.

Equation (1) states that the optimization will seek to activate synchronization variables Z_{rsb}^{ij}—as many as possible. Constraints determine that variables Z_{rsb}^{ij} only take the value 1 if the corresponding transfer is synchronized. In Equations (2) and (3), the arrival time of trip r of line i to transfer zone b is A^i_{rb}, and the arrival time of trip s of line j to transfer zone b is A^j_{sb}. For an interpretation of constraint (2), consider that the limit time $A^i_r + d^{ij}_b + W^{ij}_b$ defines the maximum time passengers are willing to wait for a transfer between trip r of line i and trip s of line j at transfer zone b. Whenever the arrival time of trip s of line j does not surpass that limit, the right-hand side of Equation (2) is greater (or equal) to 1, so this is the only case when Z^{ij}_{rsb} (the synchronization variable) is allowed to take the value 1. Furthermore, it is also necessary for passengers alighting from trip r of line i to walk to the transfer point (arriving at time $A^i_{rb} + d^{ij}_b$) before trip s of line j arrives (at time A^j_{sb}). Otherwise, passengers will not complete the transfer on time. This second condition, when met, also allows Z^{ij}_{rsb} to be set to 1, as the right term of constraints in Equation (3) is positive. To date, there is a potential issue when non-synchronized trips lead to negative values on the right term of Equation (3), which produces unfeasible constraints. The formulation only needs one value, $(A^i_{rb} + d^{ij}_b + W^{ij}_b) - A^j_{sb}$ or $A^j_{sb} - (A^i_{rb} + d^{ij}_b)$, to be lower than zero, so the synchronization variable Z^{ij}_{rsb} is deactivated. Thus, it is enough to introduce a constant value M, large enough to guarantee that both Equations (2) and (3) are always feasible. In a real solver implementation, considering large values of M might cause numerical stability problems. Thus, appropriate and relatively low values for M are computed as the maximum value within the union of sets $\{(H_i(j) + (f_i(j) - 1) \times \Delta X^j + TT^j_b) - (TT^i_b + d^{ij}_b + W^{ij}_b)\}$ and $\{(H_i(i) + (f_i(i) - 1) \times \Delta X^i + TT^i_b + d^{ij}_b) - TT^j_b\}$ for all transfer zones b in B. These values of M can be easily calculated during the process of crafting the MIP formulation before using any solver implementation, since finding M is a polynomial complexity problem. Equation (4) indicates that the maximum value for the offset of each line is the minimum between T (mod F^i) and the maximum headway H_i. No constraints are defined over headways, since a fixed frequency F^i, which satisfies the bounds for headways, is assumed for all subsequent trips. Finally, Equation (5) defines that decision variables Z^{ij}_{rsb} are within the domain of binary variables.

Without losing generality, the proposed problem formulation assumes that $F^j > W^{ij}_b$, $\forall j \in I$, i.e., headways of bus lines are larger than the waiting time thresholds for users. The case where $F^j \leq W^{ij}_b$ corresponds to a scenario in which the headway of line j is lower than the time users are willing to wait; thus, all transfers with line j would be synchronized, and they would not be part of the problem to solve (i.e., those lines can be removed for the specific problem instance to solve).

4.3. Problem Variant #2: Headways Optimization

The second problem variant proposes optimizing not only the offsets of each line but also the headways of each line. Headways are allowed to vary within a fixed range of the reference value F^i. The interval for variation is defined by $F^i(1-\alpha) \leq X^i_r - X^i_{r-1} \leq F^i(1+\alpha), \forall r \in 1, 2, \ldots, f_i$. To work under the assumption of maintaining an appropriate quality of service for the transportation system, which is assumed to be fairly provided by the current situation (i.e., using the reference values for the time difference between consecutive trips of each line), small values of α are considered (see a graphical description in Figure 3).

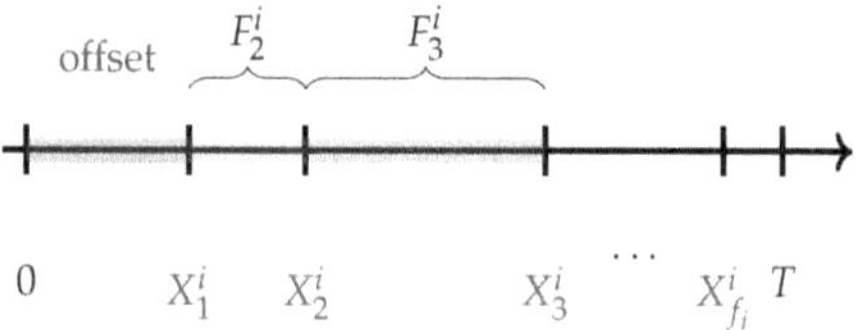

Figure 3. Graphical representation of BSP variant #2. The offset (X^i_1) and headways (F^i_r) (in blue) are the control variables, and departures are separated by variable headway values.

The mathematical model of BSP variant #2 as an MIP problem is formulated in Equations (6)–(13).

$$\text{maximize} \quad \sum_{b \in B} (\sum_{r=1}^{f_i} \sum_{s=1}^{f_j} Z_{rsb}^{ij}) \cdot \frac{P_b^{ij} \times (X_r^i - X_{r-1}^i)}{T} \tag{6}$$

$$\text{subject to} \quad Z_{rsb}^{ij} \leq 1 + \frac{(A_{rb}^i + d_b^{ij} + W_b^{ij}) - A_{sb}^j}{M}, \qquad \begin{array}{l} \forall b = <i,j,d_b^{ij}> \in B, \\ 1 \leq r \leq f_i, 1 \leq s \leq f_j, \end{array} \tag{7}$$

$$Z_{rsb}^{ij} \leq 1 + \frac{A_{sb}^j - (A_{rb}^i + d_b^{ij})}{M}, \qquad \begin{array}{l} \forall b = <i,j,d_b^{ij}> \in B, \\ 1 \leq r \leq f_i, 1 \leq s \leq f_j, \end{array} \tag{8}$$

$$\text{with } A_{rb}^i = X_r^i + TT_b^i, \ A_{sb}^j = X_s^j + TT_b^j$$

$$\sum_{s=1}^{f_j} Z_{rsb}^{ij} \leq 1, \forall i, j \in I, 1 \leq r \leq f_i \tag{9}$$

$$\max(F^i(1-\alpha), h_i) \leq X_r^i - X_{r-1}^i, \qquad \forall r \in 2, .., f_i \tag{10}$$

$$X_r^i - X_{r-1}^i \leq \min(F^i(1+\alpha), H_i), \qquad \forall r \in 2, .., f_i \tag{11}$$

$$T - H_i \leq X_{f_i}^i \leq T, \forall i \in I \tag{12}$$

$$Z_{rsb}^{ij} \in \{0,1\}, 0 \leq X_1^i \leq \min(H_i, mod(T, F^i)) \tag{13}$$

In Equations (6)–(13), X_r^i is the time of departure of trip r of line i, and F^i is the reference value for the time difference between consecutive trips of line i, as defined in problem variant #1. In turn, $\alpha \in [0,1]$ is the coefficient used for defining the allowed deviation of times between buses from the reference value F^i. Finally, $X_{f_i}^i$ is the departing time of the last trips of the line i.

The objective function is Equation (6), i.e., maximizing the number of successful transfers completed in every transfer zone in the planning period. In this case, the demand is split considering the (flexible) time between consecutive buses of line i (i.e., $X_r^i - X_{r-1}^i$). Equations (7)–(13) formulate the problem constraints. Equations (7) and (8) define a synchronization, as in the previous problem formulation.

Equation (9) guarantees that every trip r of line i synchronizes with, at most, one trip s of line j. The last point is necessary because headways are part of the control variables in this case. In the previous model, headways were known in advance, so they could be pre-filtered in the case where a headway was lower than the maximum time passengers are willing to wait at some stop. Conversely, in this model, Equation (9) prevents us from counting such synchronizations more than once. Constraints in Equations (10) and (11) impose the limits for headways, according to the allowed variation α. Constraints in Equation (12) guarantee that the last trip is within the specified maximum range. Finally, Equation (13) defines the domain for decision variables Z_{rsb}^{ij}.

The formulation in Equations (6)–(13) intuitively extends the previous to incorporate headways as control variables, but it has the drawback of being quadratic in its objective function due to the product of $Z_{rsb}^{ij}(X_r^i - X_{r-1}^i)$. However, it is linearized by a change of variables and additional constraints. Let y_{rsb}^{ij} be as $Z_{rsb}^{ij}(X_r^i - X_{r-1}^i)$ in (6), so the objective turns out to be $\frac{1}{T} \sum_{b \in B} \sum_{r=1}^{f_i} \sum_{s=1}^{f_j} y_{rsb}^{ij} \cdot P_b^{ij}$, which is linear. Moreover, two equations per each y_{rsb}^{ij} variable must be included: $y_{rsb}^{ij} \leq (X_r^i - X_{r-1}^i)$ and $y_{rsb}^{ij} \leq H_i \cdot Z_{rsb}^{ij}$. Within a maximization problem, variables y_{rsb}^{ij} will take a value as high as possible. H_i is an upper bound for $(X_r^i - X_{r-1}^i)$ because of Equation (11), so whenever $Z_{rsb}^{ij} = 1$, the second equation results,

$y_{rsb}^{ij} \le H_i$, and it is the first equation that guarantees y's value to be $(X_r^i - X_{r-1}^i)$ at most, which is then exactly the value that the variable y_{rsb}^{ij} will take. Conversely, when $Z_{rsb}^{ij} = 0$, the second equation forces y_{rsb}^{ij} to be 0. This behavior replicates that of $Z_{rsb}^{ij}(X_r^i - X_{r-1}^i)$'s product, so the change of variables is equivalent and linear.

Table 1. Summary of the proposed BSP variants.

Problem	Decision Variables	Model (Objective, Constraints)
variant #1	X_1^i (offset)	Equations (1)–(5)
variant #2	X_1^i (offset), X_s^i (headways, within range)	Equations (6)–(13)

Table 1 summarizes both problem variants. Variables X_1^i (i.e., offsets) are the control variables in both proposed problem variants, while the latter adds new control variables X_r^i with $r > 1$. The difference is semantic rather than substantive, and we refer to them as X_r^i in general. The group of variables Z_{rsb}^{ij} are auxiliary in both proposed problem variants. They are used to identify synchronizations as an outcome of control variable X_r^i values. These variables are only meaningful when stated as Boolean variables, a condition that must be explicitly set in the formulation since it is not achieved otherwise. Finally, problem variant #2 introduces y_{rsb}^{ij} variables, which are also auxiliary and number as many as those of Z_{rsb}^{ij}. They are used to obtain to a purely linear alternative formulation for the second problem variant.

Unlike Z_{rsb}^{ij} variables, the domain of X_r^i and y_{rsb}^{ij} is that of the real numbers, since their values are a consequence of active constraints in an optimization process. However, whenever parameters that define an instance are integers, optimal X_1^i values Equations (1)–(5) are to be integers as well, as they are the result of pushing the objective function variables against integer constraints. The situation is not so for Equations (6)–(13), because α values greater than 0 generally lead to non-integer bounds in both proposed problem variants.

5. Exact and Evolutionary Computation Methods for the BSP

This section presents the exact and evolutionary computation approaches developed to address the BSP, considering extended transfer zones.

5.1. Mathematical Programming

The exact method for solving the formulated model was developed by combining several tools. Each instance in the input dataset was imported into MATLAB matrices, just as it was for each solution (externally found). In this way, results could be easily analyzed, debugged, and post-processed. The software version of MATLAB is R2015a-8.5.0.

Some parameters of the models formulated in both proposed problem variants are to be computed during the MIP construction itself. The most notorious is the value of the parameter M, which has to be large enough to prevent (2), (3) or (7), (8) from being unfeasible but not so large to lead to numerical issues. Because of this, we decided not to use AMPL or other algebraic modeling language to feed the solver. Instead, we developed a C++ program to read instances in the input dataset and convert them to CPLEX LP-format.

IBM(R) ILOG(R) CPLEX(R) Interactive Optimizer 12.6.3.0 was used as the optimization tool. The default GAP tolerance for the MIP solver is 0.01%. This corresponds to the relative distance between the best integer solution found and the best upper bound estimated up to that moment: $(f(x) - bestBound)/f(x)$, where x is a solution, $f(x)$ is its objective function value, and $bestBound$ is the lowest upper bound found for the optimum value.

5.2. Evolutionary Algorithm

The EA was developed using the Malva library (github.com/themalvaproject, accessed on 15 June 2021) in the C++ programming language. The main implementation details are described in the next subsections.

Solution Encoding

For both problem variants, candidate solutions are represented using integer vectors. Next, both representations are explained.

Problem variant #1: offset optimization. For problem variant #1, each integer value in the solution representation indicates the offset for each bus line, i.e., the time elapsed between time zero (start of the planning period) and the time when the first trip of the line departs. A candidate solution of the BSP is represented by a vector $X = \{X_1^1, X_1^2, \ldots X_1^n\}$ (n is the number of lines in the instance), $X_0^i \in \mathbf{Z}^+$, and $0 \le X_0^i \le T \mod H^i$. Figure 4 describes the solution representation for a problem instance with n bus lines.

X_1^1	X_1^2	X_1^3	$\cdots$	X_1^{n-1}	X_1^n

Figure 4. Solution representation for BSP variant #1.

Problem variant 2: headways optimization. For problem variant #2, integer values in the solution representation represent, for each bus line, the offset (in minutes) and the subsequent headway values (also in minutes). A candidate solution of the BSP is represented by a vector $X = \{X_1^1, X_2^1, \cdots, X_{f_1}^1, X_1^2, X_2^2 \cdots, X_{f_2}^2, \cdots, X_1^n, \cdots, X_1^n, \cdots, X_{f_n}^n\}$, where $X_0^i \in \mathbf{Z}^+ \cup 0$, and $X_{k_i}^i \in \mathbf{Z}^+$, with $k_i \in \{1, \cdots, f_i\}$. Figure 5 describes the solution representation for a problem instance with n bus lines (offset values for each line are marked in blue font).

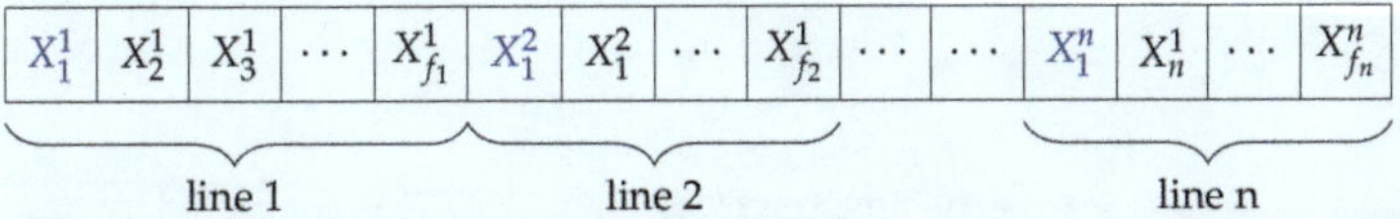

Figure 5. Solution representation for BSP variant #2.

Evolution model. The EA follows the $(\mu + \lambda)$ evolution model [20]. λ offsprings are generated from μ parents, and all compete among themselves to select the solutions to include in the population in a new generation. The $(\mu + \lambda)$ evolution model computed better and more diverse solutions than a standard generational model in configuration experiments.

Initialization operator. The initialization generates random solutions, selecting integer values within the corresponding ranges and taking into account the problem constraints. The random initialization operator is conceived to generate an appropriate diversity for the evolutionary process.

Selection operator. A tournament method is used for selecting individuals during the evolutionary search. A tournament size of three individuals was used (just one individual survives). The tournament operator outperformed the standard proportional selection in configuration experiments, providing a proper selection pressure for candidate solutions during the evolutionary process.

Recombination operator. An ad hoc variant of the well-known two-point crossover operator was designed. Crossover points are randomly selected in $[1, n-1]$, and information from both parents is exchanged between the crossover points. The main idea is to maintain those features of bus lines synchronized in the parent to be used as relevant information for offspring generation. The recombination probability is p_R. A correction procedure is applied to guarantee the feasibility of the generated solutions by properly shifting conflicting offset and/or headway values.

Mutation operator. An ad hoc variant of Gaussian mutation is applied. Selected position(s) in an individual are changed according to a Gaussian distribution and considering the limits (minimum and maximum) defined for both offsets and headways of each line. The mutation probability is p_M.

6. Experimental Evaluation

The experimental analysis of the exact and evolutionary methods for the BSP is reported in this section.

6.1. Methodology

This subsection describes the methodology applied for the experimental evaluation of the proposed methods to solve the BSP.

6.1.1. Problem Instances

The experimental evaluation considered realistic problem instances, generated using real information from the case study, i.e., the STM in Montevideo, Uruguay.

Several data sources were considered to gather information to build the BSP instances. All the information about the bus network (lines, routes, real timetables, bus stops locations, etc.) was retrieved from the National Open Data Catalog (catalogodatos.gub.uy, accessed on 15 June 2021). The real information about transfer demands was provided by the city administration. All the gathered data were analyzed applying urban data analysis [7]. Several elements define the scenario and the problem instances built:

- The starting time was set to 12:00 p.m., and the final time was set to 2:00 p.m., including one of the two peak hours occurring in a working day for the public transportation system in Montevideo [21].
- The transfer demand function (P) is generated from real information registered by the smart cards of the STM.
- The transfer zones are selected from the pairs of bus stops with the largest demand of registered transfers for the period; all pairs are candidates to be selected in the created BSP instances; the number of considered transfer zones is 170.
- The bus lines are those connecting the considered transfer zones; a total number of 250 lines are considered.
- The time traveling function (TT) is computed by processing the real GPS data from the operating vehicles for each line;
- The walking time function is defined by the product of the estimated pedestrian speed (6 km/h) and the distance between bus stops in each transfer zone; the distance is computed using geospatial analysis.
- The maximum waiting time (W) is λH; five values of λ are considered ($\lambda \in [0.3, 0.5, 0.7, 0.9]$ to define BSP instances with different tolerance levels.

A total number of 75 BSP instances are defined, of three different dimensions (30, 70, and 110 transfer zones), using the real information of buses operating in the STM of Montevideo, Uruguay. Problem instances are named as [NP].[NL].[λ].[id]: NP $= n$ the number of transfer zones, NL $= m$ the number of bus lines, λ the tolerance to define the maximum waiting time (percentage), and id is a number to differentiate instances with the same values of the other parameters. Problem instances are publicly available at www.fing.edu.uy/inco/grupos/cecal/hpc/bus-sync/, accessed on 15 June 2021.

6.1.2. Execution Platform

Experiments were carried out on a Quad-core Xeon E5430 at 2.66GHz, 8 GB RAM, from the high-performance computing infrastructure of the National Supercomputing Center (Cluster-UY), Uruguay [22].

6.1.3. Baseline Solution for the Comparison

A significant solution to be used as baseline for the comparison is the real timetable used in the transportation system of Montevideo. This solution determines the current level of service regarding both direct trips and transfers. The real timetable does not apply an explicit synchronization approach for transfers. The first trip of each line departs when the planning period starts, and subsequent trips depart according to the real predefined headways for each line. The comparison with the real timetable allows determining the advantages of the explicit optimization approaches proposed for the studied BSP variants.

6.1.4. Statistical Analysis and Metrics

The effectiveness of the proposed EA is assessed via proper statistical analysis. For every problem instance and experiment, 30 independent executions (i.e., using a different seed for the random numbers generator) were performed. Then, a three-step analysis is applied to study the fitness results distributions:

1. *Normality check.* The Shapiro–Wilk statistical test is applied to determine if the results distribution is modeled by a normal distribution, i.e., computing the likeliness of the underlying randomness to be normally distributed.
2. *Mean rank comparison* (for parametric configuration analysis). The Friedman rank statistical test is applied to detect/analyze the differences in the distributions of fitness values across multiple executions.
3. *Pairwise comparison of distributions.* The Kruskal–Wallis non-parametric test is applied to determine if the differences between two parameter configurations have statistic significance.
4. *Boxplots* are used for results visualization and graphical comparison. Relevant order statistics are computed and reported: first quartile (Q1), third quartile (Q3), median, minimum, and maximum values, since the Shapiro–Wilk test confirmed that the results do not follow a normal distribution. The interquartile range (IQR) is used as a measure of statistical dispersion, as usual for non-normal distributions.

Several metrics are used for evaluating the exact and the evolutionary approach: (i) the number of successfully synchronized trips for passengers (i.e., the objective function defined in Equations (1) and (6)), (ii) the improvements over the real solution, and (iii) the average waiting for transfers in a transfer zone.

6.1.5. Parameter Setting

Exact Mathematical Programming. The default set of parameters of CPLEX was used to find exact solutions, except for the timeout parameter, which was set to 7200 seconds. Variant #2 with $\alpha = 0$ is equivalent to variant #1, where headways are fixed and equal to F^i. However, problem formulations are not equal nor are the proposed methods to solve them. Variant #2 implies significantly more variables than variant #1, so it is worth checking that algorithms are also able to rapidly find solutions when $\alpha = 0$. Such instances were solved within a few seconds, so the timeout limit only applies for $\alpha = 0.3$. It was verified, though, that the performance of the more general variant #2 is quite good for very low values of α, comparable to the performance of the much simpler version #1. Runtimes degrade for higher values of α, and in fact, the best solutions found for $\alpha = 0.3$ were not proven to be optimal according to the defined tolerance for the MIP solver (0.01%). There is room to explore changes in parameters in order to improve performance, which is appointed as a future line of work.

Evolutionary algorithm. Since EAs are non-deterministic, parameter configuration analysis is mandatory to find the proper combination of parameter values to compute the best results. Studied parameters included population size (ps), recombination probability (p_R), and mutation probability (p_M). Experiments were performed on five small problem instances with different features to avoid bias. Candidate values considered for each parameter were $ps \in \{15, 25, 50\}$, $p_R \in \{0.5, 0.75, 0.9\}$, and $p_M \in \{0.001, 0.01, 0.1\}$.

A summary of the results obtained in the analysis of the population size is presented in Table 2. Results of median and best fitness values are reported for each population size. The Friedman rank test is applied to analyze the results distribution.

Table 2. Analysis of the population size.

	Average Fitness				
ps	30.37.30.1	30.37.50.1	30.37.70.1	30.37.90.1	30.37.100.1
15	251.96	267.15	286.97	295.06	296.51
25	253.78	266.50	288.01	295.45	297.48
50	253.70	271.89	287.39	295.37	297.19
	Friedman Rank (p-Value $< 10^{-3}$)				
ps	30.37.30.1	30.37.50.1	30.37.70.1	30.37.90.1	30.37.100.1
15	3	2	3	3	3
25	1	3	1	1	1
50	2	1	2	2	2
	Best Fitness				
ps	30.37.30.1	30.37.50.1	30.37.70.1	30.37.90.1	30.37.100.1
15	260.71	276.41	291.07	296.69	298.31
25	260.94	275.33	291.72	297.95	298.46
50	262.18	273.62	290.06	296.09	298.22
	Friedman Rank (p-Value $< 10^{-3}$)				
ps	30.37.30.1	30.37.50.1	30.37.70.1	30.37.90.1	30.37.100.1
15	3	1	2	2	2
25	2	2	1	1	1
50	1	3	3	3	3

Results in Table 2 demonstrate that the best results were computed using a population size of 25 individuals. Using a larger population size turned to be counterproductive for the EA, since suboptimal solutions dominate quickly and results do not improve in the long term.

Table 3 reports the best fitness values and the Friedman ranks of the nine configurations (C1,...,C9) considered in the analysis of the recombination and mutation probabilities. The p-value of the Friedman rank was below 10^{-3}, thus assuring statistical significance of the results. In turn, the boxplots in Figure 6 graphically compare the median, best, worst, Q1, and Q3 values obtained for each studied configuration in the problem instance, which is representative of the results computed for other instances too. According to the reported metrics, the best results were obtained with configuration C2, i.e., setting $PS = 25$, $p_R = 0.5$, and $p_M = 0.005$. Configuration C2 systematically obtained the best Friedman ranking in all but one of the studied problem instances.

Table 3. Analysis of the recombination and mutation probabilities in the proposed EA.

			Fitness				
id	p_R	p_M	30.37.30.1	30.37.50.1	30.37.70.1	30.37.90.1	30.37.100.1
C1	0.5	0.010	253.18	263.90	285.82	292.03	294.97
C2	0.5	0.005	263.30	274.21	293.15	298.19	298.79
C3	0.5	0.001	253.99	267.70	287.67	295.56	297.06
C4	0.75	0.010	250.67	268.65	286.70	293.12	294.50
C5	0.75	0.005	262.91	276.65	291.69	296.48	298.25
C6	0.75	0.001	250.32	270.05	284.09	292.87	294.89
C7	0.9	0.010	255.12	269.66	289.43	293.91	296.06
C8	0.9	0.005	260.94	275.33	291.72	297.95	298.46
C9	0.9	0.001	256.80	269.87	288.57	296.71	297.45

Table 3. *Cont.*

			Friedman Rank					
id	p_R	p_M	30.37.30.1	30.37.50.1	30.37.70.1	30.37.90.1	30.37.100.1	*avg.*
C1	0.50	0.01	7	9	8	9	7	8.0
C2	0.50	0.005	1	3	1	1	1	1.4
C3	0.50	0.001	6	8	6	5	5	6.0
C4	0.75	0.01	8	7	7	7	9	7.6
C5	0.75	0.005	2	1	3	4	3	2.6
C6	0.75	0.001	9	4	9	8	8	7.6
C7	0.90	0.01	5	6	4	6	6	5.4
C8	0.90	0.005	3	2	. 2	2	2	2.2
C9	0.90	0.001	4	5	5	3	4	4.2

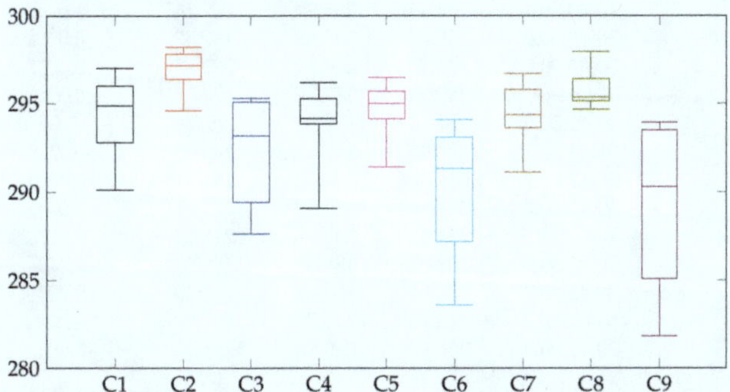

Figure 6. Boxplot analysis of parameter configurations for a representative BSP instance.

6.2. Numerical Results

This subsection reports the numerical results of the proposed methods for the BSP and the comparison with the baseline real timetable for both studied problem variants.

6.2.1. Problem Variant #1: Offset Optimization

Table 4 presents the objective function values achieved by the exact and the EA for the studied BSP instances for variant #1 (offset optimization). The real timetable is reported as baseline for the comparison. Column 'EA vs. exact' compares the results of both proposed approaches (a negative percentage value means a smaller function value for the EA solution). Relative improvements over the real timetable (in percentage values) are also reported. Column 'EA vs. real' reports the relative improvement over the real timetable for the EA solution, and column 'exact vs. real' reports the relative improvement over the real timetable for the exact solution.

Results in Table 4 demonstrate that both exact and EA significantly outperform the baseline real solution in all problem instances. EA improved the real solution up to 66,3% in instance 40.37.30.1, and the exact method improved over the real solution up to 66.6% in instance 70.63.30.2. When comparing the EA and the exact solutions, the evolutionary approach demonstrated to be highly efficient at solving the problem: it computed the optimal solution in 54 out of the 75 problem instances studied. In all other instances, the distance to the optimum was below 1%. Overall, improvements over the real timetable were 24.5% (on average) for EA and 24.6% (on average) for the exact solution.

Table 4. Summary of results: exact and EA for BSP variant #1.

Scenario	EA	Exact	Real	EA vs. Exact	EA vs. Real	Exact vs. Real
30.37.100.1	265.76	265.76	245.33	0.00%	8.3%	8.3%
30.37.30.1	171.12	171.12	102.88	0.00%	66.3%	66.3%
30.37.50.1	203.52	203.58	165.28	−0.03%	23.1%	23.2%
30.37.70.1	252.22	252.22	219.60	0.00%	14.9%	14.9%
30.37.90.1	260.68	260.68	245.33	0.00%	6.3%	6.3%
30.40.100.0	205.68	205.68	187.57	0.00%	9.7%	9.7%
30.40.100.4	223.17	223.17	204.49	0.00%	9.1%	9.1%
30.40.30.0	131.76	132.13	82.83	−0.28%	59.1%	59.5%
30.40.30.4	143.98	143.98	98.36	0.00%	46.4%	46.4%
30.40.50.0	162.15	162.17	122.31	−0.01%	32.6%	32.6%
30.40.50.4	169.66	169.66	123.83	0.00%	37.0%	37.0%
30.40.70.0	195.54	195.54	166.10	0.00%	17.7%	17.7%
30.40.70.4	208.99	208.99	179.18	0.00%	16.6%	16.6%
30.40.90.0	205.68	205.68	187.57	0.00%	9.7%	9.7%
30.40.90.4	223.12	223.12	202.69	0.00%	10.1%	10.1%
30.41.100.2	247.59	247.59	224.50	0.00%	10.3%	10.3%
30.41.30.2	154.11	154.20	95.89	−0.06%	60.7%	60.8%
30.41.50.2	186.95	187.35	145.39	−0.21%	28.6%	28.9%
30.41.70.2	236.50	236.50	195.47	0.00%	21.0%	21.0%
30.41.90.2	247.59	247.59	224.05	0.00%	10.5%	10.5%
30.42.100.3	231.13	231.13	211.67	0.00%	9.2%	9.2%
30.42.30.3	138.83	138.83	93.27	0.00%	48.8%	48.8%
30.42.50.3	168.64	168.64	142.38	0.00%	18.4%	18.4%
30.42.70.3	212.78	212.78	187.80	0.00%	13.3%	13.3%
30.42.90.3	230.98	230.98	211.12	0.00%	9.4%	9.4%
70.60.100.1	540.41	540.41	492.20	0.00%	9.8%	9.8%
70.60.30.1	332.11	333.02	211.56	−0.27%	57.0%	57.4%
70.60.50.1	404.30	406.61	310.96	−0.57%	30.0%	30.8%
70.60.70.1	509.60	509.60	435.28	0.00%	17.1%	17.1%
70.60.90.1	539.53	539.53	491.70	0.00%	9.7%	9.7%
70.62.100.3	522.34	522.34	479.58	0.00%	8.9%	8.9%
70.62.30.3	306.14	309.02	207.12	−0.93%	47.8%	49.2%
70.62.50.3	388.99	389.69	321.41	−0.18%	21.0%	21.2%
70.62.70.3	486.82	486.82	417.53	0.00%	16.6%	16.6%
70.62.90.3	521.82	521.82	478.46	0.00%	9.1%	9.1%
70.63.100.2	525.90	525.90	478.87	0.00%	9.8%	9.8%
70.63.30.2	332.06	334.45	200.77	−0.55%	65.7%	66.6%
70.63.50.2	405.90	405.90	316.53	0.00%	28.2%	28.2%
70.63.70.2	497.84	497.84	427.21	0.00%	16.5%	16.5%
70.63.90.2	525.90	525.90	477.09	0.00%	10.2%	10.2%
70.67.100.0	487.42	487.42	440.27	0.00%	10.7%	10.7%
70.67.30.0	304.60	304.85	203.81	−0.08%	49.5%	49.6%
70.67.50.0	374.91	374.98	289.39	−0.02%	29.6%	29.6%
70.67.70.0	459.73	459.93	386.96	−0.04%	18.8%	18.9%
70.67.90.0	487.42	487.42	440.27	0.00%	10.7%	10.7%
70.69.100.4	505.15	505.15	456.10	0.00%	10.8%	10.8%
70.69.30.4	322.57	322.57	209.04	0.00%	54.3%	54.3%
70.69.50.4	385.63	385.63	295.16	0.00%	30.7%	30.7%
70.69.70.4	475.36	475.36	408.63	0.00%	16.3%	16.3%
70.69.90.4	504.20	504.20	454.31	0.00%	11.0%	11.0%

Table 4. *Cont.*

Scenario	EA	Exact	Real	EA vs. Exact	EA vs. Real	Exact vs. Real
110.76.100.2	777.19	777.19	706.18	0.00%	10.1%	10.1%
110.76.30.2	487.71	489.13	298.20	−0.29%	63.6%	64.0%
110.76.50.2	591.50	591.50	465.58	0.00%	27.0%	27.0%
110.76.70.2	731.20	731.20	631.71	0.00%	15.7%	15.7%
110.76.90.2	777.06	777.06	703.90	0.00%	10.4%	10.4%
110.78.100.0	806.58	806.58	728.94	0.00%	10.7%	10.7%
110.78.100.1	826.45	826.45	752.16	0.00%	9.9%	9.9%
110.78.100.3	803.13	803.13	731.61	0.00%	9.8%	9.8%
110.78.30.0	487.92	490.34	315.02	−0.49%	54.9%	55.7%
110.78.30.1	507.28	511.37	311.43	−0.80%	62.9%	64.2%
110.78.30.3	499.39	499.39	307.45	0.00%	62.4%	62.4%
110.78.50.0	606.89	610.36	461.39	−0.57%	31.5%	32.3%
110.78.50.1	623.89	623.89	467.20	0.00%	33.5%	33.5%
110.78.50.3	605.72	609.60	472.20	−0.64%	28.3%	29.1%
110.78.70.0	753.93	754.26	648.61	−0.04%	16.2%	16.3%
110.78.70.1	775.14	775.14	659.51	0.00%	17.5%	17.5%
110.78.70.3	753.11	753.76	635.61	−0.09%	18.5%	18.6%
110.78.90.0	805.77	805.77	727.38	0.00%	10.8%	10.8%
110.78.90.1	824.47	824.47	751.09	0.00%	9.8%	9.8%
110.78.90.3	802.42	802.42	729.99	0.00%	9.9%	9.9%
110.83.100.4	779.09	779.09	713.79	0.00%	9.1%	9.1%
110.83.30.4	501.43	501.43	305.51	0.00%	64.1%	64.1%
110.83.50.4	593.89	593.89	464.88	0.00%	27.8%	27.8%
110.83.70.4	730.68	730.68	635.29	0.00%	15.0%	15.0%
110.83.90.4	778.09	778.09	711.49	0.00%	9.4%	9.4%

The average improvements of both exact and EA over the baseline solutions, grouped by tolerance and scenario size, are reported in Table 5. Regarding tolerance levels, improvements of EA and the exact algorithm over the real timetable were up to 57.56% and 57.96%, respectively, in scenarios with $\lambda = 30$, which pose the tighter bounds for user waiting times. In less restrictive scenarios, both studied methods improved over 9.74% in regard to the real timetable. Regarding the scenario dimension, both methods computed robust solutions that improve between 23.88% (EA, for scenarios with 30 transfer zones) and 25.72% (exact, for scenarios with 110 transfer zones).

Table 5. Exact and EA improvements over the baseline real solutions, grouped by tolerance and problem dimension (number of transfer zones).

EA Over Real			Exact Over Real			EA vs. Exact		
λ	Δ	Instances	λ	Δ	Instances	λ	Δ	Instances
30	57.56%	15	30	57.96%	15	30	−0.25%	15
50	28.49%	15	50	28.68%	15	50	−0.15%	15
70	16.79%	15	70	16.80%	15	70	−0.01%	15
90	9.79%	15	90	9.79%	15	90	0.00%	15
100	9.74%	15	100	9.74%	15	100	0.00%	15

EA Over Real			Exact Over Real			EA vs. Exact		
N	Δ	Instances	N	Δ	Instances	N	Δ	Instances
30	23.88%	25	30	23.92%	25	30	−0.02%	25
70	23.99%	25	70	24.15%	25	70	−0.11%	25
110	25.55%	25	110	25.72%	25	110	−0.12%	25

The reported improvements on the objective function values grouped by tolerance and problem dimension indicate that for all sizes, the improvements over the real timetable increase as user tolerance decreases. This is a relevant result, which demonstrates that the optimization methods properly scale when the complexity of the problem increases, providing solutions with a better quality of service. Improvements also slightly increase for larger instance dimensions.

6.2.2. Problem Variant #2: Offset and Headways Optimization

Problem variant #2 imposes a significantly harder challenge for the studied optimization algorithms. Control variables, originally spanning a sole variable per line (offsets), also incorporate several more variables per line (headways). In addition, as mentioned in Section 4.3, in the exact model, the number of variables increases not only because of control variables (i.e., offsets and headways) but also because of auxiliary variables (i.e., synchronizations and the objective's contributions), half of which are Boolean. In turn, for the EA, the number of variables increased more than 20 times, and a correction procedure is needed to repair non-feasible solutions generated by the evolutionary operators. Both facts result in a larger and more complex search space for optimization.

Table 6 reports the results of the exact algorithm for BSP variant #2. Each scenario has an associated row in the table. Columns *#vars*, *#varsCtl*, and *#varsBool*, respectively, correspond to the total number of variables, the number of control variables (all of which are real variables), and the number of Boolean variables. The column named *best-sol* reports the value of the objective function for the best solution found before timeout. Column *GAP* corresponds to the estimated worst-case gap to the optimum, that is, the relative gap between the reported solution and the lowest upper bound for the optimum estimated up to that moment. Finally, column *time to solution* shows the time required for the solver to find that solution.

Table 6. Summary of results of the exact algorithm for BSP variant #2.

Scenario	#vars	#varsCtl	#varsBool	Best-Sol	GAP	Time to Solution
30.40.100.0	7762	410	3676	240.48	15.01%	6533
30.40.30.0	7762	410	3676	231.80	15.36%	7015
30.40.50.0	7762	410	3676	236.45	12.78%	6952
30.40.70.0	7762	410	3676	239.78	13.84%	6487
30.40.90.0	7762	410	3676	240.31	13.58%	6502
30.40.100.4	8086	418	3834	261.02	11.26%	1914
30.40.30.4	8086	418	3834	246.10	16.24%	6822
30.40.50.4	8086	418	3834	251.91	14.88%	328
30.40.70.4	8086	418	3834	258.56	12.69%	640
30.40.90.4	8086	418	3834	261.22	11.48%	272
30.42.100.3	9053	481	4286	266.78	13.77%	6544
30.42.30.3	9053	481	4286	250.58	14.47%	6912
30.42.50.3	9053	481	4286	261.86	12.01%	6606
30.42.70.3	9053	481	4286	265.36	12.92%	319
30.42.90.3	9053	481	4286	266.51	14.40%	196
30.37.100.1	9107	435	4336	301.73	16.91%	6456
30.37.30.1	9107	435	4336	283.66	19.13%	6562
30.37.50.1	9107	435	4336	294.14	18.31%	7061
30.37.70.1	9107	435	4336	300.95	16.84%	6692
30.37.90.1	9107	435	4336	299.54	15.92%	6553
30.41.100.2	9435	467	4484	279.61	19.24%	4747
30.41.30.2	9435	467	4484	256.80	26.11%	375
30.41.50.2	9435	467	4484	271.55	20.20%	7157
30.41.70.2	9435	467	4484	278.92	18.63%	267
30.41.90.2	9435	467	4484	277.95	19.53%	2493

Table 6. *Cont.*

Scenario	#vars	#varsCtl	#varsBool	Best-Sol	GAP	Time to Solution
70.67.100.0	18466	692	8887	560.70	17.97%	1972
70.67.30.0	18466	692	8887	504.97	27.64%	7153
70.67.50.0	18466	692	8887	533.73	21.65%	7194
70.67.70.0	18466	692	8887	542.97	22.29%	4448
70.67.90.0	18466	692	8887	553.31	19.78%	7134
70.69.100.4	19408	720	9344	579.77	17.29%	6601
70.69.30.4	19408	720	9344	509.99	29.70%	4178
70.69.50.4	19408	720	9344	534.51	25.21%	3671
70.69.70.4	19408	720	9344	566.38	19.77%	4382
70.69.90.4	19408	720	9344	581.45	17.21%	6980
70.60.100.1	19469	659	9405	609.27	21.51%	7087
70.60.30.1	19469	659	9405	539.19	33.65%	7189
70.60.50.1	19469	659	9405	570.25	27.76%	7119
70.60.70.1	19469	659	9405	602.28	21.85%	7062
70.60.90.1	19469	659	9405	614.67	20.26%	7098
70.62.100.3	19608	648	9480	605.51	16.38%	7084
70.62.30.3	19608	648	9480	535.45	27.84%	7120
70.62.50.3	19608	648	9480	573.81	21.20%	7148
70.62.70.3	19608	648	9480	592.85	18.51%	7033
70.62.90.3	19608	648	9480	602.33	16.96%	7139
70.63.100.2	19667	653	9507	598.86	20.99%	7049
70.63.30.2	19667	653	9507	514.17	37.05%	7152
70.63.50.2	19667	653	9507	554.54	28.91%	7192
70.63.70.2	19667	653	9507	594.11	21.18%	7160
70.63.90.2	19667	653	9507	594.78	21.66%	6979
110.78.100.0	29561	763	14399	891.64	25.00%	7194
110.78.30.0	29561	763	14399	804.94	35.38%	7154
110.78.50.0	29561	763	14399	811.56	34.99%	7177
110.78.70.0	29561	763	14399	896.75	22.85%	7155
110.78.90.0	29561	763	14399	902.26	23.58%	7175
110.78.100.1	29861	799	14531	936.73	21.63%	7164
110.78.30.1	29861	799	14531	791.59	40.92%	7159
110.78.50.1	29861	799	14531	848.34	32.66%	7142
110.78.70.1	29861	799	14531	916.96	23.12%	7166
110.78.90.1	29861	799	14531	920.41	23.52%	7163
110.78.100.3	30376	810	14783	924.06	21.21%	7159
110.78.30.3	30376	810	14783	807.20	35.26%	7174
110.78.50.3	30376	810	14783	857.82	28.76%	7152
110.78.70.3	30376	810	14783	904.65	23.12%	7180
110.78.90.3	30376	810	14783	920.32	21.23%	7142
110.76.100.2	30558	772	14893	886.30	21.99%	7158
110.76.30.2	30558	772	14893	752.96	40.05%	7172
110.76.50.2	30558	772	14893	789.53	34.76%	7116
110.76.70.2	30558	772	14893	852.75	25.33%	7174
110.76.90.2	30558	772	14893	870.84	24.08%	7176
110.83.100.4	30762	858	14952	887.73	20.74%	7139
110.83.30.4	30762	858	14952	744.44	40.45%	7181
110.83.50.4	30762	858	14952	816.41	29.22%	7112
110.83.70.4	30762	858	14952	852.20	24.63%	7054
110.83.90.4	30762	858	14952	887.83	20.57%	7154

Since no instance was solved to optimality when $\alpha = 0.3$, the overall runtime was 7200 s for all instances. For example, in instance 30.40.50.4, the best solution was found in 328 seconds, and it was not improved in the remaining 6872 seconds. The rest of the processing was spent to improve the gap to the lower bound. GAPs reported in Table 6 are always best-gaps, i.e., the lowest gap, the last reported in logs before timeout.

The remarkable facts of the results in Table 6 are: (i) control variables only represent between 2.5% and 5.3% of all variables, meaning that most resources in the exact approach are put into the auxiliary variables needed to reach the linear mixed integer programming formulation; (ii) the total number of variables is highly correlated with both the number of lines and bus stops; and (iii) although gap and time to solution are both correlated with the number of control and total variables, those correlations are higher for the total number of variables. The last two observations arise from computing Pearson's linear correlation coefficient over the whole set of instances used as a dataset: number of bus stops, number of lines, *#vars*, *#varsCtl*, *GAP*, and *time to solution*, as in Table 6.

The most notorious difference between variant #1 and variant #2 comes from the value of α, not from the higher number of variables. The number and type of variables in Equations (6)–(13) are not affected by α, but CPLEX time-to-optimal is in the order of the second for $\alpha = 0$ (no headway tolerance at all). Conversely, when $\alpha = 0.3$, none of the solutions is proven optimal within the 2 h timeout period, and only in 7 out of the 75 instances could the best solution be found before 20 min. Thus, the increased size and complexity of the search space coming from higher values of α have a much higher impact in the performance than the number of variables.

The objective function values achieved by exact and EA for the studied problem instances in BSP variant #2 are reported in Table 7. The real timetable is reported as baseline for the comparison, and columns 'EA vs. exact', 'EA vs. real', and 'exact vs. real' summarize the comparison, as in Table 4.

Table 7. Summary of results: exact and EA for BSP variant #2.

Scenario	*EA*	*Exact*	*Real*	*EA vs. Exact*	*EA vs. Real*	*Exact vs. Real*
30.37.100.1	298.79	301.73	245.33	−0.98%	21.8%	23.0%
30.37.30.1	282.68	283.66	102.88	−0.35%	155.9%	175.7%
30.37.50.1	294.14	294.14	165.28	0.00%	66.6%	78.0%
30.37.70.1	300.95	300.95	219.60	0.00%	33.5%	37.0%
30.37.90.1	298.19	299.54	245.33	−0.45%	21.5%	22.1%
30.40.100.0	240.48	240.48	187.57	0.00%	26.8%	28.2%
30.40.100.4	261.02	261.02	204.49	0.00%	26.8%	27.6%
30.40.30.0	223.96	231.80	82.83	−3.38%	152.8%	179.9%
30.40.30.4	234.73	246.10	98.36	−4.62%	133.2%	150.2%
30.40.50.0	233.81	236.45	122.31	−1.12%	80.1%	93.3%
30.40.50.4	247.48	251.91	123.83	−1.76%	92.7%	103.4%
30.40.70.0	239.44	239.78	166.10	−0.14%	40.2%	44.4%
30.40.70.4	258.56	258.56	179.18	0.00%	41.1%	44.3%
30.40.90.0	240.31	240.31	187.57	0.00%	25.7%	28.1%
30.40.90.4	261.22	261.22	202.69	0.00%	27.6%	28.9%
30.41.100.2	279.61	279.61	224.50	0.00%	23.2%	24.5%
30.41.30.2	256.52	256.80	95.89	−0.11%	150.6%	167.8%
30.41.50.2	265.18	271.55	145.39	−2.35%	76.2%	86.8%
30.41.70.2	283.15	278.92	195.47	1.52%	37.7%	42.7%
30.41.90.2	275.52	277.95	224.05	−0.87%	23.0%	24.1%
30.42.100.3	263.26	266.78	211.67	−1.32%	24.4%	26.0%
30.42.30.3	242.43	250.58	93.27	−3.25%	148.0%	168.7%
30.42.50.3	254.03	261.86	142.38	−2.99%	74.1%	83.9%
30.42.70.3	265.36	265.36	187.80	0.00%	37.6%	41.3%
30.42.90.3	266.51	266.51	211.12	0.00%	24.6%	26.2%

Table 7. *Cont.*

Scenario	EA	Exact	Real	EA vs. Exact	EA vs. Real	Exact vs. Real
70.60.100.1	609.27	609.27	492.20	0.00%	22.9%	23.8%
70.60.30.1	539.19	539.19	211.56	0.00%	135.9%	154.9%
70.60.50.1	570.25	570.25	310.96	0.00%	70.8%	83.4%
70.60.70.1	602.28	602.28	435.28	0.00%	32.9%	38.4%
70.60.90.1	614.67	614.67	491.70	0.00%	21.4%	25.0%
70.62.100.3	605.51	605.51	479.58	0.00%	24.6%	26.3%
70.62.30.3	527.03	535.45	207.12	−1.57%	135.6%	158.5%
70.62.50.3	563.79	573.81	321.41	−1.75%	64.0%	78.5%
70.62.70.3	592.85	592.85	417.53	0.00%	34.3%	42.0%
70.62.90.3	602.33	602.33	478.46	0.00%	23.4%	25.9%
70.63.100.2	598.86	598.86	478.87	0.00%	23.8%	25.1%
70.63.30.2	514.17	514.17	200.77	0.00%	148.7%	156.1%
70.63.50.2	542.788	554.54	316.53	−2.12%	62.9%	75.2%
70.63.70.2	594.11	594.11	427.21	0.00%	37.0%	39.1%
70.63.90.2	594.7	594.78	477.09	0.00%	22.7%	24.7%
70.67.100.0	560.70	560.70	440.27	0.00%	26.2%	27.4%
70.67.30.0	489.28	504.97	203.81	−3.11%	130.4%	147.8%
70.67.50.0	518.43	533.73	289.39	−2.87%	69.1%	84.4%
70.67.70.0	542.97	542.97	386.96	0.00%	39.1%	40.3%
70.67.90.0	552.94	553.31	440.27	−0.07%	25.6%	25.7%
70.69.100.4	579.77	579.77	456.10	0.00%	26.4%	27.1%
70.69.30.4	483.19	509.99	209.04	−5.26%	122.1%	144.0%
70.69.50.4	534.51	534.51	295.16	0.00%	69.7%	81.1%
70.69.70.4	566.38	566.38	408.63	0.00%	35.1%	38.6%
70.69.90.4	581.45	581.45	454.31	0.00%	26.0%	28.0%
110.76.100.2	886.30	886.30	706.18	0.00%	25.5%	25.5%
110.76.30.2	712.33	752.96	298.20	−5.40%	138.9%	152.5%
110.76.50.2	766.31	789.53	465.58	−2.94%	64.6%	69.6%
110.76.70.2	824.29	852.75	631.71	−3.34%	30.5%	35.0%
110.76.90.2	867.47	870.84	703.90	−0.39%	23.2%	23.7%
110.78.100.0	891.64	891.64	728.94	0.00%	22.3%	22.3%
110.78.100.1	936.73	936.73	752.16	0.00%	24.5%	24.5%
110.78.100.3	919.12	924.06	731.61	−0.53%	25.6%	26.3%
110.78.30.0	745.54	804.94	315.02	−7.38%	136.7%	155.5%
110.78.30.1	761.42	791.59	311.43	−3.81%	144.5%	154.2%
110.78.30.3	744.56	807.20	307.45	−7.76%	142.2%	162.5%
110.78.50.0	800.06	811.56	461.39	−1.42%	73.4%	75.9%
110.78.50.1	798.96	848.34	467.20	−5.82%	71.0%	81.6%
110.78.50.3	840.01	857.82	472.20	−2.08%	77.9%	81.7%
110.78.70.0	887.16	896.75	648.61	−1.07%	36.8%	38.3%
110.78.70.1	919.96	916.96	659.51	−0.33%	39.5%	39.0%
110.78.70.3	870.30	904.65	635.61	−3.80%	36.9%	42.3%
110.78.90.0	898.51	902.26	727.38	−0.42%	23.5%	24.0%
110.78.90.1	919.59	920.41	751.09	−0.09%	42.4%	22.5%
110.78.90.3	920.32	920.32	729.99	0.00%	26.1%	26.1%
110.83.100.4	887.73	887.73	713.79	0.00%	24.4%	24.4%
110.83.30.4	715.95	744.44	305.51	−3.83%	134.3%	143.7%
110.83.50.4	861.29	816.41	464.88	−6.75%	63.8%	75.6%
110.83.70.4	829.03	852.20	635.29	−2.72%	30.5%	34.1%
110.83.90.4	879.41	887.83	711.49	−0.95%	23.6%	24.8%

Results in Table 7 demonstrate that both exact and EA significantly outperform the baseline real solution in all BSP instances. EA improved over the real solution up to 155.9% in instance 30.37.30.1, and the exact method improved over the real solution up to 179.9% in instance 30.40.30.0. The comparative analysis of EA and the exact solutions indicates that the evolutionary approach demonstrated to be highly efficient at solving the problem: on average, the distance to the exact solution was below 1.32%. Overall, EA improved

63.4% (on average) over the real timetable, and the exact method improved 66.2% (on average) over the real timetable.

The average improvements of exact and EA over the baseline real solution, grouped by tolerance and instance size for BSP variant #2, are reported in Table 8. Regarding tolerance levels, EA improved up to 150.95%, and the exact method improved up to 158.13% over the real timetable. The best improvements were computed for λ = 30, which pose the tighter bounds for user waiting times. In less restrictive scenarios, improvements over the real timetable were over 25.23% for both methods. Regarding scenario dimension, both methods computed robust solutions that improve between 58.50% (EA, instances with 110 transfer zones) and 70.25% (exact, instances with 30 transfer zones) over the real solution.

Table 8. Exact and EA improvements over baseline real solution, grouped by tolerance and problem dimension, for BSP variant #2.

EA Over Real			*Exact Over Real*			*EA vs. Exact*		
λ	Δ	*Instances*	λ	Δ	*Instances*	λ	Δ	*Instances*
30	150.95%	15	30	158.13%	15	30	−3.32%	15
50	78.06%	15	50	82.16%	15	50	−2.26%	15
70	38.95%	15	70	39.79%	15	70	−0.61%	15
90	25.05%	15	90	25.32%	15	90	−0.22%	15
100	25.23%	15	100	25.47%	15	100	−0.19%	15
EA Over Real			*Exact Over Real*			*EA vs. Exact*		
N	Δ	*Instances*	N	Δ	*Instances*	N	Δ	*Instances*
30	68.28%	25	30	70.25%	25	30	−0.89%	25
70	63.37%	25	70	64.84%	25	70	−0.67%	25
110	58.50%	25	110	63.43%	25	110	−2.41%	25

The reported improvements on the objective function values grouped by tolerance and problem dimension demonstrate that for all BSP instances, the improvements over the real timetable increase as user tolerance decreases. Similar to problem variant #1, the optimization algorithms properly scale with the complexity of the problem, computing better solutions for tighter instances.

The proposed EA computed the same solution as the exact method in 33 problem instances (10 with 30 transfer zones, 18 with 70 transfer zones, and 5 with 110 transfer zones). Furthermore, in another 14 instances, the difference was below 1%, and in two problem instances, the EA computed a better solution than the exact method.

6.2.3. Quality of Service Results

Table 9 reports the average objective values (normalized by the number of transfer zones) for all scenarios, grouped by tolerance. Results show that both EA and the exact methods significantly improve over the baseline real solution. For BSP variant #1, the largest difference is 1.72 (4.68 − 2.97) between the exact approach and the real solution, when low user tolerance is considered (λ = 30). Similar to the results reported in Table 5, the graphic demonstrates that improvements of the optimization methods are better in tighter BSP instances. For BSP variant #2, the best improvement of the exact method was 4.45, and the best improvement of EA was 4.45, both when λ = 30. Results are graphically presented in Figure 7 for BSP variant #1.

Table 9. Average objective values (normalized by the number of transfer zones), grouped by tolerance, for both problem variants.

				Problem Variant #1			
λ	*EA*	*Exact*	*Real*	*EA/Real*	*Exact/Real*	*EA − Real*	*Exact − Real*
30	4.67	4.68	2.97	1.57	1.58	1.70	1.72
50	5.68	5.69	4.43	1.28	1.28	1.25	1.26
70	7.04	7.04	6.03	1.17	1.17	1.01	1.01
90	7.47	7.47	6.81	1.10	1.10	0.66	0.66
100	7.36	7.36	6.61	1.11	1.11	0.75	0.75
				Problem Variant #2			
λ	*EA*	*Exact*	*Real*	*EA/Real*	*Exact/Real*	*EA − Real*	*Exact − Real*
30	7.42	7.66	2.97	2.50	2.58	4.45	4.70
50	7.88	8.06	4.43	1.78	1.82	3.45	3.63
70	8.38	8.43	6.03	1.39	1.40	2.35	2.40
90	8.51	8.52	6.81	1.25	1.25	1.70	1.72
100	8.57	8.59	6.61	1.30	1.30	1.96	1.98

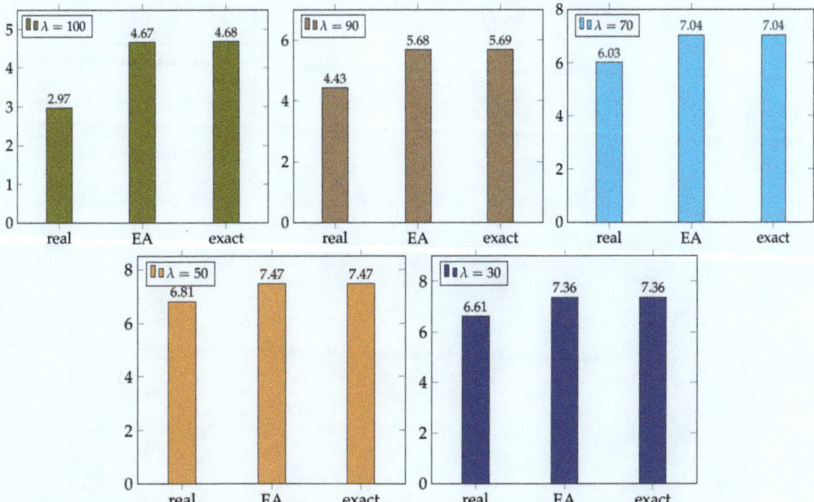

Figure 7. Comparison of normalized objective function values, grouped by tolerance, for BSP variant #1.

Three quality of service metrics (r, l_s, and l_n) are considered in the results reported for the computed solutions in Table 10, grouped by dimension (NP) and tolerance (λ). Metric r is the ratio between the average waiting time computed by the proposed timetable schedules and the threshold value that passengers are willing to wait for a transfer in each transfer zone (W_b^{ij}). The r metric evaluates the number of successful synchronized trips, which are represented by $r \leq 1.0$, whereas unsuccessful synchronizations are represented by $r > 1.0$. Metric r also allows evaluating how far from acceptable (i.e., synchronized) the trips of two lines are, considering the deviation from the ratio defined by the threshold value passengers are willing to wait (1.0). In turn, metric l_s is the average number of successfully synchronized lines, whereas metric l_n is the average number of not synchronized lines. All metrics are reported for each instance class regarding dimension and tolerance.

Table 10. Average synchronizations for the studied methods.

		Problem Variant #1					
m	*λ*	*Real*		*EA*		*Exact*	
	r	l_s/l_n	*r*	l_s/l_n	*r*	l_s/l_n	
30	100	0.45	22/0	0.44	22/0	0.44	22/0
30	90	0.47	21/1	0.46	22/0	0.47	22/0
30	70	0.59	19/3	0.54	21/1	0.53	22/0
30	50	0.85	12/10	0.76	16/6	0.75	18/4
30	30	1.33	3/19	1.15	10/12	1.12	11/11
70	100	0.46	39/0	0.44	39/0	0.44	39/0
70	90	0.47	39/0	0.46	39/0	0.47	39/0
70	70	0.59	33/6	0.54	39/0	0.52	39/0
70	50	0.85	24/14	0.74	30/9	0.73	32/6
70	30	1.35	5/33	1.14	18/21	1.14	20/19
110	100	0.48	49/0	0.45	49/0	0.45	49/0
110	90	0.49	47/2	0.48	49/0	0.46	49/0
110	70	0.61	40/9	0.57	43/6	0.53	49/0
110	50	0.87	28/20	0.74	33/16	0.72	40/9
110	30	1.38	7/42	1.14	21/28	1.10	24/23
		Problem Variant #2					
m	*λ*	*Real*		*EA*		*Exact*	
	r	l_s/l_n	*r*	l_s/l_n	*r*	l_s/l_n	
30	100	0.45	22/0	0.44	22/0	0.44	22/0
30	90	0.47	22/0	0.46	22/0	0.47	22/0
30	70	0.59	22/0	0.54	22/0	0.53	22/0
30	50	0.85	16/6	0.76	19/3	0.75	19/3
30	30	1.33	3/19	1.15	5/16	1.12	7/15
70	100	0.46	39/0	0.44	39/0	0.44	39/0
70	90	0.47	39/0	0.46	39/0	0.47	39/0
70	70	0.59	38/1	0.54	39/0	0.52	39/0
70	50	0.85	28/10	0.74	35/3	0.73	35/4
70	30	1.35	5/33	1.14	18/21	1.14	20/19
110	100	0.48	49/0	0.45	49/0	0.45	49/0
110	90	0.49	49/0	0.48	49/0	0.46	49/0
110	70	0.61	46/2	0.57	48/1	0.53	49/0
110	50	0.87	34/14	0.74	41/8	0.72	44/5
110	30	1.38	7/42	1.14	21/28	1.10	22/27

Results in Table 10 indicate that both studied optimization methods significantly improved the quality of service when compared with the real solution. Both the exact method and the EA computed lower values of the waiting time metric for all instance classes. The best improvements of the studied methods over the baseline real solution occurred for problem instances with NP = 70/110 and λ = 30, where both exact and EA computed solutions with a higher number of synchronized lines. In turn, the largest number of synchronized lines were computed by the exact method in problem instances with NP = 110 and λ = 30. Furthermore, better waiting time values were obtained by both exact and EA in problem instances with lower waiting time tolerance from users, with respect to the baseline real solution.

A comparison of waiting time values for transfers of all lines (normalized by W_b) is presented in the histograms in Figure 8. Results correspond to scenario 70.63.30.2, which is representative of the results computed for other studied BSP instances. Normalized average waiting time results are reported for the baseline real solution, the EA solution for BSP variant #1 (α = 0.0), and the exact solution for BSP variant #2 (α = 0.3). Values of the normalized average waiting time over 1.0 mean that for a number of lines in the reported scenario, the optimization methods were not able to find a combination of offset and headway values that allows lowering the waiting time for transfers under W_b. This occurs as a direct consequence of the inter-related design of bus lines and the behavior of passengers in each scenario, since when adjusting offsets and headways to achieve successful transfers between line i and line j in transfer zone b (i.e., for those transfers avg(wait) < W_b), those lines are not synchronized (at least for some trips) with other lines on other transfer zones $b*$ (i.e., for those transfers, avg(wait) > W_{b*}). The histograms in Figure 8 report both successful synchronizations (avg(wait) < W_b, on the left side of each histogram) and non-successful synchronizations (avg(wait) > W_b, on the right side of each histogram).

Results reported in the histograms in Figure 8 demonstrate that the exact solution effectively reduces the waiting time for a significant percentage of lines in the studied instance. The exact solution has more lines with an average waiting time lower or equal to 1 (i.e., 23 in the exact solution vs. only 6 in the real solution). Furthermore, five lines in the exact solution have an average waiting time less than 0.5 of W_b, whereas there is none in the real solution. Regarding larger waiting times, only 3 lines in the exact solution have more than 1.5 of the maximum value for a successful synchronization, whereas in 13 lines of the real solution, passengers must wait 1.5 more than expected to perform a multi-leg trip. The comparative results demonstrate that the solution found by the exact method synchronizes a larger number of lines, thus improving the QoS. Solutions computed by the proposed EA have a similar behavior regarding waiting times, as reported in Table 10. Similar results were obtained for other studied BSP instances, and even better results were achieved in the most restrictive instances ($\lambda = 30$).

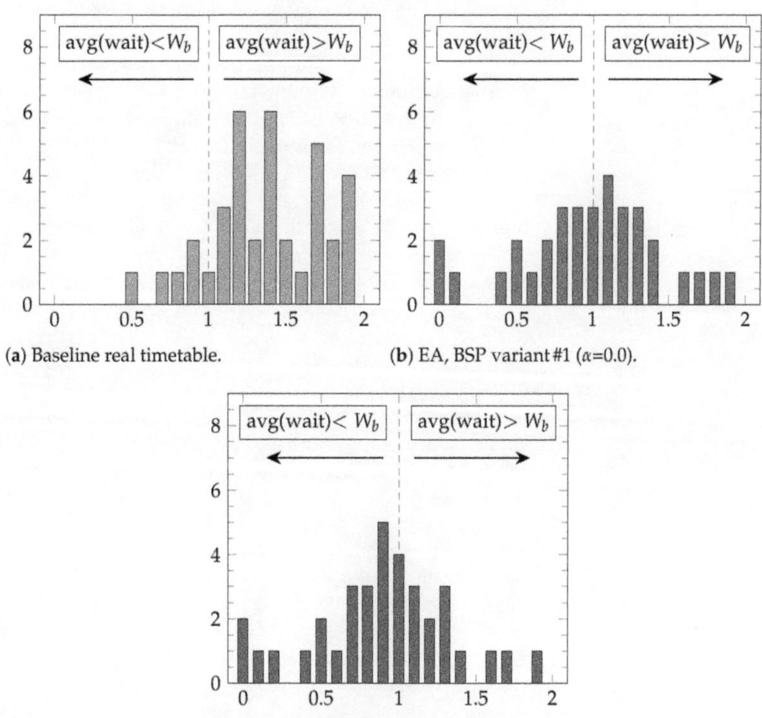

Figure 8. Comparative results of waiting times for the real solution, EA for BSP variant #2 ($\alpha = 0.0$), and exact solution for BSP variant #2 ($\alpha = 0.3$) in scenario 70.63.100.2.

Finally, regarding execution time, both the exact and evolutionary approaches were able to compute accurate solutions in very short execution times. For problem variant #1, the execution times were less than 10 seconds. For problem variant #2, the time limit of the exact method was two hours, and the EA found the best solutions in less than five minutes.

7. Concluding Remarks

This article studied the bus timetabling synchronization problem. The problem model extends existing ones by defining extended transfer zones for every pair of bus stops. In turn, an improved model was included to account for the transfer demands of passengers, and two variants of the problem were addressed: optimizing the offset for each line, and optimizing the offset and headways for each line. For the second problem variant, the optimization domain was restricted to a certain deviation on the real headways defined for the case study to provide a proper quality of service to users.

An exact MILP approach and an EA were proposed to solve the problem. Both methods were evaluated for a real case study for the public transportation system in Montevideo, Uruguay. A total number of 75 scenarios were defined using real data about lines, headways, and transfer demands, gathered from the intelligent transportation system of Montevideo. Results were compared with the real timetable currently implemented by the city administration.

The empirical evaluation indicated that both proposed methods are able to compute solutions that are significantly improved over the real current timetable in Montevideo. The exact method computed the optimal solution for all instances in problem variant #1, improving successful synchronizations up to 66.6% (24.6% on average) over the real timetable. The EA was highly efficient for this problem variant, too, improving the synchronizations up to 66.3% (24.5% on average) over the current timetable. Problem variant #2 poses a significantly harder challenge for optimization, as the number of variables significantly increases for each considered scenario. Despite this, the exact method was able to compute an accurate solution, which provides significant improvements on the number of successfully synchronized trips over the real timetable. Improvements up to 179.9% (66.2% on average) for the exact methods and up to 174.8% (66.2% on average) for the EA. The proposed optimization methods also improved relevant quality of service metrics, such as the average waiting times for transfers, especially in tight problem instances, which reduced up to 57.8% for BSP variant #1 and up to 158.3% for BSP variant #2.

Both methods were efficient regarding computing times. The exact method provided accurate objective function values within less than a second for problem variant #1. In turn, the proposed EA is useful for solving larger problem instances of problem variant #2 in reduced execution times.

Future lines of research are related to addressing other BSP variants, e.g., explicitly focusing on the perceived waiting times for users, and modeling the real demand for direct trips from real data on ticket sales provided by the transportation system. Multi-objective versions of the problem can be devised, too, by including the explicit optimization of other relevant functions, such as the operation cost of the bus fleet and other quality of service metrics.

Author Contributions: Conceptualization, S.N.; methodology, S.N. and C.R.; software, S.N. and C.R.; validation, S.N. and C.R.; formal analysis, S.N. and C.R.; investigation, S.N. and C.R.; writing—original draft preparation, S.N. and C.R.; writing—review and editing, S.N. and C.R.; supervision, S.N.; project administration, S.N. All authors have read and agreed to the published version of the manuscript.

Funding: This research received no external funding.

Institutional Review Board Statement: Not applicable.

Informed Consent Statement: Not applicable.

Acknowledgments: The work of S. Nesmachnow and C. Risso is partly supported by ANII and PEDECIBA, Uruguay. Authors would like to thank Jonathan Muraña and Renzo Massobrio for assistance in data processing for scenarios generation.

Conflicts of Interest: The authors declare no conflict of interest.

References

1. Grava, S. *Urban Transportation Systems*; McGraw-Hill: New York, NY, USA, 2002.
2. Nesmachnow, S.; Baña, S.; Massobrio, R. A distributed platform for big data analysis in smart cities: combining Intelligent Transportation Systems and socioeconomic data for Montevideo, Uruguay. *EAI Endorsed Trans. Smart Cities* **2017**, *2*, 1–18. [CrossRef]
3. Ceder, A.; Wilson, N. Bus network design. *Transp. Res. Part B Methodol.* **1986**, *20*, 331–344. [CrossRef]
4. Hipogrosso, S.; Nesmachnow, S. Analysis of Sustainable Public Transportation and Mobility Recommendations for Montevideo and Parque Rodó Neighborhood. *Smart Cities* **2020**, *3*, 479–510. [CrossRef]
5. Ceder, A.; Tal, O. Timetable Synchronization for Buses. In *Lecture Notes in Economics and Mathematical Systems*; Springer: Berlin/Heidelberg, Germany, 1999; pp. 245–258.
6. Risso, C.; Nesmachnow, S. Designing a Backbone Trunk for the Public Transportation Network in Montevideo, Uruguay. In *Smart Cities*; Springer: Berlin/Heidelberg, Germany, 2020; pp. 228–243.
7. Massobrio, R.; Nesmachnow, S. Urban Mobility Data Analysis for Public Transportation Systems: A Case Study in Montevideo, Uruguay. *Appl. Sci.* **2020**, *10*, 5400. [CrossRef]
8. Nesmachnow, S.; Muraña, J.; Risso, C. Exact and Metaheuristic Approach for Bus Timetable Synchronization to Maximize Transfers. In *Smart Cities*; Springer: Berlin/Heidelberg, Germany, 2021; pp. 183–198.
9. Eranki, A. A Model to Create Bus Timetables to Attain Maximum Synchronization Considering Waiting Times at Transfer Stops. Master's Thesis, University of South Florida, Tampa, FL, USA, 2004.
10. Ibarra-Rojas, O.; Rios-Solis, Y. Synchronization of bus timetabling. *Transp. Res. Part Methodol.* **2012**, *46*, 599–614. [CrossRef]
11. Ibarra-Rojas, O.; López-Irarragorri, F.; Rios-Solis, Y. Multiperiod Bus Timetabling. *Transp. Sci.* **2016**, *50*, 805–822. [CrossRef]
12. Saharidis, G.; Dimitropoulos, C.; Skordilis, E. Minimizing waiting times at transitional nodes for public bus transportation in Greece. *Oper. Res.* **2013**, *14*, 341–359. [CrossRef]
13. Salicrú, M.; Fleurent, C.; Armengol, J. Timetable-based operation in urban transport: Run-time optimisation and improvements in the operating process. *Transp. Res. Part A Policy Pract.* **2011**, *45*, 721–740. [CrossRef]
14. Daduna, J.; Voß, S. Practical Experiences in Schedule Synchronization. In *Lecture Notes in Economics and Mathematical Systems*; Springer: Berlin/Heidelberg, Germany, 1995; Volume 430, pp. 39–55.
15. Ceder, A.; Golany, B.; Tal, O. Creating bus timetables with maximal synchronization. *Transp. Res. Part Policy Pract.* **2001**, *35*, 913–928. [CrossRef]
16. Fleurent, C.; Lessard, R.; Séguin, L. Transit timetable synchronization: Evaluation and optimization. In Proceedings of the 9th International Conference on Computer-aided Scheduling of Public Transport, San Diego, CA, USA, 9–11 August 2004.
17. Shafahi, Y.; Khani, A. A practical model for transfer optimization in a transit network: Model formulations and solutions. *Transp. Res. Part A Policy Pract.* **2010**, *44*, 377–389. [CrossRef]
18. Abdolmaleki, M.; Masoud, N.; Yin, Y. Transit timetable synchronization for transfer time minimization. *Transp. Res. Part B Methodol.* **2020**, *131*, 143–159. [CrossRef]
19. Nesmachnow, S.; Muraña, J.; Goñi, G.; Massobrio, R.; Tchernykh, A. Evolutionary Approach for Bus Synchronization. In *High Performance Computing*; Springer: Berlin/Heidelberg, Germany, 2020; pp. 320–336.
20. Bäck, T.; Fogel, D.; Michalewicz, Z. (Eds.) *Handbook of Evolutionary Computation*; Oxford University Press: Oxford, UK, 1997.
21. Massobrio, R.; Nesmachnow, S. Urban Data Analysis for the Public Transportation System of Montevideo, Uruguay. In *Smart Cities*; Springer: Cham, Switzerland, 2020; pp. 199–214.
22. Nesmachnow, S.; Iturriaga, S. Cluster-UY: Collaborative Scientific High Performance Computing in Uruguay. In *Communications in Computer and Information Science*; Springer: Berlin/Heidelberg, Germany, 2019; pp. 188–202.

applied
sciences

MDPI

Review

Smart Cities' Applications to Facilitate the Mobility of Older Adults: A Systematic Review of the Literature

Nelson Pacheco Rocha [1,*], Rute Bastardo [2], João Pavão [3], Gonçalo Santinha [4], Mário Rodrigues [5], Carlos Rodrigues [4], Alexandra Queirós [6] and Ana Dias [7]

[1] IEETA-Institute of Electronics and Informatics Engineering of Aveiro, Department of Medical Sciences, University of Aveiro, 3810-193 Aveiro, Portugal
[2] UNIDCOM, Science and Technology School, University of Trás-os-Montes and Alto Douro, Quinta de Prado, 5001-801 Vila Real, Portugal; rutebastardo@gmail.com
[3] INESC-TEC, Science and Technology School, University of Trás-os-Montes and Alto Douro, Quinta de Prado, 5001-801 Vila Real, Portugal; jpavao@utad.pt
[4] GOVCOPP-Governance, Competitiveness and Public Policies, Department of Social, Political and Territorial Sciences, University of Aveiro, 3810-193 Aveiro, Portugal; g.santinha@ua.pt (G.S.); cjose@ua.pt (C.R.)
[5] IEETA-Institute of Electronics and Informatics Engineering of Aveiro, Águeda School of Technology and Management, University of Aveiro, 3750-127 Águeda, Portugal; mjfr@ua.pt
[6] IEETA-Institute of Electronics and Informatics Engineering of Aveiro, Health Sciences School, University of Aveiro, 3810-193 Aveiro, Portugal; alexandra@ua.pt
[7] GOVCOPP-Governance, Competitiveness and Public Policies, Economics, Management, Industrial Engineering and Tourism, University of Aveiro, 3810-193 Aveiro, Portugal; anadias@ua.pt
* Correspondence: npr@ua.pt; Tel.: +351-234-247-292

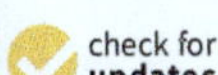
check for
updates

Citation: Rocha, N.P.; Bastardo, R.; Pavão, J.; Santinha, G.; Rodrigues, M.; Rodrigues, C.; Queirós, A.; Dias, A. Smart Cities' Applications to Facilitate the Mobility of Older Adults: A Systematic Review of the Literature. *Appl. Sci.* **2021**, *11*, 6395. https://doi.org/10.3390/app11146395

Academic Editors: Luis Hernández-Callejo, Víctor Alonso Gómez, Sergio Nesmachnow, Vicente Leite, Javier Prieto and Ângela Ferreira

Received: 17 June 2021
Accepted: 9 July 2021
Published: 11 July 2021

Publisher's Note: MDPI stays neutral with regard to jurisdictional claims in published maps and institutional affiliations.

Abstract: This study aimed to identify: (i) the relevant applications based on information technologies and requiring smart cities' infrastructure to facilitate the mobility of older adults in URBAN SPACES; (ii) the type of data being used by the proposed applications; (iii) the maturity level of these applications; and (iv) the barriers TO their dissemination. An electronic search was conducted on Web of Science, Scopus, and IEEE Xplore databases, combining relevant keywords. Then, titles and abstracts were screened against inclusion and exclusion criteria, and the full texts of the eligible articles were retrieved and screened for inclusion. A total of 28 articles were included. These articles report smart cities' applications to facilitate the mobility of older adults using different types of sensing devices. The number of included articles is reduced when compared with the total number of articles related to smart cities, which means that the mobility of older adults it is still a not significant topic within the research on smart cities'. Although most of the included studies aimed the implementation of specific applications, these were still in an early stage of development, without the assessment of potential end-users. This is an important research gap since it makes difficult the creation of market-oriented solutions. Another research gap is the integration of knowledge generated by other research topics related to smart cities and smart mobility. Consequently, important issues (e.g., user privacy, data standardization and integration, Internet of Things implementation, and sensors' characteristics) were poorly addressed by the included studies.

Keywords: smart city; older adults; mobility; systematic review

1. Introduction

Population ageing is taking place worldwide. In most countries, people live longer and, overall, in better health conditions than before. Coping with such a demographic shift has become a major issue for public policies and has gained attention from international institutions [1,2].

Currently, the ideal political paradigm considered for the care of older adults is that they should continue living in the community rather than being forced to move to residential care units because of their cognitive and physical limitations. Though not recent, it is within this context that the concepts such as ageing in place [3] or active ageing [4,5] have been developed. According to these concepts, the maintenance of patterns of activities and values typical of middle age can optimize opportunities for social participation, health conditions, and the safety of the individuals as they age [5–8]. This depends not only on the characteristics of the individuals but also on environmental factors [9], such as, for instance, urban infrastructures and services.

The Ageing in Cities Report from the Organisation of Economic Cooperation and Development (OECD) states that "designing policies that address ageing issues requires a deep understanding of local circumstances, including communities' economic assets, history and culture. (...) Cities need to pay more attention to local circumstances to understand ageing, and its impact. They are especially well-equipped to address the issue, given their long experience of working with local communities and profound understanding of local problems" [10] p. 18.

The World Health Organization (WHO) age-friendly cities and communities' approach [11] is in line with this concern. Following the idea that an age-friendly city is as a place where older adults are actively involved, valued, and supported with infrastructure and services that accommodate their needs, the WHO produced a guide identifying the key characteristics of an age-friendly environment in terms of service provision (e.g., healthcare services or transportation), built environment (e.g., housing, outdoor spaces, or buildings), and social aspects (e.g., civic, or social participation) [12]. Based on the experience, the WHO provides guidance [13] to ensure that research and initiatives held at country and regional levels on topics relevant to healthy ageing can be widely shared.

Although ageing societies pose diverse challenges, they also provide a large set of opportunities that society can benefit from [10]. Such opportunities include new developments in technology and innovation. In fact, due to technological advances in the last couple of decades, the use of smart technologies is increasingly considered an important means to promote active ageing [1,14]. Therefore, many researchers aimed to develop new services based on Information Technologies (IT), such as Ambient Assisted Living (AAL) [15–17], to enable older adults to achieve their full potential in terms of physical, social, and mental wellbeing.

Additionally, due to the technological developments, during the last few years the smart city paradigm has been the object of great attention from different sectors: scientific journals publish specific issues on this topic, local governments fight to label their city as such, firms advertise smart cities' solutions, and international and national programs aim to promote adequate implementations [18]. In this respect, several initiatives of the European Commission, namely the Digital Agenda (one of the seven pillars of the Europe 2020 Strategy) and the Smart Cities and Communities initiative, aimed at bringing together cities, industry, and citizens through more sustainable integrated solutions [19].

1.1. Objective and Contributions

Systematic reviews allow us not only to answer clearly formulated questions, using systematic methods, but also to critically evaluate and synthesize results from multiple studies, thus consolidating knowledge and identifying gaps in a given research field.

Concretely, the systematic review reported in this article aimed to identify relevant IT-based applications requiring smart cities' infrastructure to facilitate the mobility of older adults in the urban space and, consequently, their participation and inclusion in the community as full citizens.

To the best of our knowledge, the systematic review literature reported by this article constitutes the first review that covers the mobility of older adults supported by smart cities' infrastructure. It aims to make the following contributions in terms of smart city applications to facilitate the mobility of older adults: (i) review of the main recently published research; (ii) identification and discussion of relevant applications; (iii) typification of the applications being developed; (iv) identification of current approaches to use sensors and smart city data to support the mobility of older adults; (v) analysis of the maturity level of the reported applications; (vi) discussion about the main results and contributions of the current research; (vii) identification of the barriers for the dissemination of the identified applications; and (viii) identification of research gaps. These contributions might be useful to inform smart cities' stakeholders about state-of-the-art solutions with impacts on active ageing and the gaps of the current research.

1.2. Smart Cities and Smart Mobility

A set of characteristics has been identified as relevant in the context of smart cities [20,21]: smart governance, smart economy, smart environment, smart people, smart living, and smart mobility.

Smart mobility includes local, national, and international accessibility, and the availability of communication infrastructure or sustainable and safe transport systems and is aligned with the United Nations (UN) Sustainable Development Goals [22,23]. Smart mobility is often seen as related to the use of IT to adequately orchestrate services designed to improve urban mobility [24]. In this respect, a wide range of information services can be foreseen, such as intelligent transportation systems [25–28] or algorithms to infer mobility patterns [29,30]. These information services might contribute to the reduction of air and noise pollution, traffic congestion, and travel costs, while increasing individuals' safety [24,31,32].

Moreover, smart mobility might facilitate older adults' activities and participation, in line with the goals of active ageing and age-friendly cities and communities' approaches [13]. In its baseline report for the decade of healthy ageing, the WHO states that "engagement of older people and municipalities can steer the use of digital technology to support enabling environments—and reduce the digital divide between older and younger people" [1] p. 67, pointing out Chicago as a good example of a city in which the labels of Age-Friendly and Smart City have been brought together.

1.3. An Overview of Related Reviews

Several articles published in scientific journals [33–65] reported different types of reviews, including systematic literature reviews, related to various aspects of smart cities' implementation, including data analytics [33–37], systems architectures [38], data security [39–41], data security and Internet of Things (IoT) [42–44], IoT [45], ontologies [46], healthcare [47–49], energy efficiency [50], citizenship [51], smart city indicators [52], mobility [53–58] and age-friendly initiatives [59–65].

Looking specifically for reviews related to mobility or age-friendly initiatives (Table 1), is possible to conclude that the state-of-art studies did not aim to analyze relevant IT-based applications requiring smart cities' infrastructures to facilitate the mobility of older adults.

In turn, concerning age-friendly initiatives, the state-of-the-art reviews aimed to study aspects as the impact of smart cities in different countries [60–62], barriers to the implementation of age-friendly concepts in smart cities [59], the role of IT and the barriers and stressors for age and disability-friendly communities [65], and the impact of IT on the quality of life of older adults [63]. Finally, one article [64] is a narrative review on the role of mobility digital ecosystems for age-friendly urban public transport. However, this narrative review (not a systematic review), instead of performing a quantitative evidence-based analysis of empirical studies, aimed to present a discussion on how transport technology and transport mobility practices might contribute to the promotion of mobility rights of older adults and age-friendly mobility [64].

The article is outlined as follows. In Section 2, the proposed research questions and methods are detailed. The results are presented in Section 3. Finally, the discussion and answers to the research questions are presented in Section 4, and general conclusions are presented in Section 5.

Table 1. State-of-the-art reviews related to smart city mobility and age-friendly initiatives.

Topic	Reference	Title	Year
Mobility	[53]	*Emerging big data sources for public transport planning: a systematic review on current state of art and future research directions.*	2019
	[54]	*Smart parking: a literature review from the technological perspective.*	2019
	[55]	*The quality of smart mobility: a systematic review.*	2020
	[56]	*A systematic review of urban navigation systems for visually impaired people.*	2021
	[57]	*Enabling technologies for urban smart mobility: recent trends, opportunities and challenges.*	2021
	[58]	*Barriers and risks of Mobility-as-a-Service (MaaS) adoption in cities: a systematic review of the literature.*	2021
Age-friendly initiatives	[59]	*Implementation of age-friendly initiatives in smart cities: probing the barriers through a systematic review.*	2020
	[60]	*Smart and age-friendly cities in Romania: an overview of public policy and practice.*	2020
	[61]	*Smart and age-friendly cities in Russia: an exploratory study of attitudes, perceptions, quality of life and health information needs.*	2020
	[62]	*Smart and age-friendly communities in Poland: an analysis of institutional and individual conditions for a new concept of smart development of ageing communities.*	2020
	[63]	*Quality of life framework for personalised ageing: a systematic review of ICT solutions.*	2020
	[64]	*The role of mobility digital ecosystems for age-friendly urban public transport: a narrative literature review.*	2020
	[65]	*Use of connected technologies to assess barriers and stressors for age and disability-friendly communities.*	2021

2. Materials and Methods

This systematic review followed the guidelines of the Preferred Reporting Items for Systematic Reviews and Meta-Analyses (PRISMA) [66]. To achieve its objectives the following research questions were formulated:

- RQ1—What are the relevant smart city IT-based applications to facilitate the mobility of older adults?
- RQ2—What type of data (i.e., sensors and city data) are being used to support the proposed applications?
- RQ3—What are the maturity levels of the solutions being reported?
- RQ4—What are the barriers for the dissemination of the solutions that were identified?

Boolean queries were prepared to include all the articles that have in their titles, abstract or keywords a reference to smart city (i.e., one of the following expressions 'Smart City', 'Smartcity', 'Smart-city', 'Smart Cities', 'Smartcities' or 'Smart-cities') together with at least one of the following terms: 'Elderly', 'Older', 'Senior', 'Ageing' or 'Aging'. The resources considered to be searched were two general databases, Web of Science and Scopus, and one specific technological database, IEEE Xplore. The literature search was concluded in April 2021.

As inclusion criterion, the authors aimed to include all the articles that reported evidence of explicit use of IT-based application requiring smart cities' infrastructures to facilitate the mobility of older adults.

Considering the exclusion criteria, the authors aimed to exclude all the articles not published in English, without abstract or without access to full text. Furthermore, the authors also aimed to exclude all the articles that reported overviews, reviews and applications that do not explicitly require smart cities' infrastructures, or that were not relevant for the specific aim of this study.

The selection of the articles was performed in three steps:

- First, the authors removed the duplicates and the articles without abstract.
- The abstracts of the retrieved articles were screened against inclusion and exclusion criteria to exclude non relevant articles. When from the title and abstract of an article it was not possible to take a decision, the article was considered for the next step (i.e., full text screening).
- The full texts of the eligible articles were retrieved and screened for inclusion.

Then, the full texts of the included articles were analyzed and classified using a synthesis process based on the method proposed by Ghapanchi and Aurum [67] (i.e., terms and definitions used in the included articles were identified to create a primary list of application domains, which was afterwards refined by further analyses).

Data from each included article were extracted using a standardized form including: (i) article authors, title, and year of publication; (ii) aim of the study being reported; (iii) details of the applications being reported; (iv) details of the applied research methods; (v) methods applied for the assessment of the proposed applications; (vi) results; and (vii) author's interpretations.

In all the steps the articles were reviewed by at least two authors and any disagreement was discussed and resolved by consensus.

3. Results

Figure 1 presents the PRISMA flowchart of this systematic review.

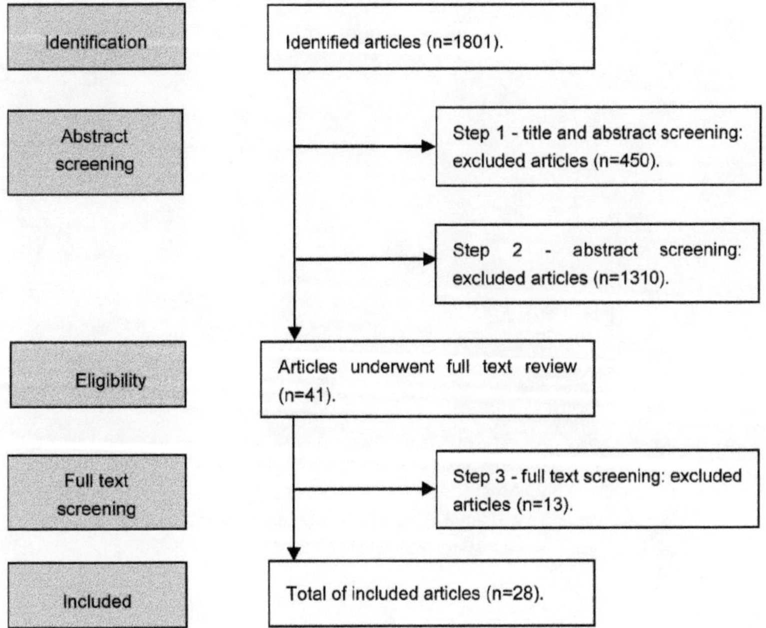

Figure 1. Study flowchart.

A total of 1801 articles were retrieved from the initial search on Web of Science (718 articles), Scopus (869 article) and IEEE Xplore (214 articles).

The initial step of the screening phase (i.e., step 1) yielded 1351 articles by removing the duplicates (348 articles) or the articles without abstracts (102 articles).

Based on abstracts (i.e., step 2), 390 articles were removed since they were not published in English, or they were overviews or reviews, editorials, prefaces, and announcements of special issues, workshops, or books. Moreover, 920 articles were removed because they did not report evidence of the explicit use of IT-based applications requiring smart cities' infrastructures to facilitate the mobility of older adults.

Finally, the full texts of the remaining 41 articles were screened (i.e., step 3) and 13 articles were excluded because they did not meet the inclusion criteria. Therefore, 28 articles [68–95] were included in this systematic review.

Of these 28 articles, some reported the same research projects: articles [79,80,85,87,88] were related to a Portuguese funded project called SmartWalk; articles [73,77] were related to a European funded project called City4Age; and articles [78,93] reported on the same application.

Considering the articles included for this systematic review, 18 were published in conference proceedings [68–70,72–74,79–87,89,91,92] and ten were published in scientific journals [71,75–78,88,90,93–95].

In terms of publication years, the included articles were published between 2013 (i.e., two articles [68,69]) and 2021 (i.e., one article [95]). The diagram in Figure 2 demonstrates that the trend of the development of IT-based applications requiring smart cities' infrastructures to facilitate the mobility of older adults is increasing, and more than two-thirds of the articles (i.e., 19 articles [77–95]) were published in the last four years.

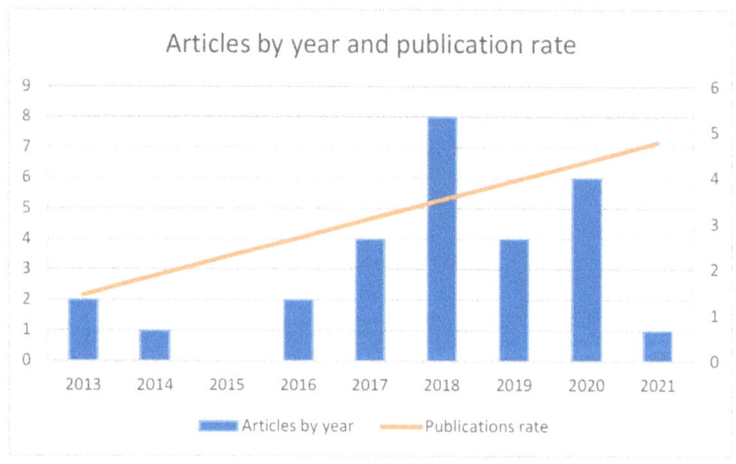

Figure 2. Articles by year and publication rate (calculated using RMS Lest Square Fit).

Figure 3 reveals the country of origin for all primary studies. Europe (in particular, Portugal, six articles [79,80,85,87,88], and Spain, five articles [68,73,75–77]) contributed the largest number of articles (i.e., 20 articles [68–73,75–80,83–85,87–89,92,93]). In turn, Asia contributed five articles [74,81,82,90,91], North America (i.e., United States of America) contributed two articles [86,94] and South America (i.e., Brazil) contributed one article [95].

3.1. Quality of the Retrieved Studies

In addition to general inclusion and exclusion criteria, it is considered critical to assess the quality of primary studies. Therefore, the 28 articles were evaluated against a set of study quality assessment questions listed in Table 2, which were adopted and adjusted from other studies [96–98]. Two of the authors independently answered the questions according to binary scale (i.e., 1 for Yes or 0 for No) and any disagreements were resolved through discussion until consensus was reached.

Table 2. Study quality assessment questions.

#	Study Quality Assessment Questions
Q1	Are the objectives and aims of the study clearly identified?
Q2	Is the context of the study clearly stated?
Q3	Does the research methods support the aims of the study?
Q4	Has the study an adequate description of the technologies being used?
Q5	Is there a clear statement of the findings?
Q6	Are limitations of the study discussed explicitly?

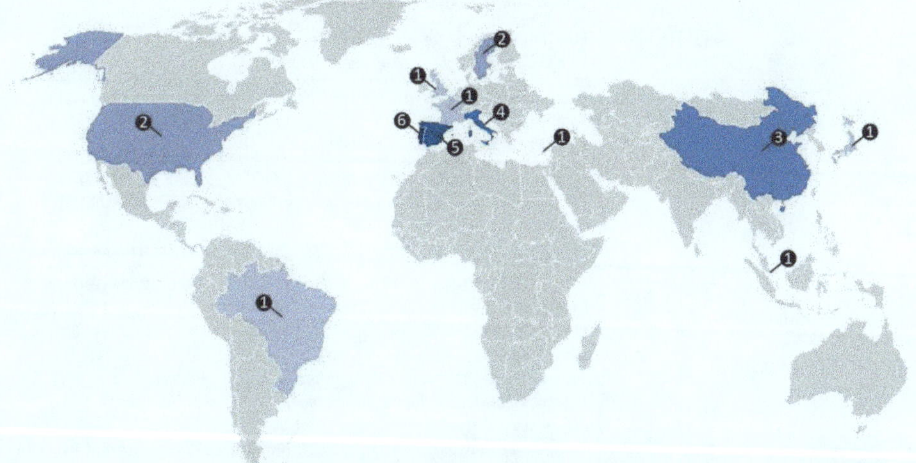

Figure 3. Distribution of the selected articles by country.

As can be seen in Figure 4, all studies state the aims and objectives of the conducted research (Q1). In turn, only three articles explicitly included the limitations of the respective studies (Q6). Additionally, only nine articles conveniently described the research methods (Q3). The results for the remainder three questions varied from 23 (Q2 and Q5) to 19 (Q4).

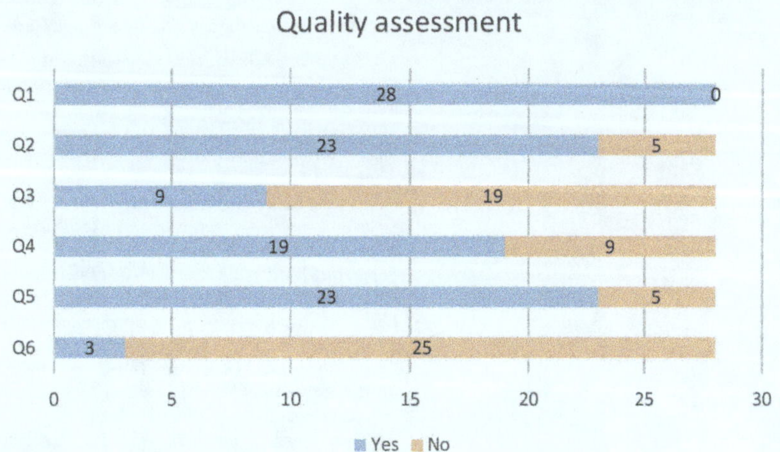

Figure 4. Results of the quality assessment of the included articles.

3.2. Application Domains

Based on the analysis of the included articles, the following application domains were considered: (i) requirements and development platforms; (ii) accessibility; (iii) localization; (iv) mobility assistance; (v) health conditions monitoring; (vi) promotion of healthy lifestyles; and (vii) data analytics. Table 3 presents the classification of the 28 included articles.

Table 3. Classification of the included articles.

Application Domains	Reference	Title	Year
Requirements and development platforms	[68]	*Towards ambient assisted cities and citizens.*	2013
	[76]	*GAWA—manager for accessibility wayfinding apps.*	2017
	[89]	*Age and the city: the case of smart mobility.*	2020
Accessibility	[71]	*A service-oriented approach to crowdsensing for accessible smart mobility scenarios.*	2016
	[83]	*IoE accessible bus stop: an initial concept.*	2018
Localization	[72]	*A genetic-based localization algorithm for elderly people in smart cities.*	2016
	[81]	*OmimamoriNet: an outdoor positioning system based on Wi-SUN FAN network.*	2018
	[82]	*GPS trajectories based personalized safe geofence for elders with dementia.*	2018
Mobility assistance	[69]	*DisAssist: An Internet of Things and Mobile Communications platform for Disabled Parking Space Management.*	2013
	[70]	*On combining crowdsourcing, sensing and open data for an accessible smart city.*	2014
	[86]	*Smart mobility for seniors through the urban connector.*	2019
	[91]	*Beyond walking: improving urban mobility equity in the age of information.*	2020
	[95]	*Towards a collaborative model to assist people with disabilities and the elderly people in smart assistive cities.*	2021
Health conditions monitoring	[73]	*An architecture for combining open data with sensors' data for effective prevention of MCI and frailty in elderly people.*	2017
	[77]	*Definition of technological solutions based on the internet of things and smart cities paradigms for active and healthy ageing through cocreation.*	2018
	[84]	*Remote and non-invasive monitoring of elderly in a smart city context.*	2018
	[94]	*Wearable biosensor and hotspot analysis-based framework to detect stress hotspots for advancing elderly's mobility.*	2020
Promotion of healthy lifestyles	[75]	*Healthy routes in the smart city: a context-aware mobile recommender.*	2017
	[78]	*The Smart City Active Mobile Phone Intervention (SCAMPI) study to promote physical activity through active transportation in healthy adults: a study protocol for a randomized controlled trial.*	2018
	[79]	*Meet Smartwalk, smart cities for active seniors.*	2018
	[80]	*Smartwalk: personas and scenarios definition and functional requirements.*	2018
	[85]	*Customized walk paths for the elderly.*	2019
	[87]	*Smartwalk mobile—a context-aware m-health app for promoting physical activity among the elderly.*	2019
	[88]	*Supporting better physical activity in a smart city: a framework for suggesting and supervising walking paths.*	2019
	[93]	*User perception of a smartphone app to promote physical activity through active transportation: inductive qualitative content analysis within the Smart City Active Mobile Phone Intervention (SCAMPI) study.*	2020
Data analytics	[74]	*Identifying points of interest for elderly in Singapore through mobile crowdsensing.*	2017
	[90]	*Analysis of the temporal characteristics of the elderly traveling by bus using smart card data.*	2020
	[92]	*Classification of users' transportation modalities in real conditions.*	2020

3.2.1. Requirements and Development Platforms

Three articles were focused on the mobility requirements of older adults and IT platforms to help the development of new applications [68,76,89].

Sourbati [89] discussed the older adults' mobility practice in smart city environments as a phenomenon at the intersection of age, IT, and data, and aimed to study transport mobilities of older adults and mobility practices interactions with smart cities' IT services. The analysis highlighted age-bias in inherited transport systems, gaps in available data about older adults' mobility practices and IT use, and opportunities for more inclusive (and sustainable) smart transport. Moreover, based on the discussion of the older adults' mobility practices, the author presented the features of a journey planning application [89].

In turn, Lopez-de-Ipina et al. [68] proposed a proof-of-concept platform for ambient assisted cities able to sense the city to promote the creation of an ecosystem of user-centric applications to help older adults in their daily activities, considering their disabilities. This platform was developed within a research project, and the authors reported a work-in-progress and explained how they brought together the achievements of earlier works.

Finally, Rodriguez-Sánchez and Martinez-Romo [76] presented a platform to provide a universal and accessible solution to manage wayfinding applications that focuses on individuals with disabilities in outdoor and indoor environments. The platform is composed by different subsystems to perform user profiles management, accessibility data management, routes management and accessible application generation. The Assisted Multimodal Wayfinding Application was presented a case study: it can be installed on older adults' smartphones or wearable devices and combines text, maps, auditory, augmented reality, and tactile feedback to provide indoor and outdoor guidance considering personal factors (e.g., disabilities, impairments, or languages) [76].

The authors reported an accessibility guidelines validation and a prototype evaluation [76]. In terms of accessibility guidelines, the authors followed the Web Content Accessibility Guidelines (WCAG), version 2.0, and obtained the maximum level. In turn, the prototype was evaluated by a group of 20 participants (11 males and nine females, aged from 20 to 75), which were divided into the following categories: users without a disability (eight), blind users (three), limited vision users (four), deaf users (three) and wheelchair users (two). This evaluation was composed by three experiments: one for testing the application using Android devices, another for testing the application using iOS devices and the third one for testing the web interface. The evaluation methods include direct observation and questionnaires [76]. According to the results, the application could be used for supporting daily living activities [76].

3.2.2. Accessibility

Two articles [71,83] were considered as proposing applications to improve the accessibility of urban spaces.

Rodrigues et al. [83] proposed a smart bus stop, which would integrate all the features from the existing bus stops, as well as intelligent features to allow its adaptation to different users' needs. The authors reported an initial stage of the project and only the conceptual architecture was presented.

In turn, Mirri et al. [71] presented the design of an infrastructure called Smart Mobility for All (SMAll) aiming to support the mobility of impaired individuals within urban environments. Since a mobile user can be at the same time a consumer and a provider of the sensing services, the authors proposed both participatory sensing (i.e., mobile users actively engage in sensing activities by manually determining how, when, what, and where to sense) and opportunistic sensing (i.e., fully automated sensing activities without the involvement of the users) to collect that about the accessibility conditions of the urban spaces exploiting, for example, Radio-Frequency Identification (RFID) and Global Positioning System (GPS). Moreover, the crowdsensing data were aggregated with other data sources including official data about the transport infrastructure (e.g., static features and real-time information versus planned timetables). To prove the effectiveness of the approach, the authors considered two scenarios illustrating urban accessibility issues involving a wheelchair user and an older adult.

3.2.3. Localization

Three articles [72,81,82] were considered within the scope of applications to determine the localization of older adults.

Considering that the city should provide outdoor monitoring for older adults while promoting their ability to perform leisure activities, the final aim of the study reported by Liouane et al. [72] was to ensure a localization service for older adults that should be provided in time and when it is necessary. According to the authors, they took advantage of the IoT paradigm for collecting heterogeneous data that are broadcast in the city. The assessment of the location performance was performed through a set of experimental simulations using the MATLAB environment.

Chen et al. [81] presented an outdoor positioning system that might be used to protect older adults, via estimating their locations in a city. Particularly, locations were estimated using machine learning algorithms from the data measured at multiple base-stations that are densely deployed over a city to construct an ad hoc network. A prototype system consisting of nine base-stations was deployed on a university campus to conduct an experiment to validate the proposed approach in terms of the performance of the communication networks and machine learning algorithms.

Lin et al. [82] argued that wandering is among the most problematic, dangerous, and frequent behaviors of individuals with dementia and that frequent wanders are more vulnerable to experiencing adverse events than the healthy ones, ranging from falling, getting lost, elopement or boundary transgression to emotional distress. To minimize wandering-related adverse consequences, the authors proposed a geofence based in a virtual boundary delineated around an area of interest that can be created with a variety of different technologies, such as WiFi, cellular mobile, RFID and GPS. Moreover, a data mining-based approach was used to construct a personalized safe geofence by mining older adults' historical GPS trajectories. In terms of validation, an open dataset of older adults' GPS traces was used and, according to the authors, the qualitative results showed that the method is workable for constructing personalized safe geofences by mining older adults' GPS trajectories [82].

3.2.4. Mobility Assistance

Considering the articles related to mobility assistance, Lambrinos and Dosis [69] focused on an application to enhance the parking experience from the perspective of impaired individuals, while four articles [70,86,91,95] reported studies aiming to support older adults when moving in the city.

Lambrinos and Dosis [69] proposed a smart parking management system called DisAssist which takes advantage of the capabilities of mobile devices to allow users to find, reserve and access real-time parking availability information obtained via machine-to-machine communications.

Mirri et al. [70] proposed a system that exploits real time data provided by bus operating companies combined with data produced by sensors and crowdsourcing data to provide routes tailored to older adults' specific needs. The users might access two mobile applications: mobile Pervasive Accessibility Social Sensing (mPASS) and WhenMyBus. The first collects data about urban accessibility and provides older adults with personalized and accessible pedestrian paths and maps, while the second aims to support older adults who travel by bus in the city by providing real time information about transport availability and accessibility facilities.

The important features of these applications are the integration of data produced by sensors (i.e., gyroscope, accelerometer, and GPS) and data gathered via crowdsourcing by users, together with official accessibility reviews conducted by experts. The authors argued that although any instance of the crowdsourced and sensed data may be unreliable, aggregating a large amount of information related to the urban area makes the data more trustworthy (i.e., an error made by a single sensor, or a single user, become less significant as the volume of data increases). Additionally, the data quality might be increased when considering their aggregation with accessibility reviews conducted by experts.

Badii et al. [91] described a concept based on various technologies, including location-based services, augmented reality, and crowdsourcing. Specifically, a smartphone application prototype was developed to support wheelchair users with a customized information service. The important features of the system are path navigation and public participation. Path navigation is focused on individuals who find moving difficult (especially those who use wheelchairs) and aimed to provide an effective and accurate map information service and travel route planning, with two path navigation modes (i.e., wheelchair and walking). In turn, public participation allows users to take photos and upload the improper public facilities around them, such as a broken blind path, a broken elevator, and a damaged or blocked ramp. The positioning information of the photos can be uploaded to the urban management platform, providing data for urban management and maintenance.

Vargas-Acosta et al. [86] presented the Urban Connector, a mobile application that provides relevant information about city services to assist older adults during their travels within a city and to mitigate their risks (e.g., being caught in traffic congestion, getting lost, or being involved in a crash). Based on the recommendations followed in the design of the Urban Connector, an Android-native prototype has been developed, which was tested for a period of 30 days with a group of 38 older adults as early adopters.

Finally, Matos et al. [95] proposed the SafeFollowing application to provide collaborative support for individuals with disabilities and older adults in adverse situations from qualified agents and volunteers. In concrete, the SafeFollowing mobile application allows users to ask for assistance by sending a notification to agents and volunteers who are nearby. The SafeFollowing evaluation consisted of two scenarios (i.e., a situation in which an older adult receives special care provided by an agent, and a situation in which an agent helps a wheelchair user). After performing tests on the above SafeFollowing scenarios, a questionnaire based on the Technology Acceptance Model was used to gather the opinions of the participants.

3.2.5. Health Conditions Monitoring

Four articles [73,77,84,94] described solutions aiming to monitor the health conditions of older adults when moving in the urban spaces.

Lee et al. [94] proposed a wearable biosensor (i.e., a smart watch) and hotspot analysis-based framework to continuously monitor older adults' interactions with the built environment aiming to support interventions to minimize stressful interactions. To test the proposed framework, stress hotspots were detected based on the data related to 30 older adults, which were collected during two weeks of their daily trips. The detected stress hotspots were then investigated by site inspections and interviews with subjects. According to the authors, the proposed sensing framework strengthens the smart city paradigm and can be a basis for optimizing interventions to improve the mobility of older adults.

Through highlighting the importance of integrating social resources into the core of what a smart city is, the ideas presented in [73,77], which took part in the same project (City4Age), were related to the use technologies for the prevention of mild cognitive impairment and frailty in older adults.

Open data services available in the smart city technological infrastructure together with data collected by wearables, smart phones and sensors were used to track the users within the city, including visited points of interest and some activities performed, such as visiting a family member, and their usage of public transportation (e.g., buses or railway trains). The authors claimed that this tracking is useful to build a comprehensive and predictive picture of older adults' wellbeing and health conditions, aiming to sustain better health outcomes and deliver early interventions to anticipate needs [73,77].

Medrano-Gil et al. [77] reported the cocreation framework used for the Madrid City4Age pilot, which involved informal interviews with representatives of public transport organizations and a manager from a private older adults' care service, as well as structured interviews with a manager of day centers and home care services and the people responsible for social services of a municipality of fifty thousand inhabitants. In turn, [73] described the architecture supporting an innovative approach for interfacing open data from smart cities (e.g., public transportation or weather conditions) with data acquired (e.g., walking or enter home) via multiple sensors, namely GPS, Bluetooth Low Energy (BLE) beacons, smartphones, and smart wristbands.

A similar solution was presented in article [84], which described a smartphone-based prototype for remote and non-invasive older adults' monitoring. In the described implementation, a smartwatch with a heart rate monitor and a tri-axial accelerometer was used to determine if a dangerous situation is occurring (e.g., abrupt changes of the heart rate or a sudden acceleration, followed by a state of quiet that might be a sign of fall or fainting).

3.2.6. Promotion of Healthy Lifestyles

Eight articles focused the promotion of healthy lifestyles [75,78–80,85,87,88,93].

Article [75] reported a context-aware recommender application that offers personalized recommendations of exercise routes to older adults according to their health conditions and real-time information from a smart city. The application has predefined routes and recommends the best route based on a memory-based method that employs neighborhood search (i.e., to determine groups of similar individuals) and information acquired from the smart city infrastructure such as air quality, ultraviolet radiation, wind speed, temperature, or precipitation. The older adults can then select the best course according to their profile (e.g., age, effort, or distance) and can inform the application about unexpected situations that could affect other users. Moreover, they can also propose new routes. The application was configured for two cities and first tested with simulated users whose age distribution and medical statistics followed the reports of the WHO and the World Heart Federation. Later, an experiment was conducted to verify that the quality of the recommendations provided by the system were similar to those obtained by simulation.

A similar approach was followed by the SmartWalk project that was described by five articles [79,80,85,87,88]. The objective of the SmartWalk project was to implement a platform to promote physical activity by older adults, through the suggestion of routes that met both the individual requirements in terms of physical activity and personal preferences. The choice of routes is made by a healthcare professional that considers the individual health conditions. The vision and general architecture of the SmartWalk project was presented in two articles [79,88], while a set of personas and scenarios that were developed to help the systematization of the system functional requirements was presented in [80]. Moreover, the algorithm to determine the routes was described in [85] and the SmartWalk application was presented in [87].

Finally, articles [78,93] were related to the project Smart City Active Mobile Phone Intervention (SCAMPI), which evaluates an application to promote physical activity together with data acquisition related to behavior, mode of travel, duration, and speed. The application collects data in real time on location and travel speed using GPS. Moreover, accelerometers are used to provide an objective assessment of physical activity.

Ek et al. [78] presented the design of a two-arm parallel randomized controlled trial to assess the application promoting physical activity selected for SCAMPI trial in terms of monitoring behavior change towards active transport. The primary outcome is moderate-to-vigorous intensity physical activity, while secondary outcomes include time spent in active transportation, perceptions about active transportation and health related quality of life. In turn, Lindqvist et al. [93] presented a qualitative study that aimed to examine the acceptance and user experience of the referred application promoting physical activity.

3.2.7. Data Analytics

Concerning the application of data analytics, three articles [74,90,92] were identified. The studies aimed to process data to determine activities and behaviors of older adults in the urban space.

The study presented in [74] adopts a smartphone-based mobile application as a tool to better understand the daily activity of older adults. Using the smartphone sensor information collected through the mobile application, the authors intended to prove that it is possible to identify the points of interest of older adults in Singapore.

In turn, in the study reported by [90], a large volume of smart card data was used to determine the public transport travel behavior of older adults in Beijing, China, that were classified as long, medium, and short bus-travelers considering their objective mobility characteristics. According to the authors, the findings might be used to tailored mobility policies of the cities.

The research reported in [92] aimed to understand the older adults' means of traveling considering contextual data and data coming from the smartphones. The correct classification of transportation can be also used for providing suggestions in the context of public or private transportation.

3.3. Types of Data Being Used

Since smart city technological platforms aim to promote automated and intelligent processes based on the analysis of vast quantities of data, data gathering is an important issue for the proposed applications. According to the included studies, the data acquired from the smart city infrastructure were complemented with continuous data gathered by personal sensors that are gradually being pushed into the market (Table 4).

Table 4. Types of data being used by the proposed applications.

Types of Sensors	NotSpecified	Location	Movement and Speed	Using Transport	Physiological Parameters	Points of Interest	Presence of Vehicles	Air Quality	Weather Conditions	Illumination
Smartphone's sensors										
Unspecified	[68]	[92]	[73,77,92]		[93]					
Gyroscopes and accelerometers			[70,71]							
Global Positioning System		[79,80,82,87,88]	[70,71,73,77,93]	[73,77]		[73,74,76,77]				
Bluetooth Low Energy beacons			[73,77]	[73,77]		[73,76,77]				
Wearables										
Body Area Network		[72]			[73,79,80,87,88]					
Smartwatch					[84,94]					
Sensors of deployed in the cities										
Unspecified	[68]		[70]					[75,80,88]	[75,80,88]	[80]
Magnetic sensors							[69]			
Radio-Frequency Identification		[71,82]								
QR-Code						[76]				

As shown in Figure 5, 18 of the examined articles indicated the use of personal sensors or sensors from the smart city infrastructures to gather different types of data. Sixteen articles referred to the use of personal sensors (e.g., sensors deployed in the older adults' smartphones or wearables), and nine articles referred to the use of sensors deployed in the city. Only seven articles referred to the use of both personal sensors and sensors from the smart city structure.

Looking at the technological details presented by the articles, ten articles considered the concept of IoT, although just two articles referred to the protocols being used. Mechanisms to aggregate data from different sensors and to guarantee the quality of these data were presented in two articles, while five articles mentioned the need for data interoperability, although just one referred to how interoperability can be achieved.

In turn, 16 articles referred to concerns related to the privacy, integrity and confidentiality of the data gathered by personal sensors, but only one described the implementation of mechanisms for data protection.

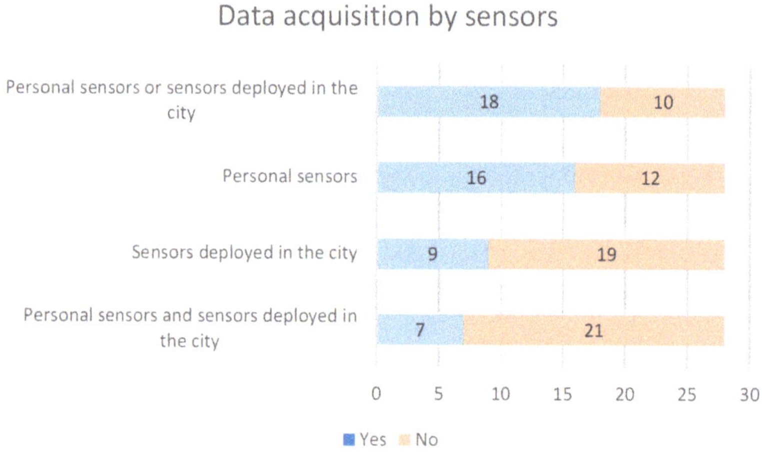

Data acquisition by sensors

Figure 5. Number of articles reporting data gathered by personal sensors and sensors deployed in the city.

In addition to the data gathered by the sensors, several articles reported the integration of open data from the smart city (Table 5), namely publicly available accessibility information [68], transport infrastructures [68–71,73,77,78], real-time data about public transport [70,71] and points of interest in the city [76].

Table 5. Open data from the smart city.

Type of Data	References
Accessibility of the city	[68]
Transport infrastructures	
Roads	[85]
Parking	[69,71]
Public transport	[68,70,71,73,77]
Scheduled data (e.g., buses)	[70,71]
Shared transport	[71]
Real-time transport data	
Real-time data of public transport	[70,71]
Touristic features	[76]

Moreover, seven articles [70,71,74,75,86,91,95] reported the implementation of crowdsourcing mechanisms (Table 6).

Table 6. Articles reporting crowdsourcing.

References	Aims
[70]	To determine urban accessibility (i.e., barriers and facilities).
[71]	To determine urban accessibility (i.e., barriers and facilities).
[74]	To identify the points of interest among the older adults.
[75]	To determine the route's status.
[86]	To identify recommended places.
[91]	To determine problems of walking environments.
[95]	To support the coverage of individuals with disabilities.

3.4. Maturity Level

In terms of the development of applications, it is important distinguish the different development phases, each one with a different maturity level (e.g., requirements, analysis, design and implementation, testing, and evolution or maintenance). As can be seen in Table 7 and Figure 6, a significant percentage of the included studies aiming to develop smart cities' applications to support the mobility of older adults were still in an early development phase (i.e., description of concepts or architectures). In turn, 11 articles [72,74,76,81,82,85,86,90,92,93,95] reported proof-of concept prototypes. Four of these articles [76,86,93,95] reported the participation of older adults for the qualitative evaluation of the prototypes, although the measured outcomes were poorly described, and the number of participants was small. Finally, one article [94] reported the assessment of prototypes in a real-life scenario.

Table 7. Maturity level of the proposed solutions.

Maturity Level	References
Concepts for further development	[77,78,80,83,91]
Architectures	[68–71,73,75,79,84,87–89]
Proof-of-concept prototypes	[72,74,76,81,82,85,86,90,92,93,95]
Assessments in real-life scenarios	[94]

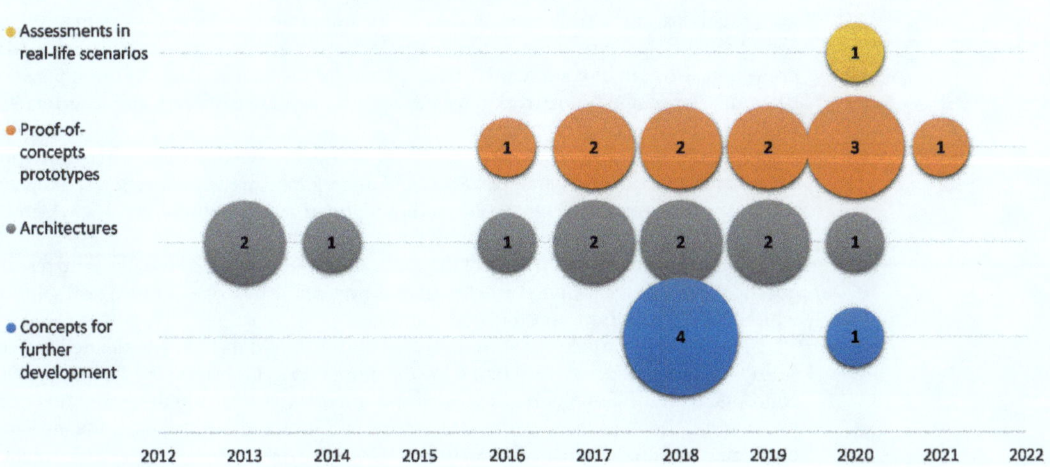

Figure 6. Number of articles by maturity level and years.

In the study reported in article [94], which aimed to develop a wearable biosensor and hotspot analysis-based framework to continuously monitor the older adults' stressful interactions with the built environment, it was conducted a pilot. Thirty older adults participated in various types of data collection, including controlled route and daily trips, so that stress hotspots could be identified. As a result, the authors concluded that hotspot analysis with wearable biosensors can detect spatiotemporal stressful interactions between the older adults and the built environment.

4. Discussion

Looking for the first research question (i.e., the relevant smart city applications to facilitate the mobility of older adults), seven application domains emerged from this systematic review: (i) requirements and development platforms (three articles); (ii) accessibility (two articles); (iii) localization (three articles); (iv) mobility assistance (five articles); (v) health conditions monitoring (four articles); (vi) promotion of healthy lifestyles (eight articles); and (vii) data analytics (three articles).

These results are in line with current concerns related to the improvement of environmental factors [9], including urban infrastructure and services, to facilitate active ageing [10,11] and the promotion of the wellbeing of older adults [99], and show that smart cities, if well-equipped to address the needs of their older adults, might facilitate their mobility [10,11].

In terms of the type of data being used by the proposed applications (i.e., the second research question), the different articles reported the use of data acquired from the smart city infrastructure, such as air quality, weather conditions, or light conditions and open data available in the smart city databases, namely static data (e.g., city accessibility, transport infrastructures and touristic points of interest), and real-time data about public transport. On the other hand, sensors deployed in the older adults' smartphones and wearables were used to acquire continuous monitoring data to determine localization, movement, speed, points of interest, or utilization of public transport, as well as to monitor physiological parameters. Moreover, one fourth of the articles reported the implementation of crowdsensing mechanisms.

Using sensors to constantly monitor individuals has the potential to put them at risk. Therefore, secure data transmission and the guarantee that the stored data would only be accessed by individuals who are authorized are important requirements [39–41]. If these requirements are not satisfied, data such as the location of an individual at a given time might be used for nefarious purposes (e.g., to know when and for how long the individuals are out of their homes). Despite the importance of data privacy, integrity, and confidentiality, just one of the included articles described mechanisms to guarantee data protection. This can be considered a major barrier for the dissemination of the applications being developed and future work must pay special attention to privacy and security issues when using emerging sensor technologies.

Considering the need to use huge amounts of data from different sources, interoperability and data quality, standardization and aggregation are key aspects [46]. However, these aspects are almost absent in the studies reported by the included articles, which also negatively impacts the dissemination of the applications.

Concerning the maturity level of the applications being reported (i.e., the third research question), most of the included articles generally tended to describe technological solutions, which were still far from consolidated solutions.

Twenty-seven articles proposed concepts for further development, defined architectures, or presented prototypes that were developed to demonstrate the feasibility of the concepts. Four articles reported the participation of older adults in the evaluation of the prototypes, although the experimental setups had limitations in terms of the measured outcomes, and the number of participants. Only one article reported the assessment of the proposed application in a real-life scenario. It is worth emphasizing the importance of going beyond technological determinism and, accordingly, to consider the impact on the target users.

This low maturity level is an important drawback when comparing the included studies with similar research using different approaches. For instance, in terms of the monitoring of health conditions and the promotion of healthy lifestyles, the scientific literature reports relevant research studies based on robust methods and involving a significant number of participants to assess the impact of mobile health applications (e.g., [100,101]).

Therefore, concerning the fourth research question (i.e., what are the barriers for the dissemination of the solutions that were identified?), it is possible to conclude that the lack of assessment in real-life scenarios, together with a set of unsolved aspects related to the data being used (i.e., data privacy, integrity, and confidentiality, data interoperability, data aggregation, and data quality) constitute major barriers for the dissemination of smart cities' applications to facilitate the mobility of older adults.

5. Conclusions

The study reported by this article aimed to review smart city applications to facilitate the mobility of older adults. Relevant application domains were identified, including requirements and development platforms, accessibility, localization, mobility assistance, health conditions monitoring, promotion of healthy lifestyles, and data analytics.

The results show that there is an ongoing effort to take advantage of the smart cities' paradigm to make cities more age-friendly by facilitating the mobility of older adults, namely by using a diversity of sensing data provided by a broad range of sensors. However, issues such as user privacy, data standardization and integration, and sensors' characteristics were poorly addressed. The results also show that there is a lack of maturity of the developed applications, which constitutes a major barrier to their dissemination. Moreover, it is foreseen that the number of articles related to the topic will increase in the future, since the research effort increased over the years: the oldest articles that were included were published in 2013 and more than two-thirds of the included articles were published in the last four years.

According to the study protocol (e.g., search keywords and inclusion and exclusion criteria) this review only considered the research related to smart cities' applications specifically focused on facilitating the mobility of older adults. However, older adults constitute a heterogeneous population group with different needs, life experiences, expectations, and personal factors, which means that, when needed, there are older adults able to use applications developed for the general population. In this respect, one of the limitations of this study is related to the fact that there are other smart cities' applications and services developed for the general population that can be used by older adults to facilitate their mobility (e.g., pedestrian mobility [102], multi-modal routes' support [103], smart parking [54], traffic management [104], assistance to drivers [105,106], intelligent transport systems [25,105,106], or mobility as a service [38,58]). Moreover, it should be noted that this review did not consider other possible studies aiming to facilitate the mobility of older adults without being supported in smart city infrastructures, nor articles whose primary focus was not the use of IT (e.g., [107]). Additionally, the review did not consider articles published after March 2021.

Moreover, is always possible to point out limitations about both the chosen keywords and the databases that were used in the research. Likewise, since most articles were published in conference proceedings, there will certainly be similar articles that have not been included because they were presented in non-indexed conferences. It should also be noted that the grey literature was not considered in this review and that this can be seen as a gap of some significance.

Despite these limitations, in methodological terms, the authors tried to follow rigorous procedures for the articles' selection and data extraction, so that the results are relevant for identifying IT-based applications requiring smart cities' infrastructures to facilitate the mobility of older adults, which may contribute to future developments.

Based on the findings of this systematic literature review, it is possible to conclude that most of the identified studies supported older adults' requirements, which means that there is a trend in the research of practical solutions. However, only one article reported an experimental set-up to assess the proposal solution in a real-life scenario. The remainder articles proposed concepts, architectures, and proof-of-concept prototypes. Four proof-of-concept prototypes were validated by older adults, but the respective experimental designs exhibit limitations.

The technological solutions should respond to user needs and not the other way around. As such, it is possible to conclude that the assessment of the smart cities' applications should be emphasized, as well as the use of robust methodological approaches. Robust evidence is required to show that new developments are valid, reliable, cost-effective, and able to make a difference. Collecting this evidence requires considerable resources to integrate new applications into real life conditions, to be used by many users for long periods of time [108].

Since this review identified a research trend to develop practical solutions and a considerable investment is being made to bring together smart city stakeholders, including industry, to create market-ready solutions [19], the low maturity level of the developments is a major gap on the current research related to IT-based applications requiring smart cities' infrastructures to facilitate the mobility of older adults.

Together with sustainable and environmentally friendly mobility, mobility-as-a-service, traffic management, namely using different data analytics techniques [33–37], and autonomous vehicles are important concerns in the current smart mobility research [57]. Surprisingly, these topics are almost absent in the set of the included studies. Moreover, other relevant topics of the smart cities' research, such as user privacy, data standardization and integration, IoT implementation, and sensors' characteristics, are poorly addressed by the included studies. Therefore, according to the results, another research gap emerged, which is related to the difficulty of incorporating into the specific topic of this systematic review, the knowledge generated by the research related to other topics related to smart cities. Since smart cities are digital ecosystems resulting from a combination of business models and innovation to transform the cities' processes, structures, and strategies [38], researchers should consider these ecosystems to promote a comprehensive view and to take advantage of all relevant smart cities' developments.

After this systematic review, it is possible to conclude that some attention should be given to the fact that the total number of included articles is not very representative within the total number of articles related to smart cities. Furthermore, most of the included articles were published in conference proceedings. Therefore, the implementation of smart cities is still not largely imbued within the active ageing domain, as can be seen in other topics (e.g., AAL [15–17]). Therefore, IT-based applications requiring smart cities' infrastructure to facilitate the mobility of older adults still represent a relevant research opportunity.

Author Contributions: Conceptualization, N.P.R.; writing—original draft preparation, N.P.R.; writing—review and editing, N.P.R., R.B., J.P., G.S., M.R., C.R., A.Q. and A.D.; investigation, N.P.R., R.B., J.P., G.S., M.R., C.R., A.Q. and A.D. All authors have read and agreed to the published version of the manuscript.

Funding: This research was partially funded by Fundação para a Ciência e a Tecnologia in the scope of the project UID/CEC/00127/2021.

Institutional Review Board Statement: Not applicable.

Informed Consent Statement: Not applicable.

Data Availability Statement: Not applicable.

Conflicts of Interest: N.P.R., M.R., and A.Q. were authors of some included studies, but they did not analyze these studies. The authors declare no other conflict of interest.

References

1. World Health Organization. *Decade of Healthy Ageing: Baseline Report*; WHO: Geneva, Switzerland, 2020.
2. World Health Organization. *Global Report on Ageism*; WHO: Geneva, Switzerland, 2021.
3. Sixsmith, A.; Sixsmith, J. Ageing in place in the United Kingdom. *Ageing Int.* **2008**, *32*, 219–235. [CrossRef]
4. World Health Organization. *Active Ageing: A Policy Framework*; WHO: Geneva, Switzerland, 2002.
5. World Health Organization. *A Glossary of Terms for Community Health Care and Services for Older Persons*; WHO: Geneva, Switzerland, 2004.
6. Havighurst, R.J. Successful aging. *Process Aging Soc. Psychol. Perspect.* **1963**, *1*, 299–320.
7. Rowe, J.W.; Kahn, R.L. Successful Aging. *Gerontologist* **1997**, *37*, 433–440. [CrossRef]
8. Cosco, T.D.; Prina, A.M.; Perales, J.; Stephan, B.C.M.; Brayne, C. Operational definitions of successful aging: A systematic review. *Int. Psychogeriatr.* **2014**, *26*, 373. [CrossRef]
9. Annear, M.; Keeling, S.; Wilkinson, T.I.M.; Cushman, G.; Gidlow, B.O.B.; Hopkins, H. Environmental influences on healthy and active ageing: A systematic review. *Ageing Soc.* **2014**, *34*, 590–622. [CrossRef]
10. OECD. *Ageing in Cities*; OECD Publishing: Paris, France, 2015.
11. World Health Organization. *Global Age-Friendly Cities: A Guide*; World Health Organization: Geneva, Switzerland, 2007.

12. Buffel, T.; Phillipson, C. Can global cities be 'age-friendly cities'? Urban development and ageing populations. *Cities* **2016**, *55*, 94–100. [CrossRef]
13. World Health Organization. *Age-Friendly Environments in Europe*; A handbook of domains for policy action; WHO Regional Office for Europe: Copenhagen, Denmark, 2017.
14. Van Hoof, J.; Kazak, J.K.; Perek-Białas, J.M.; Peek, S. The challenges of urban ageing: Making cities age-friendly in Europe. *Int. J. Environ. Res. Public Health* **2018**, *15*, 2473. [CrossRef] [PubMed]
15. Dobre, C.; Mavromoustakis, C.X.; Garcia, N.M.; Mastorakis, G.; Goleva, R.I. Introduction to the AAL and ELE Systems. In *Ambient Assisted Living and Enhanced Living Environments*; Elsevier: Amsterdam, The Netherlands, 2017; pp. 1–16.
16. Van Grootven, B.; van Achterberg, T. The European Union's Ambient and Assisted Living Joint Programme: An evaluation of its impact on population health and well-being. *Health Inform. J.* **2019**, *25*, 27–40. [CrossRef] [PubMed]
17. Stara, V.; Rossi, L.; Borrelli, G. Medical and para-medical personnel' perspectives on home health care technology. *Informatics* **2017**, *4*, 14. [CrossRef]
18. Santinha, G.; Dias, A.; Rodrigues, M.; Queirós, A.; Rodrigues, C.; Rocha, N.P. How Do Smart Cities Impact on Sustainable Urban Growth and on Opportunities for Entrepreneurship? Evidence from Portugal: The Case of Águeda. In *New Paths of Entrepreneurship Development*; Springer: Berlin/Heidelberg, Germany, 2019; pp. 31–53.
19. Allwinkle, S.; Cruickshank, P. Creating smarter cities: An overview. *J. Urban Technol.* **2011**, *18*, 1–16. [CrossRef]
20. Marcos-Pablos, S.; García-Peñalvo, F.J. Technological ecosystems in care and assistance: A systematic literature review. *Sensors* **2019**, *19*, 708. [CrossRef]
21. Vanolo, A. Smartmentality: The smart city as disciplinary strategy. *Urban Stud.* **2014**, *51*, 883–898. [CrossRef]
22. Papa, E.; Lauwers, D. Smart mobility: Opportunity or threat to innovate places and cities. In Proceedings of the 20th international conference on urban planning and regional development in the information society (REAL CORP 2015), Gent, Belgium, 5–7 May 2015; pp. 543–550.
23. Assembly, G. *United Nations: Transforming Our World: The 2030 Agenda for Sustainable Development*; UN: New York, NY, USA, 2015.
24. Benevolo, C.; Dameri, R.P.; D'auria, B. Smart mobility in smart city. In *Empowering Organizations*; Springer: Berlin/Heidelberg, Germany, 2016; pp. 13–28.
25. Mangiaracina, R.; Perego, A.; Salvadori, G.; Tumino, A. A comprehensive view of intelligent transport systems for urban smart mobility. *Int. J. Logist. Res. Appl.* **2017**, *20*, 39–52. [CrossRef]
26. Sumalee, A.; Ho, H.W. Smarter and more connected: Future intelligent transportation system. *Iatss Res.* **2018**, *42*, 67–71. [CrossRef]
27. Pauer, G. Development potentials and strategic objectives of intelligent transport systems improving road safety. *Transp. Telecommun. J.* **2017**, *18*, 15–24. [CrossRef]
28. Reagan, I.J.; Cicchino, J.B.; Kerfoot, L.B.; Weast, R.A. Crash avoidance and driver assistance technologies—Are they used? *Transp. Res. Part F Traffic Psychol. Behav.* **2018**, *52*, 176–190. [CrossRef]
29. Tosi, D.; Marzorati, S. Big data from cellular networks: Real mobility scenarios for future smart cities. In Proceedings of the 2016 IEEE second international conference on big data computing service and applications (BigDataService), Oxford, UK, 29 March–1 April 2016; pp. 131–141.
30. Calabrese, F.; Colonna, M.; Lovisolo, P.; Parata, D.; Ratti, C. Real-time urban monitoring using cell phones: A case study in Rome. *IEEE Trans. Intell. Transp. Syst.* **2010**, *12*, 141–151. [CrossRef]
31. Frank, L.; Kavage, S.; Litman, T. *Promoting Public Health through Smart Growth: Building Healthier Communities through Transportation and Land Use Policies and Practices*; Smart Grow BC: Vancouver, BC, Canada, 2006.
32. Bencardino, M.; Greco, I. Smart communities. Social innovation at the service of the smart cities. *TeMA-J. Land Use Mobil. Environ.* **2014**. [CrossRef]
33. Moustaka, V.; Vakali, A.; Anthopoulos, L.G. A systematic review for smart city data analytics. *ACM Comput. Surv.* **2018**, *51*, 1–41. [CrossRef]
34. Brohi, S.N.; Bamiah, M.; Brohi, M.N. Big data in Smart Cities: A systematic mapping review. *J. Eng. Sci. Technol.* **2018**, *13*, 2246–2270.
35. Souza, J.T.d.; Francisco, A.C.d.; Piekarski, C.M.; Prado, G.F.d. Data mining and machine learning to promote smart cities: A systematic review from 2000 to 2018. *Sustainability* **2019**, *11*, 1077. [CrossRef]
36. Soomro, K.; Bhutta, M.N.M.; Khan, Z.; Tahir, M.A. Smart city big data analytics: An advanced review. *Wiley Interdiscip. Rev. Data Min. Knowl. Discov.* **2019**, *9*, e1319. [CrossRef]
37. Kong, L.; Liu, Z.; Wu, J. A systematic review of big data-based urban sustainability research: State-of-the-science and future directions. *J. Clean. Prod.* **2020**, *273*, 123142. [CrossRef]
38. Anthony Jnr, B. Managing digital transformation of smart cities through enterprise architecture–A review and research agenda. *Enterp. Inf. Syst.* **2020**, *15*, 1–33. [CrossRef]
39. Ahmed, S.; Shah, M.A.; Wakil, K. Blockchain as a Trust Builder in the Smart City Domain: A Systematic Literature Review. *IEEE Access* **2020**, *8*, 92977–92985. [CrossRef]
40. Laufs, J.; Borrion, H.; Bradford, B. Security and the smart city: A systematic review. *Sustain. Cities Soc.* **2020**, *55*, 102023. [CrossRef]
41. Machin, J.; Batista, E.; Martínez-Ballesté, A.; Solanas, A. Privacy and Security in Cognitive Cities: A Systematic Review. *Appl. Sci.* **2021**, *11*, 4471. [CrossRef]

42. Liao, B.; Ali, Y.; Nazir, S.; He, L.; Khan, H.U. Security analysis of IoT devices by using mobile computing: A systematic literature review. *IEEE Access* **2020**, *8*, 120331–120350. [CrossRef]
43. Yu, Z.; Song, L.; Jiang, L.; Sharafi, O.K. Systematic literature review on the security challenges of blockchain in IoT-based smart cities. *Kybernetes* **2021**. ahead-of-print. [CrossRef]
44. Lee, E.; Seo, Y.-D.; Oh, S.-R.; Kim, Y.-G. A Survey on Standards for Interoperability and Security in the Internet of Things. *IEEE Commun. Surv. Tutor.* **2021**, *23*, 1020–1047. [CrossRef]
45. Hajjaji, Y.; Boulila, W.; Farah, I.R.; Romdhani, I.; Hussain, A. Big data and IoT-based applications in smart environments: A systematic review. *Comput. Sci. Rev.* **2021**, *39*, 100318. [CrossRef]
46. De Nicola, A.; Villani, M.L. Smart City Ontologies and Their Applications: A Systematic Literature Review. *Sustainability* **2021**, *13*, 5578. [CrossRef]
47. Pacheco Rocha, N.; Dias, A.; Santinha, G.; Rodrigues, M.; Queirós, A.; Rodrigues, C. Smart cities and healthcare: A systematic review. *Technologies* **2019**, *7*, 58. [CrossRef]
48. Buttazzoni, A.; Veenhof, M.; Minaker, L. Smart City and High-Tech Urban Interventions Targeting Human Health: An Equity-Focused Systematic Review. *Int. J. Environ. Res. Public Health* **2020**, *17*, 2325. [CrossRef]
49. da Rosa Tavares, J.E.; Victória Barbosa, J.L. Ubiquitous healthcare on smart environments: A systematic mapping study. *J. Ambient Intell. Smart Environ.* **2020**, *12*, 1–17. [CrossRef]
50. Kim, H.; Choi, H.; Kang, H.; An, J.; Yeom, S.; Hong, T. A systematic review of the smart energy conservation system: From smart homes to sustainable smart cities. *Renew. Sustain. Energy Rev.* **2021**, *140*, 110755. [CrossRef]
51. Clarinval, A.; Simonofski, A.; Vanderose, B.; Dumas, B. Public displays and citizen participation: A systematic literature review and research agenda. *Transform. Gov. People Process Policy* **2020**, *15*, 1–35. [CrossRef]
52. Purnomo, F.; Prabowo, H. Smart city indicators: A systematic literature review. *J. Telecommun. Electron. Comput. Eng.* **2016**, *8*, 161–164.
53. Zannat, K.E.; Choudhury, C.F. Emerging big data sources for public transport planning: A systematic review on current state of art and future research directions. *J. Indian Inst. Sci.* **2019**, *99*, 601–619. [CrossRef]
54. Barriga, J.J.; Sulca, J.; León, J.L.; Ulloa, A.; Portero, D.; Andrade, R.; Yoo, S.G. Smart parking: A literature review from the technological perspective. *Appl. Sci.* **2019**, *9*, 4569. [CrossRef]
55. Nagy, S.; Csiszár, C. The Quality of Smart Mobility: A Systematic Review. *Sci. J. Sil. Univ. Technol.* **2020**, *109*, 117–127.
56. El-Taher, F.E.-Z.; Taha, A.; Courtney, J.; Mckeever, S. A systematic review of urban navigation systems for visually impaired people. *Sensors* **2021**, *21*, 3103. [CrossRef] [PubMed]
57. Paiva, S.; Ahad, M.A.; Tripathi, G.; Feroz, N.; Casalino, G. Enabling Technologies for Urban Smart Mobility: Recent Trends, Opportunities and Challenges. *Sensors* **2021**, *21*, 2143. [CrossRef] [PubMed]
58. Butler, L.; Yigitcanlar, T.; Paz, A. Barriers and risks of Mobility-as-a-Service (MaaS) adoption in cities: A systematic review of the literature. *Cities* **2020**, *109*, 103036. [CrossRef]
59. Torku, A.; Chan, A.P.C.; Yung, E.H.K. Implementation of age-friendly initiatives in smart cities: Probing the barriers through a systematic review. *Built Environ. Proj. Asset Manag.* **2020**. ahead-of-print. [CrossRef]
60. Ivan, L.; Beu, D.; van Hoof, J. Smart and Age-Friendly Cities in Romania: An Overview of Public Policy and Practice. *Int. J. Environ. Res. Public Health* **2020**, *17*, 5202. [CrossRef] [PubMed]
61. Ziganshina, L.E.; Yudina, E.V.; Talipova, L.I.; Sharafutdinova, G.N.; Khairullin, R.N. Smart and age-friendly cities in Russia: An exploratory study of attitudes, perceptions, quality of life and health information needs. *Int. J. Environ. Res. Public Health* **2020**, *17*, 9212. [CrossRef]
62. Podgórniak-Krzykacz, A.; Przywojska, J.; Wiktorowicz, J. Smart and Age-Friendly Communities in Poland. An Analysis of Institutional and Individual Conditions for a New Concept of Smart Development of Ageing Communities. *Energies* **2020**, *13*, 2268. [CrossRef]
63. Baraković, S.; Baraković Husić, J.; van Hoof, J.; Krejcar, O.; Maresova, P.; Akhtar, Z.; Melero, F.J. Quality of life framework for personalised ageing: A systematic review of ICT solutions. *Int. J. Environ. Res. Public Health* **2020**, *17*, 2940. [CrossRef] [PubMed]
64. Loos, E.; Sourbati, M.; Behrendt, F. The Role of Mobility Digital Ecosystems for Age-Friendly Urban Public Transport: A Narrative Literature Review. *Int. J. Environ. Res. Public Health* **2020**, *17*, 7465. [CrossRef]
65. Zanwar, P.; Kim, J.; Kim, J.; Manser, M.; Ham, Y.; Chaspari, T.; Ahn, C.R. Use of Connected Technologies to Assess Barriers and Stressors for Age and Disability-Friendly Communities. *Front. Public Health* **2021**, *9*, 578832. [CrossRef]
66. Moher, D.; Liberati, A.; Tetzlaff, J.; Altman, D.G.; Group, P. Preferred reporting items for systematic reviews and meta-analyses: The PRISMA statement. *Ann. Intern. Med.* **2009**, *151*, 264–269. [CrossRef] [PubMed]
67. Ghapanchi, A.H.; Aurum, A. Antecedents to IT personnel's intentions to leave: A systematic literature review. *J. Syst. Softw.* **2011**, *84*, 238–249. [CrossRef]
68. López-de-Ipiña, D.; Klein, B.; Vanhecke, S.; Pérez-Velasco, J. Towards ambient assisted cities and citizens. In Proceedings of the 2013 27th International Conference on Advanced Information Networking and Applications Workshops, Barcelona, Spain, 25–28 March 2013; pp. 1343–1348.
69. Lambrinos, L.; Dosis, A. DisAssist: An internet of things and mobile communications platform for disabled parking space management. In Proceedings of the 2013 IEEE Global Communications Conference (GLOBECOM), Atlanta, GA, USA, 9–13 December 2013; pp. 2810–2815.

70. Mirri, S.; Prandi, C.; Salomoni, P.; Callegati, F.; Campi, A. On combining crowdsourcing, sensing and open data for an accessible smart city. In Proceedings of the 2014 Eighth International Conference on Next Generation Mobile Apps, Services and Technologies, Oxford, UK, 10–12 September 2014; pp. 294–299.

71. Mirri, S.; Prandi, C.; Salomoni, P.; Callegati, F.; Melis, A.; Prandini, M. A service-oriented approach to crowdsensing for accessible smart mobility scenarios. *Mob. Inf. Syst.* **2016**, *2016*, 2821680. [CrossRef]

72. Liouane, Z.; Lemlouma, T.; Roose, P.; Weis, F.; Messaoud, H. A genetic-based localization algorithm for elderly people in smart cities. In Proceedings of the Proceedings of the 14th ACM International Symposium on Mobility Management and Wireless Access, Valletta, Malta, 2–6 November 2016; pp. 83–89.

73. de los Ríos Pérez, S.; Castrillo, M.P.; Abril-Jiménez, P.; Montalvá-Colomer, J.B.; Cabrera-Umpiérrez, M.F.; Waldmeyer, M.T.A. An architecture for combining open-data with sensors' data for effective prevention of MCI and frailty in elderly people. In Proceedings of the 2017 IEEE 3rd International Forum on Research and Technologies for Society and Industry (RTSI), Modena, Italy, 11–13 September 2017; pp. 1–5.

74. Hasala, M.S.; Pik Lik Lau, B.; Kadaba, V.S.; Balasubramaniam, T.; Yuen, C.; Yuen, B.; Nayak, R. Identifying points of interest for elderly in Singapore through mobile crowdsensing. In *Proceedings of the 6th International Conference on Smart Cities and Green ICT Systems, Volume 1*; SCITEPRESS, Science and Technology Publications: Setúbal, Portugal, 2017; pp. 60–66.

75. Casino, F.; Patsakis, C.; Batista, E.; Borràs, F.; Martínez-Ballesté, A. Healthy routes in the smart city: A context-aware mobile recommender. *IEEE Softw.* **2017**, *34*, 42–47. [CrossRef]

76. Rodriguez-Sánchez, M.C.; Martinez-Romo, J. GAWA–Manager for accessibility Wayfinding apps. *Int. J. Inf. Manage.* **2017**, *37*, 505–519. [CrossRef]

77. Medrano-Gil, A.M.; de los Ríos Pérez, S.; Fico, G.; Montalvá Colomer, J.B.; Cea Sáncez, G.; Cabrera-Umpierrez, M.F.; Arredondo Waldmeyer, M.T. Definition of technological solutions based on the internet of things and smart cities paradigms for active and healthy ageing through cocreation. *Wirel. Commun. Mob. Comput.* **2018**, *2018*, 1949835. [CrossRef]

78. Ek, A.; Alexandrou, C.; Nyström, C.D.; Direito, A.; Eriksson, U.; Hammar, U.; Henriksson, P.; Maddison, R.; Lagerros, Y.T.; Löf, M. The Smart City Active Mobile Phone Intervention (SCAMPI) study to promote physical activity through active transportation in healthy adults: A study protocol for a randomised controlled trial. *BMC Public Health* **2018**, *18*, 1–11. [CrossRef] [PubMed]

79. Rodrigues, M.; Santos, R.; Queirós, A.; Silva, A.G.; Amaral, J.; Gonçalves, L.J.; Pereira, A.; da Rocha, N.P. Meet smartwalk, smart cities for active seniors. In Proceedings of the 2018 2nd International Conference on Technology and Innovation in Sports, Health and Wellbeing (TISHW), Thessaloniki, Greece, 20–22 June 2018; pp. 1–7.

80. Queirós, A.; Silva, A.G.; Simões, P.; Santos, C.; Martins, C.; da Rocha, N.P.; Rodrigues, M. Smartwalk: Personas and scenarios definition and functional requirements. In Proceedings of the 2018 2nd International Conference on Technology and Innovation in Sports, Health and Wellbeing (TISHW), Thessaloniki, Greece, 20–22 June 2018; pp. 1–7.

81. Chen, Y.; Sakamura, M.; Nakazawa, J.; Yonezawa, T.; Tsuge, A.; Hamada, Y. OmimamoriNet: An Outdoor Positioning System Based on Wi-SUN FAN Network. In Proceedings of the 2018 Eleventh International Conference on Mobile Computing and Ubiquitous Network (ICMU), Auckland, New Zealand, 5–8 October 2018; pp. 1–6.

82. Lin, Q.; Liu, X.; Wang, W. GPS Trajectories Based Personalized Safe Geofence for Elders with Dementia. In Proceedings of the 2018 IEEE SmartWorld, Ubiquitous Intelligence & Computing, Advanced & Trusted Computing, Scalable Computing & Communications, Cloud & Big Data Computing, Internet of People and Smart City Innovation (SmartWorld/SCALCOM/UIC/ATC/CBDCom/IOP/SCI), Guangzhou, China, 8–12 October 2018; pp. 505–514.

83. Rodrigues, J.M.F.; Martins, M.; Sousa, N.; Rosa, M. IoE accessible bus stop: An initial concept. In Proceedings of the Proceedings of the 8th International Conference on Software Development and Technologies for Enhancing Accessibility and Fighting Info-exclusion, Thessaloniki, Greece, 20–22 June 2018; pp. 137–143.

84. Bellagente, P.; Crema, C.; Depari, A.; Ferrari, P.; Flammini, A.; Lanfranchi, G.; Lenzi, G.; Maddiona, M.; Rinaldi, S.; Sisinni, E. Remote and non-invasive monitoring of elderly in a smart city context. In Proceedings of the 2018 IEEE sensors applications symposium (SAS), Seoul, Korea, 12–14 March 2018; pp. 1–6.

85. Amaral, J.; Rodrigues, M.; Gonçalves, L.J.; Teixeira, C. Customized Walk Paths for the Elderly. In Proceedings of the International Conference on Information Technology & Systems, Quito, Ecuador, 6–8 February 2019; Springer: Berlin/Heidelberg, Germany, 2019; pp. 754–763.

86. Vargas-Acosta, R.A.; Becerra, D.L.; Gurbuz, O.; Villanueva-Rosales, N.; Nunez-Mchiri, G.G.; Cheu, R.L. Smart Mobility for Seniors through the Urban Connector. In Proceedings of the 2019 IEEE International Smart Cities Conference (ISC2), Casablanca, Morocco, 14–17 October 2019; pp. 241–246.

87. Bastos, D.; Ribeiro, J.; Silva, F.; Rodrigues, M.; Santos, R.; Martins, C.; Rocha, N.; Pereira, A. Smartwalk mobile–a context-aware m-health app for promoting physical activity among the elderly. In Proceedings of the World Conference on Information Systems and Technologies; Springer: Berlin/Heidelberg, Germany, 2019; pp. 829–838.

88. Rodrigues, M.; Santos, R.; Queirós, A.; Silva, A.; Amaral, J.; Simoes, P.; Gonçalves, J.; Martins, C.; Pereira, A.; da Rocha, N.P. Supporting Better Physical Activity in a Smart City: A Framework for Suggesting and Supervising Walking Paths. *Adv. Sci. Technol. Eng. Syst. J.* **2019**, *4*, 404–413. [CrossRef]

89. Sourbati, M. Age and the City: The case of smart mobility. In Proceedings of the International Conference on Human-Computer Interaction, Copenhagen, Denmark, 19–24 July 2020; Springer: Berlin/Heidelberg, Germany, 2020; pp. 312–326.

90. Shi, Z.; Pun-Cheng, L.S.C.; Liu, X.; Lai, J.; Tong, C.; Zhang, A.; Zhang, M.; Shi, W. Analysis of the Temporal Characteristics of the Elderly Traveling by Bus Using Smart Card Data. *ISPRS Int. J. Geo-Inf.* **2020**, *9*, 751. [CrossRef]
91. An, D.; Wang, J.; Wang, P.; Yang, Y.; Pu, Y.; Ke, H.; Chen, Y. Beyond Walking: Improving Urban Mobility Equity in the Age of Information. In Proceedings of the International Conference on Applied Human Factors and Ergonomics, San Diego, CA, USA, 16–20 July 2020; Springer: Berlin/Heidelberg, Germany, 2020; pp. 204–209.
92. Badii, C.; Difino, A.; Paoli, I.; Paolucci, M.; Nesi, P. Classification of Users' Transportation Modalities in Real Conditions. In Proceedings of the DMSVIVA, Pittsburgh, PA, USA, 7–8 July 2020; pp. 74–81.
93. Lindqvist, A.-K.; Rutberg, S.; Söderström, E.; Ek, A.; Alexandrou, C.; Maddison, R.; Löf, M. User Perception of a Smartphone App to Promote Physical Activity Through Active Transportation: Inductive Qualitative Content Analysis Within the Smart City Active Mobile Phone Intervention (SCAMPI) Study. *JMIR mHealth uHealth* **2020**, *8*, e19380. [CrossRef]
94. Lee, G.; Choi, B.; Ahn, C.R.; Lee, S. Wearable Biosensor and Hotspot Analysis–Based Framework to Detect Stress Hotspots for Advancing Elderly's Mobility. *J. Manag. Eng.* **2020**, *36*, 4020010. [CrossRef]
95. Matos, C.M.; Matter, V.K.; Martins, M.G.; da Rosa Tavares, J.E.; Wolf, A.S.; Buttenbender, P.C.; Barbosa, J.L.V. Towards a Collaborative Model to Assist People with Disabilities and the Elderly People in Smart Assistive Cities. *JUCS J. Univers. Comput. Sci.* **2021**, *27*, 65. [CrossRef]
96. Shahin, M.; Liang, P.; Babar, M.A. A systematic review of software architecture visualization techniques. *J. Syst. Softw.* **2014**, *94*, 161–185. [CrossRef]
97. Dybå, T.; Dingsøyr, T. Empirical studies of agile software development: A systematic review. *Inf. Softw. Technol.* **2008**, *50*, 833–859. [CrossRef]
98. Yang, L.; Zhang, H.; Shen, H.; Huang, X.; Zhou, X.; Rong, G.; Shao, D. Quality Assessment in Systematic Literature Reviews: A Software Engineering Perspective. *Inf. Softw. Technol.* **2020**, *130*, 106397. [CrossRef]
99. Connelly, K.; Mokhtari, M.; Falk, T.H. Approaches to understanding the impact of technologies for aging in place: A mini-review. *Gerontology* **2014**, *60*, 282–288. [CrossRef] [PubMed]
100. Mao, Y.; Lin, W.; Wen, J.; Chen, G. Impact and efficacy of mobile health intervention in the management of diabetes and hypertension: A systematic review and meta-analysis. *BMJ Open Diabetes Res. Care* **2020**, *8*, e001225. [CrossRef] [PubMed]
101. Tong, H.L.; Quiroz, J.C.; Kocaballi, A.B.; Fat, S.C.; Dao, K.P.; Gehringer, H.; Chow, C.K.; Laranjo, L. Personalized mobile technologies for lifestyle behavior change: A systematic review, meta-analysis, and meta-regression. *Prev. Med.* **2021**, *24*, 106532. [CrossRef] [PubMed]
102. Carter, E.; Adam, P.; Tsakis, D.; Shaw, S.; Watson, R.; Ryan, P. Enhancing pedestrian mobility in smart cities using big data. *J. Manag. Anal.* **2020**, *7*, 173–188. [CrossRef]
103. Rocha, N.P.; Santinha, G.; Rodrigues, M.; Rodrigues, C.; Queirós, A.; Dias, A. Mobility Assistants to Support Multi-Modal Routes in Smart Cities: A Scoping Review. *J. Digit. Sci.* **2021**, *3*, 26–40. [CrossRef]
104. Skabardonis, A. Traffic management strategies for urban networks: Smart city mobility technologies. In *Transportation, Land Use, and Environmental Planning*; Elsevier: Amsterdam, The Netherlands, 2020.
105. Arena, F.; Pau, G.; Severino, A. An overview on the current status and future perspectives of smart cars. *Infrastructures* **2020**, *5*, 53. [CrossRef]
106. Mchergui, A.; Moulahi, T.; Othman, M.T.; Nasri, S. Enhancing VANETs broadcasting performance with mobility prediction for smart road. *Wirel. Pers. Commun.* **2020**, *112*, 1629–1641. [CrossRef]
107. Phannil, N.; Jettanasen, C. Design of a Personal Mobility Device for Elderly Users. *J. Healthc. Eng.* **2021**, *2021*, 8817115. [CrossRef]
108. Rashidi, P.; Mihailidis, A. A survey on ambient-assisted living tools for older adults. *IEEE J. Biomed. Health Inform.* **2012**, *17*, 579–590. [CrossRef] [PubMed]

 *applied sciences*

Article

Active Safety System for Urban Environments with Detecting Harmful Pedestrian Movement Patterns Using Computational Intelligence [†]

Juan Chavat [1,*] , Sergio Nesmachnow [1] , Andrei Tchernykh [2,3,4] and Vladimir Shepelev [3]

[1] Facultad de Ingeniería, Universidad de la República, 11200 Montevideo, Uruguay; sergion@fing.edu.uy
[2] CICESE Research Center, 22860 Ensenada, BC, Mexico; chernykh@cicese.mx
[3] Automobile Transport, South Ural State University, 454080 Chelyabinsk, Russia; shepelevvd@susu.ru
[4] Ivannikov Institute for System Programming, 109240 Moscow, Russia
* Correspondence: juan.pablo.chavat@fing.edu.uy
[†] This is extend version of conference paper Computational Intelligence for Detecting Pedestrian Movement Patterns, in Ibero-American Congress of Smart Cities, Soria, Spain, 26–27 September 2018.

Received: 17 November; Accepted: 9 December 2020; Published: 17 December 2020

Abstract: This article presents a system for detecting pedestrian movement patterns in urban environments, by applying computational intelligence methods for image processing and pattern detection. The proposed system is capable of processing multiple images and video sources in real-time. Furthermore, it has a flexible design, as it is based on a pipes and filters architecture that makes it easy to evaluate different computational intelligence techniques to address the subproblems involved in each stage of the process. Two main stages are implemented in the proposed system: the first stage is in charge of extracting relevant features of the processed images, by applying image processing and object tracking, and the second stage is responsible for the patterns detection. The experimental analysis of the proposed system was performed over more than 1450 problem instances, using PETS09-S2L1 videos, and the results were compared with part of the Multiple Object Tracking Challenge benchmark results. Experiments covered the two main stages of the system. Results indicate that the proposed system is competitive yet simpler than other similar software methods. Overall, this article provides the theoretical frame and a proof of concept needed for the implementation of a real-time system that takes as input a group of image sequences, extracts relevant features, and detects a set of predefined patterns. The proposed implementation is a reliable proof of the viability of building pedestrian movement pattern detection systems.

Keywords: computational intelligence; image processing; pedestrian movement patterns; surveillance cameras

---

1. Introduction

Nowadays, there is a growing trend of installing security and surveillance cameras, with the main goal of increasing security in public spaces, streets, offices, and private facilities [1]. Standard systems that handle traditional security cameras do not use automatic methods for detecting incidences in real-time. These surveillance systems are mainly based on operational centers that are in charge of receiving, storing, and analyzing the images and take actions when certain events happen. Operational centers needed to process images from security cameras are operated by technical professionals that play the role of visualizing agents. The job of each visualizing agent consists in constantly observing an (often large) number of images from different sources (security cameras), detecting and generating alerts in case an event of interest is observed [2]. Because of the high personnel costs, most operational centers assign to each visualizing agent a significantly large number of image sources, which could exceed their capacity. Therefore, two phenomena occur, which reduce the surveillance capabilities of the operational center. On the one hand, a degradation of the global attention or the visualizing agents, which are forced to just pay attention to a reduced number of image sources at a time, ignoring events from the rest of the sources. On the other hand, the fact that visualizing agents are human can cause them to feel bored and/or fatigue due to the monotony of the task. It frequently causes poor results.

This article presents an approach applying computational intelligence to overcome the attentional problem of human visualizing agents that works in operational centers, making use of different techniques for image processing and pattern detection. A system capable of processing in real-time multiple image/video sources is proposed to help human visualizing agents in the process of identifying pedestrian movement patterns.

The system focuses on harmful patterns in pedestrian movements, considering common situations for pedestrian safety as defined by the World Health Organization [3], including patterns that can cause a risk of harm (robbery, agglomeration, prowling, etc.) or have negative impacts on pedestrian safety (running, sudden direction changes, walking in forbidden areas). Furthermore, the proposed system is flexible to include/detect different pedestrian movement patterns, which are relevant for each studied scenario.

The proposed system is based on a *filter and pipes* architecture that makes it easy to exchange and evaluate different computational intelligence techniques in each stage of the process. The system is comprised of two main stages. In the first stage, image processing and filtering techniques are applied to images generated from multiple sources, extracting relevant features of images and discarding the ones that are not of interest, according to pre-loaded rules. This stage allows visualizing agents to focus their attention efficiently, on important images. The second stage is responsible for the detection of patterns, taking into account typical situations arising in surveillance and security that are worth identifying (e.g., people running, agglomerations, prowling). The system architecture and design allow extending its capabilities without significant effort, as it is easily adaptable for detection of different types of events of interest and different computational intelligence methods can be incorporated easily in the flow of the system.

This article is an extended version of our conference article "Computational intelligence for detecting pedestrian movement patterns" [4] presented at Iberoamerican Congress on Smart Cities. The main contributions of the research reported in this article are multi-fold. First, we extend our understanding of what and why existing techniques and methods are applied in the areas of image processing and pattern detection. Secondly, we propose and study a system that conveniently combines the surveyed techniques and methods to be capable of processing multiple sources of images in real-time and detect potential events of interest. Specific computational intelligence methods (Hungarian algorithm, Kalman filters, classification) are applied to handle the particularities of pedestrian movements. Finally, to demonstrate the practical applicability of the proposed scheme and study its properties, we conduct experiments over several benchmark videos from PETS09-S2L1 [5] and the comparative analysis of results with state-of-the-art methods from the Multiple Object Tracking (MOT) Challenge competition. The insights from our study contribute to understanding the design methodology, the state-of-the-art of related works, architecture, and components of the proposed system. We proposed to develop a system based on simple image processing and computational intelligence techniques, to provide an efficient solution to be applied in real-time and realistic scenarios. Thus, we focused on the utilization of free software libraries and standard methods for pattern detection, which have been properly extended, configured, and integrated into the proposed system.

The rest of the article is structured as follows. Section 2 contains a brief theoretical introduction to image processing and pattern detection. A review of related work on recognition and pattern detection and tracking on surveillance systems is presented in Section 3. Section 4 presents the general architecture and design of the system. The main implementation details of the proposed system are described in Section 5. Sample results from the evaluation are presented in Section 6. Finally, Section 7 presents the conclusions and the main directions for future work.

2. Methodology: Image Processing and Pattern Detection

This section presents a brief introduction to image processing and pattern detection for surveillance systems.

2.1. Image Processing

Image processing is defined as the process of applying computational techniques in order to modify, improve, change the appearance, or obtain information from images [6]. A standard image processing flow includes five steps (capture, preprocessing, segmentation, feature extraction, and object identification), where the output of each phase is the input of the next one:

1. *Capture* consists in acquiring raw images from a source (e.g., surveillance cameras). In this step, noise and other types of degradation such as blurring, high contrast of the scene, and others, are always added to the image, depending on the specific device used [7].
2. *Preprocessing* applies techniques to remove or reduce the information in those images that are not of interest for solving the problem. The main goal of the preprocessing step is to improve those characteristics of the image that are important for solving the problem (e.g., contour and shine), by using mathematical tools.
3. *Segmentation* consists of splitting each image into regions that represent different objects or background, based on contour, connectivity, or in pixel based characteristics (e.g., shades of gray, textures, gradient magnitude). Some authors state that segmentation algorithms focus on two relevant properties of images: discontinuity and similarity [8], while others add a third property: connectivity [9]. The output of this step is a binary representation of the original image.
4. *Features extraction* consists in finding, selecting, and extracting relevant features of an image, which allow identifying objects of interest for the problem. The features identified in the image must satisfy three properties: robustness, discrimination, and invariance [6].
5. *Object identification* is responsible for categorizing the set of features extracted in the previous step. In this step, different decision models are applied, such as supervised classifiers.

2.2. Pattern Detection

Pattern detection is the study of how computer programs can observe a context, learn, and classify patterns of interest, allowing to make intelligent decisions [10]. A pattern detection system consists of procedures that partition the universe of classes in such a way as to be able to assign elements to classes depending on a set of characteristics of each studied element (the characteristics pattern). On the one hand, when patterns are unknown a priori, the process is called *pattern recognition*; on the other hand, when the patterns are known, the process is called *pattern matching*. The pattern detection process usually consists of three stages: segmentation, features selection, and extraction and classification. The main details of each stage are commented next:

1. *Segmentation* in pattern detection systems is similar to the segmentation applied in image processing systems. The goal of the segmentation stage is to simplify the input, resulting in a set of information that is easier to process.
2. *Feature extraction* takes as input the result of the segmentation stage. The input is processed to extract relevant information about specific objects, remove redundant/irrelevant information, to reduce the dimension of the problem. Quantitative (e.g., speed, distance) and qualitative (e.g., occupation, sex) features are extracted and used to build a vector of features. The goal is to select a subset of features (from the original set) in order to optimize a predefined target function. Feature selection can be done by applying statistical techniques. This procedure usually requires

a deep knowledge of the problem. Selection feature methods consist of three components: at least one evaluation criterion, a procedure or search algorithm and a stop criterion [11].

3. *Classification* processes features a vector to assign features to specific classes, according to predefined metrics. Classification methods (i.e., *classifiers*) depend on the nature of the problem and, in general, their performance also depends on the quality and number of extracted features. There are two main groups of classifiers: *supervised*, and *unsupervised* [12]. Supervised classifiers are based on a set of elements, known as *training data*, for which the class they belong is previously known by the classifier [13]. Some typical supervised methods are Bayesian, Support Vector Machine (SVM), *k*-nearest neighbors (*k*-NN) and neural networks, among others [14]. Unsupervised classifiers tries to discover the classes of a given problem from a set of elements of which the classes are unknown [13]. The number of classes to be discovered by an unsupervised classifier can be known in advance or not, depending exclusively on the datasets handled. Some typical unsupervised methods are Simple Link, ISODATA and *k*-means, among others [14].

2.3. Methodology Applied in the Proposed System

The proposed system consists of two stages for detecting pedestrian movement patterns. In the first stage, it receives images from different sources and applies image processing techniques to extract a set of features from the scene to detect objects of interest (e.g., pedestrians) and ignore the rest of the image. In a second stage, a specific module applies pattern analysis techniques to detect patterns fulfilled by objects of interest detected in the previous stage. The main details of the proposed system are presented in Section 4.

3. Related Works

Surveillance systems and pedestrian movement detection have been the subject of several articles in the related literature.

The survey by Valera and Velastin [15] highlighted the growing importance of intelligent surveillance systems with distributed architecture. Several important issues were identified, including: detection and recognition of moving objects; tracking and detection of anomalous movement patterns in video surveillance, behavior analysis; and recovering/storing images. Existent systems were classified in three categories, according to their *generation*: (i) a *first generation* of analog Closed Circuit TeleVision (CCTV) systems that performed fairly good but were not easy to distribute and store; (ii) a *second generation* of systems that combined computer vision technologies with CCTV for automation, which allowed increasing the surveillance efficiency by providing accurate event detection; and (iii) a *third generation* of automatic systems that cover large areas by a distributed deploy, combining sensors, robust tracking algorithms, and optimized algorithms for managing large volumes of information. Taking into account the classification by Valera and Velastin, the system proposed in our research is within the third generation, as complex pattern detection methods are included and parallel computing techniques are applied in order to deal with multiple distributed sources of data.

Piccardi [16] presented a review of background subtraction methods, describing the main features of seven algorithms and analyzing their performance (processing speed, memory utilization and accuracy). The studied methods included: Running Gaussian Average (RGA), Temporal median filter, Mixture of Gaussians (MOG), Kernel Density Estimation (KDE), Sequential Kernel Density approximation, and Concurrence of image variations. A performance analysis was reported, based on the complexity of applying the methods over each pixel (for the case of processing speed and memory utilization) and on categorizing into limited, intermediate or high (L, M, H) accuracy provided. Analysis showed that Running Gaussian Average obtained the best processing speed and the lower memory utilization, while Mixture of Gaussians and Kernel Density Estimation were the best methods regarding accuracy. Table 1 summarizes the main features (time and space complexity, and accuracy) of the methods studied by Piccardi.

Table 1. Performance of the background subtraction methods (adapted from the work of Piccardi [16]).

Method	Speed	Memory	Accuracy
Running Gaussian Average	$O(1)$	$O(1)$	L/M
Temporal Median Filter	$O(n_s)$	$O(n_s)$	L/M
Mixture of Gaussians	$O(m)$	$O(m)$	H
Kernel Density Estimation	$O(n)$	$O(n)$	H
Sequential Kernel Density Approximation	$O(m+1)$	$O(m)$	M/H
Concurrence of Image Variations	$O(8n/N^2)$	$O(8nK/8N^2)$	M
Eigen-backgrounds	$O(M)$	$O(n)$	M

Lopez [17] presented a system capable of detecting apparent movement on images (caused by camera movements). Two types of methods were reviewed: those that detect movement from the difference of consecutive images and those that detect the movement from a single image (e.g., optical flow methods, occasionally used for background subtraction). The author concluded that methods for detecting from a single image have a higher computational cost and introduce noise to the process. Thus, global alignment methods using points with correspondence were used. The system proposed by Lopez obtains an aligned image without apparent movement and both original and aligned images are sent to a segmentation module composed of three layers that apply background subtraction, labeling, and process grouping. The output of the segmentation module results in a set of regions of interest (*blobs*). Blobs are sent to the tracking module that applies filters, including the algorithm by Stauffer and Grimson [18] and also Kalman filters [19] to decide if apparent movement is detected or not. Results close to 90% were achieved without using tracking and almost 100% using tracking.

Regarding pedestrian detection/tracking, Lefloch [20] presented a system for counting people based on detecting the moment in which a person crosses a previously established virtual line. The author proposed counting people on images captured with a camera placed on the roof, in order to eliminate occlusion problems. The proposed system first applies background subtraction to obtain a binary image, which is used to determine which pixels belong to the bottom and to the front of the area that has movement. After that, morphological operations (e.g., erosion, dilatation, opening, and closure) were applied to eliminate noise and also small, isolated areas that exhibit minimal movement. The resulting image is sent to a stage of detection and classification of blobs, which detects contiguous pixels and calculates its bounding box. Bounding boxes that do not meet certain criteria (e.g., wide-high ratio and area) are discarded and those that potentially contain people are identified.

Rodriguez et al. [21] addressed the problem of detecting and tracking people in very dense crowds. Specific problems arise in this scenario, such as occlusion and change of shape and location of people, which pose big challenges for detection. The proposed approach is based on detecting heads, in order to mitigate occlusion problems when detecting the whole body. Background subtraction and segmentation techniques are not useful in this type of scenario, because they cannot isolate people. Rodriguez et al. applied an object detector, trained to detect human heads, and density estimation algorithms, which provide information about the number of persons within a region (but not their locations). The detector was applied to all regions of the image, generating a map that contains scores that indicate the possible presence of people. The map of scores was combined with data obtained by the density estimation algorithms to obtain accurate detection results.

Dollar et al. [22] analyzed the best approaches for people detection. A set of evaluations were performed for varying scenarios and data sets, studying the performance and limitations of different approaches. Authors concluded that people detectors are far from perfection, even under the most favorable conditions. Significant performance degradation was detected when working with images of people whose area is below 30 pixels or when occlusion is greater than 35%. Several areas were identified to improve people detection, including dealing with images between 30 and 80 pixels, improving performance against occlusions, using movement patterns, and using temporary and context information together with monitoring information.

Leach et al. [23] studied the problem of detecting subtle behavior anomalies in surveillance videos. A decision-making process was presented, taking into account *social signals*, i.e., information about the scenario and social context. Social context is based on the premise that two individuals who share a high degree of information in their trajectories (direction, speed, proximity, and trajectory overlap) have similarities and a 'social dependency'. Four scenarios were considered: traffic, with a high number of trajectories; idle, with stationary trajectories; convergence and divergence, with separate or join trajectories; and general, not classifiable in the previous ones. Pedestrian detection [24] and a follow-up Tracking-Learning-Detection method [25] were applied. The process was focused on detecting human heads to reduce the problems generated by the occlusion. Experiments were performed on data from PETS 2007 and Oxford datasets. Effective results were reported, for example, true positive rate 0.78 and false positive rate 0.19, improving 0.13 in comparison with methods that do not take into account the social context. These results suggest that inferring social connections between people helps improve decision making in the detection process.

Cho and Kang [26] presented an abnormal behavior detection system based on hybrid agents for complex scenarios studying group interaction and individual behavior. In the proposed system, agents are categorized into static and dynamic. Static agents are located at fixed points, in order to compute temporal movement information (e.g., speed and direction) of the objects belonging to the background of the image, considering the optical flow variation. Dynamic agents are assigned to moving objects and they move according to the optical flow. Dynamic agents are responsible for calculating the information about social interaction between neighbors, using a Social interaction Force Magnitude (SFM) model and the potential energy of interaction. The proposed system divides each video into non-overlapping blocks and places a static agent and a dynamic agent in the center of each block of the initial frame. Then, the optical flow field between pairs of consecutive frames is computed, extracting information from the behavior of individuals and groups, using the static and dynamic agents. Feature descriptors are created from the extracted information of each block and then mapped to codewords by a clustering process. Codewords from all the blocks plus the bag-of-word method are used to create feature word vectors. The feature word vectors are then processed by a trained SVM classifier and determine the abnormality of the scene. The system was implemented in Visual C++, using OpenCV and LIBSVM libraries. The reported results allowed concluding that the proposed system outperformed the SFM method over PETS 2009 and UCSD datasets. In addition, the authors argued that static agents help to detect quick movements, movements in restricted areas and abnormal behaviors of individual subjects, while dynamic agents help to detect disordered movements (such as panic situations) and separation movements, which are poorly detected by simple agent systems.

Zhu et al. [27] presented a method for detecting abnormal events in crowded scenes based on modeling motion and context information. Low- or medium-level visual information from surrounding local regions was used as context information. The proposed method focuses on modeling the activity of a crowd using context and motion information, considering that all objects or regions with detected motion provide context information. The method extracts dense trajectories and applies noise filters to filter non-typical trajectories. After that, Multiscale Block–Local Binary Pattern coding on Three Orthogonal Plans (MB-LBP-TOP) was used to encode local context information, and the Multiscale Histogram of Frequency Coefficient (MHFC) feature descriptor was used to describe the motion information and extract dense trajectories. Authors claimed that extracting dense trajectories, preferably with noise screening, using MHFC descriptors has two benefits: (i) the method achieves independence of the processed data and, (ii) it is able to better capture the motion characteristics of trajectories. Based on context patterns and motion information, the model is trained using sparse coding and sparse reconstruction cost is applied to classify events into normal and abnormal. The system was implemented in MATLAB and the experimental analysis was performed over UCSD and Subway datasets, obtaining a processing time of 3.7 and 4.2 s/frame, respectively. The authors concluded that the proposed system computed better results than previously developed methods, arguing that improvements are due to the fact that the proposed model does not take into account only the movement of the crowd, but it also uses context information contributed by surrounding objects. False alarms were detected in case of great perspective distortion or when the size of the objects vary significantly. This issue could be solved using different scales of objects in the training phase.

The analysis of related works indicates that there is still room to contribute regarding efficient systems for detecting pedestrian movement patterns applying computational intelligence techniques.

4. The Proposed Detection System

This section describes the architecture and design features of the proposed system for detecting pedestrian movement patterns.

4.1. Architecture and Design

The proposed system was designed to be able to collect and process images generated from different data sources. In turn, a specific architecture was conceived to be flexible enough to allow replacing or adding new algorithms without significant effort. To assure efficiency, the concurrent processing of multiple data sources must be supported. The processes applied on the images are independent of each other and they adapt correctly to a chain pattern [15,20]. Furthermore, all the applied algorithms are able to take as input the output of their previous immediate(s). The aforementioned schema is modeled by the pipes and filters architecture [28].

Taking into account the previous comments and the review of related works, an architecture based on *pipes* and *filters* was proposed for the developed system. The system consists of two main modules. The *Recognition and Tracking* module is responsible for detecting and monitoring pedestrians (objects of interest); the *Pattern Detection* module analyzes the results of the first module to detect patterns based on (recent) historical information. Both modules support multiple concurrent executions using separate processes. The Pattern Detection module supports processing multiple unrelated data sources. The system also includes three auxiliary modules: *control panel, instance launcher* and *events generator*. The Control Panel module graphically shows the received events and sends command orders (which refers to requested actions) to the Instance Launcher. The Instance Launcher is responsible for the creation of Recognition and Tracking module instances, as separate processes. Finally, the Events Generator receives the information of detected patterns and injects them back to the Control Panel. The system modules and the exchanged information are described in Figure 1.

Advanced Message Queuing Protocol (AMQP) protocol is used for communications between modules. AMQP is an open and secure protocol that guarantees delivery on time (or the consequent expiration), uniqueness, and correct ordering of messages, and also data integrity. The following subsections describe each module of the system.

4.2. Recognition and Tracking Module

The Recognition and Tracking module consists of four stages, arranged in pipes and filters. The first filter receives raw images and applies background subtraction, resulting in a binary image. The second filter takes binary images, detects blobs (set of adjacent pixels that belongs to the front of the image) and transfers the set of blobs to the blobs filter, which discards those blobs that do not contain objects of interest and adds spatial information to those relevant blobs. The last stage takes the information of the objects of interest in the image space and associates each one of them to the position of previously detected objects; thus, calculating the movement of each object. The different stages that compose the tracking and recognizing module are presented in Figure 2.

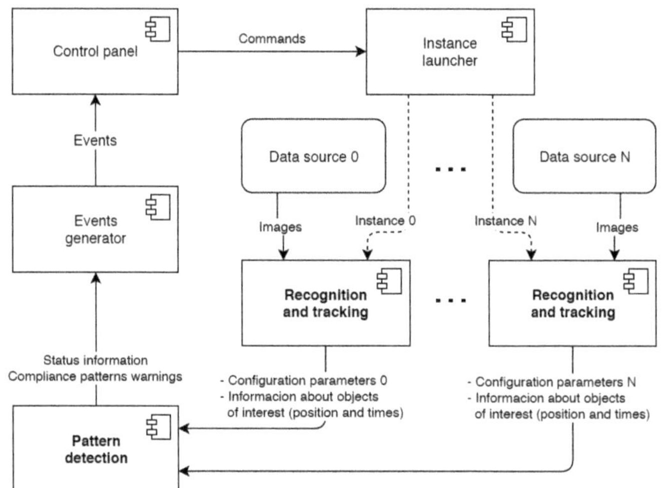

Figure 1. Diagram of the architecture of the proposed system.

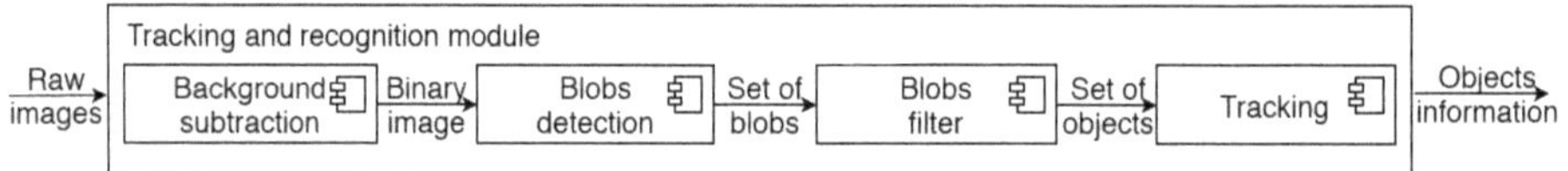

Figure 2. Diagram of components of the Recognition and Tracking module.

4.3. Pattern Detection Module

The Pattern Detection module receives information about objects of interest from multiple instances of the Recognition and Tracking module. The information is stored in a repository that contains the recent history for each object. Periodically, the module processes the last entries to identify a set of features, called *primitives*, which represent basic characteristics of the objects. Characteristics depend on the movement speed, direction, or another attribute of the object. A sequence of primitives plus a set of associated values of properties defines a *pattern*.

Single and multi-target primitives are used. Single-target primitives take into account only one object of interest, ignoring the rest of the objects in the scene, and multi-target primitives takes into account multiple objects in the scene. For example, single-target primitives can determine if a person is standing, walking, or running, depending only on the speed of movement of that person. On the other hand, a multi-target primitive can detect an agglomeration depending on the position of a group of persons for a period of time.

A specific method for pattern detection, based on previous work [29] is implemented. The proposed method for pattern detection takes into account the 'proximity' between an identified sequence of primitives and a set of previously established patterns. Proximity is evaluated using an error function that applies the concept of temporal distance, i.e., the total time of primitives within a sequence that are not included in the reference pattern.

Reference patterns are integrated into the system dynamically. For each primitive that integrates a pattern, a quantifier and a value are defined. For example, a primitive that takes into account the movement speed is fulfilled within a pattern if a pedestrian walks with a movement speed greater or equal to (*quantifier*) 5 km per hour (*value*).

The way patterns are defined and integrated into the system makes them generic, i.e., agnostic of specific information about the data sources and the scenes processed. For instance, an 'agglomeration' is defined by two relevant parameters: the number of people and the time that these people stay within a radius of less than a defined parameter distance. Parameters that define a given pattern keep unchanged along with the different data sources or the scenes, but can be modified according to specific needs.

4.4. Auxiliary Modules

Three auxiliary modules allow simplifying the operation of the main modules of the system and displaying results.

The Instance Launcher module starts instances of the pattern detection, control panel, and events generator modules. After that, it waits for the arrival of command orders. For instance, a command order can require to attend a new source of data, which causes a new instance of the Recognition and Tracking module to be launched. The Control Panel module consists of a web service and a web interface that allows final users to start new processing instances and visualize partial and final results. The Event Generator module stays idle while waiting for results generated by the patterns detector. Generated results, when available, are sent to the event generator. Based on the results received, the event generator generates web events that are sent to all web users using the control panel.

5. Implementation

This section describes the main decisions about technologies and algorithms taken during the implementation of the system.

5.1. Technology Selection

The search of technologies for implementing the system was based on a set of predefined conditions related to the main requirements of a pedestrian movement patterns detection system, including: (i) using a cross-platform programming language; (ii) develop over a programming language without technical complexities (not hard to type, has an automatic memory handler, etc.) and having a broad and active community; (iii) using free libraries and preferably open source; (iv) achieving good performance on all tasks covered by the system: image processing, pattern detection, message passing and management, etc.

After a literature and technology research [30,31], a group of configurations were selected for a deeper study: Matlab, OpenCV over C/C++, and OpenCV over Python. Both OpenCV and Python are free and open source. In addition, Python is a dynamically typed language and counts with an automatic memory manager. Python has a wide variety of free and open source scientific libraries and the community is broad and active. Regarding performance, Python is also an efficient option. The study allowed to conclude that the best choice for implementing the system was using OpenCV library (version 3.0.0 was selected) over Python (version 3.4.3).

5.2. Communication between Modules

The AMQP implementation from RabbitMQ [32] is used for the communication between modules. RabbitMQ was thought to support parallelism and be robust for managing messages. For the connection between Python and the RabbirMQ service, the pika library was used.

In AMQP, *exchange* elements provide the message delivery service, according to instructions about how and where to send them. There are four types of exchange elements: direct, topic, fanout, or header. All data in RabbitMQ are in JSON format, a standard, language independent, and simple format for data exchange.

An exchange of type 'direct' was defined between Recognition and Tracking and Pattern Detection modules. Each instance of the Recognition and Tracking module generates messages that are addressed to a unique queue attended by an instance of the Pattern Detection module. Messages exchanged between the two main modules are of two types: *configuration*, used to attend the instance of recognition and tracking that sends the message; and *data*, which contains precise information about objects of interest. The Pattern Detection module sends its results to an exchange of type topic. Each message contains a key that indicates the message type: commands, state information, and matched patterns warnings. The events generator module binds the exchange with a queue to receive the three types of messages, while the instance launcher module binds the exchange to receive just command messages.

5.3. Recognition and Tracking Module

The main implementation details of the four stages of the Recognition and Tracking module are described next.

5.3.1. Background Subtraction

Algorithm 1 presents the steps followed by the component in charge of performing background subtraction. The background is subtracted directly from the raw image. First, background subtraction transforms the raw image to an image in grayscale (line 2 in Algorithm 1). The grayscale image allows processing less information and results in a lower processing time. After that, the grayscale image is blurred (line 3). Blurring is a technique for reducing the noise presented in the image [33]. Blur operations are made by the application of filters. In the proposed system, a Gaussian filter is applied. A Gaussian filter applies a convolution in each point of the image using a Gaussian kernel and then returns the summation as the final result. The implementation of the Gaussian filter used in the system is included in the OpenCV library. After blurring the image, the background is subtracted (line 4). Two different methods were integrated into the system for this purpose: Improved Mixture of Gaussians (MOG) [34] and k-NN [35]. MOG and k-NN are robust methods that provide support for handling dynamic backgrounds too. Both methods are included in OpenCV and a parameter in the configuration of each instance of the Recognition and Tracking module indicates which method is applied.

Algorithm 1 Background subtraction steps

  1: frame ← raw image
  2: grey image ← BGRToGrey(frame)
  3: blurred image ← GaussianBlur(grey image)
  4: binary image ← BackgroundSubtractor(blurred image)
  5: binary image without noise ← MorphologicalOperations(binary image)
  6: output ← binary image without noise

A binary image is obtained after background subtraction. In the binary image, some elements are detected incomplete or are too close to others, generating a not-desired union of blobs. Morphological operations are applied (line 5) to mitigate these problems. A variety of morphological operations are included in OpenCV: *erosion* allows separating elements that appear together by small contact areas; *dilatation* allows joining nearby elements by applying edge thickening; *opening* consists in applying first erosion and then dilatation; and *closing* is the result of applying first dilatation and then erosion. The result obtained after applying morphological operations is an image with less noise and better identified elements. In the proposed system, two operations are applied: erosion and dilatation. A sample of the processing is presented in Figure 3.

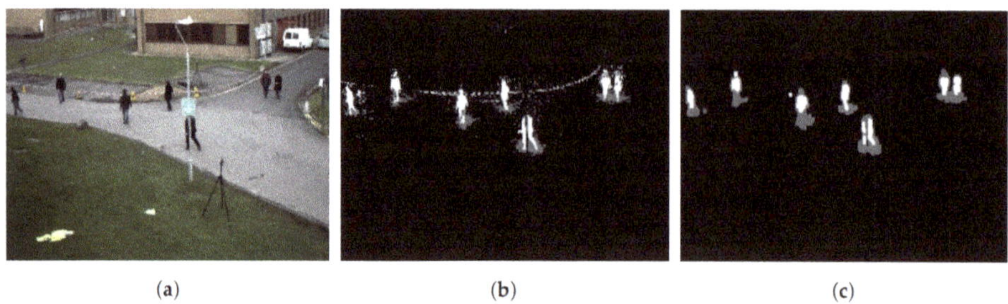

(a)   (b)   (c)

Figure 3. Background subtraction steps. (**a**) Raw image. (**b**) k-NN before morphological operations. (**c**) k-NN after morphological operations.

5.3.2. Blobs Detection

The proposed system includes two methods for processing the binary images for detecting blobs: *simple blob detector* (SBD) and *blob detection based on bounding boxes* (BBBD). SBD is a basic implementation of a blob extractor provided by OpenCV [36]. BBBD is a specific method implemented as part of the reported research. It operates in two phases: the first phase consists of detecting the contour of elements and the second phase performs a search of the minimum rectangles that contains the detected contours (the *bounding boxes*). Both methods return a set of rectangles that contains the blobs detected.

5.3.3. Blobs Classification

Blobs classification takes the set of rectangles as input and classifies them into *useful blobs*, i.e., those containing objects of interest, or as *not useful blobs* when not. Not useful blobs are discarded. This stage takes as input the set of blobs and not the entire image and avoids processing image areas without particular interest, therefore, resulting in a better performance. Three different techniques are implemented, which can be applied isolated or in combination with each other, to improve the results of the classification:

- *Aspect ratio* (AR) classifies blobs based on the relation (ratio) between their width and height. If the ratio is close to the average value of the objects of interest, AR indicates that the blob contains at least one object of interest. The major benefit of AR is its low computational cost. However, it tends to be inaccurate because the reference aspect ratio often varies significantly for different data sources. AR is not useful for discarding blobs (i.e., the fact that a blob fulfills the relation does not mean that it contains an object of interest) and wrongly discards blobs that do not comply with the established aspect ratio criterion due to the fact that they contain multiple objects of interest (e.g., objects close enough one of each other that conforms a unique blob).
- *Computational intelligence* uses the *default people detector*, a pre-trained learning algorithm included in OpenCV. Dalal and Triggs [37] demonstrated that a combination of Histograms of Oriented Gradients (HOG) for feature extraction, and Support Vector Machine (SVM) for the classification of the feature vectors, allows obtaining accurate detection results. The method is based on moving a gridded window all over the image, extracting the vectors of features (using HOG) and classifying them (using SVM) to decide if the image contains a person. Considering that the proposed system studies the movement of persons, the blob classification technique processes just those areas where movement was detected. Thus, the default people detector algorithm is applied just over each detected blob, reducing the computational cost of the processing. This method returns a set of rectangles that contains persons, some of them overlapped. To reduce and unify the number of rectangles, the *Non-Maximum Suppression* algorithm [38] is used.
- *Aspect ratio frequency* filters blobs depending on the frequency that similar blobs were filtered by computational intelligence algorithms. In this way it is possible to simulate a behavior close to the computational intelligence algorithms without having to execute them in each iteration. Aspect ratio frequency can be applied in two different ways: (i) directly filtering those blobs in which the frequency is greater than a threshold, or (ii) categorizing the blobs frequencies into three different levels and depending on that level, discarding the blob, requesting a computational intelligence method to process the blob, or keeping the blob.

5.3.4. Tracking

This stage determines the one-to-one correspondence between detected objects of interest in the current and previous frames. Specific techniques are applied to handle pedestrian movements, which have special features. Since pedestrians are autonomous and can make irregular movements, they differ from other pattern recognition systems, like the ones used to process traffic videos, where vehicles tend to move in more regular patterns and follow normalized rules.

A specific variant of the Hungarian algorithm [39] was developed for this purpose. The Hungarian algorithm receives as input a set of blobs, a set of objects of interest, and a cost function, and returns a correspondence between both input sets that optimizes the defined cost function. The original Hungarian algorithm only accepts inputs of the same size, thus the result is always surjective. A modified implementation was developed to allow the system to handle a different number of blobs than the number of objects of interest. This way, it is possible to process those cases where the number of blobs detected is lower than the objects of interest in the previous frame, or vice versa. In addition, the modified version declares invalid all correspondences in which the cost is greater than a certain threshold, assuming that the blobs do not correspond to the tracked objects.

The cost function used in the proposed system has three components, weighted according to specific parameters in the instance file configuration: (i) the distance between the position of an object in the previous frame and the current position of the blob; (ii) the distance between the predicted position of the object for the current frame and the current position of the blob; and (iii) the difference between the colors of the blob that contained the object in the previous frame and the color of the blob in the current frame.

The position of a blob is not always accurately adjusted to the shape of the objects. As a consequence, the raw trace of an object can suffer zig-zag movements, making it difficult to track the object and detect movement patterns. Kalman filters [19] are applied to avoid the zig-zag effect and to predict the next probable position of each object.

The Kalman filters method keeps the state of the objects, updating it in each frame based on a prediction and correction model (considering position, speed and acceleration for each person). The prediction uses a matrix, known as *transition matrix*, to predict the next state. The correction uses the information from the prediction plus a matrix, known as *noise matrix*, to calculate a correction of the predicted state. In addition, a *gain* parameter is used to smooth the trace, avoiding the zig-zag effect. Empirical results demonstrated that the smoothness achieved by a Kalman filters method is not enough to achieve an accurate tracking, as occlusions of objects are too frequent. In order to mitigate this problem, the system applies an extra smoothing process using Kalman filters smoothing in *fixed-lag smoother* mode. In a given time (state t), this method uses the data of the last $N - 1$ Kalman filter iterations to improve state $t - N$. The cost of the extra processing method is a delay in the detection of patterns in the order of the time needed to process the last N iterations. The implementation for Kalman filter smoothing provided by the FilterPy library was used.

Two structures were implemented to store information of different objects and their tracking, and to resolve occlusions: *tracklets*, associated to a unique object, to store and update the relevant tracking information (position, color, frame when it appears, last frame when its object was not occluded, etc.) and *groups*, used to store tracklets and associate blobs to frames. Ideally, each group should have a single tracklet and a single blob associated. This is the case when the blob represents the object that is tracked by the tracklet. However, a single blob can be associated to a group with many tracklets. This is the case when a multiple occlusion occurs, i.e., multiple objects are close enough to form a single blob. In turn, when many blobs and many tracklets are associated to a single group, it is the case in which multiple objects where previously in occlusion and are splitting up at this moment. In this last case, the system divides the group in order to obtain a single blob per group.

Tracklets are updated or removed in each iteration of the tracking algorithm, depending on: (i) the groups they belong to, (ii) how long the tracklets have belonged to the group, or (iii) how long the *tracklet* has been in the system. On the other hand, a *tracklet* can be removed from the system due to three reasons: (i) the object tracked by the tracklet is the result of a ghost blob, i.e., a newly created blob resulting from noise in the cameras or in previous steps; (ii) low tracking confidence of the object, which happens when the tracklet has not been associated to a one-to-one group for a certain time; and (iii) because the object definitely disappears from the scene.

Three levels are considered for tracklet information updating: (i) *correction with maximum confidence*, when a tracklet is associated one-to-one to a group, the blob of the group represents the tracked object and the tracklet is updated with the information of position and appearance of the blob; (ii) *correction with minimum confidence*, when an object suffers multiple occlusion for a certain time, the tracklet is not associated one-to-one to a group, the predicted position is no longer trustworthy and the tracklet is updated with the position of the blob that represents the occlusion; and (iii) *prediction only*, when a tracklet was recently associated one-to-one to a group, it is assumed that the predicted position is reliable and no correction is made (e.g., when objects are occluded by a short time or two paths cross each other); this level makes it possible to keep tracking positions of the objects even when there is no blob assigned in the current iteration. The pseudocode in Algorithm 2 synthesizes how the system tracks the objects, frame by frame.

Algorithm 2 starts applying the Hungarian method to determine the correspondence between the blobs of the current frame and the objects detected in the previous frames (line 1). Then, it iterates over the blobs detected for the current frame (line 2) and, if the blob is in correspondence with a tracklet, this is added to the group of the tracklet (lines 3–4), otherwise a new tracklet is created, including the blob and a new group containing the tracklet (lines 6–7), because it means that the blob represents a new object in the system. The next step is iterating over the groups that have no blobs assigned (line 10). For each of these groups, the system looks for the closest blob that overlaps the tracklets of the group; if such a blob does not exist, the system looks for the closest blob. In the case that a blob was found, all the tracklets of the group are moved to the group that contains the blob in question (lines 11). Then, each blob is processed in accordance with the number of blobs assigned to it (lines 13–36). Three different situations are possible:

- The group has no assigned blobs. If the tracklet was recently updated with a blob position, then it will be updated in the current frame with the prediction values (line 17), otherwise, the tracklet is removed from the system, assuming that the objects that were tracked disappeared from the scene (line 19).
- The group has a single assigned blob. If the group has just one tracklet (the ideal case), the tracklet is updated with the blob data (line 24). Otherwise, if the group has more than one tracklet, it is the case where many objects are occluded in one blob. In this case, the system first deletes the tracklets with ghost blobs and then deletes all tracklets, except the most reliable, updating it with the blob data (line 26).
- The group has more than one assigned blob. This is the case where objects that were occluded are splitting up. To avoid any loss of information between iterations, each blob is assigned to a new group and the Hungarian algorithm is executed (up to two times, using different weights) between the tracklets of the current group and the blobs associated to it (line 30). After applying the Hungarian algorithm, if there are still many tracklets assigned to a single blob in a group, it is a situation where occlusions and/or ghost blobs happen. In this case, the method in line 26 is applied again.

Algorithm 2 Tracking steps

 1: calculate assignations between tracklets and blobs using Hungarian algorithm
 2: **for each** blob in the system **do**
 3:     **if** assigned to a tracklet **then**
 4:         add blob to the group of the tracklet
 5:     **else**
 6:         create tracklet containing blob
 7:         create group containing the just created tracklet
 8:     **end if**
 9: **end for**
10: **for each** group without assigned blobs **do**
11:     move tracklets to the group of the closest blob, or the group of the blob that contains it
12: **end for**

Algorithm 2 *Cont.*

13: **for each** group in the system **do**
14:     **if** has no blobs assigned **then**
15:         **for each** tracklet in the group **do**
16:             **if** object disappears for a moment **then**
17:                 update position with prediction
18:             **else**
19:                 remove the tracklet
20:             **end if**
21:         **end for**
22:     **else if** has one blob **then**
23:         **if** has one tracklet **then**
24:             update it with the blob information
25:         **else if** has many tracklets **then**
26:             resolves with one-blob-to–many algorithm
27:         **end if**
28:     **else**
29:         **if** has more tracklets than blobs **then**
30:             resolves with many–to–many algorithm
31:         **end if**
32:     **end if**
33:     **if** the group has no tracklets **then**
34:         remove the group from the system
35:     **end if**
36: **end for**

5.4. Pattern Detection Module

The Pattern Detection module is capable of processing multiple data sources concurrently. The module consists of two stages:

1. The first stage receives messages from multiple instances of the Recognition and Tracking module and routes them depending on the source identifier. Two types of messages exist. The first type of message is the least common (but always necessary) and correspond to the ones used when processing requests by new instances of recognition and tracking. When a request arrives, the Pattern Detection module creates the structures to handle data from the new data source identified in the message, and configuration values in the message are applied to process data from the respective data source. The second type of message is used for communicating data of detected objects. When these data messages arrive, they are routed to the structures previously created to handle the data source.

2. The second stage receives data of the detected objects and has the patterns definition, the recent history of primitives fulfilled by each detected object, and all the logic needed to check patterns compliance. Patterns are defined as a sequence of primitives, defined by a 'primitive type', an 'event type' for each type of primitive, and a 'quantifier' that defines how the values of the met primitives are compared with that required by the pattern. For example, the primitives defined in the implemented system are SPEED, DIRECTION, and AGGLOMERATION. Each primitive type has events, for example, for the SPEED primitive, three events are possible: WALKING, RUNNING and STOPPED. The quantifiers defined in the proposed system are LE–lesser or equal, GE–greater or equal, AX–approximate, EQ–equal and NM–irrelevant value.

6. Validation and Results

This section presents validation results of the proposed system for detecting pedestrian movement patterns. Validation is performed for a case study in Montevideo, Uruguay, relying on security patterns, which are the most relevant for Ministerio del Interior, the ministry in charge of the public security.

6.1. Recognition and Tracking Module

The validation of the Recognition and Tracking module was performed using a video from scenario S2.L1 of the PETS09 dataset. PETS09 videos were recorded using a fixed camera installed at a higher height than the head of the people, at a rate of 7 FPS with a resolution of 768×576 pixels and natural light [5]. After 1:54 min of the video, 19 people get in and out of the scene and walk around, generating multiple occlusions among them and with objects of the scene.

A good performance of the module is characterized by an accurate tracking, where data are processed and sent to the Pattern Detection module in real-time (i.e., in less than a second). Thus, the metrics used in the experimental analysis focus on the final result of the module and not in partial filter results. In order to evaluate the processing efficiency, the time required to process each frame was collected and used to calculate the average and maximum processing time per frame. The MOTChallenge benchmark, a unified evaluation platform created by Leal-Taixé et al. [40], is used to evaluate the tracking accuracy. Experiments were executed on Xeon Gold 6138 processors with 128 GB of RAM, from the National Supercomputing Center (Cluster-UY), Uruguay [41].

The MOTChallenge benchmark consists of three components: (i) a public dataset including its own and well known videos (some of them including ground truth information, like PETS09-S2L1, which is used for the evaluation of the proposed system); (ii) a centralized evaluation method that allows the comparison of results obtained with different methods; and (iii) an infrastructure that makes the crowdsourcing of new data possible, new evaluation methods, and new notations (i.e., ground truth).

The evaluation method included in MOTChallenge provides several metrics. One of them, Multiple Object Tracking Accuracy (MOTA) was applied to evaluate the tracking accuracy of the proposed system. MOTA is a percentage that combines three indicators: false positives, false negatives, and identity changes of the tracked persons. The greater the MOTA value is, the more accurate is the tracking of the persons. Furthermore, the average and maximum difference between the number of persons in each frame (from the ground truth) and the detected tracklets and blobs in each frame are evaluated.

The system has a set of parameters that determine how accurate the module performs in a given scene. Thus, previously to the evaluation, a set of experiments was performed to find the best parameter configuration. Since the module has a pipes and filters architecture, the performance of each filter depends only on its configuration values, so finding the best configuration values for each filter results in the best configuration for the entire module. Taking into account these considerations, different values for 40 parameters were studied. A total number of 1458 experiments were performed. For each executed block, three configurations were selected to process the next block. To find the best performing configuration, the following three criteria were taken into account: (i) higher MOTA value (M-MOTA), (ii) lower average difference in the person counting (M-Count), and (iii) from the ones with higher MOTA value, the one with lower average processing time per frame (M-MOTA-TC). Full details of configuration experiments and best values are reported in the complete execution plan available in the project website fing.edu.uy/inco/grupos/cecal/hpc/APMP.

The baseline for the comparison was the manual configuration obtained by empirical tuning and used during the development of the system. The three configurations that achieved the best results in the configuration experiments used k-NN and were able to improve the MOTA value up to 14.8% and the person counting up to 34%. In addition, the highest MOTA value obtained by the system (52.7) was higher than the average MOTA value (36.6) obtained by algorithms in the 2D MOT 2015 benchmark [42]. However, it is worth noting that results for the proposed system correspond to an evaluation performed on a single scene, while results for the 2D MOT 2015 benchmark correspond to averages obtained from applying different algorithms to multiple scenes.

Regarding the other metrics, the obtained average processing time per frame was similar for all three configurations, between 0.024 and 0.046 s, being the blobs classification the filter that requires the most processing time. The maximum processing times per frame are in the range of 0.058 and 0.103 s. The complete average and maximum processing times for each stage are reported in Tables 2 and 3, respectively. An important result for the proposed system is that the average processing time is lower than 0.05 s for all cases. This value indicates that the proposed system allows processing in real-time a 20 FPS data source, which is twice the number of frames required for detecting pedestrian movements [43].

Table 2. Average execution time (in seconds) of the evaluation using the three best configurations.

Configuration	Background Subtraction	Blobs Detection	Blobs Classification	Tracking	Total
M-MOTA	0.00357	0.00051	0.03894	0.00294	0.04595
M-Count	0.00355	0.00053	0.03902	0.00296	0.04606
M-MOTA-TC	0.00356	0.00049	0.01712	0.00265	0.02381

Table 3. Maximum execution time (in seconds) of the evaluation using the three best configurations.

Configuration	Background Subtraction	Blobs Detection	Blobs Classification	Tracking	Total
M-MOTA	0.00552	0.00094	0.09065	0.00657	0.10369
M-Count	0.00559	0.00080	0.08779	0.00562	0.09980
M-MOTA-TC	0.00578	0.00091	0.04628	0.00556	0.05853

Execution time is not strongly linked to the resolution of the processed images. Image resolution only affects the first stage (background subtraction) and the second stage (transformation of the binary image to a set of blobs) of the Recognition and Tracking module. The third and subsequent stages use only the blobs and not the entire image. The two first stages have very low processing times compared with the following stages (one order the magnitude lower, as reported in Tables 2 and 3, so the impact of changing the resolution is not high).

Regarding the goal of person counting, the differences were always below four for two of the best configurations and below five for the other. From a total of 795 frames, no differences were registered in 533 frames for the best configuration. The system was robust, as the worst configuration also had no differences for 433 frames. The aforementioned values of the person counting metric are rather good, especially taking into account that multiple occlusions occur in the videos.

Results in Tables 4 and 5 highlight the accuracy of the proposed system. Table 4 reports the comparison of the number of blobs and number of trackings between the proposed system and ground truth (GT) in MOTChallenge. The obtained results suggest that the proposed system provides an accurate and robust method for recognition and tracking.

Table 4. Results of the differences in people counting for the experiments of the three best configurations.

Best Three Configurations	# Blobs vs. GT			# Trackings vs. GT		
	Mean	Min.	Max.	Mean	Min.	Max.
M-MOTA	0.67	0	4	0.43	0	3
M-Conteo	0.67	0	4	0.33	0	3
M-MOTA-TC	1.14	0	5	0.52	0	4

Table 5. Results of the Multiple Object Tracking (MOT) Challenge for the experiments of the three best configurations.

		Best Three Configurations		
		M-MOTA	**M-Count**	**M-MOTA-TC**
	Recall	78.6	78.9	75.3
	Precisión	76.3	74.5	76.5
	FAF	1.31	1.44	1.24
	MT	10	10	8
	PT	9	9	11
Metrics on MOT Challenge	ML	0	0	0
	FP	1040	1148	986
	FN	910	898	1053
	IDs	63	55	58
	FRA	253	235	234
	MOTA	52.7	50.7	50.8
	MOTP	64.2	64.1	64.3

Sample results for the counting difference for people for each best configuration are presented in the histograms in Figure 4, compared with the baseline GT. Values for counting using the number of blobs (# blobs vs. GT) and using the number of trackings (# tracking vs. GT) are reported. These results are representative of the behavior of the proposed system for other metrics and indicators.

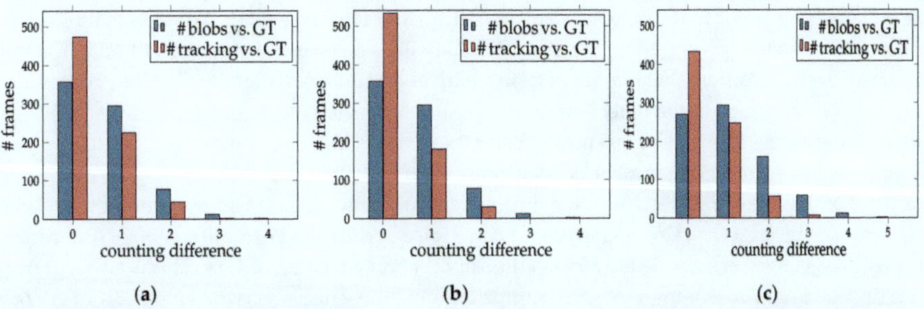

Figure 4. Histogram of counting differences for people, for each best configuration. (**a**) Higher Multiple Object Tracking Accuracy (M-MOTA). (**b**) Lower average difference in the person counting (M-Count). (**c**) The one with lower average processing time per frame (M-MOTA-TC).

Histograms in Figure 4 indicate that the tree configurations have accurate values (differences near to zero). M-Count computed the best results (more than 500 frames with zero differences), M-MOTA is the second best and M-MOTA-TC computed the worst results. Values for counting using the number of blobs are accurate, but lower than using the number of tracking, for the three studied configurations.

Table 5 reports the results of the metrics included in MOTChallenge for the three best configurations. *Recall* corresponds to the percentage of persons detected (higher is better), *precision* is the percentage of persons correctly detected (higher is better), *FAF* measure the false alarms per frame (lower is better), *MT*, *PT* and *ML* are the number of ground truth trajectories that are tracked for at least 80 percent, between 20 and 80 percent, and less than 20 percent of its existence, respectively; *FP* and *FN* are false positives and negatives, respectively; *IDs* counts the changes of identity; *FRA* (for fragmentations) counts the interrupted or restarted trackings (lower is better); *MOTA* was previously introduced; and *MOTP* is the difference between the positions of the ground truth and correctly predicted positions (lower is better).

The scalability of the Recognition and Tracking module was validated regarding the number of sources (videos) that it can handle and process simultaneously. The proposed architecture (described in Figure 1) is robust and scalable, being able to work concurrently with many sources thanks to its parallel/distributed design that considers many instances of the Recognition and Tracking module.

Regarding the number of tracked objects (persons) in a single video, experiments were performed over sparse and medium-density scenes. However, the Recognition and Tracking module applies a parallel approach for blob classification, controlled by a specific parameter (PERSON_DETECTION_PARALLEL_MODE). Using this feature, the module is able to process many objects in the same scene and reducing the overall execution time in up to 20%.

6.2. Pattern Detection Module

The validation of the Pattern Detection module was performed over a recorded video that is representative of the current reality for recognition, tracking, and detection in nowadays surveillance systems in smart cities, such as Montevideo. The recorded video has a duration of 2:51 min, a resolution of 800 × 600 pixels and was recorded in natural light. In the video, nine people walk around and get in and out of the scene occasionally. Five events occur in the video, which can be detected by the four pre-loaded patterns in the proposed system: two agglomerations, two street robberies and one 90-degree turn. The Recognition and Tracking module was configured with the three best parameters values. No significant differences were found in the results between the executions that vary the configuration of the first module.

Experiments were oriented to evaluate the capability of the system to detect predefined patterns exactly, at the moment they occur. When the system detects an event that in fact occurred, a true positive is produced. In this case, the closer to the start of the event it is detected, the more accurate the system is. The evaluation also takes into account the number of false positives (i.e., when the system detects an event that did now occur in reality) and false negatives (i.e., events that occurred but were not detected) produced by the system during the processing. Thus, in addition to the timing accuracy of event detection, the number of false positives and false negatives during the processing was evaluated. The accuracy of the detection of true positive cases is evaluated considering the difference between the moment that an event is notified and the real starting time of the event. For false positives and false negatives, only the number of occurrences is taken into account.

Three of the five events were notified during the processing. All of them correspond to real events. Three true positives, two false negatives, and no false positive event notifications were recorded. True positive events were notified in a mean time of 8.3 s and a median of 4.0 s. This is an accurate time for event detection, which allows to take appropriate and timely decisions when a street robbery occurs.

As an example, the scene where the first street robbery occurs is shown in Figure 5. Blobs are represented by the violet rectangles and tracking is represented by the continuous lines. The robbery is detected in the leftmost blob: one person rapidly approaches another, steals the bag and runs fast, as represented by the violet line. This event was correctly detected by the proposed system.

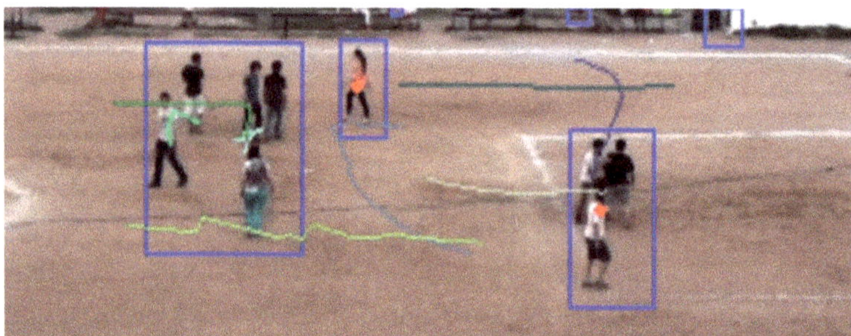

Figure 5. Scene where the first street robbery occurs.

The first not reported event was the second street robbery. This event was not detected due to an incorrect resolution of an occlusion. This case can be observed in Figure 6. The second not reported event corresponded to a 90-degree turn. From the empirical evaluation, it was observed that the detection patterns module has a high sensitivity to small variations of consecutive positions of objects. The fact that the system detects a sequence of small turns instead of a single turn suggests that, in order to detect the event correctly, it is necessary to use a longer history of the last positions of the person who turns.

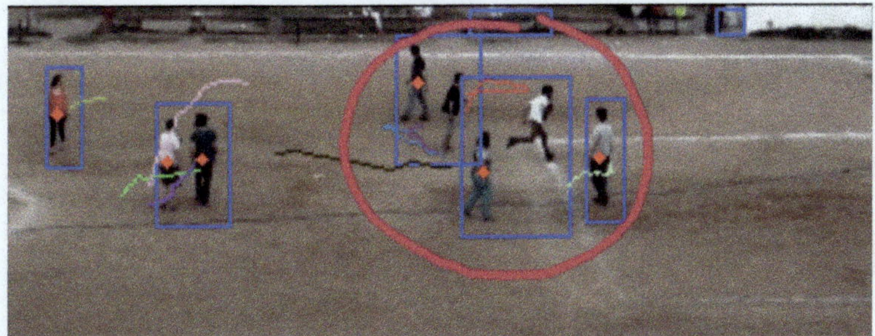

Figure 6. Scene where the second street robbery occurs (it was not detected by the system).

Several additional experiments were performed to evaluate the impact of relevant parameters of the captured videos: contrast and brightness. Real surveillance videos were modified varying the contrast and brightness of the captured scenes. Obtained results allowed concluding that contrast does not significantly impact the efficacy of pattern detection (variation on the prediction capabilities under 5% in the experiments performed). On the other hand, the study of the impact of brightness showed that it affects the successful prediction rate of the proposed system, especially in scenes/videos with poor light conditions (the successful prediction rate is below 50%). These results indicate that further enhancements are needed to properly work on hazy scenes (such as foggy/rainy days). Regarding approaches to overcome issues that emerge under poor illumination conditions, enhancements on two methods already included in the system can be developed: Kalman filters can be applied to predict values of pixels (or groups of pixels) to reduce the variation of illumination, and HOG can be applied to exploit information of the gradient of the images using local histograms that can be grouped in blocks to improve detection and tracking, by normalizing the final representation of each frame to make it less sensitive to illumination changes and distortion. Other methods, such as similarity techniques between frames, can also be applied to mitigate the impact of sudden illumination changes. These enhancements are proposed as one of the main lines for future work.

7. Conclusions and Future Work

This article presented a system for detecting pedestrian movement patterns, based on computational intelligence for image processing and pattern detection.

The proposed system is capable of processing in real-time multiple image/video sources. An architecture based on pipes and filters is used to allow an easy evaluation of different computational intelligence techniques in each stage of the processing. Two main stages are identified in the system, focusing on extracting relevant features of the processed images (implemented in the Recognition and Tracking module) and detecting movement patterns (implemented in the Pattern Detection module). Several techniques are applied for image processing and pattern detection. The proposed implementation fulfills important requirements for a pedestrian movement patterns detection system: it is cross-platform, open-source, and efficient.

The experimental analysis performed over more than 1450 problem instances covers the two main stages of the system. The system was evaluated using PETS09-S2L1 videos and the results were compared with part of the MOTChallenge benchmark results. Results suggest that the proposed system is competitive, yet simpler, than other similar software methods.

The main lines for future work include studying other algorithms for person detection that maintain the low processing cost and possibly improve the dynamical adaptation of blobs aspect ratios; and studying the auto-adaptation of parameters for different scenarios. In turn, to better handle scenes from foggy/rainy days, including support for scenes with different light conditions, and the integration of a specific module to automatically extract/learn the sequence of primitives from particular videos. Filters and estimators (e.g., velocity, depth) can be integrated to deal with the detection of people using moving cameras and other problems that emerge when background subtraction is not directly applicable. The integration of such modules is easily supported by the flexible pipes and filters architecture. Finally, the experimental analysis should be extended to account for the performance of the proposed system to deal with crowded scenarios. Further details about the proposed system are available on the project website fing.edu.uy/inco/grupos/cecal/hpc/APMP.

Author Contributions: J.C.: conceptualization, methodology, software, validation, research, writing; S.N.: conceptualization, methodology, software, validation, research, writing, supervision; A.T.: conceptualization, methodology, research, writing; V.S.: conceptualization, methodology, research, writing. All authors have read and agreed to the published version of the manuscript.

Funding: This research received no external funding.

Acknowledgments: Part of the research was developed in project "Algoritmos de inteligencia computacional para detección de patrones de movimiento de personas" by J.C., J. Gómez and I. Silveira (advisor, S.N.).

Conflicts of Interest: The authors declare no conflict of interest.

References

1. La Vigne, N.; Lowry, S.; Markman, J.; Dwyer, A. Evaluating the Use of Public Surveillance Cameras for Crime Control and Prevention. Technical Report. Urban Institute, Justice Policy Center, 2011. Available online: https://www.urban.org/research (accessed on 1 October 2018).
2. Kruegle, H. *CCTV Surveillance, Second Edition: Video Practices and Technology*; Butterworth-Heinemann: Newton, MA, USA, 2006.
3. World Health Organization. *Pedestrian Safety: A Road Safety Manual for Decision-Makers and Practitioners*; World Health Organization: Geneva, Switzerland; Foundation for the Automobile and Society: London, UK; Global Road Safety Partnership: Geneva, Switzerland; World Bank: Geneva, Switzerland, 2013.
4. Chavat, J.; Nesmachnow, S. Computational Intelligence for Detecting Pedestrian Movement Patterns. In *Smart Cities*; Springer International Publishing: Berlin/Heidelberg, Germany, 2019; pp. 148–163.
5. University of Reading. Performance Evaluation of Tracking and Surveillance. Available online: www.cvg.reading.ac.uk/PETS2009 (accessed on 26 October 2019).
6. Gonzalez, R.; Woods, R. *Digital Image Processing*; Pearson Education: London, UK, 2008.
7. Ramírez, B. *Procesamiento Digital de Imágenes: Fundamentos de la Imagen Digital*; Universidad Nacional Autónoma de México: Mexico City, Mexico, 2006.
8. Martín, M. *Técnicas Clásicas de Segmentación de Imagen*; Universidad de Valladolid: Valladolid, Spain, 2002.
9. Muñoz, J. Segmentación de Imágenes. 2009. Available online: www.lcc.uma.es/~munozp/documentos/procesamiento_de_imagenes/temas/pi_cap6.pdf (accessed on 26 October 2019).
10. Jain, A.; Duin, R.; Mao, J. Statistical pattern recognition: A review. *IEEE Trans. Pattern Anal. Mach. Intell.* **2000**, *22*, 4–37. [CrossRef]
11. Ramírez, D. Desarrollo de un méTodo de Procesamiento de Imágenes para la Discriminación de Maleza en Cultivos de Maíz. Master's Thesis, Universidad Autónoma de Querétaro, Santiago de Querétaro, Mexico, 2012.
12. Webb, A. *Statistical Pattern Recognition*; John Wiley & Sons: Hoboken, NJ, USA, 2003.
13. Corso, C. *Aplicación de Algoritmos de Clasificación Supervisada usando Weka*; Facultad Regional Córdoba, Universidad Tecnológica Nacional: Buenos Aires, Argentina, 2009.
14. Carrasco, J.; Martínez, J. Reconocimiento de patrones. *Komput. Sapiens* **2011**, *2*, 5–9.
15. Valera, M.; Velastin, S. Intelligent distributed surveillance systems: A review. *IEE Proc. Image Signal Process.* **2005**, *152*, 192–204. [CrossRef]

16. Piccardi, M. Background subtraction techniques: A review. In Proceedings of the IEEE International Conference on Systems, Man and Cybernetics, The Hague, The Netherlands, 10–13 October 2004; pp. 3099–3104.

17. López, H. Detección y Seguimiento de Objetos con Cámaras en Movimiento. Bachelor's Thesis, Universidad Autónoma de Madrid, Madrid, Spain, 2011.

18. Stauffer, C.; Grimson, E. Adaptive background mixture models for real-time tracking. In Proceedings of the 1999 IEEE Computer Society Conference on Computer Vision and Pattern Recognition, Fort Collins, CO, USA, 23–25 June 1999; Volume 2.

19. Kalman, R. A new approach to linear filtering and prediction problems. *J. Basic Eng.* **1960**, *82*, 35–45. [CrossRef]

20. Lefloch, D. Real-Time People Counting System Using Video Camera. Master's Thesis, Université de Bourgogne, Dijon, France, 2007.

21. Rodriguez, M.; Laptev, I.; Sivic, J.; Audibert, J. Density-aware person detection and tracking in crowds. In Proceedings of the IEEE International Conference on Computer Vision, Barcelona, Spain, 7–13 November 2011; pp. 2423–2430.

22. Dollar, P.; Wojek, C.; Schiele, B.; Perona, P. Pedestrian Detection: An Evaluation of the State of the Art. *IEEE Trans. Pattern Anal. Mach. Intell.* **2012**, *34*, 743–761. [CrossRef] [PubMed]

23. Leach, M.; Sparks, E.; Robertson, N. Contextual anomaly detection in crowded surveillance scenes. *Pattern Recognit. Lett.* **2014**, *44*, 71–79. [CrossRef]

24. Felzenszwalb, P.; Girshick, R.; McAllester, D.; Ramanan, D. Object Detection with Discriminatively Trained Part-Based Models. *IEEE Trans. Pattern Anal. Mach. Intell.* **2010**, *32*, 1627–1645. [CrossRef] [PubMed]

25. Kalal, Z.; Matas, J.; Mikolajczyk, K. Online learning of robust object detectors during unstable tracking. In Proceedings of the 2009 IEEE 12th International Conference on Computer Vision Workshops, Kyoto, Japan, 27 September–4 October 2009; pp. 1417–1424.

26. Cho, S.; Kang, H. Abnormal behavior detection using hybrid agents in crowded scenes. *Pattern Recognit. Lett.* **2014**, *44*, 64–70. [CrossRef]

27. Zhu, X.; Jin, X.; Zhang, X.; Li, C.; He, F.; Wang, L. Context-aware local abnormality detection in crowded scene. *Sci. China Inf. Sci.* **2015**, *58*, 1–11. [CrossRef]

28. Garlan, D.; Shaw, M. An Introduction to Software Architecture. In *Advances in Software Engineering and Knowledge Engineering*; Ambriola, V., Tortora, G., Eds.; World Scientific Publishing: Singapore, 1993.

29. Van Huis, J.; Bouma, H.; Baan, J.; Burghouts, G.; Eendebak, P.; den Hollander, R.; Dijk, J.; van Rest, J. Track-based event recognition in a realistic crowded environment. In Proceedings of the SPIE–The International Society for Optical Engineering, Amsterdam, The Netherlands, 22–25 September 2014; Volume 9253, p. 92530E.

30. Mallick, S. Learn OpenCV (C++/Python). 2016. Available online: www.learnopencv.com/ (accessed on 26 October 2019).

31. Coelho, L. Why Python is Better than Matlab for Scientific Software. 2013. Available online: metarabbit.wordpress.com/2013/10/18/ (accessed on 26 October 2019).

32. Pivotal Software Inc. RabbitMQ–Messaging that just Works. Available online: www.rabbitmq.com/ (accessed on 26 October 2019).

33. Nixon, M.; Aguado, A. *Feature Extraction and Image Processing*; Academic Press: Amsterdam, The Netherlands, 2008.

34. Zivkovic, Z. Improved adaptive Gaussian mixture model for background subtraction. In Proceedings of the 17th International Conference on Pattern Recognition, 2004, ICPR 2004, Cambridge, UK, 23 August 2004; Volume 2, pp. 28–31.

35. Zivkovic, Z.; Van Der Heijden, F. Efficient adaptive density estimation per image pixel for the task of background subtraction. *Pattern Recognit. Lett.* **2006**, *27*, 773–780. [CrossRef]

36. OpenCV Developers Team. OpenCV Simple Blob Detector. Available online: https://docs.opencv.org/3.4/d0/d7a/classcv_1_1SimpleBlobDetector.html (accessed on 26 October 2019).

37. Dalal, N.; Triggs, B. Histograms of oriented gradients for human detection. In Proceedings of the IEEE Conference on Computer Vision and Pattern Recognition, San Diego, CA, USA, 20–26 June 2005; pp. 886–893.

38. Rosebrock, A. Non-Maximum Suppression for Object Detection in Python. 2014. Available online: www.pyimagesearch.com/2014/11/17 (accessed on 26 October 2019).

39. Kuhn, H. The Hungarian method for the assignment problem. *Nav. Res. Logist. Q.* **1955**, *2*, 83–97. [CrossRef]

Appl. Sci. **2020**, *10*, 9021

40. Leal-Taixé, L.; Milan, A.; Reid, I.; Roth, S.; Schindler, K. MOTChallenge: Multiple Object Tracking Benchmark. Available online: motchallenge.net/ (accessed on 26 October 2019).
41. Nesmachnow, S.; Iturriaga, S. Cluster-UY: Collaborative Scientific High Performance Computing in Uruguay. In Proceedings of the 10th International Conference on Supercomputing in Mexico, Monterrey, Mexico, 25–29 March 2019; Volume 1151; pp. 188–202.
42. Leal-Taixé, L.; Milan, A.; Reid, I.; Roth, S.; Schindler, K. MOTChallenge Results 2D. Available online: motchallenge.net/results/2D_MOT_2015/ (accessed on 26 October 2019).
43. Honovich, J. Frame Rate Guide for Video Surveillance. 2014. Available online: ipvm.com/reports/frame-rate-surveillance-guide (accessed on 26 October 2019).

Publisher's Note: MDPI stays neutral with regard to jurisdictional claims in published maps and institutional affiliations.

applied sciences

Article

Environmental Footprint Assessment of a Cleanup at Hypothetical Contaminated Site

Muhammad Azhar Ali Khan [1], Zakria Qadir [2],[*], Muhammad Asad [1], Abbas Z. Kouzani [3] and M. A. Parvez Mahmud [3]

1 Mechanical Engineering Department, College of Engineering, Prince Mohammad bin Fahd University, Al-Khobar 31952, Saudi Arabia; mkhan6@pmu.edu.sa (M.A.A.K.); masad@pmu.edu.sa (M.A.)
2 School of Computing Engineering and Mathematics, Western Sydney University, Locked Bag 1797, Penrith 2751, Australia
3 School of Engineering, Deakin University, Geelong 3216, Australia; abbas.kouzani@deakin.edu.au (A.Z.K.); m.a.mahmud@deakin.edu.au (M.A.P.M.)
* Correspondence: z.qadir@westernsydney.edu.au

Abstract: Contaminated site management is currently a critical problem area all over the world, which opens a wide discussion in the areas of policy, research and practice at national and international levels. Conventional site management and remediation techniques are often aimed at reducing the contaminant levels to an acceptable level in a short period of time at low cost. Owing to the fact that the conventional approach may not be sustainable as it overlooks many ancillary environmental effects, there is an immense need of "sustainable" or "green" approaches. Green approaches address environmental, social and economic impacts throughout the remediation process and are capable of conserving the natural resources and protecting air, water and soil quality through reduced emissions and other waste burdens. This paper presents a methodology to quantify the environmental footprint of a cleanup for a hypothetical contaminated site by using the US Environmental Protection Agency's (EPA) Spreadsheet for Environmental Footprint Assessment (SEFA). The hypothetical contaminated site is selected from a metropolitan city of Pakistan and the environmental footprint of the cleanup is analyzed under three different scenarios: cleanup without any renewable energy sources at all, cleanup with a small share of renewable energy sources, and cleanup with a large share of renewable energy sources. It is concluded that integration of renewable energy sources into the remedial system design is a promising idea which can reduce CO_2, NOx, SOx, PM and HAP emissions up to 68%.

Keywords: environmental footprint; cleanup; green remediation; renewable energy sources

Citation: Khan, M.A.A.; Qadir, Z.; Asad, M.; Kouzani, A.Z.; Parvez Mahmud, M.A. Environmental Footprint Assessment of a Cleanup at Hypothetical Contaminated Site. *Appl. Sci.* **2021**, *11*, 4907. https://doi.org/10.3390/app11114907

Academic Editor: Luis Hernández-Callejo

Received: 6 April 2021
Accepted: 15 May 2021
Published: 26 May 2021

Publisher's Note: MDPI stays neutral with regard to jurisdictional claims in published maps and institutional affiliations.

1. Introduction

Over the last few years, a rapid increase is observed in the awareness and dialogue about the environment in general and, in particular, about the issues, such as sustainability, recycling, greenhouse gas (GHGs) emissions, and a greener world [1–4]. In almost all spheres of life, people take a keen interest in understanding how goods are produced, how they are delivered and, eventually, how it all impacts the environment [5]. Within this context, the restoration of contaminated and toxic places is now also identified as a critical problem, which opens a wide discussion in the areas of policy, research, and practice at national and international levels [6–11]. At this point, the following question arises: "Is there any environmentally friendly way to clean the environment?" The US Environmental Protection Agency (EPA) tries to answer this question by adding a new phrase to today's environmental lexicon: Green Remediation [12,13].

Conventional and Green remediation are in contrast to each other in many perspectives. A conventional site remediation approach is normally based on (a) the effectiveness and appropriateness of the particular remediation method to meet the remedial goals; (b) ease of implementation; (c) remediation costs; and (d) remediation timeframe [14].

Green remediation, on the other hand, employs the idea of protecting human health and environment while minimizing the environmental side effects. It asks for (a) efficient use of natural resources and energy; (b) reduction in the negative impacts on the environment; (c) minimization or elimination of pollution at its source; and (d) reduction in waste to greatest possible extent [14].

Owing to the fact that the conventional approaches may not be sustainable as they overlook many ancillary environmental effects, there is an immense need of "Sustainable" or "Green" approaches which address environmental, social and economic aspects throughout the remediation process. One way of making the remediation process green could be analyzing the extent to which it is impacting the environment. Such analyses in a green remediation setting are often termed Environmental Footprint Analysis. Primarily, an environmental footprint highlights the aspects of a cleanup that dominates the footprint and then provides the opportunity to improve the remedy efficiency and effectiveness by implying a range of alternatives to the existing remedial system.

Incorporating renewable energy sources into a cleanup activity can offer increased sustainability and long-term cost savings [15]. The use of renewable energy in remedial system designs is not a new idea, and it is already observed at various sites in the world. Amanda [15] identifies solar, wind, landfill gas, and biodiesels as the possible options that can be integrated into a remedy. The Saint St. Croix Alumina site in the Virgin Islands uses wind-driven turbine compressors (WDTC) to drive hydraulic oil "skimmer" pumps to recover free product (oil) from ground water. The system does not produce electricity to power pumps; instead, it uses compressed air generated by WDTCs. PV arrays and wind-driven electricity generators are also employed to power submersible pumps for oil, groundwater and petroleum hydrocarbon recovery [15]. Another application of renewable energy in remedial systems is the use of solar energy to pump water into and circulate through a bioreactor installed at the Altus Air Force Base in Oklahoma, to remove TCE from ground water. This project is found to be cost effective in terms of avoiding construction of a power transmission line from the utility grid to the bioreactor's remote location [16]. The BP Paulsboro, a former petroleum and specialty-chemical storage facility, is being remediated by a large onsite pump and treat (P&T) system which is empowered by solar energy since 2003. Almost 20–25% electricity requirements are met by solar energy and for the rest the system relies on electricity from the utility grid. A reduction in the emissions of CO_2 by 571,000 pounds per year, sulfur dioxide (SO_2) by 1600 pounds per year, and nitrogen oxide (NOx) by 1100 pounds per year is expected from this hybrid remedial system design [17].

The EPA's Spreadsheet for Environmental Footprint Assessment (SEFA) is among the key tools available for such investigations. However, very limited published literature is available about using SEFA for environmental impact assessment of remediation of contaminated sites. Marco et al. [18] study the environmental impact of the remediation of an aquifer below an industrial site in the Bologna area. Three proposed systems were investigated for environmental impact using two environmental footprint analysis tools, i.e., SiteWise^{TM} and SEFA. The three solutions studied and compared for environmental footprint are (i) groundwater extraction system, treatment and reinjection, (ii) reductive bioremediation, and (iii) in situ chemical oxidation (ISCO). Based on the results obtained from both tools, bioremediation is found to be the appropriate remediation with minimum GHG and lower levels of environmental impacts. In addition, the higher environmental impacts caused by ISCO is due to frequent multiple injection events with Potassium Permanganate, in contrast with the general single injection performed with bioremediation.

In this paper, the environmental footprint of a cleanup at a hypothetical contaminated site is studied by using EPA's Spreadsheet for Environmental Footprint Assessment (SEFA). The underlying idea is to develop a methodology for analyzing the environmental impact of a cleanup of any contaminated site using this easily accessible tool. Therefore, a hypothetical contaminated site is selected from a metropolitan city in Pakistan and the environmental footprint of the cleanup is analyzed under three different scenarios: cleanup without any renewable energy sources at all, cleanup with a small share of renewable energy sources, and cleanup with a large share of renewable energy sources.

2. Core Elements of Green Remediation

Green remediation aims to minimize the energy and environmental footprint of a site remediation and revitalization. A set of core elements is made by the US EPA [19] as shown in Figure 1, which actually describes the potential areas that can reduce the environmental footprint of a site cleanup. The details of these core elements are provided in the following sub-sections.

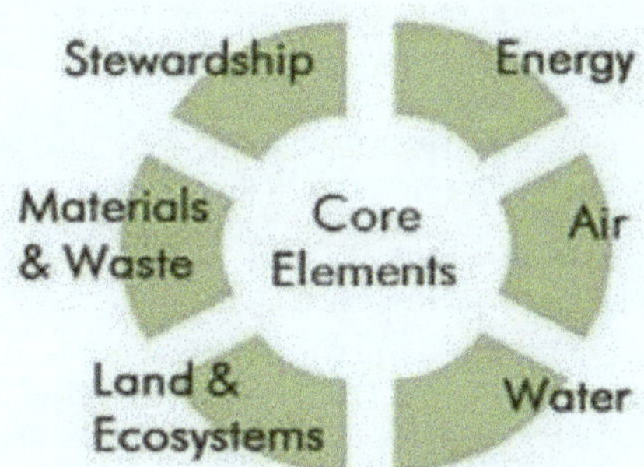

Figure 1. Core elements of green remediation [18].

2.1. Energy

The energy requirement of the treatment system is extremely important to analyze in terms of green remediation. This element emphasizes the use of passive energy sources in order to meet all remediation objectives. In addition to that, energy efficient equipment should be used and maintained at peak performance to maximize efficiency. Moreover, periodical evaluation and optimization of energy efficiency of equipment with high energy demand can significantly reduce the energy consumption. Besides all this, the integration of renewable energy systems can replace or at least offset the electricity requirements otherwise met by the utility grid.

2.2. Air

This element is mostly concerned with the air emissions caused by different types of fuel in any onsite or offsite operation of a cleanup. It emphasizes to minimize the use of heavy equipment requiring high amounts of fuels and to use cleaner fuels for the operation of these equipment. It also takes into account the reduction of toxic and priority pollutants, such as ozone, particulate matter, carbon monoxide, nitrogen dioxide, sulfur dioxide, and lead, with the minimization of dust export of the contaminants.

2.3. Water

Water requirement and the impacts on water resources is also a key component of green remediation by minimizing the freshwater use and maximize the water reuse during daily operations and treatments processes. The treated water can be reclaimed for beneficial use such as irrigation. Moreover, the nearby water bodies should be prevented from impacts such as nutrient loading.

2.4. Land and Ecosystems

In terms of land and ecosystems impacts, the minimum invasive in situ technologies should be used and passive energy technologies like bioremediation and phytoremediation should be selected as primary remedies where possible and effective. This component also calls for the minimization of soil and habitat disturbance and reduction in noise and lighting disturbance.

2.5. Materials and Wastes

For green remediation, the selected technologies should be capable of generating minimum waste. Re-use and recycling of material generated at or removed from the site should be promoted. A major concern in this component is the minimization of natural resource extraction and disposal. If feasible, passive sampling devices should be used that produce minimum waste.

2.6. Stewardship

The stewardship goals are usually long-term such as reduction of CO_2, N_2O, CH_4, and other greenhouse gases emissions contributing to climate change, integration of an adaptive management approach into controls for a site, installation of renewable energy systems for cleanup and future activities on redeveloped land, and community involvement to increase public acceptance and awareness of long-term activities and restrictions.

3. Environmental Footprint of a Cleanup Project

The term "footprint" refers to the quantification of a specific parameter that has been assigned a particular meaning. For example, in terms of "carbon footprint", it is the quantification of carbon dioxide (and other GHGs) emitted into the air by a particular activity, facility, or individual. Green remediation should be analyzed in detail to closely examine the components of the remedial system and to identify the large contributors to the environmental footprint. The purpose, limitations, value, and the level of effort and cost for an environmental footprint assessment are discussed in the following subsections.

3.1. Purpose

The first and foremost purpose of environmental footprint analysis is to facilitate the implementation of EPA's principles for greener cleanups [8]. By doing this analysis, the quantification of metrics for a cleanup can be done and a set of technical suggestions can be made on the approaches to reduce the footprint of a remedial system.

3.2. Limitations

The environmental footprint assessment is not intended to be a detailed life-cycle analysis (LCA). It uses a suitable number of green remediation metrics to represent the core elements of green remediation but limits the number of metrics to streamline the footprint analysis process. It is also limited in a sense that it is not a mandatory requirement of EPA but it is just intended to support the remedial process and to reduce the environmental impact of a cleanup activity.

3.3. Value

The environmental footprint assessment can be considered as a valuable component in a cleanup because it can quantify the footprint reductions of a cleanup project. The dominant aspects of the footprint can be highlighted and thus strategies can be adapted to reduce their contributions in the footprint. Based on its results, it also provides the opportunity to improve the remedy efficiency and effectiveness which is usually a missing element in a more conventional evaluation.

3.4. Level of Effort and Cost

According to the US EPA, the environmental footprint analysis adds negligible amount to the level of effort or cost for an overall remediation and a fraction of any particular remedial activity, such as a remedy design or an optimization evaluation. EPA expects to have an addition of 10% to the level of effort or cost of an optimization evaluation, or less than 5 percent to the level of effort or cost of a remedial design [20]. Footprint evaluation mainly focuses on green remediation metrics and does not quantify the cleanup costs. Since the cleanup costs are directly related to core elements of greener cleanups, the cost savings can be expected over the life of a cleanup project. However, these reductions are project-specific and are highly related to the location and time span of a remedial operation.

4. Method and Material

4.1. Site Selection

The Sustainable Development Policy Institute (SDPI) in collaboration with the Blacksmith Institute (BSI), USA, carried out a Global Inventory Project (GIP) for the mapping of chemical contaminated sites in Pakistan, to improve public health and environment in and around the public site area. SDPI identified a total of 31 contaminated sites in Pakistan [21], the distribution of which is shown in Figure 2.

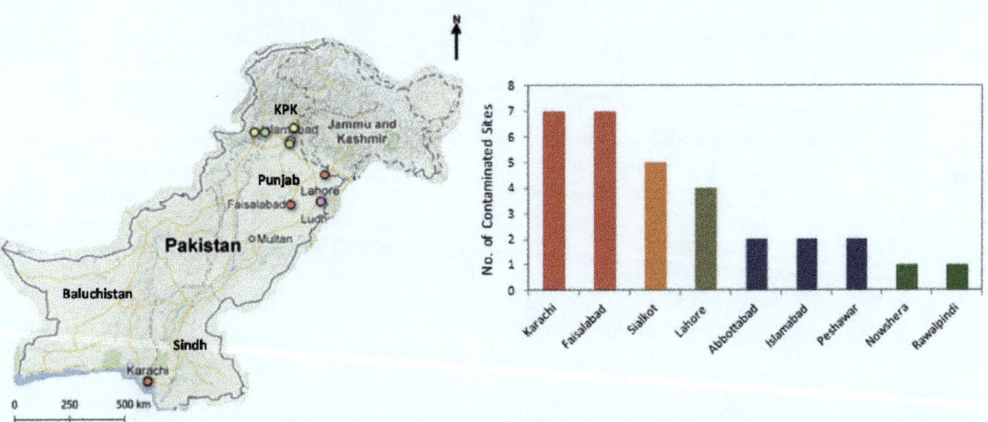

Figure 2. Distribution of contaminated sites in Pakistan.

In general, the two provinces Khyber Pakhtunkhwa (KPK) and Punjab are found to have maximum concentration of contaminated sites, as shown in the map of Pakistan. Since Karachi and Faisalabad are considered to be the industrial hubs of Pakistan, each of them contains seven contaminated sites. For the purpose of this study, a contaminated site is selected from Karachi. The site is potentially contaminated due to a number of industries in the vicinity. Moreover, the site is near to the creek. Thus, it is expected to have both groundwater and surface water pollution.

4.2. Hypothetical Situation

A pump and treat system are under design to treat LNAPL (petroleum product) contamination caused by a spill from an underground tank. An oil/water separation technique will be used for cleanup. The clean water is then discharged to the creek. A schematic of the LNAPL release and subsequent migration and the remedial system is presented in Figure 3a,b, respectively.

The construction of the remedial system will include:

i.   Ten 6-inch extraction wells, each to 60 feet deep with 20-foot screens;
ii.   3000 feet of 6-inch HDPE piping with electrical conduit and wiring;
iii.   80 ft × 100 ft building that is 30 feet high;
iv.   200 ft × 200 ft reinforced concrete pad and containment area (20,000 ft^3 of concrete).

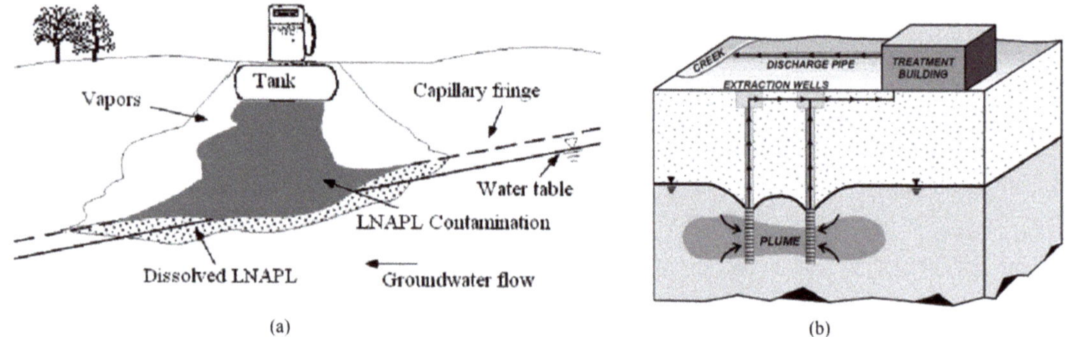

(a)　　　　　　　　　　　　　　　(b)

Figure 3. (**a**) LNAPL release and subsequent migration [22] and (**b**) pump and treat remedial system.

The data required to analyze the environmental footprint of this remedial system will include the details of materials used, wastes generated, water consumption, energy consumption and air emissions. The largest contributor to refined materials is expected to be the building steel (292,000 lbs) over a 30-year period. The largest contributor for unrefined materials is expected to be the aggregate in the concrete for the building foundation (about 1200 tons). No specific appreciable non-hazardous waste streams have been identified. The dewatered sludge is expected to be 2600 tons. The project team has chosen a percent-based screening limit of 1 percent for refined and unrefined materials and magnitude-based limits of 1000 lbs for refined materials and 1 ton for unrefined materials and wastes. Table 1 lists all the data which are provided to the EPA's Spreadsheet for Environmental Footprint Assessment.

4.3. Scenarios for Analysis

The environmental footprint of the cleanup at the hypothetical site is discussed in Sections 4.1 and 4.2 and is analyzed under three different scenarios.

4.3.1. Cleanup without Any Renewable Energy Resources at All

In this scenario, it is assumed that all of the energy used in the remedial systems, which is estimated to be 33,000,000 kWh, is taken from the grid for which the fuel mix is already described in Table 1. Neither the renewable energy is generated onsite nor it is voluntarily purchased from a renewable energy producer. It is important to note that the fuel mix for grid electricity in Pakistan is dominated by conventional fossil fuels such as oil and natural gas.

4.3.2. Cleanup with a Small Share of Renewable Energy Sources

For this case, it is assumed that the energy demands in remedial operations are met by employing a small share of renewable energy. This means that 70% of the electricity is taken from grid while 30% from renewable energy resources. Out of this 30%, 20% electricity is generated onsite using renewable energy sources and 10% is voluntarily purchased from a renewable energy producer.

4.3.3. Cleanup with a Large Share of Renewable Energy Sources

This scenario assumes to have a large share of renewable energy resources in order to meet the energy demands of remedial operations. In this case, only 30% electricity is taken from grid while 70% from renewable energy resources. Out of this 70%, 50% electricity is generated onsite using renewable energy sources and 20% is voluntarily purchased from a renewable energy producer.

272

Table 1. Manual data input to EPA's spreadsheet for environmental footprint assessment of a cleanup.

Material and Use	Units	Quantity	Conversion Factor to lbs	% Recycled or Reused Content	Quantity (lbs) Virgin	Quantity (lbs) Recycled
Refined Materials						
Well-PV casing and grout					0	0
Wells-Screen					0	0
Piping and Conduit	ft	3000	7.5	0%	22,500	0
Building Steel	ft^3	240,000	1	55%	108,000	132,000
Concrete Reinforcing Steel	ft^2	40,000	1.3	55%	23,400	28,600
Cement Portion of Concrete	ft^3	20,000	22	20%	352,000	88,000
Process Equipments					0	0
Process Controls					0	0
					0	0
Unrefined Materials						
Well-Sand Pack					0	0
Aggregate for Concrete	ft^3	20,000	0.0575	0%	1150	0
Waste Disposal (tons)						
Hazardous Waste						
2600 tons of Hazardous Waste in the form of Sludge						2600

Water Resource	Description of Quality of Water Used	Volume Used (1000 gallons)	Uses	Fate of Used Water
Water Usage				
Extracted groundwater #1				
Location:	Shallow Aquifer, Marginal Quality	11,000,000	Treatment	Creek
Aquifer:				

Participant	Crew Size	Number of Days Worked	Hours Worked Per Day	Total Hours Worked	Number of Roundtrips to Site
Labor, Mobilizations, Mileage, and Fuel					
SGS Pakistan	20	90	8	14,400	100

Roundtrip Miles to Site	Mode of Transport.	Fuel Type	Total Miles	Fuel Usage Rate	Total Fuel Used (gal)
7.2	Bus	Diesel	720	96	8

Table 1. *Cont.*

Material and Use	Units	Quantity	Conversion Factor to lbs	% Recycled or Reused Content	Quantity (lbs)	
					Virgin	Recycled
On-Site Equipment Use, Mobilization, and Fuel Usage						
Equipment Type *		HP	Load Factor	Equip. Fuel Type	Units of Fuel Used per Hour	Total Hours Operated
Drilling-medium rig (150 HP)		150	1%	Diesel	0.05625	320
Gallons of Fuel Used On-Site	Number of Roundtrips to Site	Roundtrip Miles to Site	Total Miles Transported	Transport Fuel Type	Fuel Usage Rate	Total Fuel Used for Transport (gal)
18	1	7.2	7.2	Diesel	6	1.2
On-Site Electricity Use						
Equipment Type	HP	% Full Load	Efficiency (%)	Electrical Rating (kW)	Hours Used	Energy Used (kWh)
Six 0.75 hp extraction pump	4.5	80%	65%	4.131692308	1800	7437.046154
Two 1 hp discharge pumps	2	80%	75%	1.591466667	1800	2864.64
On-Site Natural Gas Use						
Equipment Type		Power Rating (btu/hr)	Efficiency	Total Hours Used	Btus of Gas Required	Total ccf Used
Building Heat		200,000	80%	2000	500,000,000	4854.368932
Materials Use (including Potable Water) and Transportation						
Material Type or Public Water			Unit	Quantity	Tons	Default One-Way Miles
Cement			dry-lb	440,000	220	500
Steel			lb	292,000	146	500
HDPE			lb	22,500	11.25	500
Concrete			lb	2,300,000	1150	25
Site-Spec. One-Way Distance (miles) *	Number of One-way Trips to Site	Mode of Transport.		Fuel Type	Fuel Usage Rate (gptm or mpg)	Total Fuel Used (gallons)
25	1	Truck (mpg)		Diesel	6	4.2
25	1	Truck (mpg)		Diesel	6	4.2
25	1	Truck (mpg)		Diesel	6	4.2
25	1	Truck (mpg)		Diesel	6	4.2

Table 1. *Cont.*

Material and Use	Units	Quantity	Conversion Factor to lbs	% Recycled or Reused Content	Quantity (lbs)	
					Virgin	Recycled
Waste Transportation and Disposal						
	Waste Destination		Unit	Quantity	Tons	Default One-Way Miles
	Hazardous waste landfill		tons	2600	2600	500
Site-Spec. One-Way Distance (miles)	Number of One-way Trips to Site		Mode of Transport.	Fuel Type	Fuel Use Rate (gptm or mpg)	Total Fuel Use (gallons)
30	1		Truck (mpg)	Diesel	6	5
Fuel Mix for Grid Electricity						
Type				% of Total Used		
Conventional Energy						
Coal				0%		
Natural Gas				27%		
Oil				34%		
Nuclear				6%		
Biomass				0%		
Geothermal				0%		
Hydro				33%		
Solar				0%		
Wind				0%		
Other (enter information below)				0%		
Total				100%		

4.4. Green Remediation Metrics

The green remediation metrics are mainly based on five out of six core elements. These metrics provide an opportunity to the remedial team to change their strategies for environmental benefit. The details of these green remediation metrics are presented in Table 2.

Table 2. Green remediation metrics (modified from [20]).

Core Element		Metric	Unit of Measure
Materials and Waste	M&W-1	Refined materials used on site	tons
	M&W-2	Percent of refined materials from recycled or waste material	percent
	M&W-3	Unrefined materials used on site	tons
	M&W-4	percent of unrefined materials from recycled or waste material	pecent
	M&W-5	Onsite hazardous waste generated	tons
	M&W-6	Onsite non-hazardous waste generated	tons
	M&W-7	Percent of total potential onsite waste that is recycled or reused	percent
Water		Onsite water use (by source)	
	W-1	Source: Groundwater, Purpose: Treatment, Fate: Creek	millions of gallons
Energy	E-1	Total energy use	MMBtu
	E-2	Total energy voluntarily derived from renewable resources	
	E-2A	- Onsite generation or use and biodiesel use	MMBtu
	E-2B	- Voluntary purchase of renewable electricity	MWh
	E-2C	- Voluntary purchase of RECs	MWh
Air	A-1	Onsite Nox, Sox, and PM emissions	lbs
	A-2	Onsite HAP emissions	lbs
	A-3	Total Nox, Sox, and PM emissions	lbs
	A-4	Total HAP emissions	lbs
	A-5	Total GHG emissions	tons CO_2-e
Land and Ecosystem		Qualitative Description	

4.4.1. Materials and Waste Metrics

The material metrics takes into account the total amount of materials used onsite and the percentage of those materials that are produced from recycled material, reused material, or waste material. The waste metrics consider the total amount of waste generated on site and the percentage of total potential onsite waste that is recycled or reused [19].

4.4.2. Water Metrics

According to [8], the water metrics include the source and amount of water used onsite, and the fate of water after use. The sources of water could be public potable water supply, ground water from a local aquifer, surface water, and reclaimed water, etc. In a remedial system, water is mainly used in equipment decontamination, treatment, injection for plume migration, and chemical blending, etc. In terms of fate of used water, it can be discharged to groundwater and fresh surface water, used in irrigation and industrial processes, or reused in a public or domestic water supply.

4.4.3. Energy Metrics

The energy metrics are mainly focused on the total amount of energy used in the remedial operation (both onsite and offsite). Moreover, they also take into account the total amount of renewable energy used in a remedial operation, such as onsite generation and use of renewable energy and use of biodiesel, voluntary purchase of renewable electricity and renewable energy certificates.

4.4.4. Air Metrics

The air metrics consider emissions of GHGs, nitrogen oxides (NOx), sulfur oxides (SOx), particulate matter less than 10 microns in size (PM10), and hazardous air pollutants (HAPs) [21].

4.4.5. Land and Ecosystem

This metric is about a qualitative description of potential disturbance in land and ecosystem which will be caused by the employed remediation technique.

4.5. Footprint Methodology

The footprint methodology is actually based on seven-steps as shown in Figure 4. The evaluation of footprint begins with setting goals and scope of the analysis followed by gathering the information of the remedial system to be footprinted. Based on this information, the quantification of onsite materials metrics, waste metrics and water metrics is carried out. The materials, waste and water information, and other remedy information, then helps quantify the energy and air metrics. A qualitative description of the ecosystems, which will be disturbed by the implementation of remedial system, is also included in the analysis. Finally, the results are presented and analyzed for the identification of large contributors to the metrics and for the opportunities to reduce overall footprint of the remedial system.

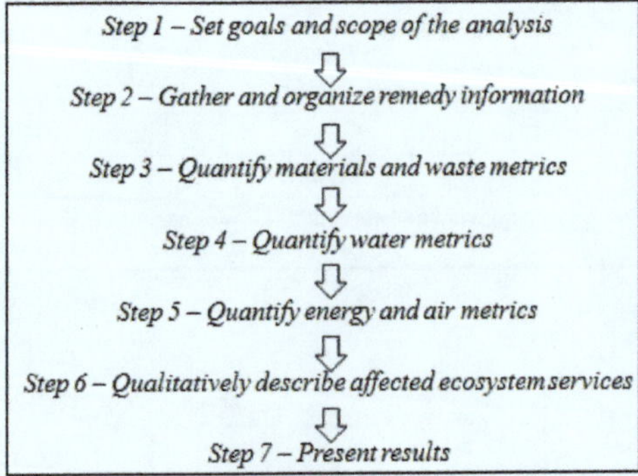

Figure 4. Steps in footprint methodology (modified from [8]).

5. Results and Discussions

In Table 3, the summary of environmental footprint of the cleanup under all three scenarios is presented. Figures 5–7 represent the contributions of different elements of the cleanup in environmental footprint in terms of total energy, CO_2 emissions, and NOx, SOx, PM, and HAP emissions, respectively. Table 4 lists the reductions in CO_2, HAP, and NOx, SOx, and PM emissions as a result of renewable energy integration into the remedial system as discussed above in Section 4.3.

Table 3. Summary of the environmental footprint of cleanup under all scenarios.

Core Element		Metric	Unit of Measure	Scenario 1	Scenario 2	Scenario 3
Materials and Waste	M&W-1	Refined materials used on site	tons	377	377	377
	M&W-2	Percent of refined materials from recycled or waste material	percent	33	33	33
	M&W-3	Unrefined materials used on site	tons	1150	1150	1150
	M&W-4	percent of unrefined materials from recycled or waste material	pecent	0	0	0
	M&W-5	Onsite hazardous waste generated	tons	2600	2600	2600
	M&W-6	Onsite non-hazardous waste generated	tons	0	0	0
	M&W-7	Percent of total potential onsite waste that is recycled or reused	percent	0	0	0
Water		Onsite water use (by source)				
	W-1	Source: Groundwater, Purpose: Treatment, Fate: Creek	millions of gallons	110,000	110,000	110,000
Energy	E-1	Total energy use	MMBtu	408,551	309,966	182,275
	E-2	Total energy voluntarily derived from renewable resources				
	E-2A	- Onsite generation or use and biodiesel use	MMBtu	0	22,526	56,315
	E-2B	- Voluntary purchase of renewable electricity	MWh	0	3300	6600
	E-2C	- Voluntary purchase of RECs	MWh	0	0	0
Air	A-1	Onsite Nox, Sox, and PM emissions	lbs	55	55	55
	A-2	Onsite HAP emissions	lbs	0	0	0
	A-3	Total NOx, SOx, and PM emissions	lbs	219,726	155,511	69,891
	A-4	Total HAP emissions	lbs	1119	801	377
	A-5	Total GHG emissions	tons CO_2-e	41,741,757	29,609,307	13,432,707
Land and Ecosystems		Land and Ecosystem will be disturbed in terms of hazardous waste disposal offsite, depending upon the technique to be used in its disposal.				

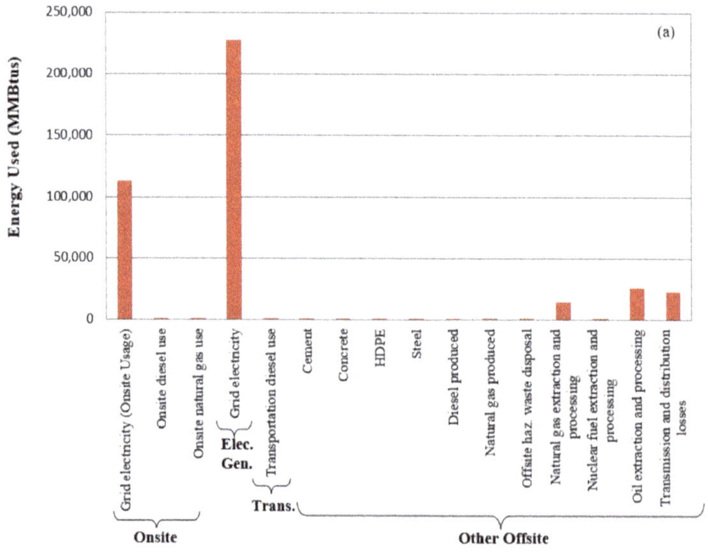

Figure 5. *Cont.*

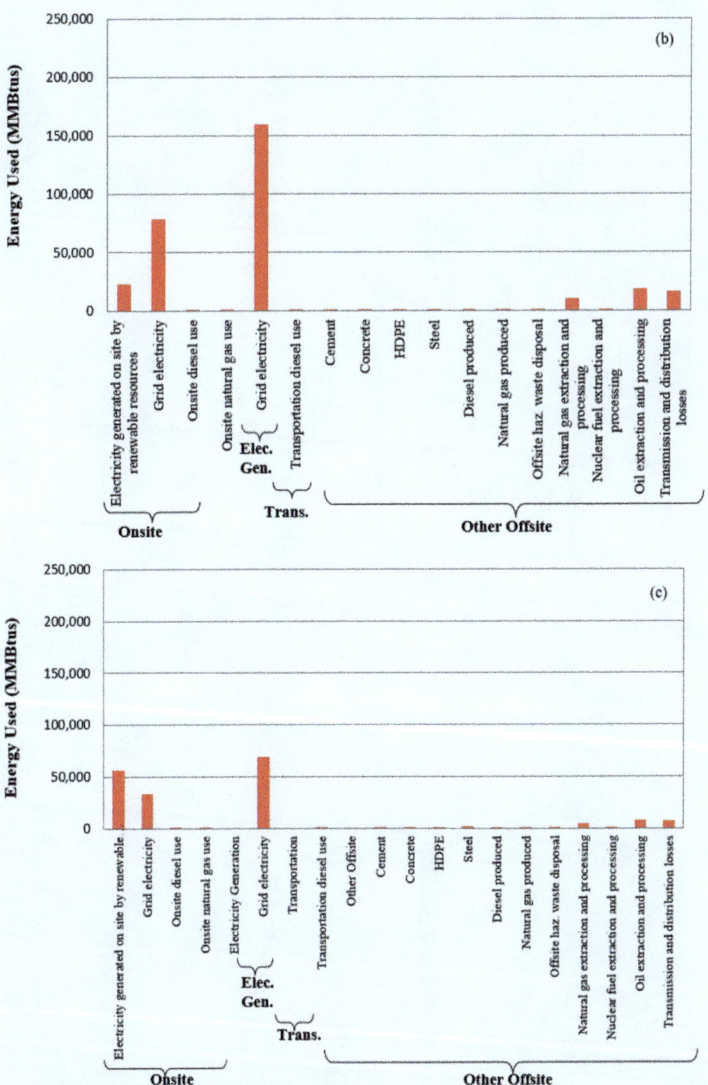

Figure 5. Contributors to the total energy used metric; (**a**) Scenario 1, (**b**) Scenario 2 and (**c**) Scenario 3.

From Figures 5–7, it can be observed that the greatest contributor to the environmental footprint is the grid electricity used in remedial operations. One reason of this high contribution is the fuel mix of grid electricity in Pakistan. As discussed earlier, Pakistan mostly relies on conventional fossil fuels such as oil and natural gas for electricity generation. SEFA not only takes into account the amount of fossil fuels used in the electricity generation but also considers the energy which is put into the extraction of these fossil fuels. Thus, the higher the use of fossil fuels, the greater will be their contribution in the overall footprint of the remedy. Integration of onsite renewable energy in cleanup helps in the reduction of electricity usage from grid and consequently lowers the energy demand for fuel extraction and electricity transmission as shown in Figure 5b,c.

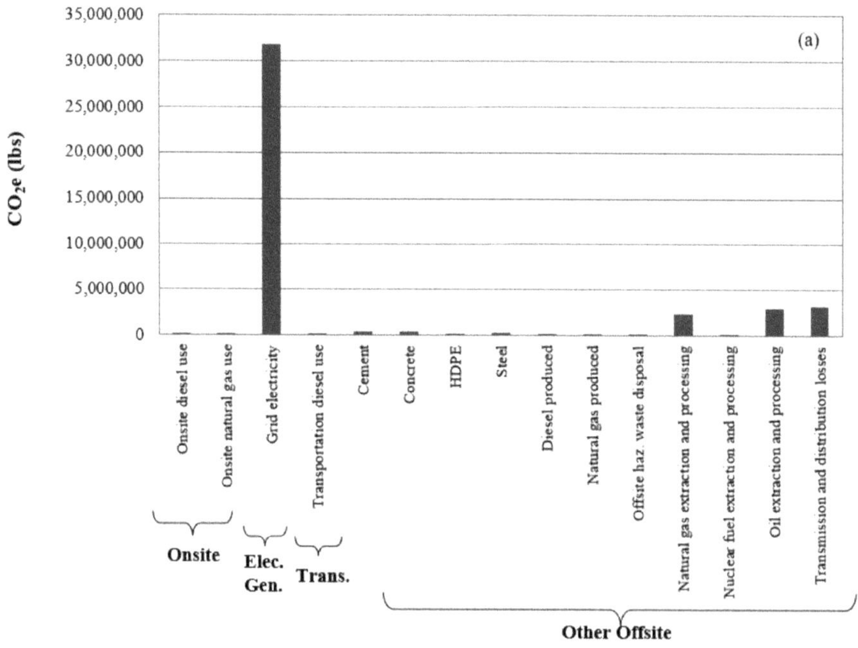

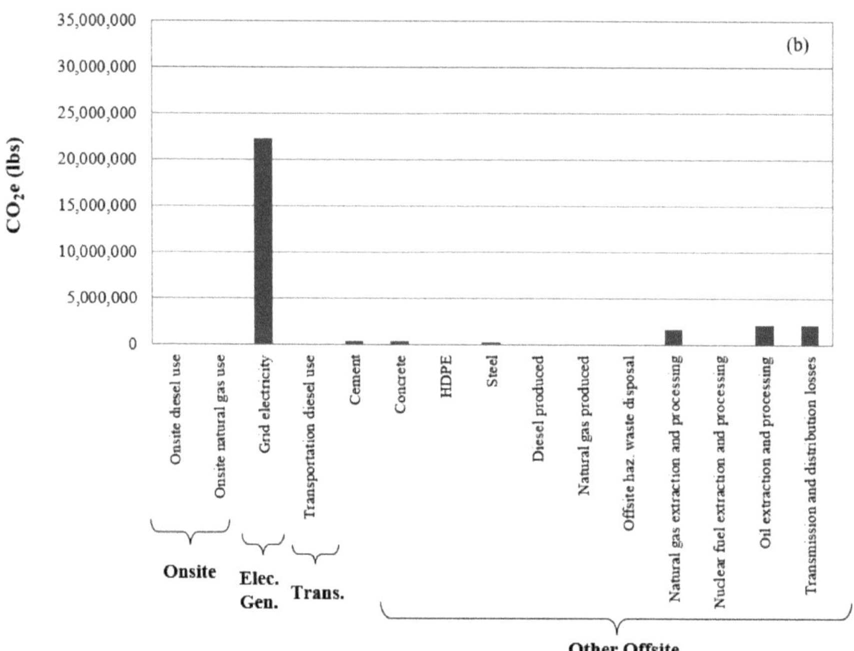

Figure 6. *Cont.*

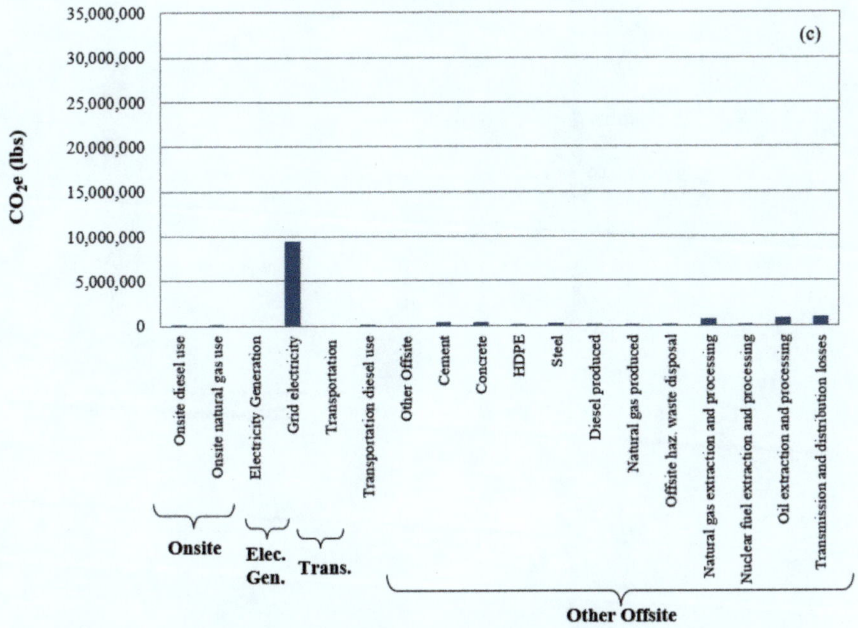

Figure 6. Contributors to the CO2 emissions; (**a**) Scenario 1, (**b**) Scenario 2 and (**c**) Scenario 3.

Table 4. Reduction in emissions via renewable energy integration in the remedial system.

Scenario	CO₂ Emissions		Total HAP Emission		Total NOx, SOx, PM Emissions	
	tons	% Reduction	lbs	% Reduction	lbs	% Reduction
1	41,741,757	-	1119	-	219,726	-
2	29,609,307	29	801	28	155,511	29
3	13,432,707	68	377	66	69,891	68

The CO_2 emissions are also high in the first scenario, as shown in Figure 6a, owing to the fact that only grid electricity is used in the remediation process. The overall behavior of CO_2 emissions is nearly the same as of Figure 5, and it is found to have almost negligible emissions in terms of fossil fuel extraction and electricity transmission when renewable energy is incorporated into the remedial system.

The NOx, SOx, HAP and PM emissions are presented in terms of onsite operations, electricity generation, transportation and other offsite operations. It can be observed that the electricity generation and other offsite operations are major contributors in these emissions. The SOx emissions are found to have maximum contribution among all NOx, SOx, PM and HAP. This may be due to the fact that SOx will be emitted when the fuel contains sulphur such as coal and oil. Gasoline is extracted from oil and metals are extracted from ore [23]. For the remedial system, the main sources of SOx emissions are the extraction of fossil fuels for grid electricity, fuels used in transportation, extraction of metals because building steel is used in a very large quantity (292,000 lbs), which makes it a dominant contributor in this category [24–26]. Although NOx emissions are quite less when compared to SOx, and, primarily, they are only due to burning of fuels in motor vehicles and electric utilities, they cannot be ignored in the overall environmental footprint analysis.

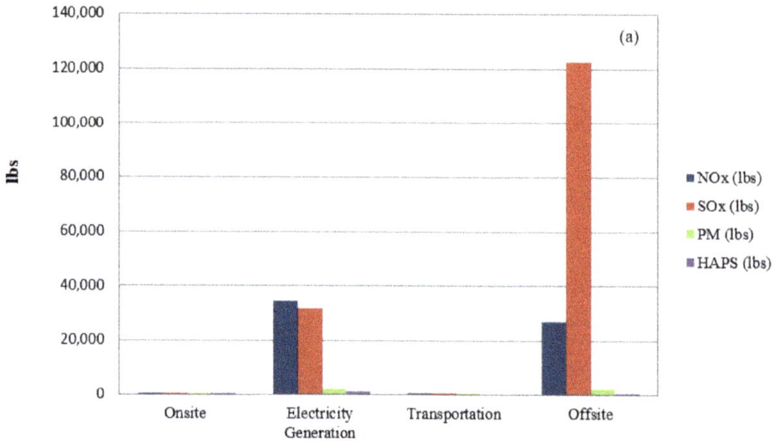

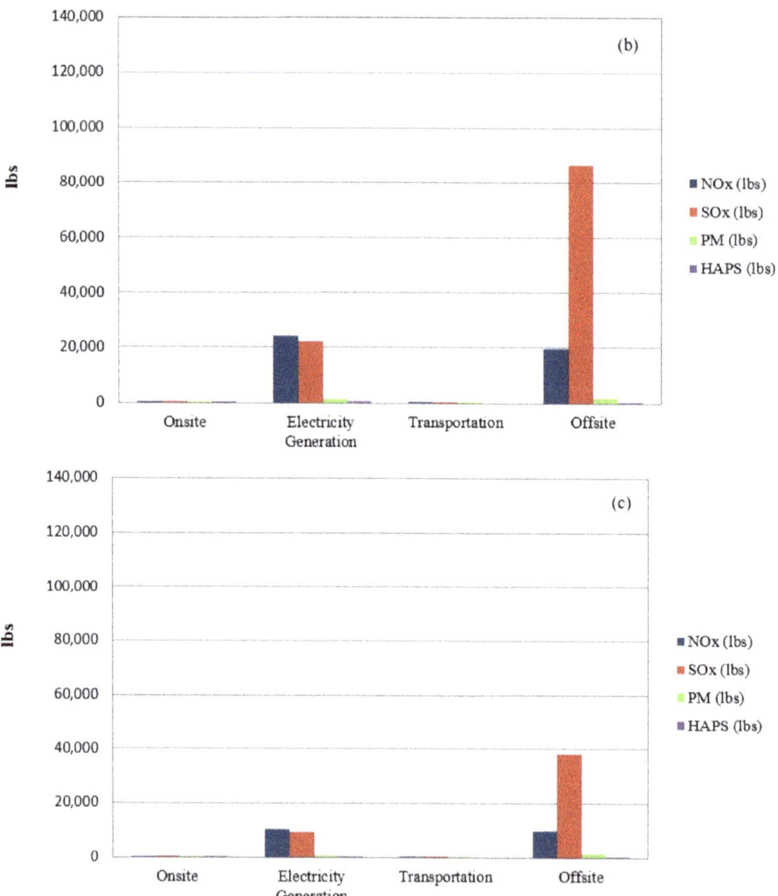

Figure 7. Contributors to the total NOx, SOx, PM and HAP emissions; (**a**) Scenario 1, (**b**) Scenario 2 and (**c**) Scenario 3.

Based on the above findings, it is highly recommended to incorporate renewable energy sources not only into the remediation activities but also into the fuel mix of grid electricity in Pakistan. Figures 8 and 9 demonstrate the solar and wind energy potential of Pakistan, respectively, which is quite promising for reducing the overall footprint of any activity in general and, in particular, the remediation system under consideration. For example, the annual average mean daily solar radiation in Karachi is in between 5.1–5.4 kWh/m^2 [27], and the wind power class for Karachi is from "Fair" to "Excellent", with a wind speed between 6.2 and 7.8 m/s [28], depending on the region. Thus, installation of renewable energy systems onsite and purchase of electricity from local renewable energy producers for the remediation of contaminated sites in Karachi is a promising idea which can lead to a reduction in the CO_2, NOx, SOx, PM and HAP emissions even beyond 68%.

The research methodology used in this study has limitations. The most significant is that it is a single case study on a hypothetical contaminated site and uses a particular data set for analysis. Therefore, it has potential limitations for systematic generalization. Moreover, the results obtained are based on the share of renewable energy resources in each cleanup scenario and could lead to a completely different set of results if the use of renewable energy resources is increased for more sustainability or decreased due to their limited availability.

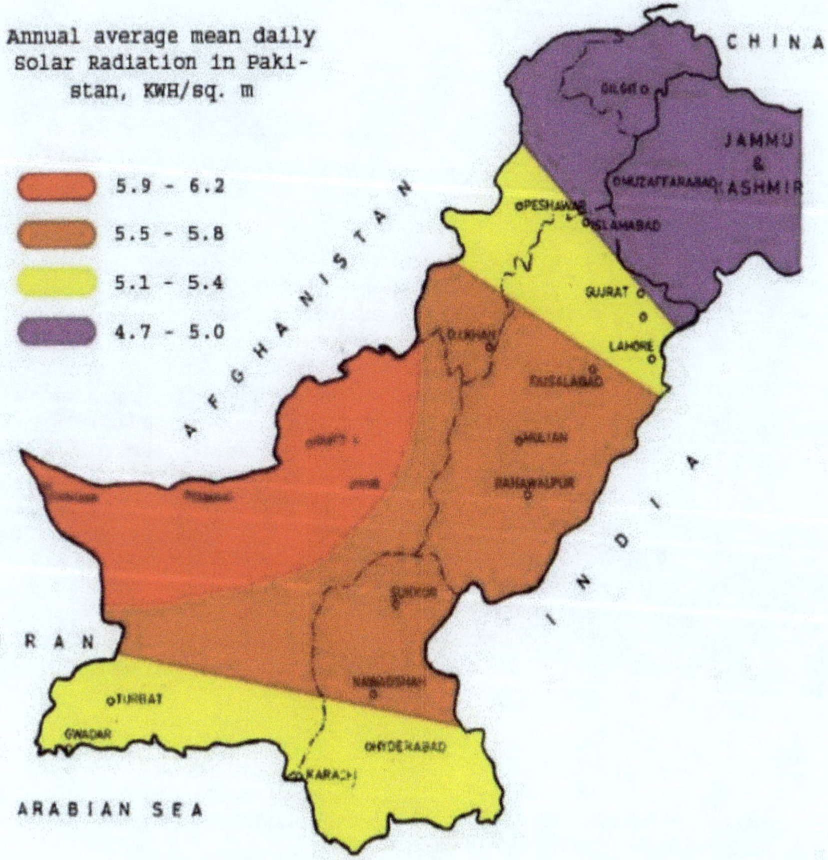

Figure 8. Solar energy potential of Pakistan [27].

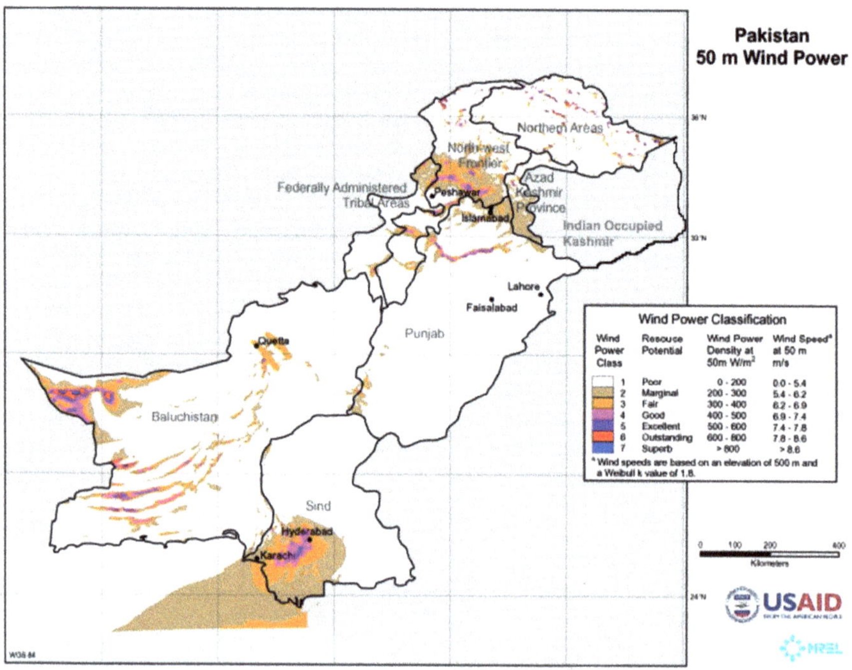

Figure 9. Wind energy potential of Pakistan [28].

6. Conclusions and Future Work

The environmental footprint of a cleanup at a hypothetical contaminated site is analyzed by using EPA's Spreadsheet for Environmental Footprint Assessment (SEFA). The effect of renewable energy integration into the remedial system is studied by considering three scenarios: cleanup without any renewable energy sources at all, cleanup with a small share of renewable energy sources, and cleanup with a large share of renewable energy sources. It is observed that the greatest contributor in the overall environmental footprint for this cleanup activity is grid electricity due to the fuel mix used in Pakistan. However, integration of renewable energy systems onsite and voluntary purchase of renewable energy can reduce the CO_2, NO_X, SOx, PM and HAP emissions up to 29% if done on a small scale, and up to 68% if done on a large scale.

The study presented here can be extended further by taking an actual contaminated site and applying this methodology to analyze the environmental footprint of remediation with possible shares of renewable energy resources. Owing to abundant solar resources and excellent wind power class in Pakistan, there is an immense need to incorporate these renewable energy sources in the remediation and to accurately predict the realistic reduction in CO_2, NO_X, SOx, PM and HAP emissions, thereby promoting green remediation across the country.

Author Contributions: Conceptualization, M.A.A.K., Z.Q., and M.A.; methodology M.A.A.K., M.A. and soft-ware, M.A.A.K., Z.Q.; validation, M.A.; formal analysis, M.A., Z.Q. investigation, M.A.A.K., Z.Q. and M.A. resources, A.Z.K.; data curation, M.A.P.M.; writing—original draft preparation, M.A.A.K., Z.Q. and M.A. funding acquisition, M.A.P.M. All authors have read and agreed to the published version of the manuscript.

Funding: This research received no external funding.

Institutional Review Board Statement: Not applicable.

Informed Consent Statement: Not applicable.

Data Availability Statement: Not applicable.

Conflicts of Interest: The authors declare no conflict of interest.

Nomenclature

Symbols	Definition
kW	Kilo Watt
kWh	Kilo Watt Hour
HP	Horsepower
Btu	British thermal unit
gptm	Gallons per ton-mile
mpg	Miles per gallon
gal	Gallon
MMBtu	Metric Million British Themral Unit
MWh	Mega Watt Hour
lbs	Pound

References

1. Islam, M.; Sahadath, M. Prediction of Potential Offsite TEDE, Excess Cancer Risk, Dominant Exposure Pathways and Activity Concentration for Hypothetical Onsite Soil Contamination At the Proposed Rooppur Nuclear Power Plant. *J. Nucl. Eng. Radiat. Sci.* **2020**. [CrossRef]
2. Harwell, M.C.; Jackson, C.; Kravitz, M.; Lynch, K.; Tomasula, J.; Neale, A.; Mahoney, M.; Pachon, C.; Scheuermann, K.; Grissom, G.; et al. Ecosystem services consideration in the remediation process for contaminated sites. *J. Environ. Manag.* **2021**, *285*, 112102. [CrossRef] [PubMed]
3. Hu, G.; Liu, H.; Rana, A.; Li, J.; Bikass, S.; Hewage, K.; Sadiq, R. Life cycle assessment of low-temperature thermal desorption-based technologies for drill cuttings treatment. *J. Hazard. Mater.* **2021**, *401*, 123865. [CrossRef] [PubMed]
4. Carriger, J.F.; Parker, R.A. Conceptual Bayesian networks for contaminated site ecological risk assessment and remediation support. *J. Environ. Manag.* **2021**, *278*, 111478. [CrossRef] [PubMed]
5. Brown, J.B.; Conder, J.M.; Arblaster, J.A.; Higgins, C.P. Assessing human health risks from per-and polyfluoroalkyl substance (PFAS)-impacted vegetable consumption: A tiered modeling approach. *Environ. Sci. Technol.* **2020**, *54*, 15202–15214. [CrossRef] [PubMed]
6. Koiwanit, J.; Filimonau, V. Carbon footprint assessment of home-stays in Thailand. Resources. *Conserv. Recycl.* **2021**, *164*, 105123. [CrossRef]
7. Morgan, D.R.; Styles, D.; Lane, E.T. Thirsty work: Assessing the environmental footprint of craft beer. *Sustain. Prod. Consum.* **2021**, *27*, 242–253. [CrossRef]
8. Valencia, G.; Fontalvo, A.; Forero, J.D. Optimization of waste heat recovery in internal combustion engine using a dual-loop organic Rankine cycle: Thermo-economic and environmental footprint analysis. *Appl. Therm. Eng.* **2021**, *182*, 116109. [CrossRef]
9. Bartie, N.J.; Cobos-Becerra, Y.L.; Fröhling, M.; Schlatmann, R.; Reuter, M.A. The resources, exergetic and environmental footprint of the silicon photovoltaic circular economy: Assessment and opportunities. Resources. *Conserv. Recycl.* **2021**, *169*, 105516. [CrossRef]
10. Mayer, M.J.; Szilágyi, A.; Gróf, G. Ecodesign of ground-mounted photovoltaic power plants: Economic and environmental multi-objective optimization. *J. Clean. Prod.* **2021**, *278*, 123934. [CrossRef]
11. Tennison, I.; Roschnik, S.; Ashby, B.; Boyd, R.; Hamilton, I.; Oreszczyn, T.; Owen, A.; Romanello, M.; Ruyssevelt, P.; Sherman, J.D.; et al. Health care's response to climate change: A carbon footprint assessment of the NHS in England. *Lancet Planet. Health* **2021**, *5*, e84–e92. [CrossRef]
12. Saget, S.; Costa, M.; Santos, C.S.; Vasconcelos, M.W.; Gibbons, J.; Styles, D.; Williams, M. Substitution of beef with pea protein reduces the environmental footprint of meat balls whilst supporting health and climate stabilisation goals. *J. Clean. Prod.* **2021**, *28*, 126447. [CrossRef]
13. Young, R.J.; Howard, R. *A Clean, Green Remediation*; TechLaw, Inc.: USA. Available online: http://www.alterecho.com/artfiledownload/2/CleanGreenRemdiation.pdf (accessed on 12 December 2012).
14. Reddy, K.R.; Adams, J.A. Towards Green and Sustainable Remediation of Contaminated Site. In Proceedings of the 6th International Congress on Environmental Geotechnics, New Delhi, India, 8–12 November 2010; pp. 1222–1227.
15. Dellens, A.D. *Green Remediation and the Use of Renewable Energy Sources for Remediation Projects*; Case Western Reserve University; U.S. Environmental Protection Agency (USEPA), August 2007. Available online: https://clu-in.org/download/studentpapers/Green-Remediation-Renewables-A-Dellens.pdf (accessed on 24 May 2021).
16. U.S. Environmental Protection Agency. Technology News and Trends. May 2007. Available online: http://www.epa.gov/tio/download/newsltrs/tnandt0507.pdf (accessed on 4 January 2013).

17. BP Global Press Release. BP Builds, Operates Largest Solar Power System on East Coast. 22 April 2003. Available online: http://www.bp.com/genericarticle.do?categoryId=2012968&contentId=2001546 (accessed on 6 January 2013).
18. Pagano, M. A Low Impact Technology Chemical Oxidation, Bioremediation and Groundwater Reinjection Analysed with SiteWiseTM and SEFA. Relazione tecnico-scientifica in attesa di pubblicazione agli Atti del Convegno Ecomondo 2018 edito da Maggioli Editore. 2018, p. 11. Available online: https://www.ecosurvey.it/wp-content/uploads/2018/11/Ecomondo-2018 _Paper_impacts.pdf (accessed on 24 May 2021).
19. US Environmental Protection Agency (USEPA). *Green Remediation: Incorporating Sustainable Environmental Practices into Remediation of Contaminated Sites*; US Environmental Protection Agency, April 2008; EPA 542-R-08-002. Available online: https://www.epa. gov/sites/production/files/2015-04/documents/green-remediation-primer.pdf (accessed on 24 May 2021).
20. US Environmental Protection Agency (USEPA). *Methodology for Understanding and Reducing a Project's Environmental Footprint*; US Environmental Protection Agency, February 2012; EPA 542-R-12-002. Available online: www.cluin.org/greenremediation/ methodology (accessed on 24 May 2021).
21. Sustainable Development Policy Institute (SDPI). Mapping of Chemical Contaminated Sites in Pakistan. 2010. Available online: http://www.sdpi.org/research_programme/researchproject-18-12-31.html (accessed on 25 December 2012).
22. Meegoda, J.N.; Hu, L. A Review of Centrifugal Testing of Gasoline Contamination and Remediation. *Int. J. Environ. Res. Public Health* **2011**, *8*, 3496–3513. [CrossRef]
23. US Environmental Protection Agency (USEPA). Air Pollution Control Orientation Course. January 2010. Available online: http://www.epa.gov/apti/course422/ap5.html (accessed on 6 January 2013).
24. Munawar, H.S.; Khan, S.I.; Qadir, Z.; Kouzani, A.Z.; Mahmud, M.A.P. Insight into the Impact ofCOVID-19 on Australian Transportation Sector: An Economic and Community-Based Perspective. *Sustainability* **2021**, *13*, 1276. [CrossRef]
25. Labianca, C.; De Gisi, S.; Picardi, F.; Todaro, F.; Notarnicola, M. Remediation of a Petroleum Hydrocarbon-Contaminated Site by Soil Vapor Extraction: A Full-Scale Case Study. *Appl. Sci.* **2020**, *10*, 4261. [CrossRef]
26. Farhat, S.K.; Adamson, D.T.; Gavaskar, A.R.; Lee, S.A.; Falta, R.W.; Newell, C.J. Vertical Discretization Impact in Numerical Modeling of Matrix Diffusion in Contaminated Groundwater. *Groundw. Monit. Remediat.* **2020**, *40*, 52–64. [CrossRef]
27. Alternate Energy (AET) Pakistan. Solar Energy Potential in Pakistan. 2012. Available online: http://aetisb.com/?page_id=256 (accessed on 6 January 2013).
28. Global Energy Network Institute (GENI). Wind Energy Potential in Pakistan. September 2009. Available online: http://www.geni. org/globalenergy/library/renewable-energy-resources/world/asia/wind-asia/wind-pakistan.shtml (accessed on 6 January 2013).

applied sciences

MDPI

Case Report

Post-Flood Risk Management and Resilience Building Practices: A Case Study

Hafiz Suliman Munawar [1], Sara Imran Khan [2], Numera Anum [3], Zakria Qadir [4,*], Abbas Z. Kouzani [5] and M. A. Parvez Mahmud [5]

1 School of Built Environment, University of New South Wales, Kensington, Sydney, NSW 2052, Australia; h.munawar@unsw.edu.au
2 Faculty of Chemical Engineering, University of New South Wales, Kensington, Sydney, NSW 2052, Australia; saraimrankhan17@gmail.com
3 Department of Urban and Regional Planning, National Institute of Transportation, National University of Science and Technology, Islamabad 44000, Pakistan; numerasyed@gmail.com
4 School of Computing Engineering and Mathematics, Western Sydney University, Locked Bag 1797, Penrith, NSW 2751, Australia
5 School of Engineering, Deakin University, Geelong, VIC 3216, Australia; abbas.kouzani@deakin.edu.au (A.Z.K.); m.a.mahmud@deakin.edu.au (M.A.P.M.)
* Correspondence: z.qadir@westernsydney.edu.au

Abstract: The study was conducted to assess the post 2010 flood risk management and resilience-building practices in District Layyah, Pakistan. Exploratory research was applied to gain knowledge of flood risk management to embed the disaster risk reduction, mitigation, and adaptation strategies at the local government and community level. Around 200 questionnaires were collected from the four devastated areas/union councils. Primary data from the field uncovered flood risk management practices by organizations, local government, and the community. It highlights resilience-building practices undertaken by the community through rehabilitation, community participation, and local indigenous practices. The role of the District Layyah's local government and organizations to mitigate the 2010 flood and their contribution towards flood resilience in affected communities was investigated, as no comparable studies were carried out in the riverine belt of District Layyah previously. Moreover, the tangible and non-tangible measures to lessen the vulnerability to floods and improve flood risk governance at a local level were identified. This study makes a valuable contribution in strengthening the resilience building of vulnerable communities by recommending few changes in existing practices concerning flood risk at a local level.

Keywords: flood; integrated flood risk management; resilience building; emergency planning; disaster risk reduction; community participation; disaster preparedness

Citation: Munawar, H.S.; Khan, S.I.; Anum, N.; Qadir, Z.; Kouzani, A.Z.; Parvez Mahmud, M.A. Post-Flood Risk Management and Resilience Building Practices: A Case Study. *Appl. Sci.* **2021**, *11*, 4823. https://doi.org/10.3390/app11114823

Academic Editors: Luis Hernández-Callejo, Víctor Alonso Gómez, Sergio Nesmachnow, Vicente Leite, Javier Prieto and Ângela Ferreira

Received: 30 March 2021
Accepted: 19 May 2021
Published: 24 May 2021

Publisher's Note: MDPI stays neutral with regard to jurisdictional claims in published maps and institutional affiliations.

1. Introduction

An extreme weather event is defined as a phenomenon that leads to injuries, health disparities, or loss of human lives. Besides this, the disaster also causes damage to the properties, economy, and ecosystem. These losses are such that the community cannot cope with them using their present resources [1,2]. Over the last decades, Pakistan has suffered different natural hazards, which cost numerous lives and a considerable effect on the economy. Such as the 2005 massive earthquake in Kashmir taking the lives of more than 75,000 people [3]. In 2010, Pakistan was struck with the worst flood in its history, with a death toll of 1800 and a total of about 21 million affected. The 2010 flood in Pakistan was highly devastating and is recalled as the 'super flood' by the Government of Punjab. Followed by the floods in 2013 and 2014, which took the lives of 178 and 367 people, respectively [4,5]. The vulnerabilities in the system for disaster management, its prevention, and socio-culture issues aggravate the effects of these events resulting in larger shocks. Hence, its management calls for coordination of government institutions, private sector, community, national and international organizations during, pre and post-disaster situation [3,4]. Amongst all-weather events, floods are most common. There is a trend of increasing urbanization and it is estimated that 61% of the world population will live in cities by 2030 [5]. The persistent urban growth and shortage of diligent planning boost the cities' vulnerability to flooding disasters [6]. So, understanding flood risk is very crucial to manage and make timely decisions to address flood challenges. The European Flood Directive (2007) has defined flood risk as an association of the likelihood of flood disaster and its possible detrimental impacts on the environment, health, culture, and economic life of human beings [7]. Floods are a bare opportunity for a city to get rightful economic and social advantages, and the relative potency of institutions that take part in flood risk management can be scrutinized against their area.

In developing countries like Pakistan reactive practices have been adopted to mitigate flood hazards and using ad hoc structural inventions for diverting flood risks rather than reducing them. These practices are deemed to be effective partially. Risk, hazard, vulnerability, exposure, resistance, resilience, coping, and adaptive capacity of communities; form an ally to describe flood risk management. Flood risk management needs a strategy that can stabilize sustainability with present needs. So, integrated flood risk management includes a combination of both structural and non-structural measures [8–11]. River defenses such as the construction of levees, dams, dikes, embankments, and channelization are included in structural measures. While nonstructural measures reduce the impacts of floods and the vulnerability of people and communities with the help of early flood warning systems, flood emergency response, disaster risk reduction (DRR), and evacuation schemes.

There is a need to improve existing structural measures by increasing the water storage capacity. Pakistan's existing reservoir capacity is 9% which is quite lower than the world average, i.e., 40%. The current storage capacity is 22.8 bcm which is expected to reduce by 57% by 2025 [12–14]. A lot of emphases have been laid on developing small dams which can provide irrigation facilities to small-scale farmers. The country has heavily invested in its water infrastructure, however, due to lack of management, much of the infrastructure is decaying and needs an asset management plan [15–17]. The budget allocated for the maintenance and repair of the infrastructure is very limited. These strategic structures are susceptible to unforeseen events and damages. The deteriorating conditions of the watercourses, gates, and drainage outlets result in low performance and water losses. The government needs to pay attention to the delayed maintenance, and a shortage of rehabilitation [18–20].

A holistic framework is needed to address the needs, develop policies to enhance capacity building in flood management. However, a lack of coordination between different departments at the provincial and federal levels will also limit the effective implementation of flood management approaches [21–24]. Moreover, it is of extreme importance to focus on the resilience for unforeseen events involving important decision making, innovative and technological solutions for addressing community problems, achieve speedy recovery to normalcy, and adaptation to changing environment. The resilience formalizations bring along added value as it is based on the knowledge of both the normal functioning of a system as well as its functioning in problematic scenarios [12]. However, due to the increase in flood events, recognition that flooding cannot always be prevented, and increased uncertainty of weather events, a shift towards flood resilience has been achieved within the traditional Flood Risk Management (FRM). The concept of resilience has gained greater popularity, thus leading to a plethora of definitions, measurements, and applications. Therefore, the formulation of the resilience framework requires the integration of the FRM approaches (i.e., top-down and bottom-up) along with a more interdependent holistic approach [13].

Generally, for flood risk management three different types of conceptual resilience frameworks have been reported in literature including; engineering, ecological, and social–ecological frameworks. The application of engineering resilience has benefitted the design and functionality of both the technological systems and the flood resilient technological frameworks. Whereas socio-ecological resilience displays a broader perspective as it is based on human interaction with natural systems. However, the design of the socio-ecological resilience framework in flood risk management is based on an adaptive approach that entails a set of short-term measures and monitoring criteria. Moreover, it is well established in literature that the utilization of ecological and socio-ecological resilience frameworks in the developing and developed country can help in providing guidance for the formulation of a more resilient flood risk management framework that deals with flood protection, prevention, and preparedness [14].

To sum up, a resilient flood risk management strategy focuses on reducing the impacts of floods via efficient warning systems, evacuation plans, building regulations in flood-prone areas, and the applications of spatial planning [15,16]. Hampering of floods is difficult, but damages and exposure of flooding to risk-lain communities can be minimized with flood risk management [15,17–20]. Additionally, there has been a change from structural flood protection measures towards resilient or integrated flood risk management over the past ten years as many studies have revealed that communities adapted to shocks and disturbances rather than resistance are more durable [21]. Forrest and coworkers have shown the importance of time and place in understanding the ways in which a contribution can be made to the local level flood resilience by the civil society members. It is recommended for the policymakers to critically recognize the significance of both time-related variations for the involvement of citizens and the place-based capacities in local flood-resilience policies [22]. The findings of the semi-structured qualitative interviews with the flood affectees of both urban and rural areas of northern England have been reported elsewhere [23]. Through the utilization of the proposed 'conscious community' concept, a complex relationship between attachment to the communal identities, locality, and the local networks was revealed in the study. The concept of 'conscious community' builds on the conceptualization of community as an inevitable linkage of the cultural, spatial, and social elements of the local community. For the local communities, a shift to the localized and community-based approach to the FRM was variable. Whereas the flood experience in the context of the urban area emerged as an important factor in the construction of community and generating responses after flooding. Yet, a better understanding of the local community construction is highly necessitated to prepare the local communities for flooding [23].

Furthermore, the EU Directive (2007/EC/60) proclaimed that flood control plans demand the identification of tangible and non-tangible measures which are efficient enough to lessen the vulnerability to floods and improve flood risk governance [17]. For understanding flood challenges due to the intensity of flooding impacts, there is a need to enhance the policies, technical knowledge, and strategies [24,25]. It, therefore, calls for the systematic assessment at a local level and the use of more featured datasets and analysis of the flood risk areas. Hence, the objective of the paper is to assess the post 2010 flood risk management and resilience-building practices in the selected case study. The study also provides some observations and draws conclusions about coping strategies against the flood challenges. Moreover, it also elaborates the community awareness and level of preparedness based on the review of the flood risk management in the riverine area of district Layyah.

2. Literature Review

Secondary research included a review and analysis of the existing literature on floods, disaster risk reduction, community resilience, annual reports prepared by government departments and organizations, and the updates on the flood relief and the response of 2010 floods in district Layyah. A detailed literature review was carried out for this study, and a VOSVIEWER analysis was conducted based on the most used keywords in this research area, as shown in Figure 1. Popular and widely used search engines were opted to retrieve research articles for the current study, such as Scopus, Google Scholar, Science

Direct, Elsevier, Springer, ACM, and MDPI. A set of queries were formulated to perform an exhaustive search for a maximum number of research articles relevant to the area of interest. It was found that articles retrieved for this research revolved around the same keywords used in this study. The articles were screened based on (1) publication period: 2010–2021 (2) no duplicates (3) document selection based on research article and book chapter, and (4) published in English language only. The detailed screening process is shown in Figure 2.

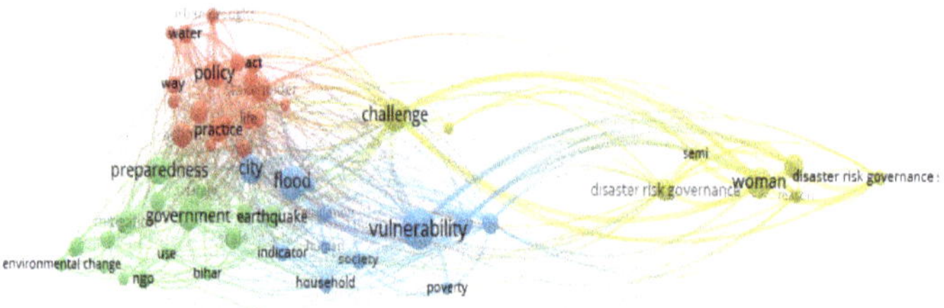

Figure 1. Keywords found in the articles retrieved for this research.

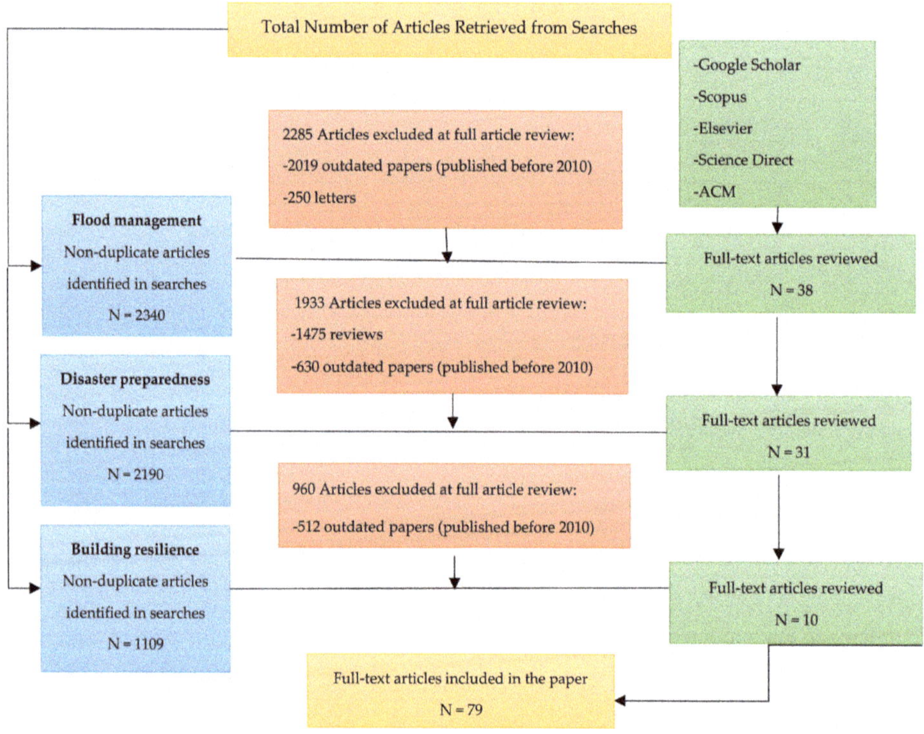

Figure 2. The Detailed Screening Process.

Secondary data was collected through content analysis of government policies, schemes, and reports prepared by various organizations at the local, national, and international levels. Organizations' reports and Government documents were utilized to get descriptive data of flood-affected communities in the case study area. Precisely, this literature review was focused on answering the following research questions:

Q1. What are the existing methods and technologies for flood disaster management in Pakistan?

Q2. What is the Perception of the community about the disaster response and recovery during past flood events?

3. Research Methodology

The mixed-method research approach incorporates the characteristics of quantitative and qualitative research to give aggregate results obtained from complicated research questions [26]. A mixed-method research design has been used for this study because it is descriptive as well as exploratory in nature. Descriptive research is applied to describe the major gaps in the flood response plans of the Government and other non-governmental humanitarian sectors during and post 2010 flood events in District Layyah and how did they play their role [22]. The exploratory dimension of research relates to gaining knowledge of flood risk management and explore how to embed the Disaster Risk Reduction, mitigation, and adaptation strategies at the local government and community development planning level. It also helps to investigate the structures if they are resilient against floods.

A questionnaire survey technique has been employed and 200 samples were drawn from residents of the four most devastated Union Councils of Layyah District. The survey was conducted in 2013. The survey team comprised three members, including the researcher and two volunteers from a local NGO (IDSP). Each questionnaire respondent was the head of his/her family. The sample drawn from each Union Council was 50 each (randomly selected from the list of the residents within each union council and was ensured that the selected respondents were victims of the 2010 floods), as it is stated by Sudman and Glenn [27,28], that for each subgroup at least 20 to 50 sample size is sufficient for analysis. The major issue associated with the sample size is the selection of number (size) to perform the data analysis adequately and precisely. While using descriptive statistics (e.g., mean, frequencies), then nearly any sample size will serve well. Generally, a sample size of 200–500 is a good size sample when we need to perform multiple regressions, analysis of covariance, or log-linear analysis. This type of analysis is often applied to perform more rigorous state impact evaluations. Therefore, the selection of the sample size should be made appropriately for the planned analysis. Particularly, for the comparative analysis of subgroups (e.g., an evaluation of program participants with nonparticipants), an adjustment in the sample size might be required. It is clearly established by Sudman (1976), that a minimum of 100 samples is required for each major group or subgroup in the sample and for each minor subgroup sample size of 20 to 50 elements is obligatory (Sudman 1976, Israel 1992). Based on this observation, we set the sample size of all small union councils to 50, which is sufficient for analysis by keeping in mind the targeted population. However, there are certain limitations of using a sample of the population for a questionnaire-based survey, such as (1) failure in the identification of the target population (2) consideration of inadequate survey population (3) the bias due to non-responses (4) bias due to respondents (5) questionnaire related problems in terms of content and wording, (6) questions should not be misleading, double-barreled, or ambiguous, rather they should be direct and clearly relevant to the survey objectives, (7) processing errors (8) misinterpretation of the results in relevance to the conducted survey (9) time period bias.

Among the 200-sample size, there were 140 male respondents and 60 female respondents. The questionnaires for the community consisted of both open-ended and closed-ended questions to help in determining the socio-economic status of the local people, their concerns about the existing situation, their level of preparedness, and construction patterns of their structures while living in flood-prone areas. The first section of interview questions was related to the background information of the participants such as gender, age, locality, highest education level, family system, household size, number of bread earners, and occupation [24,25]. After gathering the background information, all the questions that were asked by the participants were based on existing flood risk management practices, rehabilitation process, community participation, local indigenous practices, regain in livelihood, infrastructure resilience, and level of satisfaction against emergency relief in 2010 floods. After gathering the results for the top research questions, the data was cross-tabulated, and results were filtered. The statistical significance of the data was evaluated based on sample size, its accuracy within confidence level, and margin of error [26]. To analyze the satisfaction level of the respondents, Yeh's satisfaction index (YSI) method was used on the collected datasets. It is given by the equation below:

$$YSI = Satisfied - Dissatisfied / total\ respondents \times 100$$

The obtained numerical values may be zero, positive or negative. Zero indicates that there is no satisfaction. The satisfaction level may be minimum, moderate, or strong if the result is under 25%, 50%, or 75%. While if the values fall under -25%, -50%, or -75% then dissatisfaction level is minimum, moderate, or strong [26].

Calculating resilience index: Four resilience components, i.e., social, economic, physical, and institutional resilience was measured by taking percentages of all selected variables to avoid the normalization process. Weights were assigned to each variable using the subjective method [27] The component resilience index (CRI) was calculated by taking an average of the respective variable resilience index.

In Punjab Pakistan, a union council is an area within a district consisting of one or more revenue census villages, census blocks, or revenue estates. These are divided into rural and urban areas in a way the population remains the same with each area [27]. There were 14 Union Councils of a riverine belt which got affected in 2010 flood, and the selected Union Councils were the most damaged ones as expressed by an official from one of the National Organizations, which is a member of District Disaster Management Authority (DDMA). Figure 3 below shows the four selected union councils, including Lohanch Nashaib, Bakhri Ahmed Khan, Kotla Haji Shah, and Sahu Wala. The former three are part of Tehsil Layyah, while the latter one is included in Tehsil Karor. Twelve sample villages were selected among the most vulnerable villages where government and NGOs had implemented various risk reduction, relief, early recovery, and response plans, as mentioned in Table 1. The villages included are ThallaInayat Wala (Sahu Wala), Arif Abad (Sahu Wala), Qazi Wala (Sahu Wala), Khokkar Wala (Kotla Haji Shah), GurmaniSikandary (Kotla Haji Shah), Tali Musa Khan (Kotla Haji Shah), Mancharay Mohana (Lohanch Nashaib), Chandia Wala (Lohanch Nashaib), Siraye Janube (Lohanch Nashaib), Majhi (Bakhri Ahmed Khan), Shah Wala (Bakhri Ahmed Khan) and ChahKhoay Wala (Bakhri Ahmed Khan).

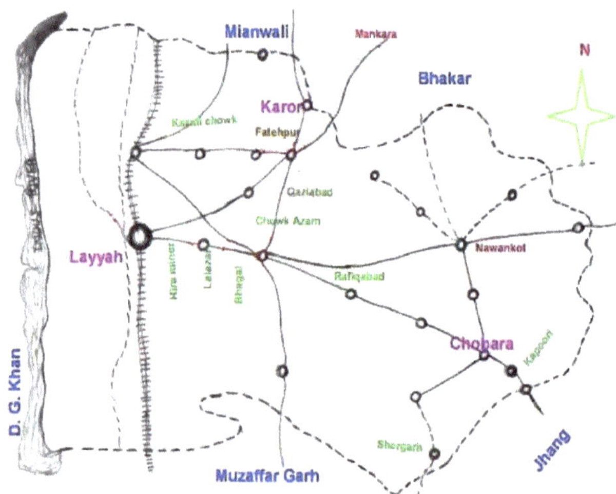

Figure 3. Case study areas within a riverine belt of district Layyah. Reprinted from the ref. [28].

Table 1. Sample Villages in selected union councils from district Layyah.

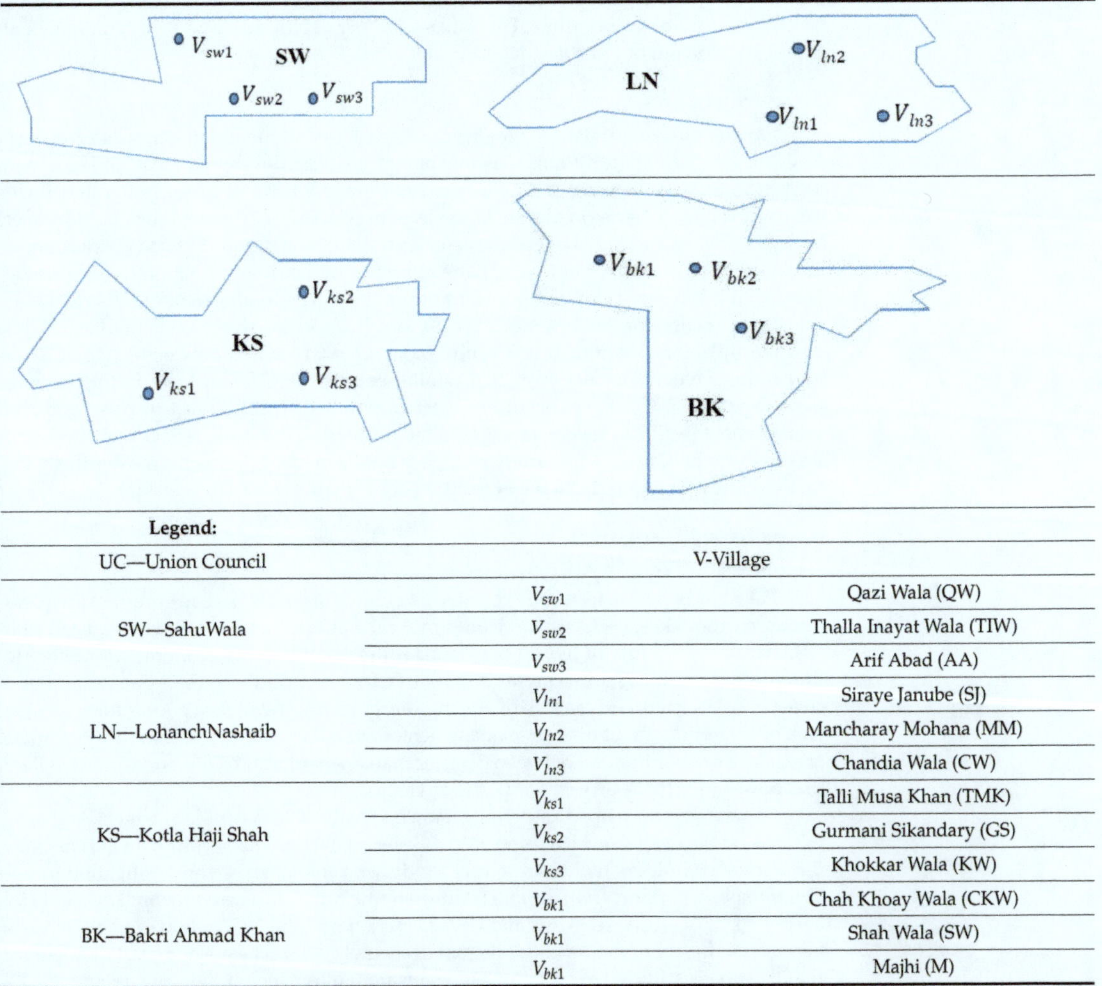

Legend:		
UC—Union Council		V-Village
SW—SahuWala	V_{sw1}	Qazi Wala (QW)
	V_{sw2}	Thalla Inayat Wala (TIW)
	V_{sw3}	Arif Abad (AA)
LN—LohanchNashaib	V_{ln1}	Siraye Janube (SJ)
	V_{ln2}	Mancharay Mohana (MM)
	V_{ln3}	Chandia Wala (CW)
KS—Kotla Haji Shah	V_{ks1}	Talli Musa Khan (TMK)
	V_{ks2}	Gurmani Sikandary (GS)
	V_{ks3}	Khokkar Wala (KW)
BK—Bakri Ahmad Khan	V_{bk1}	Chah Khoay Wala (CKW)
	V_{bk1}	Shah Wala (SW)
	V_{bk1}	Majhi (M)

4. Floods 2010 in Pakistan

According to [29], Pakistan has had seven major flood events which affected approximately 40 million people since 1973. Pakistan experienced the utmost cruel flood in August 2010 in the history of the country. Floods are a recurring natural hazard and among the population influenced by natural hazards; 90% of it is affected by floods in Pakistan [30–32]. The 2010 flood was highly devastating and is recalled as the 'super flood' by the Government of Punjab. The flood accompanied the annual monsoon season and reached unusual levels in the history of the Indus River system in Pakistan. Moreover, 78 districts were heavily affected by the floods. In Punjab, 200 villages, 1800 casualties, 500,000 homes, 1.7 million acres of cultivated land, and billion dollars' value of livestock and seasonal crops were destroyed [29]. Also, it has been identified that people residing in rural areas, associated with the agricultural sector, and having low income are the most vulnerable group having the low-income status, minimal access to education, health services, water supply, and sanitation [27]. In particular, the homes, infrastructure, transport system, and schools of this vulnerable group were destroyed due to the 2010 flood. In district Layyah,

172,607 people were affected by the 2010 flood, and 6459 houses were fully damaged [33]. Hence, The Human Development Report (2009) reveals that Pakistan's HDI was 0.572 thus, placing it at a rank of 141 while, Pakistan's GDP per capita was $955 hence, ranking it at 132nd position out of 182 countries [29].

5. Case Study Area

For the current study, villages were selected from four union councils of district Layyah, located in South Punjab, as it is one of the least developed areas and has been subject to annual flooding due to its proximity to the bank of River Indus that flows nearby. Flood is a common feature in the lives of residents of district Layyah. Moreover, between Chenab and Indus rivers, the associated districts are prone to flood every summer season. From (February-September), the rainy period is usually 7.5 months long and at least with a sliding 31-day rainfall of 13mm rainfall. The total area of the district taken into consideration for this case study is 6291 km^2 [East-West], which is (55 mi) East-Width, and (45 mi) North-South. It is a sandy block of land located between Chenab River and Indus River. Seasonal Monsoon rainfall coupled with other local factors such as poor drainage, poverty, negligence of local government institutions towards flood risk management results in severe damages to the properties, livestock, crops, livelihood, and loss of lives. Enhancing community resilience has been identified as the most influencing factor of disaster risk reduction by the 2009 Global Platform for Disaster Risk.

6. Results and Discussions

6.1. Literature Review Results

Pakistan is a disaster-prone country as evident from its disaster profile. Frequent floods, earthquakes, cyclones, landslides make the affected region vulnerable. Each year Pakistan faces the loss of lives, flora, fauna, infrastructure, and economic instability in the affected region. Furthermore, different factors contribute to the vulnerability and severity of disasters such as lack of effective early warning systems, poor infrastructures, lack of awareness about disaster management, poor communities living in disaster-prone areas, lack of coordination between disaster management authorities and limited skilled manpower to provide an early response to victims.

Pakistan has developed an institute known as Natural Disaster Management Authority (NDMA) at the federal level to respond to disaster scenarios. The NDMA plan outlines how different stakeholders will work jointly for disaster management and mitigation [8–10]. The monetary support from donors is directed towards disaster reduction. Different UN agencies such as JICA, ADB, EU and USAID work with local NGOs, however, NDMA has been set up for coordination with different agencies [21–23]. The major devastation during the 2010 floods attracted a large number of aids from international countries, relief agencies and local civil societies which helped NDMA being the central institute to carry out relief activities [24–26].

At the provincial and local level, NDMA constituted a Provincial Disaster Management Authority to manage disaster events and help people to recover from the loss of property, food, and their livelihood. PDMA provides a platform to bring all provincial organization together to provide timely disaster response and mitigate flood risk. The limitation of PDMA is the lack of risk management and making communities resilient towards extreme weather events such as floods. They focus more on response with little importance to long-term strategies for making communities more adaptive to floods. Agricultural losses are not addressed by PDMA, indicating that there are no plans for the management of framing communities to save their crops and reduce losses. Furthermore, disaster management authorities at the district level are not fully function and are equipped to assist communities during a disaster event.

Ashfaq et al. [27] surveyed the household vulnerability and resilience to flood disasters in two districts within Khyber Pakhtunkhwa after the 2010 floods. Data were collected from 600 households through interviews. Variability and resilience indices were calculated based on the selected variables for community households in the district of Charsadda and Nowshera. Higher vulnerability index along with lower resilience index indicated that communities in Nowshera were more vulnerable to flood disasters as compared to those in Charsadda. The study suggested that the local authorities need to create awareness about flood mitigation and prevention practices among the district and emphasize strengthening social, physical, and economic resilience. Educating communities about flood disaster management and effective zoning strategies to prevent houses from being built in flood-prone areas will reduce casualties. Through awareness about infrastructure material, people can move from traditional materials such as clay to more resistant and durable material, i.e., concrete or bricks.

Baig et al. [28] investigated community needs and perception about the rehabilitation process and the support from the government and NGOs during the 2010 floods in Swat, Pakistan. No humanitarian aid reached the community after the first ten days as the main road was washed away. When the roads were partially opened, a few NGOs were observed distributing water disinfectants while the community was waiting for food and shelter. After the 2010 floods, the drinking water quality deteriorated and affected sanitation systems in many districts of Pakistan. In Swat and Sukkur high microbial load was observed in collected water samples. Federal level NDMA and Provincial level PDMAs collaborated with all the NGOs to enhance the humanitarian aid work. Around 36 NGOs were registered to help the communities. During the initial three months, NGOs focused on providing clean drinking water, setting up latrines, hygiene awareness, etc., after the emergency response period was over, attention was laid on mid to long-term development interventions. As per the reported activities carried out by the NGO, it was assumed that all field activities have been carried out smoothly; however, the on-ground reality was quite different, and people were not satisfied with the relief response by the government and NGOs.

Pakistan's measures on flood forecasting, early warning systems, and evacuation plans are not developed yet. A lot of work needs to be done in this space to develop effective systems for flood management. Lack of disaster preparedness and management leads to huge losses and damage to lives, infrastructure, and livestock [21]. The government needs to improve flood warning and response systems across the country. Implementation of cutting-edge technology and equipment which can assist in decision making, creating awareness among the community, developing flood maps, and evacuation plans are highly necessitated. Satellite imagery and big data analysis will help authorities in flood predictions. Important consideration must be given to wetland restoration, and cleaning river basins will also facilitate the flood retentions [22–24]. Installation of radar will improve gathering information of real-time data on precipitation, accuracy in forecasting, and improved planning. Other environmental factors such as soil characteristics, infiltration rates, and vegetative cover are important to comprehend catchment response to flood vents. Urban planning and land use also play a vital role tool [25–27].

6.2. Survey Results

After data collection from the community through a questionnaire survey, it was analyzed by comparing results from the four localities. The results assessed flood risk management and resilience-building practices in district Layyah. Using the statistical findings, we can draw conclusions and recommendations for future advancement in flood risk management and resilience-building of local communities.

6.2.1. Reflection on the Socioeconomic Status of a Case Study Community

The socioeconomic data of community gave the insight into gender, age, locality, highest education level, family system, household size, number of bread earners, and occupation. The age of respondents was categorized into five groups and the respondents in the age group of 36–40 years dominated the other age groups. Both male and female respondents were asked questions while working in the fields as the data was collected at the time of sugar-cane crop harvesting in the district Layyah. It was observed that people of all age groups worked in the fields on daily wages. The results showed that 74% of the respondents had a joint system in their families, and only 26% had a single-family. It was observed in a field survey that people lived close to each other, and communities had a strong connection with each other in all the four localities of the riverine area in district Layyah. The adaptive capacity of a community lies in living in sheer vicinity as [34] claimed that there are dense networks of concern and association within the village periphery. Most people live in a combined family system, each household in the family shares courtyard and baths while having separate kitchens. They share happiness and sorrows, which makes them socially resilient to any stress or pressure, and hazard. Many examples boost the influence of social capital as a principle for resilient communities against disaster [35–37]. According to [38], coordinated actions provide materialist and emotional aids, and endangered communities are ought to work together manageably to figure out problems [34]. Moreover, in the community resilience model by [39], social capital addresses three elements that are (i) community participation (ii) social bonds (iii) social support. The socioeconomic status of people living in LN was found comparatively low as 74% population belonged to the laborer occupation. During a field survey in one of the villages named MM, an old man reported the 2010 flood to be a major disaster as he lost his home. One local NGO helped him to rebuild his house, but the help was limited. He said, "I live in a tent with my wife after the 2010 flood, whereas the NGO helped to build only one room, bathroom, and kitchen. Since my son got married, I have been living in a tent under severe environmental conditions". Both males and females work in flood-affected areas to fulfill their basic needs (Figure 4). During the infield survey, it was observed that females are the major contributors to family income in riverine areas. When asked by one female respondent in union council LN about her earning source, she said, "All women from our locality work from morning to evening in fields. We get up early in the morning, and after leaving our children for school, we go to work. As it is sugarcane season, so nowadays we go for cutting off this crop". Seasonal crop harvesting was the major occupation of people living close to the riverine belt. Infield survey, the education level in 4 union councils was found varyingly. During a survey in LN, SW, and KHS, respondents highlighted a lack of education facilities in their areas for their daughters, especially as there was only a primary school in each union council for girls, although there was a high school in each union council for boys only. One of the female respondents from village MM in union council LN asked for a permanent place in Layyah city so their children's education would not be affected by annual flooding. As she mentioned, schools remain closed when the flood hits the area, and school buildings get deteriorated each year. Many respondents in union council SW and KHS criticize the unjust educational facilities, as they want higher education in their areas. When asked for a reason behind criticism, one of the female respondents from KHS who was carrying out her bachelor's degree briefed that, female and male students have to travel daily to the city for higher secondary education and to attend university. Child labor was observed in all localities, but in LN higher rate of child labor was observed after floods. These results are in line with the survey conducted to capture the Bangladesh flood event in 2005 through a semi-structured survey. It was found that the effects of flood were more prevalent in low-income areas, people who lacked ownership on land and suffered economic inequality. An adaptive strategy to overcome income inequality was income diversification. Similarly, Ruffat et al. [35] reviewed and analyzed the drivers of social vulnerability in almost 70 case studies based on flood disasters. The most frequently occurring drivers for social vulnerability were found to be demographics, coping capacity, health, land tenure, risk perception, and socioeconomic status. The influence of each indicator is variable depending on the disaster stage and the setting of the country.

Figure 4. An old woman making mat with date leaves in union council S.

6.2.2. Emergency Relief Measures in the Case Study Area

Figure 5 depicts the flood risk management practices in the district. When the flood hit the riverine area in 2010, emergency shelter and rescue were the priority of the government and NGOs. Almost 47% of flood-affected communities reported in favor of the government in providing them emergency shelter. While 24% of respondents claimed that their relatives gave emergency shelter when the flood came, and 18% got emergency shelter provided by NGOs. When asked about relief camps, it was found that the government declared few schools, colleges, and district Layyah jail as camps for flood-affected communities. After the 2010 flood, the government has started to do this practice of shifting people to different schools of Layyah city annually (Figure 6). In KHS, 23% of people got emergency shelter, while in LN 17% of the people received emergency shelter. In relief camps, 50% of the population got the food from the local, national and international organizations.

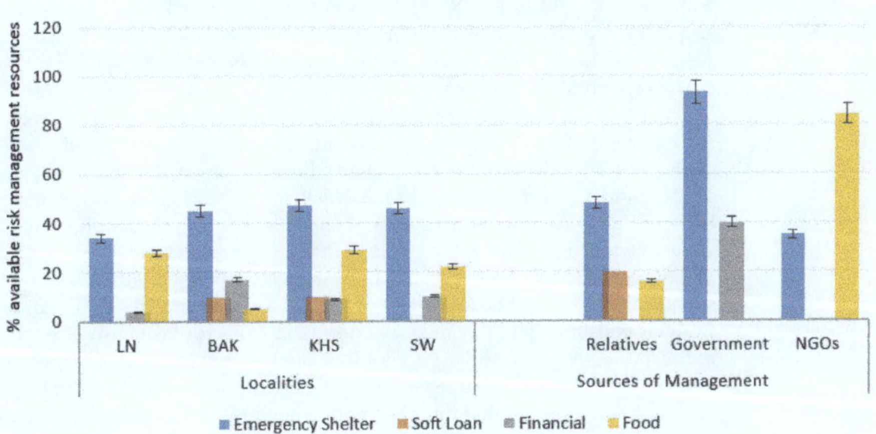

Figure 5. Flood risk management in district Layyah.

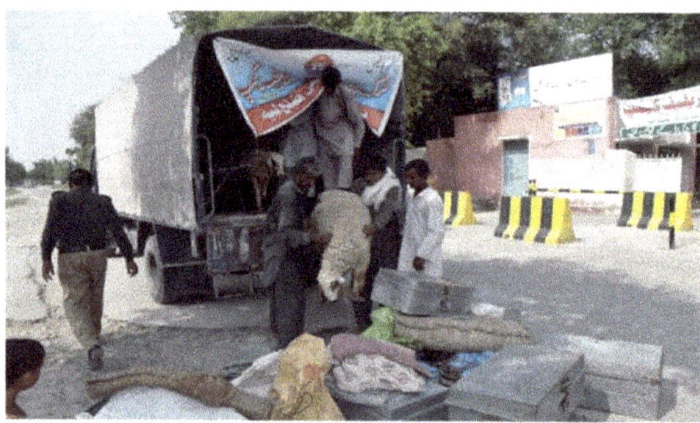

Figure 6. Rescue operation (2010) by district Police Layyah: Transferring flood affected family and its assets in Government College for Women Layyah.

The provision of food was mostly carried out by various organizations, as 42% of respondents claimed they received cooked and non-cooked food items from the different organizations and 8% said they received food from relatives in the city. In contrast, the government did not play a role in providing food in relief camps. Around 15% of people claimed to receive food in KHS, while only 2% of respondents in BAK got food. When the level of flood water became low, people went back to their homes and started living there in deteriorated houses and tents provided by the organizations. The government had started a scheme to provide financial assistance to the flood-affected community. A cash transfer scheme named Watan Cards were distributed by the government as compensation for flood victims. Debit cards carrying compensation money were distributed to transfer money in the future, depending on the need of the people [36]. The system faced some technical and administrative difficulties such as delays in money transfers, lack of monitoring corruption and black marketing, and including the name of people in beneficiaries list based on political influence. Some female-headed households were not eligible for Watan cards as they did not have identification cards [37,38].

Around 9% of respondents got these cards in BAK as its local representatives have strong support from the government. Contrary to it, in LN only 2% of people got Watan cards. Watan cards were issued to married men/women, through which they received 20 thousand rupees (equivalent to 165 AUD). In LN, respondents were not satisfied with financial help. The cards were given to only a few deserving people, while the majority were given to those who had support from the government and political references. Only 10% of respondents got a soft loan in BAK and KHS from their relatives. One of the male respondents in KHS said he got a soft loan of five thousand rupees (equivalent to 41.26 AUD) from his cousin to get food and other non-food items for his family. On other hand, one old respondent in LN said: "I do not like to take anything from anybody; flood-hit not only our homes but also our self-respect. I felt ashamed of getting any help from relatives". Hence, it is concluded that there is a gap in flood risk management concerning the integration of structural and non-structural measures in district Layyah. These responses are in line with the study conducted by Farman Ullah et al. [39] who determined the flood risk perception and its determinants in rural households of two communities in Khyber Pakhtunkhwa province. A survey was carried out in two districts through a questionnaire from 382 respondents. It was found that around 50% of respondents received high-risk perception based on education, flood experience, location of the house near stream bank, and the distance from the river. Also, respondents in flash flood-prone regions experienced a lower risk than those in the riverain flood-prone areas.

6.2.3. Recovery from Disaster

Rehabilitation

Figure 7 shows the pattern and sources of recovery from disaster in flood-hit localities. As the 2010 flood caused havoc in the riverine area, houses were destroyed, almost 74% population reported that they had to rebuild their houses. In BAK, 80% of respondents rebuilt their houses. Moreover, 26% of the population repaired houses with a maximum number of respondents from SW. In the case study area, katcha and pakka houses (mud or adobe houses) were seen. People use mud, bamboo, sheets made of bamboo sticks, wood, and bricks as a building material for the construction of their houses. Besides, 48% of respondents reported that Non-government and International organizations helped them to rebuild their houses. In LN, 70% reported that their houses were rebuilt by NGOs and 20% in SW got help from the government to rebuild their houses. In KHS, 64% population informed that they helped themselves to repair and rebuild their houses by taking soft loans from relatives or by selling their livestock. It can be concluded that the government and organizations responsible for post-disaster recovery did not satisfy the community in KHS and SW. Figure 8a below depicts the repairing of a deteriorated house in 2010 and after floods in KQN (one of the villages of LN). While Figure 8b depicts the rebuilt houses by the Integrated Development Support Program (IDSP), which is one of the local NGOs working in district Layyah. Hence it can be concluded that NGOs (IDSP and Awami Development Authority) and self-help of the community were the two prominent factors in rehabilitation with minimum contribution from the local government after the 2010 floods in district Layyah.

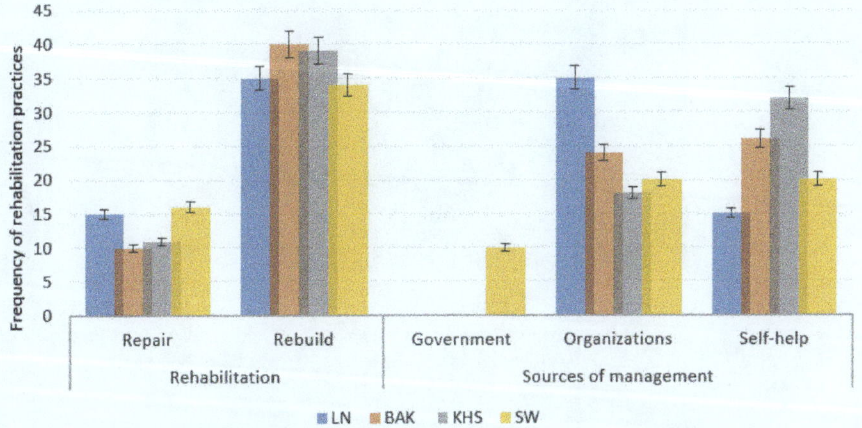

Figure 7. Rehabilitation practices and sources in flood-affected localities.

Community Participation

Figure 9 shows the community participation to get their lives back to a normal routine after the flood when they reached their communities from flood relief camps. 56% of the whole community responded that they gathered materials from their deteriorated buildings and helped themselves to fix their houses, while 70% population of LN reported similar findings. In BAK, the maximum population reported that they did nothing to help themselves as their locality was fully devastated and waited for assistance from organizations and government, while 22% of the community did not respond. As concluded by [40], community flood disaster risk management relies upon the participation and active role of the stakeholders in each locality. Moreover, The Global Platform for Disaster Risk Reduction (2009) stated that the best way to reduce the disaster risks is to strengthen the communities around the world and increase their resilience so that they can combat adverse situations [17,41,42]. Similarly, the Hyogo Framework (2005–2015) perceived the necessity of producing resilient communities plus established methods to build it by (1) Preparing policies that are integrated with disaster prevention, preparedness, mitigation, and reduction of vulnerability (2) Increasing the local capacity to build hazard resilience (3) Introducing the attributes of risk reduction, emergency preparedness, response, recovery, and reconstruction into the policies for the disaster management [43,44]. Figure 10 illustrates that people live in a hazardous area despite the river erosion.

(a)

(b)

Figure 8. (a): A repaired house by an old respondent in LN, district Layyah. **(b)**: Rebuilt houses in a row for one of the villages of union council, LN, district Layyah.

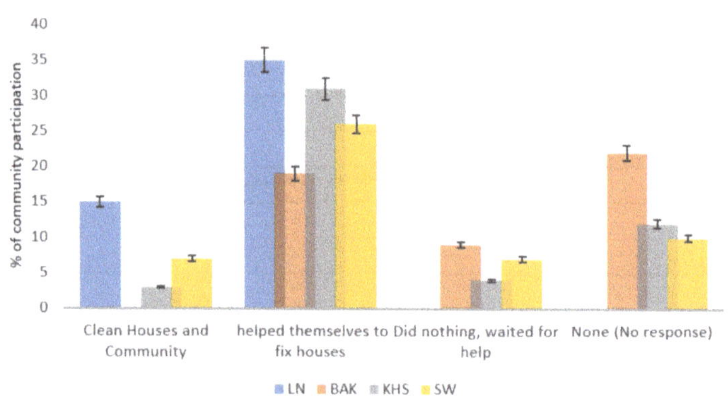

Figure 9. Community participation in flood-affected community.

Figure 10. People living with the highest risk of soil erosion in River Indus, district Layyah.

Local Indigenous Practices

Figure 11 shows the local indigenous practices taken during and post-flood conditions, along with the knowledge gained by the community from the 2010 flood event in district Layyah. 36% of people said that they started using a tractor tube for crossing, rescuing their family and other assets by themselves. 5% of the population reported that they use Sandhari for crossing flood water to reach a rescue point in case of emergency. Sandhari is a type of outfit made with goatskin which is used by the cobblers of the locality. 37% of the community reported that they started a practice of raising the platform of their houses by 4 to 5 feet after the 2010 flood disaster. This practice was seen in four localities during the field survey. When asked about the knowledge gained after the 2010 flood, 22% of people said that they have learned to evacuate their localities as soon as possible after an early warning was received. One of the male respondents in LN said: "We did not consider the 2010 flood warning, and when a flood hit our area it was havoc that destroyed us". Moreover, it was concluded that resilience is not only the capacity of the system to return to its original state but to do advancement in it by learning from past experiences and adaptation [43]. Mustafa [44] surveyed to gather information from the farmers in Pind Patekhan flood plains. To prevent water from entering the house, most of the farmers-built houses on the elevated mounds of muds. This indigenous practice was adopted by most of the locals. However, some of the tenant farmer houses lacked such features, either due to lack of funds or influence on the landlord to lend them government-owned bulldozers. The tenant farmers were in worse condition, as they were unsure of the future and it is not feasible to spend money on building elevated mounds.

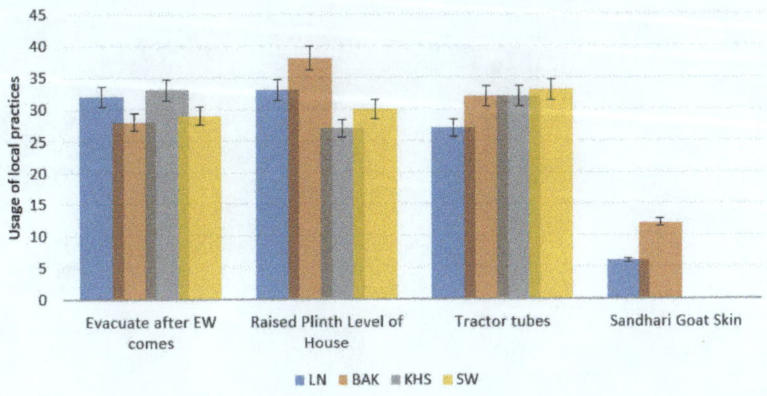

Figure 11. Local indigenous practices for flood risk management in affected union councils of district Layyah in 2010 floods.

Livelihood Regain

The results in Figure 12 showed that 53% of the population from district Layyah regained their livelihood by cutting vegetables in fields located in their respective union councils on a daily wage of 300 to 400 rupees (equivalent to 2.48 to 3.30AUD). As the flood victims had to shift in relief camps in city Layyah from their communities. The males had started working in brick kilns and in block making local industries on daily wages during their stay in relief camps so that they could get enough savings to feed their children after leaving flood relief camps in Layyah city. The major occupation of males and females in flood-affected communities is mentioned in Section 6.1 above. As shown in Figure 13 men, women, and children are harvesting sugar-cane crops near the river, while in Figure 14, an old man is cutting the crop. Holling defined resilience as the capacity of the system to endure tough situations and manage the changes and perturbations such that the relationship between the state and the community remains undisturbed [15].

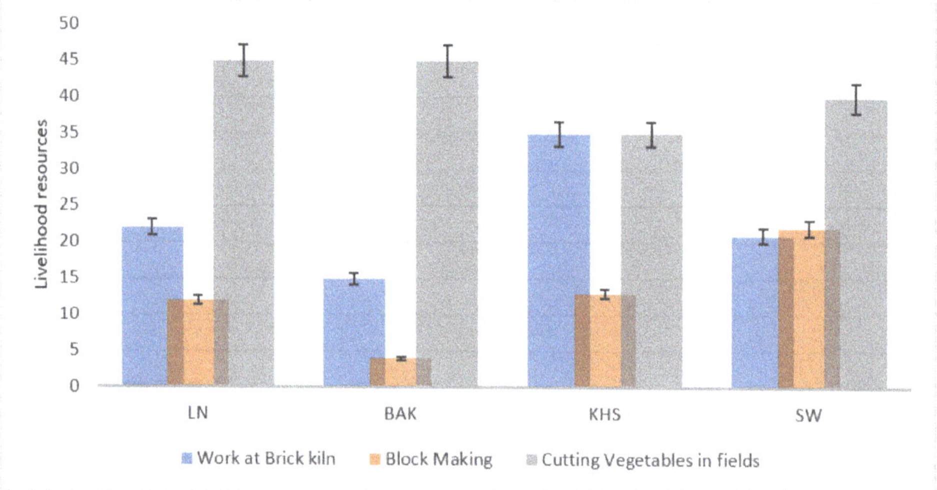

Figure 12. Major livelihood sources in flood-affected areas of district Layyah.

Figure 13. Men, women and children are cutting sugar-cane crop near the river in LN, district Layyah.

Figure 14. A man cutting crop on daily wages in SW, District Layyah.

6.2.4. Infrastructure Resilience

Figure 15 below shows the infrastructure protection practices in the case study area. Two practices were found to be common that was protection bund around houses and raised platform of houses. In SW, 75% population reported that they had protection bund/embankment around their community, while in LN, 81% of respondents reported that they had raised platforms of their houses. Even though this practice is not sustainable, however, contributes towards infrastructure resilience in localities of riverine area. Figure 16 shows the practice and construction of houses at a raised level which is considered flood resilient in the community. As resilience is also defined by Bruijn and Klijn in terms of flood risk management, and it focuses on the minimization of impacts by living with floods instead of fighting with them [15]. One form of resilience is 'process-related resilience. This is developed utilizing a long and continuous process of learning about increasing the capacity for managing disasters. This is considered highly important at the level of communities as the mitigation and relief process largely rely upon the resilience of the affected populations. Moreover, the hazard mitigation strategies and increased capacity are related to the awareness and resilience of the people [45]. Ahmed and Afzal [46] reviewed the advanced adaptation measures for improving building resilience against floods. Four mitigation strategies were widely practiced and supported by the government i.e., using reinforced material for building houses, building houses on the elevated ground floor, strengthening the foundation, and precautionary savings to overcome future uncertainties. The government needs to focus on developing strategies for adapting advanced measures, spatial planning, and improved practices for infrastructure building.

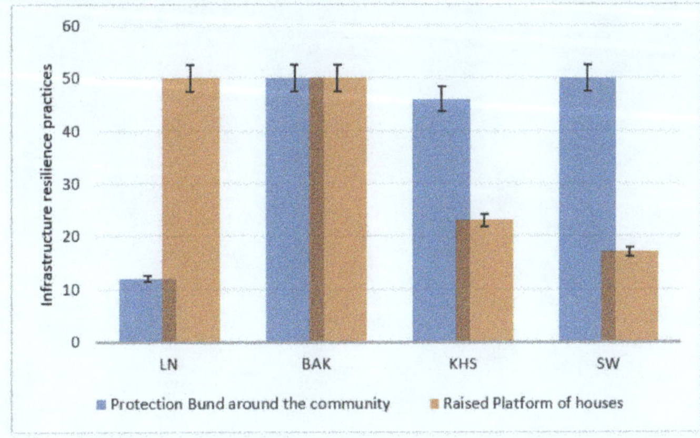

Figure 15. Infrastructure resilience practices in flood-affected communities in district Layyah.

Figure 16. Practice of raised plinth level in flood-prone communities of LN, district Layyah.

6.2.5. Comparison of Resilience Indices

Based on the collected data resilience, indices were calculated of the households across the four severely affected communities of LN, BAK, KHS, and SW (Figure 17). Four components of resilience i.e., social, physical, economic, and institutional resilience were calculated. Social resilience depicted the ability of the community to deal with the flood risk and was based on different social variables such as education, past flood experience, and the social network of the community. The economic resilience looked at the livelihood sources within the community. While institutional resilience considered the support provided by the government and NGOs, recovery, rehabilitation, and restoring livelihood. Physical or infrastructure resilience was based on variables such as the raised platform of houses and flood protection bund around the community. From the findings, it was revealed that the BAK, KHS, and SW were less vulnerable to floods, as higher infrastructure and social resilience were observed for these communities. Also, these communities were economically resilient as compared to LN. LN was found to be socially resilient, having a high index of 0.5.

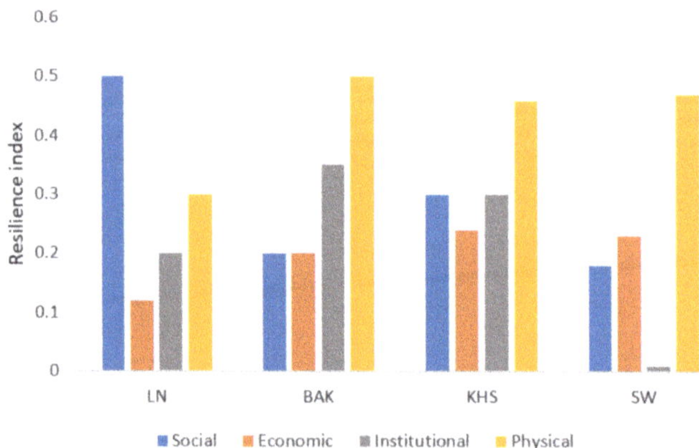

Figure 17. Comparison of resilience indices among the districts.

6.2.6. Satisfaction Level of Communities towards Government and Organization Role

Comparison of Satisfaction Level against Emergency Relief in 2010 Flood between Government and NGOs

The emergency services and response plans are the variables for the institutional Dimension of community resilience, as stated by [43,45]. Table 2 below represents the findings of a comparison of satisfaction level between government and NGOs against emergency relief received by the affected community of 2010 flood in four localities of district Layyah using Yeh's Index of Satisfaction [47]. The results show that respondents had a very low satisfaction level in SW (−1.56) and KHS (−1.76) for the government. This indicates that these two areas did not get any emergency relief from the government. In SW and KHS, flood victims were overlooked due to the negligence of local government and local representatives of areas. As SW is in Tehsil Karor and KHS is adjacent to it, so both union councils could not get emergency relief as the focus of government and NGOs were Tehsil Layyah being the main city in tehsil Layyah. Although in LN and BAK, the values of the index were 1.54 and 1.62, which showed that the satisfaction level was very high due to political biases. In these two localities, the community that experienced the 2010 flood reported vast destruction. LN is the nearest union council from the city of Layyah. While in BAK, political persons had a strong connection with district disaster management authority (DDMA) Layyah.

Table 2. Comparison of satisfaction level against emergency relief in 2010 floods between Government and NGOs.

	Localities	For Government		For NGOs	
		Value of Index	Satisfaction Level	Value of Index	Satisfaction Level
Factor: Emergency Relief	LN	1.54	Very High	1.70	Very High
	BAK	1.62	Very High	−1.70	Very Low
	KHS	−1.72	Very Low	−1.72	Very Low
	SW	−1.56	Very Low	−0.54	Very Low

On the other side, the satisfaction level for NGOs in BAK, KHS, and SW was very low as the value of the index is below 0.2 in these three localities, which indicated that organizations did not play any part in relief works such as evacuation, rescue, and first aid. While in LN, the affected community responded positively against NGOs role in emergency relief activities during and after the 2010 flood event [48].

Comparison of Satisfaction Level against Response after 2010 Flood between Government and NGOs

Table 3 below depicts the comparison of results for the satisfaction level of response plans between government and NGOs in four localities of district Layyah. The findings show that values of the index were negative in LN (−1.72), KHS (−0.28), and SW (−1.32), indicating a very low satisfaction level of respondents for the government. Lack of political support and coordination between local government, NGOs, and international donors were the major reasons for low satisfaction within the community. The satisfaction level for the government was very high, with the value of the index being positive (1.78) in BAK. As the local representatives of this union council were active and had connections with disaster managing departments in district Layyah local government. So, political influence was the main reason for the high satisfaction level of the affected community. During the field survey, the community had shown favorable behavior for government long-term assistance in BAK, which included the issuance of Watan cards, installation of hand pumps, and flood protection bunds around the union council [49].

Table 3. Comparison of satisfaction level against flood response received after the 2010 floods between Government and NGOs.

	Localities	For Government		For NGOs	
		Value of Index	Satisfaction Level	Value of Index	Satisfaction Level
Factor: Response Plans	LN	−1.72	Very Low	0.00	Acceptable (Very low)
	BAK	1.78	Very High	−1.70	Very low
	KHS	−0.28	Very Low	−0.44	Very Low
	SW	−1.32	Very Low	−0.62	Very Low

Contrary to this, the satisfaction level for NGOs in three localities was found very low as the index values were negative. For LN, the value of the index is 0.0, which is acceptable according to YIS. People criticized the NGOs support after the 2010 flood event. Many respondents informed that organizations helped them rebuilt their houses. However, after reconstructing their houses, there was no monitoring and evaluation system as the houses get cracked and were in critical conditions.

7. Conclusions

From the finding of the systematic literature review, it can be concluded that in Pakistan, disaster preparedness is higher for floods than other disaster events. The authorities continuously monitor and provide information on any emergency crisis. However, these authorities lack the technical skill and equipment to deal with large-scale disaster events such as the 2010 floods. There is a need to develop the capacity to respond to the emergency crisis, effectively relocate, rescue, and rehabilitate the victims in the affected regions. It has been observed that delayed response of government and NGOs during crisis further deteriorated the condition of the victims. Lack of clean drinking water, food, and shelter increased the suffering of the community. Despite having frequent floods proper awareness and education programs are not given to the people to prepare themselves for the disaster. Lack of coordination between government and NGOs is evident. The reported humanitarian relief work carried out is different from what is being carried out. The community was not satisfied with the relief response, reconstruction of houses by NGOs, and financial assistance provided by the government. The authorities at the district and sub-district level are unable to implement medium to long-term emergency plans and interventions to mitigate the floods [47–49].

It can be concluded from the case study that the socioeconomic status of people living in the district of Layyah was found to be low and most people worked as laborers and on the farms. After the floods, 74% population reported that they had to rebuild their house, and the support was mostly provided by NGOs instead of the Government. The local indigenous practices such as using tractor tubes and Sandhari for crossing over floodwaters were used. Houses were built on raised platforms and protection bund/embankments were built around their community as a protection against floods. The satisfaction level of the community for flood response by the Government and NGO was found to be high in LN and BAK, while very low satisfaction level was evident in SW and KHS. The flood risk management can be upgraded by integrating community resilience. The BAK, KHS, and SW community can fundamentally restore their living status and curtail long-run vulnerability using their local resources in addition to government and NGOs support. These three communities were considered as social and infrastructural resilient. The concept of community resilience to disasters has gained significant impact over the last decade, which could help policymakers and practitioners to identify the strengths and vulnerabilities of populations endangered by floods [34]. Although LN is not infrastructurally and economically resilient, however, people of LN were found to be socially resilient. There was no planning and coordination between Government and NGOs for flood risk management, due to which the satisfaction level of affected communities in all four UCs towards the government's emergency relief and rescue was found to be very low. In BAK, the satisfaction level was found to be very high due to the government's long-term assistance as compared to the other three localities which depict the bias of government officials. While people were not satisfied with organizations long term assistance, NGOs helped in reconstruction and resilience, the building of community, but there was no monitoring and evaluation system. The national and international NGOs come forward with projects only during a disaster event and leave right after the completion of their projects [50–52].

8. Recommendations

The most important measure of increasing flood resilience in a community is to increase the level of awareness and train the locals so that they are well prepared to deal with floods. Some of the preparations include a supply of sandbags, building the houses on an elevated platform, avoiding storage of food in basements, evacuation plans, being able to recognize the early warnings, and knowing the emergency procedures [53]. NGOs have limited time and budget so the Government should focus on the training and mock drills to the flood-prone community for first aid and emergency response, including evacuation. The study by [17] contributes towards tackling challenges and adds details about the opportunities and ways to promote resilience and truly bring it into practice. The findings revealed that the participation of all stakeholders and communities enhance the resilience against flood. They are well-prepared, have better awareness, and are quite knowledgeable about the risks, and respond in case of a flood [54,55]. Among the plausible explanation for these findings are resilience indicators in the context of flood risk management. The utilization of the different tools for flood management, like the management plans and the early warning systems, can act as catalysts towards increasing awareness and preparedness. Similarly, risk communication during an event falls under the domain of risk communication and perception. Moreover, institutional cooperation and coordination, preparation of emergency services, and spatial planning are primary indicators for the resilience of policies and institutions [17,56].

There is a need to build a proper channel of communication between line departments, NGOs, and the community. Due to a lack of education and awareness, people cannot differentiate between the understanding of their rights and obligation [57]. People give less time for training and awareness programs, so NGOs provides them with a daily wage to compensate for their loss. For example, the Bangladesh Red Crescent Society in Bangladesh has a community-based disaster preparedness scheme in coastal areas of the Cox Bazar district [58–60]. The objective of the program is to strengthen the self-help capability of vulnerable communities and the major initiative is the establishment of Village Disaster Preparedness Committees. Similarly, after the 1998 floods, Care Bangladesh took many community-based initiatives [61]. This includes "floodproofing" by food for work to support the community. The works include plinth level upraising of the house for five-year flood stratum, schools to 20-year flood level, and village level safety plans that enable to meet emergency relief services [47,59–61]. Furthermore, policymaking and risk analysis help in understanding and predicting the responses of the public to flood hazards by improving communication among the locals, the professionals, and all the decision-makers [62]. Communication includes spreading awareness among people and to increase their preparedness for dealing with disasters. Awareness and communication should mainly include correct and up to date information about the risks of floods during the crisis, announcing alerts, and making decisions during emergencies as it is clear that the perception of risk stems from communication about the risks and determines how the locals will apprehend these risks [63–65]. The government should build flood protection bunds, and NGOs should focus on gender sensitivity issues for DRR training in the local community. Flash floods are the main cause of flooding in district Layyah. There is a need to focus on community awareness and periodic sessions on WASH (Water Sanitation and Hygiene), DRR (Disaster Risk Reduction), environment, and health. There is also a need for updated flood maps of the riverine area by a special task committee [65–70]. In Manipur and Rajasthan states, legislation for flood plain zoning has been ordained with teamwork of the National Natural Resources Management System (NNRMS) in1999 by the Ministry of Water Resources. To enable appropriate flood zoning, the readiness of survey maps on a large scale is required. So, about 55,000 km^2 flood plains were surveyed, and 570 maps were composed. The government should take steps to lessen the soil erosion in the riverine delta. The process of soil erosion is still ongoing in two union councils of the district named LN and BAK. Provincial Disaster Management Authority (PDMA) did not appoint any representative in the district since 2011 [71,72]. To make DRR effective government should adopt a strategy that can give a permanent solution like the regulation of Lala Kareek with fixed equipment. Need assessment and resource planning are very crucial. Government and organizations should focus on preparedness, early recovery, and livelihood sources of the affected community. Modern cutting-edge technologies including image processing, machine learning, and AI can be used to deal with disaster situations and save disastrous victims [73–75]. A community should follow the seasonal calendar so that crop damages can be reduced. The government should appoint proper staff for the management of relief camps instead of patwari (local representative) and teachers. Contingency planning technique at the district level might improve the adequacy of emergency relief but, it also displays considerable needs in terms of efficiency in response plans and flood risk management. There must be a contingency plan at the union council and village level for flood risk management, and guidelines must be followed by community and line departments.

Author Contributions: Conceptualization, H.S.M., N.A., Z.Q. and S.I.K.; methodology H.S.M., N.A. and S.I.K. and software, H.S.M., Z.Q.; validation, S.I.K.; formal analysis, N.A., Z.Q. investigation, H.S.M., Z.Q. and N.A. resources, A.Z.K.; data curation, M.A.P.M.; writing—original draft preparation, H.S.M., N.A. and S.I.K. funding acquisition, M.A.P.M. All authors have read and agreed to the published version of the manuscript.

Funding: This research received no external funding.

Institutional Review Board Statement: Not applicable.

Informed Consent Statement: Not applicable.

Data Availability Statement: Not applicable.

Conflicts of Interest: The authors declare no conflict of interest.

References

1. UNISDR. *UNISDR Terminology on Disaster Risk Reduction*; United Nations International Strategy for Disaster Reduction (UNISDR): Geneva, Switzerland, 2009.
2. Modgil, S.; Singh, R.K.; Foropon, C. Quality management in humanitarian operations and disaster relief management: A review and future research directions. *Ann. Oper. Res.* **2020**, 1–54. [CrossRef]
3. Quarantelli, E. Ten Criteria for Evaluating the Management of Community. *Disasters* **1997**, *21*, 39–56. [CrossRef] [PubMed]
4. Cheema, A.R.; Mehmood, A.; Imran, M. Learning from The Past: Analysis Of Disaster Management Structures, Policies And Institutions In Pakistan. *Disaster Prev. Manag.* **2016**, *25*, 449–463. [CrossRef]
5. Prasad, N.; Ranghieri, F.; Shah, F.; Trohanis, Z.; Kessler, E.; Sinha, R. *Climate Resilient Cities: A Primer on Reducing Vulnerabilities to Disasters*; World Bank Publications: Washington, DC, USA, 2019.
6. Restemeyer, B.; Woltjer, J.; Van Den Brink, M. A Strategy-Based Framework for Assessing the Flood Resilience of Cities–A Hamburg Case Study. *Plan. Theory Pract.* **2015**, *16*, 45–62. [CrossRef]
7. Velasco, M.; Cabello, À.; Russo, B. Flood damage assessment in urban areas. Application to the Raval district of Barcelona using synthetic depth damage curves. *Urban Water J.* **2014**, *13*, 426–440. [CrossRef]
8. Jha, A.K.; Bloch, R.; Lamond, J. *Cities and Flooding: A Guide to Integrated Urban Flood Risk Management for the 21st Century*; World Bank Publications: Washington, DC, USA, 2012.
9. Schneidergruber, M.; Cierna, M.; Jones, T. *Living with Floods: Achieving Ecologically Sustainable Flood Management in Europe*; WWF European Policy: Brussels, Belgium, 2004.
10. Bănică, A.; Kourtit, K.; Nijkamp, P. Natural disasters as a development opportunity: A spatial economic resilience interpretation. *Rev. Reg. Res.* **2020**, *40*, 223–249. [CrossRef]

11. Chan, N.W.; Ab Ghani, A.; Samat, N.; Hasan, N.N.N.; Tan, M.L. Integrating Structural and Non-Structural Flood Management Measures For Greater Effectiveness in Flood Loss Reduction in the Kelantan River Basin Malaysia. In Proceedings of the AICCE'19, Penang, Malaysia, 21–22 August 2019; Springer International Publishing: Cham, Switzerland, 2020.

12. Di Nardo, M.; Clericuzio, M.; Murino, T.; Madonna, M. An Adaptive Resilience Approach for a High Capacity Railway. *Int. Rev. Civ. Eng. IRECE* **2020**, *11*, 98. [CrossRef]

13. McClymont, K.; Morrison, D.; Beevers, L.; Carmen, E. Flood resilience: A systematic review. *J. Environ. Plan. Manag.* **2020**, *63*, 1151–1176. [CrossRef]

14. Zevenbergen, C.; Gersonius, B.; Radhakrishan, M. *Flood Resilience*; The Royal Society Publishing: London, UK, 2020.

15. Vis, M.; Klijn, F.; De Bruijn, K.; Van Buuren, M. Resilience strategies for flood risk management in the Netherlands. *Int. J. River Basin Manag.* **2003**, *1*, 33–40. [CrossRef]

16. Rubinato, M.; Nichols, A.; Peng, Y.; Zhang, J.-M.; Lashford, C.; Cai, Y.-P.; Lin, P.-Z.; Tait, S. Urban and river flooding: Comparison of flood risk management approaches in the UK and China and an assessment of future knowledge needs. *Water Sci. Eng.* **2019**, *12*, 274–283. [CrossRef]

17. Schelfaut, K.; Pannemans, B.; van der Craats, I.; Krywkow, J.; Mysiak, J.; Cools, J. Bringing flood resilience into practice: The FREEMAN project. *Environ. Sci. Policy* **2011**, *14*, 825–833. [CrossRef]

18. Bharwani, S.; Magnuszewski, P.; Sendzimir, J.; Stein, C.; Downing, T.E. Vulnerability, Adaptation and Resilience: Progress Toward Incorporating VAR Concepts into Adaptive Water Resource Management. In *Report of Newater Projects–News Approaches to Adaptive Water Research Management under Uncertainty*; Stockholm Environment Institute: Stockholm, Sweden, 2008.

19. Krywkow, J.; Filatova, T.; Van Der Veen, A. Flood Risk Perceptions in The Dutch Province of Zeeland: Does the Public Still Support Current Policies? Floodrisk 2008. In Proceedings of the European Conference on Flood Risk Management, Oxford, UK, 30 September–2 October 2008.

20. Martinez, M.B.R.; Akib, S. Innovative Techniques in The Context of Actions for Flood Risk Management: A Review. *Eng* **2021**, *2*, 1–11. [CrossRef]

21. Liao, K.-H. A Theory on Urban Resilience to Floods-A Basis for Alternative Planning Practices. *Ecol. Soc.* **2012**, *17*. [CrossRef]

22. Forrest, S.; Trell, E.; Woltjer, J. Civil society contributions to local level flood resilience: Before, during and after the 2015 Boxing Day floods in the Upper Calder Valley. *Trans. Inst. Br. Geogr.* **2019**, *44*, 422–436. [CrossRef]

23. Coates, T. Understanding Local Community Construction Through Flooding: The 'Conscious Community'and The Possibilities for Locally Based Communal Action. *Geo Geogr. Environ.* **2015**, *2*, 55–68. [CrossRef]

24. Djordjević, S.; Butler, D.; Gourbesville, P.; Mark, O.; Pasche, E. New policies to deal with climate change and other drivers impacting on resilience to flooding in urban areas: The CORFU approach. *Environ. Sci. Policy* **2011**, *14*, 864–873. [CrossRef]

25. Thaler, T.; Nordbeck, R.; Löschner, L.; Seher, W. Cooperation in flood risk management: Understanding the role of strategic planning in two Austrian policy instruments. *Environ. Sci. Policy* **2020**, *114*, 170–177. [CrossRef]

26. Lingard, L.; Albert, M.; Levinson, W. Grounded theory, mixed methods, and action research. *BMJ* **2008**, *337*, a567. [CrossRef]

27. Shah, A.A.; Ye, J.; Abid, M.; Khan, J.; Amir, S.M. Flood hazards: Household vulnerability and resilience in disaster-prone districts of Khyber Pakhtunkhwa province, Pakistan. *Nat. Hazards* **2018**, *93*, 147–165. [CrossRef]

28. Hussain, M.; Anwar, S.A.; Sehar, S.; Zia, A.; Kamran, M.; Mehmood, S.; Ali, Z. Incidence of plant-parasitic nematodes associated with okra in district Layyah of the Punjab, Pakistan. *Pakistan J. Zool.* **2015**, *47*, 3.

29. Baig, S.A.; Xu, X.; Navedullah, M.N.; Khan, Z.U. Pakistan's Drinking Water and Environmental Sanitation Status In Post 2010 Flood Scenario: Humanitarian Response And Community Needs. *J. Appl. Sci. Environ. Sanit.* **2012**, *7*, 49–54.

30. Glenn, D. *Determining Sample Size. A Series of the Program Evaluation and Organizational Development*; University of Florida: Gainesville, FL, USA, 1992.

31. Haider, N. *Living with Disasters: Disaster Profiling of Districts of Pakistan*; National Disaster Management Authority Islamabad: Islamabad, Pakistan, 2006.

32. Tariq, M.A.U.R.; Van de Giesen, N. Floods and flood management in Pakistan. *Phys. Chem. Earth Parts A/B/C* **2012**, *47*, 11–20. [CrossRef]

33. Munawar, H.S. Flood Disaster Management. *Mach. Vis. Insp. Syst.* **2020**, *1*, 115–146.

34. Khan, D.L.N.M.A. *District Disaster Management Plan*; Indian Chemical Society: Gujarat, India, 2020; p. 76.

35. Walters, P. The problem of community resilience in two flooded cities: Dhaka 1998 and Brisbane 2011. *Habitat Int.* **2015**, *50*, 51–56. [CrossRef]

36. Aldrich, D.P. *Building Resilience: Social Capital in Post-Disaster Recovery*; University of Chicago Press: Chicago, IL, USA, 2012.

37. Bankoff, G. Dangers to going it alone: Social capital and the origins of community resilience in the Philippines. *Contin. Chang.* **2007**, *22*, 327–355. [CrossRef]

38. Patterson, O.; Weil, F.; Patel, K. The Role of Community in Disaster Response: Conceptual Models. *Popul. Res. Policy Rev.* **2010**, *29*, 127–141. [CrossRef]

39. Putnam, R.D. *Bowling Alone: The Collapse and Revival of American Community*; Simon And Schuster: New York, NY, USA, 2000; p. 541. ISBN 0-684-83283-6.

40. Norris, F.H.; Stevens, S.P.; Pfefferbaum, B.; Wyche, K.F.; Pfefferbaum, R.L. Community Resilience as a Metaphor, Theory, Set of Capacities, and Strategy for Disaster Readiness. *Am. J. Community Psychol.* **2008**, *41*, 127–150. [CrossRef]

41. Li, W.; Xu, B.; Wen, J. Scenario-based community flood risk assessment: A case study of Taining county town, Fujian province, China. *Nat. Hazards* **2016**, *82*, 193–208. [CrossRef]
42. Munawar, H.S.; Hammad, A.; Ullah, F.; Ali, T.H. After the Flood: A Novel Application of Image Processing and Machine Learning For Post-Flood Disaster Management. In Proceedings of the International Conference on Sustainable Development in Civil Engineering (ICSDC 2019), Jamshoro, Pakistan, 5–7 December 2019.
43. Noonan, D.S.; Liu, X. Heading for the Hills? Effects of Community Flood Management on Local Adaptation to Flood Risks. *South. Econ. J.* **2019**, *86*, 800–822. [CrossRef]
44. Mustafa, D. To Each According to his Power? Participation, Access, and Vulnerability in Irrigation and Flood Management in Pakistan. *Environ. Plan D Soc. Space* **2002**, *20*, 737–752. [CrossRef]
45. Ahmad, D.; Afzal, M. Flood hazards and factors influencing household flood perception and mitigation strategies in Pakistan. *Environ. Sci. Pollut Res.* **2020**, *27*, 15375–15387. [CrossRef] [PubMed]
46. Cutter, S.L.; Barnes, L.; Berry, M.; Burton, C.; Evans, E.; Tate, E.; Webb, J. A place-based model for understanding community resilience to natural disasters. *Glob. Environ. Chang.* **2008**, *18*, 598–606. [CrossRef]
47. Islam, E.; Wahab, H.A.; Benson, O.G. Structural and operational factors as determinant of meaningful community participation in sustainable disaster recovery programs: The case of Bangladesh. *Int. J. Disaster Risk Reduct.* **2020**, *50*, 101710. [CrossRef]
48. Tierney, K.; Bruneau, M. Conceptualizing and Measuring Resilience: A Key to Disaster Loss Reduction. *TR News* **2007**, *17*, 14–15.
49. Yeh, S.H. *Homes for the People: A Study of Tenants' Views on Public Housing in Singapore. 1972*; US Government Printing Office: Washington, DC, USA, 1972.
50. Yodmani, S. *Disaster Risk Management and Vulnerability Reduction*; Springer International Publishing: New York, NY, USA, 2001.
51. Mohapatra, P.K.; Singh, R.D. Flood Management in India. *Nat. Hazards* **2003**, *28*, 131–143. [CrossRef]
52. Ullah, F.; Saqib, S.E.; Ahmad, M.M.; Fadlallah, M.A. Flood risk perception and its determinants among rural households in two communities in Khyber Pakhtunkhwa, Pakistan. *Nat. Hazards* **2020**, *104*, 225–247. [CrossRef]
53. Rufat, S.; Tate, E.; Burton, C.G.; Maroof, A.S. Social vulnerability to floods: Review of case studies and implications for measurement. *Int. J. Disaster Risk Reduct.* **2015**, *14*, 470–486. [CrossRef]
54. Schilling, J.; Vivekananda, J.; Khan, M.A.; Pandey, N. Vulnerability to Environmental Risks and Effects on Community Resilience in Mid-West Nepal and South-East Pakistan. *Environ. Nat. Resour. Res.* **2013**, *3*, 27. [CrossRef]
55. Siddiqi, A. The Emerging Social Contract: State-Citizen Interaction after the Floods of 2010 and 2011 in Southern Sindh, Pakistan. *IDS Bull.* **2013**, *44*, 94–102. [CrossRef]
56. Akbar, M.S.; Aldrich, D.P. Social capital's role in recovery: Evidence from communities affected by the 2010 Pakistan floods. *Disasters* **2018**, *42*, 475–497. [CrossRef]
57. Kurosaki, T.; Khan, H.; Shah, M.K.; Tahir, M. *Household-Level Recovery after Floods in a Developing Country: Further Evidence from Khyber Pakhtunkhwa, Pakistan*; Institute of Economic Research: Stockholm, Sweden, 2012.
58. Ahmad, M.I.; Ma, H. An investigation of the targeting and allocation of post-flood disaster aid for rehabilitation in Punjab, Pakistan. *Int. J. Disaster Risk Reduct.* **2020**, *44*, 101402. [CrossRef]
59. Azad, A. *Ready or Not: Pakistan's Resilience to Disasters One Year on from the Floods*; Oxfam: Oxford, UK, 2011; Volume 150.
60. Kosec, K.; Khan, H.; Taffesse, A.S.; Tadesse, F. Promoting aspirations for resilience among the rural poor. *Resil. Food Nutr. Secur.* **2014**, *11*, 91.
61. Ali, R.A.; Mannakkara, S.; Wilkinson, S. Factors affecting successful transition between post-disaster recovery phases: A case study of 2010 floods in Sindh, Pakistan. *Int. J. Disaster Resil. Built Environ.* **2020**, *11*. [CrossRef]
62. Hussain, M.; Butt, A.R.; Uzma, F.; Ahmed, R.; Irshad, S.; Rehman, A.; Yousaf, B. A comprehensive review of climate change impacts, adaptation, and mitigation on environmental and natural calamities in Pakistan. *Environ. Monit. Assess.* **2019**, *192*, 48. [CrossRef] [PubMed]
63. Cancan, M.; Mahsud, M.; Ullah, S.; Mahsud, Z. National space legislation: A dire need for Pakistan. *J. Stat. Manag. Syst.* **2020**, 1–11. [CrossRef]
64. Munawar, H.S. COVID-19 Community Mobility Reports. *IJWMT* **2020**, 33. Available online: https://arxiv.org/abs/2004.04145 (accessed on 12 November 2020).
65. Munawar, H.S. *Applications of Leaky-Wave Antennas: A Review*; MECS Press: Hong Kong, China, 2020.
66. Munawar, H.S.; Awan, A.A.; Khalid, U.; Maqsood, A. Revolutionizing Telemedicine by Instilling, H.265. *Int. J. Image Graph. Signal Process.* **2017**, *9*, 20–27. [CrossRef]
67. Munawar, H.S.; Awan, A.A.; Maqsood, A.; Khalid, U. Reinventing Radiology in Modern ERA. *I J. Wirel. Microw. Technol.* **2020**, *4*, 34–38.
68. Munawar, H.S.; Khalid, U.; Jilani, R.; Maqsood, A. Version Management by Time Based Approach in Modern Era. *Int. J. Educ. Manag. Eng.* **2017**, *4*, 13–20. [CrossRef]
69. Munawar, H.S. An Overview of Reconfigurable Antennas for Wireless Body Area Networks and Possible Future Prospects. *Int. J. Wirel. Microw. Technol.* **2020**, *10*, 1–8. [CrossRef]
70. Munawar, H.S.; Qayyum, S.; Ullah, F.; Sepasgozar, S. Big Data and Its Applications in Smart Real Estate and the Disaster Management Life Cycle: A Systematic Analysis. *Big Data Cogn. Comput.* **2020**, *4*, 4. [CrossRef]

71. Qadir, Z.; Al-Turjman, F.; Khan, M.A.; Nesimoglu, T. ZIGBEE Based Time and Energy Efficient Smart Parking System Using IOT. In Proceedings of the 2018 18th Mediterranean Microwave Symposium (MMS), Istanbul, Turkey, 31 October–2 November 2018; pp. 295–298. [CrossRef]
72. Qadir, Z.; Tafadzwa, V.; Rashid, H.; Batunlu, C. Smart Solar Micro-Grid Using ZigBee and Related Security Challenges. In Proceedings of the 2018 18th Mediterranean Microwave Symposium (MMS), Istanbul, Turkey, 31 October–2 November 2018; pp. 299–302. [CrossRef]
73. Munawar, H.S.; Zhang, J.; Li, H.; Mo, D.; Chang, L. *Mining Multispectral Aerial Images for Automatic Detection of Strategic Bridge Locations for Disaster Relief Missions*; Springer International Publishing: Cham, Switzerland, 2019; pp. 189–200.
74. Qadir, Z.; Ullah, F.; Munawar, H.S.; Al-Turjman, F. Addressing disasters in smart cities through UAVs path planning and 5G communications: A systematic review. *Comput. Commun.* **2021**, *168*, 114–135. [CrossRef]
75. Qadir, Z.; Khan, S.I.; Khalaji, E.; Munawar, H.S.; Al-Turjman, F.; Mahmud, M.P.; Kouzani, A.Z.; Le, K. Predicting the energy output of hybrid PV–wind renewable energy system using feature selection technique for smart grids. *Energy Rep.* **2021**. [CrossRef]

 *applied sciences*

Article

The Feasibility and Environmental Impact of Sustainable Public Transportation: A PV Supplied Electric Bus Network

Zakariya Dalala *, Omar Al Banna and Osama Saadeh

Energy Engineering Department, German Jordanian University, P.O. Box 35247, Amman 11180, Jordan; o.albanna@gju.edu.jo (O.A.B.); osama.saadeh@gju.edu.jo (O.S.)
* Correspondence: zakariya.dalalah@gju.edu.jo

Received: 1 April 2020; Accepted: 5 June 2020; Published: 8 June 2020

Featured Application: Transforming traditional public transportation systems into zero carbon sustainable systems.

Abstract: Limited fuel resources and the huge negative impact on the environment from using fossil fuels have led to an urgency to utilize the most energy efficient solutions for public transportation. Environmentally sustainable solutions can deliver the same benefits of traditional systems, but without the negative impacts. The Bus Rapid Transit Project of Amman (Amman BRT) is used as a case study. Proposed measures include using electric buses instead of diesel ones, and installing elevated photovoltaic systems above buses parking and routes, in addition to using LED street lighting. The feasibility study of applying the proposed measures on the Amman BRT project showed that only 7.1 years is needed to payback the incremental investment throughout this transformation. Capital expenditure (CAPEX) is higher than the baseline buses, while operational expenditure (OPEX) is much lower, resulting in a 32% lower total cost of ownership (TCO). In addition, greenhouse gas (GHG) emissions are reduced by 27,203.68 metric ton of CO_2 per year and 408,055.26 metric tons for the 15-year lifetime of the project.

Keywords: electric buses; public transportation; PV supplied sustainable transportation

---

1. Introduction

Implementing environmentally friendly systems and using renewable energy are not luxuries anymore. They are needed to sustain the planet's limited resources and maintain the environment. Transportation is one of the main sectors that depletes traditional fuel reserves, in addition to damaging the environment.

As per the World Bank Group database and Energy Information Administration (EIA) statistics, overall worldwide energy consumption is continuously increasing. The 2016, the International Energy Outlook predicted that the annual transportation sector energy consumption was increasing at an average rate of 1.4%, from 104 quadrillion British thermal units (Btu) in 2012 to 155 quadrillion Btu in 2040 [1–3].

As for the transportation sector, the Energy Information Administration (EIA) has released data showing that the transportation of people and goods accounts for approximately 25 percent of global energy consumption [2]. Passenger transportation, in particular, using light-duty vehicles, accounts for most of the transportation energy consumption. Therefore, establishing reputable public transportation systems is necessary, and using clean energy instead of fossil fuels within the nominated system is of extreme importance.

Depletion of non-renewable energy sources, such as fossil fuels, is not the only problem associated with using such fuels for transportation. Production of remarkable amounts of dreadful emissions that hurt the environment and inhabitants' health is a problem of equal significance. Twenty percent of global CO_2 emissions are caused by the transportation sector.

Streets occupy a large area of public land, which offers a unique opportunity to implement green technology fundamentals and elements that improve communities, the environment, and health. One such technology is using photovoltaics (PV) to generate the required energy to power the electric buses. PV generated energy is sustainable, as it is harvested directly from the sun. As PV technology installation investment costs have decreased, global deployment has rapidly increased. PV systems may be installed near the load, with no need for a fuel source or transmission lines, in essence promoting sustainable local generation. This is very attractive for public transportation, as the energy is consumed locally at the route. The drawback of PV energy is that it is dependent on environmental factors. Power generation depends on solar radiation and temperature. As these are random in nature, and cannot be controlled, the availability of solar generated energy is not as dependable as traditional fuel sources. In addition, PV systems do not generate electricity at night, and generation levels differ by season. To combat this effect, storage systems and grid integration schemes may be employed [4].

In this study, a net-metering scheme is assumed. In essence, this scheme uses a bidirectional energy meter. All energy generated by the PV system that is not consumed by the local load is injected into the grid, and the net consumption is billed at the end of the contract-defined period. Therefore, the PV system may be designed to produce enough energy during the day to cover the complete consumption, storing the night-required energy on the electric grid.

The authors in [5] discuss electric buses as a city transportation system with the main target of reducing the carbon footprint. However, the paper does not discuss the bus electrical source. While the authors in [6] focus on the optimization of public transportation electric buses' employment, targeting the minimum capital cost of the fleet, the charging demand and stations, and solving the scheduling problem of engaged electric buses. Hence, it focuses on optimizing the charging systems, with no analysis of the environmental effects of traditional buses and electric buses.

In [7], the total cost of ownership (TCO) for different electric buses and diesel buses is studied and outlines some real cases in different cities for using electric buses in public transportation. However, the paper does not discuss the implementation of renewable energy to supply the buses.

Regarding photovoltaic systems, much research discusses the feasibility of deployment for different uses and configurations. The authors of [8] present computer based design of off-grid solar photovoltaic systems, where the author describes each step in the design procedure that includes electrical load calculations, photovoltaic panel sizing and selection, storage system sizing and selection, inverter sizing, and specification in addition to the selection of the charging controller. The authors in [9] discuss the design of on grid PV systems. In [4], the author provides a comprehensive guide to solar photovoltaic systems, starting from the planning stage, design, specifications selection, and control technologies for photovoltaic off-grid system components, and ending with the implementation phase.

The research available in the literature focuses on electrical buses or on PV components and systems, but does not address implementing such systems together, to elevate the environmental and financial burdens of large-scale public transportation systems.

In this work, sustainable solutions are proposed and renewable energy systems are utilized to transform the transportation infrastructure and streets to environmentally friendly systems by utilizing available spaces to generate clean energy, in an innovative solution to reach the concept of the so-called "green streets." The main transportation project in Jordan, the Bus Rapid Transit Project of Amman, "Amman BRT," was used as the case study. Photovoltaic systems "PV" are a mature economic technology that uses solar energy to produce electricity. Abundance research, installations, projects, manufacturers, and interested associations around the world helped in making PV dependable and fully reliant technology. Therefore, PV is elevated as a main part of the green street methodology adapted in this manuscript, driven by huge incentives from governmental and financial donor programs. Electric vehicles' (EVs) proven reliability and has increased in market penetration recently due to the fast and diverse advancements in the field. Utilization of EVs is essential in modern green transportation systems.

2. Materials and Methods

2.1. Amman BRT

The Jordanian Ministry of transportation reports that 46% of energy used in Jordan is for the transportation sector [10]. As Jordan imports most of its energy needs, this is a major burden on the economy and national security.

In Amman, Jordan, the current public transportation networks consists of private taxis, shared taxis, buses, and mini-buses, which are managed by the public sector and 14,000 small operators. The supply of public transport is insufficient to cope with demographic and urban growth. This situation strongly constrains both inhabitants' mobility and economic growth, in addition to being harmful to the environment.

Amman Bus Rapid Transit project (Amman BRT) is owned by the Greater Amman Municipality (GAM), and aims to provide a quality, economic, car-competitive mass transit system that will attract the entire spectrum of Amman citizens, including car owners [11]. With around 25 km total length of networks, and dedicated bus lanes separated from other regular traffic, Amman BRT is expected to reduce the distance travelled using private vehicles by 85 million km per year, and to be used by 142 million passengers each year [12].

Routes are divided into two lines and will serve major destinations in the city. The first stage of Amman BRT will include 100 buses. Line 1 will employ 60 buses that travel 12 day-tours an hour in addition to 3 night-tours per hour. Line 2 will employ 40 buses that will travel 20 day-tours per hour and 5 night-tours per hour [13]; routes are shown in the Figure 1. The current design makes use of traditional diesel buses.

Figure 1. Amman Bus rapid transit (BRT) routs.

2.2. Transportation Network Energy Demand

Jordan employs a net-metering scheme for connecting PV generation to the grid. This allows one total renewable energy self-consumption, without the need to install any storage systems. All excess energy generated during the day, or during high generation summer months is injected to the grid and credited in kWh. At the end of every billing year, the net generation and consumption are calculated, and only excess energy drawn is billed. In essence, the electric grid is used as a free storage sink, but any excess energy at the end of the billing cycle is forfeit to the grid operator. Therefore, it is very important to correctly size the designed PV system, such that the annual generated energy matches actual annual energy demand. The first step in designing the PV system for AMMAN BRT is to determine the annual energy needs. Generated electricity shall be used to cover required energy for charging the electric buses, for charging LED streets lights, and for the service buildings' electricity.

To properly characterize the load demand, the electric buses that will be deployed must be selected, and transportation network routes, distance, and operation must be well defined. For proper comparison between business as usual and the proposed upgrade, buses of the same size and passenger capacity are selected. The 18-m BYD K11 electric bus was selected to replace the diesel buses. It is worth mentioning that smaller electric buses (12 m) are widely used in many public transportation projects, but the larger size 18-m electric buses are new to the market. The bus has a 652 kWh battery that requires three and half hours to charge, which is enough to run the bus throughout the day for the distance required by operation plans [14].

Although the capital expenditure (CAPEX) of the electrical buses is higher than that for diesel buses, the total cost of ownership (TCO) is very different for the two options, since the operational cost for electric buses is much lower due to less energy consumption, less moving parts, no oil change, and longer life span [15,16]. The total demand of the proposed BRT project along with all services was projected to be 32 GWh, as shown in Table 1.

Table 1. Transportation network energy demand.

Electric Energy Consumer	Energy Per Year
Electric Buses	17,313,480 kWh/year
Street Lighting	800,230 kWh/year
Stations and Service Buildings	10,908,570 kWh/year
Total Energy Demand	29,022,280 kWh/year
Transformers Losses	1%
Cables Losses	1%
Additional Capacity to Cover Yield Losses	8%
Design Grand Total	32,000,000 kWh/year

2.3. PV System Design

2.3.1. Shading Analysis

For an accurate PV system design, the proposed system is simulated using PVSYST [17]. Simulation results are only as accurate as the input data. Data related to solar radiation in Amman, including horizontal global irradiation, diffused irradiation, ambient temperature, and global incident radiation, were obtained from the Meteonorm database [18].

Due to the route being within city limits, mostly in business districts, the crucial input data is the shading information. Shading data wrtr obtained by actual readings using SunEye equipment. This enabled capturing the losses due to shading and high rising objects along the zones.

As the routes cross many different areas with different topography and buildings distribution, shading at each point on the route is not like any other points. This is the main reason for using the zoning system and calculating the shading for each zone as an average. The routes are divided into fifteen zones, as shown in Figure 2. The start and end of every zone were determined based on direction and length. This determines the studying each zone as a separate project for designing the PV systems.

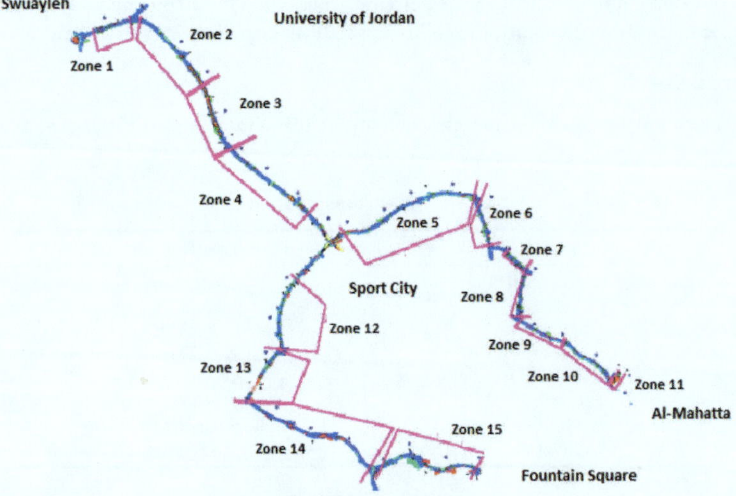

Figure 2. Zones of Amman BRT route.

According to this configuration, line 1 has zones 1 through 11, while line 2 includes zones 12 through 15. Some parts of the corridors were eliminated because of inconvenient conditions, such as intersections, bridges, tunnels, and many existing high buildings.

For each zone along the routes, several points were selected for data measurement, one point representing the worst possible shading in that zone where high buildings or obstacles exist. Another point represents the lowest possible shading in that zone. The remaining points cover the normal or average shading in the zone. The average shading for all points will be considered as the shading factor of this zone. Selection of highest and lowest shading points took place based on actual site visits.

Zone 2 was chosen to validate this approach, as it has the highest shading due to many existing commercial buildings in that zone. The shading analysis first started using 5 points (highest shade, lowest shade and 3 average points). Then, a new session for shading analysis started using 19 points; the difference in overall average annual solar access was 2%. This percentage is acceptable, especially knowing that zone 2 has the highest shading in comparison with the other zones, which means this difference will be lower over other zones. Figure 3 shows the location of data sampling in Zone 2.

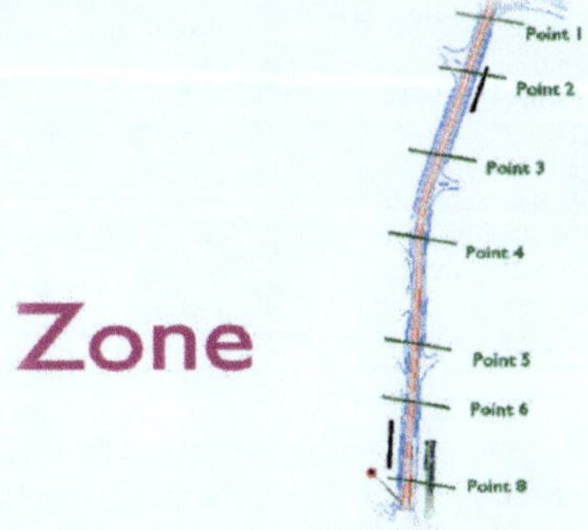

Figure 3. Location of data samples in zone 2.

Readings for shading and solar access were at 1.5–1.7 m elevation. This formulates an extra indirect safety factor, since the installation of PV panels will be much higher than this elevation. Thus, PV panels' access to solar radiation will be higher than what is used in the calculations. Therefore, results and outcomes of this study regarding electric generation from photovoltaic system will include a safety factor that affirms, in reality, the ability to produce more electricity than proposed.

2.3.2. Electrical Design

The geolocation specifications, electrical components, and designed tilt angle are depicted in Table 2.

Table 2. PV system design parameters.

Parameter	Specification
Geographical Site	Amman
Latitude	32.02° N
Longitude	35.85° E
Altitude	1022 m
Horizontal global irradiation	2063 kWh/m^2
Tilt	10°
Modules	Jinko Solar-JKM265P-60
Inverter	SMA Solar Technology-STP 60-10

The PV system was designed for each zone, taking into consideration length, shading, and mutual shading along the route. PVSYST was used to simulate the PV system and determine system capacity in each zone. The designed PV system comprises elevated PV modules along the path of the route with a tilt angle of 10° and an azimuth angle that corresponds to the route path. The azimuth angle of different PV sections is not optimum; however, it matches the route direction to take the overall scenery into consideration and to generate shading over the routes and pedestrian paths. Table 3 summarizes the resulting PV system design for each zone, where P_{nom} is the nominal power for PV system in the specific zone (in kw peak).

Table 3. Electrical design summary.

Zone No.	Equivalent Length (m)	Width (m)	Area (m^2)	Azimuth Ang.	Losses Due to Directions (from Optimum)	Far Shading Losses	Pnom (kWp)	Net Production (MWh/Yr)
1	345	9	3105	−19	−5.10%	−8.50%	501	843
2	1846	9	16,614	44	−6.90%	−30.70%	2688	3409
3	1388	9	12,492	62	−8.60%	−6%	2017	3395
4	1973	9	17,757	37	−6.40%	−5.60%	2874	4957
5	2974	9	26,766	−13	−5.30%	−21.50%	4250	6256
6	863	9	7767	67	−9.10%	−20.80%	1256	1772
7	655	9	5895	43	−6.90%	−11.30%	951	1533
8	893	9	8037	−87	−11.40%	−23.20%	1300	1756
9	1007	9	9063	16	−5.40%	−3.80%	1463	2616
10	1240	9	11,160	37	−6.40%	−10.80%	1804	2939
11	Mahatta Parking		33,120	38	−6.50%	−4.80%	5358	9292
12	1434	9	12,906	−68	−9.30%	−20.40%	2087	2967
13	1164	9	10,476	−55	−8.00%	−10%	1694	2745
14	2991	9	26,919	26	−5.80%	−4.40%	4358	7645
15	2365	9	21,285	−10	−5.20%	−7.10%	3446	5948
Total	21,138		223,362				36,047	58,073

As can be seen from Table 3, if the whole route is covered by PV systems, the generation greatly exceeds the system energy requirements. Some of the zones' yields are less than others due to shading, or direction, leading to zone-specific payback periods. Therefore, which zone to use must be carefully evaluated.

2.3.3. Economic Analysis

The payback period and net present value are the main aspects of the feasibility study in this research. Payback period is the time required to earn back the amount invested in an asset from its net cash flows. It is a direct method to evaluate the risk associated with a proposed project. An investment with a shorter payback period is considered to be better, since the investor's initial outlay is at risk for a shorter period of time. The payback period is expressed in years and fractions of years [19].

To make this method more accurate, one should convert all values to the present values, getting the net present values before applying the payback formulas. Furthermore, all capital costs and running costs shall be taken into consideration within the calculations. Total cost of ownership (TCO) will be highlighted as an indicator.

By identifying the number of years needed to payback the incremental investment in this transition, this research will highlight the economic feasibility of implementing the renewable system upgrade for the Amman BRT project. Therefore, the aim is to calculate the increase of capital cost, and the savings in the operational cost in order to calculate the payback period.

In order to calculate the net present value (NPV) for capital and operational amounts, the discount rate should be defined. The discount rate or real interest rate is dependent on the interest percentage and inflation factor. Fisher equation is used to calculate the discount rate.

2.3.4. Environmental Analysis

The approach to calculate CO_2 emissions is by multiplying estimated fuel volume by a default CO_2 emission factor. Table 4 shows road transportation default CO_2 emission factors for different fuel types [20].

Table 4. Transportation fuel default CO_2 emission factors.

Fuel Type	Emission Factor (kg/TJ)
Motor Gasoline	69,300
Gas/ Diesel Oil	74,100
Liquefied Petroleum Gases	63,100
Kerosene	71,900
Lubricants	73,300
Compressed Natural Gas	56,100
Liquefied Natural Gas	56,100

For the indirect emissions, the most significant source is electricity generation using fossil fuels in upstream processes, such as extraction, transport, refinery, and generation. However, here, as solar PV is proposed to produce the required energy for the system, the indirect emissions will only be taken into consideration for the baseline case. According to United Nations Framework Convention on Climate Change (UNFCCC), the default emission factor for upstream emissions (well-to-tank) is 16.7 (t CO_2e/TJ). For electricity generation emissions, according to the US Environmental Protection Agency (EPA), the US annual non-base load CO_2 output emission rate that is considered as an accurate "emission factor" for electrical consumption can be calculated using the rate (7.03×10^{-4} metric tons CO_2/kWh) [20]. Therefore, CO_2 emissions will be increased around 22.5–23% via indirect upstream emissions [21].

This research considers both direct and indirect (WTT, TTW) emissions caused by using the baseline buses in the business-as-usual scenario and electricity from the grid, focusing on CO_2 as a major pollutant. Then, it provides the difference in emissions upon implementing the proposed upgrade. Table 5 shows the emissions produced in the business-as-usual scenario.

Table 5. Environmental impact of baseline buses.

Item	Emission
Emission Factor of Diesel	74,100 kg/TJ
Quantity of Consumed Diesel	5,818,447 L
Diesel Calorific Value	43 MJ/kg = 0.000043 TJ/kg
Direct CO_2 Emission	15,424,713 kg CO_2
WTT (Well-to-Tank)–Indirect Emissions	23% including black carbon
Total Carbon Emissions Due to Diesel (Well-to-Wheel)	18,972,398 kg CO_2

3. Results

3.1. Overall System Design

As stated in the previous section, it is very important to properly select the zones that will be used for the PV systems. Several factors must be used in determining the optimal combination of zones. The zone score should reflect the priority of implementation. The main parameter to focus on is the required energy (E_R), where the optimum value is that the generated energy exactly matches the projected load demand. Higher or lower energy generation will be penalized equally, since the system is not paid back for any additional energy generated as per the net metering regulations. Minimizing the losses (L) is very important to optimizing the infrastructure utilization. Specific yield (SC) quantifies the ratio of the energy generated to the installed capacity, and it is affected by various design parameters, such as the height of the installation, inclination angle, and projected dust accumulation. But it is confined with the constraints of the site nature, and thus lower weight is assigned but is considered valuable in the score equation. System capacity (SC) is considered as the lowest weighted factor as it influences the initial investment only. Changing the factors might change the priorities of implementation and changing the weights might influence the zone score as well, but that would not change the order of zones significantly. Regardless of the order, it is a guide for the designer to focus on certain zones to meet the projected load demand. A factor-weighted equation is used to score the different combinations in Equation (1).

$$\text{Score} = 10\% \text{ SC} + 40\% \text{ } E_R + 30\% \text{ L} + 20\% \text{ SY,} \tag{1}$$

SC: system capacity factor, the value will be 1 if it meets or exceeds the required capacity.

RE: required energy, the value will be 1 if it exactly meets the required Energy, while it will be less for scenarios which provide more or less energy than the required energy.

OL: overall average loses due to shading and direction. The value will be 1 for the lowest losses scenario, while it will be less for scenarios with higher losses.

SY: specific yield which is the energy to power ratio. The value will be 1 for the highest ratio scenario, while it will be less for scenarios with smaller ratio values.

The weights for the different factors represent site and project specific priorities. The weight may be changed based on the priorities that best serve implementation.

As quite a large number of combinations may be used, six scenarios are defined as summarized in Table 6. The scenarios follow two main themes: utilizing the lowest shading-induced losses zones or utilizing main areas close to the major load demand. The first scenario considers all zones along the routes. Hence, it is a very big area and possibly able to contain large PV systems with more energy generation capacity than needed. Although a huge amount of energy can be generated in this scenario, losses due to direction and shading are tremendous. In scenario two, efficiency is increased by eliminating PV systems in zones with shading losses higher than 20%. In defining scenarios, shading losses were more important than losses due to the direction of the installed PV system. It is noticed that in this scenario, zones 2, 5, 6, 8, and 12 were excluded from the calculation. Therefore, the energy produced overall is less than in scenario one, but still covers more than the required energy for the project. In the third scenario, only zones that have 10% shading losses or lower are considered. In this case, zones 1, 3, 4, 9, 11, 14, and 15 are still within the group. As a result, the efficiency will be higher due to removing zones with higher losses. Zones such as 2, 5, 6, 7, 8, 10, 12, and 13 were not included in the calculation in this scenario. Less energy is generated in the third scenario (34,696.00 MWH/Year), but it still covers the needs of the proposed systems. Scenario 4 is tackling the highest efficiency by incorporating the lowest losses zones. This includes zones 4, 9, 11, and 14. Despite the efficiency being at higher limits in this scenario, it excludes most of the zones from the calculations. The result is a lower amount of energy that could be insufficient to cover the project needs. Scenario five is selecting zones which are close to parking areas. These areas are large and close to the energy consuming systems. In scenario 6, zones next to Mahatta parking only are considered. That is because Mahatta parking is a potential area for BRT buses to park overnight. Therefore, most of the consumption is expected to be there. It can be seen that energy produced is less than what the green streets model requires. The score for each scenario is depicted in Figure 4 by utilizing Equation (1), and the results of each zone's specifications in Table 3.

Table 6. Scenario definitions.

Scenario Number	Scenario Definition	Equivalent Length (m)	Area (m^2)	Pnom (kWp)	Net Production (MWH/Yr)
1	Using all zones.	21,138	223,362	36,047	58,073
2	all zones with shading losses below 10%	13,128	151,272	24,466	41,913
3	all zones with shading losses below 6%	10,069	123,741	20,017	34.696
4	only zones next to parking areas	5971	86,859	14,053	24,510
5	only zones next to main bus parking	5684	84,276	13,555	22,339
6	all zones with shading losses below 10%	2247	53,343	8625	14,847

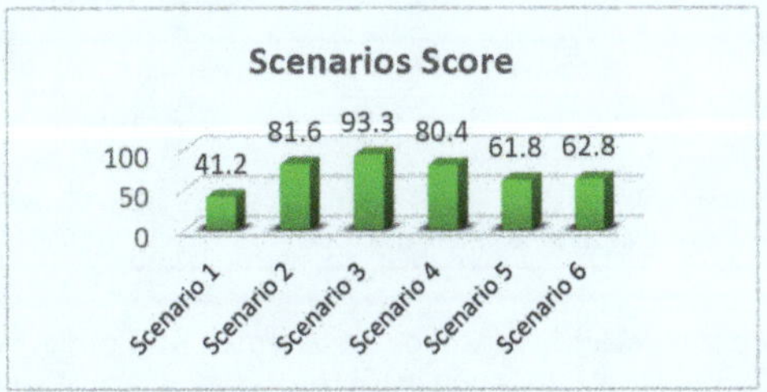

Figure 4. Scenario scores.

Based on these results, scenario 3 has the highest score of (93.3/100), making this scenario the best candidate. Scenario 3 meets capacity requirements; has lower losses than scenarios 1, 2, 5, and 6; and has a high specific yield. Scenario 3 suggests installing PV panels with an overall capacity of 20,017.00 kWp in order to generate 34,696,000.00 kWh per year. This energy is slightly above the required energy needed to run the green street model (32,000,000.00 kWh/Year); therefore, this scenario covers and exceeds energy demand with a slight indirect safety factor. Table 7 lists the details of scenario 3 design.

Table 7. Details of scenario number 3.

Zone No.	Equivalent Length (m)	Width (m)	Area (m^2)	Azimuth Ang.	Losses Due to Directions (From Optimum)	Far Shading Losses	Pnom Kwp	Net Production (MWh/Yr)
1	345	9	3105	−19	−5.10%	−8.50%	501	843
3	1388	9	12,492	62	−8.60%	−6%	2017	3395
4	1973	9	17,757	37	−6.40%	−5.60%	2874	4957
9	1007	9	9063	16	−5.40%	−3.80%	1463	2616
11	Mahatta Parking		33,120	38	−6.50%	−4.80%	5358	9292
14	2991	9	26,919	26	−5.80%	−4.40%	4358	7645
15	2365	9	21,285	−10	−5.20%	−7.10%	3446	5948
Total	10,069		123741				20,017	34,696

3.2. Environmental Analysis

One of the main targets of this research is to reduce green house gas (GHG) emissions from transportation. This is based on overall CO_2 emissions. Transportation equipment produces direct greenhouse gas emissions of carbon dioxide (CO_2), methane (CH4), and nitrous oxide (N2O) from the combustion of various fuel types, and several other pollutants, such as carbon monoxide (CO), non-methane volatile organic compounds (NMVOCs), sulfur dioxide (SO_2), particulate matter (PM), and oxides of nitrate (NOx), which cause or contribute to local or regional air pollution. However, UNFCCC advises that N2O emissions from transportation are very limited, and CH4 is a major component in case of gas engines. Hence, this study included only CO_2 emissions [20].

Table 8 summarizes the environmental impact of electrical consumption in the bassline scenario. Table 9 summarizes the overall environmental impact of the traditional baseline without the proposed upgrade. On the other hand, the proposed approach offers a reduction of emissions, by using electric buses instead of diesel buses and using PVs to produce electricity from solar energy to cover the required electric demand. As long as the scenario that is chosen for implementation provides all the required energy, all GHG emission from the baseline will be eliminated.

Table 8. Environmental impacts of baseline electrical consumption.

Item	Valu
Electric Consumption in Service Buildings	10,908,570 kWh/Year
Electric Consumption due to Streets Lighting	800,230 kWh/Year
Annual CO_2 Emissions due to Electric Consumption	8,231,286 kg CO_2/Year

Table 9. Overall environmental impact of the baseline.

Summary for Emissions-Baseline	
Total Annual Emissions due to Diesel Buses	18,972,398 kg CO_2
Total Annual Emissions due to Electricity	8,231,286 kg CO_2
Total Annual Emissions	27,203,684 kg CO_2
Total Emissions during Project's Lifetime (15 Years)	408,060 ton CO_2

Deploying the PV-supplied electric buses for the Amman BRT project will result in eliminating the release of 27,204 metric tons a year and 408,060 metric tons of CO_2 during the 15-year lifetime of the project, which has a tremendous influence on the project's carbon footprint.

The social cost of carbon is a measure of the economic harm from those impacts, expressed as the dollar value of the total damages from emitting one ton of carbon dioxide into the atmosphere. The current estimate of the social cost of carbon is roughly $40 per ton [22].

3.3. Economic Analysis

All amounts will be rounded to present worth for proper comparison. The capital expenditure (CAPEX) for the baseline buses over 15 years lifespan is summarized in the following investment timeline in Figure 5:

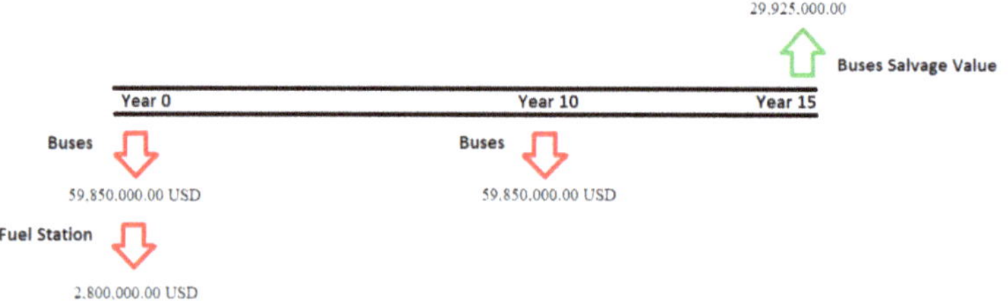

Figure 5. Investment timeline for baseline buses over 15 years.

CAPEX for the renewable energy model is higher than the baseline case. This is due to the more expensive electric buses and the installation of photovoltaic systems. The salvage values of all components within the alternative system are taken into consideration.

The following investment timeline depicted in Figure 6 shows the capital expenditure (CAPEX) of the PV supplied electric bus upgrade over 15 years:

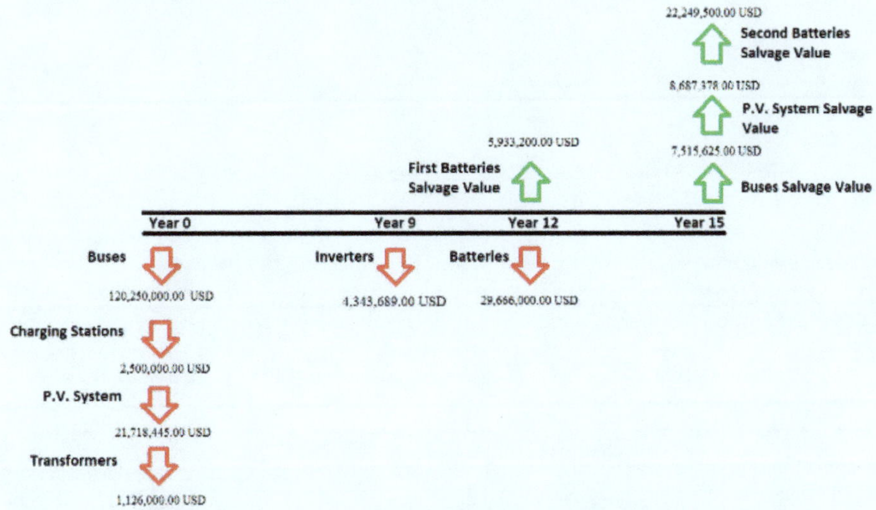

Figure 6. Investment timeline for PV electric buses over 15 years.

PV system inverters have an expected lifetime of 8–10 years. Therefore, for this study, inverters are replaced after 9 years of operation. In general, inverters share 20% of the PV system's capital cost.

The operating expenses (OPEX) for the baseline case include fuel, maintenance, lubricants, fuel station maintenance, electricity for streets lighting, and electricity for service buildings and stations, in addition to the expected increment in diesel and grid electricity prices. These factors are summarized in Figure 7.

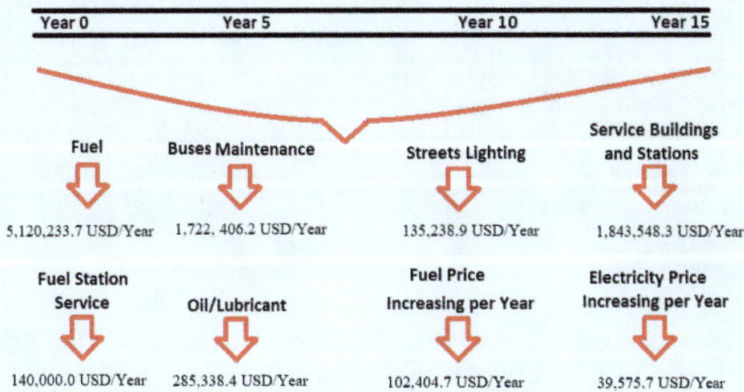

Figure 7. Operational costs for baseline buses.

On the other hand, OPEX for the PV electric bus model is much lower due to use of electric buses and electricity generated from the PV system. Savings from social cost and carbon trade of emissions are included. Servicing of a PV system is included as 3% of the capital cost per year for general maintenance, and 400 USD/MWp per month for panel cleaning. Figure 8 shows the OPEX of the PV electric bus model.

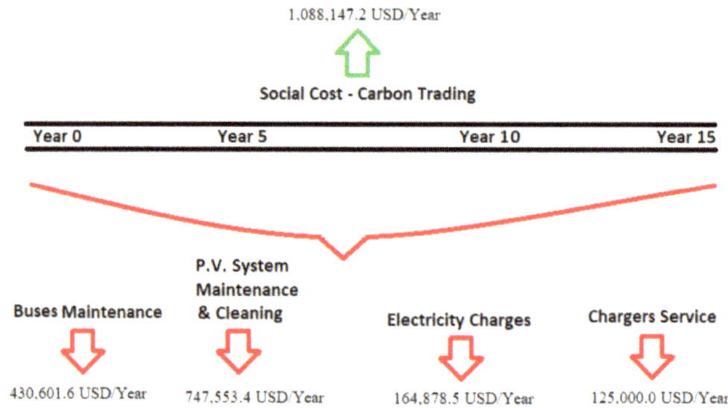

Figure 8. Operational costs for PV electric bus model.

The nominal interest rate of 7% is used in the calculations; 3.1% is the average inflation rate in Jordan. Assigning these rates in the net present value formulas and applying the same on the CAPEX and OPEX of both baseline and PV electric bus model results in the following NPV values in Table 10.

Table 10. Capital expenditure (CAPEX) and operational expenditure (OPEX) for the baseline and green streets model in net present value.

CAPEX (present worth)	15 Years	Baseline Buses	86,765,290 USD
		PV powered Electric Buses	141,892,633 USD
Incremental Capital Cost			55,127,343 USD
OPEX (present worth)	15 Years	Baseline Buses	105,862,835 USD
		PV powered Electric Buses	4,283,408 USD
Savings in Operational Cost			101,579,427 USD
Payback Period (Years)			7.1

4. Discussion

To be able to properly design a PV system to meet the energy needs of transforming the traditional system to a more sustainable system, several scenarios were proposed for the PV system design. The scenarios were based on what areas to use, and analyzed based on energy yield, using actual shading data that were acquired from the physical location. A grading system was proposed, based on these results. The equation proposed uses weights that represent project priorities. In this case study, it was important to produce the required energy, with minimal losses, and a small overhead to make use of the connecting net metering scheme. If a project has different priorities, such as limitless energy injection profit, then the priorities and weight can be changed to reflect that specific scenario. The goal is to use this type of information to provide the designer or decision maker with additional information for better design decisions.

Scenario 3 has the highest score and meets capacity requirements, with lower losses and a higher specific yield than scenarios 1, 2, 5, and 6. This scenario proposes installing PV panels with an overall capacity of 20,017.00 kWp in order to generate 34,696,000.00 kWh per year. As the generated energy is slightly above the requirement, it provides a safety margin. This is important, as the design is based on a simulation; and as the system will be connected using a net-metering scheme, excess energy is forfeit and energy deficit is billed.

Deploying this sustainable upgrade will result in eliminating the release of 27,204 metric tons of CO_2 a year. As the project has an expected lifetime of 15 years; this results in 408,060 metric tons of CO_2 emission reduction. Regardless of what scenario is chosen, as long as all energy required from the upgrade is supplied through a renewable energy sources, then all emissions are eliminated.

After converting the capital investments for the baseline buses and PV-supplied electric buses to net present value (NPV), and using the same approach for OPEX, it was found that the NPV for the incremental investment is 55,127,343 USD, while savings in operational costs are 101,579,427 USD over 15 years. With those results, the payback period is around seven years. For such an important project and the huge impact on the environment and society, seven years for returning the initial investment is very good, bearing in mind the calculations were based on a 15 year project lifetime. Much of the installed equipment will last for much longer.

The total cost of ownership (TCO) for the baseline buses and proposed model is another valid comparison. It was found that the TCO for the business-as-usual scenario is much higher than the TCO of the PV electric bus model by about 32%. TCO for business-as-usual scenario in NPV is 192,628,125 USD over 15 years, while TCO for PV electric bus model in NPV is 146,176,041 USD over the same time period.

5. Conclusions

The use of sustainable and renewable energy concepts for mass public transportation was proposed and analyzed in this paper. Electric buses are used instead of diesel buses, and elevating the photovoltaic system above the bus routes is used to charge the buses' batteries. The concept was applied to the Amman BRT project as a case study under investigation.

Converting conventional transportation systems into clean and green systems requires careful design and projecting multiple scenarios that vary by nature and impact. PV system yield along the whole route was studied and zoned, and several design scenarios were proposed. The most feasible and practical scenario for PV system implementation is the one that excludes zones with high shading losses, due to high raised objects, and generates energy that covers the load demand with an acceptable cap over. Elevated PV panels are available on the market, easy to install, with high efficiency and low cost, easy to maintain, and give the opportunity to benefit horizon above buses and routes. Moreover, they provide attractive shading that does eliminate direct sunrays but not the light from reaching buses cabins and then reduce AC loads.

The PV-supplied electric bus upgrade is a better alternative, with about 46,452,084 USD NPV savings over 15 years compared to the baseline case. The payback period for the proposed scenario was found to be 7.1 years. Payback period calculations included the incremental cost by calculating CAPEX and OPEX for the two cases.

The negative environmental effects of using diesel as a fuel for the buses were eliminated. The overall result is eliminating the release of 27,204 metric tons of CO_2 per year, the equivalent of 408,060 metric tons in the 15 year lifetime of the project. In conclusion, the proposed upgrade is economically feasible and environmentally viable.

The green streets fundamentals and models are economically feasible and environmentally viable. With the negative effects of emissions on the planet and life, converting to green streets is substantial and not optional anymore. Implementing green streets solutions in cities will reduce emissions and pollution, enhance the quality of life, save money, and attain awareness for people towards sustainable transportation means and renewable energy systems.

Author Contributions: Conceptualization, Z.D. and O.S.; methodology, Z.D.; software, Z.D. and O.A.B.; validation, Z.D., O.S., and O.A.B.; formal analysis, O.A.B.; investigation, O.A.B.; resources, Z.D.; data curation, O.A.B. and O.S.; writing—original draft preparation, O.A.B.; writing—review and editing, Z.D. and O.S.; visualization, O.A.B.; supervision, Z.D.; project administration, O.S.; funding acquisition, Z.D. All authors have read and agreed to the published version of the manuscript.

Funding: This research received no external funding.

Acknowledgments: This work is supported by the deanship of scientific research at the German Jordanian University.

References

1. EIA, U.S. Energy Information Administration. International Energy Outlook 2016. Available online: https://www.eia.gov/outlooks/ieo/pdf/0484(2016).pdf (accessed on 3 January 2020).
2. World Bank. World Bank Collection of Development Indicators. 2014. Available online: http://documents.worldbank.org/curated/en/752121468182353172/World-development-indicators-2014 (accessed on 3 January 2020).
3. International Energy Agency. Energy USe. 2014. Available online: https://data.worldbank.org (accessed on 5 December 2018).
4. Mohanty, P.; Muneer, T.; Kolhe, M. *Solar Photovoltaic System Applications*; Springer International Publishing: Cham, Switzerland, 2016.
5. Pihlatie, M.; Kukkonen, S.; Halmeaho, T.; Karvonen, V.; Nylund, N.-O. Fully electric city buses: The viable option. In Proceedings of the IEEE International Electric Vehicle Conference, IEVC 2014 IEEE Institute of Electrical and Electronic Engineers, Florence, Italy, 17–19 December 2014. [CrossRef]
6. Chao, Z.; Chen, X. Optimizing Battery Electric Bus Transit Vehicle Scheduling with Battery Exchanging: Model and Case Study. *Procedia-Soc. Behav. Sci.* **2013**, *96*, 2725–2736. [CrossRef]
7. Landerl, P. Status and Future Perspectives of Electric Buses in Urban Public Transport. Internationale Energiewirtschaftstagung- Klimaziele 2050. 15–17 February 2017. Available online: https://docplayer.net/58141648-Status-and-future-perspectives-of-electric-buses-in-urban-public-transport.html (accessed on 4 January 2020).
8. Egbivwie, E.A.; Ughwumiakpor, V.O.; Ayinuola, A.P. Computer Aided Off-Grid Solar PV System Design. *Int. J. Sci. Res. Publ.* **2017**, *7*, 310–317.
9. Abuelrub, A.; Saadeh, O.; Al-Masri, H.M.K. Scenario Aggregation-Based Grid-Connected Photovoltaic Plant Design. *Sustainability* **2018**, *10*, 1275. [CrossRef]
10. MEMR, Ministry of Energy and Mineral Resources of Jordan. *Transport Energy Survey*; Ministry of Energy and Mineral Resources of Jordan: Amman, Jordan, 2012.
11. GAM. *Brief BRT Report*; GAM: Amman, Jordan, 2005.
12. GAM. *Design of Infrastructure and Operations Planning for a Bus Rapid (BRT) System in Amman*; GAM: Amman, Jordan, 2010.
13. GAM. *Amman BRT—Demand & Operations Report*; GAM: Amman, Jordan, 2010.
14. BYD. 2018. Available online: http://en.byd.com/usa/about/ (accessed on 5 December 2018).
15. Field, K. BYD Wins Contract to Supply 64 Electric Buses for Medellín, Colombia. 2019. Available online: https://cleantechnica.com (accessed on 3 January 2019).
16. Imam, R.; Jamrah, A. Energy Consumption and Environmental Impacts of Bus Rapid Transit (BRT) Systems. *Jordan J. Civ. Eng.* **2012**, *159*, 1–12.
17. Pvsyst. 2018. Available online: www.pvsyst.com (accessed on 5 December 2018).
18. Meteonorm. Available online: https://meteonorm.com/en/ (accessed on 3 January 2018).
19. Bragg, S. Payback Method|Payback Period Formula. Accounting Tools. 2018. Available online: https://www.accountingtools.com/articles/2017/5/17/payback-method-payback-period-formula (accessed on 5 December 2018).
20. IPCC. *Chapter 3: Mobile Combustion*; IPCC: Geneva, Switzerland, 2006.
21. UNFCCC. *Upstream Leakage Emissions Associated—Clean Development Mechanism*; UNFCCC: Rio de Janeiro, Brazil; New York, NY, USA, 2014.
22. Environmental Defence Fund. 2019. Available online: www.edf.org (accessed on 3 December 2019).

Article

Evaluation of Artificial Intelligence-Based Models for Classifying Defective Photovoltaic Cells

Álvaro Pérez-Romero [1], Héctor Felipe Mateo-Romero [2], Sara Gallardo-Saavedra [3], Víctor Alonso-Gómez [3], María del Carmen Alonso-García [4] and Luis Hernández-Callejo [3,*]

[1] Department of Mathematics, Statistics and Computing, Universidad de Cantabria, Av. de los Castros, s/n, 39005 Santander, Spain; alvaro.perezr@alumnos.unican.es
[2] Artificial Intelligence Department, Universidad Politécnica de Madrid, Pº Juan XXIII 11, 28031 Madrid, Spain; h.mateo@alumnos.upm.es
[3] Department of Applied Physics, School of Engineering of the Forestry, Agronomic and Bioenergy Industry, Campus Duques de Soria, 42004 Soria, Spain; sara.gallardo@uva.es (S.G.-S.); victor.alonso.gomez@uva.es (V.A.-G.)
[4] Photovoltaic Solar Energy Unit, Centro de Investigaciones Energéticas, Energy Department, Medioambientales y Tecnológicas (CIEMAT), 28040 Madrid, Spain; carmen.alonso@ciemat.es
*   Correspondence: luis.hernandez.callejo@uva.es; Tel.: +34-975129418

Abstract: Solar Photovoltaic (*PV*) energy has experienced an important growth and prospect during the last decade due to the constant development of the technology and its high reliability, together with a drastic reduction in costs. This fact has favored both its large-scale implementation and small-scale Distributed Generation (*DG*). *PV* systems integrated into local distribution systems are considered to be one of the keys to a sustainable future built environment in Smart Cities (*SC*). Advanced Operation and Maintenance (O&M) of solar PV plants is necessary. Powerful and accurate data are usually obtained on-site by means of current-voltage (I-V) curves or electroluminescence (*EL*) images, with new equipment and methodologies recently proposed. In this work, authors present a comparison between five *AI*-based models to classify *PV* solar cells according to their state, using *EL* images at the *PV* solar cell level, while the cell I-V curves are used in the training phase to be able to classify the cells based on its production efficiency. This automatic classification of defective cells enormously facilitates the identification of defects for *PV* plant operators, decreasing the human labor and optimizing the defect location. In addition, this work presents a methodology for the selection of important variables for the training of a defective cell classifier.

Keywords: photovoltaic cell defect; classifier; artificial intelligence

Citation: Pérez-Romero, Á.; Mateo-Romero, H.F.; Gallardo-Saavedra, S.; Alonso-Gómez, V.; Alonso-García, M.d.C.; Hernández-Callejo, L. Evaluation of Artificial Intelligence-Based Models for Classifying Defective Photovoltaic Cells. *Appl. Sci.* **2021**, *11*, 4226. https://doi.org/10.3390/app11094226

Academic Editor: Francesco Calise

Received: 15 April 2021
Accepted: 30 April 2021
Published: 6 May 2021

1. Introduction

During the last decade, worldwide installation of renewable generation plants has considerably increased. Among renewables, photovoltaic (*PV*) solar plants have been the most interesting in recent years, and it seems that they will be the most installed in the following years [1,2]. During 2019, the last analyzed year in the Global Status Report [3], 201 GW of renewable power capacity were installed in the World; 115 GW of Solar *PV* capacity, corresponding to more than 57% of total renewable additions. The solar PV cumulative installed capacity was raised to 633.7 GW by the end of 2019.

The reason for the spectacular growth and prospect of this energy source lies in the constant development of the technology and its high reliability. This has made possible a drastic reduction in costs, which has favored both its large-scale implementation and small-scale Distributed Generation (*DG*). Many countries have already begun to review their climate and energy policies. Innovation in sustainable energy supply is, thus, crucial for providing reliable and clean energy sources and improving the quality of life on this planet. To achieve this goal, the idea of smart energy buildings or energy-neutral buildings has been launched. The main objective of an energy regulation of a building is to maintain internal thermal comfort, as well as minimizing energy consumption [4]. *PV* systems and electric vehicles (*EVs*) integrated into local distribution systems are considered to be two of the keys to a sustainable future built environment in Smart Cities (*SC*) [5]. The term "smart energy city" has arisen in parallel with these developments, notably since at least the turn of the 2010s in connection with the energy-relevant components of smart cities [6]. In this scenario, ensuring energy production is a key factor in warranting plant profitability, and this has forced the design of increasingly intelligent and advanced operation and maintenance (*O&M*) strategies.

The *O&M* of *PV* solar plants is critical for industry players. The development of new equipment and methodologies for its application in *PV* solar plants is necessary, and in this sense, research and industry in this area are rapidly evolving [7]. Some researchers have worked with real data from *PV* solar plants, and they have obtained important results showing defects in them [8]. The *PV* inverter is a critical element in *PV* solar plants, but so is the *PV* solar module. A *PV* solar module is made up of *PV* solar cells, and these cells can have different problems or defects, as shown in [9]. Although the knowledge at the *PV* solar cell level is interesting for operators and maintainers of *PV* solar plants, the measurements should be taken in the *PV* solar module.

The detection of faults in *PV* solar modules is subject to obtaining field data, as currently, monitoring does not reach the module level in *PV* plants. Powerful and accurate data can be obtained by current-voltage (I-V) curves, both from the *PV* solar cell (when the cell contacts can be accessed) or at the *PV* solar module level [9,10]. Nowadays, I-V tracers present some drawbacks, such as being only for the string level, working offline, or being expensive. However, advanced instruments recently developed allow automating the online module I-V curves acquisition without requiring the *PV* plant to be shut down [11]. Additionally, it is also possible to perform online dark I-V curves of modules in *PV* plants without the need to disconnect them from the string. For this, a combination of electronic boards in the *PV* modules and a bidirectional inverter is required, as presented in [12].

Another way to obtain field data is by taking images. Classically, thermographic imaging (*IRT*) has been used for early detection of hot spots [13], and in recent years this *IRT* method has been carried out using drone flight [14,15]. One of the most promising maintenance techniques is the study of electroluminescence (EL) images as a complement of *IRT* analysis. However, its high cost has prevented its use regularly up to date. The authors in [16] proposed a maintenance methodology to perform on-site EL inspections as efficiently as possible using a bidirectional inverter.

All these inspection techniques make it possible to locate and identify the defects present in *PV* modules. However, nowadays, solar plants are becoming larger, and the manual treatment of all the data obtained through the previous techniques presented can be a very expensive and time-consuming task. Artificial Intelligence (*AI*) is already being applied in *PV* solar plants. *AI* application has long focused on energy production forecasting issues. The authors in [17] developed a solution that provides electricity production based on historical and current available solar radiation data in real-time. Other authors have presented a taxonomy study, showing a process to divide and classify the different forecasting methods, and the authors also presented the trends in *AI* applied to generation forecasting in solar *PV* plants [18]. The use of artificial neural networks (*ANN*) has been successful in the last decade; some authors use *ANNs* together with climatic variables to forecast generation in *PV* solar plants [19], while others use Support Vector Machine (*SVM*) together with an optimization of the internal parameters of the model [20].

ANNs have also been used for other tasks, such as for the detection of problems in energy production, as is the case in [21], where the authors use radial basis function (*RBF*) to detect this type of failure in production. A similar goal is sought in [22], where an *SVM*-based model was employed for describing a failure diagnosis method that uses the linear relationship between the solar radiation and the power generation graphs. This research studies the following failure types: inverter failures, communication errors, sensor failures, junction box errors, and junction box fires. The model classifies string and inverter failures. However, in actual *PV* plants, each inverter can cover thousands of modules, and therefore important failure information can be lost during classification.

Therefore, it is possible to affirm that the use of *AI* is common in *PV* solar plants. Research has studied its application in energy production forecasting issues or for the detection of problems in energy production. However, it has been highlighted how the detection of defects using inverter-level information can be imprecise. In this work, the authors present a comparison between five *AI*-based models to classify *PV* solar cells according to their state based on EL images. The five well-known models used for classification have been: k-nearest neighbors (*KNN*), *SVM*, Random Forests (*RF*), Multilayer Perceptron (*MLP*), and Convolutional Neural Networks (*CNNs*). This automatic classification of defective cells enormously facilitates the identification of defects in a precise way for *PV* plant operators, decreasing the human labor and optimizing the defect location. For this, the authors used an ad-hoc *PV* solar module manufactured in a special way since *PV* solar cells have their back contacts accessible, allowing their total characterization [9]. With this manufactured module, it was possible to obtain each cell I-V curve, in addition to EL images, so it was feasible to label each cell (group 1: good, group 2: fair, and group 3: bad) based on its production efficiency. This allowed an accurate classification for model training. The study presents a novel method for the labeling of cells based on their production efficiency, and this was possible due to the customized *PV* solar module, which clearly differentiates this research. The classifications discussed in this document are an extension of the previous work "*Photovoltaic cell defect classifier: a model comparison*" presented at the Smart Cities—III Iberoamerican Congress on Smart Cities (ICSC-CITIES 2020) [23]. The document is structured as follows: Section 2 presents the materials and methodology used, Section 3 shows the results, and Section 4 contains conclusions and future work proposals.

2. Materials and Methods

This section is intended to explain the materials used, as well as the methodology followed to validate the classifier.

2.1. Materials

A 60-cells polycrystalline module composed of cells with and without defects was used. The front and back views of the module are presented in Figure 1a,b respectively. The module was ad-hoc manufactured with all cells accessible from the backside of the module. Regarding the cell labeling, numbers from 0 to 59 have been used to identify the cell, as detailed in Figure 2. Additionally, the four corner cells have been labeled both in Figure 1a,b to facilitate understanding.

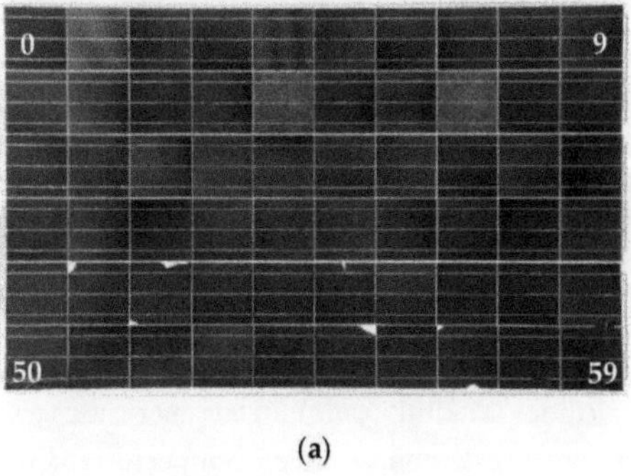

(a)

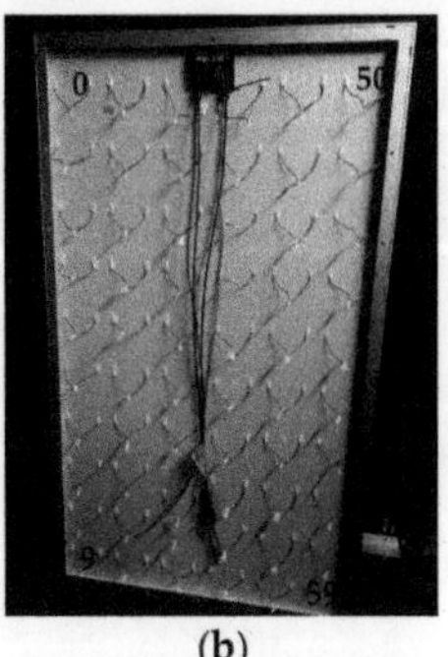

(b)

Figure 1. (a) Front view of the module. (b) Back view of the module.

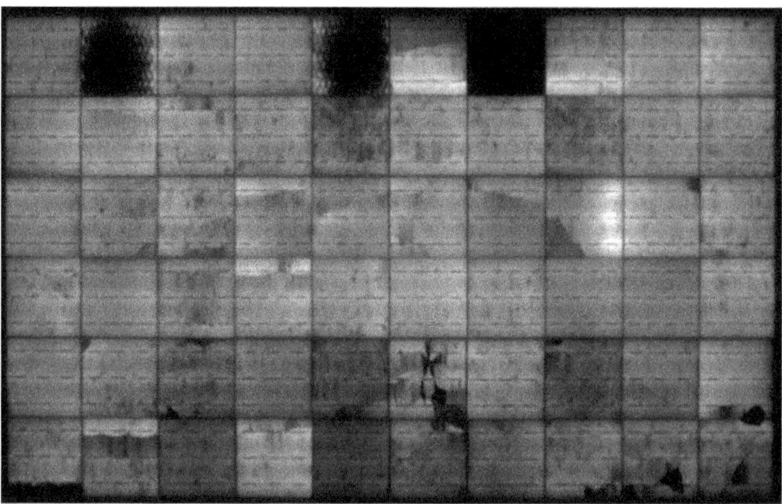

Figure 2. EL image and cell numbering for model training.

The first string (first and second columns in the back view) contains manufacturing defects, the central string (third and fourth columns) contains soldering faults, while the third string (fifth and sixth columns in the back view) contains breaking deficiencies. The low-efficiency defects (cells 1 and 4) and medium efficiency cells (cells 14 and 17) were due to manufacturing problems, with an efficiency of 9% and 16.4% approximately, but they did correspond with breaking or short-circuited cells. Short-circuit cell (cell 6) has been generated by extending the cell connection tabs beyond the ordinary placement, short-circuiting the cell. In order to simulate the bad soldering defects, buses from the back of some cells were left without soldering, either one bus (cell 22) or two buses (cell 34). Three buses have not been left without soldering in any case as it would have meant that this cell would not be series connected as the rest. In cell 38, all tabs were lose (without soldering), although they made contact allowing module production. The cell with only 1 cm welded (cell 27) was used to simulate bad soldering, in which only 1 cm of the bus was welded instead of the typical 15 cm being welded. The third-string contains some cracked cells without cell area decrease (cell 50 and 51), with cell area decrease (cell 41, 42, 55, and 57) or a combination of both in the same cell (cell 45). When a piece of broken cell was placed on top of another cell (cells 49, 58, and 59), it generates partial shading, simulating the important aspect of permanent bird droppings. These types of defects were analyzed as they ordinarily appear in commercial modules in operation, either during manufacturing, transport, or operation. However, commercial modules are not accessible at the cell level. That is why an ad hoc module was manufactured for defect characterization.

The nominal characteristics of a standard module of this type are: nominal power (P) 250 W, efficiency 15.35%, maximum power point current (Impp) 8.45 A, maximum power point voltage (Vmpp) 29.53 V, short circuit current (Isc) 8.91 A and open-circuit voltage (Voc) 37.6 V. Having 60 cells in series, the nominal values of a healthy cell were considered: nominal power 4.17 W, Impp 8.45 A, Vmpp 0.49 V, Isc 8.91 A, and Voc 0.63 V.

I-V curve measurement (at the cell level) and EL images were carried out at the CIEMAT's facilities (Madrid, Spain). In summary, the facilities used were the following:

- The indoor measurements have been performed in the commercial system Pasan SunSim 3 CM, which consisted of a light pulse solar simulator class AAA according to IEC 60904-9 standard, which can perform I-V curve measurements at Standard Test conditions;

- The *EL* and indoor *IRT* tests were simultaneously performed with the *EL* in this chamber. The module was fed with a Delta power supply SM 70-22. A Fluke 189 multimeter connected to module terminals allowing it to register the exact module voltage. *EL* and *IR* images were captured with a PCO 1300 and a FLIR SC 640 camera, respectively.

In this way, the information of the *EL* image and I-V curve of each *PV* solar cell was obtained. In the following Figure 2, an *EL* image of the measurement module and the numbering of the cells is shown.

This information could serve to validate the different models. In the training phase, the cells used were labeled according to their potency (measured through the I-V curve) and with the following criteria:

- Group 1: Power $\geq$ 95%;
- Group 2: 80% $\leq$ Power < 95%;
- Group 3: Power < 80%.

According to the measured I-V curves, the classification of the cells in the three proposed groups was as follows:

- Group 1: 0, 2, 3, 5, 7, 8, 9, 10, 11, 12, 13, 15, 16, 18, 19, 20, 21, 22, 23, 24, 25, 28, 30, 31, 33, and 53;
- Group 2: 14, 17, 26, 27, 29, 32, 34, 35, 36, 37, 38, 39, 40, 41, 42, 43, 44, 46, 47, 48, 51, 52, 54, 55, 56, 57, and 58;
- Group 3: 1, 4, 6, 45, 49, 50, and 59.

For the test phase, the classifiers only used the *EL* images, which were perfectly applicable in the field in common *PV* modules. I-V curves were only used in the model training in order to classify the cells based on their production efficiency, and this was possible thanks to the customized *PV* solar module, which clearly differentiates this research from previous work.

2.2. Methods

In this section, the methodology followed for defective cell classification is presented. Firstly, the pretreatment of the *EL* images was explained. Secondly, Self-Organizing Maps (*SOM*) were used to observe the similarities detected between cells and groups. Thirdly, different variables were proposed, and applying the method combining *KN* and *SVM*, the variables with the best performance were selected. Finally, four classifiers were developed and tested using those variables: *KNN*, *SVM*, *RF*, and *MLP*. Besides, a fifth independent classifier, using *CNN*, was proposed. More explicit information about the well-known models used (*KNN*, *SVM*, *RF*, *MLP*, and *CNN*) can be found in the following outstanding references [24–28].

According to the pretreatment of *EL* images for feeding the classifiers, since the *EL* image was taken from the entire module, as showed in Figure 2, the first thing was to cut out the cells for individual assessment. For this, a 115 × 115-pixel cropping window was established, which offered the best adjustment by superimposing it on each cell. It was necessary to zoom to the maximum to adjust the window optimally, thus ensuring that all the cropped images were aligned and thus avoiding including the black margins that separate (due to the natural structure of the panel) the cells. After that, with the objective of improving the edges, the images were loaded into Python, reducing each side by 1 pixel, finally resulting in 60 images of size 113 × 113.

Given the classification method proposed in this work, in which the cells were grouped according to the power measured individually in each one (groups 1, 2, or 3, as previously detailed), it was not possible to appreciate visible features (see Figure 2) that characterized the elements of the different groups, except for the black spots in group 3.

To demonstrate the correctness of this classification, the authors proposed using *SOM* to be able to observe the similarities detected between cells. *SOM* is a type of unsupervised neural network, and its purpose is to reduce the dimensionality of the data preserving the topological properties of it [29]. It is made up of only 2 layers, the input and the output layers. The input layer has a number N of neurons equivalent to the dimension of the data. The output layer represents a two-dimensional array of neurons, each with an assigned N-dimension weight. Each time the *SOM* map algorithm is executed, the weights are randomly initialized (once initialized, they are organized and distributed on the map based on their proximity), and they compete with each other each time data is entered into the input layer. The neuron whose weight most closely resembles the input information is the winner, which causes an update in the value of its weight, as well as that of its neighbors. Therefore, the algorithm consists of iterating enough times and each time choosing random data in the training sample to progressively update the map until it is molded to the structure of the starting information. In this study, a *SOM* network of 20 rows and 20 columns was proposed, using Euclidean distance as the standard distance, and the results are shown in the results section.

Therefore, it was necessary to decide which variables were the representative ones of the 60 members. Each image (in grayscale) was a matrix of 113 rows by 113 columns, where each coordinate or pixel had values from 0 (black) to 255 (white). Since each matrix had 12,769 coordinates or pixels that take 256 possible values, then every image could also be seen as a flattened matrix, that is, a vector $C_i = (p_{i,0}, \ldots, p_{i,12768}) \in \mathbb{R}^{12769}, 0 \leq i \leq 59$, where $p_{i,0}, \ldots, p_{i,112}$ are the values of the pixels that correspond to the first row of the ith matrix, $p_{i,113}, \ldots, p_{i,226}$ are the pixels correspond to the second row of the ith matrix, and so on. From this information, 28 variable candidates were calculated, as can be observed in Figure 3.

$$C_i = (p_{i,0}, \ldots, p_{i,12768}) \in \mathbb{R}^{12769}, \; p_{i,j} \in \{0, 1, \ldots, 255\}, \; 0 \leq i \leq 59, \; 0 \leq j \leq 12768$$

- 1st to 9th variable $\longrightarrow$ $\mathrm{card}(\{p_{i,j} : 25(k-1) \leq p_{i,j} < 25k, 0 \leq j \leq 12768\}), \; 1 \leq k \leq 9$
- 10th variable $\longrightarrow$ $\mathrm{card}(\{p_{i,j} : 225 \leq p_{i,j} < 256, 0 \leq j \leq 12768\})$
- 11th to 19th variable $\longrightarrow$ $P_{10 \cdot (k-10)}(C_i), \; 11 \leq k \leq 19$
- 20th variable $\longrightarrow$ $P_{99}(C_i) - P_1(C_i)$
- 21st variable $\longrightarrow$ $\sum_{j=0}^{12768} p_{i,j}$
- 22nd variable $\longrightarrow$ $mean(C_i)$
- 23rd variable $\longrightarrow$ $var(C_i)$
- 24th variable $\longrightarrow$ $mode(C_i)$
- 25th variable $\longrightarrow$ $energy(C_i)$
- 26th variable $\longrightarrow$ $entropy(C_i)$
- 27th variable $\longrightarrow$ $kurtosis(C_i)$
- 28th variable $\longrightarrow$ $skewness(C_i)$

Figure 3. Pseudocode of the procedure for selecting the most significant variables.

Within the set of variables, the first 10 represent the number of pixels of any cell with values comprised in bands of length 25, except for the tenth variable that represents the number of pixels between 225 and 255. The next 9 variables represent the 9 percentiles ranging from the tenth percentile to the ninetieth percentile. The last 9 represent the range, the global sum of the 12,769 pixels, the mean, the variance, the mode, the energy, the entropy, the kurtosis, and the statistical skewness.

Before deciding the variables with which to start the study, some requirements that influence their choice had to be taken into account. Mainly, it was sought to obtain the best global results, but it was also used to minimize the error in the classification of the elements related to group 3, that is, to avoid classifying elements from group 1 or 2 into group 3 and, vice versa. Since there could also be the possibility that there existed more than one variable that returned similar values for cells of opposite groups and therefore caused noise in the information, it was necessary to find and leave aside those variables.

Given the difficulty of knowing which variables offer optimal results, the chosen strategy consisted of repeating the following method successively, previously standardizing all the data (60 cells explained in 28 variables) in mean 0 and variance 1. For this, 55 random cells were chosen 20,000 times.

In the following paragraphs, the method is explained, and the proposed methodology for the selection of main variables are shown in Figure 4.

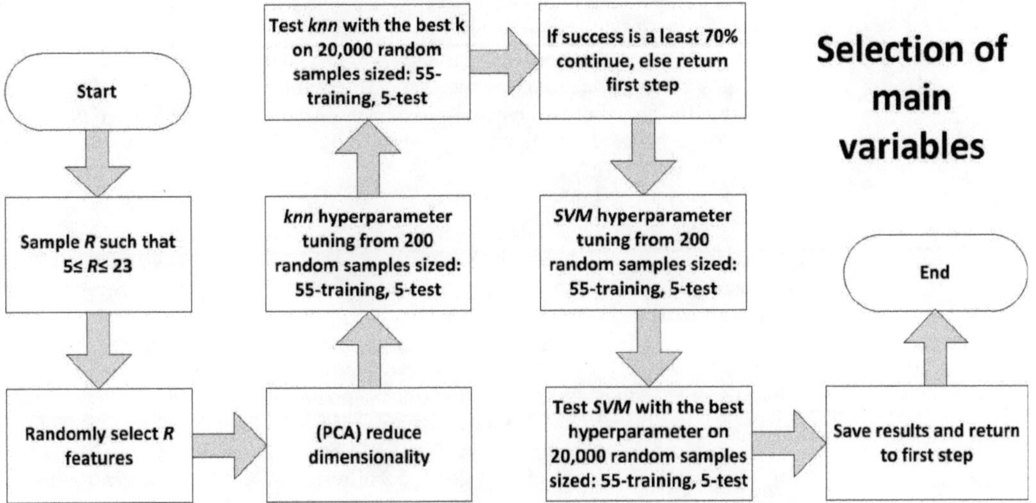

Figure 4. Proposed methodology for the selection of main variables.

1. Sample R such that $5 \leq R \leq 23$ and randomly select R features: A random number R is obtained between 5 and 23. Next, R variables are chosen at random from among the 28 possible ones;
2. Principal Component Analysis (*PCA*) to Reduce dimensionality: Subsequently, principal component analysis is applied, saving the first variables that explain more than 99.5% of the variance of the data. Therefore, from now on, we work with 60 individuals explained in at most R new variables;
3. *KNN* hyperparameter tuning from 200 random samples sized 55-training and 5-test, test *KNN* with the best K on 20,000 random samples sized 55-training and 5-test: As we were interested in obtaining a good classification, the optimal number of neighbors k was sought, choosing between 1 and 10, starting from the one that offers the best results when applied in 200 random samples of size 55-training, 5-test. Once k has been obtained, the percentage of success with *KNN* is now estimated from 20,000 random samples of size 55-training, 5-test. Finally, the proportion of bad classifications related to group 3 is noted. If the percentage of success with *KNN* is less strict than 70%, step 1 becomes:
4. *SVM* hyperparameter tuning from 200 random samples sized 55-training, 5-test: Now, exceeding 70% of success with *KNN*, *SVM* is applied taking into account the following parameters:

    ○   Core: function in charge of transporting the data to a higher dimension where a better separation of the same can be achieved. Sigmoidal, polynomial, and Gaussian functions and the linear core were taken into account for the experiment;

    ○   Penalty parameter C: it is an indicator of the error that one is willing to tolerate. The values for C of 10, 50, 75, and 100 were taken into account for the experiment;

    ○   Gamma: indicates how far the points are taken into account when drawing up the separating boundary. The gamma values of $1, 0.8, 0.6, 0.4, 0.1, 0.01$, and 0.001 were taken into account for the experiment;

    ○   Degree: degree of the function in the polynomial nucleus. Grades $1, 2, 3$, and 4 were taken into account for the experiment;

    ○   Based on these parameters, a search was done among all the possible combinations in order to calculate which one of them offered the best results applied to 200 random samples of size 55-training, 5-test. GridSearchCV was used to perform the above task.

5. Test *SVM* with the best hyperparameters on 20,000 random samples sized 55-training 5-test: Once the ideal combination has been obtained, the efficacy of *SVM* is estimated running the supervised method based on these parameters and applied to 20,000 random samples of size 55-training, 5-test. Hit and misclassification ratios related to group 3 are saved;

6. Save results and return to first step: Back to step 1.

For sufficiently wide data collection, it was necessary to run the previous process in Python for around 40 h to obtain 1800 iterations, of which 250 corresponded to those cases where *KNN* and *SVM* were calculated at the same time (since *KNN* achieved more than 70% of success).

The data available was 60 cells, which supposed very little information with which to carry out the study. This influenced the search for the best parameters for *KNN* and *SVM*, since *Cross Validation* was not applied (the best parameters would hardly be obtained). Instead, a search based on 200 random samples of size 55-train, 5-test was applied.

Considering the results obtained, it could be concluded that the variables 4, 5, 8, 10, 13, 17, 19, 20, and 24 offered good global results as well as low error rates when making classifications related to group 3. Hence, those variables have been selected as representative ones, and it reduced the number of them from 9 to 7 by applying *PCA* (saving the first variables that explain more than 99.5% of the variance of the data). Afterward, groups of 55 cells were made again to train each model, and it was validated with the 4 remaining classifiers in each case. The classifiers to be tested were the following: *KNN*, *SVM*, *RF*, and *MLP*. On the other hand, *CNN* was also tested. Nevertheless, this classifier did not follow the same strategy as the four algorithms mentioned before (it directly used the $60, 113 \times 113$ matrix as input data). These models were chosen and compared since they are the most used in classification [30–32].

Below are included some details of the architecture of the models used. *KNN* and *SVM* architectures have already been described in the previous paragraphs, in which the representative variables selection was detailed. The number of neighbors considered in *KNN* was equal to the number of them obtained in the previous algorithm when the representative variables were selected. The same occurred with the hyperparameters obtained in *SVM*.

According to *RF*, each classifier was built based on 500 trees. Additionally, *hyperparameter tuning* was applied, combining the following parameters:

1. Maximum depth: represents the maximum number of levels allowed in each decision tree. The values 20, 40, 60, 80, and 100 were taken into account;

2. Minimum points per node: this is the minimum number of data allowed in each partition. The values $1, 2, 3, 4$, and 5 were taken into account.

3. Maximum variables: indicates the maximum number of variables (chosen at random) that are taken into consideration when partitioning a node. Usually, $\sqrt{n}$ is used as a standard parameter, where n is the number of total variables, but $\sqrt{n} - 1$, $\sqrt{n}$, and $\sqrt{n} + 1$ were taken into account.

In the case of *MLP*, the model was built from an input layer made up of 7 neurons (coinciding with the dimensionality of the data), a first hidden layer made up of 128 neurons, a second hidden layer made up of 64 neurons, and an output layer made up of only

3 neurons (matching the number of classifications). The neural network created was dense, a network formed by neurons that were each connected to all possible neurons belonging to contiguous layers. The activator used in the process was the rectifier or *ReLU* activator, except in the last layer where the *softmax* function was used. In addition, *hyperparameter tuning* was applied, taking into account the following parameters:

1. *Epochs*: indicates the number of times that the neural network reads the data from the training sample in order to adjust to them (translated into a successive update of its parameters). The values 25, 50, 75, 100, 150, and 200 were taken into account;
2. *Batches*: indicates the speed with which the network parameters are updated as the epochs progress. The values 15, 25, 50, 75, 100, 150, and 200 were taken into account.

For the validation of the different models and obtaining the results, the methodology used with each of the 4 classifiers is shown in Figure 5.

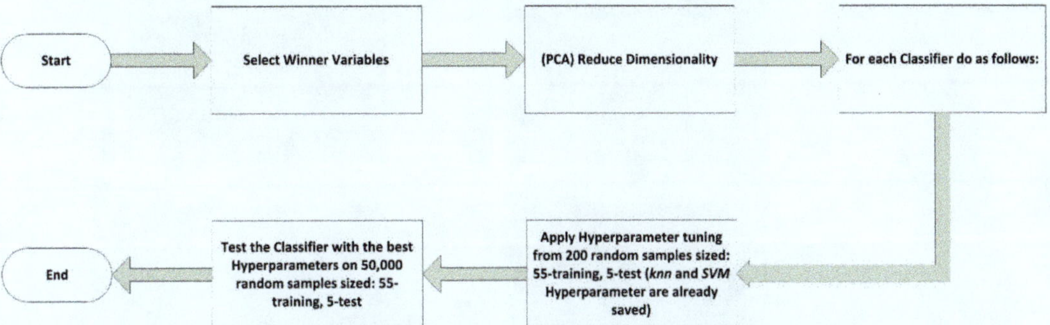

Figure 5. Methodology used with each of the 4 classifiers.

In the case of *CNN*, it was built with a similar approach to the *MLP*. In order to find enough patterns to solve this problem, we have used a convolutional layer with 64 filters and a kernel size of 3 × 3. As to reduce the dimensionality, we also used a maxpool layer. Finally, a dense layer of 128 was introduced. In this architecture, we did not use dynamic parameter optimization. The networks need a high number of epochs to train (around 1000). We use the Nadam Optimizer [33] since it is the best in the tests that we have executed. We also exploit early-stopping, stopping the training when we do not have obtained better results in a certain number of epochs.

For the validation of the different models and obtaining the results, the methodology used with each of the five classifiers was as follows:

1. For *KNN* and *SVM*:

    a. Test the classifier with the best hyperparameters (previously calculated) on 50,000 random samples sized: 55-training, 5-test.

2. For *RF* and *MLP*:

    a. Apply hyperparameter tuning from 200 random samples sized: 55-training and 5-test.

    b. Test the classifier with the best hyperparameters on 50,000 random samples sized: 55-training and 5-test.

3. For *CNN*:

    a. Test the classifier with the best hyperparameters (manually settled) on 100 random samples sized: 55-training and 5-test.

3. Results and Discussion

This section provides a concise description of the experimental results, their interpretation, as well as the discussion of the results and how they can be interpreted.

3.1. Justification of the Correct Initial Power Rating

As already mentioned, *SOM* was used to have some idea of whether the power classification was correct.

Next, Figure 6 shows four maps obtained considering the data with the representative variables previously obtained. Each pixel represented one of the 400 possible output neurons. Nearby pixels with dark values reflect proximity to each other, while nearby pixels with the contrast between light and dark representing distance. The elements of group 1 are represented in red, the elements of group 2 in green, and the elements of group 3 in blue.

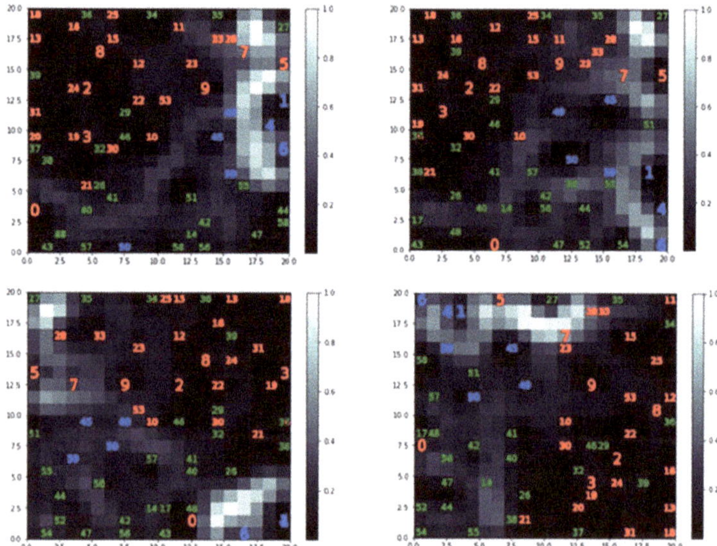

Figure 6. Result when applying *SOM* to the available data [23].

Observing the results, it was possible to determine the existence of a certain grouping between the elements of the same color, and therefore of the same group. It was possible to conclude that the classification based on power was correct. Within group 3, it was deduced by the white border that the cells that are most distinguished from the rest of the groups were number 1, 4, and 6. Furthermore, comparing group 1 and 2, it could be concluded that it was easier to make a mistake when classifying cells from group 2 (green) in group 1 (red) than otherwise. This was due to the fact that some green points were mixed within the main mass of red points, which did not happen in the green group, where its elements had hardly any red elements inside them, except for element 0. Similarly, when classifying elements of group 3, it was possible to make a certain mistake and to identify them as elements of group 2, or vice versa, due to their greater closeness (compared to group 1). This could be verified by observing the results of the classifications, which will be shown later.

3.2. Classification of Variables

As already mentioned, the detection of the most important variables for the training of the models is crucial. To do this, it was necessary to run the previous process in Python for 40 h to obtain 1800 iterations, of which 250 iterations correspond to those cases where *KNN* and *SVM* were calculated at the same time.

Next, Table 1 shows the best results obtained according to different criteria, such as the success in the classification with *KNN* and *SVM* (columns 1 and 2), and proportion of bad classifications caused between groups 1 and 2 with respect to group 3, using *KNN* and *SVM* respectively (columns 4 and 5). Therefore, the value 0.9331 in row 5 and column 5 was interpreted by associating 93.331% to the percentage of bad classifications (applying *SVM* with the indicated variables) that were not related to group 3, to which only 6.669% corresponded. Therefore, a high proportion represented a smaller error in the classification of elements related to group 3. The third column is the sum of the fourth and fifth columns.

Table 1. Relationship between success in classification and variables used.

	KNN Success	SVM Success	KNN + SVM Success Group 3	KNN Success Group 3	SVM Success Group 3	Variables
Top 1 *KNN* success	0.745	0.756	1.425	0.734	0.691	2, 4, 5, 7, 10, 13, 16, 17, and 22
Top 1 *SVM* success	0.7077	0.7628	1.5973	0.6975	0.8998	4, 5, 8, 10, 13, 17, 19, 20, and 24
Top 1 *KNN* + *SVM* success group 3	0.7005	0.7187	1.7646	0.9093	0.8552	2, 4, 6, 8, 10, 20, 23, and 24
Top 1 *KNN* success group 3	0.7095	0.7187	1.7646	0.9093	0.8552	2, 4, 6, 8, 10, 20, 23, and 24
Top 1 *SVM* success group 3	0.7099	0.7401	1.6371	0.704	0.9331	2, 4, 5, 7, 8, 9, 10, 12, 15, 16, 17, 18, 19, 20, and 21

In addition, the frequency of appearance of each of the 28 variables in all the iterations (1800 in total) was studied based on two criteria:

- Success with *KNN* greater than 68.5% (seventy-fifth percentile);
- The proportion of bad classifications not related to group 3 higher than 79.4% (eighty-fifth percentile).

From the first criteria, it was deduced that good candidate variables were 2, 4, 5, and 7 whereas bad candidate variables were 1, 3, 6, 26, 27, and 28. From the second criteria, it was concluded that good candidate variables were 2, 4, 6, 23, and 28 whereas bad candidate variables were 1, 7, 25, and 27. Comparing the results obtained, it was possible to conclude that there was a relationship between the importance of a variable and its frequency of appearance following the two criteria mentioned.

The same type of criteria could be considered for *SVM*. However, it was possible that given the low number of iterations (250), the results were not entirely reliable.

3.3. Convergence and Results of the Models

An important aspect when working with AI is the convergence of the model. Next, Figure 7 shows the behavior of the hit obtained with each of the classifiers, depending on the number of iterations performed. In the case of *MLP*, 50,000 iterations were not reached due to the high computational cost, although there was no loss of efficiency, as was well observed in all models since there was some convergence at a lower number of iterations.

In the *CNN* model compilation appeared a critical problem. The amount of time needed to finish one iteration was extremely high. This was vital in order to decide the number of iterations. The authors finally decided to use 100 iterations of 1000 epochs. With this number of iterations, some convergence was obtained.

From the results shown in Figure 7, the most successful model from the first fourth classifiers (*KNN*, *SVM*, *RF*, and *MLP*) was *SVM*, closely followed by *MLP* and *RF*. The worst result was obtained by *KNN*; however, the success rate was 70.61%, and it was possible to consider it as being of high value. On the other hand, it could be seen that the fifth classifier, *CNN*, achieved higher success than the rest, exceeding 80%.

Table 2 shows the results (percentage of success) of the five models used once the cells used for the validation phase were classified. The time (hours) required are also shown. With regard to time, the first column of times shows the time needed to locate the ideal parameters (*hyperparameter tuning*), while the second column of times indicates the time needed for classifier training.

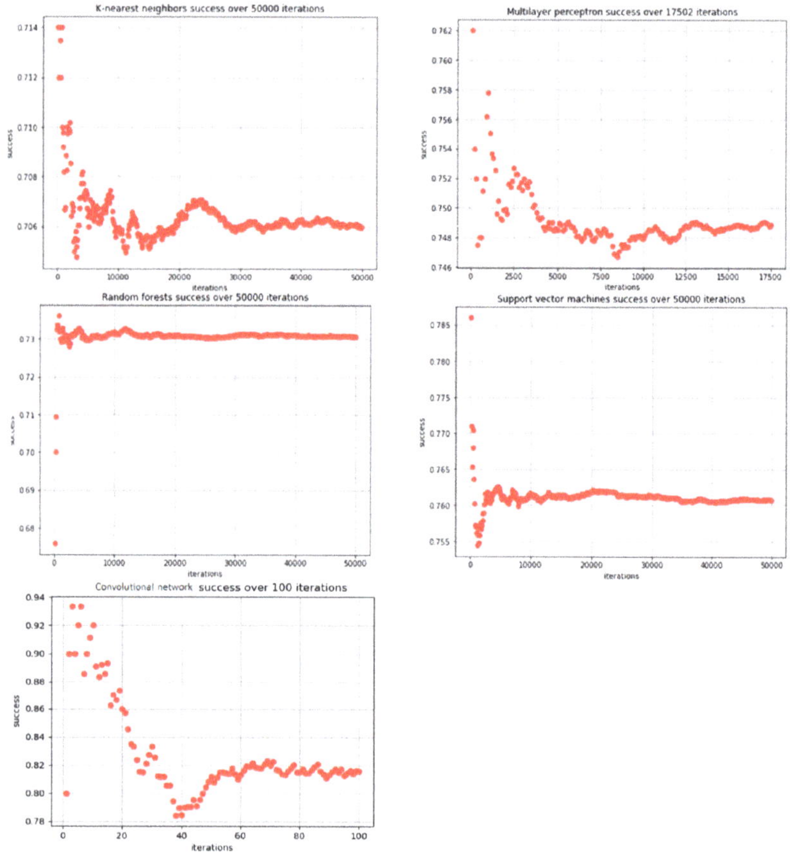

Figure 7. Comparison of classification success versus the number of iterations for the five models [23] and authors elaboration.

Table 2. Classification results in the five models.

	Classification Success	Time Spent on Hyperparameter Tuning (hr)	Time Spent Testing 50,000 (17,502 with *MLP*/100 *CNN*) Samples (hr)
KNN	0.7061	0.0017	0.0267
SVM	0.7607	0.0223	0.0196
RF	0.7308	2.1552	8.8071
MLP	0.7488	0.9633	15.5442
CNN	0.8160		100

Regarding the time spent in locating the main parameters, the fastest model was *KNN*, with *SVM* following and with a time close to 1 h, *MLP*. *RF* presented the worst time to locate these parameters, requiring more than 2 h.

Regarding the time spent on training, *SVM* was the fastest, closely followed by *KNN*. At the other extreme, *RF* required almost 9 h, followed by *MLP* with 15.54 h, and *CNN* with 100 h.

Observing results presented in Table 2, it could be concluded that *CNN* presented the highest percentage of success, with 81.6% but was also extremely slow (around 100 h).

SVM was the model that presented the second-highest efficiency (76.07%), and it was the model with reasonably lower times (search for parameters and training).

One of the main goals of sorting was to detect bad cells (group 3). In this sense, Table 3 shows the results of the classification of cells in group 3. In the same way, Table 3 shows the results of the misclassification between groups.

Table 3. Group 3 classification results and results of misclassification between groups.

	Success on Group 3	1 Missclassified as 2	1 Missclassified as 3	2 Missclassified as 1	2 Missclassified as 3	3 Missclassified as 1	3 Missclassified as 2
KNN	0.7002	0.1017	0	0.5985	0.0117	0.0579	0.2302
SVM	0.9055	0.1377	0.0077	0.7678	0.0142	0.0044	0.0752
RF	0.7746	0.2525	0	0.5221	0.0142	0.0173	0.1939
MLP	0.8224	0.2166	0.0018	0.6058	0.0687	0.0073	0.0998
CNN	0.6956	0.2282	0.0217	0.3586	0.0865	0.0543	0.25

Focusing only on the classification of group 3, *SVM* obtains success results of 90.55%, while *MLP* obtains 82.24%, *RF* obtains 77.46%, *KNN* 70.02%, and the worst result is for *CNN* with 69.56%.

It could also be observed that there was hardly any confusion between cells of group 1 with group 3 or cells of group 2 with group 3. In the first case, *KNN* and *RF* did not present confusion (0%), while *MLP, SVM,* and *CNN* presented 0.18%, 0.77%, and 2.17%, respectively. In the second case, *KNN, SVM,* and *RF* presented a value below 1.5%, while *MLP* presented a value of 6.87% and CNN the highest value, 8.65%.

Greater confusion appears between groups 1 and 2. As can be seen, the misclassification of cells in group 1 as group 2 varied between 10.17% for *KNN* and 25.25% for *RF*. In the case of misclassification of cells in group 2 to group 1 was when the *CNN* performed better than the other algorithms, with 35.86% for *CNN*, reaching 76.78% in the case of *SVM*. This inaccuracy was due to the similarity between some cells, as can be seen in Figure 6.

4. Conclusions

The work presents different solar *PV* cell defect classifiers, using five different classifier models, *KNN, SVM, RF, MLP,* and *CNN*. The classification was carried out based on EL images and I-V curves, all of them at the solar *PV* cell level. For all cases, good classification was obtained, and the differences between the proposed models were analyzed. *CNN* presented the highest percentage of success, with 81.6%, but it was also extremely slow (around 100 h). *SVM* was the model that presented the second-highest efficiency (76.07%), and it was the model with a reasonably short computation time.

The classifiers' biggest application is in defective solar *PV* cells classification. Furthermore, this group of cells is the one of greatest interest since it is the group that contains cells with almost zero electrical production. The work also presents a method to be used to select variables of interest, which will serve to train the different models of the classifiers. This process is essential since the use of variables without relevance can cause noise in training and the consequent obtaining of bad results in the classification.

The study has focused on *PV* solar cells from a single *PV* solar module. As future work, it is proposed that the dataset should be extended, and the evaluation of how this extension improves the results in each of the five models should be evaluated. We will also work on collecting data from individual cells (isolated, which are not part of a module) to expand the dataset in the future. We are also going to explore the option of generating synthetic data with *GAN* networks. The authors will also work on collecting data from isolated individual PV cells (which are not part of a module) to expand the dataset in the future. We are also going to explore the option of generating synthetic data with *GAN* networks. The authors will also extend this work, applying the classification to *PV* solar cells to other modules. Testing *IRT* image-based classifiers and *IRT* and *EL* image classifiers together are also proposed as future works. This work is of interest as it is known that both techniques are complementary in certain aspects. Another application of *AI* in which authors are working is the estimation of the I-V curve from *EL* images and *IRT*, at the level of the solar *PV* module.

Author Contributions: Conceptualization, L.H.-C., S.G.-S., and V.A.-G.; methodology, Á.P.-R., L.H.-C., S.G.-S., and V.A.-G.; software, Á.P.-R. and H.F.M.-R.; validation, L.H.-C., S.G.-S., V.A.-G., and M.d.C.A.-G.; formal analysis., L.H.-C. and M.d.C.A.-G.; investigation, L.H.-C., S.G.-S., V.A.-G. and M.d.C.A.-G.; resources, Á.P.-R., L.H.-C., S.G.-S., V.A.-G., and M.d.C.A.-G.; data curation, Á.P.-R., S.G.-S., and H.F.M.-R.; writing—original draft preparation, Á.P.-R., L.H.-C., and S.G.-S.; writing—review and editing, L.H.-C. and M.d.C.A.-G.; visualization, L.H.-C. and M.d.C.A.-G.; supervision, L.H.-C., and M.d.C.A.-G.; project administration, L.H.-C.; funding acquisition, L.H.-C., S.G.-S., and V.A.-G. All authors have read and agreed to the published version of the manuscript.

Funding: This research was funded by the "Ministerio de Industria, Economía y Competitividad" grant number "RTC-2017-6712-3" with name "Desarrollo de herramientas Optimizadas de operaCión y manTenimientO pRedictivo de Plantas fotovoltaicas—DOCTOR-PV" and by the University of Valladolid 2018 pre-doctoral research contract from the budget application 180113-541A.2.01e691.

Institutional Review Board Statement: Not applicable.

Informed Consent Statement: Not applicable.

Data Availability Statement: The data is not provided.

Conflicts of Interest: The authors declare no conflict of interest.

References

1. Scholten, D.; Bazilian, M.; Overland, I.; Westphal, K. The geopolitics of renewables: New board, new game. *Energy Policy* **2020**, *138*, 111059. [CrossRef]
2. Gugler, K.; Haxhimusa, A.; Liebensteiner, M.; Schindler, N. Investment opportunities, uncertainty, and renewables in European electricity markets. *Energy Econ.* **2020**, *85*, 104575. [CrossRef]
3. REN21. *Renewables 2020 Global Status Report*; REN21 Secretariat: Paris, France, 2020; ISBN 978-3-948393-00-7. Available online: http://www.ren21.net/gsr-2020/ (accessed on 15 January 2021).
4. Behzadi, A.; Arabkoohsar, A. Feasibility study of a smart building energy system comprising solar PV/T panels and a heat storage unit. *Energy* **2020**, *210*, 118528. [CrossRef]
5. Fachrizal, R.; Ramadhani, U.H.; Munkhammar, J.; Widén, J. Combined PV–EV hosting capacity assessment for a residential LV distribution grid with smart EV charging and PV curtailment. *Sustain. Energy Grids Netw.* **2021**, *26*, 100445. [CrossRef]
6. Thornbush, M.; Golubchikov, O. Smart energy cities: The evolution of the city-energy-sustainability nexus. *Environ. Dev.* **2021**, *100626*, 100626. [CrossRef]
7. Hernández-Callejo, L.; Gallardo-Saavedra, S.; Alonso-Gómez, V. A review of photovoltaic systems: Design, operation and maintenance. *Sol. Energy* **2019**, *188*, 426–440. [CrossRef]
8. Gallardo-Saavedra, S.; Hernández-Callejo, L.; Duque-Pérez, O. Quantitative failure rates and modes analysis in photovoltaic plants. *Energy* **2019**, *183*, 825–836. [CrossRef]
9. Gallardo-Saavedra, S.; Hernández-Callejo, L.; Alonso-García, M.D.C.; Santos, J.D.; Morales-Aragonés, J.I.; Alonso-Gómez, V.; Moretón-Fernández, Á.; González-Rebollo, M.A.; Martínez-Sacristán, O. Nondestructive characterization of solar PV cells defects by means of electroluminescence, infrared thermography, I–V curves and visual tests: Experimental study and comparison. *Energy* **2020**, *205*, 117930. [CrossRef]
10. Blakesley, J.C.; Castro, F.A.; Koutsourakis, G.; Laudani, A.; Lozito, G.M.; Fulginei, F.R. Towards non-destructive individual cell I-V characteristic curve extraction from photovoltaic module measurements. *Sol. Energy* **2020**, *202*, 342–357. [CrossRef]
11. Morales-Aragonés, J.; Gallardo-Saavedra, S.; Alonso-Gómez, V.; Sánchez-Pacheco, F.; González, M.; Martínez, O.; Muñoz-García, M.; Alonso-García, M.; Hernández-Callejo, L. Low-cost electronics for online i-v tracing at photovoltaic module level: Development of two strategies and comparison between them. *Electronics* **2021**, *10*, 671. [CrossRef]
12. Morales-Aragonés, J.; Alonso-García, M.; Gallardo-Saavedra, S.; Alonso-Gómez, V.; Balenzategui, J.; Redondo-Plaza, A.; Hernández-Callejo, L. Online distributed measurement of dark i-v curves in photovoltaic plants. *Appl. Sci.* **2021**, *11*, 1924. [CrossRef]
13. Jordan, D.C.; Silverman, T.J.; Wohlgemuth, J.H.; Kurtz, S.R.; VanSant, K.T. Photovoltaic failure and degradation modes. *Prog. Photovolt. Res. Appl.* **2017**, *25*, 318–326. [CrossRef]
14. Gallardo-Saavedra, S.; Hernández-Callejo, L.; Duque-Perez, O. Technological review of the instrumentation used in aerial thermographic inspection of photovoltaic plants. *Renew. Sustain. Energy Rev.* **2018**, *93*, 566–579. [CrossRef]
15. Gallardo-Saavedra, S.; Hernandez-Callejo, L.; Duque-Perez, O.; Hermandez, L. Image resolution influence in aerial thermographic inspections of photovoltaic plants. *IEEE Trans. Ind. Inform.* **2018**, *14*, 5678–5686. [CrossRef]

16. Ballestín-Fuertes, J.; Muñoz-Cruzado-Alba, J.; Sanz-Osorio, J.F.; Hernández-Callejo, L.; Alonso-Gómez, V.; Morales-Aragones, J.I.; Gallardo-Saavedra, S.; Martínez-Sacristan, O.; Moretón-Fernández, Á. Novel utility-scale photovoltaic plant electroluminescence maintenance technique by means of bidirectional power inverter controller. *Appl. Sci.* **2020**, *10*, 3084. [CrossRef]
17. Gligor, A.; Dumitru, C.-D.; Grif, H.-S. Artificial intelligence solution for managing a photovoltaic energy production unit. *Procedia Manuf.* **2018**, *22*, 626–633. [CrossRef]
18. Wang, H.; Liu, Y.; Zhou, B.; Li, C.; Cao, G.; Voropai, N.; Barakhtenko, E. Taxonomy research of artificial intelligence for deterministic solar power forecasting. *Energy Convers. Manag.* **2020**, *214*, 112909. [CrossRef]
19. Kayri, I.; Gencoglu, M.T. Predicting power production from a photovoltaic panel through artificial neural networks using atmospheric indicators. *Neural Comput. Appl.* **2017**, *31*, 3573–3586. [CrossRef]
20. Li, L.-L.; Wen, S.-Y.; Tseng, M.-L.; Chiu, A.S.F. Photovoltaic array prediction on short-term output power method in centralized power generation system. *Ann. Oper. Res.* **2018**, *290*, 243–263. [CrossRef]
21. Hussain, M.; Dhimish, M.; Titarenko, S.; Mather, P. Artificial neural network based photovoltaic fault detection algorithm integrating two bi-directional input parameters. *Renew. Energy* **2020**, *155*, 1272–1292. [CrossRef]
22. Cho, K.-H.; Jo, H.-C.; Kim, E.-S.; Park, H.-A.; Park, J.H. Failure diagnosis method of photovoltaic generator using support vector machine. *J. Electr. Eng. Technol.* **2020**, *15*, 1669–1680. [CrossRef]
23. Pérez-Romero, Á.; Hernández-Callejo, L.; Gallardo-Saavedra, S.; Alonso-Gómez, V.; Alonso-García, M.d.C.; Mateo-Romero, H.F. Photovoltaic cell defect classifier: A model comparison. In Proceedings of the III Ibero-American Conference on Smart Cities, San José, Costa Rica, 9–11 November 2020; pp. 257–273.
24. Bishop, C.M. Pattern recognition and machine learning. In *Information Science and Statistics*; Springer: New York, NY, USA, 2006; pp. 21–24. ISBN 9780387310732.
25. Kohonen, T. Self-organized formation of topologically correct feature maps. *Biol. Cybern.* **1982**, *43*, 59–69. [CrossRef]
26. Kohonen, T. *Self-Organization and Associative Memory*, 2nd ed.; Springer: Berlin/Heidelberg, Germany, 1988; ISBN 978-3540183143.
27. Rumelhart, D.E.; Hinton, G.E.; Williams, R.J. Learning representations by back-propagating errors. *Nature* **1986**, *323*, 533–536. [CrossRef]
28. Silverman, B.W.; Jones, M.C.; Fix, E. An important contribution to nonparametric discriminant analysis and density estimation: Commentary on fix and hodges (1951). *Int. Stat. Rev.* **1989**, *57*, 233. [CrossRef]
29. Kohonen, T. *Self-Organizing Maps*; Springer Series in Information Sciences: Berlin/Heidelberg, Germany, 2001; Volume 30, ISBN 978-3-540-67921-9.
30. Kothari, S.; Oh, H. Neural networks for pattern recognition. *Adv. Comput.* **1993**, *37*, 119–166. [CrossRef]
31. Clancey, W.J. Heuristic classification. *Artif. Intell.* **1985**, *27*, 289–350. [CrossRef]
32. Rodrigues, M.A. Invariants for Pattern Recognition and Classification. In *Ensemble Learning*; World Scientific: Singapore, 2000; Volume 42.
33. Dozat, T. *Incorporating Nesterov Momentum into Adam*; ICLR Work: Paris, France, 2016; pp. 1–4.

Review

Development of a Preliminary-Risk-Based Flood Management Approach to Address the Spatiotemporal Distribution of Risk under the Kaldor–Hicks Compensation Principle

Muhammad Atiq Ur Rehman Tariq [1,2,]*, **Rashid Farooq [3,4]** and **Nick van de Giesen [5,]***

[1] College of Engineering & Science, Victoria University, Melbourne, VIC 8001, Australia
[2] Institute for Sustainable Industries & Livable Cities, Victoria University, P.O. Box 14428, Melbourne, VIC 8001, Australia
[3] Department of Civil Engineering, International Islamic University, Islamabad 44000, Pakistan; rashidmeo50@gmail.com
[4] Department of Civil Engineering, Lakehead University, Thunder Bay, ON P7B 5E1, Canada
[5] Water Resources Management Section, Faculty of Civil Engineering and Geosciences, Delft University of Technology, 2600 GA Delft, The Netherlands
* Correspondence: atiq.tariq@yahoo.com (M.A.U.R.T.); n.c.vandegiesen@tudelft.nl (N.v.d.G.)

Received: 8 November 2020; Accepted: 14 December 2020; Published: 17 December 2020

Abstract: All over the world, probability-based flood protection designs are the ones most commonly used. Different return-period design floods are standard criteria for designing structural measures. Recently, risk-based flood management has received a significant appraisal, but the fixed return period is still the de facto standard for flood management designs due to the absence of a robust framework for risk-based flood management. The objective of this paper is to discuss the economics and criteria of project appraisal, as well as to recommend the most suitable approach for a risk-based project feasibility evaluation. When it comes to flood management, decision-makers, who are generally politicians, have to prioritize the allocation of resources to different civic welfare projects. This research provides a connection between engineering, economics, and management. Taking account of socioeconomic and environmental constraints, several measures can be employed in a floodplain. The Kaldor–Hicks compensation principle provides the basis for a risk-based feasibility analysis. Floods should be managed in a way that reduces the damage from minimum investments to ensure maximum output from floodplain land use. Specifically, marginal losses due to flood damage and the expense of flood management must be minimized. This point of minimum expenses is known as the "optimum risk point" or "optimal state". This optimal state can be estimated using a risk-based assessment. Internal rate of return, net present value, and benefit–cost ratio are indicators that describe the feasibility of a project. However, considering expected annual damage is strongly recommended for flood management to ensure a simultaneous envisage of the performance of land-use practices and flood measures. Flood management ratios can be used to describe the current ratio of expected annual damage to the expected annual damage at the optimal risk point. Further development of the approach may replace probability-based standards at the national level.

Keywords: cost–benefit analysis; multicriteria analysis; comparative risk assessment; expected annual damages; economic rent; internal rate of return; present value; optimal risk point; flood management ratio

1. Introduction

Some research studies support the notion that the severity and frequency of floods have increased in recent years [1]. Flood situations are likely to worsen due to climate change, land subsidence, urbanization, and population growth. Floods will become more frequent, prevalent, and serious in the future due to demography, lifestyle, and climate trends [2]. Greater flood damage demands an effective decision-making system for flood management and makes a revision of existing strategies a priority. Apollonio, C. et al. [3] emphasized a detailed economic analysis of flood events before the decision-makers deploy flood mitigation options.

As to flooding, it is not a new issue, and neither are the solutions being proposed. However, with the ever-increasing awareness of environmental concerns, living standards, and evolved technical expertise, the expectations of having a more efficient flood management approach is natural and justifiable. As a result, several flood management options, theories, approaches, measures, plans, and strategies are emerging. Some examples are flood protection, evacuation, environmentally friendly solutions, renaturalization of rivers, foolproof structural measures, nonstructural measures, soft measures, hard measures, room for the river, flood fighting, the resilient approach, sustainable flood management, the integrated approach, risk-based flood management, living with floods, nature-based solutions, economical solutions, spatial planning modifications, flood risk zoning, coping with floods, floodplain restoration, the no-adverse-impact approach, catchment management, community cooperation, compensation, dikes, floodwalls, warning systems, encroachment control, flood insurance, floodproofing, flow diversion, groundwater recharge, public awareness, relief efforts, river improvement, storage and retention, and sustainable drainage systems.

Flood management practices have evolved on the basis of the severity of floods and the socioeconomic aspects of a region. A single solution cannot be extended to all flood situations because the situations are not the same everywhere [4]. Climate variability, land use, and land cover are among the significant parameters affecting the regime of flooding [5]. Solutions have to be determined by the socioeconomic aspects of a country and flood behavior [6]. However, probability-based flood design prevails worldwide. For example, the 1/100-year flood is the de facto flood risk management norm in the US [7–9]. The 1/100-year flood is considered the base-flood for all structural and nonstructural designs. The UK's typical design tolerance for urban areas is 1% and 0.5% probability of floods [10,11]. The Netherlands protects its population by 53 dike rings with a design probability of 1/1250 for river floods and up to 1/10,000 probability for the coasts of South Holland [12,13]. Flood safety standards along the coast are based on a 100-year flood return period in Germany and Austria [14]. Flood protection safety standards along the main courses of rivers differ greatly and range from 1/30 years to 1/1000 years along the upper Rhine [15]. In France, only one return period is calculated for flood hazards [16]. A 100-year flood is a widely accepted design standard for flood protection measures in many parts of the world [17].

Risk-based assessment methods were introduced during the 1990s in the field of flood management, and they provide a clear focus on flood impacts and the reduction of all possible flood-related risks [18]. However, three decades on, the probability-based flood protection design has become predominant due to the absence of rigid and well-modeled risk-based flood management design criteria. Other than the differences in development approaches and design criteria, the associated regulations and guidelines are not uniform for risk-based flood management throughout the world.

1.1. Research Questions

Decision-makers face the following questions when choosing their flood options and deciding on the extent to which resources will be deployed for flood management while bearing in mind other civic welfare priorities:

1. How many types of resources need to be deployed for flood management while there are also other priorities of civic welfare public works?

2. How can the best value of resources be obtained in a floodplain, including the cost of land-use practices and flood management measures?

The research presented in this manuscript addresses the fundamental questions facing decision-makers when choosing flood management measures. The article lays out a provisional risk-based approach based on the Kaldor–Hicks compensation principle to address the economic justifications of risk-based flood management.

1.2. Limitations and Scope

The research presented covers the development of a provisional basis for choosing appropriate flood management measures and the limits of financial resources for risk-based flood management options. The implementation strategies, technical designs, cost–benefit analyses, and discussions on possible flood management options are beyond the scope of this research. Floods take several forms: pluvial, fluvial, urban, flash, tidal, and dam breaks. However, for this research, only fluvial floods are considered for the development of a decision support system for flood management.

1.3. Flood Management Plans

Governments incur considerable expenditure in connection with plans for flood management. The intensity of flooding problems and resource availability govern the extent of resource allocation for flood projects. Flood management projects or plans are the practical steps of a flood management strategy consisting of one or more measures planned according to the nature of flood issues. Plans are transformed into projects using resources, while a combination of measures may be implemented throughout a strategy's lifetime [19,20]. The following examples demonstrate different strategies:

- **Flood control/flood mitigation** tends to limit flood distribution through structural measures.
- The adjustment or modification of human activities to minimize flood losses is obtained by **adaptation**.
- Flood management based on **resilience** is an approach that emphasizes structural floodplain operations to ensure that after a flood occurs, the system can recover.
- **Integrated flood management** ensures that flood management is appropriately linked to other river functions and floodplain operations. There is multidisciplinary knowledge involved.
- **Sustainable flood management** ensures that measures that will not cause serious complications in the future are chosen. A fair valuation of social, economic, and environmental assets is required for this approach.
- The **no adverse impact** strategy considers plans that do not shift or increase flood hazards to neighboring floodplains or areas.
- **Floodplain restoration** assumes that a natural floodplain has better flood-handling capacities.

The modern flood management concepts above have been evaluated over time. A brief history of different flood management eras will shed light on the development of the different concepts over time.

1.4. Four Generations of Flood Management

The first human response to floods or any natural hazard was to adapt. Human lifestyle and activities were adjusted to ensure minimum interference to waterways [6]. This indigenous era of flood adaptations started in the 19th century and included growing flood-resistant crops, relocating dwellings, and early embankments.

Growing urbanization and greater economic exposure in floodplains heralded the second generation of flood management, which was mainly characterized by structural measures to control and defend against floods. Flood control remained the dominant approach for the late 19th century and most of the 20th century. Initially, flood protection measures were designed to ensure that structures or measures must follow a minimum standard; this design approach was known as the element-design approach. For example, flood protection structures were designed in light of historical flood experiences [21,22]. Despite involving analyses of previous floods, they were not optimized in terms of a cost-to-benefit ratio. This approach is still taken in situations where high accuracy is not required [23]. However, the realization dawned that structures should be designed where there is a known probability of occurrence. As a result, probability-based safety standards were developed and implemented by prioritizing safety in the field of hazard and disaster management while controlling the return period of the flood [18]. A probability-based standard can be applied only when the risk falls below an arbitrary level, and these methods lack the ability to combat severe floods [4].

Although second-generation flood control was primarily an engineering approach, it produced more societal confidence in flood management. Nevertheless, there was a range of drawbacks, including growing exposure in a floodplain, detrimental environmental effects, and catastrophes once the flow exceeds the design flood. Initially, nonstructural measures were introduced to support the structural measures, but later on, the nonstructural possibilities were also presented as an alternative to conventional engineering solutions in the third generation of flood management. The main focus was on reducing the susceptibility to flooding or the exposure in the floodplain. Floodproofing, early warning systems, evacuation plans, public awareness, and flood zoning were introduced.

Substantial analyses of the strengths and failures of nonstructural approaches and the recurrence of severe and extremely devastating floods have culminated in a variety of strategies (rather than a single strategy) based on a more holistic approach (fourth generation) to address floods and the root causes of flooding [24]. The approach emphasizes the inclusion of all possible measures and adaptations at the catchment, river, and floodplain levels. Flood insurance and resilience capacities are recognized as flood management options. The involvement of multiple options at different stages of flood generation and flooding need criteria that would be capable of expressing the efficiencies of spending and the benefits thus obtained. Probability-based design standards failed to support the holistic approach. Risk-based assessment gained popularity due to its ability to assess and design measures under diverse situations and options. Risk-based criteria are also helpful when choosing the appropriate flood management strategy [25].

2. Risk Redistribution under the Kaldor–Hicks Compensation Principle

Holistic flood management involves a large-scale redistribution of flood risk over spatial and temporal frames. A risk-based assessment can evaluate and compare the outcomes, but the redistribution of risks involves different benefits and different losses. The Brownfield redevelopment principle is an example that combines different economic parameters under the concept of "enhancing public benefit" [26]. It is also true that the aim of all the standards developed so far is to evaluate the proportion of the costs and benefits of measures [27].

Losses/gains to small business setups involve small-scale studies, whereas such gains/losses might not be that important at a regional level due to redistributions of gains and losses to other small business setups. In this case, a business setup loses its production and another setup may have gains in its production to meet the market requirements. Thus, the loss to an individual is transferred in the shape of gains to another while having no impact at the regional level. The gains and losses of individual households, industries, or businesses are analyzed under "financial analysis", whereas the macroeconomic effects at the regional level are covered under "economic analysis" [28]. Economic analysis is carried out to assess the feasibility of a flood management scheme. The Kaldor–Hicks compensation principle justifies redistribution and provides a criterion for combining the redistributed losses and benefits. The Kaldor–Hicks compensation principle states, "A redistribution of risk is efficient if it enables the gainers to compensate the losers, whether or not they do so" [29]. Flood management measures require significant resources, and sometimes, these measures just shift the flood spatially instead of suppressing the flooding. In such situations, just focusing on the loss reductions at a particular area of floodplain proves quite illusive. Kaldor proposed the principle of potential compensations in 1939. The main opposition that Kaldor faced about the principle was of an ethical nature, where marginal costs were ignored and the same weightage was given to all individuals. Hicks supported the proposed approach to aggregate welfare analysis proposed by Kaldor, which had benefits [30]. The Kaldor–Hicks compensation principle emphasizes the holistic consideration of all impacts. The principle of compensation provides for a rule that benefits (gains in human well-being) should exceed the cost (losses in human well-being) of projects or plans to be approved [29].

Structural measures are changing the pattern of floods. Such a shift of flow pattern brings a redistribution of risk within the floodplain. Normally, such a shift of flow results in the redistribution of flood risk from densely populated and intensively industrialized areas to rural areas characterized by sparse population and agricultural activities. Such redistributions are best handled by the Kaldor–Hicks compensation principle, which ignores the marginal costs and benefits at the regional level. The overall reduction in risk needs to be evaluated in totality. However, this shift in flooding does not protect against extreme and catastrophic events, which can be covered by insurance to support floodplain activities with lower costs [31].

2.1. Risk-Based Flood Management

Management and coordination of measures/actions that are used to reduce flood problems can be considered to be flood management [32]. An effective flood strategy reduces risk by reducing hazards (by reducing the probability and/or intensity of flooding) or vulnerability (by reducing the susceptibility and/or exposure). The measures are selected and designed to achieve the desired results within the economic, social, and environmental limitations. Evaluation criteria need to be clear, transparent, and object-oriented and include social appraisal and economic and environmental evaluation due to nonstructural and structural technical requirements. The reliability of evaluation depends on the correct assessment of the inputs, such as flood frequency, the vulnerability of the exposed asset, and other risk factors [3].

"Flood management" refers to the overall process involved in mitigating the extent of flooding and the resulting damage by flooding [18], whereas "risk-based flood management" is the combination of all actions that aim to meliorate overall activities in a floodplain [33]. Risk-based management is geared towards evaluating the projects for reducing, but not necessarily eliminating, the flood risk [34]. In practical terms, the chance of flooding can never be eliminated. However, the consequences of flooding can be mitigated by appropriate behaviors and actions [2]. The basic principle of flood risk management should be "adjusting from both ends to achieve moderation" [35]. Vulnerability and/or hazard parameters can be adjusted to improve the floodplain functions. The flood management options available can be defined as those that reduce the challenge and those that enhance the individuals' and society's ability to cope with the flood. Typically, a mixture of methods would be the most effective management technique.

The most important task regarding flood management is to select the most effective and suitable measure from direct (flood abatement, flood control, and flood alleviation) and indirect measures (involving structural and nonstructural measures) [17,36]. A risk-based assessment enables policy-makers to focus on the outcomes and, thus, results in a set of the most suitable flood management measures. Risk-based flood management provides a rationale to spend resources on flood management options [3]. Resources can be spent in proportion to the risk involved, whereas the risk arises due to the combined environmental, social, and economic impacts of flooding. Hence, risk-based flood management facilitates the effective selection of different options. The research conducted by Pezzoli A., et al. [37] discussed the meteorological, hydraulic, and statistical fields of knowledge. In another similar attempt, a multivariate statistical model was set up to represent a typical probability distribution of measured river or sea level data at multiple locations by R. Lamb et al. [38] for the development of a risk-based assessment methodology for the UK. A proposed stepwise framework to evaluate the efficiencies of the flood management scheme is presented in Figure 1. The proposed framework is fitted into the recommended context of ISO Guide 31000 [39] to ensure the standardization of the approach.

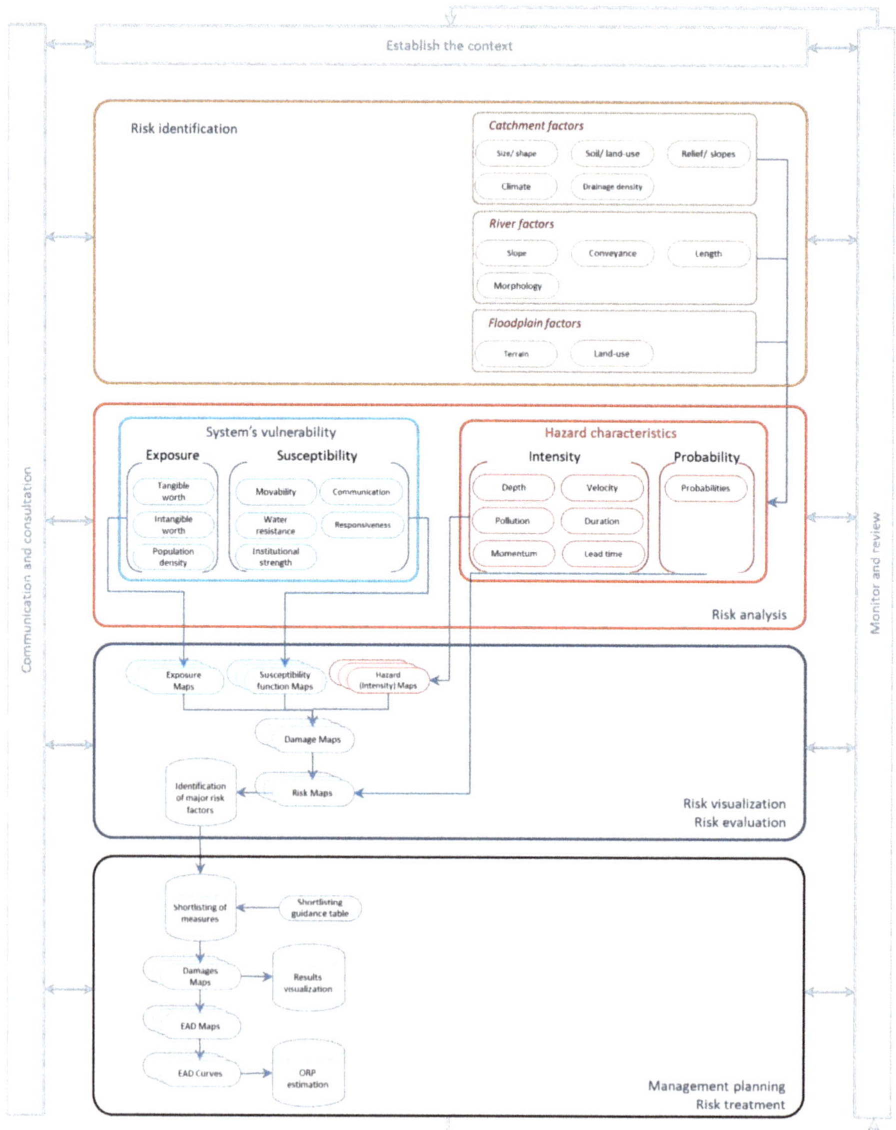

Figure 1. The proposed basic framework for risk-based flood management fitted into the context of ISO Guide 31000 [39].

2.2. Classifying Suitable Measures

A detailed risk assessment of all possible flood management measures is not economically feasible. Using available data, a competent professional with knowledge of the flood management domain can discard certain flood management options without analyzing each of them in depth [40]. Similarly, there are certain measures that are preferable under certain prevailing conditions [41]. For example, floodproofing of individual areas within the floodplain has proved economically more feasible than flood protection by constructing dikes under conditions where the floodplain is not densely populated [36]. A few measures are highly counterproductive in certain situations. One example is constructing a dike in mountainous areas, where flood levels fluctuate more rapidly and the chances of dike failure are high. In this case, the dike prevents the floodwaters from receding once the flood peak has passed. Which flood measures are selected depends on the available resources, technical limitations, and the effectiveness of an individual measure. The following are common examples of flood management measures, with a brief description:

- **Reservoirs** reduce the flood peaks holding the surplus flows.
- **Rain harvesting** also works in the same way as reservoirs but on a smaller scale. The stored water is used mainly for domestic purposes.
- **Dikes, levees, floodwalls, and other barriers** suppress the flood hazard from reaching settlements, thus avoiding the risk.
- **Channel improvements** facilitate the convenience of flow within a channel and reduce the stage against a high discharge by widening, deepening, smoothing, or removing curves from a channel.
- **A diversion (or bypass)** of the flow capacity of a channel is increased by diverting a part of the flow to follow a different flow path and rejoining the main stream at a downstream point, where appropriate.
- **Insurance and relief** are approaches to make the system more robust by increasing the recovery speed and extent of recovery. These options also play an indirect role in controlling exposure and susceptibility within a floodplain.
- **Flood warnings, rescue, and pre-emptive evacuation** are capacity-building measures within the floodplain that enhances the capacity of settlements to react against floods by moving out or removing the assets from a floodplain in time.
- **Public awareness** is a multidimensional tool that helps us to react against floods in many ways because of the flexibility and effectiveness of allowing settlements to understand flood issues, management options, and response options.
- **Flood zoning, encroachment control, and implementing building codes** are effective tools to control the vulnerability of floodplains in an effective way.

The selection of a suitable measure is carried out by considering the appraisal requirements and the contribution of the measures in achieving the appraisal objectives. Different flood management options should be evaluated one by one by approximation. Developing options correctly can save time and resources. The following steps are recommended to develop the most suitable combination for appraisal needs.

1. Identification of a wide range of options by considering the topographical, environmental, social, cost-effective constraints.
2. Screening out of impractical and infeasible options.
3. Shortlisting of options that will most probably achieve the best use of resources.
4. Development of a wide range of options by combining different measures.

Table 1 provides a rule of thumb guideline example to shortlist the measures under the predominant conditions of the floodplain. The data are compiled based on an extensive literature review and field experts' opinions. This guidance is tentative and based on a range of typical benefits, including reduced costs and additional socioenvironmental benefits.

Table 1. A practical example of flood management options and priorities for different characteristics of the floodplain.

Parameter	Characteristics	Suitable Measures
Terrain	Hilly	Rain harvesting Reforestation Storage and retention Encroachment control Land use adaptation
	Flat	Groundwater recharging Dikes, Floodwalls
Type of flood	Flash	Rain harvesting Reforestation Soil conservation Storage and retention Encroachment control Land use adaptation
	Fluvial	Storage and retention Dikes, Floodwalls Evacuation
Population density	Dense	Dikes, Floodwalls Flow diversion River conveyance Encroachment control Building codes
	Sparse	Encroachment control Building codes Rescue Evacuation Floodproofing Flood insurance
Capitals	High	Encroachment control Flood insurance
	Low	Land use adaptation
Responsiveness	Responsive	Encroachment control Building codes Rescue Evacuation Land use adaptation Floodproofing Public awareness Flood insurance
	Reluctant	Rescue

3. Comparison of Mutually Exclusive Alternatives

The ultimate purpose of risk analysis is to enable decision-makers to select the best available option and envisage risk reduction after a proposed project is implemented [17,25]. There are several financial techniques to support the appraisal process (e.g., cost–benefit analysis, multicriteria analysis) [6].

3.1. Opportunity Cost

Resources at the national level are limited and have different uses. The use of resources for one purpose or project implies the foregone benefit from another purpose or project that could be achieved by using the same resources. Since the objective is to maximize returns, the alternative project that is relevant here is one with the maximum output or benefit. The opportunity cost of using resources in any project is, therefore, equal to the maximum value of output that could have been produced elsewhere with the same resources. For example, the opportunity cost of money invested by a farmer is the annual interest that the money would have received if the farmer had lent it to someone or deposited it in a bank. Similarly, the cost of producing a single unit of hydroelectricity is the cost of producing a single unit of thermal energy.

More importantly, this concept can be used to determine the value of unpriced assets. Thus, the opportunity cost of preserving the land for a national park can be estimated by deriving the income from other land uses that had to be abandoned or forfeited for the sake of conservation. This technique is also used to evaluate the benefits of preservation that are not valued on their own.

3.2. Feasibility Analysis

Damage can be divided into tangible and intangible damage. Intangibles are the losses that cannot be expressed in monetary terms with conventional approaches. Multicriteria analysis (MCA) is a common tool to assess intangible losses [42]. However, for a risk-based design, all possible damage due to floods are converted into tangible losses and evaluated through a cost–benefit analysis (CBA) to achieve maximum efficiency of the flood management approach.

3.2.1. Cost–Benefit Analysis

Most of the projects need to be financially attractive to be sanctioned for financing. CBA is performed to determine the adequacy of a project to assess its ability to meet the objectives. The analysis is helpful for designing the best combinations of flood measures [43–45]. While comparing the costs and benefits of a project, the costs for flood management has two components: the capital costs of measures and their maintenance cost. The costs also include the reduction in productivity within the floodplain, which might have been there in the absence of flooding or flood management measures as these costs are invested in before the flood occurs. Therefore, discount rates must be considered by assuming the service life cycle of the measure (see Section 3.4).

Benefits can be defined as the damage that is avoided due to flood management measures. This way, the benefits include a reduction in losses due to a flood management project [43]. Risk-based flood management demands that all costs and all benefits must be accounted for. The costs and benefits must be converted into monetary terms to perform a CBA.

3.2.2. Valuation of Intangible Assets

As mentioned above, the costs, as well as the benefits, consist of tangibles and intangible components. CBA becomes difficult if these intangible costs and benefits are not converted into tangibles [46]. Some advanced and systematic approaches are required to handle the intangibles. Alternatively, CBA can be expanded using a nested multicriteria framework to consider intangible factors along with tangible factors [47]. MCA covers multiple aspects of gains and losses and handles these separately. Sometimes, weights are assigned to different criteria to determine the results of the MCA, but these are always highly controversial and questionable. MCA provides an elaborated and detailed picture of the problem [47] but, at the same time, lacks the ability to facilitate policy-makers choosing between the options as an apple-to-apple comparison is not carried out with this approach. CBA is, therefore, recommended to allow decision-makers to make a "like for like" comparison. Zeleňáková and Zvijáková [41] introduced a qualitative approach to cover multiple parameters in a risk assessment.

When costs and benefits are compared against different projects, a comparative risk assessment (CRA) is obtained. CRA is used to identify the most effective flood management scheme in the context of appropriate flood management options. As a governing principle, the lowest risk is the most desirable [44]. Therefore, utmost efforts should be made to ensure that the maximum possible number of alternative projects or policies are analyzed under CRA [29]. CRA is used to obtain the "optimal risk point" (ORP; discussed under Section 4.2) for the design of "master flood action plans" and the most appropriate land use planning in the floodplains (see Figure 2). It is important to note that CRA for an individual project is not possible. Consequently, isolated evaluation of projects without considering the optimum risk status of the overall strategy could be misleading.

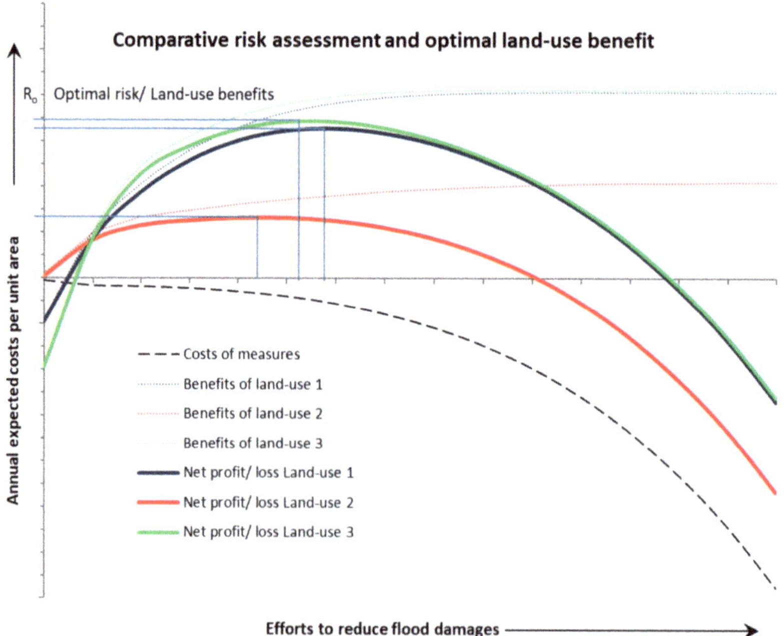

Figure 2. The graph representing the comparative risk analysis of different strategies and the identification of optimal risk for different land-use schemes in a floodplain, along with the net profit of typical land uses.

3.3. Economic Efficiency Indicators

The outcomes of a CBA are communicated using economic efficiency indicators; these economics efficiency indicators are broadly divided into two classes. The first group of indicators expresses the ratio between inputs and outputs, while the other class of indicators considers the difference in costs and benefits. A few of the most common indicators are described in the following sections.

3.3.1. Benefit–Cost Ratio (BC Ratio)

The benefit–cost ratio can be considered as the most common indicator that compares the investments and the benefits. The decision-making process can have high confidence in this indicator due to its ability to provide the gain in terms of investment percentage [48]. An example of the popularity of this indicator is that the flood defense programs in the United Kingdom must have a BC ratio of more than 8 to receive public funding [25].

3.3.2. Internal Rate of Return (IRR)

Another important term used to evaluate the feasibility of a project is the "internal rate of return" (IRR). IRR evaluates the project through its life of cash flows. If net present value (NPV) is equal to zero, IRR is equal to the discount rate. IRR is the discount rate at which the cost of the project leads to the benefit of the project [8,49]. The World Bank uses IRR as a qualifying indicator for financing projects [50].

3.3.3. Economic Rent (*ER*)

An increase in land-use efficiency due to reduced flooding is a benefit of flood management. Economic rent (*ER*) considers the efficiency of land uses in a floodplain. *ER* is defined as the net annual revenue associated with a resource that can be land use for a floodplain [51,52]. *ER* is an indicator that combines the effectiveness of alternatives to achieve the economic objectives of floodplain management.

The average annual net return, discounted to its present value, can be considered the *ER* of land use in a floodplain. If "R_n" is the annual net return of a unit of land area in year "n", assuming a constant discount rate "r", then the "*ER*" in time "t" will be the average discounted annual net return. This relationship can be expressed using the following formula:

$$ER = \sum_{n=0}^{n=t} \frac{R_n}{t(1+r)^n} \tag{1}$$

It is also possible to define the net annual return per unit of land as the difference of annual gross returns to total costs (excluding land rent), which represents economic efficiency [51]. In the case of a floodplain, the annual cost of the measure, including capital and maintenance costs, and the expected residual damage reduce the ER; these are called combined deductions. Therefore, ORP can also be defined as a state of flood management when the risk of *ER* reduction is minimal.

"Location benefits" (R_{loc}) and "intensifying benefits" (R_{int}) are the two parts of *ER* (please refer to Equation (2)). R_{loc} is connected with the availability of land for intensive economic uses, e.g., shifting from agriculture to industrial land use [53], whereas, R_{int} is the value of increasing land use such as changing from lower- to higher-value crop cultivation or, alternatively, producing higher-yield crops or intensifying agricultural economic activities [53].

$$\Delta R_n = \Delta R_{loc} + \Delta R_{int} \tag{2}$$

The negative impacts of floods are reduced with improved flood management, which results in an increase in both components by higher flows of investments. Although several factors depend on an increase in location benefits and intensification benefits, the availability of investments is the most important factor by itself. Investment project planning in the floodplain is subject to a precise evaluation of *ER*.

3.4. Discounting Procedure

Although floods cause damage to life and infrastructure, in order to reduce this damage, flood measures also require substantial investment. Investments in flood measures are ideally low and paid in advance, while the cost of flood damage is high and, with some probability, may occur in the future [54]. By reducing losses every year, flood measures provide advantages throughout their service life. Flood measures need capital investment at the start and some maintenance costs annually. Such capital investments, if deposited in a bank, could produce some annual return (equal to the compound interest or discount rate "i"). Therefore, it is not possible to consider future returns from flood measures equal to those obtained at the present time. These two sets of unevenly distributed advantages and costs are handled by the concept of present value (*PV*).

PV is a basic concept of economics that accounts for the time value of money as opposed to the future value concept. The Kaldor–Hicks compensation principle is a criterion that helps in this analysis by measuring the NPV over the expected life of the project [55]. In all analyses that compare investments and results, the discount rate is considered. Typically, the discount rate used is set by the federal government and is much higher in developing nations than in developed nations. It is possible to find the present value "*PV*" of investment "*P*" with "*i*" interest rate over time "*t*" (in years) using Equation (3).

$$PV = \frac{P}{(1+i)^t} \tag{3}$$

The costs and benefits are spread over different timestamps. Therefore, the rent on investments cannot be ignored. It is logical to compare the difference in investments and benefits in terms of present values. The *BC* ratio and net benefits (*NBs*) of a project can be calculated using the following formulae (refer to the formulas in Equations (4) and (5)) after the *PV* of benefits and costs have been estimated [29]:

$$BC\ ratio = \frac{PV_b}{PV_c} \tag{4}$$

$$NB = PV_b - PV_c \tag{5}$$

If the net benefits are positive, then the project is cost-efficient, and the *BC* ratio is greater than 1. Kaldor and Hicks had a simple criterion, *NPV* > 0, for determining whether a project is economically feasible over time [55]. The necessary condition for the adoption of a project is that discounted benefits "PV_b" should exceed discounted costs "PV_c" (see Equation (6)) or NPV should be greater than zero [8,56] (see Equation (7)).

$$PV_b > PV_c \tag{6}$$

$$NPV > 0 \tag{7}$$

As mentioned above, a project is economically feasible if the benefits are more than the costs, but the priority of projects is required in case of limited resource availability. This prioritization of projects may help in phase-wise execution of the projects or may help in selecting which projects to reject based on a cutline criterion. Nevertheless, the priority ranking of projects is always based on BC ratios. Additionally, the projects exhibiting the highest net benefits could be the preferred projects when reasonable resources become available or the availability of resources is not restrained [43].

4. Decision Support System

The aim of flood management can be to maximize net benefits within floodplains, rather than aiming solely at minimizing flood damage.

4.1. Expected Annual Damage (EAD)

Risk-based assessment methods assist decision-makers in envisaging the spatial and temporal risk distributions of floodplains. A probabilistic event [57] is flood damage, and flood intensity varies greatly over the floodplain. To date, the spatial distribution of risks and the benefits of flood mitigation measures have seldom been taken into account [58]. The average level of damage that could be expected annually due to the probabilistic nature of floods and flood losses is the expected annual damage (*EAD*). De Risi et al. [59] used an approach to evaluate the feasibility of life cycle cost (LCC) and return on investment (ROI) analysis strategies and found that by reducing expected annual damage, the projects produced net welfare gains. Damage curves and maps of *EAD* distribution provide an extensive illustration of the distribution of risk. Pellicani, R. et al. [60] used risk curves on a basin and subbasin basis to represent the temporal variability of flood risk for decision making. To discover the lateral trends in risk along the river, *EAD* can also be calculated across the flow. Therefore, decisions based on *EAD* have benefits over conventional approaches. *EAD* provides the risk estimate and shows how far from achieving the optimal risk point ("ORP") the current flood management practice is.

In the 1990s, to develop a hydroeconomic model for *EAD* estimation [57], the US Army Corps of Engineers used the relationships between damage, frequency, discharge, and stage. If flow conditions and societal vulnerability remain unchanged, these correlations function well. Structural measures substantially alter the flow regime and flood behavior. Once the flow regime is changed, relationships previously developed are no longer applicable. Therefore, a hydroeconomic model cannot be used under such conditions. In order to handle such changes appropriately, detailed analysis using GIS data and a 2D hydrodynamic model is required. Based on the actual conditions of the floodplain, the method supports *EAD* calculation. In addition to their potential benefits and proposed uses, the concepts of *EAD*, damage curve, and *EAD* distribution map have been introduced (see Equations (8) and (9)).

$$EAD = \sum_{i=0}^{i=\infty} D_i \times \Delta P_i \tag{8}$$

$$D_i = \frac{D_{P_{i-1}} + D_{P_i}}{2} \tag{9}$$

It is important to note that the term "expected" is used instead of "average". The reason is that the damages or losses are probabilistic in nature and are calculated using the exceedance of probability fittings to estimate the hazard and the resulting damage [61].

4.2. Optimum State and Optimal Risk Point

Flood management must have the ultimate purpose of reducing risk. In such situations, risk cannot be entirely nullified. Flood measures reduce flood losses, but, at the same time, these measures incur high capital and maintenance costs for operational maintenance. This means that the risk cannot be reduced beyond a certain threshold in a floodplain. At this threshold, the combined annual expected losses due to flood damage and the cost of measures are reduced to a minimum.

Measures taken in a floodplain generally have high BC ratios in the initial stages. For measures taken after that, the BC ratio decreases and may reach 1 or even lower for subsequent measures. The stage in which marginal measurement costs and marginal benefits are equal reflects the "optimum state" (see Figure 3) [62,63].

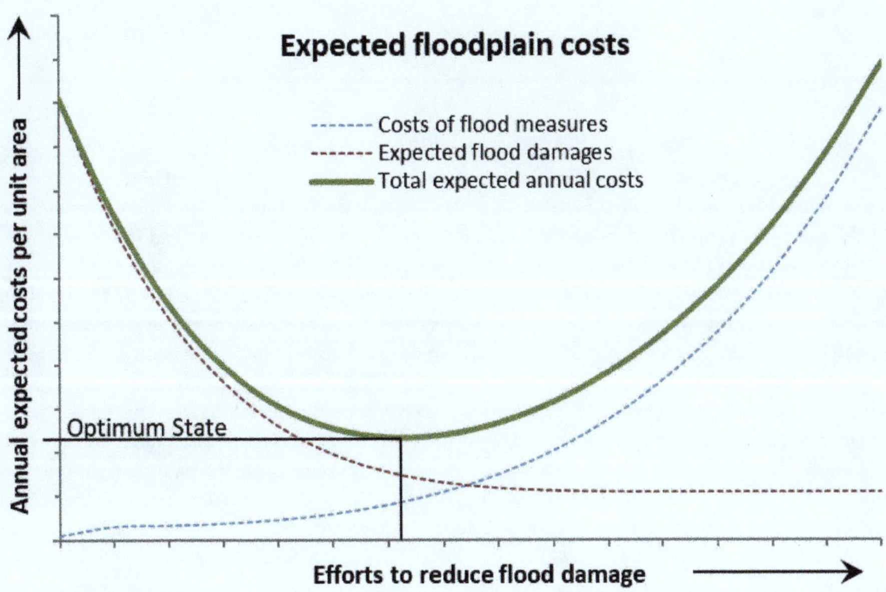

Figure 3. Graphs representing the increasing costs of flood measures and reduction in flood damages, along with the variation in total costs within the floodplain, which is minimum at the optimum state (OS).

Various flood management schemes may show different optimum states. Under all favorable combinations of flood measures, the optimum state with the lowest risk (expected costs) is known as the "optimal risk point" (ORP). As to how different strategies (consisting of appropriate measures) are compared, the ORP concept is explained in Figure 4. It also compares the feasibility of the land-use practices proposed and supports the idea that only when ORP is achieved in a floodplain can maximum land-use benefits be obtained.

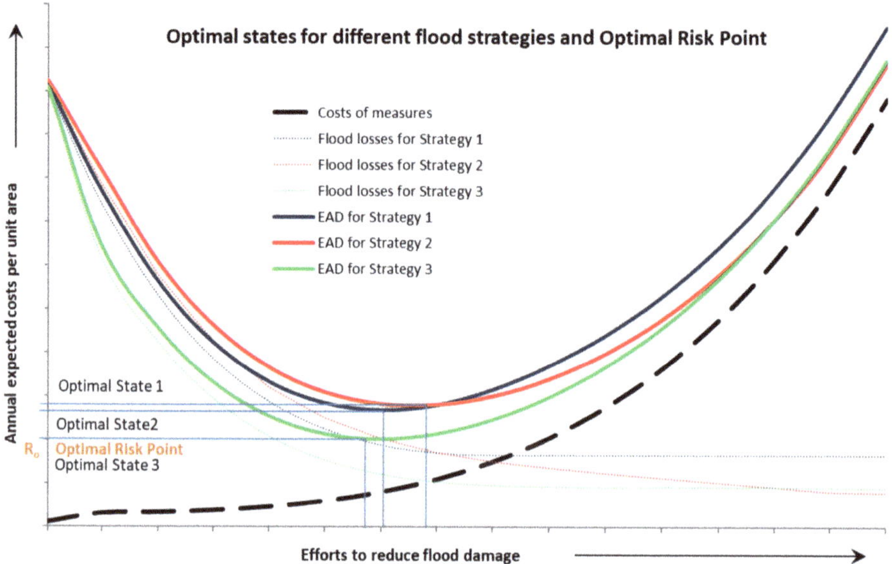

Figure 4. An illustration of three examples of land use and the establishment of the optimal risk point (ORP) in a floodplain using comparative risk analysis of different strategies.

4.3. Options under Constrained Resources

The benefit of a probability-based approach is that it is possible to establish uniform standards nationwide. So far, risk-based approaches have failed to provide standards that could be practiced across a nation uniformly. In terms of risk-based standards, the ORP concept allows flood management to be described. The present or proposed flood management status can be described in a floodplain as a ratio or percentage of ORP. The "flood management ratio" (FM ratio) reflects the ratio of present *EAD* and *EAD* at ORP. Theoretically, the FM ratio for any flood management is justified at 1 or 100% except where a shortage of resources are the main constraint and stepwise phases are planned to achieve ORP.

The use of common economic efficiency factors, as described in the above sections, may not bring up the most efficient flood management plans. As mentioned in Section 3.3, economic indicators can be classified into two distinct groups. Some indicators compare the ratios of input–output and others compare the differences in input–output. The input–output ratio is represented most effectively by the BC ratio. The difference in cost–benefit, ER, IRR, NPV, and *EAD* are based on the difference in input–output (or sometimes, the rate of difference). These indicators (apart from *EAD*) are intended to assess the benefits of an investment. These can also be used for performance assessment of investments made on flood measures. Nevertheless, the use of these indicators for flood management planning is not recommended. These indicators could be misleading and may derail the planning techniques from achieving ORP. The maximum possible ER of an available resource in a floodplain can only be attained at ORP (see Figure 5). The most efficient line of action should be determined or established upon using the proposed ORP standards. However, priorities can be assigned to individual measures within the master plan for situations where there are budget constraints, with the help of common indicators (normally the BC ratio) that are appropriate to the situation and the constraints.

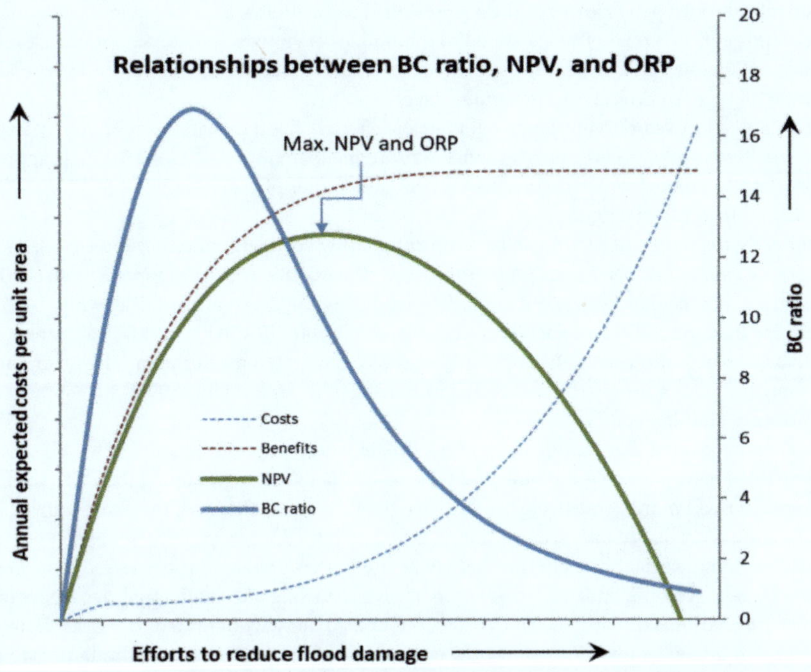

Figure 5. Illustration of flood management costs, benefits, benefit–cost ratio (BC ratio), net present value (NPV), and ORP concepts and their relative variations as efforts to reduce flood damage increase.

Flood-related damage and measure-related costs and benefits can be represented using *EAD* (see Section 4.1). The calculation of *EAD* involves multiple cycles of analysis, including hydrodynamic and economic modeling. Different scenarios of flood management options on proposed land use may come up with different *EAD*s, giving multiple optimal points against different measures or combinations of measures. Multiple iterations of analyses are required to be carried out in a floodplain under several scenarios to optimize flood management. The economic efficiencies of proposed measures can be reassessed at the planning stage to allocate priorities to individual projects in the context of situational constraints and preferences. The outcomes of risk analysis can be expressed using the BC ratio, NPV, ER, IRR, and *EAD* to facilitate policy-makers who are finalizing flood management arrangements.

5. Conclusions and Recommendations

Flood management practices have evolved. Risk-based flood management is now generally recognized around the world. Nevertheless, design procedures and criteria follow the probability-based design of flood management schemes. An estimate of risk reduction of a proposed flood management project can be made using risk analysis to help decision-makers select the most viable option. The Kaldor–Hicks compensation principle acts as the ruling principle for floodplain management systems. It establishes the governing rule that benefits should exceed the cost before policies and projects are sanctioned. The research also recommends that objective flood management should be the "welfare of society" by assigning the same weightage to all individuals.

For risk-based flood management, CBA is recommended to determine the efficiency of a project and define the best composition of a project. Costs can be estimated for the service life using discounting procedures, and the benefits can be equated to the damage avoided. The benefits of a project are, therefore, equal to the difference in damage due to the project. Intangible losses must be included in the analysis. In the analysis, social and environmental assets are as important as economic assets in order to obtain reliable outcomes. However, the conversion of an intangible component into scalar monetary value remains controversial. The exclusion of intangibles is, however, a worse option. Developing intangible-to-tangible value criteria at the national level can reduce disagreements. Such criteria could be used for all types of projects at the national level as well.

CRA is an effective option to single out the most effective flood management option by analyzing the risks of alternative projects or policies while considering that the lowest risk is the most appropriate solution. CRA is a useful tool to determine ORP for any master plan consisting of flood management measures and land use in a floodplain.

Normally, the results of a CBA can be represented using economic efficiency indicators, and these indicators are used to find the most adequate measure. The BC ratio is used to express, with confidence, the proportion of what has been spent and achieved during the decision appraisal, while IRR is used to determine the return of the project throughout its service life. IRR, NPV, and BC ratio are common indicators to express the feasibility or priority of projects within a master plan. However, for flood management, *EAD* is strongly recommended to envisage the performance of land-use practices and flood measures simultaneously.

It can be concluded that the efficiency of floodplain land use is at its maximum at ORP under the prevailing flood management techniques and their costs. Therefore, determining the ORP could be the start point for flood management planning. Once the ORP is determined, the short-term planning measures should consist of only those projects that match the long-term strategy of achieving ORP. Nevertheless, common established indicators can be used to assign the priorities to shortlisted projects using the BC ratio (recommended). However, extensive modeling efforts and analyses are required to determine the ORP. It is difficult to achieve the ORP immediately with limited resources. In that case, the flood FM ratio can be used to describe the extent of flood management in a standardized way for stepwise planning to obtain the ORP on a national level. The FM ratio guides us in identifying the present divergence from achieving the ORP within a floodplain.

It is also important to mention here that, once established, the ORP must be re-evaluated periodically. The research presented here is among the initial attempts to develop a formal risk-based flood management standard that can be established on a nationwide basis. Several potential modifications and improvements to the presented methodology are possible. Further research, case studies, and discussions are required before the proposed approach is established and recognized as a standard approach at a national level. The authors expect that further development of the approach may replace probability-based standards.

Author Contributions: Conceptualization, M.A.U.R.T. and N.v.d.G.; Data curation, M.A.U.R.T.; Formal analysis, M.A.U.R.T. and R.F.; Investigation, M.A.U.R.T. and R.F.; Methodology, M.A.U.R.T. and R.F.; Software, M.A.U.R.T. and R.F.; Supervision, M.A.U.R.T. and N.v.d.G.; Validation, M.A.U.R.T.; Visualization, M.A.U.R.T. and R.F.; Writing—Original draft, M.A.U.R.T. and R.F.; Writing—Review & editing, M.A.U.R.T. All authors have read and agreed to the published version of the manuscript.

Funding: This research received no external funding.

Conflicts of Interest: The authors declare no conflict of interest.

References

1. Aronica, G.T.; Candela, A.; Fabio, P.; Santoro, M. Estimation of flood inundation probabilities using global hazard indexes based on hydrodynamic variables. *Phys. Chem. Earth Parts A B C* **2012**, 119–129. Available online: http://www.sciencedirect.com/science/article/pii/S1474706511000568 (accessed on 15 October 2020). [CrossRef]

2. Borrows, P.; De Bruin, D. The management of riverine flood risk. *Irrig. Drain.* **2006**, *55*, S151–S157. [CrossRef]

3. Apollonio, C.; Bruno, M.F.; Iemmolo, G.; Molfetta, M.G.; Pellicani, R. Flood Risk Evaluation in Ungauged Coastal Areas: The Case Study of Ippocampo (Southern Italy). *Water* **2020**, *12*, 1466. [CrossRef]

4. APFM. *Risk Sharing in Flood Management: A Tool for Integrated Flood Management*; Associated Programme on Flood Management: Geneva, Switzerland, 2009.

5. Blöschl, G.; Ardoin-Bardin, S.; Bonell, M.; Dorninger, M.; Goodrich, D.; Gutknecht, D.; Matamoros, D.; Merz, B.; Shand, P.; Szolgay, J. At what scales do climate variability and land cover change impact on flooding and low flows? *Hydrol. Process.* **2007**, *21*, 1241–1247. [CrossRef]

6. Green, C.H.; Parker, D.J.; Tunstall, S.M.; Berga, L. *Assessment of Flood Control and Management Options*; World Commission on Dams: Cape Town, South Africa, 2000.

7. FEMA. *Flood Insurance Manual*; FEMA: Washington, DC, USA, 2008.

8. Tudose, N.C.; Ungurean, C.; Davidescu, Ș.; Clinciu, I.; Marin, M.; Nita, M.D.; Adorjani, A.; Davidescu, A. Torrential flood risk assessment and environmentally friendly solutions for small catchments located in the Romania Natura 2000 sites Ciucas, Postavaru and Piatra Mare. *Sci. Total Environ.* **2020**, *698*, 134271. [CrossRef] [PubMed]

9. Qiang, Y. Flood exposure of critical infrastructures in the United States. *Int. J. Disaster Risk Reduct.* **2019**, *39*, 101240. [CrossRef]

10. Agency, E. *Understanding Flood Risk Using Our Flood Map*; Identifying and Understanding Flood Risk in England & Wales; Environment Agency: Bristol, UK, 2006. Available online: www.environment-agency.gov.uk/floodmap (accessed on 11 October 2020).

11. Prime, T.; Brown, J.M.; Plater, A.J. Flood inundation uncertainty: The case of a 0.5% annual probability flood event. *Environ. Sci. Policy* **2016**, *59*, 1–9. [CrossRef]

12. Merz, B.; Thieken, A.H.; Gocht, M. Flood risk mapping at the local scale: Concepts and challenges. In *Flood Risk Management in Europe*; Begum, S., Ed.; Springer: Berlin/Heidelberg, Germany, 2007; pp. 231–251.

13. Jonkman, S.N.; Voortman, H.G.; Klerk, W.J.; Van Vuren, S. Developments in the management of flood defences and hydraulic infrastructure in the Netherlands. *Struct. Infrastruct. Eng.* **2018**, *14*, 895–910. [CrossRef]

14. Nordbeck, R.; Steurer, R.; Löschner, L. The future orientation of Austria's flood policies: From flood control to anticipatory flood risk management. *J. Environ. Plan. Manag.* **2019**, *62*, 1864–1885. [CrossRef]

15. Linde, A.H.T.; Bubeck, P.; Dekkers, J.E.C.; De Moel, H.; Aerts, J.C.J.H. Future flood risk estimates along the river Rhine. *Nat. Hazards Earth Syst. Sci.* **2011**, *11*, 459–473. [CrossRef]

16. De Moel, H.; Van Alphen, J.; Aerts, J.C.J.H. Flood maps in Europe—Methods, availability and use. *Nat. Hazards Earth Syst. Sci.* **2009**, *9*, 289–301. [CrossRef]

17. Tariq, M.A.U.R.; Hoes, O.; Ashraf, M.T. Risk-Based Design of Dike Elevation Employing Alternative Enumeration. *J. Water Resour. Plan. Manag.* **2014**, *140*, 05014002. [CrossRef]

18. Van Duivendijk, J. *Assessment of Flood Control and Management Options*; World Commission on Dams: Cape Town, South Africa, 1999.

19. Halcrow; Euroconsult; NDC. *Capacity Building for Integrated River Management and Subprojects Implementation Design Criteria and Methodology: Package B*; Federal Flood Commission: Islamabad, Pakistan, 2001.

20. Halcrow. *River Dove Strategy Scoping Report*; Environment Agency: Bristol, UK, 2004.

21. Andjelkovic, I. *Guidelines on Non-Structural Measures in Urban Flood Management*; UNESCO, International Hydrological Programme: Paris, France, 2001.

22. ASFPM. *Reducing Flood Losses: Is the 1% Chance (100-Year) Flood Standard Sufficient?* Report of the 2004 Assembly of the Gilbert F. White National Flood Policy Forum; National Academies Keck Center: Washington, DC, USA, 2004.

23. Nathwani, J.S.; Lind, N.C.; Pandey, M.D. *Affordable Safety by Choice: The Life Quality Method*; Institute for Risk Research: Waterloo, ON, Canada, 1997.

24. Rusman, B.; Istijono, B.; Ophyandri, T.; Junaidi, A. Social, Economic and Environmental Perspectives of Flood Assessment on Delta Lowland. *Int. J. Civ. Eng. Technol.* **2017**, *8*, 966–978.

25. Hino, M.; Hall, J.W. Real Options Analysis of Adaptation to Changing Flood Risk: Structural and Nonstructural Measures. *ASCE ASME J. Risk Uncertain. Eng. Syst. Part A Civ. Eng.* **2017**, *3*, 04017005. [CrossRef]

26. Song, Y.; Kirkwood, N.; Maksimović, Č.; Zheng, X.; O'Connor, D.; Jin, Y.; Hou, D. Nature based solutions for contaminated land remediation and brownfield redevelopment in cities: A review. *Sci. Total Environ.* **2019**, *663*, 568–579. [CrossRef] [PubMed]

27. Hoes, O.; Schuurmans, W. Flood standards or risk analyses for polder management in the Netherlands. *Irrig. Drain.* **2006**, *55*, S113–S119. [CrossRef]
28. Door, K.; de Bruijn, M.; van Beek, E. *Resilience and Flood Risk Management: A Systems Approach Applied to Lowland Rivers*; Technische Universiteit Delft: Delft, The Netherlands, 2005.
29. Pearce, D.; Atkinson, G.; Mourato, S. *Cost-Benefit Analysis and the Environment: Recent Developments*; Organisation for Economic Co-Operation and Development: Paris, France, 2006.
30. Martin, S. The Potential Compensation Principle and Constant Marginal Utility of Income. *Jpn. Econ. Rev.* **2019**, *70*, 383–393. [CrossRef]
31. Erdlenbruch, K.; Thoyer, S.; Grelot, F.; Kast, R.; Enjolras, G. Risk-sharing policies in the context of the French Flood Prevention Action Programmes. *J. Environ. Manag.* **2009**, *91*, 363–369. [CrossRef]
32. Riquet, M.D.; Paudyal, D.G.; Dolcemascolo, M.G.; Abo, M.F.J. *Integrated Flood Risk Management in Asia 2*; The Asian Disaster Reduction Center (ADRC): Bangkok, Thailand, 2005.
33. Tariq, M.A.U.R.; van de Giesen, N.; Hoes, O. Development of expected annual damage curves and maps as a basic tool for the risk-based designing of structural and non-structural measures. In *Floods in 3D: Processes, PatternsPredictions, Proceedings of the EGU Leonardo Conference, Bratislava, Slovakia, 23–25 November*; Szolgay, J., Danáčová, M., Hlavčová, K., Kohnová, S., Pišteková, V., Eds.; Slovak University of Technology in Bratislava: Bratislava, Slovakia, 2011.
34. Pilon, P.J.; Davis, D.A.; Halliday, R.A.; Paulson, R. *Guidelines for Reducing Flood Losses*; Inter-Agency Secretariat of the International Strategy for Disaster Reduction (UN/ISDR): Geneva, Switzerland, 2003.
35. Cheng, X. Recent progress in flood management in China. *Irrig. Drain.* **2006**, *55*, S75–S82. [CrossRef]
36. Abbas, A.; Amjath-Babu, T.S.; Kächele, H.; Usman, M.; Iqbal, M.A.; Arshad, M.; Shahid, M.A.; Müller, K. Sustainable survival under climatic extremes: Linking flood risk mitigation and coping with flood damages in rural Pakistan. *Environ. Sci. Pollut. Res.* **2018**, *25*, 32491–32505. [CrossRef]
37. Pezzoli, A.; Cartacho, D.L.; Arasaki, E.; Alfredini, P.; Sakai, R.D.O. Extreme Events Assessment Methodology Coupling Rainfall and Tidal Levels in the Coastal Flood Plain of the Sao Paulo North Coast (Brazil) for Engineering Projects Purposes. *J. Clim. Weather Forecast.* **2013**, *1*. [CrossRef]
38. Lamb, R.; Keef, C.; Tawn, J.; Laeger, S.; Meadowcroft, I.; Surendran, S.; Dunning, P.; Batstone, C. A new method to assess the risk of local and widespread flooding on rivers and coasts. *J. Flood Risk Manag.* **2010**, *3*, 323–336. [CrossRef]
39. Purdy, G. ISO 31000:2009-Setting a New Standard for Risk Management. *Risk Anal.* **2010**, *30*, 881–886. [CrossRef] [PubMed]
40. Ahmad, S.; Simonovic, S.P. Integration of heuristic knowledge with analytical tools for the selection of flood damage reduction measures. *Can. J. Civ. Eng.* **2001**, *28*, 208–221. [CrossRef]
41. Zeleňáková, M.; Zvijáková, L. Environmental impact assessment of structural flood mitigation measures: A case study in Šiba, Slovakia. *Environ. Earth Sci.* **2016**, *75*. [CrossRef]
42. Dassanayake, D.R.; Burzel, A.; Oumeraci, H. Coastal Flood Risk: The Importance of Intangible Losses and Their Integration. *Coast. Eng. Proc.* **2012**, *1*. [CrossRef]
43. Medina, D. *Benefit-Cost Analysis of Flood Protection Measures*; Metropolitan Water Reclamation District of Greater Chicago: Chicago, IL, USA, 2006.
44. Azmeri, A.; Yunita, H.; Safrida, S.; Satria, I.; Jemi, F.Z. Physical vulnerability to flood inundation: As the mitigation strategies design. *J. Water Land Dev.* **2020**, *46*, 20–28. [CrossRef]
45. Pesaro, G.; Mendoza, M.; Minucci, G.; Menoni, S. Cost-benefit analysis for non-structural flood risk mitigation measures: Insights and lessons learnt from a real case study. In *Safety and Reliability—Safe Societies in a Changing World*; Haugen, S., Barros, A., van Gulijk, C., Kongsvik, T., Vinnem, J.E., Eds.; CRC Press: Boca Raton, FL, USA, 2018; Chapter 14; pp. 109–118. [CrossRef]
46. Kron, W. Flood Risk = Hazard X Exposure X Vulnerability. In Proceedings of the 2nd International Symposium on Flood Defence, Beijing, China, 10–13 September 2002; pp. 82–97.
47. Tung, Y.-K. Risk-based design of flood defense systems. In Proceedings of the 2nd International Symposium on Flood Defence, Beijing, China, 10–13 September 2002.
48. Messner, F.; Penning-Rowsell, E.; Green, C.; Meyer, V.; Tunstall, S.; van der Veen, A. *Evaluating Flood Damages: Guidance and Recommendations on Principles and Methods*; FLOODsite Consortium: Wallingford, UK, 2007.

49. Yi, C.-S.; Lee, J.-H.; Shim, M.-P. GIS-based distributed technique for assessing economic loss from flood damage: Pre-feasibility study for the Anyang Stream Basin in Korea. *Nat. Hazards* **2010**, *55*, 251–272. [CrossRef]

50. Tan, J.-P.; Anderson, J.R.; Belli, P.; Barnum, H.N.; Dixon, J.A. *Economic Analysis of Investment Operations*; The Word Bank: Washington, DC, USA, 2001; Volume 8. [CrossRef]

51. Weisz, R.N.; Day, J.C. A regional planning approach to the floodplain management problem. *Ann. Reg. Sci.* **1975**, *9*, 80–92. [CrossRef]

52. Wandji, Y.D.F.; Bhattacharyya, S.C. Evaluation of economic rent from hydroelectric power developments: Evidence from Cameroon. *J. Energy Dev.* **2017**, *42*, 239–270.

53. USACE. *Engineering and Design: Risk-Based Analysis for Flood Damage Reduction Studies*; U.S. Army Corps of Engineers: Washington, DC, USA, 1996.

54. Bakkensen, L.A.; Ma, L. Sorting over flood risk and implications for policy reform. *J. Environ. Econ. Manag.* **2020**, *104*, 102362. [CrossRef]

55. Heydt, G.T. The Probabilistic Evaluation of Net Present Value of Electric Power Distribution Systems Based on the Kaldor–Hicks Compensation Principle. *IEEE Trans. Power Syst.* **2017**, *33*, 4488–4495. [CrossRef]

56. Feng, J.; Tang, W.; Liu, T.; Wang, N.; Tian, Y.; Shi, J.; Zhao, M. A novel method to evaluate the well pattern infilling potential for water-flooding reservoirs. In *Proceedings of the International Field Exploration and Development Conference, Xi'an, China, 18–20 September 2018*; Springer Series in Geomechanics and Geoengineering; Lin, J., Ed.; Springer: Singapore, 2020; pp. 643–650. [CrossRef]

57. Yoe, C. *Framework for Estimating National Economic Development Benefits and Other Beneficial Effects of Flood Warning and Preparedness Systems*; U.S. Army Corps of Engineers: Washington, DC, USA, 1994; Available online: https://apps.dtic.mil/dtic/tr/fulltext/u2/a281145.pdf (accessed on 15 October 2020).

58. Meyer, V.; Haase, D.; Scheuer, S. A multicriteria flood risk assessment and mapping approach. In *Flood Risk Management: Research and Practice*; Samuels, P., Huntington, S., Allsop, W., Harrop, J., Eds.; Taylor & Francis Group: London, UK, 2009.

59. De Risi, R.; De Paola, F.; Turpie, J.K.; Kroeger, T. Life Cycle Cost and Return on Investment as complementary decision variables for urban flood risk management in developing countries. *Int. J. Disaster Risk Reduct.* **2018**, *28*, 88–106. [CrossRef]

60. Pellicani, R.; Parisi, A.; Iemmolo, G.; Apollonio, C. Economic Risk Evaluation in Urban Flooding and Instability-Prone Areas: The Case Study of San Giovanni Rotondo (Southern Italy). *Geosciences* **2018**, *8*, 112. [CrossRef]

61. USACE. *Engineering and Design—Hydrologic Frequency Analysis*; U.S. Army Corps of Engineers: Washington, DC, USA, 1993.

62. Park, C. *Natural Hazards*; Lancaster University: Lancaster, UK, 1999.

63. Jonkman, S.N.; Kok, M.; Van Ledden, M.; Vrijling, J. Risk-based design of flood defence systems: A preliminary analysis of the optimal protection level for the New Orleans metropolitan area. *J. Flood Risk Manag.* **2009**, *2*, 170–181. [CrossRef]

Publisher's Note: MDPI stays neutral with regard to jurisdictional claims in published maps and institutional affiliations.

applied sciences

Review

A Critical Review of Flood Risk Management and the Selection of Suitable Measures

Muhammad Atiq Ur Rehman Tariq [1,*], **Rashid Farooq [2,3]** and **Nick van de Giesen [4,*]**

1   College of Engineering & Science, Victoria University, Melbourne, VIC 8001, Australia
2   Department of Civil Engineering, International Islamic University, Islamabad 44000, Pakistan;
    rashidmeo50@gmail.com
3   Department of Civil Engineering, Lakehead University, Thunder Bay, ON P7B 5E1, Canada
4   Water Resources Management Section, Faculty of Civil Engineering and Geosciences,
    Delft University of Technology, 2600 GA Delft, The Netherlands
*   Correspondence: atiq.tariq@yahoo.com (M.A.U.R.T.); n.c.vandegiesen@tudelft.nl (N.v.d.G.)

Received: 23 November 2020; Accepted: 5 December 2020; Published: 7 December 2020

Abstract: Modern-day flood management has evolved into a variety of flood management alternatives. The selection of appropriate flood measures is crucial under a variety of flood management practices, approaches, and assessment criteria. Many leading countries appraise the significance of risk-based flood management, but the fixed return period is still the de facto standard of flood management practices. Several measures, approaches, and design criteria have been developed over time. Understanding their role, significance, and correlation toward risk-based flood management is crucial for integrating them into a plan for a floodplain. The direct impacts of a flood are caused by direct contact with the flood, while indirect impacts occur as a result of the interruptions and disruptions of the socio-economical aspects. To proceed with a risk-based flood management approach, the fundamental requirement is to understand the risk dynamics of a floodplain and to identify the principal parameter that should primarily be addressed so as to reduce the risk. Risk is a potential loss that may arise from a hazard. On the one hand, exposure and susceptibility of the vulnerable system, and on the other, the intensity and probability of the hazard, are the parameters that can be used to quantitatively determine risk. The selection of suitable measures for a flood management scheme requires a firm apprehension of the risk mechanism. Under socio-economic and environmental constraints, several measures can be employed at the catchments, channels, and floodplains. The effectiveness of flood measures depends on the floodplain characteristics and supporting measures.

Keywords: flood risk management; measures; approaches; design criteria; plans; schemes; hazard; vulnerability; risk; expected annual damages; direct and indirect flood losses

---

1. Introduction

Floods can be devastating, and they are frequently compared to all other natural disasters worldwide, because they have the highest number of casualties and the highest economic losses. The worldwide average direct annual cost of natural disasters between 2000 and 2012 was roughly $100 billion [1]. The following facts are presented based on flood-associated studies [2]. One-third of natural catastrophes can be considered to be floods, one of every ten fatalities due to natural disasters are flood-related, and flooding is responsible for one-third of overall economic loss ($250 billion worldwide over the last fifteen years alone). A stark 95–97% of fatalities from natural disasters are caused by floods in developing countries. At the same time, 90% of natural disasters include flooding, causing a $6 billion annual loss in the world economy. The impact of floods can vary throughout the world due to geographical, agricultural, and economic reasons. The Nile, the Mekong, and the lower Indus flood plains are used for regular agricultural needs, while the risk of floods is largely overlooked by the growing population and industrial activities of both developed and developing countries [3,4]. However, some research studies support the severity of floods, and in recent years, the frequency of floods has increased [5]. Flood situations tend to worsen with climate change, land subsidence, urbanization, and population growth.

The problem and its solutions are nothing new, but the demand for further efficient and effective flood management is quite obvious with the ever-growing environmental awareness, higher living standards, and technological expertise. Moreover, management of floods is an obtuse, complex phenomena that should be periodically revised roughly every 30–50 years [6]. A variety of new strategies, concepts, and initiatives have consequently been proposed, namely: flood protection, evacuation, environmentally-friendly solutions, the denaturalization of rivers, foolproof structural measures, non-structural measures, soft measures, hard measures, room for the river, flood fighting, resilient approach, sustainable flood management, integrated approach, risk-based flood management, living with floods, nature-based solutions, economical solutions, spatial planning modifications, flood risk zoning, coping with floods, floodplain restoration, non-adverse impact approaches, catchment management, community cooperation, compensation, dykes, floodwalls, warning system, encroachment control, flood insurance, floodproofing, flow diversion, groundwater recharge, public awareness, relief efforts, river improvement, storage and retention, and sustainable drainage systems, just to name a few.

1.1. Research Question

In the presence of numerous flood management approaches, measures, criteria, plans, and strategies, it is difficult to choose the most appropriate flood management practice; however, choosing the most efficient and appropriate strategies is the most critical step toward flood management. With the increasing number of new concepts and terminologies, the following main concern arises:

How can an effective and efficient flood management practice be developed under the prevailing conditions of a floodplain?

The following sub-questions are formulated to build the investigation structure of the main research question:

1. How can an appropriate measure/scheme of measures for flood management be chosen?

    a. What are the basic elements of flood management?
    b. What is the role of different measures in flood risk reduction?

2. What is risk-based flood management, and where does it stand when there are so many modern concepts, theories, approaches, and strategies?

    a. Among so many understandings and definitions of risk, which one is best, and why should it be considered for flood management?
    b. How can different flood management measures and approaches be compared while they have different targets and different outcomes?
    c. Is a detailed assessment of all possible combinations of measures required in order to design a more effective scheme for a floodplain?

The research presented through this manuscript addresses the basic questions that decision-makers have while choosing flood management measures. Furthermore, fundamental principles of flood management have also been identified in order to clarify essential theories, techniques, and solutions for flood prevention. Herein, specific guidelines are discussed for an optimal flood management approach.

1.2. Limitations and Scope

This research paper covers the development of a general framework for flood management options. The implementation strategies, technical designs, cost–benefits analysis, and discussions on possible flood management options are beyond the scope of this research. Several types of floods exist, such as pluvial, fluvial, urban, flash, tidal, and dam breaks. However, for this research, only fluvial flood is considered for the formation of a decision supporting system in managing floods. The research strongly recommends considering the environmental and social effects of flood management aspects, and assumes that intangibles related to these two aspects are converted into tangibles for risk evaluation. However, the negative environmental impacts of the structural measures are not considered while shortlisting the preferred flood management options.

2. Flood Management Practices

Risk-based flood management is in the initial evolving stages. The comprehension, practices, and criteria are maturing, with slight variations around the globe. Despite differences in the formation of strategies, the rules and obligations involved with stormwater maps are often varied around the globe. From region to region, procedural information and mapping methods often differ, mainly because of local knowledge and problems. Thus, in several countries, flood maps but with different features have been developed.

2.1. Worldwide Practices

In the USA, the Federal Emergency Management Agency (FEMA) identifies flood hazard areas, maps flood hazard areas, sets flood insurance rates, covers the risk, establishes design requirements for floodplain development, and provides funding for mitigation projects [7]. The 1/100-year flood is the de facto flood risk management norm in the USA. Local governments prohibit construction in a floodway of a 1/100-year flood zone. In addition, the ground floor of new or significantly renovated buildings in a 1/100-year flood zone must be elevated if a levee is not provided. Similarly, for both river and sea flooding, flood maps are developed in the U.K. "River or sea flooding was mapped for 1 percent and 0.5 percent probability floods, respectively, not considering defenses because if an extreme event flood occurs, defenses can be overtopped or fail" [8]. The Netherlands has a long history of flood management. At the moment, about 55% of the Netherlands' total surface area is vulnerable to flooding, and about 25% is below the mean sea level [9]. The country follows different standards, including the highest safety standard (1/10,000-year flood) [10]. At the same time, Germany's Parliament adopted the "Flood Control Act" in July 2004 as a consequence of severe flood damage in August 2002; however, no formal and uniform nomenclature or accepted flood management practices are available [11]. Flood safety standards along the coast are based on a 100-year flood return period. Flood protection safety standards along the main courses of the river differ greatly, and range from 1/30 years to 1/1000 years along the upper Rhine [12]. Approximately one-fourth of the total French population is exposed to floods [13]. In France, only one return period is calculated for flood hazard [10]. The Federal Flood Commission (FFC) is responsible for all flood-related issues in Pakistan [14]. A 1/100-year flood is a widely accepted design standard of flood protection measures. However, flood management is dominated by political influence.

2.2. Measures, Approaches, and Practices

The most crucial step for flood management is to select effective and suitable measures, such as direct measures (flood abatement, flood control, and flood alleviation) and indirect measures, both structural and non-structural. Flood control activities can conventionally be categorized as recovery measures, flood alleviation, flood control, and flood relief. These activities include structural and non-structural approaches, or contribute to the reduction of floodplain activity. The risk-based assessment even impacts measures and approaches of flood management. A conceptual correlation between the different measures, plans/projects, design criteria, and approaches are depicted in Figure 1. This section discusses the role of these components and illustrates the significance of implementing practices related to risk-based management of floods.

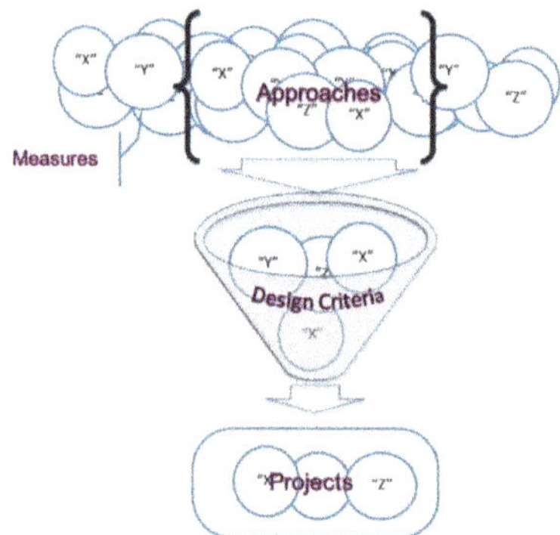

Figure 1. Schematics of flood measures, approaches, design criteria, and projects.

2.2.1. Measures

Understanding the role of numerous measures is a requirement for designing an efficient approach. Flood measures can broadly be divided into three classes, namely: thwarting flood peak formation (flood abatement), regulating the inundation (flood control), and reducing adverse effects (flood alleviation) [15]. When selecting a flood management measure, it is important to consider what the targeted risk parameter is. Table 1 describes the various flood management options/measures related to target risk factors. Understanding the role of a selected flood management measure in reducing risk by reducing one or more components of risk helps facilitate choosing an appropriate flood management option under site constraints. The constraints could be economic, social, topographical, or environmental.

Often, it is crucial to verify that measures include some structural activity that may alter a river's flow. Non-structural measures to control floods involve practices to mitigate or alleviate damage caused by floods without constructing a flow-changing infrastructure [16]. Although non-structural measures were initially suggested as part of an integrated floodplain management plan [17], the non-structural possibilities were also presented as an alternative to conventional engineering solutions.

Some practices limit risk by decreasing the chance and/or magnitude, although other measures aim to decrease susceptibility and/or damage by reducing vulnerability. Focusing on the mechanism of risk, measures are being divided into two main categories—direct and indirect measures. It is necessary to remember which criteria a flood control measure aims at during the selection of a measure. The adoption of structural measures and risk mechanics together with the target risk parameters is critical for achieving desired outcomes in the consideration of flood management solutions based on classification. Considering flood management options based on classification, the involvement of structural measures and risk mechanics, along with the target risk parameters, is important in order to obtain desirable results.

Table 1. Flood management measures and target risk parameters.

Class	Measure	Target Parameter	
Abatement	Rain harvesting	Hazard	Probability
	Reforestation	Hazard	Probability
	Soil conservation	Hazard	Intensity
	Groundwater recharging	Hazard	Probability
Control	Storage and retention	Hazard	Probability
	Dykes, Floodwalls	Hazard	Intensity
	Flow diversion	Hazard	Probability
	River re-profiling	Hazard	Intensity
	River conveyance	Hazard	Intensity
Alleviation	Encroachment control	Vulnerability	Exposure
	Building codes	Vulnerability	Exposure, Susceptibility
	Rescue	Vulnerability	Exposure
	Evacuation	Vulnerability	Exposure
	Land-use adaptation	Vulnerability	Susceptibility
	Floodproofing	Vulnerability	Susceptibility
	Public awareness	Vulnerability	Susceptibility *, Exposure *
Recovery	Flood insurance	Vulnerability	Exposure, Susceptibility
	Relief efforts	Vulnerability	Exposure °, Susceptibility
	Compensation	Vulnerability	Exposure °, Susceptibility °

* Public awareness plays an indirect role as it provokes other direct measures. ° Play negative role instead, or at least no direct role in reducing risk.

Some measures strive to improve society's ability to rebuild after floods. Examples include flood insurance, rehabilitation, reimbursement, and voluntary assistance, which can be categorized as recovery/indirect intervention, as they do not explicitly mitigate losses [18]. Relief efforts do less to mitigate potential casualties in floods [19], and likewise, the damage can also be raised by compensating.

2.2.2. Approaches

An approach or strategy is comprised of motivations and primacies that aim to select the most suitable measures. Depending on the specific economic, environmental, social, hydrological, and geological conditions, the most appropriate approach will vary accordingly [19].

The risk-based strategy not only enables analyzing and designing a measure, but also identifies the most suitable strategy in terms of the effectiveness of flood management according to the local demands. Flood management approaches, based on flooding risk mechanics, may be classified into two groups. Direct approaches strive to reduce the losses, while indirect approaches tend to recover from these losses rapidly. Direct approaches can be further distinguished in relation to those measures that minimize the vulnerability of floodplain inhabitants with those that minimize hazards [20].

- Direct approaches

  - Hazard-focused
  - Vulnerability-focused

- Indirect approaches

  - Resilience-based

The approaches for hazard management are based on theory "to keep floods away from people". Important examples are flood management and flood mitigation. The approach itself seeks to minimize floods by adopting structural measures. These strategic measures must be comprehensive and manage to reduce the flood occurrence events' probability if installed at upper catchments. Such measures may otherwise be locally applied to minimize the magnitude with which a flood affects an area [21].

The "lives with floods" theory has been adopted as an adapted indicator of risk management [22]. The components of preparedness (event-based response) and adaptation (societal adjustment) [22] minimize susceptibility. Annual flood events are anticipated and expected in highly adapted societies, while such severe events may still lead to losses and destruction [20]. Flood zoning restricts vulnerability by minimizing both susceptibility (flood proofing) and exposure (encroachment control) [23]. Encroachment control is another option based on the concept of "restraining people from the flood" by limiting exposure within a floodplain.

Examples of indirect measures include rehabilitation, compensation, flood insurance, and relief. In general, resilience-based flood management is related to indirect approaches. The persisting philosophy is "accept floods and recover afterward", while K.M. de Bruijn and Beek [24] define resilience as "the ability of a system to return to its equilibrium after a reaction to a disturbance". In order to support swift rescue after a flood subsides, a resilient mechanism needs to be developed. Indirect approaches substantially decrease indirect losses and impact direct damage indirectly.

Other concepts exist in addition to the above-stated approaches, like "integrated flood management", "no adverse impact approach", "floodplain restoration", "sustainable flood management", and so on. These options are exclusive in nature and may be selected along with other methods. Integrated flood management, for instance, avoids discrete perspectives for flood control measures [25] and does not exclude or limit a particular measure. In general, the integration of approaches and methods will bring about the most appropriate management strategy [20], and also involves complementary options [26].

2.2.3. Design Criteria

Developing an effective strategy is necessary for the enhanced effectiveness of various measures to reduce hazards (by reducing the probability and/or intensity of floods) or vulnerability (by reducing susceptibility or exposure) by adjusting one or more risk parameters. The measures are selected and designed in order to achieve the desired results within economic, social, and environmental limitations. Evaluation criteria need to be object-oriented, transparent, and clear, including economic, social appraisal, and environmental evaluation due to non-structural and structural technical requirements. Risk-based assessment methods were introduced during the 1990s in the area of flood management, and provide a clear focus on flood impacts by reducing all possible flood-related risks [16].

2.2.4. Plans and Schemes

The severity of a flood issue and the availability of resources dictates the allocation of resources for projects. A risk-based assessment enables policymakers to focus on the outcome, thus resulting in choosing a set of the most suitable flood management measures. Risk-based analysis aids in assessing and designing measures, as well as choosing an appropriate flood management strategy. Risk-based management is designed to evaluate the projects for minimizing, but not exactly eliminating, flood risk [27].

The primary goal of all of the standards that have so far been established is weighing the benefits and costs of the measures [28]. For a risk-based design, all of the possible damages due to floods are converted into tangible losses and evaluated through cost–benefit analysis in order to achieve the maximum efficiency of the flood management approach.

Small business loss is considered in small scale studies because, owing to financial transfers, such damage may not be severe at a regional level. Production in the area may be taken over by non-flooded firms, which will not cause any production losses at this stage. Loss to individual families, companies, or manufacturing plants is referred to as "financial injury", while the macroeconomic consequences are referred to as "economic damage" for a region or country [24].

3. Risk-Based Flood Management

"Flood management" refers to the general method involved in flood prevention and subsequent flood loss [16], whereas "risk-based flood management" is the combination of all of the actions that aim at the amelioration of the overall activities in a floodplain. Risk-based flood management provides a rationale to spend resources on flood management options. Resources can be spent proportional to the risk involved. Risk arises because of the combined environmental, social, and economic impacts of flooding. As stated in the preceding section, these impacts can be positive or negative. Thus, risk-based flood management facilitates the effective selection of different options.

Human society must adopt a risk management strategy in order to coexist harmoniously with flood events. In fact, it will never be possible to eliminate the risk of floods. Nevertheless, appropriate actions and behaviors can mitigate the consequences of flooding [29]. Complete abandonment of floodplains or full flood control could be extreme reactions to flooding, but these reactions could either be practically impossible or emphatically uneconomical. The basic principle of flood risk management should be "adjusting from both ends to achieve moderation" [30]. Vulnerability and/or hazard parameters can be adjusted to improve the floodplains' functions. A proposed framework is explained in Figure 2 to explore all available options in a systematic way to reduce the flood risk.

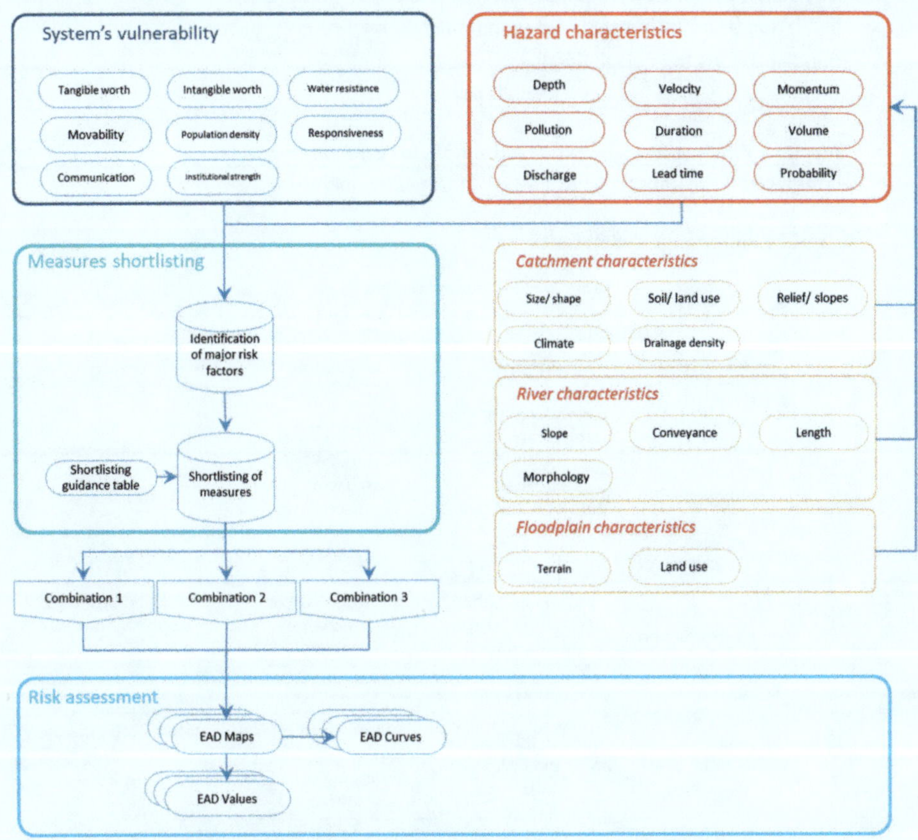

Figure 2. A proposed basic framework for risk-based flood management.

The advantage of a risk-based strategy is that it trades outcomes, which separates it from other design or decision-making methods [31]. Risk-based land-use planning in a floodplain is capable of providing a non-structural means of lowering flood losses and harmonizing floodplain activities. There is a strong need for more comprehensive studies on flood vulnerability [11]. Therefore, recent research indicates a risk-based approach to flood management [10,32–34]. Risk-based management counts every aspect of risk (flood impacts) and also considers all options to manage it the most efficiently [35]. In this way, risk-based assessment not only leads to an optimum flood management plan, but also creates an understanding of the mechanism generating the risk.

3.1. Flood Impacts

Flood impacts may include damage, disruptions, restructuring, compensation, and management expenses. There are a significant number of positive and adverse consequences of floods [24]. These consequences can be classified according to the following criteria:

1. **Type of impact:** Whether flooding is causing a positive impact or a negative
2. **Connection with the flood:**

   a. *Direct:* Whether there was a physical connection with the flood? [36,37]

   b. *Indirect:* In case the damage occurs spatially and temporally out of a flood event [36,38]

      i. *Primary:* Not direct, but still occurred within the floodplain

      ii. *Secondary:* The impact is not within the floodplain

   c. *Induced:* Efforts to manage floods

3. Capability of expression in monetary terms [39,40]

   a. *Tangible:* Whether it can be expressed simply in currency terms

   b. *Intangible:* If the impacts have a social and/or emotional value

Practical examples are provided in Table 2.

Table 2. Flood impact classification.

		Tangible		Intangible	
		Positive	**Negative**	**Positive**	**Negative**
Direct		High nutrient water to crops reducing water and fertilizer costs, flushing of salt from the land's surface	Capital loss (houses, crops, cars, factory buildings), deposition of pollution and debris or salts	Increased biodiversity, archaeological discoveries	Victims, ecosystems, pollution, monuments, culture loss
Indirect	**Primary**	Replenishing lakes and ponds for fishing production	Production losses, income loss, theft, and robbery during an evacuation	Groundwater recharge	Social disruption, emotional damage
	Secondary	Increase in production and sales of competitors in an outside area	Production losses for supplier from outside the flooded area, unemployment, inflation	Occlusion of seawater intrusion in estuaries and coastal areas, aquifer recharging of outside areas	Transmission of vector-borne diseases to an outside area, migration
Induced		Increased business & production for relief & rehabilitation, insurance business	Costs for relief aid, flood protection measures, and all management costs	Raised patriotism & regional cooperation in relief and rehabilitation	Evacuation stress Land-use restrictions/regulations

3.2. Flood Risk: Definitions and Parameters

The main objective of this research is to provide a basic framework to develop a risk-based flood management strategy for a floodplain. A first task is to put together various views of what constitutes "risk" in general, and flood risk in particular. The research also discusses how flood risk evolves, what the causes of this development are, how they can be affected so as to mitigate the risks (by what points of attack and by what measures), and how these can be combined. A detailed explanation of risk-based management is carried out in order for a better understanding of the approach. In addition, for clear comprehension and to avoid ambiguity and misconceiving, the basic terms involved have been defined here. Although there is a general agreement on most of the definitions, some terms are used for different meanings by some scientists.

3.3. Risk Perception

The term risk has been defined and understood in many instances to have somewhat similar meanings. Emphasis has been placed on various elements when being defined by different scientists, as follows:

- Risk = hazard × vulnerability [41]
- Risk = impact of hazard × elements at risk × vulnerability of elements at risk [42,43]
- Risk = hazard × vulnerability × value (of the threatened area)/preparedness [44–48]
- Risk = probability × consequences [49,50]
- Risk = hazard × consequences [51–53]

Overall, the term "risk" is somehow is related with the vulnerability, hazard, impacts values, elements at risk, preparedness of the people, results or the consequences, and the chances of all of these happening (these factors have also been differently defined). Risk may be generally envisioned as "an estimate of potential consequences associated with a hazard". The following section provides further understanding of risk, and proves how various definitions of risk are representing the same concept, just in different terms.

3.4. Risk Parameters

Risk-based assessment evaluates flood measurements based on potential impacts, while reducing negative impacts of floods, such as socioeconomic and environmental factors. This method provides a logical ground to shortlist and finalize flood management options. Understanding risk concepts, elaboration of the role of the river process, and societal activities should be considered during the assessment. Risk can be broadly defined as "an estimation of expected results connected to a hazard".

$$\text{Risk (\$ year}^{-1}) = \text{Probability (year}^{-1}) \times \text{Consequences (\$)} \tag{1}$$

Also, the risk is a function of how vulnerability interacts with hazard or vice versa.

$$\text{Risk (\$ year}^{-1}) = \text{Hazard (m year}^{-1}) \times \text{Vulnerability (\$m}^{-1}) \tag{2}$$

The above-stated term indicates that the hazard itself is not the only cause for risk induction. Everyday vulnerability and hazards shape trends of rising risk, which can end in a catastrophe arising from an exceptionally natural incident [54]. Therefore, two independent variables are considered for risk analysis—hazard and vulnerability. In flood management, the term "hazard" indicates the occurrence of a high-water-level event with a given probability of exceedance [55]. Thus, the hazard may be classified by its intensity and probability.

$$\text{Hazard (m year}^{-1}) = \text{Probability (P) (year}^{-1}) \times \text{Intensity (I) (m)} \tag{3}$$

Probability can be defined as the chance of a hazard to occur that can be defined annually for a flood event. While some negative intensity features (depth, velocity, etc.) of hazards cause damage, vulnerability is the outcome of susceptibility and exposure.

$$\text{Vulnerability (\$m}^{-1}) = \text{Susceptibility (S) (m}^{-1}) \times \text{Exposure (E) (\$)} \tag{4}$$

where people and value currently exist in an under-threat region is known as exposure. Susceptibility is generally defined as the relative loss function [56,57]. The ability to recognize the losses from the hazard is susceptibility. The extent of loss is based on the susceptibility of vulnerable items and life, and indeed the magnitude of hazard. In order to comprehend the role of both sides, risk in terms of vulnerability and hazard needs to be articulated, as illustrated in Equation (2). Thus, we get Equation (5) by substituting vulnerability and hazard by their components in Equation (2). Likewise, expressing probability alone will result in Equation (6).

$$\text{Risk} = [\underline{P \times I}] \ (\text{m year}^{-1}) \times [\underline{S \times E}] \ (\$\text{m}^{-1}) \tag{5}$$

$$\text{Risk} = P \times [\underline{I \times S \times E}] \ (\$ \ \text{year}^{-1}) \tag{6}$$

If the chance of the mutual interaction of vulnerability and hazard is considered separately, the consequences are definitely as a result of a hazard's intensity, and the total exposures of assets and how much they accept the damage from the hazard (the susceptibility). It can be observed that Equation (6) is an elaborated expression of Equation (2). A graphical demonstration is shown in Figure 3 below.

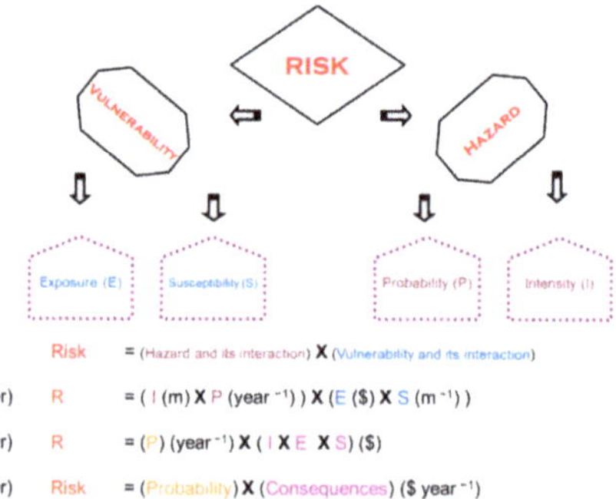

Figure 3. Graphical explanation of risk into its components and demonstration equivalency of different definitions of risk (source [18]).

3.5. Scheme of Measures

Flood management options can be defined as those that reduce challenges and those that enhance the individual and societal ability to cope with a flood. Typically, a mixture of methods would be the most effective management technique. There is an inversely proportional impact of probability on vulnerability. Normally, floods that occur more frequently are more adequately responded to, and therefore cause less harm. Quite rarely, (extreme) floods can cause significant harm. In areas that are located away from the river, these types of losses primarily occur. Because of more exposure and higher susceptibility, vulnerability rises as people who live far from the river do not expect a flood. The risk of occurrence primarily defines the magnitude of the protection measures initiated in a floodplain. Flood-suffering areas also implement infrastructure steps, such as floodproofing, etc. Flood interventions are highly dependent on risk factors, as well as technological, financial, and social constraints, for their efficacy and suitability.

Therefore, comparisons are made between alternatives, and evaluations are usually comparative. Therefore, the advantages and drawbacks of one alternative exist only in comparison with those of another. Identifying the potential role of one strategy in managing flood hazards includes identifying the potential roles of all other flood hazard management strategies at the same time. Compared with the best choices, the solution found through the evaluation process cannot be any better. Therefore, it is important to allow a sufficient range of options if the best choice among them is to be sought. While some guidelines have been suggested to maximize the probability of choosing the best alternative among those considered, the most effective method is to include the public in defining both important problems and potential solutions early on.

3.6. Assorting Suitable Measures

Detailed risk assessment of all possible flood management measures is not economically feasible. A competent professional with knowledge of the domain of flood management while using available data can discard certain flood management options without going into a deep analysis [58]. Similarly, certain measures are preferable under certain prevailing conditions. For example, floodproofing of individual areas within the floodplain has been proven to be more economically feasible compared with flood protection by constructing dykes under conditions where the floodplain is not densely populated. Some measures are highly counterproductive in certain situations. One example is constructing a dyke in a mountainous area, where flood levels fluctuate more rapidly and chances of dyke failure are high. In this case, the dyke prevents the floodwaters from receding once the flood peak has passed. The selection of a suitable measure is done by considering the appraisal needs and the contribution of the measures in achieving appraisal objectives.

Different flood management options should be evaluated one by one by approximation. Developing options correctly can save time and resources. The following steps are recommended in order to develop the most suitable combination suited to appraisal needs:

1. Identification of a wide range of options considering the topographical, environmental, social, and cost-effective constraints
2. Screening out impractical and infeasible options
3. Shortlisting options that most probably can achieve the best use of resources
4. Development of a wide range of options by combining different combinations of measures

Figure 4 provides a rule-of-thumb guideline example in order to shortlist the measures under the predominant conditions of the floodplain. The data are compiled based on extensive literature review [2,5,16,19,29,33,59–64] and field experts' opinions. Plenty of case studies were observed and their preferred approaches were noted. Figure 4 is developed after detailed brainstorming sessions with field experts. This guidance is tentative and is based on a range of typical benefits, including reduced costs and additional socio-environmental benefits.

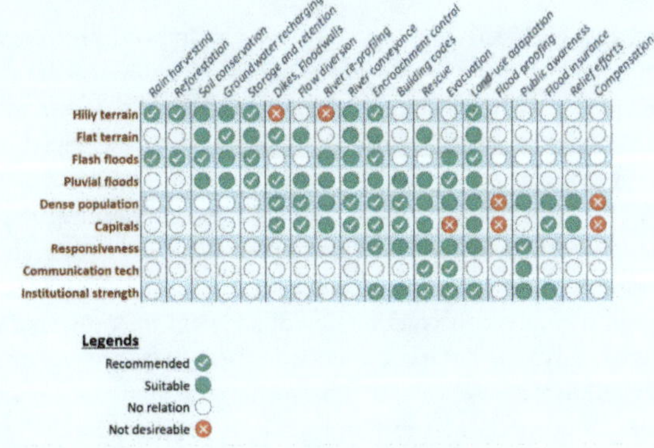

Figure 4. A practical example of flood management option priorities about different characteristics of the floodplain.

3.6.1. Suitable Measures at Catchment

Catchment management can be an effective option with or without the support of other measures. Some land-use practices can increase the tendency of rapid and high peak runoff, especially if the soil surface is left bare for long periods of the year, or periods when heavy drainage is more likely [59,65]. Not only land use, but also installing meteorological and hydrometric telemetric stations in the basin, is important for an effective flood forecast-warning system. It is important to record meteorological and hydrometric data in real-time [66]. In most cases, there is ample time to identify and notify the local authorities of an oncoming significant flood event [60].

Agricultural activities play a significant role in the generation of total volume and peak flow. For example, livestock above certain densities can lead to soil compaction (which increases rapid surface runoff) and the loss of water to evapotranspiration is minimized by reducing biomass, thereby increasing the runoff [67]. The way the catchment's surface is controlled by tillage will influence rapid runoff for a given land-use [68]. Through evapotranspiration, trees can increase water loss and increase infiltration via their root networks, both reducing overland flow generation [69]. Watershed protection steps can also be seen in watershed planning from the perspective of soil conservation, economic and agricultural aspects, and recreational concerns [70].

In the case of hill torrents/flash floods, the most effective measures lie with catchment management. A sudden increase in torrent kinetic energy may occur in the high regions. Technical interventions, such as tanks and inhibitory dams, can be designed for flood interception before the floods start heading to the low regions [66]. Steps taken at the catchment may help to de-synchronize the accumulation of water at points where the torrent's kinetic energy is increased. Furthermore, forest conservation measures should be considered, as they play a significant role in the prevention of flood phenomena [66].

3.6.2. Suitable Measures at the River

There are fewer options that can be taken at the river to interrupt the floods, such as shielding banks from erosion and increasing the conveyance of the channel [66]. The rate of flow conveyance, an inverse function of the attenuation rate in the absence of tributary inflows, depends on river and floodplain roughness; by increasing river roughness, the flow tends to deepen in the river and more easily cause flooding. As floodplain flows are generally shallow, they appear to be sluggish, and so attenuation will increase [59,71]. Excavation should be easily feasible; that is, no bedrock should be found within the diversion excavation depth. Another important consideration is that either a natural body of water should be available to dispose of the diverted water, or enough gradient should be available to drop the water back into the river without having a significant backwater effect [58].

3.6.3. Suitable Measures at Floodplain

Densely populated and high economic growth areas have been protected by building dykes. The construction and maintenance cost of dykes can be justified for the level of protection they provide to the floodplain [72]. Klijn et al. [73] investigated the enhanced exposure and increased vulnerability of The Netherlands' defense system and found that the risk was increased because of increased vulnerability. However, floodproofing or relocation can be considered if it is a small town or if there are only a few houses in the region to be covered from potential flooding. Relocation is an option only when damage is high and there is no economic reason for any other choices. Medium–low economic growth limits the deployment of costly defense systems to high-risk areas [62]. In addition, in a region with a clay soil type, the construction of a dyke is not the preferred choice [58]. Because of environmental concerns, a dyke is not recommended where there are wetlands or marshes in the region, because they dry out in the absence of regular recharging by natural floods [58].

Water storage, artificial groundwater recharging basins, and wetlands have multiple benefits in addition to flood attenuation. Artificial groundwater recharging is one of the effective options to handle flash flooding [66]. However, many considerations are required in order to implement the option. For a single location to be protected from flooding, reservoirs are not an option, unless the project can be economically justified by considering additional uses, such as hydropower generation, water supply, recreation, etc. Most essential of all, a suitable place should be available for building a reservoir [58]. Farmers' willingness to allow flooding to their land is very necessary for the option of the retention basin. Water-logging (high groundwater levels) in the region should not be serious. There should be no historical site of significance, or it should be possible to protect the site from the damage caused by managed flooding by building a dyke around it or by any other steps [58].

3.6.4. Interconnected Suitability of Measures

Flood management measures are strongly connected, and their performance is affected when applied in combination. For example, if the majority of the watershed is barren ground, dredging should not be considered, as surface runoff would carry tons of sediment. Water-logging should not be serious, as there will be a lack of additional capacity due to increased groundwater flow into the system. Dredging is not an option if the river has a serious sedimentation problem. Preferably, the additional river capacity needed to handle damaging floods should be equal to or less than 50% of the current river capacity [58].

In addition, few measures are alternative and exclusive to each other. For example, Hsieh et al. [61] evaluated three mitigation steps for flood-prone lowlands with a high population density along the Keelung River in Taiwan, including dykes, a drainage channel, and a storage reservoir. A diversion channel was proven to be more feasible than the other two flood mitigation options because of the high population and economic activity concentration in the floodplain. Where budgets are too poor to support defenses, successful, controlled realignment, or planned destruction will occur, with habitat recreation where necessary [62]. Dixon et al. [59] found that for smaller drainage basins, usually less than 100 km^2, the restoration of the river channel morphology and floodplain forest can be the basis of flood risk mitigation.

3.7. Software Tools

Several tools/software programs have been developed to carry out flood simulations in order to estimate flood damage/risk. The methodology/approaches of these tools vary at a large scale, extending from hydrodynamic modeling to statistical, artificial intelligence-based, and even stage-damage or discharge damage empirical models. The majority of tools are hydrodynamic models that simulate the flood wave in 1D, 2D, and 3D [74,75]. These include models that have compatibility with geographic information systems (GIS) [76]. A few tools have additional extensions/modules to perform statistical analysis and damage calculations as well. The inclusion of GIS capabilities has become more popular in recent years in hydrodynamic models. However, damage estimation tools recognize GIS capabilities on a larger scale. Most common software takes basic hydrologic, hydraulic, and topographical inputs to simulate flooding, while others estimate the flood damage considering the depth of flood and land-uses parameters in an empirical way. Table 3 lists a few software tools, along with their capability for GIS and country of origin.

Table 3. Popular software tools, their compatibility to geographic information systems (GIS), and their country of origin.

Software	GIS-Based	Country
ANUFLOOD	No	Australia
ESTDAM	No	UK
FAT	Yes	Czech Republic
FDAM	Yes	Japan
FLODSIM	No	South Africa
FloodAUS	Yes	Australia
HAZUS-MH	Yes	USA
HEC-FDA	No	USA
HIS-SSM	Yes	Netherlands
HWSCalc	No	Germany
TEWA	Yes	South Africa
MDSF	Yes	UK

3.8. Practical Examples

A few case studies are already being carried out under the defined framework. However, there are some good examples that can explain parts of the framework with real examples of flood management. Klijn, F. et al. [73] explained various concepts of risk-based flood management and provided initial suggestions to approach towards risk reduction. Two case studies have been referred to in order to demonstrate flood adaptations (vulnerability adjustments). Jukrkorn, N. [77] performed an ex-post analysis of a 2011 flood in central Thailand, and discussed the selection of various flood measures and their integration to a well-connected system. Agent-based modeling was performed by Yared, A. A. et al. [78], where they studied the impact of hazard reduction and vulnerability control in Sint Maarten. Another case study conducted by Rogger, M. et al. [79] explained land use effects in catchment on flood management with supporting options. Serre, D. et al. [80] explained the effectiveness of structural and non-structural measures regarding a few historic floods in Europe and the USA.

4. Conclusions and Recommendations

The concept of risk-based flood management is not new and has evolved over time. Risk-based flood management is replacing the probability-based (return period) approach, but the design standards, implementation procedures, and detailed protocols are yet to be determined. Risk-based flood management is popular and has received broader appraisal in many parts of the world, yet the approach needs a few finer refinements before being accepted as a standard approach at national levels.

Comprehending the role of risk parameters is important for the scrutiny of flood management schemes. A careful valuation of the economic, social, and environmental assets is indispensable in order to develop a balanced flood management approach. Understanding risk phenomenon is very important in order to devise the most effective solutions under the socioeconomic and environmental constraints. The basis of effective and efficient risk reduction measures lies in risk assessment, which considers the different aspects of flood risk, e.g., hydrological, hydraulic, economic, social, and ecological aspects. The presented risk definition and mechanism provide a standardized approach and establish common grounds to compare different flood management options.

Many options and measures exist for flood management. Their effectiveness, suitability, and role within a plan are highly associated with the characteristics of a floodplain and the combination of measures selected. Every floodplain has its own circumstances and needs a tailor-made solution based on thorough analysis. The initial shortlisting of measures should be done considered alongside other models suitability matched to the floodplain, as well as the combined efficiency of the measures themselves. However, the final scheme of measures must undergo a detailed risk assessment.

Author Contributions: Conceptualization, M.A.U.R.T. and N.v.d.G.; Data curation, M.A.U.R.T.; Formal analysis, M.A.U.R.T. and R.F.; Investigation, M.A.U.R.T. and R.F.; Methodology, M.A.U.R.T. and R.F.; Software, M.A.U.R.T. and R.F.; Supervision, M.A.U.R.T. and N.v.d.G.; Validation, M.A.U.R.T.; Visualization, M.A.U.R.T. and R.F.; Writing—original draft, M.A.U.R.T. and R.F.; Writing—review & editing, M.A.U.R.T. All authors have read and agreed to the published version of the manuscript.

Funding: This research received no external funding.

Conflicts of Interest: The authors declare no conflict of interest.

References

1. Kousky, C. Who holds on to their flood insurance. *Common Resour.* **2015**.
2. Loster, T. Flood trends and Global Change. In Proceedings of the EuroConference on Global Change and Catastrophe Risk Management: Flood Risks in Europe, Laxenburg, Austria, 6–9 June 1999.
3. Hassan, F.; Hamdan, M.A.; Flower, R.J.; Shallaly, N.; Ebrahem, E. Holocene alluvial history and archaeological significance of the Nile floodplain in the Saqqara-Memphis region, Egypt. *Quat. Sci. Rev.* **2017**, *176*, 51–70. [CrossRef]

4. Manning, J.G.; Ludlow, F.; Stine, A.R.; Boos, W.R.; Sigl, M.; Marlon, J.R. Volcanic suppression of Nile summer flooding triggers revolt and constrains interstate conflict in ancient Egypt. *Nat. Commun.* **2017**, *8*, 900. [CrossRef] [PubMed]

5. Aronica, G.T.; Candela, A.; Fabio, P.; Santoro, M. Estimation of flood inundation probabilities using global hazard indexes based on hydrodynamic variables. *Phys. Chem. Earth Parts A/B/C* **2012**, *42*, 119–129. Available online: http://www.sciencedirect.com/science/article/pii/S1474706511000568 (accessed on 15 June 2020). [CrossRef]

6. Plate, E.J. Flood risk and flood management. *J. Hydrol.* **2002**, *267*, 2–11. Available online: http://www.sciencedirect.com/science/article/B6V6C-46HBKF8-4/2/93f2763765cd0bad006fed4cfc6fa5c1 (accessed on 15 June 2020). [CrossRef]

7. FEMA. *Flood Insurance Manual*; FEMA: Washington, DC, USA, 2008.

8. Agency, E. *Understanding Flood Risk Using Our Flood Map*; Identifying and understanding Flood Risk in England & Wales; Environment Agency: Bristol, UK, 2006. Available online: www.environment-agency.gov.uk/floodmap (accessed on 15 June 2020).

9. De Moel, H.; Aerts, J.C.; Koomen, E. Development of flood exposure in the Netherlands during the 20th and 21st century. *Glob. Environ. Chang.* **2011**, *21*, 620–627. [CrossRef]

10. De Moel, H.; Van Alphen, J.; Aerts, J.C.J.H. Flood maps in Europe—Methods, availability and use. *Nat. Hazards Earth Syst. Sci.* **2009**, *9*, 289–301. [CrossRef]

11. Merz, B.; Thieken, A.H.; Gocht, M. Flood risk mapping at the local scale: Concepts and challenges. In *Flood Risk Management in Europe*; Begum, S., Ed.; Springer: Berlin/Heidelberg, Germany, 2007; pp. 231–251.

12. Linde, A.H.T.; Bubeck, P.; Dekkers, J.E.C.; De Moel, H.; Aerts, J.C.J.H. Future flood risk estimates along the river Rhine. *Nat. Hazards Earth Syst. Sci.* **2011**, *11*, 459–473. [CrossRef]

13. Cabal, A.; Erlich, M. Flood risk management approaches and tools for mitigation strategies of coastal submersions and preparedness of crisis management in France. *Int. J. River Basin Manag.* **2018**, *16*, 353–369. [CrossRef]

14. Tariq, M.A.U.R.; Van De Giesen, N. Floods and flood management in Pakistan. *Phys. Chem. Earth Parts A/B/C* **2012**, *47–48*, 11–20. [CrossRef]

15. De Bruijn, K.; Green, C.; Johnson, C.; McFadden, L. Evolving concepts in flood risk management: Searching for a common language. In *Flood Risk Management in Europe*; Springer: Berlin/Heidelberg, Germany, 2007; pp. 61–75.

16. Van Duivendijk, J. *Assessment of Flood Control and Management Options*; World Commission on Dams: Cape Town, South Africa, 1999.

17. White, G.F. *White, Natural Hazards, Local, National, Global*; Oxford University Press: New York, NY, USA, 1974.

18. Tariq, M.A.U.R.; Hoes, O.; Van De Giesen, N. Development of a risk-based framework to integrate flood insurance. *J. Flood Risk Manag.* **2013**, *7*, 291–307. [CrossRef]

19. Andjelkovic, I. *Guidelines on Non-Structural Measures in Urban Flood Management*; UNESCO, International Hydrological Programme: Paris, France, 2001.

20. Green, C.; Nicholls, R.; Johnson, C. *Climate Change Adaptation: A Framework for Analysis and Decision-Making in the Face of Risks and Uncertainties*; FHRC, Middlesex University: Middlesex, UK, 2000.

21. Tucci, C.E.M. *Urban Flood Management*; World Meteorological Organization: Porto Alegre, Brazil, 2007.

22. Genovese, E. *A Methodological Approach to Land Use-Based Flood Damage Assessment in Urban Areas: Prague Case Study*; Institute for Environment and Sustainability: Ispra, Italy, 2006.

23. Tariq, M.A.U.R. Risk-based flood zoning employing expected annual damages: The Chenab River case study. *Stoch. Environ. Res. Risk Assess.* **2013**, *27*, 1957–1966. [CrossRef]

24. Door, K.; de Bruijn, M.; van Beek, E. *Resilience and Flood Risk Management: A Systems Approach Applied to Lowland Rivers*; Technische Universiteit Delft: Delft, The Netherlands, 2005.

25. APFM. *Integrated Flood Management Concept Paper*; World Meteorological Organization: Geneva, Switzerland, 2009.

26. APFM. *Economic Aspects of Integrated Flood Management*; Associated Programme on Flood Management: Geneva, Switzerland, 2007; ISBN 92-63-11010-7.

27. Pilon, P.J.; Davis, D.A.; Halliday, R.A.; Paulson, R. *Guidelines for Reducing Flood Losses*; Inter-Agency Secretariat of the International Strategy for Disaster Reduction (UN/ISDR): Geneva, Switzerland, 2003.

28. Hoes, O.; Schuurmans, W. Flood standards or risk analyses for polder management in the Netherlands. *Irrig. Drain.* **2006**, *55*, S113–S119. [CrossRef]
29. Borrows, P.; De Bruin, D. The management of riverine flood risk. *Irrig. Drain.* **2006**, *55*, S151–S157. [CrossRef]
30. Cheng, X. Recent progress in flood management in China. *Irrig. Drain.* **2006**, *55*, S75–S82. [CrossRef]
31. Sayers, P.B.; Gouldby, B.P.; Simm, J.D.; Meadowcroft, I.; Hall, J. *Risk, Performance and Uncertainty in Flood and Coastal Defence: A Review*; HR Wallingford Report SR 587; Environment Agency DEFRA/EA R&D Technical Report FD2302/TR1; HR Wallingford: Wallingford, UK, 2002.
32. Hooijer, A.; Klijn, F.; Pedroli, G.B.M.; Van Os, A. Towards sustainable flood risk management in the Rhine and Meuse river basins: Synopsis of the findings of IRMA-SPONGE. *River Res. Appl.* **2004**, *20*, 343–357. [CrossRef]
33. Van Alphen, J.; Lodder, Q. Integrated flood management: Experiences of 13 countries with their implementation and day-to-day management. *Irrig. Drain.* **2006**, *55*, S159–S171. [CrossRef]
34. Petrow, T.; Merz, B.; Lindenschmidt, K.-E.; Thieken, A.H. Aspects of seasonality and flood generating circulation patterns in a mountainous catchment in south-eastern Germany. *Hydrol. Earth Syst. Sci.* **2007**, *11*, 1455–1468. [CrossRef]
35. Hooijer, A.; Klijn, F.; Kwadijk, J. (Eds.) *Towards Sustainable Flood Risk Management in the Rhine and Meuse River Basins*; Main Results IRMA Sponge Res. Proj. NCR-Publication 18-2002; NCR: Delft, The Netherlands, 2002.
36. Merz, B.; Kreibich, H.; Thieken, A.; Schmidtke, R. Estimation uncertainty of direct monetary flood damage to buildings. *Nat. Hazards Earth Syst. Sci.* **2004**, *4*, 153–163. [CrossRef]
37. Nascimento, N. Flood-damage curves: Methodological development for the Brazilian context. *Water Pract. Technol.* **2006**, *1*. [CrossRef]
38. Oosterhaven, J.; Többen, J. Wider economic impacts of heavy flooding in Germany: A non-linear programming approach. *Spat. Econ. Anal.* **2017**, *12*, 404–428. [CrossRef]
39. Smith, K.; Ward, R. *Floods: Physical Processes and Human Impacts*; John Wiley and Sons: Hoboken, NJ, USA, 1998.
40. Hudson, P.; Botzen, W.J.W.; Poussin, J.; Aerts, J.C.J.H. Impacts of Flooding and Flood Preparedness on Subjective Well-Being: A Monetisation of the Tangible and Intangible Impacts. *J. Happiness Stud.* **2017**, *20*, 665–682. [CrossRef]
41. UNDHA. *Internationally Agreed Glossary of Basic Terms Related to Disaster Management*; United Nations Department of Humanitarian Affairs: Geneva, Switzerland, 1992.
42. Blong, R. Volcanic hazards risk assessment. In *Monitoring and Mitigation of Volcano Hazards*; Scarpa, R., Tilling, R.I., Eds.; Springer: Berlin/Heidelberg, Germany; New York, NY, USA, 1996; p. 23.
43. Santos, P.P. A comprehensive approach to understanding flood risk drivers at the municipal level. *J. Environ. Manag.* **2020**, *260*, 110127. [CrossRef] [PubMed]
44. La Cruz-Reyna, S.D. Long-term probabilistic analysis of future explosive Eruptions. In *Monitoring and Mitigation of Volcano Hazards*; Scarpa, R., Tilling, R.I., Eds.; Springer: Berlin/Heidelberg, Germany; New York, NY, USA, 1996.
45. Kron, W.; Eichner, J.; Kundzewicz, Z. Reduction of flood risk in Europe—Reflections from a reinsurance perspective. *J. Hydrol.* **2019**, *576*, 197–209. [CrossRef]
46. Crichton, D. The risk triangle. In *Natural Disaster Management*; Ingleton, J., Ed.; Tudor Rose: London, UK, 1999; p. 2.
47. Granger, K.; Jones, T.; Leiba, M.; Scott, G. *Community Risk in Cairns: A Multi-Hazard Risk Assessment*; AGSO (Australian Geological Survey Organisation): Canberra, Australia, 1999.
48. Smith, K. *Environmental Hazards: Assessing Risk and Reducing Disaster*; Routledge: London, UK, 1996.
49. Helm, P. Integrated Risk Management for Natural and Technological Disasters. *Tephra* **1996**, *9*, 15.
50. Sayers, P.B.; Gouldby, B.P.; Simm, J.D.; Meadowcroft, I.; Hall, J. Risk, Performance and Uncertainty in Flood and Coastal Defence—A Review R&D Technical Report FD2302/TR1. Defra/Environment Agency. 2003. Available online: www.defra.gov.uk/environ/fcd/research (accessed on 15 June 2020).
51. Kron, W. Flood Risk = Hazard X exposure X vulnerability. *J. Lake Sci.* **2003**, *15*, 185–204. [CrossRef]
52. Ciullo, A.; Viglione, A.; Castellarin, A.; Crisci, M.; Di Baldassarre, G. Socio-hydrological modelling of flood-risk dynamics: Comparing the resilience of green and technological systems. *Hydrol. Sci. J.* **2017**, *62*, 880–891. [CrossRef]

53. Knighton, J.; Steinschneider, S.; Walter, T. A Vulnerability-Based, Bottom-up Assessment of Future Riverine Flood Risk Using a Modified Peaks-Over-Threshold Approach and a Physically Based Hydrologic Model. *Water Resour. Res.* **2017**, *53*, 10043–10064. [CrossRef]

54. Pelling, M.; Maskrey, A.; Ruiz, P.; Hall, L. *Reducing Disaster Risk: A Challenge for Development. United Nations Development Programme, Bureau for Crisis Prevention and Recovery*; John S. Swift Co., Inc.: New York, NY, USA, 2004.

55. Kron, A. Flood damage estimation and flood risk mapping. In *Advances in Urban Flood Management*; Ashley, R., Garvin, S., Pasche, E., Vassilopoulos, A., Zevenbergen, C., Eds.; Taylor & Francis/Balkema: Leiden, The Netherlands, 2007; pp. 213–235.

56. Merz, B.; Kreibich, H.; Schwarze, R.; Thieken, A. Review article "Assessment of economic flood damage". *Nat. Hazards Earth Syst. Sci.* **2010**, *10*, 1697–1724. [CrossRef]

57. Messner, F.; Penning-Rowsell, E.; Green, C.; Meyer, V.; Tunstall, S.; van der Veen, A. Evaluating flood damages: Guidance and recommendations on principles and methods. *FLOODsite Consort.* **2007**, 189.

58. Ahmad, S.; Simonovic, S.P. Integration of heuristic knowledge with analytical tools for the selection of flood damage reduction measures. *Can. J. Civ. Eng.* **2001**, *28*, 208–221. [CrossRef]

59. Dixon, S.J.; Sear, D.A.; Odoni, N.A.; Sykes, T.; Lane, S.N. The effects of river restoration on catchment scale flood risk and flood hydrology. *Earth Surf. Process. Landforms* **2016**, *41*, 997–1008. [CrossRef]

60. Kourgialas, N.N.; Karatzas, G.P.; Nikolaidis, N.P. Development of a thresholds approach for real-time flash flood prediction in complex geomorphological river basins. *Hydrol. Process.* **2011**, *26*, 1478–1494. [CrossRef]

61. Hsieh, L.-S.; Hsu, M.-H.; Li, M.-H. An Assessment of Structural Measures for Flood-prone Lowlands with High Population Density along the Keelung River in Taiwan. *Nat. Hazards* **2006**, *37*, 133–152. [CrossRef]

62. Dawson, R.; Ball, T.; Werritty, J.; Werritty, A.; Hall, J.W.; Roche, N. Assessing the effectiveness of non-structural flood management measures in the Thames Estuary under conditions of socio-economic and environmental change. *Glob. Environ. Chang.* **2011**, *21*, 628–646. [CrossRef]

63. Van Duivendijk, J. The systematic approach to flooding problems. *Irrig. Drain.* **2006**, *55*, S55–S74. [CrossRef]

64. Rikz, R.W.S. *Atlas of Flood Maps; Examples from 19 European Countries, USA and Japan*; WL Delft Hydraulics: Delft, The Netherlands, 2007.

65. Evans, R.; Boardman, J. Curtailment of muddy floods in the Sompting catchment, South Downs, West Sussex, southern England. *Soil Use Manag.* **2003**, *19*, 223–231. [CrossRef]

66. Kourgialas, N.N.; Karatzas, G.P. Flood management and a GIS modelling method to assess flood-hazard areas—A case study. *Hydrol. Sci. J.* **2011**, *56*, 212–225. [CrossRef]

67. Orr, H.G.; Carling, P.A. Hydro-climatic and land use changes in the River Lune catchment, North West England, implications for catchment management. *River Res. Appl.* **2006**, *22*, 239–255. [CrossRef]

68. Holman, I.P.; Hollis, J.M.; Bramley, M.E.; Thompson, T.R.E. The contribution of soil structural degradation to catchment flooding: A preliminary investigation of the 2000 floods in England and Wales. *Hydrol. Earth Syst. Sci.* **2003**, *7*, 755–766. [CrossRef]

69. Marc, V.; Robinson, M. The long-term water balance (1972–2004) of upland forestry and grassland at Plynlimon, mid-Wales. *Hydrol. Earth Syst. Sci.* **2007**, *11*, 44–60. [CrossRef]

70. Yazdi, J.; Neyshabouri, S.A.A.S.; Niksokhan, M.H.; Sheshangosht, S.; Elmi, M. Optimal prioritisation of watershed management measures for flood risk mitigation on a watershed scale. *J. Flood Risk Manag.* **2013**, *6*, 372–384. [CrossRef]

71. Dixon, S.J.; Sear, D.A.; Nislow, K.H. A conceptual model of riparian forest restoration for natural flood management. *Water Environ. J.* **2018**, *33*, 329–341. [CrossRef]

72. Tariq, M.A.U.R.; Hoes, O.; Ashraf, M.T. Risk-Based Design of Dike Elevation Employing Alternative Enumeration. *J. Water Resour. Plan. Manag.* **2014**, *140*, 05014002. [CrossRef]

73. Klijn, F.; Kreibich, H.; De Moel, H.; Penning-Rowsell, E.C. Adaptive flood risk management planning based on a comprehensive flood risk conceptualisation. *Mitig. Adapt. Strat. Glob. Chang.* **2015**, *20*, 845–864. [CrossRef]

74. Huang, J. Numerical Study on the Impact of Revising the Rainstorm Intensity Formula on the Storm Flood Disaster Prediction in Huinan, Pudong District. *J. Civ. Eng. Constr.* **2018**, *7*, 19. [CrossRef]

75. Macchione, F.; Costabile, P.; Costanzo, C.; De Lorenzo, G.; Razdar, B. Dam breach modelling: Influence on downstream water levels and a proposal of a physically based module for flood propagation software. *J. Hydroinform.* **2015**, *18*, 615–633. [CrossRef]

76. Albano, R.; Mancusi, L.; Sole, A.; Adamowski, J. FloodRisk: A collaborative, free and open-source software for flood risk analysis. *Geomat. Nat. Hazards Risk* **2017**, *8*, 1812–1832. [CrossRef]

77. Jukrkorn, N.; Sachdev, H.; Panya, O. Community-based flood risk management: Lessons learned from the 2011 flood in central Thailand. *Flood Recovery Innov. Response IV* **2014**, *184*, 75–86. [CrossRef]

78. Abebe, Y.A.; Ghorbani, A.; Nikolic, I.; Vojinovic, Z.; Sanchez, A. Flood risk management in Sint Maarten—A coupled agent-based and flood modelling method. *J. Environ. Manag.* **2019**, *248*, 109317. [CrossRef]

79. Rogger, M.; Agnoletti, M.; Alaoui, A.; Bathurst, J.C.; Bodner, G.; Borga, M.; Chaplot, V.; Gallart, F.; Glatzel, G.; Hall, J.; et al. Land use change impacts on floods at the catchment scale: Challenges and opportunities for future research. *Water Resour. Res.* **2017**, *53*, 5209–5219. [CrossRef] [PubMed]

80. Serre, D.; Barroca, B.; Diab, Y. Urban flood mitigation: Sustainable options. *Sustain. City* **2010**, *129*, 299–309. [CrossRef]

Publisher's Note: MDPI stays neutral with regard to jurisdictional claims in published maps and institutional affiliations.

applied sciences

Article

Measuring the Lighting Quality in Academic Institutions: The UPM Faculty of Aerospace Engineering (Spain)

María Zamarreño-Suárez [1], Daniel Alcala-Gonzalez [2], Daniel Alfonso-Corcuera [1] and Santiago Pindado [1],*

[1] Instituto Universitario de Microgravedad "Ignacio Da Riva" (IDR/UPM), ETSI Aeronáutica y del Espacio, Universidad Politécnica de Madrid, Pza. del Cardenal Cisneros 3, 28040 Madrid, Spain; maria.zamarreno.suarez@alumnos.upm.es (M.Z.-S.); daniel.alfonso.corcuera@alumnos.upm.es (D.A.-C.)

[2] Departamento de Ingeniería Civil: Hidráulica y Ordenación del Territorio, ETSI Civil, Universidad Politécnica de Madrid, Alfonso XII 3, 28014 Madrid, Spain; d.alcalag@upm.es

* Correspondence: santiago.pindado@upm.es

Received: 21 October 2020; Accepted: 19 November 2020; Published: 24 November 2020

Abstract: This article analyzes the current status of the lighting quality at the *Escuela Técnica Superior de Ingeniería Aeronáutica y del Espacio* (ETSIAE), the aerospace engineering faculty at the *Universidad Politécnica de Madrid* (UPM), and evaluates possible improvement actions based on the use of DIALux® lighting simulation software together with measured data. The results show rather low levels of measured illuminance on classroom desks and blackboards in one of the buildings comprising the faculty. The improvements proposed (a new coat of paint on the walls and replacement of luminaires) were simulated in four individual classrooms representing all rooms in two of the ETSIAE buildings (where the lower illuminance levels were measured). In order to study these improvements, the current situation of the four selected classrooms was simulated using DIALux® and fine-tuning attenuation of the luminaires to take into account their wear and tear. The correlation between the DIALux® simulation and the test results was analyzed with quite good results. The results clearly reveal a need to fully replace the classroom lighting systems in ETSIAE building A (the oldest building, dating back to 1955). According to the results from the selected classrooms, the average lighting over the desks can be greatly improved by using LED technology in order to meet UNE 12464-1 standard (that is, 500 lx, from an initial situation with much lower illuminance values: 129 lx to 295 lx). This article represents an innovative way to perform lighting improvement projects as real measured lighting data is used as initial input for the lighting simulations.

Keywords: DIALux; simulation; lighting quality; teaching institution; classroom lighting

1. Introduction

The importance of proper lighting conditions (i.e., visual comfort) at academic institutions has been emphasized at all levels, from elementary, primary, and secondary education schools [1–6] to university faculties [7–16]. In general, it can be said that comfort in academic environments has been widely analyzed as it has important effects on learning. On this subject and attending to the available literature, three main different aspects can be underlined in first place: thermal comfort, noise comfort, and visual comfort [12,14,17]. After them, other less relevant aspects such as wall colors, furniture, spatial arrangements, ventilation, etc. can also be added when analyzing the aforementioned comfort in academic institutions [4,10,12,13,15]. Visual comfort mainly depends on lighting. However, it is somehow affected by thermal and noise levels [14,17].

In the available literature, some examples were found of lighting analysis in classrooms and other academic environments [8,18–23]. Some of these focused mainly on energy costs and efficiency [18,20,21,23], as lighting represents about 20% of the world's demand for electrical energy production [23]. On the other hand, many analyses on visual comfort and the effects of lighting seem to be based on surveys/interviews rather than measuring the illuminance levels [5,6,11,12,15]. Even those works that reflect lighting measurements include surveys [13,16,24,25], adding subjective (and statistical) factors to the study.

Many of the studies on lighting similar to the present one found in the available literature were conducted using DIALux® [8,19–23], although it seems that there are many possibilities when it comes to lighting simulation tools [26–29]. However, leaving aside some works related to validation of lighting analysis software [30–33], there seems to be a lack of studies evaluating the accuracy in the prediction of actual lighting results. Finally, it is worth mentioning the quite high accuracy in artificial lighting simulations (0.5% average error) reported by Maamari et al. [33].

This paper discusses the results of research conducted on the current lighting in the classrooms of the *Universidad Politécnica de Madrid* (UPM) Faculty of Aerospace Engineering (*Escuela Técnica Superior de Ingeniería Aeronáutica y del Espacio* (ETSIAE), see Section 2) without daylight (that is, during evening/night periods and with the window blinds lowered) and analyzes different proposals for improving it. To this end, illuminance was measured on student desks and blackboards in 39 classrooms, including those used for exams. This study excludes other areas of the faculty with special characteristics, such as the library, computer rooms, laboratories, and other common areas. It was considered relevant to focus on this group of classrooms, where students and instructors spend most of their time and where lighting quality is most crucial. Measurements were carried out by measuring the illuminance levels on the selected surfaces (blackboards and desks) with a lux meter. This is a quite direct and simple technique that allowed a first evaluation of a quite relevant problem. However, it is somehow limited when compared to High Dynamic Range (HDR) imaging technology [24,25,34,35], as it does not give other interesting variables such as glare.

In addition, it should also be said that the selection of illuminance as a lighting-quality variable has some drawbacks, as it might not correlate well with perceived brightness or visual comfort. Cuttle [36] suggested the use of the Perceived Adequacy of Illumination (PAI) as an alternative to minimum illuminance levels in standards, this concept of lighting being quantified by Mean Room Surface Exitance (MRSE) as a control variable. Besides, some studies show limitations of illuminance as the proper variable to define the adequacy of indoor lighting [37,38] and others show new lighting-quality metrics being required if health is taken into account when designing work environments [39], whereas other studies propose that the proper illumination on artwork needs to be specifically defined [40,41].

The aim of the present work is to propose a lighting analysis technique that may ensure the compliance of lighting upgrade projects with UNE 12464-1 standard, which regulates lighting projects in working environments. Taking into account that, based on this standard, visual quality is only quantitatively ensured by

- the illuminance value,
- the color rendering index,
- the illuminance uniformity index, and
- the Universal Glare Rating value,

illuminance was selected as a control variable in the present work.

Leaving aside the illuminance measurements taken in the classrooms, the most original contribution from the present work is the DIALux® calibration/tuning carried out by comparison between the data from the simulation of the present situation and the measurements. After this calibration and validation, three different lighting improvement actions were studied. To the best of the authors' knowledge, no study of this type had ever been conducted at ETSIAE or any other faculties at the *Universidad Politécnica de Madrid* (UPM). It is worth emphasizing the lack of renovation with regard to lighting quality in the oldest building at ETSIA (building A, dating back to 1955), despite the importance of lighting maintenance as one of the categories which influence building environment condition [42].

The paper is organized as follows: Section 2 describes the instrumentation and methodology used in the measurement campaign and subsequent simulations. Section 3 discusses the results and shows the actual current situation of the classroom lighting, along with the expected situation after implementing the proposed improvements. Finally, Section 4 summarizes the conclusions.

2. Materials and Methods (Testing and Simulation Methodologies)

The *Escuela Técnica Superior de Ingeniería Aeronáutica y del Espacio* (ETSIAE) at the *Universidad Politécnica de Madrid* (UPM) is the university's aerospace engineering faculty. It was created by merging two older engineering schools within the same university: the *Escuela Técnica Superior de Ingenieros Aeronáuticos* (ETSIA) and the *Escuela Universitaria de Ingeniería Técnica Aeronáutica* (EUITA). This followed the university's implementation of the new degree programs regulated by Royal Decree 1393/2007, within the framework of the European Higher Education Area. The available data, from academic year 2016–2017, indicate a total of 3468 students. This makes the ETSIAE one of the UPM's largest faculties, with a surface area of 36,500 m^2, 240 professors and researchers, and 140 service and administrative staff.

Following the renovations undertaken in recent years, ETSIAE has four buildings dedicated to teaching, known as buildings A, B, C, and E (see Figure 1). Due to the high number of students at the school and the degree programs offered and depending on the time of year, it is normal for many students to spend all or part of their school day using only artificial lighting, with the visual fatigue that this entails. For this reason, it is essential for classroom lighting to meet regulatory requirements [43] in order to be suitable for the tasks performed at this faculty.

Figure 1. Aerial view of *Escuela Técnica Superior de Ingeniería Aeronáutica y del Espacio* (ETSIAE): there are four buildings dedicated to teaching (**A,B,C,E**), and two buildings housing the library, laboratories, offices, cafeteria, and common areas (**D,F**). Source: *Universidad Politécnica de Madrid*.

This study comprises several successive steps, starting with measurement of the present illumination levels (i.e., illuminance, E) of a significant number of ETSIAE classrooms. Four different types of classrooms were then selected as representative of the entire set, and the current situation was reproduced using DIALux® lighting simulation software. Once the current situation in the selected classrooms was properly simulated, three proposals for improvement were analyzed, using as reference the simulation of the current situation.

2.1. Testing Campaign

The first step in the proposed methodology was the design of the testing campaign, where it was necessary to define the scope of the research and to establish a schedule for measuring the lighting levels in the chosen classrooms. When creating the schedule, it was important to define the times of day during which the data should be collected, as these data might be greatly affected by sunlight. Thus, it was determined that the measurements should be taken in the most unfavorable case in terms of illumination, which is between sunset and sunrise (when there is less outside light and the classroom lighting reaches the minimum possible value). It should be emphasized that there is still academic activity at ETSIAE after sunset. It was also necessary to plan the manpower and time needed to properly conduct the testing campaign, as the measurements in a single classroom can take a significant amount of time (between 30 min and several hours). Finally, it should be highlighted that the measurements were taken in winter, when there are more hours of darkness for measuring, and with all window blinds lowered.

This study includes measurements of the lighting levels on student desks and blackboards in 39 ETSIAE classrooms (see Figure 2), distributed among buildings A, B, and E. It includes all ETSIAE classrooms dedicated to traditional teaching (i.e., not multimedia classrooms). A CEM-DT 1308 lux meter and a DEXTER CM30 distance meter were used in this testing campaign (see Figure 3). In addition to the geometry of the classrooms, the arrangement of lights, desks, blackboards, and other elements that might interfere with lighting (such as platforms, cabinets, columns, pillars, beams, etc.) were precisely identified. This was done in order to reproduce both the geometry and the lighting conditions of the classrooms as accurately as possible in the subsequent simulation. Similarly, the brand and model of each light in the classrooms were also identified.

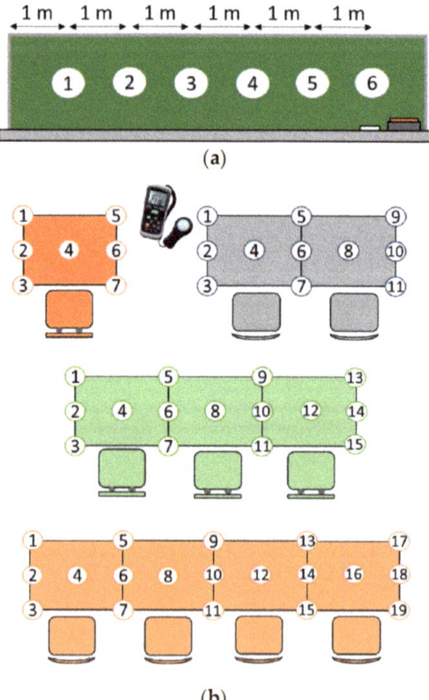

Figure 2. Diagram of the position and distribution of the points selected for illumination measurements on the blackboards (**a**) and desks (**b**) in the classrooms.

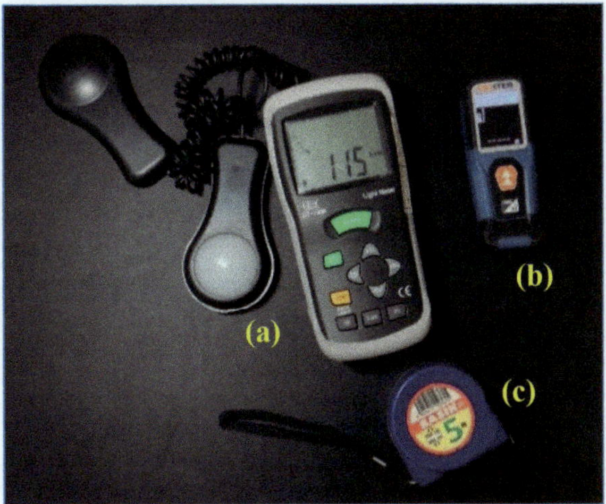

Figure 3. Equipment used in the testing campaign: (**a**) CEM-DT 1308 lux meter, (**b**) DEXTER CM30 distance meter, and (**c**) measuring tape.

Once the geometry of the classrooms was determined, the illuminance values, *E*, were measured on the surfaces of interest (blackboards, tables, and student desks). Blackboards are elements of great importance to the educational process, as students use them to take notes and to understand the professors' explanations. In fact, UNE 12464-1 gives them special importance in terms of lighting, with an average illuminance reference value of 500 lx. (This value may vary according to national regulations [20].) In order to measure the illumination distribution, it was chosen to measure points one meter apart on the average line of the blackboard (see Figure 2). These points were measured at mid-height, since direct classroom observations revealed that most of the professors' notes on the blackboard were made in this area.

With regard to student desks, it was noted that students rarely place paper in the center of the desk to write, but tend to tilt it toward one side depending on their dominant hand. Bearing this in mind, the four corners of the desk as well as the central point were defined as measurement points. For long desks with more than one seat, the part corresponding to each seat was treated as an individual desk and the two corners in contact with each other were treated as a single point (see Figure 2).

A significant number of measurement points was obtained in each classroom, ranging from 134 points in the smallest classroom to 2275 points in one of the examination rooms. The measurements were collected on cards such as the one shown in Figure 4. Each group of desks is identified in one color according to the level of lighting achieved (the closer the desks are to the dark red color, the poorer the lighting).

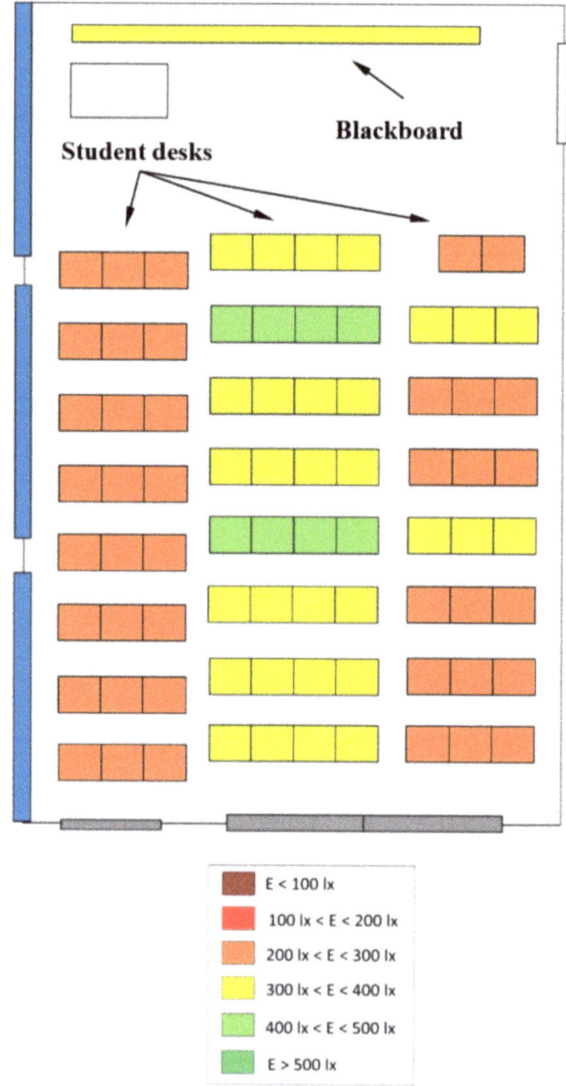

Figure 4. Illuminance measurements on the desks and blackboard in one of the classrooms studied.

2.2. Simulation

Once the rooms representative of all other ETSIAE classrooms were selected (see Section 3 of this paper), their lighting situation was represented as accurately as possible using DIALux® simulation software (see Figure 5). Thanks to the data collected in the testing campaign, the simulations were carried out taking into account precise details from

- the geometry of the classes;
- the layout of the luminaires;
- the arrangement of desks, chairs, blackboards, and elements that affect lighting; and the textures of the walls and floor.

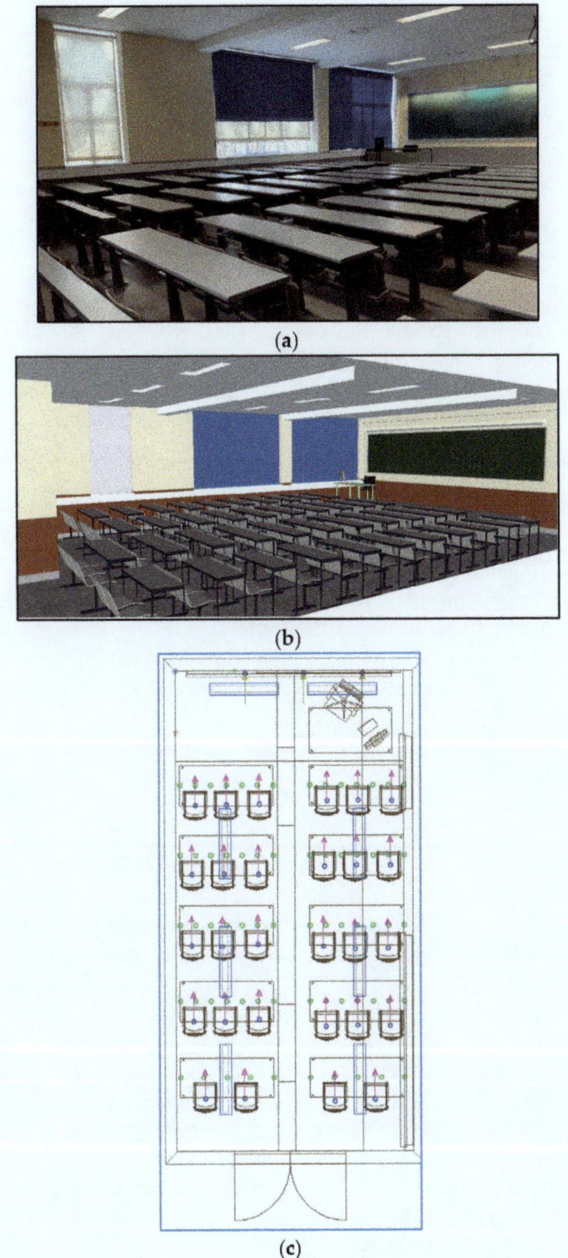

Figure 5. Comparison between an actual image of classroom A-038 (**a**) and the simulation carried out using DIALux® software (**b**), and calculation points for the illuminance and Unified Glare Rating (UGR) in the DIALux® simulation of classroom A-016 (**c**).

The geometry of the classes was reproduced using the data obtained in the test campaign. As explained in Section 2.1, with the aid of the distance meter DEXTER CM30, an accurate geometry of the classes was obtained. For the luminaires, a complete inventory of which luminaires were in use in all the classrooms was performed in the test campaign. With this information, DIALux® ready-to-use data was obtained from the manufacturers' data catalogues. This allowed to directly implement the luminaires and their performance in the simulation, with only a need to adjust their attenuation to take into account wear and tear. In the testing campaign, together with the geometry of the classrooms, an extensive set of data regarding the position and shape of all desks, chairs, blackboards, and elements that may affect lighting was measured. Using the aforementioned data, the necessary objects were created and arranged to be taken into account in the simulation. With regard to the windows, as the lighting analysis is based on the results obtained in the most unfavorable scenario (at night), no light was set to be transmitted through them, with the blinds being lowered. Finally, taking into account the importance that surface textures have in light reflection, the colors of the walls, ceiling, blackboard, chairs, and tables were reproduced in the simulation. Furthermore and in relation to the floors, as their texture was not uniform, customized textures were created with the aid of imported photos of the present floors at the simulated classrooms.

The different points of interest with regard to the lighting on student desks and blackboards as well as the Unified Glare Rating (UGR) were established in the simulations. The UGR calculation points were placed in the center of each chair where the eyes of an average-sized student would be sited.

3. Results and Discussion

The results of this study are described and discussed in this section. First, the results of the ETSIAE classroom lighting measurement campaign are analyzed. Second, the improvements applied to four representative classrooms using the DIALux® software are analyzed. This section begins with a comparison between the experimental lighting results measured on the student desks in the selected classrooms and those extracted from the direct simulation of the current situation conducted using DIALux®. An accuracy estimation of the simulated results is then established. After this calibration/tuning, three proposed improvements are studied: (1) changing the color of the walls; (2) replacing the luminaires with new ones (of the same model); and (3) replacing the luminaires with new ones but employing LED technology.

3.1. Testing Results

As mentioned above, the testing campaign included measurements from 39 classrooms in three different ETSIAE buildings (buildings A, B, and E). Table 1 shows the average illuminance values measured on student desks and blackboards in each of these classrooms, along with the surface areas of the classrooms. A first look at these results reveals rather low values in all classrooms in building A and in nearly all in building B, much lower than the 500 lx level stablished by the standards [43]. In some building A classrooms, the average lighting is below 200 lx. As expected, the situation is much better in building E.

In building A, most of the classrooms have very poor lighting, with only three rooms recording a measured illuminance, E_m, higher than 200 lx and none recording higher than 300 lx. The average illumination value on all desks measured in building A was 169 lx, well below the recommended 500 lx. In building B, better results were obtained, with 30% of the classrooms achieving the illuminance values recommended by regulations. In 50% of the classrooms, illuminance values of 300 lx to 400 lx were obtained, with only one classroom recording a value below 300 lx. Finally, all classrooms in the newest building, building E, had acceptable levels of lighting on the study desks. Based on the above results, it is clear that the improvements to be proposed should focus on increasing the average illuminance levels in the classrooms in buildings A and B while keeping the UGR glare index below the maximum level (maximum value of UGR should be below 19, according to UNE 12464-1).

Table 1. Average illuminance measured on student desks, E_{mdk}, and blackboards, E_{mbd}, in the ETSIAE classrooms studied: the table also includes the surface areas of these classrooms, A, and the number of desks, n.

Classroom	A [m²]	n	E_{mdk} [lx]	E_{mbd} [lx]	Classroom	A [m²]	n	E_{mdk} [lx]	E_{mbd} [lx]
A-003	102.57	36	133	212	A-139	399.22	276	182	107
A-004	103.5	36	133	251	A-242	455.22	325	304	183
A-005	103.53	36	133	224	BSS-02	341.3	224	740	794
A-006	114.43	42	129	236	B-003	86	23	356	443
A-013	103.53	36	138	174	B-004	86	32	336	145
A-014	68.3	24	148	244	B-005	90.21	32	307	339
A-015	114.15	42	124	205	B-202	196.49	50	662	270
A-016	31.2	10	132	112	B-205	146.91	40	759	315
A-025	103.53	36	134	224	B-305	87.40	32	336	429
A-026	103.53	36	138	224	B-306	86.94	32	293	456
A-027	104.4	37	145	195	B-319	87.40	32	338	404
A-029	74.88	19	163	271	B-324	87.40	32	350	473
A-036	105.02	36	218	439	E-004	69.44	15	688	458
A-037	66.75	19	146	203	E-005	69.76	15	744	593
A-038	111.59	42	132	246	E-006	36.80	8	987	787
A-039	31.16	10	188	142	E-104	67.52	16	907	780
A-113	105.17	24	299	373	E-105	69.76	16	913	764
A-114	104.17	24	309	336	E-107	70.08	16	759	571
A-115	114.33	39	134	206	E-303	323.13	180	875	987
A-120	102.96	37	150	188	-	-	-	-	-

3.2. DIALux® Analysis Results

As stated in the methodology described in Section 2, four model classrooms were selected as representative of the different types of poorly illuminated classrooms, three of which were in building A and one was in building B (see Figure 6):

- A-005 (medium-sized classroom model)
- A-016 (smaller classroom model)
- A-038 (large classroom model)
- B-306 (classroom with the worst lighting result in building B)

The heterogeneous classroom sizes in building A prompted the selection of a higher proportion of cases from this building. In contrast, the classrooms in building B are much more homogeneous with regard to their geometry.

To get a better idea of the current status of the illumination levels in the selected classrooms (in addition to the results from Table 1), the percentage distribution of the individual desks with regard to the measured illuminance, E_m, was plotted (see Figure 7). This histogram clearly indicates the poor lighting conditions in the building A classrooms.

Before analyzing any possible improvement, the current situation of the selected classrooms was simulated as accurately as possible using DIALux®. As shown in Figure 4, the geometry and conditions of the classrooms were simulated, including the furnishings and position of all luminaires. Finally, the attenuation of these luminaires was adjusted to bring the results of the simulation as close as possible to the measurements taken (see Table 2 for the average and maximum values of measured and simulated lighting on the desks and blackboards in the selected classrooms). Figure 8 compares the illumination from the testing campaign, E_m, and the DIALux® simulation, E_{dl}, on each individual desk. The quality of the correlation between the two results can be appreciated compared to the ideal simulation indicated by the continuous line included on the graphs.

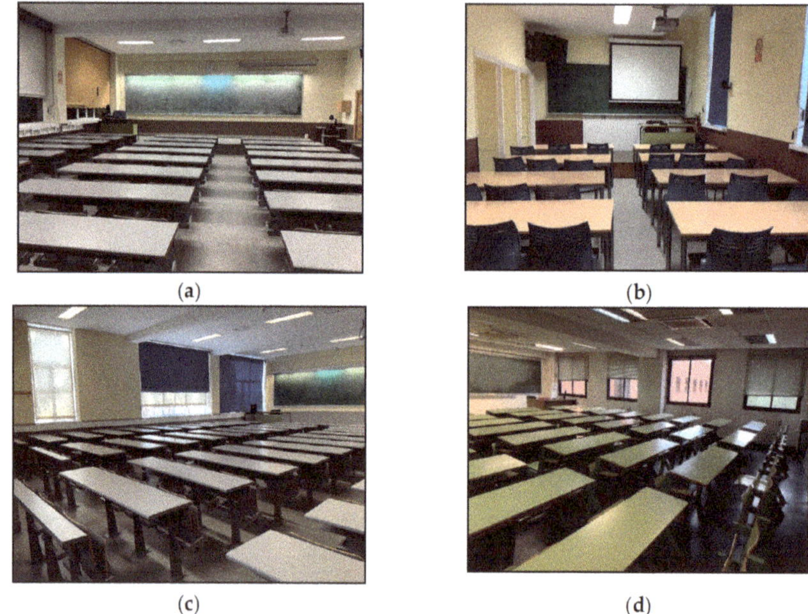

Figure 6. Pictures of classrooms A-005 (**a**), A-016 (**b**), A-038 (**c**), and B-306 (**d**) selected for this study as representative of all classrooms at the *Escuela Técnica Superior de Ingeniería Aeronáutica y del Espacio* (ETSIAE) at the *Universidad Politécnica de Madrid* (UPM).

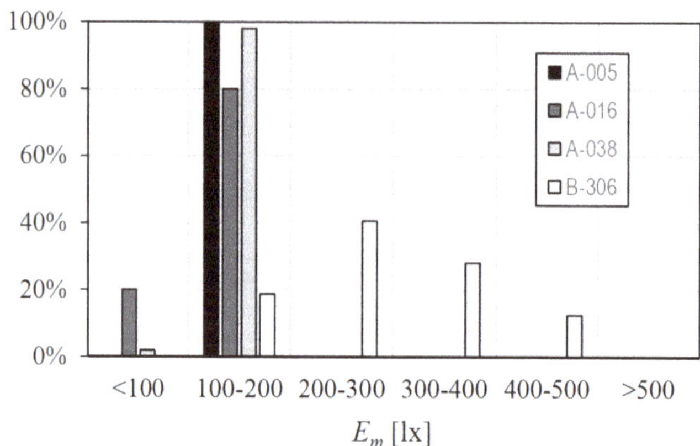

Figure 7. Percentage distribution of the individual desks with regard to the measured illuminance, E_m, in the classrooms selected for this study (A-005, A-016, A-038, and B-306).

Table 2. Average and maximum illuminance on the desks ($E_{av,dk}$ and $E_{max,dk}$) and blackboards ($E_{av,b}$ and $E_{max,b}$) in the classrooms studied in the following cases: current (measured by lux meter), current case simulation by DIALux®, case 1, case 2, and case 3. The average and maximum UGR glare values obtained in the simulations have been also included.

	Classroom	$E_{av,dk}$ [lx]	$E_{max,dk}$ [lx]	$E_{av,b}$ [lx]	$E_{max,b}$ [lx]	UGR (Average)	UGR (max)
Current Case	A-005	133	168	224	242	-	-
	A-016	112	152	112	120	-	-
	A-038	132	166	246	275	-	-
	B-306	293	478	456	509	-	-
Current Case Simulation	A-005	133	187	222	237	24	25
	A-016	135	157	113	118	18	20
	A-038	129	205	243	256	24	25
	B-306	295	509	467	501	21	23
Case 1	A-005	148	202	235	249	23	24
	A-016	152	176	126	132	17	19
	A-038	141	218	254	265	23	25
	B-306	311	531	483	520	21	23
Case 2	A-005	304	403	431	451	23	25
	A-016	213	340	354	360	23	25
	A-038	267	372	413	440	23	25
	B-306	973	1088	974	1044	22	24
Case 3	A-005	536	698	570	795	13	19
	A-016	614	686	593	696	16	19
	A-038	506	809	555	745	14	19
	B-306	564	855	527	581	15	19

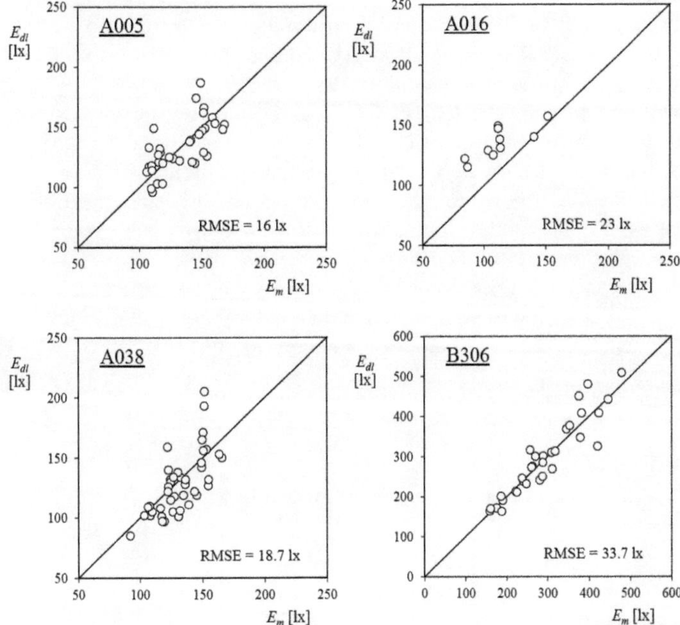

Figure 8. Illuminances calculated by DIALux®, E_{dl}, on each of the desks in classrooms A-005, A-016, A-038, and B-306, compared to the measurements taken in each case, E_m: A line indicating the ideal simulation is included on each graph. The Root Mean Square Error (RMSE) of E_{dl} in relation to E_m is included in each graph.

This correlation can also be quantified by means of the Root Mean Square Error (RMSE). For each classroom, assuming data is reasonably distributed according a Gaussian process (data from the A005, A016, and B036 classrooms are Gaussian according to the Shapiro–Wilk test with 0.05 significance level; data from A038 do not; however, this last result changes if the two points—from a 42-point sample—representing the larger difference between simulated and testing data are not taken into account), the RMSE represents the limit encompassing 66% of the differences between the simulated and measured illuminances (i.e., the limit marking 66% of the desks displaying a minor difference between those illuminances. This difference is reflected by the vertical distance between the points and the unit slope line on the graphs in Figure 8). The RMSE values obtained from the simulations are as follows: 16.0 lx (A-005), 26.3 lx (A-016), 18.7 lx (A-038), and 33.7 lx (B-306). Bearing in mind that the above data show a difference between the simulated and measured data of 3.2% to 6.7% of the reference value of the recommended illuminance level, 500 lx [43], the accuracy of the simulation performed with DIALux® was assumed to be reasonable.

After studying the reasonable suitability of DIALux® simulations, 3 different case studies were proposed in order to improve the current situation:

- Case 1: The proposal consists of painting the walls of the classrooms white. In this case, the luminaires (and lamps) and their distribution are maintained.
- Case 2: Installation of new lamps (in the luminaires), using the same model as before, should increase their luminous flux. This improvement was programmed in DIALux® by applying a 0% attenuation value to all simulated control groups.
- Case 3: Replacement of the current luminaires with a different model employing LED technology, PHILIPS SM134V PSD W20L120 1 xLED37S/840 OC, was chosen to maintain the current arrangement of the luminaires in order to simplify the installation process. Because of the project size, it was decided to choose a single luminaire model in order to achieve two fundamental advantages: to simplify maintenance and to reduce the investment by increasing the purchase volume and by possibly obtaining additional discounts. The following numbers of luminaires were used in each of the classrooms studied: 18 (A-005), 8 (A-016), 19 (A-038), and 17 (B-306). Most of the lamps currently installed (PHILIPS) have a luminous flux of 3350 lm. Since the aim is also to increase the current average illuminance value, it was decided to choose lamps with a higher luminous flux: 3700 lm. Because the ceilings in the building A classrooms are very high, it was advantageous to choose luminaires that can be hung (or installed directly on the ceiling, which is necessary in the building B classrooms due to their lower ceiling height). In addition, an "OC" diffuser was chosen. Regarding the color temperature of the luminaires, 4000 K was selected (current situation and recommended by the regulations). On the technical data sheet provided by the manufacturer, it should be emphasized that the color reproduction index is greater than or equal to 80, thus complying with European standards.

Table 2 includes the average and maximum illumination on the desks and blackboards in the selected classrooms for the cases studied, along with the maximum UGR glare values obtained in each case:

- Current Case (represented by the experimental measures)
- Current Case Simulation
- Case 1 (change of paint on the walls)
- Case 2 (replacement of the lamps using the same models)
- and Case 3 (replacement of the lamps with LED models)

In addition and due to their special importance, Figure 9 shows the percentage distributions of the desks in the selected classrooms, classified according to their illuminance in cases 1, 2, and 3 and with regard to the initial case (current case simulation). The results of this study clearly indicate the need for a change in the lighting of the ETSIAE classrooms, as a real and true approach to the minimum conditions (500 lx) is not achieved unless the luminaires and lamps are replaced. More specifically, it can be said that changing the color of the walls and replacing the existing luminaires would both have positive effects but are not sufficient to provide adequate lighting for the classrooms. Therefore, for the purposes of the case study, changing the lighting technology is considered the best option in order to meet the standard [43] in the building A and B classrooms.

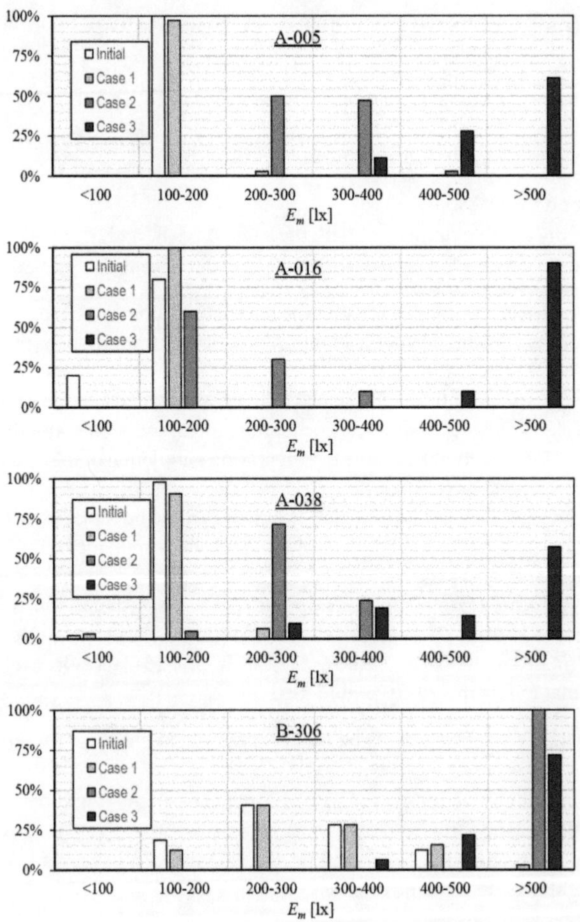

Figure 9. Distribution of desks by lighting levels and improvement cases (case 1, case 2, and case 3) from the initial illuminance levels in each of the selected classrooms.

As it can be observed in Table 2, for the option chosen as the appropriate improvement action (case 3), the UGR or glare values are below the limits established by the UNE 12464-1, which means no further action is needed to comply with the standards in terms of glare.

However, if other options were to be implemented (as in case 2 for the B-306 classroom), technologies for glare reduction should be considered and tested via simulations in order to reduce the maximum UGR value to comply with the standards (maximum UGR value of 19). Such technologies include translucent or polarizing covers or reflectors to prevent light spreading horizontally. In addition, focusing on case 2 for the B-306 classroom, this particular case could be evaluated with some of the luminaires turned off, as there is a wide margin for illuminance reduction that could lead to meeting the glare standards.

4. Conclusions

This study analyzes the quality of the lighting in the classrooms at the *Escuela Técnica Superior de Ingeniería Aeronáutica y del Espacio* (ETSIAE) at the *Universidad Politécnica de Madrid* (UPM). The main motivation for this work is the lack of lighting maintenance and improvement on ETSIAE classrooms in the last decades. The work was carried out

- by measuring the illuminance in a large number of classrooms from this faculty and
- by simulation.

The results of measurements carried out in 39 classrooms of ETSIAE revealed deficient lighting in evening/night conditions.

Three possible improvements were studied:

- white painting on the walls,
- replacement of the existing luminaires with the same models, and
- replacement of the existing luminaires with different models employing LED technology.

These improvements were analyzed using the simulation program DIALux$^{®}$ after performing a prior simulation of the current measured situation in order to calibrate/tune the attenuation of the lamps and to achieve the same lighting levels.

Analysis of the correlation between the results from the simulation and the data obtained from the measurements revealed acceptable results, showing the aforementioned method of tuning the lamp attenuation in the DIALux$^{®}$ software suitable for obtaining a simulation of the actual lighting situation for its use in improvement action evaluation.

The results clearly indicate that the current luminaires must be replaced with different ones in most of the classrooms from the oldest building of the faculty if compliance with current standards is to be ensured. In this sense, replacing the luminaires with others based on LED technology has proven to be the best option. According to the results from the selected classrooms, the average lighting over the desks can be improved from 133 lx (classroom A-005), 135 lx (classroom A-016), 129 lx (classroom A-038), and 295 lx (classroom B-306) to 536 lx, 614 lx, 506 lx, and 564 lx, respectively.

Future works that follow this study should focus on

- improving the lighting analysis by using more sophisticated tools and procedures such as High Dynamic Range (HDR) imaging technology;
- including daylight in the lighting analysis, as some classrooms might have glare problems at certain periods of the day; and
- including new lighting design, as according to the new social situation caused by covid-19, classrooms might need to combine in-class and online teaching.

Finally, the importance of this study must be emphasized as it reveals a serious problem that could also exist at other UPM schools and faculties and at other Spanish universities.

Author Contributions: Conceptualization, S.P.; methodology, S.P.; software, M.Z.-S. and D.A.-C.; validation, M.Z.-S., S.P., and D.A.-C.; formal analysis, S.P. and D.A.-C.; investigation M.Z.-S.; data curation, M.Z.-S.; writing—original draft preparation, S.P. and D.A.-G.; writing—review and editing, S.P. and D.A.-G.; supervision, S.P. All authors have read and agreed to the published version of the manuscript.

Funding: This research received no external funding.

Acknowledgments: Authors are indebted to Tania Tate for her kind help with the style of the text. Authors are also grateful to the reviewers whose wise comments helped us in improving this work.

Conflicts of Interest: The authors declare no conflict of interest.

References

1. De Giuli, V.; Zecchin, R.; Corain, L.; Salmaso, L. Measured and perceived environmental comfort: Field monitoring in an Italian school. *Appl. Ergon.* **2014**, *45*, 1035–1047. [CrossRef] [PubMed]
2. Winterbottom, M.; Wilkins, A. Lighting and discomfort in the classroom. *J. Environ. Psychol.* **2009**, *29*, 63–75. [CrossRef]
3. Mott, M.S.; Robinson, D.H.; Walden, A.; Burnette, J.; Rutherford, A.S. Illuminating the effects of dynamic lighting on student learning. *SAGE Open* **2012**, *2*. [CrossRef]
4. Samani, S.A.; Samani, S.A. The Impact of Indoor Lighting on Students' Learning Performance in Learning Environments: A knowledge internalization perspective University of Applied Sciences. *Int. J. Bus. Soc. Sci.* **2012**, *3*, 127–136.
5. Barrett, P.; Davies, F.; Zhang, Y.; Barrett, L. The impact of classroom design on pupils' learning: Final results of a holistic, multi-level analysis. *Build. Environ.* **2015**, *89*, 118–133. [CrossRef]
6. Bluyssen, P.M.; Zhang, D.; Kurvers, S.; Overtoom, M.; Ortiz-Sanchez, M. Self-reported health and comfort of school children in 54 classrooms of 21 Dutch school buildings. *Build. Environ.* **2018**, *138*, 106–123. [CrossRef]
7. Hui, S.C.M.; Cheng, K.K.Y. Analysis of Effective Lighting Systems for University Classrooms. In Proceedings of the Henan-Hong Kong Joint Symposium, Zhengzhou, China, 30 June–1 July 2008; pp. 53–64.
8. Wang, R.; He, X.Y.; Liu, S.; Liu, L.C.; Liu, B.; Wang, Z.-S.; Yang, Y.; Yu, L. Investigation and analysis on the illumination of the university classroom. In Proceedings of the 5th International Conference on Machinery, Materials and Computing Technology 2017 (ICMMCT 2017), Beijing, China, 25–26 March 2017; Volume 126, pp. 430–438.
9. Yang, Z.; Becerik-Gerber, B.; Mino, L. A study on student perceptions of higher education classrooms: Impact of classroom attributes on student satisfaction and performance. *Build. Environ.* **2013**, *70*, 171–188. [CrossRef]
10. Castilla, N.; Llinares, C.; Bravo, J.M.; Blanca, V. Subjective assessment of university classroom environment. *Build. Environ.* **2017**, *122*, 72–81. [CrossRef]
11. Castilla, N.; Llinares, C.; Bisegna, F.; Blanca-Giménez, V. Affective evaluation of the luminous environment in university classrooms. *Build. Environ.* **2018**, *58*, 52–62. [CrossRef]
12. Ricciardi, P.; Buratti, C. Environmental quality of university classrooms: Subjective and objective evaluation of the thermal, acoustic, and lighting comfort conditions. *Build. Environ.* **2018**, *127*, 23–36. [CrossRef]
13. Tahsildoost, M.; Zomorodian, Z. Indoor environment quality assessment in classrooms: An integrated approach. *J. Build. Phys.* **2018**, *42*, 1–27. [CrossRef]
14. Yang, W.; Moon, H.J. Combined effects of acoustic, thermal, and illumination conditions on the comfort of discrete senses and overall indoor environment. *Build. Environ.* **2019**, *148*, 623–633. [CrossRef]
15. Yildiz, Y.; Koçyiğit, M. Evaluation of indoor environmental conditions in university classrooms. *Proc. Inst. Civ. Eng. Energy* **2019**, *172*, 148–161. [CrossRef]
16. Leccese, F.; Salvadori, G.; Rocca, M.; Buratti, C.; Belloni, E. A method to assess lighting quality in educational rooms using analytic hierarchy process. *Build. Environ.* **2020**, *168*, 106501. [CrossRef]
17. Wu, H.; Wu, Y.; Sun, X.; Liu, J. Combined effects of acoustic, thermal, and illumination on human perception and performance: A review. *Build. Environ.* **2020**, *169*, 1–19. [CrossRef]
18. Liang, Y.; Zhang, R.; Wang, W.; Xiao, C. Design of energy saving lighting system in university classroom based on wireless sensor network. *Commun. Netw.* **2013**, *5*, 55–60. [CrossRef]
19. Dessi, V.; Fianchini, M. Lights and shadows in university classrooms. In Proceedings of the International Conference Arquitectonics Network: Architecture, Education and Society, Barcelona, Spain, 3–5 June 2015; Final papers. GIRAS. Universitat Politècnica de Catalunya. 2015; pp. 1–10.
20. Wang, S. The view of classroom light design optimization simulation. In Proceedings of the International Conference on Information Sciences, Machinery, Materials and Energy (ICISMME 2015), Chongqing, China, 11–13 April 2015; pp. 1768–1772.
21. Kamaruddin, M.A.; Arief, Y.Z.; Ahmad, M.H. Energy Analysis of Efficient Lighting System Design for Lecturing Room Using DIAlux Evo 3. *Appl. Mech. Mater.* **2016**, *818*, 174–178. [CrossRef]
22. Sielachowska, M.; Tyniecki, D.; Zajkowski, M. Measurements of the Luminance Distribution in the Classroom Using the SkyWatcher Type System. In Proceedings of the 7th Lighting Conference of the Visegrad Countries (Lumen V4), Trebic, Czech Republic, 18–20 September 2018; Institute of Electrical and Electronics Engineers (IEEE): New York, NY, USA, 2018; pp. 1–5.

23. Sathya, P.; Natarajan, R. Energy Estimation and Photometric measurements of LED lighting in Laboratory. In Proceedings of the 2014 International Conference on Advances in Electrical Engineering (ICAEE), Vellore, India, 9–11 January 2014; IEEE: New York, NY, USA, 2014; pp. 1–5.

24. Bellia, L.; Spada, G.; Pedace, A.; Fragliasso, F. Methods to evaluate lighting quality in educational environments. *Energy Procedia* **2015**, *78*, 3138–3143. [CrossRef]

25. Yacine, S.M.; Noureddine, Z.; Piga, B.E.A.; Morello, E.; Safa, D. Towards a new model of light quality assessment based on occupant satisfaction and lighting glare indices. *Energy Procedia* **2017**, *122*, 805–810. [CrossRef]

26. Ochoa, C.E.; Aries, M.B.C.; Hensen, J.J. State of the art in lighting simulation for building science: A literature review. *J. Build. Perform. Simul.* **2012**, *5*, 209–233. [CrossRef]

27. Baloch, A.A.; Shaikh, P.H.; Shaikh, F.; Leghari, Z.H.; Mirjat, N.H.; Uqaili, M.A. Simulation tools application for artificial lighting in buildings. *Renew. Sustain. Energy Rev.* **2018**, *82*, 3007–3026. [CrossRef]

28. Fernandez-Prieto, D.; Hagen, H. Visualization and analysis of lighting design alternatives in simulation software. *Appl. Mech. Mater.* **2017**, *869*, 212–225. [CrossRef]

29. Davoodi, A.; Johansson, P.; Laike, T.; Aries, M. Current Use of Lighting Simulation Tools in Sweden. In Proceedings of the 60th SIMS Conference on Simulation and Modelling SIMS 2019, Västerås, Sweden, 12–16 August 2019; Linkoping University Electronic Press: Linkoping, Sweden, 2020; pp. 206–211.

30. Houser, K.W.; Tiller, D.K.; Pasini, I.C. Toward the accuracy of lighting simulations in physically based computer graphics software. *J. Illum. Eng. Soc.* **1999**, *28*, 117–129. [CrossRef]

31. Maamari, F.; Fontoynont, M. Analytical tests for investigating the accuracy of lighting programs. *Light. Res. Technol.* **2003**, *35*, 225–239. [CrossRef]

32. Maamari, F.; Fontoynont, M.; Tsangrassoulis, A.; Marty, C.; Kopylov, E.; Sytnik, G. Reliable datasets for lighting programs validation—benchmark results. *Sol. Energy* **2005**, *79*, 213–215. [CrossRef]

33. Maamari, F.; Fontoynont, M.; Adra, N. Application of the CIE test cases to assess the accuracy of lighting computer programs. *Energy Build.* **2006**, *38*, 869–877. [CrossRef]

34. Ibañez, C.A.; Zafra, J.C.G.; Sacht, H.M. Natural and artificial lighting analysis in a classroom of technical drawing: Measurements and HDR images use. *Procedia Eng.* **2017**, *196*, 964–971. [CrossRef]

35. Pierson, C.; Wienold, J.; Jacobs, A. Luminance maps from High Dynamic Range imaging: Calibrations and adjustments for visual comfort assessment. In Proceedings of the 13th European Lighting Conference (Lux Europa 2017), Ljubjana, Slovenia, 18–20 September 2017; pp. 147–151.

36. Cuttle, C. A new direction for general lighting practice. *Light. Res. Technol.* **2013**, *45*, 22–39. [CrossRef]

37. Duff, J.; Kelly, K.; Cuttle, C. Spatial brightness, horizontal illuminance and mean room surface exitance in a lighting booth. *Light. Res. Technol.* **2017**, *49*, 5–15. [CrossRef]

38. Duff, J.; Kelly, K.; Cuttle, C. Perceived adequacy of illumination, spatial brightness, horizontal illuminance and mean room surface exitance in a small office. *Light. Res. Technol.* **2016**, *49*, 133–146. [CrossRef]

39. Hwang, T.; Kim, J.T. Effects of indoor lighting on occupants' visual comfort and eye health in a green building. *Indoor Built Environ.* **2011**, *20*, 75–90. [CrossRef]

40. Durmus, D.; Davis, W. Blur perception and visual clarity in light projection systems. *Opt. Express* **2019**, *27*, A216–A223. [CrossRef] [PubMed]

41. Smith, A.K.; Sedgewick, J.R.; Weiers, B.; Elias, L.J. Is there an artistry to lighting? The complexity of illuminating three-dimensional artworks. *Psychol. Aesthetics Creativity Arts* **2019**. [CrossRef]

42. Faqih, F.; Zayed, T.; Soliman, E. Factors and defects analysis of physical and environmental condition of buildings. *J. Build. Pathol. Rehabil.* **2020**, *5*, 1–15. [CrossRef]

43. Comité Europeo de Normalización. *UNE-EN 12464.1: Iluminación. Iluminación de los Lugares de Trabajo. Parte 1: Lugares de Trabajo en Interiores*; AENOR: Madrid, Spain, 2003.

Publisher's Note: MDPI stays neutral with regard to jurisdictional claims in published maps and institutional affiliations.

Article

A Model for Locating Tall Buildings through a Visual Analysis Approach

Mehrdad Karimimoshaver [1,*], Hatameh Hajivaliei [1], Manouchehr Shokri [2],
Shakila Khalesro [1,3], Farshid Aram [4] and Shahab Shamshirband [5,6,*]

1   Department of Architecture, Bu-Ali Sina University, Hamedan 65167-38695, Iran;
    hatameh_hajivaliei@yahoo.com (H.H.); sh.khalesro@art.uok.ac.ir (S.K.)
2   Institute of Structural Mechanics, Faculty of Civil Engineering, Bauhaus University, 99423 Weimar, Germany;
    manouchehr.shokri@uni-weimar.de
3   Department of Urbanism, University of Kurdistan, Sanandaj 66177-15175, Iran
4   Higher Technical School of Architecture, Technical University of Madrid (UPM), 28040 Madrid, Spain;
    Farshid.aram@alumnos.upm.es
5   Institute of Research and Development, Duy Tan University, Da Nang 550000, Vietnam
6   Future Technology Research Center, College of Future, National Yunlin University of Science and
    Technology 123, University Road, Section 3, Douliou, Yunlin 64002, Taiwan
*   Correspondence: mkmoshaver@basu.ac.ir (M.K.);
    shamshirbandshahaboddin@duytan.edu.vn or shamshirbands@yuntech.edu.tw (S.S.)

Received: 25 July 2020; Accepted: 25 August 2020; Published: 2 September 2020

Abstract: Tall buildings have become an integral part of cities despite all their pros and cons. Some current tall buildings have several problems because of their unsuitable location; the problems include increasing density, imposing traffic on urban thoroughfares, blocking view corridors, etc. Some of these buildings have destroyed desirable views of the city. In this research, different criteria have been chosen, such as environment, access, social-economic, land-use, and physical context. These criteria and sub-criteria are prioritized and weighted by the analytic network process (ANP) based on experts' opinions, using Super Decisions V2.8 software. On the other hand, layers corresponding to sub-criteria were made in ArcGIS 10.3 simultaneously, then via a weighted overlay (map algebra), a locating plan was created. In the next step seven hypothetical tall buildings (20 stories), in the best part of the locating plan, were considered to evaluate how much of theses hypothetical buildings would be visible (fuzzy visibility) from the street and open spaces throughout the city. These processes have been modeled by MATLAB software, and the final fuzzy visibility plan was created by ArcGIS. Fuzzy visibility results can help city managers and planners to choose which location is suitable for a tall building and how much visibility may be appropriate. The proposed model can locate tall buildings based on technical and visual criteria in the future development of the city and it can be widely used in any city as long as the criteria and weights are localized.

Keywords: tall building; locating criteria; fuzzy visibility; visual impact

---

1. Introduction

1.1. Importance of Tall Buildings and Their Various Dimensions

Tall buildings have always been one of the most critical elements of a city's visual appearance and research on the visual impact of tall buildings is underdeveloped.

Tall buildings need more attention due to the growing number of tall buildings all over the world and their high impact on the city—e.g., "they cannot be ignored" [1].

Short [2] believes that the impact of tall buildings on the city can be studied based on eight categories including context, the effect on the historic environment, the effect on the local environment, the relationship to transport, permeability, architectural quality, contributions, and sustainability. To study the tall buildings' roles in the cities, Al-Kodmany and Ali [3] consider tall buildings' relationships to the city in various dimensions including infrastructure, city skyline, place making, architectural and engineering quality, and iconic role. Another study evaluates the triple effects of tall buildings, and their impact on the city based on aesthetic, semantic, and visual criteria, and presented a framework to measure their impact.

1.2. Definition of Tall Buildings

There are different words for defining tall buildings worldwide, including high-rise, skyscraper, tower, and super tall buildings. Tall buildings' categories can be different around the world; for example, one city can consider more than 12 stories as tall buildings, but another city does not categorize 30–40 stories as tall buildings.

Tall buildings' definitions can differ due to the time and place conditions of this type. Different cities of the world have specific descriptions of tall buildings in different periods concerning urban criteria, fire stations, structure, and installation.

Al-Kodmany and Ali [2] classified these buildings as tall buildings: high-rise buildings, ultra-high-rise buildings, towers, and skyscrapers. Other definitions are based on the countries' internal regulations. For example, in Germany, some of the high-rise construction criteria are determined by the country's firefighting system's available equipment. Howeler (2003) categorized skyscrapers based on appearance into media-oriented, environmental, kinetic, etc. These buildings are defined according to each country and city's local regulations [4].

However, it seems to be an unsaid contract that tall buildings are taller than their context in their location in the world [5]. Therefore, the definition of tall buildings in this study is 20-story buildings that are considered taller than the context in Hamedan.

1.3. Tall Buildings and Visibility

Al-Kodmany and Ali [2] believe that "the visual impact of tall buildings on urban form extends far beyond their footprints". In architecture and urban studies, visibility is a criterion that deals with whether a building or an urban structure is visible from urban public spaces, to analyze its impact on the viewers [6]. This analysis calculates the visual impact of structures and buildings with the aim of controlling their impacts on the city.

London is one of the cities that has specific regulations on controlling the impact of tall buildings on urban views. The London View Management Framework (LVMF) includes strict regulations that control the possible effects of proposals for new buildings on the skyline and views in London. It proposes four kinds of views, including London Panoramas, Linear Views, River Prospect, and Townscape Views. Proposals for new buildings are assessed for their impact on the protected views within the foreground, middle ground, or background of those views [7].

Tall buildings and city skylines are essential components of these views, and in most cases, tall buildings are the most prominent parts of the city skyline [8]. Rod and Van der Meer [9] evaluated the dominance of tall buildings in new developments. They considered how much a tall building could influence visibility, using both traditional and modern methods. They believe visibility and dominance are fundamental options for locating tall buildings in cities. The question that "a building how much and how long would be visible for citizens" will be examined in analyzing visibility and dominance.

Zarghami et al [10] developed a method to study the impact of the form and the physical features of tall buildings on citizens based on Height, width, and height-to-width ratio. Karimimoahaver [11] provided and evaluated different methods of describing a building's visual impact on the city. Karimimoshaver et al. [12] examined the relationship between tall buildings' locations and the urban landscape. They evaluated the effectiveness of single and cluster tall buildings based on functional, identity, and aesthetic purposes.

Traditional visibility considers if urban elements are visible or non-visible; however, Fisher [13] does not agree with this definition. She believes, instead, that visibility is how much of a tall building is visible, and considers that as a measure using fuzzy logic [13]. In another study, Oh [14] examines the percentage of visibility and its importance as two critical things in visual analysis.

Rød and van der Meer [9] showed how a planned tall building's impact can be evaluated through GIS-based visibility and dominance analysis. To make the visibility of tall buildings more practical, they explain the dominance concept based on the distance of the building from the observation viewpoint. They believe that traditional visibility analysis calculates the relationship between buildings and the existing situation in the urban landscape, in which visual obstacles, trees or other disturbing elements are considered. However, dominance analysis adds the criterion of how a building dominates visually, taking into account the distance from the building and the degree to which the building is visible from any point of view [9].

Another study did a visibility analysis of high-rise building development, using a GIS-based map to present visual coverage and cumulative visibility of tall buildings through the study of visibility of existing single and cluster buildings in the city of Rotterdam, and the impact of those buildings on the city skyline during the time [15]. They considered a tall building's performance in the cluster, which is perceived visually instead of obtaining it individually; this offered a new way to characterize the visibility of a tall building. Czynska and Rubinowicz [16] offer the visual impact size (VIS) method to recognize tall buildings' locations and to determine how much tall buildings are visible through virtual city models.

Today visibility analysis is performed in both binary and fuzzy forms. The binary form presents the possibility of seeing the building only, while the fuzzy method determines the percentage of visibility. Since binary visibility only examines whether a building is seen or not, it can only be useful in certain cases. The purpose of visibility in this research is fuzzy visibility in order to determine the amount of visibility of tall buildings from urban spaces, and based on this to be able to measure their impact.

1.4. Locating Tall Buildings

Locating is a kind of spatial planning that locates certain activities [17]. Hence projects can exploit the identification of the most economical places, and its competitiveness should be considered as one of the critical goals. These decisions must be in line with the investors' and government's specific policies, and will primarily meet environmental needs [18].

Results from tall buildings' physical-spatial effects of Farmanieh and Kamranieh neighborhoods in the city of Tehran designate that creating tall buildings must occur based on pre-planned principles and macro-policies [19]. Azizi and Fallah [20] presented suitable areas for the tall building development by providing location criteria (social, economic, environmental, and physical) in the city of Shiraz.

Adeli and Sardarre [21] evaluated proper locations for tall buildings in the city of Qazvin by considering social, economic, environmental, and physical criteria, weighting them, and producing the corresponding layers. Sedaghati [22] studied the existing location of one of the tall building complexes in the city of Tabriz. This study evaluated the impact of distance from pollutants, faults, traffic points, highways, etc., on these buildings.

Thomasetin et al., [23] desired to turn the process of developing tall buildings into a purposeful endeavor by categorizing existing tall buildings by location (city center, middle texture, surrounding ring, next to the big ones, and close to the industrial areas) and type of accumulation (single-scattered) [23]. They concluded that urban planners must keep developing tall buildings in centralized clusters around the historic part of the city, and that constructing individual buildings near the historic center of the city should be avoided. The investigation of location methods during the last two decades in Iran reveals that existing approaches are mostly simplistic and rudimentary. Over that time period, there has not been a significant achievement in advancing the urban planning profession in Iran [24]. Another study suggested location for nine regions of the city of Mashhad by using criteria (physical-spatial, environmental) that have been extracted from the smart growth theory. This paper declares that existing tall buildings are located in inappropriate locations and defines appropriate points [25]. Salehi and Ghadiri [26] believe that the criteria required for locating and constructing tall buildings in Iran are in line with building regulations of the city of Toronto in Canada. These regulations considered "rule", "tower", and "communication with adjacent areas" in terms of shading, neighborhood, aesthetics, and urban landscape. These rules can also be used in Iranian metropolises.

Zista consulting engineering [27] considered urban planning principles, citizen orientation, and urban visualization for building locations in Tehran. Anabestani et al., [28] estimated appropriate locations for constructing tall buildings in Mashhad's 9th region. Zareian [29] studied social issues of tall buildings. He defines the most critical social problems of these buildings as increased congestion, lack of cultural compatibility, reduced neighborhood identity, difficulty in ensuring security, and weak neighborly relations.

1.5. Research Question

Previous studies that dealt with tall buildings' locations suggested appropriate locations only for specific sites and failed to provide a comprehensive model including the weight of the criteria and consideration of the two- and three-dimensional impacts of tall buildings on the city, therefore the main research question is:

What is the weight of the criteria affecting the location of tall buildings and how to create an effective combination of two-dimensional and three-dimensional criteria for locating tall buildings, as a comprehensive model?

2. Materials and Methods

The research method consists of three main stages including weighting and prioritizing criteria, applying criteria to the GIS layer, and visibility analysis.

2.1. Weighting and Prioritizing Criteria

The first stage was prioritizing criteria (Figures 1–3 and Table 1). Twenty registered and reputable experts, including twelve urban designers, five architects, and three urban planners, completed a survey through the analytic network process (ANP) technique (Figure 4). Criteria were narrowed down based on the current information and the general situation of Hamedan. The relationship between them and inhibiting sub-criteria, and their importance compared to one another, were measured. Finally, the weighting and priority of criteria were provided by using Super Decisions V2.8 software. Super Decisions software is a tool for making decisions based on multi-criteria methods including the analytic hierarchy process (AHP) and the analytic network process (ANP) [30] which is applied in this research.

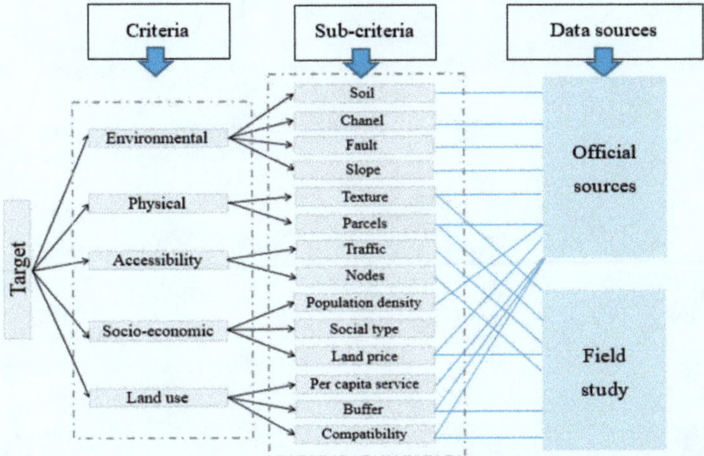

Figure 1. The structure of the criteria study including criteria (environmental, physical, accessibility, socio-economic, and land use), sub-criteria (soil, channel, fault, slope, texture, parcels, traffic, nodes, population density, social type, land price, per capita service, buffer, and compatibility), and data sources.

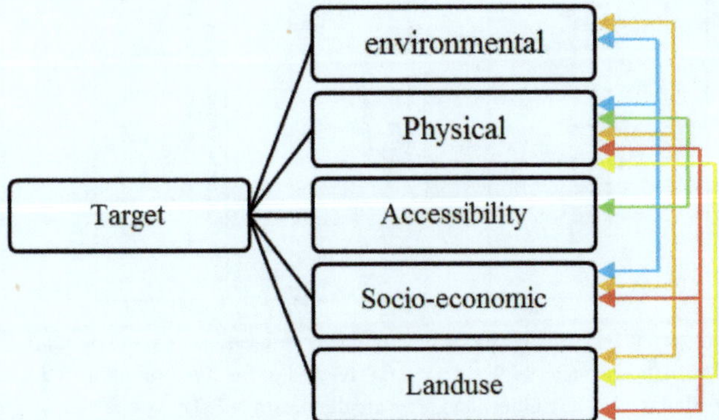

Figure 2. The network relations between the criteria, based on expert opinion. (The colors in this figure are simply due to the better readability of the relationships between the criteria and have no special meaning).

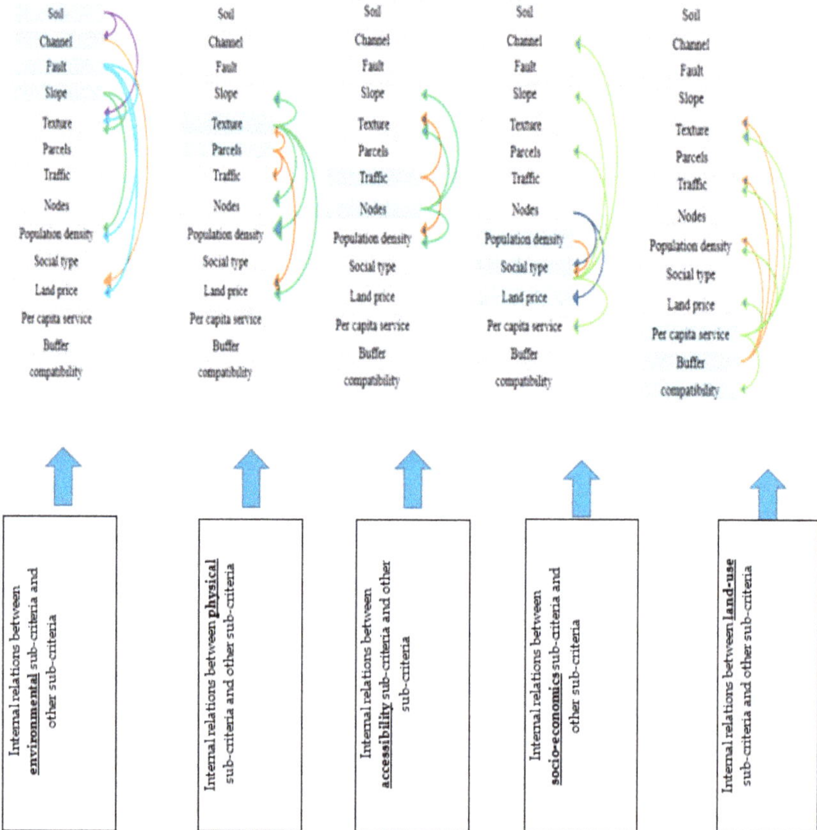

Figure 3. The internal relations between the sub-criteria, based on expert opinion. In this figure, the relationships between each of the sub-criteria (related to the five criteria introduced in Figure 2) with other sub-criteria related to other categories are demonstrated (The colors in this figure are simply due to the better readability of the relationships between the criteria and have no special meaning).

Figure 4. The result of the pairwise comparison of the criteria presented in Figure 4 from Super Decisions V2.8 software.

Table 1. The criteria and sub-criteria for locating tall buildings and their weights based on expert opinion, formulated using the ANP method and Super Decisions V2.8 software.

	Criterion	Criterion Weight	Sub-Criteria	Sub-Criteria Weight	Measure	Priority
1	Environmental (natural-geological factors)	0.04384	Slope	0.04355	0–5%	1
					5.1–10%	2
					10.1–15%	3
					>15%	4
			Channel (surface and underground sewage)	0.00062	0–25 m	3
					25–150 m	2
					>150 m	1
			Fault	0.02851	0–500 m	3
					500–1000 m	2
					>1000 m	1
			Soil type	0.00026	Clay soil	1
					Sandy soil	2
2	Physical	0.41614	Texture	0.2445	Low density	1
					Medium density	2
					dense	3
					Non-residential	4
			Land parcels	0.00328	>1000 m	1
					800–1000 m	2
					600–800 m	3
					Others	4
3	Accessibility and transportation	0.2547	Traffic	0.11677	Primary road	1
					30 m street	2
					Street	3
					Others	4
			Nodes	0.03184	short distance (first line)	3
					Medium distance (second line)	2
					Long distance (third line and so on)	1

Table 1. *Cont.*

	Criterion	Criterion Weight	Sub-Criteria	Sub-Criteria Weight	Measure	Priority
4	**Socio-economic**	0.07933	Land price	0.25698	2375–3125 U.S. dollars /m^2	4
					1625–2375 U.S. dollars/m^2	2
					875–1625 U.S. dollars/m^2	1
					250–875 U.S. dollars/m^2	3
			Social type	0.00398	Educated/ low population	4
					Educated/medium population	3
					Educated/relatively large	2
					Educated/large population	1
			Population density	0.18223	dense	4
					Medium density	3
					Medium density	2
					Low density	1
5	**Land-use**	0.20598	Buffer	0.04608	short distance	2
					Long distance	1
			Land-use compatibility	0.00446	Incompatible land-use	3
					Semi-compatible land-use	2
					Compatible land-use	1
			Per capita urban services	0.03695	Low service	4
					Medium service	3
					Acceptable service	2
					Optimal service	1

2.2. Developing GIS Layers

In the second stage, the GIS layers of Hamedan's characteristics (Figures 5–18) are applied based on the sub-criteria mentioned in Table 1. The Map Algebra tool in ArcGIS V10.3 software was applied to create the layers so that it was possible to include the weight of the criteria obtained in the first stage. Then a map (Figure 19) is provided by the ArcGIS software that shows the best locations for the proposed tall buildings in the city.

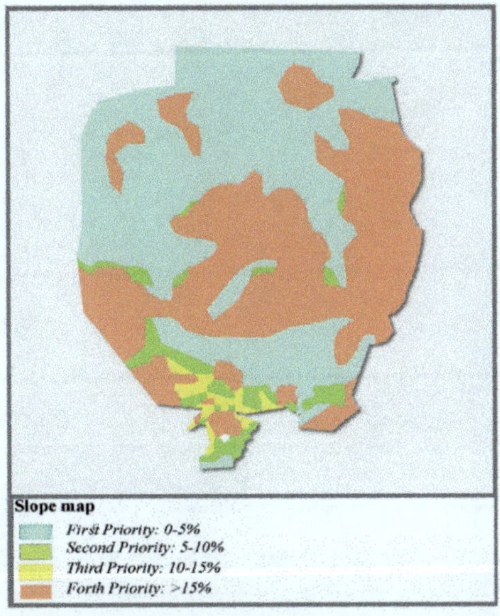

Figure 5. The slope map of Hamedan based on the sub-criteria measures in Table 1.

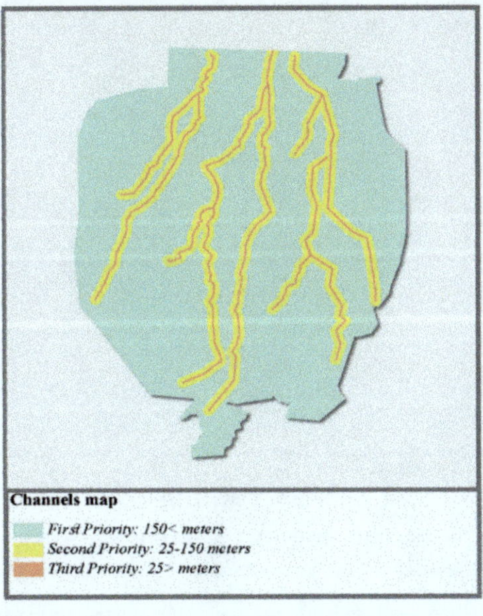

Figure 6. The channels map of Hamedan based on the sub-criteria measures in Table 1.

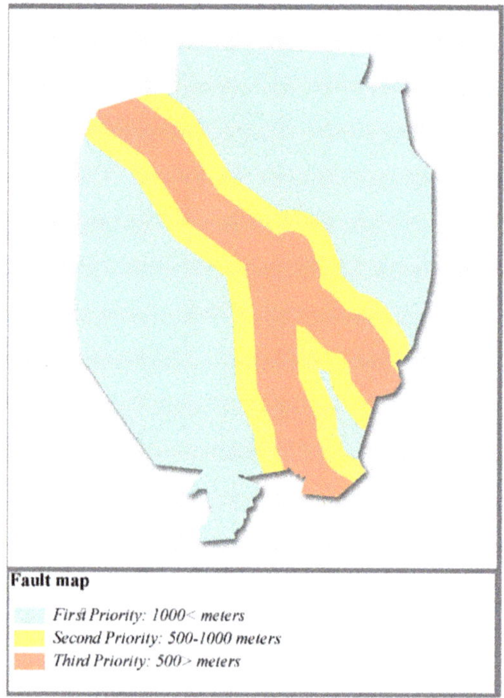

Figure 7. The fault map of Hamedan based on the sub-criteria measures in Table 1.

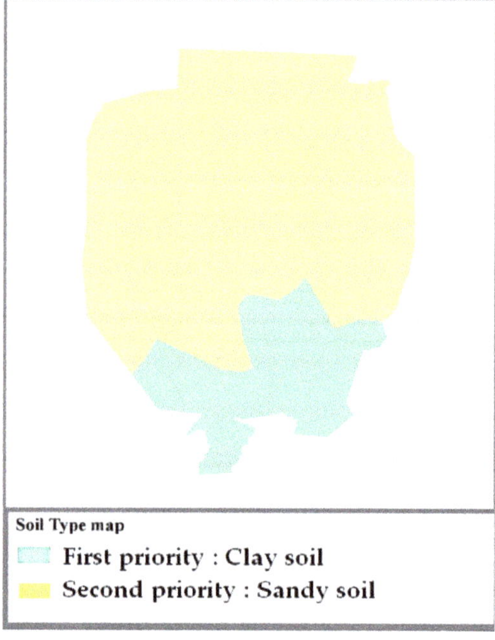

Figure 8. The soil type map of Hamedan based on the sub-criteria measures in Table 1.

Figure 9. The texture map of Hamedan based on the sub-criteria measures in Table 1.

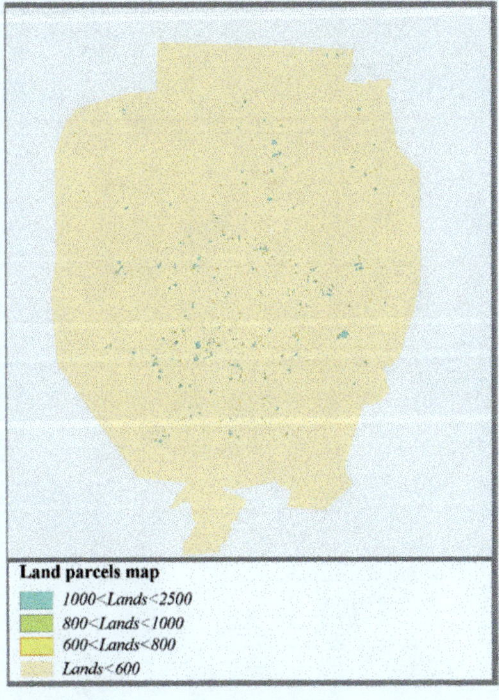

Figure 10. The land parcels map of Hamedan based on the sub-criteria measures in Table 1.

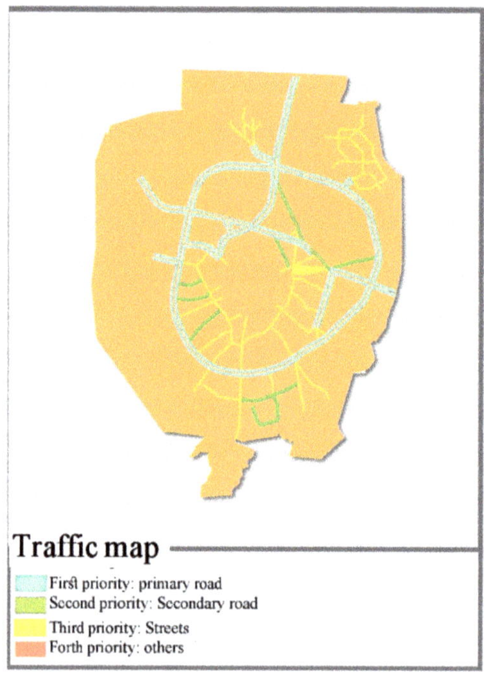

Figure 11. The traffic map of Hamedan based on the sub-criteria measures in Table 1.

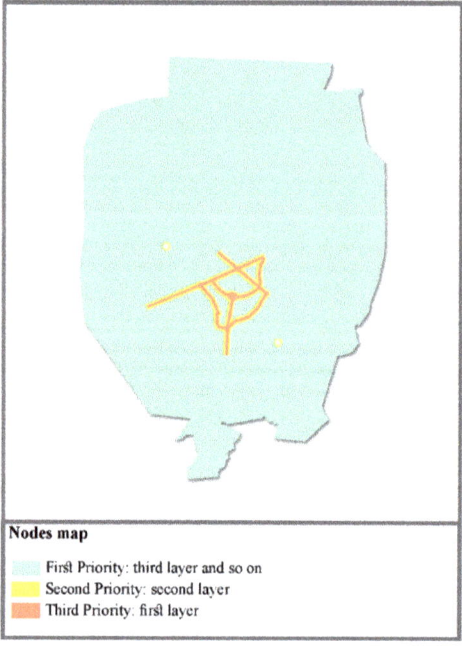

Figure 12. The nodes map of Hamedan based on the sub-criteria measures in Table 1.

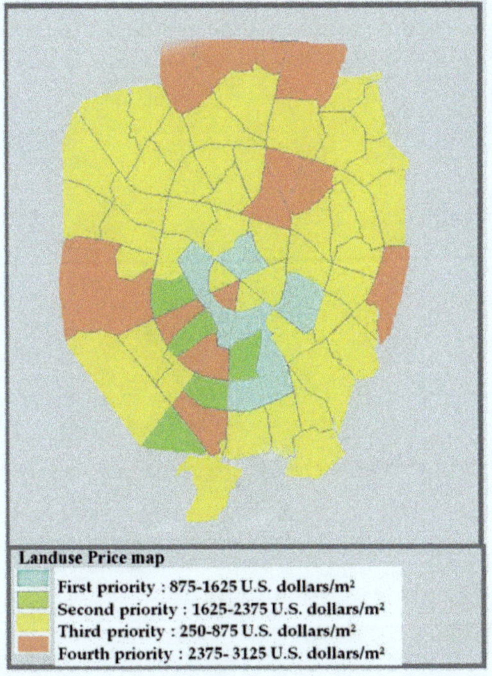

Figure 13. The land price map of Hamedan based on the sub-criteria measures in Table 1.

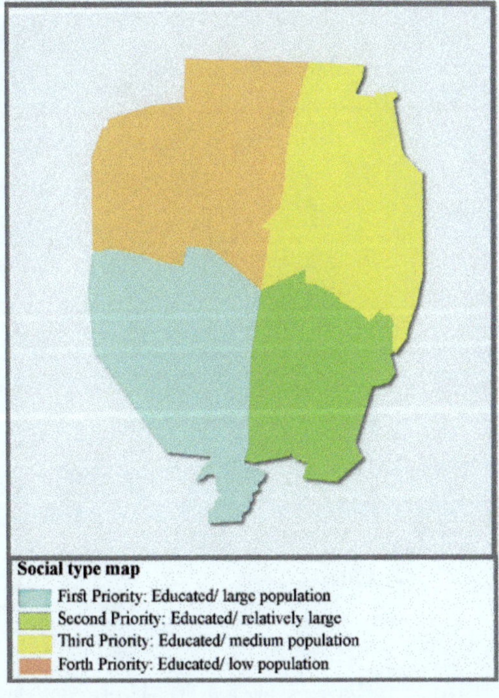

Figure 14. The social type map of Hamedan based on the sub-criteria measures in Table 1.

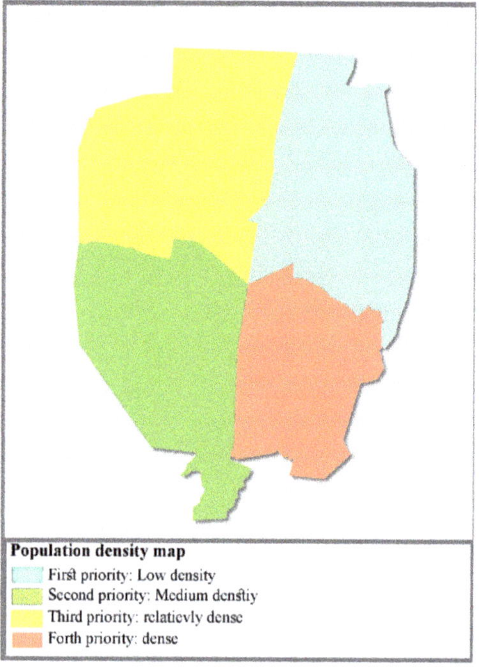

Figure 15. The population density map of Hamedan based on the sub-criteria measures in Table 1.

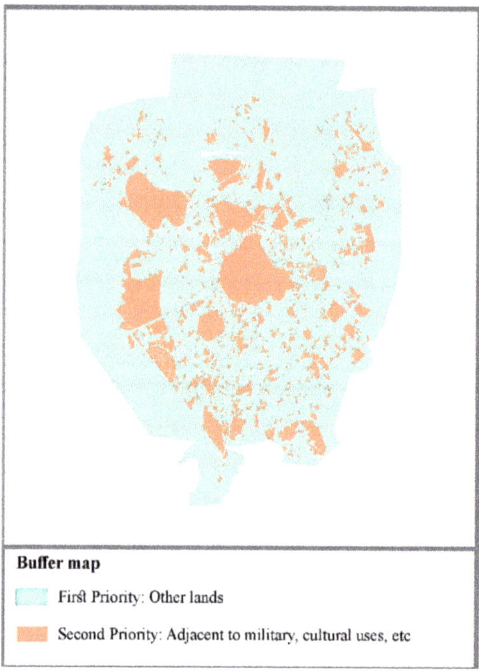

Figure 16. The buffer map of Hamedan based on the sub-criteria measures in Table 1.

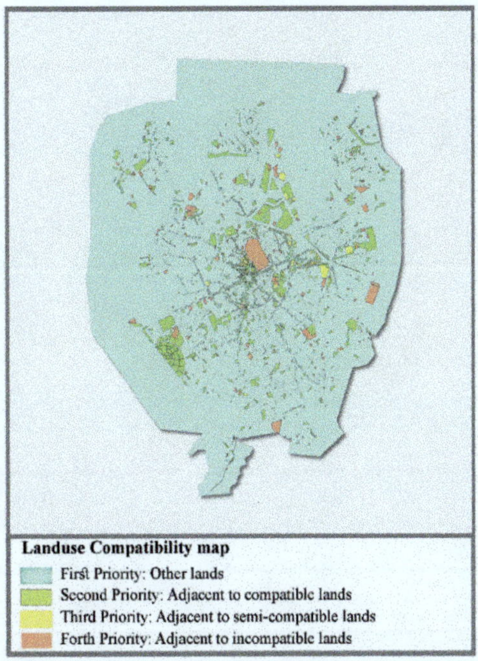

Figure 17. The land-use compatibility map of Hamedan based on the sub-criteria measures in Table 1.

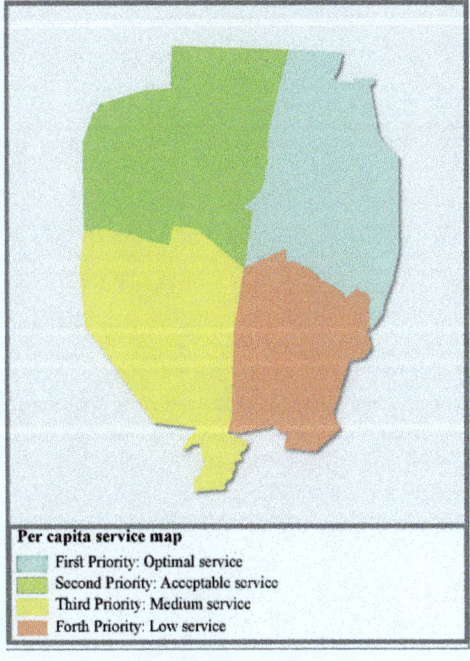

Figure 18. The per capta service map of Hamedan based on the sub-criteria measures in Table 1.

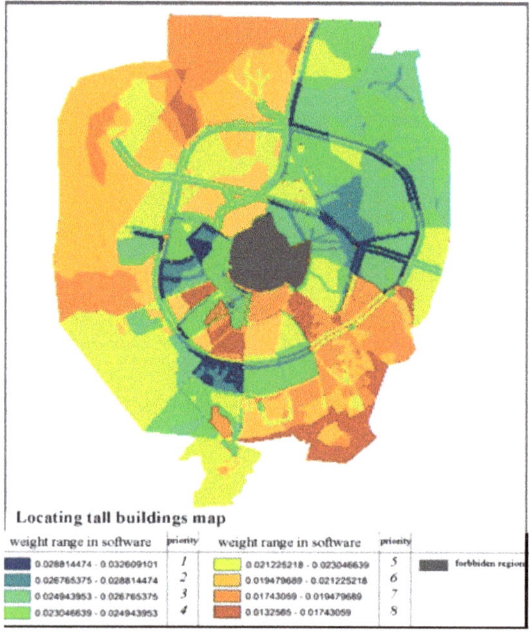

Figure 19. The best urban areas for locating tall buildings. The map shows the priority of suitable places for tall buildings on a scale from dark blue (priority 1) to dark red (priority 8).

2.3. Visibility Analysis

The last and most crucial stage was preparing a plan that addressed building visibility in future development. Then, a general algorithm was reached (Figure 20). For visibility analysis, a fuzzy model was used, which was designed to provide the numerical value of visibility for each tall building (Figure 21). Based on Figure 19, seven hypothetical 70-m-tall buildings (20 stories), with a length of 30 m and a width of 15 m, were considered. This process was programmed on Matlab software (based on the process shown in Figure 20) and outputs were imported into ArcGIS again. In this program, larger sides had 18 points (3 points in 6 rows), and smaller sides had 6 points; 48 points were defined exclusively, and the observer could see 24 points at best. The observer (height: 170 cm, radius vision: 1000 m), was located on the surrounding urban spaces (Figure 21). The purpose of doing the third stage is comparing the influence of chosen buildings and finally selecting the best place with the least negative impact on public urban spaces (Figures 22–28).

Note: all the visibility maps for tall buildings No. 1–7 are produced based on Matlab software coding and ArcGIS V10.3 outputs. In the images on the right, the blue dots show the location of seven hypothetical tall buildings, which are enlarged on the left. In these images, red spots show the highest visibility of a tall building from urban spaces (80 to 100%) and the lowest is the green color, which is less than 20% of each of these urban spaces.

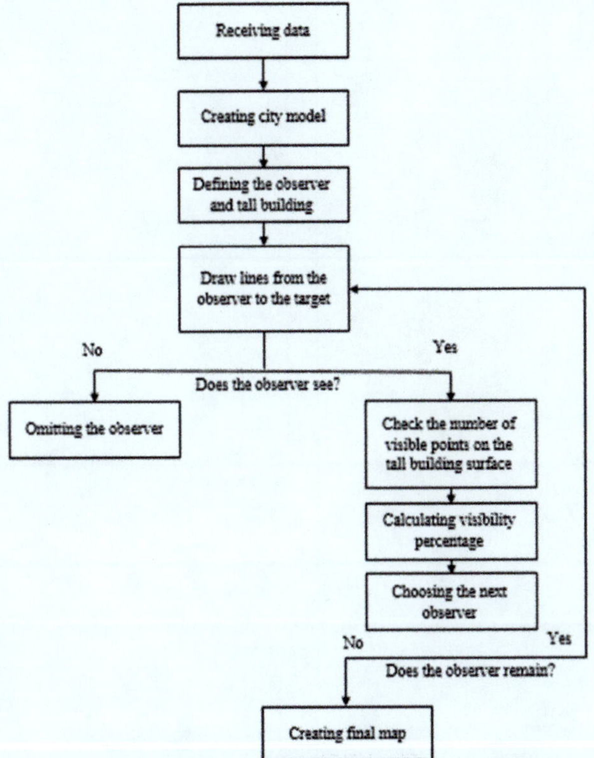

Figure 20. The fuzzy visibility's programmed process used in the Matlab software.

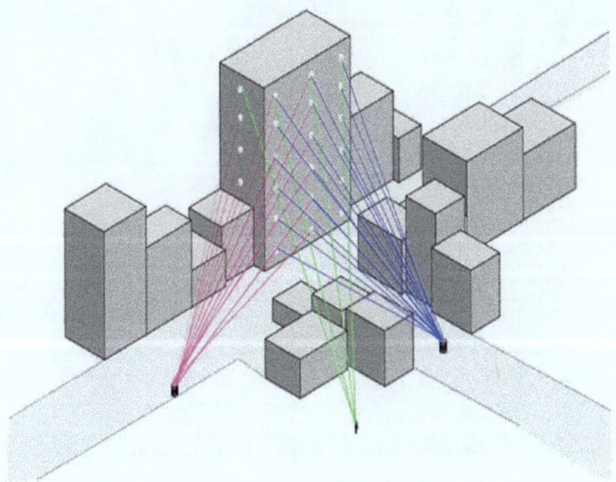

Figure 21. A 3-D image of the fuzzy visibility's programmed process. To do this, the larger side of tall buildings has 18 points (3 points in 6 rows), and the smaller side has 6 points; 48 points are defined exclusively, and the observer could see 24 points at best.

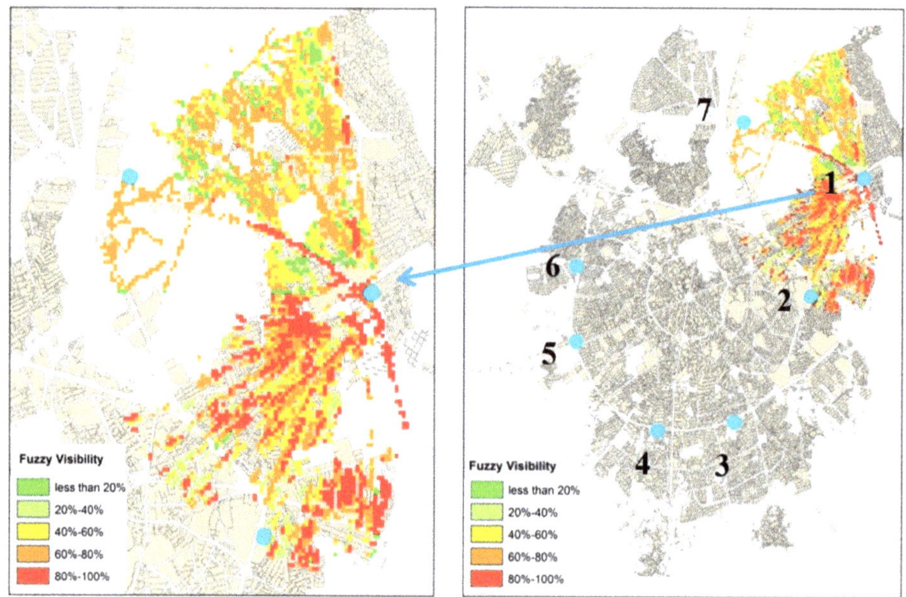

Figure 22. Visibility map of hypothetical tall building No. 1.

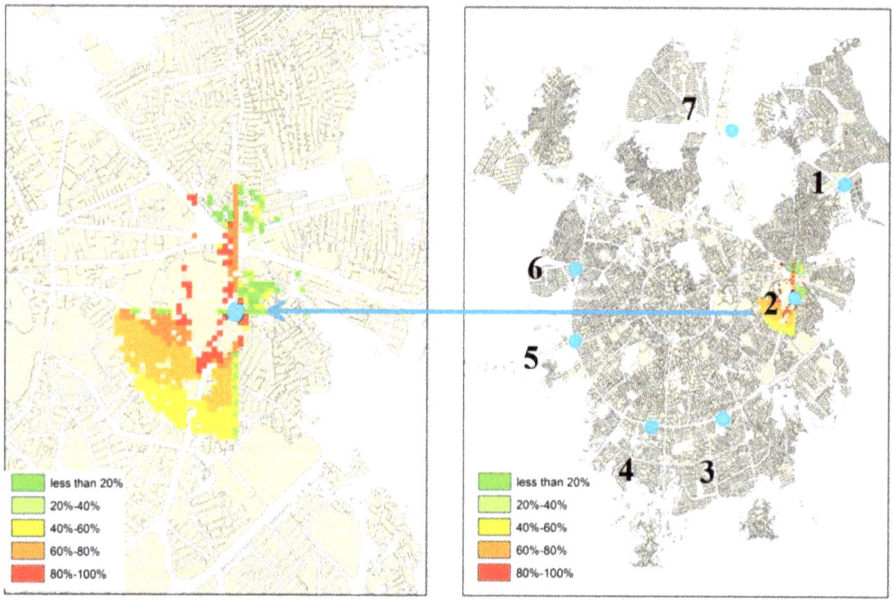

Figure 23. Visibility map of hypothetical tall building No. 2.

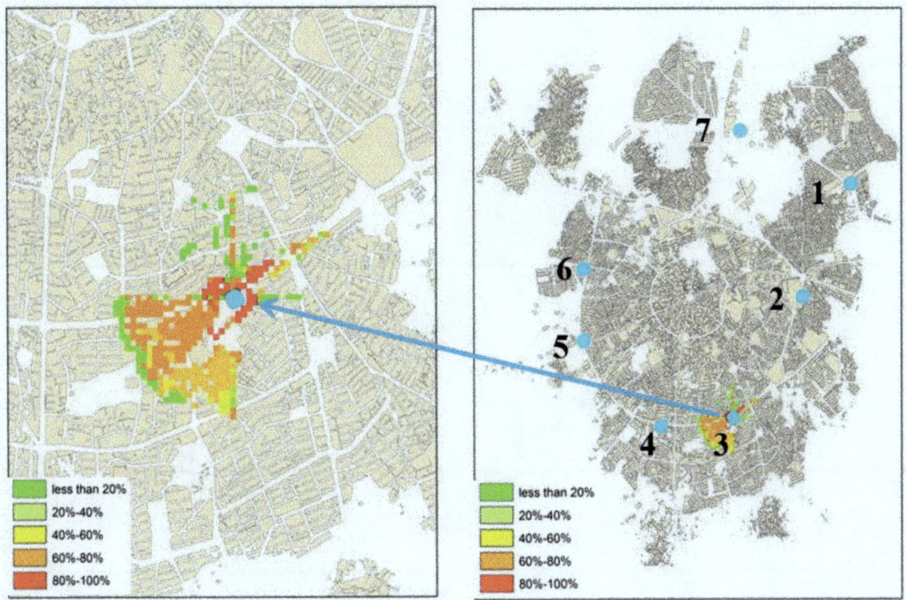

Figure 24. Visibility map of hypothetical tall building No. 3.

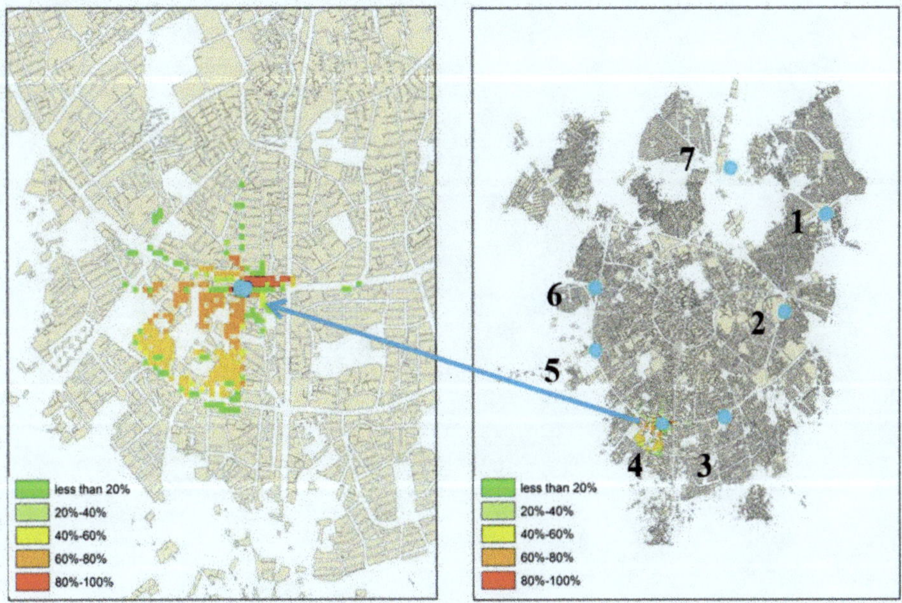

Figure 25. Visibility map of hypothetical tall building No. 4.

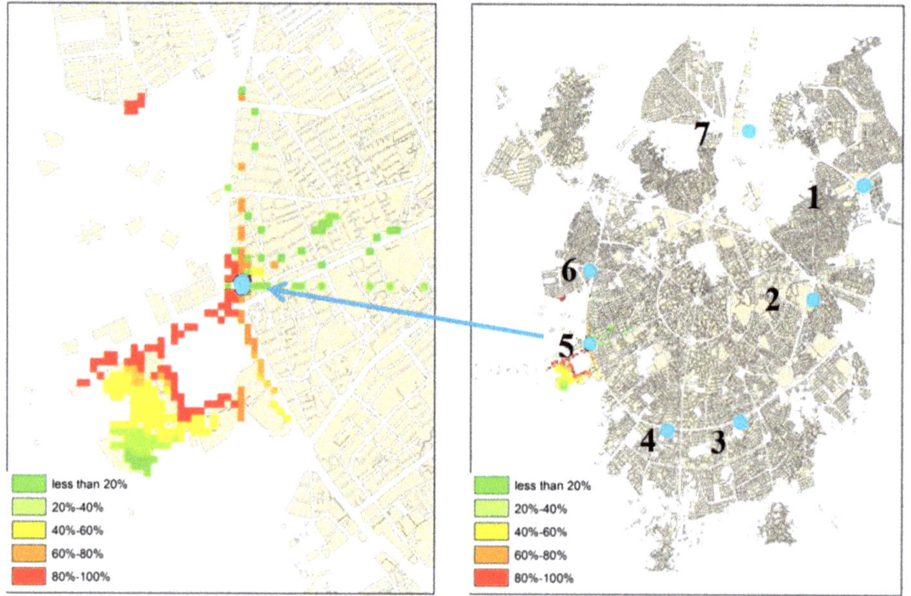

Figure 26. Visibility map of hypothetical tall building No. 5.

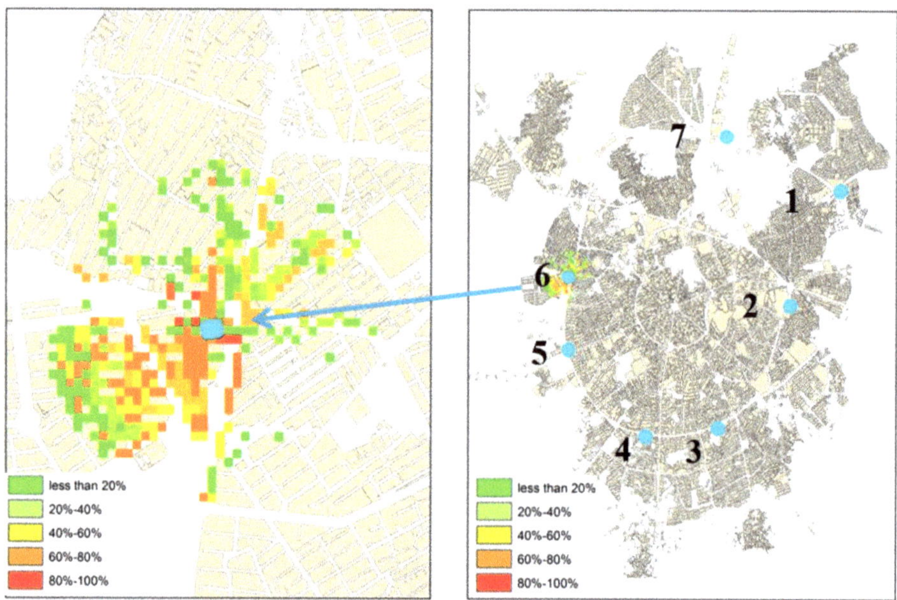

Figure 27. Visibility map of hypothetical tall building No. 6.

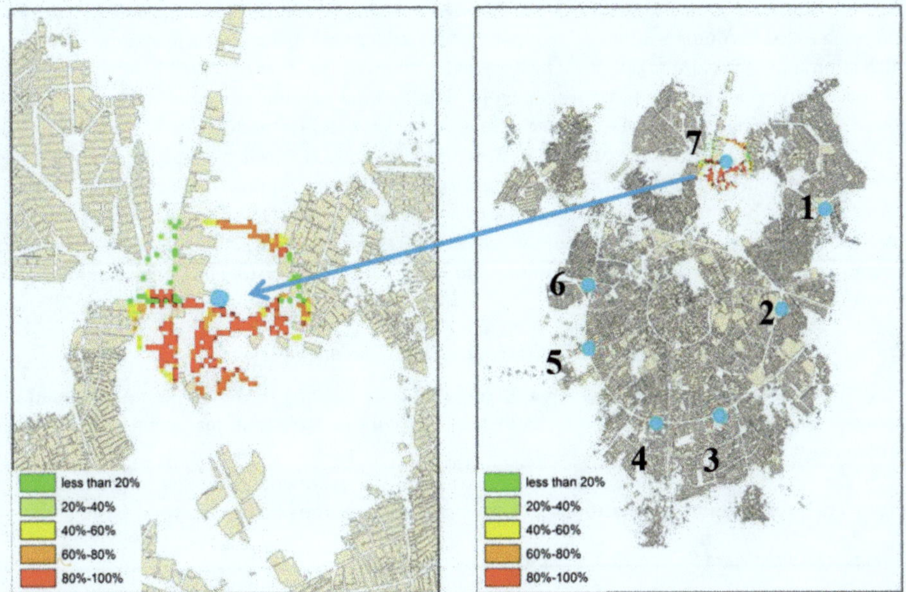

Figure 28. Visibility map of hypothetical tall building No. 7.

2.4. Case Study: City of Hamedan

The city of Hamedan is located on the west side of Iran and on the slope of Alvand Mountain. The northern side of the city's height is 1780 m, and at the end of the southern side, it reaches 1950 m; this causes a steeper slope on the northern side of the city. Hamedan's soil categories are composed of two types: clay and sandy soil. In this city, two main faults exist that begin from the southeast and continue to the center and northwest; they are named Yelfan and Kashin [31]. Five rivers have flowed through the city. However, today their surface is covered in some places, and they turned into sewers [31], which explains surface and groundwater criteria. Urban texture, based on the maps prepared by the municipality, is divided into three sections: dense, medium density, and low density, and city parcels more significant than 600 m^2 are identified and categorized by ArcGIS V10.3 software. Road maps were prepared to cover accessibility and transportation criteria, and adjacent layers of these roads have been prioritized based on the street width. A traffic node map was arranged through field study and surveys at different hours of the day. Socio-economic features were prepared by referring to the Statistics Organization, and land price information was requested from the experts of the land pricing department and field study. The calculation and prioritization of the region's per capita area in treatment, training, green space, and urban facilities provided functional and land-use maps. The detailed plan explained land-use, buffers, and their compatibility.

3. Results

The research's findings are presented in three main sections. The first section provides criteria and sub-criteria and their weights. The second section shows the appropriate regions in the city for constructing tall buildings through GIS maps. The third section clarifies proposed tall buildings' visual impacts in proper lands.

3.1. Findings of the First Section: Criteria, Sub-Criteria and Their Weight

This section specifies the essential criteria and sub-criteria based on experts' opinions for choosing proper lands as a main target (Figure 1).

The criteria weights and priorities are clarified based on the experts' viewpoints. To do this, experts are asked to compare all the related criteria and sub-criteria in a pairwise manner to determine which criteria in each comparison have priority over other criteria. Figures 2 and 3 show the criteria and sub-criteria inter-relationships respectively. The method of comparison between criteria and sub-criteria is that the criteria and sub-criteria are compared in pairs and each of them, which have more importance and priority than the other, is awarded more points, in a scale of 1 to 9, which ultimately determines how important each criterion is compared to the other (Figure 4). Then, Super Decisions V2.8 software has been used to achieve Figure 4, this image is an example of the output of this software, which shows the pairwise comparison of the criteria expressed in Figure 2. These analyses have been implemented for all the sub-criteria mentioned in Figure 3, and the results of all these analyses are presented numerically in Table 1.

3.2. Findings of the Second Section: Appropriate Locations for Tall Buildings

In this phase, essential sub-criteria (based on Table 1) for locating tall buildings are presented via GIS layers (Figures 5–19); by overlapping them, (Figure 20) the suitable areas for erecting tall buildings are identified.

In the next step, the GIS layers, shown in Figures 5–19, were overlapped together; consequently, the best areas for tall buildings locations, as shown in Figure 20, are obtained.

3.3. Findings of the Third Section: Tall Building Visual Impacts

In this phase, seven tall buildings are placed in seven more suitable zones, based on the best urban area for locating tall buildings (Figure 20), and their visual impact is determined by GIS maps. By comparing these maps and the visual impact of these buildings, the most appropriate place for tall buildings can be obtained.

Figures 23–28 show the visual impact of each of the proposed buildings. In this article, we do not intend to judge which mode is better, and we are only trying to provide a method for cities to act according to their plan and vision. Whether visibility is more or less better depends on the urban plans and visions of each city. In some cities, managers and planners need to find a way to increase the visibility of tall buildings, and in some other cities, they need to minimize the visual impact of tall buildings as much as possible.

4. Discussion

The model presented in this research can help to locate tall buildings. The method presented in this research is a combination of the two methods of planning and visual analysis. In the planning section, the criteria and sub-criteria are considered to determine the appropriate areas for the erection of tall buildings. Visual analysis is done in three dimensions based on fuzzy visibility; the positive and negative effects of tall buildings on the city are identified and help to decide on the appropriate location of the tall buildings. This model can be used for every desirable region and it suggests proper spots for all of them, provided there is localization of criteria.

In most previous researches, the influential factors in locating tall buildings have been mentioned [32,33], but the prioritization and weighting of these criteria have not been studied in detail. In this study, according to the results, land price, urban texture, population density, and traffic sub-criteria are at the highest level, and surface and groundwater sub-criteria are at the lowest level in locating tall buildings by prioritization. Other criteria like buffer, slope, per capita service, traffic nodes, fault, compatibility, and parcels are also placed among the list.

Lack of access to accurate 3D images and maps of the city was one of the main limitations of this research. Although we tried to develop a comprehensive model, further studies can examine designing adjacent sidewalks criteria, facades and their details, a building's 3D model, parametric design according to the background characteristics, etc. The effects of vegetation have not been applied in the fuzzy visibility analysis in this research. By minimizing the study area, this factor can be included.

Another thing to consider is the amount of visibility from inside of the yard. The visual analysis could accommodate different factors such as increasing legibility, making memories, and preparing maps to protect visually valuable axes.

5. Conclusions

In this research, a model is presented so that the role of the tall buildings' visual impacts in the city can be properly considered in decision-making for the location of tall buildings in the city. The central hypothesis of this research was that a set of criteria is vital in locating tall buildings.

The main question of this research was how to find a solution that is reproducible and can be used in any other city based on many heterogeneous criteria. To do this, the criteria were divided into two-dimensional criteria and three-dimensional criteria. In two-dimensional criteria, technical issues affecting the location of tall buildings were identified and weighted and prioritized based on the opinion of experts by the ANP method in Super Decisions V2.8 software. In the three-dimensional section, the fuzzy visual analysis was applied by MATLAB software considering how much of a tall building can be seen from urban spaces. In this way, this research provided a model that was able to examine a large number of heterogeneous variables affecting the location of tall buildings and to determine their weight and priority.

After evaluating the criteria, locations should be chosen that not only do not negatively impact the city, but also positively impact it. The final filter for locating tall buildings is a visual criterion that is very important in the city. In other words, in addition to evaluating technical criteria, the visual impact and its positive influence should be considered. Due to the complexity of three-dimensional analysis, the proposed model in this research is formed in two stages: the first stage is reviewing technical criteria to narrow the possible urban area for locating tall buildings, and the second stage is reviewing visual criteria in which only areas of the city that have been identified as suitable for the construction of tall buildings in terms of two dimensions are examined in the first stage.

In this study, the main limitation is the unavailability of 3D images of the city with sufficient accuracy. This limitation might change the research results. We tried to get as close as possible to the existing reality through field research. However, observing correct height codes for both the land and the building requires more accurate mapping data.

This research generally seeks to find the appropriate framework for locating tall buildings. This research's framework presented in two main steps. In the first step, technically suitable locations of the city for constructing tall buildings were recognized. In the second step, parts of the city that had both a positive visual impact on the city and positive technical criteria were identified. The method and framework used in this research can be used in all cities around the world. In fact, in each city, the criteria and sub-criteria and their weights can be reviewed based on the city's priority.

This model can be considered individually for areas considered more desirable, and offers a proper location in each zone.

The designed program can evaluate a building's visibility in future development plans if the input data is accurate enough. In this program, one of the essential requirements for preparing a precise output is the accuracy and precision of the information. In further studies, vegetation and distance impact can also be considered and can be focused on providing a set of regulations.

Author Contributions: Conceptualization, M.K.; methodology, M.K.; software, H.H., S.S., M.S. and F.A.; validation, M.K. and H.H.; formal analysis, H.H. and F.A.; investigation, M.K.; resources, S.K.; data curation, F.A.; writing—Original draft preparation, M.K. and H.H.; writing—Review and editing, M.K. and S.K.; visualization, S.S.; supervision, M.K.; funding acquisition, S.S. All authors have read and agreed to the published version of the manuscript.

Funding: We acknowledge the support of the German Research Foundation (DFG) and the Bauhaus-Universität Weimar within the Open-Access Publishing Programme.

Conflicts of Interest: The authors declare no conflict of interest.

References

1. Short, M. *Planning for Tall Buildings*; Routledge: Abingdon, UK, 2012.
2. Al-Kodmany, K.; Ali., M.M. *The Future of the City: Tall Buildings and Urban Design*; WIT Press: Chicago, IL, USA, 2013; pp. 11–44 & 68–98.
3. Karimimoshaver, M.; Winkemann, P. A framework for assessing tall buildings' impact on the city skyline: Aesthetic, visibility, and meaning dimensions. *Environ. Impact Assess. Rev.* **2018**, *73*, 164–176. [CrossRef]
4. Howeler, E. *Skyscraper: Design of the Recent Past and for the Near Future*; Thames & Hudson: London, UK, 2003.
5. Council on Tall Buildings and Urban Habitat. Ctbuh.org. 2017. Available online: http://ctbuh.org/ (accessed on 30 October 2017).
6. Karimimoshaver, M. Methods, Techniques and Tools in Urban Visual Analysis. *Bagh-e Nazar* **2014**, *11*, 73–80.
7. London View Management Framework. Available online: https://www.london.gov.uk/what-we-do/planning/implementing-london-plan/planning-guidance/london-view-management (accessed on 12 November 2017).
8. Heath, T.; Smith, S.G.; Lim, B. Tall Buildings and the Urban Skyline. *Environ. Behav.* **2000**, *32*, 541–556. [CrossRef]
9. Rød, J.K.; Van Der Meer, D. Visibility and Dominance Analysis: Assessing a High-Rise Building Project in Trondheim. *Environ. Plan. B Plan. Des.* **2009**, *36*, 698–710. [CrossRef]
10. Zarghami, E.; Karimimoshaver, M.; Ghanbaran, A.; SaadatiVaghar, P. Assessing the oppressive impact of the form of tall buildings on citizens: Height, width, and height-to-width ratio. *Environ. Impact Assess. Rev.* **2019**, *79*, 106287. [CrossRef]
11. Karimimoshaver, M. Approaches and methods in urban aesthetics. *Sci. J. NAZAR Res. Cent. (NRC) Art Archit. Urban.* **2013**, *10*, 63–72.
12. Karimimoshaver, M.; Mansouri, S.A.; Adibi, A. Relationship between the Urban Landscape and Position of Tall Buildings in the City. *Sci. J. NAZAR Res. Cent. (NRC) Art Archit. Urban.* **2010**, *7*, 89–99.
13. Fisher, P.F. Probable and fuzzy models of the viewshed operation. In *Innovations in GIS*; Worboys, M.F., Ed.; Taylor and Francis: London, UK, 1994; pp. 161–175.
14. Oh, K. Visual threshold carrying capacity (VTCC) in urban landscape management: A case study of Seoul, Korea. *Landsc. Urban Plan.* **1998**, *39*, 283–294. [CrossRef]
15. Van der Hoeven, F.; Nijhuis, S. Developing Rotterdam's skyline. *CTBUH J.* **2012**, *2*, 32–37.
16. Czyńska, K.; Rubinowicz, P. Visual impact size method in planning tall buildings. In *Education for Research, Research for Creativity*; Słyk, J., Bezerra, L., Eds.; Faculty of Architecture, Warsaw University of Technology: Warsaw, Poland, 2016; pp. 169–174.
17. Sahami, H. *Preparation and Location*; Malek-e-Ashtar Industrial University: Tehran, Iran, 2010.
18. Forghani, A.; Akhundi, A.; Sharifyazdi, M. *Industry and Service Facility Location with a Practical Approach*; Sharif University Press: Tehran, Iran, 2009.
19. Azizi, M.M. Evaluation of physical-spatial effects of tower construction in Tehran (Farmanieh-Kamranieh). *Honar-Ha-Ye-Ziba* **1999**, *4–5*, 33–46.
20. Azizi, M.M.; Fallah, E. *Determining and Applying the Criteria for Locating Tall Buildings in Metropolitan Areas (Case Study: Shiraz)*; University of Tehran: Tehran, Iran, 2008.
21. Adeli, Z.; Sardare, A. Locating of Tall buildings in Qazvin City using the hierarchical process of G.I.S., AHP. In Proceedings of the Third Conference on Urban Planning and Management, Mashhad, Iran, 20–23 April 2011; pp. 1–12.
22. Sedaghati, H. Study and Analysis of Location Criteria for Tall Buildings in Cities, Tabriz City Residential Tower. In Proceedings of the National Conference on Urban Development, Sanandaj, Iran, 15–16 October 2011.
23. Tamošaitienė, J.; Šipalis, J.; Banaitis, A.; Gaudutis, E. Complex model for the assessment of the location of high-rise buildings in the city urban structure. *Int. J. Strat. Prop. Manag.* **2013**, *17*, 93–109. [CrossRef]
24. Aminzadeh Goharrizi, B.; Roshan, M.; Badr, S. A Comparative Analysis of Site Selection Methods for New Towns in Iran. *Sci. J. NAZAR Res. Cent. (NRC) Art Archit. Urban.* **2013**, *9*, 21–32.
25. Rahnama, M.R.; Razaghian, F. Locating of High-Rise Buildings with Emphasis on Smart Growth Theory. *Geogr. Plan. Space Q. J.* **2014**, *3*, 45–64.
26. Salehi, N.; Ghadiri, H. An analysis of the design criteria for Toronto high-rise buildings for use in Iran. In Proceedings of the First National Conference on the Landscape of Natanz in the Pattern of Islamic Architecture and Urban Planning on the Horizon of 1404, Natanz, Iran, 25 February 2014.

27. Zista Consulting Engineers. *Tehran High-Rise Buildings: Criteria and Location*; Urban Planning and Processing Company Publications: Tehran, Iran, 2004.

28. Anabestani, A.; Javanshiri, M.; Anabestani, Z. Adaptive comparison of multi-criteria decision-making methods in optimal locating of tall buildings (case study: Region 9 Mashhad Municipility). *J. Spat. Plan.* **2015**, *5*, 1–24.

29. Zareian, M. Tall buildings' social issues. In Proceedings of the Sixteenth Conference on Housing Development Policies in Iran (Economy Faculty of Tehran University), Tehran, Iran, 3–4 October 2016; pp. 3–11.

30. Saaty, T.L. *Theory and Applications of the Analytic Network Process: Decision Making Benefits, Opportunities, Costs and Risks*; RWS Publications: Pittsburgh, PA, USA, 2005; ISBN 1-888603-06-2.

31. Hamadan Municipality. Statistics and Technology Center. Available online: http://www.hamedan.ir/ (accessed on 25 November 2019).

32. Dimić, V.; Milošević, M.; Milošević, D.; Stević, D. Adjustable Model of Renewable Energy Projects for Sustainable Development: A Case Study of the Nišava District in Serbia. *Sustainability* **2018**, *10*, 775. [CrossRef]

33. Faroughi, M.; Karimimoshaver, M.; Aram, F.; Solgi, E.; Mosavi, A.; Nabipour, N.; Chau, K.-W. Computational modeling of land surface temperature using remote sensing data to investigate the spatial arrangement of buildings and energy consumption relationship. *Eng. Appl. Comput. Fluid Mech.* **2020**, *14*, 254–270. [CrossRef]

applied sciences

Article

Wireless Home Energy Management System with Smart Rule-Based Controller

Hussain Shareef [1], Eslam Al-Hassan [1,*] and Reza Sirjani [2]

1. Department of Electrical Engineering, United Arab Emirates University, P.O. Box 15551, Al-Ain 15258, UAE; shareef@uaeu.ac.ae
2. Department of Electrical and Electronic Engineering, Eastern Mediterranean University, 99628 Gazimagusa, Northern Cyprus, 10 Mersin, Turkey; reza.sirjani@emu.edu.tr
* Correspondence: 201050323@uaeu.ac.ae; Tel.: +971-508380566

Received: 7 April 2020; Accepted: 2 May 2020; Published: 30 June 2020

Abstract: Despite the increasing utilization of renewable energy resources, such as solar and wind energy, most residential buildings still rely on conventional energy supply by public utility services. Such utility services often use time-of-use energy pricing, which compels residential consumers to reduce their energy usage. This paper presents a wireless home energy management (HEM) system that enables the automatic control of home appliances to reduce energy consumption to assist such energy users. The system consists of multiple smart sockets that measure the energy that is consumed by the connected appliances and are capable of implementing on/off commands. The system includes other support components for supplying data to a central controller, which utilizes a rule-based HEM algorithm. The control rules were designed, such that the lifestyle of the user would be preserved while the energy consumption and daily energy cost were reduced. The experimental results showed that the central controller could effectively receive data and control multiple devices. The system was also found to afford significant reductions of 23.5 kWh and $2.898 in the total daily energy consumption and bill of the considered household setup, respectively. The proposed HEM system promises to be particularly useful for households with a high daily energy consumption.

Keywords: home energy management; Zigbee; smart socket; monitoring; appliances scheduling

---

1. Introduction

The world still mainly relies on non-renewable resources for residential energy supply through public utility services, despite the increasing integration of renewable energy sources in the electricity generation process. In this framework, residential consumers are compelled to minimize their energy usage in order to benefit from government incentives that encourage efficient public energy utilization by adjusting the time of energy usage and efficient devices, like LED lamps [1]. However, the main obstacle to the development of such energy management systems is the user inconvenience that results from the regulated energy use. Thus, many researches have attempted to develop home energy management (HEM) systems that minimize energy utilization while maintaining the highest possible level of user comfort.

A greedy iterative algorithm was used in [2] to schedule multiple user appliances in a smart grid system. The objective was to minimize the aggregated energy usage of all the consumers, mainly while using the day-ahead hourly energy price. However, the conclusions of the authors were only based on simulations and, thus, had no practical validation. In addition, a sufficient amount of constraints, such as user preference and comfort, were not properly incorporated into the optimization problem that minimizes energy consumption.

In [3], an artificial neural network was used to establish a convenient thermal environment in a domestic building by controlling the thermal loads. In a related work, a thermal model of a two-story house while using a novel gray box modeling approach was presented in [4]. With the special focus on thermal device models, a HEM system was developed, in which machine learning (ML) methods were used in [5] to train and optimize the energy consumption. However, the application of the utilized algorithm to many energy variables encountered in practice is not feasible. In [6] and [7], an appliance modeling software and optimization methods were used to consider the realistic operation of various home devices in the simulation of an HEM controller. The main objective of the controller was to reduce the load peaks and smoothen the power consumption trend. The utilized optimization algorithm considered the time of use pricing and outdoor temperature information, together with the preferences of the user. The employed appliance models were realistic for the development of an optimized HEM controller, according to the authors. Adika and Wang in [8] developed a smart demand-side energy management and scheduling system for households with photovoltaic (PV) systems. Similarly, different types of loads are intelligently scheduled in [9] in order to minimize the electricity usage bills when considering the maximal demand limit. The scheduler computed the expected load profile and PV generation of the user based on the weather forecast for a specific day of the week. The aggregated energy consumption of all the appliances was used to minimize the energy consumption by means of an optimized schedule that utilized a PV generator whenever possible [8]. However, this simulation-based work did not consider user comfort and user override constraints and actual PV output.

Similar to [5], in the simulation in [10], three types of loads were considered in the optimization problem of energy usage. The proposed controller received the outdoor temperature, energy cost, and pricing signal as inputs to the optimization problem. A convex cost function was used as the main objective function, with the maximization of user comfort. The use of a convex objective function might not be accurate from a practical perspective, because appliance-switching operations cause discontinuity in the energy consumption pattern. In [11], smart meter data consisting of the energy consumption and pricing signal from the supplier were used to simulate a HEM system. The system performed multi-objective optimization to minimize the daily electricity cost while maximizing the user convenience. The main shortcoming of the work was the use of limited constraints (only room and water temperatures) in the optimization. In a related effort, Moghaddam et al. [12] developed a programming model for solving the mixed-objective optimization problem of energy usage and user comfort. Instead of using smart meter data, the authors assumed that the required data could be acquired from users and local energy resources, such as PV systems. Although the simulation algorithm was used to decrease the energy consumption and user satisfaction, in practice the work did not consider manual bypass by the user [12]. Recently, reinforcement learning (RL), which is a Markov decision process based technique, has been applied in the EMS problems. Q-learning RL algorithm was applied in [13] to automatically schedule appliance tasks according to the demand response signals as well as customers' requests and evaluations. However, the Q-learning RL discards the current data sample after every update, leading to the inefficient use of data. A detailed review that discusses the demand response schemes, the communication protocols, and the scheduling controller techniques, which are all required for the intelligent home EMSs, can be found in [14]. In this work, the scheduling controller techniques can be classified into rule-based, artificial intelligence (AI)-based, and optimization-based techniques.

References [15,16] bridge the gap between the simulations and practical implementation of HEM systems. The authors in [15] developed a test bed for the scheduling of lighting loads attached to a smart plug. The control scheme was based on the power consumption information received at the controller terminal, which, in turn, actuates the on/off commands for the loads. However, the work did not consider other important loads, such as thermal loads, which are major energy consumers and user comfort factors in practical households [15,17] applied a binary backtracking search algorithm to optimally schedule and control various appliances using wireless smart sockets in order to reduce customer energy consumption during peak hours on real time basis.

As an advancement on previous works, [18] presented a comprehensive HEM algorithm that considered distributed generation and a battery storage system (BSS). The aim of the controller was the optimization of the utilization schedule through battery charging and discharge to minimize the daily energy cost of the household. The main contribution of the research was its consideration of a large number of home appliances and the use of real data to simulate multiple energy usage scenarios. A practical implementation of an energy management systems consisting of a home server and Zigbee-connected home appliances was also proposed in [19]. The server acquired weather forecast information from the Internet and used the data in order to estimate the required energy generation from connected distributed generators (DGs). A user interface was established in order to display the previous load profile and the generation history of the DGs, together with weather forecast information. However, the developed test bed could not be used to monitor the information required for the maintenance of user comfort, such as room temperature, thermal appliance set points, and water temperature in the electrical heater, which can also be used to improve the efficiency of an HEM system [19]. In the meantime, several companies have designed and sell home automation systems. For example, General Electric Co. [20], Honda [21], Samsung [22], Fibaro [23], and etc. designed a system that controls home appliances using smartphones. These systems offer strong support to customers by controlling and monitoring devices with smartphone applications. However, the main disadvantage of commercial home automation systems are that the details of their software and hardware are confidential and unavailable for academic research.

The above-mentioned works only considered appliances with low energy consumption and assumed that the management system did not significantly impact the daily bill. In some, many user comfort constraints were not considered, resulting in inefficiency of the developed systems and their inability to maintain user comfort. These issues were considered in the physical system developed in the present study through the use of support circuitries that provide online information regarding inputs related to user comfort and lifestyle, such as room temperature. The main contributions of the work can be summarized, as follows:

- Design and implementation of a complete home energy management system consisting of multiple smart sockets, a programmable air conditioner (AC) remote control, and room condition monitoring nodes for the acquisition of comfort data to be inputted to the main controller. All of these components are wirelessly connected by a Zigbee communication protocol, eliminating the need for complex wiring.
- Consideration of user comfort as the first priority of the energy management process. This makes it easier to integrate the system with the daily routine of the consumer without affecting their lifestyle.
- The development of a rule-based algorithm for reducing the energy consumption within a 24-h period on a repeatable basis.

2. Proposed HEM System

The HEM system developed in this paper consists of three major modules, namely the appliance monitoring and controlling circuitries, room condition monitoring circuitry, and main scheduling terminal. These three modules are connected by a ZigBee wireless communication protocol through XBee microcontrollers that are installed in the modules. The XBee microcontrollers are configured to the API 2 communication mode, which allows for the transmission of data as packets. One of the major advantages of using such packets is that address information it can be integrated into them, enabling the main controller to identify the module sending and receiving any data. However, the data should be coded as string bytes at the transmitting end and decoded as float values at the receiving end to fit into a transmission packet. Figure 1 shows an illustration of the proposed HEM system. Further details on each module are provided in the following subsections.

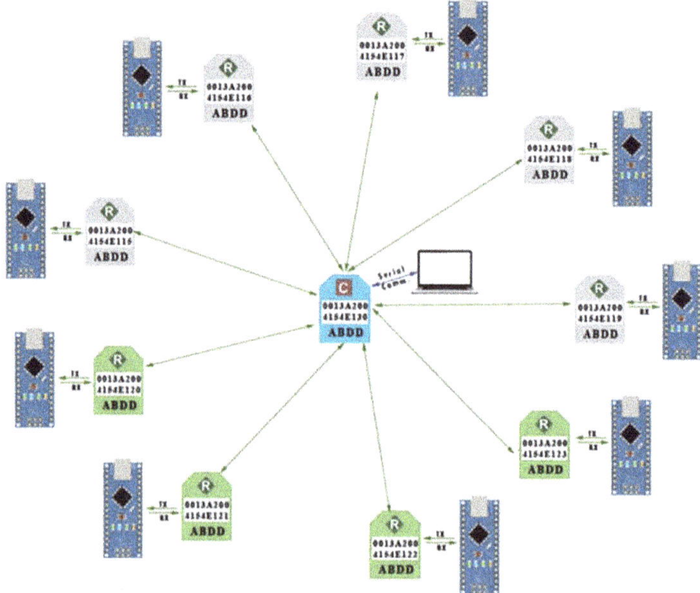

Figure 1. System configuration.

2.1. Scheduling Terminal

The configured XBee microcontroller is connected to a personal computer (PC) via a USB port and functions as a scheduling coordinator. A MATLAB program that is supported by a graphical user interface (GUI) on the main PC enables the user to input the desired comfort levels in terms of the room temperature, fridge temperature, water heater temperature, and illuminance ranges for the minimum, maximum, and normal pricing hours, respectively. The main scheduler also fetches the day-ahead pricing from the website of the utility service that are specified by the user. The day-ahead pricing trends used in the present study were obtained from [24]. The control commands are generated based on information predefined by the user, together with the power consumption information obtained from the smart sockets that are attached to the appliances, the heater outlet water temperature, and the room condition and occupancy information. The commands are in the form of binary decisions sent to specific smart plugs to turn on/off the connected appliances; dimming commands sent to Zigbee-connected dimmers to reduce/increase the light intensity of dimmable LEDs; and, temperature settings sent to Zigbee-connected infra-red (IR) remote controls to set ACs to specific temperatures. Figure 2 shows a screenshot of the scheduling coordinator's GUI.

The user is able to manually send commands to the house appliances, to adjust the desired comfort level or the power consumption of a specific appliance or to override an automatic scheduling function of a connected device, as can be seen from Figure 2. It can also be clearly seen that the GUI significantly assists the user in making their decisions by graphically showing the power consumption of any specified appliance. The user can also graphically examine the hourly pricing variation and identify the minimum and maximum pricing hours, which is beneficial to decision making.

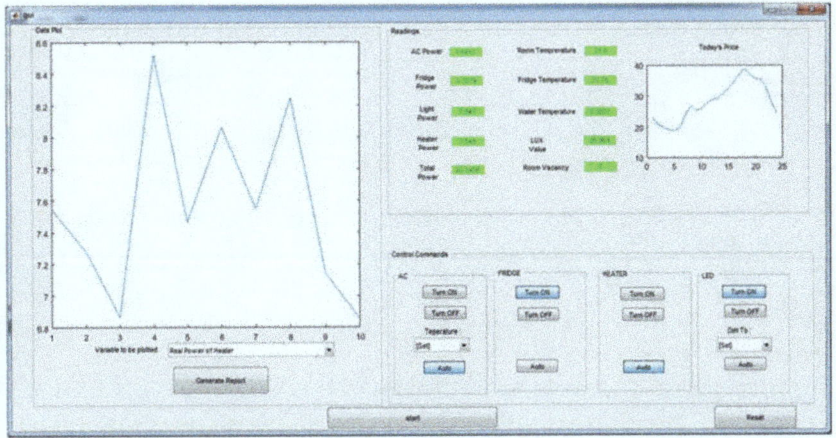

Figure 2. Graphical user interface of the proposed energy management system.

2.2. Appliance Monitoring and Controlling Modules

Many circuitries are used to facilitate the monitoring and control of the connected appliances. One circuitry monitors the power consumption and turning on/off of the connected appliances based on the commands that are received from the main controller/scheduler. It is referred to as the smart plug in the rest of this paper. The circuitries of the Zigbee-connected IR remote controls and LED light dimmers have more options for controlling the ACs and LED lights. The room condition monitoring circuits are also used to monitor the necessary inputs to the scheduling algorithm, such as the room temperature, humidity, light intensity (LUX), and CO_2 level, as well as the fridge and water heater temperatures. More details regarding these circuitries are explained in the following subsections.

2.2.1. Smart Plug/Socket

The smart plug/socket is able to monitor the power consumption of a connected appliance with a rating of up to 13 A, and turn it on/off using an integrated relay. The plug/socket determines and processes the single-phase power line voltage and current of the connected device, and then sends the captured data to a connected master node (when used) or the controller. The plug/socket uses the aforementioned wireless communication module and stores raw data provided by the microcontroller. Figure 3 shows the components (left) that were used to develop the plug/socket prototype (right).

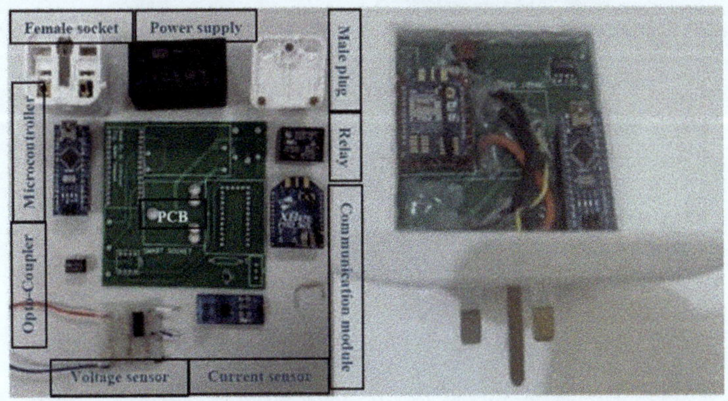

Figure 3. Smart plug/socket: components and prototype.

A printed circuit board (PCB) is used for most of the wired connections, with copper traces being used to achieve compactness and separate the high- and low-voltage connections. A two-sided PCB is used for high-voltage connection at the bottom of the system, and low-voltage connection at the top. The system components are soldered to the PCB to complete the hardware, which is placed inside an isolated box. The female socket side is used for appliance connection, while the male plug side has three pins that are inserted into the wall power socket. Interested readers can find the details of smart socket construction and software used to develop it in [25].

The following five steps can summarize the operation principle of the smart socket:

1. The male side of the smart plug is connected to the normal electricity socket to power up the internal components and prepare the smart socket for the connection of an appliance.
2. When an appliance is connected to the smart socket, the sensing modules acquire 100 readings of current and voltage and transmit the data to the microcontroller.
3. Based on the received data, the microcontroller calculates the real power, root-mean-square values of the current and voltage, complex power, and power factor.
4. The microcontroller transmits the calculated data to the main scheduling terminal at a time interval of 2 s.
5. The microcontroller receives and checks feedback data from the scheduling terminal module. If a command is received, the microcontroller sends the appropriate actuating signal to the relay of the smart socket to turn on/off the connected appliance.

2.2.2. Zigbee-Connected IR Remote Control

A learner IR remote control with Zigbee connectivity is employed for flexibility in controlling the AC units. The remote control uses an IR receiver diode to learn the preprogrammed IR commands from the original AC remote control. An ordinary IR transmitter is used to actuate the commands received from the scheduling controller terminal. Figure 4 shows the remote control circuitry.

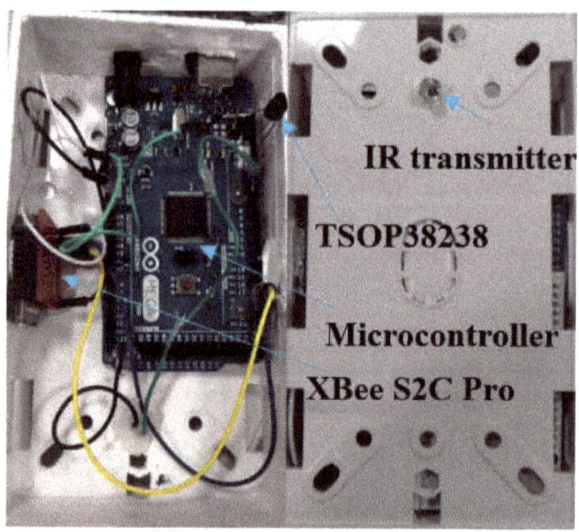

Figure 4. Zigbee-connected infra-red (IR) learner remote control.

The remote control is capable of learning the IR digital code patterns transmitted by the original AC remote control while using a TSOP decoding diode and an Arduino Mega microcontroller. The microcontroller saves the patterns as digital commands, together with various temperature settings in its internal memory. The microcontroller transmits one of the earlier-learned IR codes through the IR transmitter when a remote command is received from the scheduling controller terminal via the integrated Zigbee communication module. Thus, the Zigbee connectivity of the remote control allows for the user to control the AC units through remote commands from the scheduler algorithm, or by manual commands via the user interface. The advantage of using this Zigbee based IR report is that the controlling the high power AC units do not require contactors with high current rating, as used in [17]. The proposed circuit can be duplicated to control multiple AC units. Figure 5 illustrates the operation of the remote control.

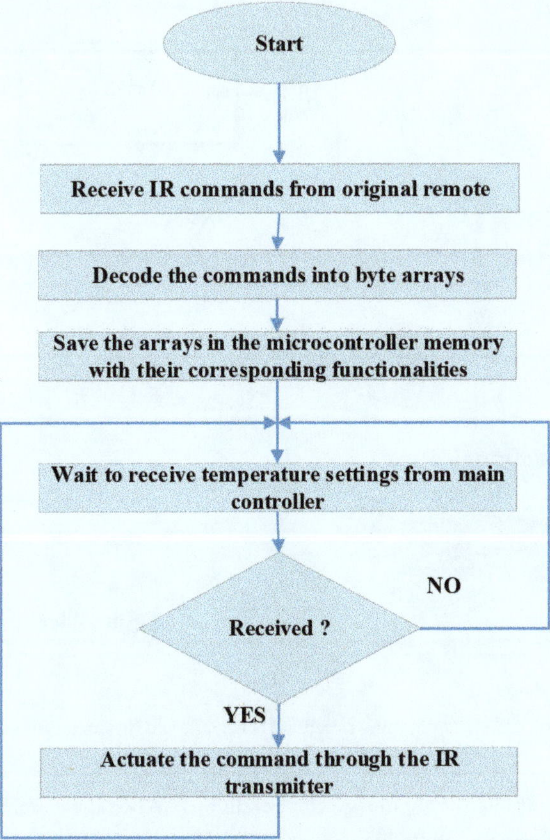

Figure 5. Operation of the IR remote control.

2.2.3. Zigbee-Connected LED Light Dimmer

Lights are the primary electrical appliances used in residential buildings and many householders neglect turning off or diming the lights when the natural lighting is adequate or a room is unoccupied. Recently, many dimmable drivers are flooding the market, offering each solutions to control individual light emitting diode (LED) lamps. Figure 6a shows such a driver circuit with PT4115. The dimming can be performed while using variable DC voltage by adjusting a variable resistor or PWM signal, for example, from a Arduino micro controller. A logic level below 0.3 V at DIM forces PT4115 (Hotchip Technology, Baoan, Shenzhen, China, 2015) to turn off the LED and the logic level at DIM must be at least 2.5 V to turn on the full LED current. The frequency of PWM dimming ranges from 100 Hz to more than 20 kHz.

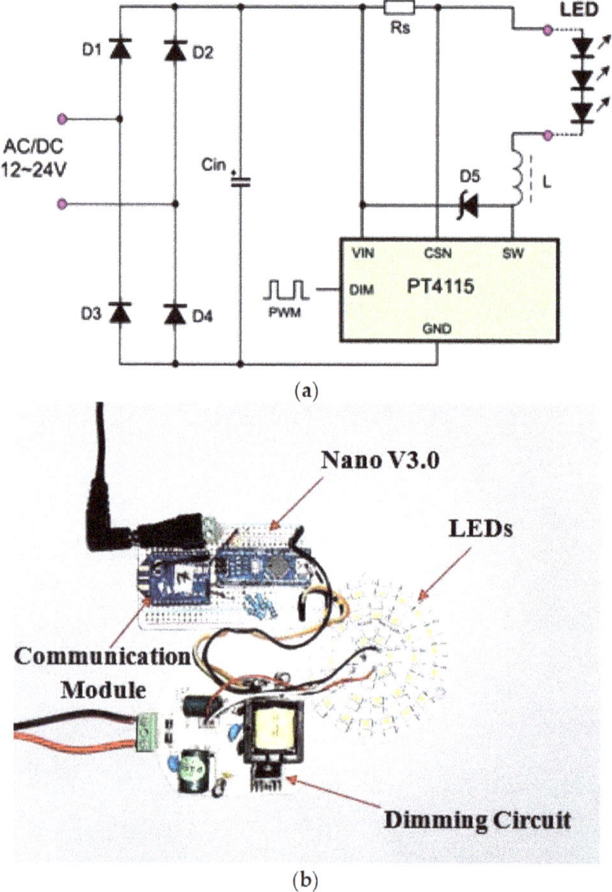

Figure 6. LED light dimming circuit: (**a**) Schematic and (**b**) Circuit connections.

The Zigbee-connected dimmer offers options for reducing the energy consumption of LED lights by decreasing the light intensity under specific conditions. When the communication module in the dimmer circuit receives a dimming command, it passes the command to the microcontroller for processing, like in the smart socket and IR remote. The microcontroller then generates a sequence of pulses that adjusts the light intensity to the level that is indicated by the remote command. Figure 6b shows the implementation of dimming circuit.

2.2.4. Room Condition Monitoring Circuit

The proposed HEM system includes a circuit for monitoring and transmitting data on the ambient conditions, such as the room temperature, humidity, illuminance, and CO_2 concentration. The circuit is also capable of detecting motion in the room through a motion sensor and is thus able to provide the main controller with the inputs required by the decision-making algorithm. Figure 7 shows the circuit and employed sensors.

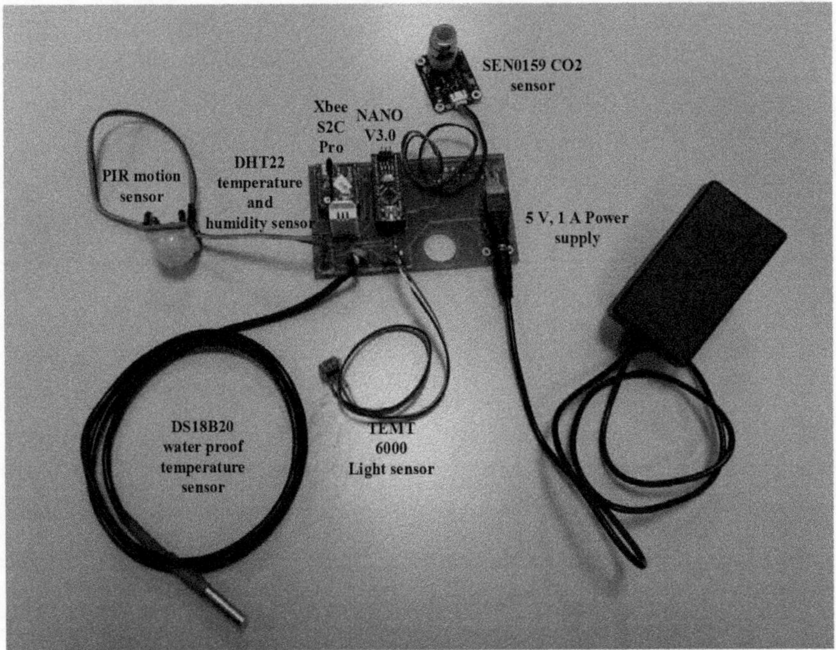

Figure 7. Room condition monitoring circuit.

Each sensor in Figure 7 has a specific purpose. For example, the room temperature and humidity are measured by the DHT22 (STmicroelectronics, Shanghai, China, 2000) sensing module, which updates its reading information every 2 s. When the condition monitoring circuit is used to monitor the internal temperature of a refrigerator or the water temperature in a water heater, the waterproof DS18B20 (Maxim Integrated, San Jose, CA, USA, 2005) temperature sensor is employed. All of the recorded data are initially compared with readings from commercial temperature and humidity loggers to ensure that all of the sensing modules are producing accurate and reliable readings for input to the algorithm. Further, the TEMT 6000 (Vishay Americas, Greenwich, CT, USA, 2004) light sensing module is used to continuously log the light intensity within the room because the user controls the light intensity inside the room during the HEM process. Here again, the readings are validated by comparison with the measurements of a commercial ST-1309 LUX (ATP Instrumentation, Ashby-de-la-Zouch, Leicestershire, UK, 2018) meter.

Two other sensors are used to identify the occupancy status of the room, namely, a passive infra-red (PIR) motion detection sensor and an SEN 1059 CO_2 sensor (DF Robot, Pudong, Shanghai China, 2018). The information acquired by these sensors are used by the algorithm to automatically control the relevant appliances. The PIR motion detection sensor continuously scans the space in front of it within its range, and then sets its output to 1 or 0, depending on whether it detects motion or not, respectively. However, because there might be the situation in which a person is within the room, but without detectable motion, a CO_2 sensing module is used for augmentation. The SEN 1059 CO_2 sensor utilizes an inverse relationship between CO_2 concentration and voltage [26].

After the development of all the necessary hardware for providing the main controller with the needed inputs, a rule-based HEM algorithm was developed and experiments were performed in order to verify the reliability of the proposed HEM system.

3. Experimental Setup and HEM Algorithm

This section consists of two parts. The first part describes the setup of the experimental hardware, which included four of the most commonly used household appliances, while the second part describes the utilized energy management algorithm and its interaction with the hardware.

3.1. Hardware

Four home appliances were selected for the experimental demonstration of the proposed HEM system. Table 1 details the appliances and their specifications. Each of the appliances in Table 1 was attached to a smart socket to monitor their energy consumption and continuously transmit the information to the main controller in the scheduling terminal. The temperatures of the first three appliances were also monitored by the appropriate sensing modules that were integrated with the condition monitoring circuit, as described above.

Table 1. Appliance details and specifications.

Appliance	Model	Power Rating (W)
Refrigerator	Super General 035H	90
Air conditioner	NIKAI NPAC12512A4	1200
Water heater	Florence FWH-50-15A	1500
LED	V-tac-22w	22

3.1.1. Heater Setup

Figure 8 shows the setup for monitoring the water heater power consumption and temperatures. The heater was attached to a smart socket and two waterproof temperature sensors were plugged into the inlet and outlet water pipes of the heater.

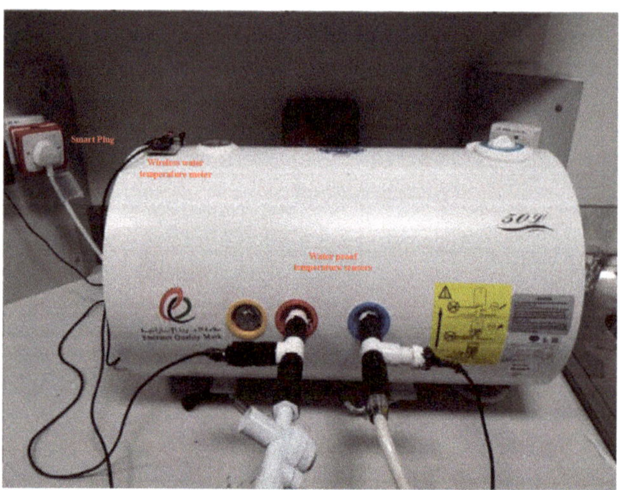

Figure 8. Experimental setup of the water heater.

The inlet and outlet water temperatures were monitored by waterproof sensors, and the temperature readings were acquired and transmitted by the microcontroller every 2 s. The smart socket to which the heater was connected measured power consumption. The heater could be turned on/off through remote commands received by the smart socket.

3.1.2. Refrigerator Setup

The working cycle of refrigerator or air conditioner refers to their operations based on which the refrigerant (in this case gas) heat exchange cycles change status, or cools down the medium. It starts at the compressor phase at which it pulls the warm refrigerant and increases its pressure and temperature. The refrigerant then travels to the condenser, where it goes through several fins with the use of a running fan that helps to release its heat to the outer environment, thus reducing its temperature. During this process, the refrigerant changes its state from the gaseous to a high-temperature liquid phase. The liquid is then passed to a valve that converts it into mist. This sudden drop of pressure results in a rapid cooling of the refrigerant, which is then passed to the evaporator coil located in front of a fan that circulates the chamber air, thus resulting in its cooling. Some heater are tuned on and the compressor is turned off during the deforest cycle to avoid ice formation and blocking the cool air flow in the chamber. This deforest cycles could last for 30 to 45 min. [26]. This clearly shows that the refrigerator could be used as a schedulable load if temperature is kept within the acceptable limits. In this study, the refrigerator was connected to another smart socket and a waterproof temperature sensor was placed inside it (Figure 9).

Figure 9. Experimental setup of the refrigerator.

The internal temperature information of the refrigerator and its power consumption were wirelessly transmitted to the main controller by Zigbee communication modules that were built into the condition monitoring circuitry and the smart plug. The refrigerator could also be turned on/off by remote commands that actuated the relay inside the smart socket. Note that the internal temperature is continuously monitored by monitoring circuit described in Section 2.2.4.

3.1.3. AC Setup

Unlike the cases of the two preceding appliances, the smart socket to which the AC was connected only monitored and transmitted the power consumption to the main controller, as shown in Figure 10. The room temperature determined the cooling performance, which was measured by a DHT22 temperature sensing module.

The AC unit was controlled through a Zigbee-connected IR remote control. All of the IR patterns that the remote controlled learned from the ordinary AC remote were stored in the memory of the mega microcontroller inside the former. The microcontroller actuated the command by releasing the corresponding IR pattern through an IR-transmitting LED bulb on the microcontroller output pin when a remote command was received by the Zigbee-connected remote control. Normally, the remote control could turn the AC unit on/off and set the AC thermostat to a specific temperature.

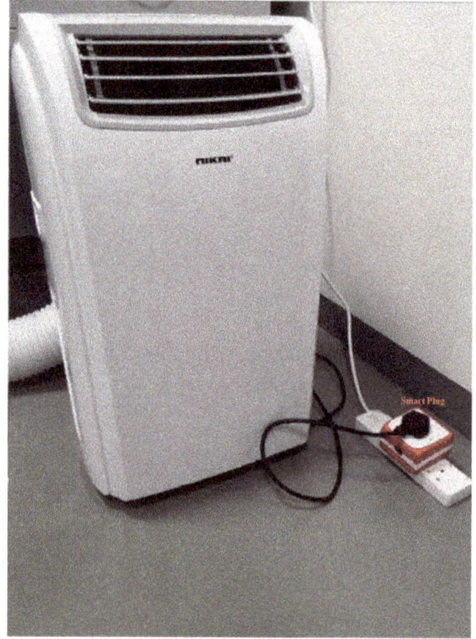

Figure 10. Experimental setup of the air conditioner (AC).

3.1.4. LED Light Setup

Similar to the case of the AC, power monitoring was the only function of the smart socket to which the LED light was connected. This was because the control task was assigned to the above-mentioned light dimming circuit. Figure 11 shows the connections for the LED light energy management. The control function of the dimming circuit was based on pulse width modulation (PWM). In the case of multiple LED light control, individual Zigbee based PWM control or single PWM generator with multiple dimmable LED driver circuits could be developed.

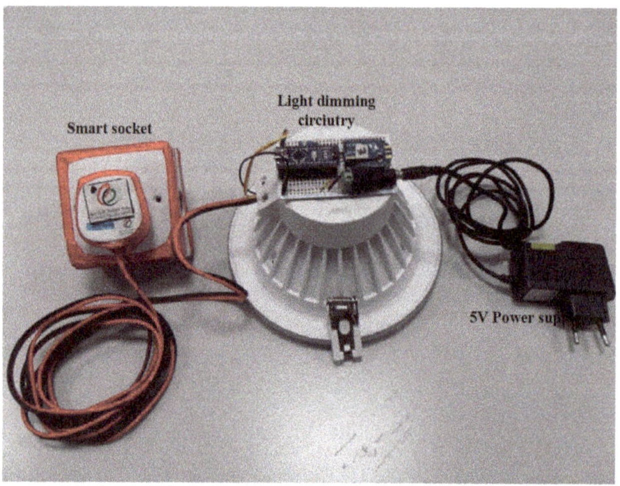

Figure 11. Experimental setup of the LED light.

Hence, when the dimming circuit received a remote command from the central controller through the Zigbee connection, the microcontroller in the dimmer analyzed the command and then actuated it by sending a PWM pattern to the connection pin of the LED driver. The LED lightning level could be varied in steps of 25% between 0% (off) and 100% (fully on).

3.1.5. Room Occupancy

Much energy is wasted through the operation of appliances in a room while no one is inside. AC units and lights should particularly be turned off when a room is unoccupied. This is the reason for employing CO_2 and motion sensors in the proposed HEM system. The use of the two sensors affords a two-fold determination of the occupancy of a room. The PIR motion sensor sends a value of 1 when motion is detected inside the room, which indicates that the appliances should be kept on for an additional 5 min. This time duration is used to address the problem that might arise from someone being in the room but without a detected motion, possibly due to an outrage of the PIR sensor trigger. During the 5-min. time gap, the CO_2 sensor reads the concentration of CO_2 in the room. If a continuous increase is detected, the central controller would maintain the occupancy status of 1, regardless of the motion sensor detection. Conversely, the central controller would wait for the next communication from the motion sensor if there is a decline in the CO_2 concentration. If it changes to 0, a vacant room decision would be taken.

An increase in CO_2 inside the room is detected by a decrease in the voltage reading of the sensor. In the present study, the CO_2 sensor was placed in the experimental room and its readings were monitored during different known occupancy statuses of the room to set an appropriate voltage threshold. A reading above the threshold indicated vacancy, and vice versa. The flowchart presented in Figure 12 describes the vacancy identification process. Figure 13 shows the CO_2 sensor readings during different occupancy statuses of the room over a 24-h period. When the CO_2 concentration in the room settles at a certain level, it is always below the set threshold, whereas the occupancy of the room always takes it above the threshold.

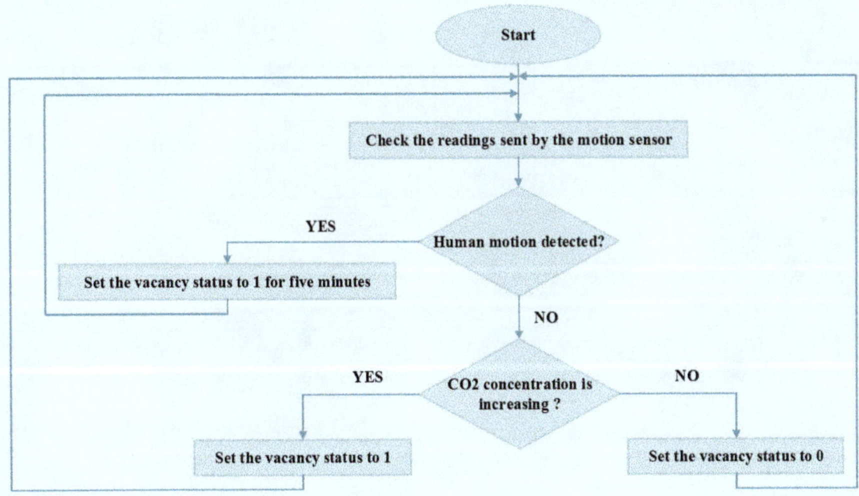

Figure 12. Room vacancy identification process.

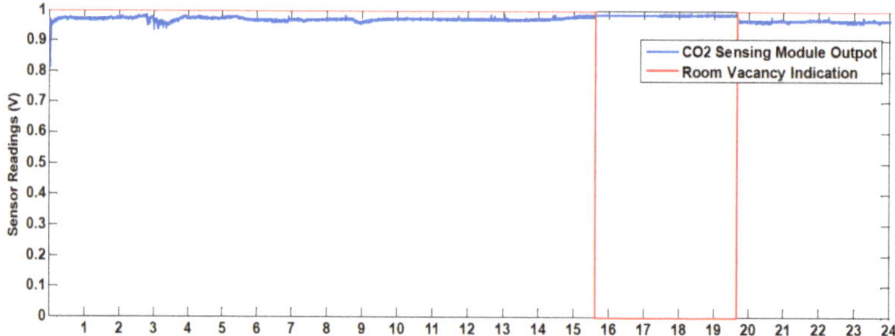

Figure 13. CO2 sensor reading with respect to room occupancy status over 24 h.

3.2. Home Appliance Usage Preference Survey

This survey aims to identify the categories of appliances that are used by the respondents, the features of household appliances, power consumption, customer behavior, and factors that can affect home electricity intake. A data collection forms were prepared on the SurveyMonkey® online platform to inquire about the different perspectives of the involved 50 participants in order achieve these objectives. The survey forms were designed with multiple choice, and Likert-type scales. The participants of the survey comprise people living in apartment building (51%), families living in landed houses (30.6%), and university students living in dormitories (18.4%) from Al Ain, UAE. The ages of the respondents ranges between 18 and 45 years. It should be noted that 96% of participants use the AC for more than 2 h per day. This could be mailing due to the hot UAE climate. Figure 14 shows some of the most important information obtained from this survey to determine the set points of the appliances and full list of questions can be seen in Appendix A.

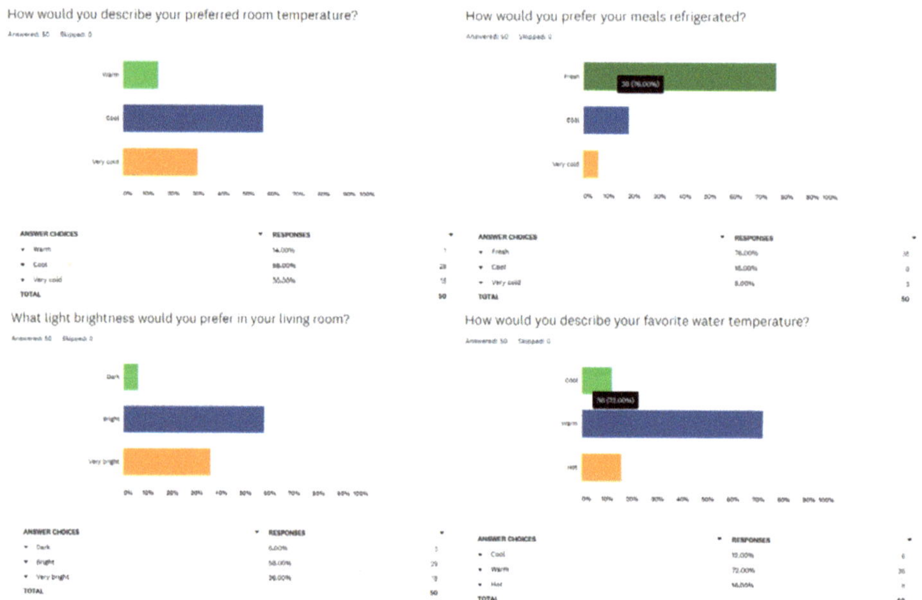

Figure 14. Results of survey questions related to thresholds.

3.3. Energy Management Algorithm

Based on the results of the survey in the previous section, Table 2 was developed to set the thresholds for the rules of the algorithm, after defining the terms in the answers and relating them with temperature and lux values [27,28]. A particular answer was chosen by the majority of the participants for each question, as can be observed from Table 2.

Table 2. Household appliance user preference survey.

Question	Choice 1	Choice 2	Choice 3
How would you describe your preferred room temperature?	Warm (25–30 °C) 7 users	Cool (20–25 °C) 28 users	Very cold (17–20 °C) 15 users
How would you prefer your groceries meals and drinks refrigerated?	Fresh (5–7 °C) 38 users	Cool (3–5 °C) 9 users	Very cold (1–3 °C) 3 users
How would you describe your favorite water temperature for bathing?	Cool (25–35 °C) 6 users	Warm (35–40 °C) 36 users	Hot (>40 °C) 8 users
What light brightness would you prefer in your living room?	Dark (<200 lux) 3 users	Bright (200–350 lux) 29 users	Very Bright (>350 lux) 18 users

A rule-based algorithm was developed to manage the usage of the household appliances discussed above while using the proposed HEM system. The algorithm operates with preset user preferences and day-ahead electricity pricing signals, as well as inputs from the different circuitries of the HEM system, enabling the most convenient and energy-efficient usage of the household appliances. This information was gained from a questionnaire in the previous section that was conducted on a representative sample of customers [29].

The predominant answers from the survey were used as the starting point for formulating the appliance schedule and light dimming rules. The algorithm also utilized online information from the room condition monitoring circuitries regarding the variations of the illuminance, room temperature, water temperature, and refrigerator temperature, as well as the variation of the time-of-use (TOU) pricing, which was obtained from a prediction algorithm that was embedded in the main controller. Thus, the room condition parameters were sent wirelessly to the main controller. The utilized circuitries, as described above, enabled measurement of the illuminance (L), internal refrigerator temperature (TREF), room temperature (TRoom), water temperature (TW), and vacancy state of the room (Vs.).

The algorithm used the preset user comfort ranges for each appliance, or the immediate previous measurements, as detailed in Equation (1).

$$
\begin{aligned}
T_{REF\_min} &\leq T_{REF\_t} \leq T_{REF\_max} \\
T_{ROOM\_min} &\leq T_{ROOM\_t} \leq T_{ROOM\_max} \\
LUX_{min} &\leq LUX_{ROOM\_t} \leq LUX_{max} \\
T_{W\_min} &\leq T_{W\_t} \leq T_{W\_max}
\end{aligned}
\tag{1}
$$

where $TREF\_t$ is the online value of the internal refrigerator temperature; $TROOM\_t$, $LUXROOM\_t$, and TW_t are the present room temperature, present illuminance, and present water temperature, respectively; and, min and max indicate the minimum and maximum comfort values of the parameters set by the user. Table 3 gives the minimum and maximum preset values of each parameter used in the present study, based on the survey results presented in Table 2.

Table 3. User comfort limits.

Parameter	Minimum Value	Maximum Value
Refrigerator temperature (°C)	5	7
Room temperature (°C)	20	25
Illuminance (lux)	200	400
Water temperature (°C)	38	42

The algorithm is designed to control the appliances such that the different parameters can only assume values within the respective preset ranges in Table 3. Each appliance schedule is determined by a set of factors, including the TOU, power consumption, room occupancy, and desired performance of the appliance, which are determined by the different circuitries of the system. Equations (2)–(5) express the schedules (S) for the LED light, AC, refrigerator, and water heater, respectively.

$$S_{LIGHT} = \begin{bmatrix} 0\%, & LUX_{ROOM\_t} = 0 \Rightarrow \text{if } v = 0 \\ 90\%, & LUX_{min} \le LUX_{ROOM\_t} \le LUX_{max} \Rightarrow \text{if } v = 1 \,\&\, t = t_{nonpeak} \\ 100\%, & 0 \le LUX_{ROOM\_t} \le LUX_{min} \Rightarrow \text{if } v = 1 \,\&\, t = t_{nonpeak} \\ 50\%, & (LUX_{max} + LUX_{min})/2 \le LUX_{ROOM\_t} \Rightarrow \text{if } v = 1 \,\&\, t = t_{peak} \end{bmatrix} \tag{2}$$

$$S_{AC} = \begin{bmatrix} 0, & T_{ROOM\_min} \le T_{ROOM\_t} \le T_{ROOM\_max} \Rightarrow \text{if } V = 0 \\ 21, & T_{ROOM\_min} \le T_{ROOM\_t} \Rightarrow \text{if } V = 1 \,\&\, t = t_{min} \\ 23, & T_{ROOM\_min} \le T_{ROOM\_t} \le T_{ROOM\_max} \Rightarrow \text{if } V = 1 \,\&\, t = t_{nonpeak} \\ 25, & T_{ROOM\_t} \le T_{ROOM\_max} \Rightarrow \text{if } V = 1 \,\&\, t = t_{peak} \\ S_{AC\_t-1}, & T_{ROOM\_min} \le T_{ROOM\_t} \le T_{ROOM\_max} \Rightarrow \text{if } V = 1 \,\&\, t = t_{nonpeak} \end{bmatrix} \tag{3}$$

$$S_{REF} = \begin{bmatrix} 1, & T_{REF\_max} \le T_{REF\_t} \\ 0, & T_{REF\_t} < T_{REF\_min} \\ 0, & T_{REF\_min} \le T_{REF\_t} \le T_{REF\_max} \,\&\, T_{ROOM\_min} \le T_{ROOM\_t} \Rightarrow \text{if } t = t_{peak} \\ S_{REF\_t-1}, & T_{REF\_min} \le T_{REF\_t} \le T_{REF\_max} \Rightarrow \text{if } t = t_{nonpeak} \end{bmatrix} \tag{4}$$

$$S_{Heater} = \begin{bmatrix} 1, & T_{Heater\_t} \le T_{Heater\_min} \\ 0, & T_{Heater\_max} \le T_{Heater\_t} \\ 0, & T_{Heater\_min} \le T_{Heater\_t} \le T_{Heater\_max} \Rightarrow \text{if } t = t_{peak} \\ S_{REF\_t-1}, & T_{Heater\_min} \le T_{Heater\_t} \le T_{Heater\_max} \Rightarrow \text{if } t = t_{nonpeak} \end{bmatrix} \tag{5}$$

As an example of the use of the rules in the above equations, the algorithm dims the light to 50% when the room is vacant, with illuminance remaining within the comfort range during the maximum pricing hours to ensure that the energy consumption is reduced without affecting user comfort. Additionally, the AC is allowed to cool the room to the lowest allowed temperature during the minimum pricing time, and to increase the compressor off period outside the minimum pricing time. In another example regarding the use of the refrigerator equation, the refrigerator is turned off when the set minimum temperature is reached, regardless of the pricing signal. Further, the heater is turned on when the present water temperature is less than the set minimum. Equation (2) expresses other lighting levels for various conditions, while Equations (3)–(5) express other scheduling rules for the other appliances. Figure 15 shows a comprehensive flowchart of the energy management algorithm.

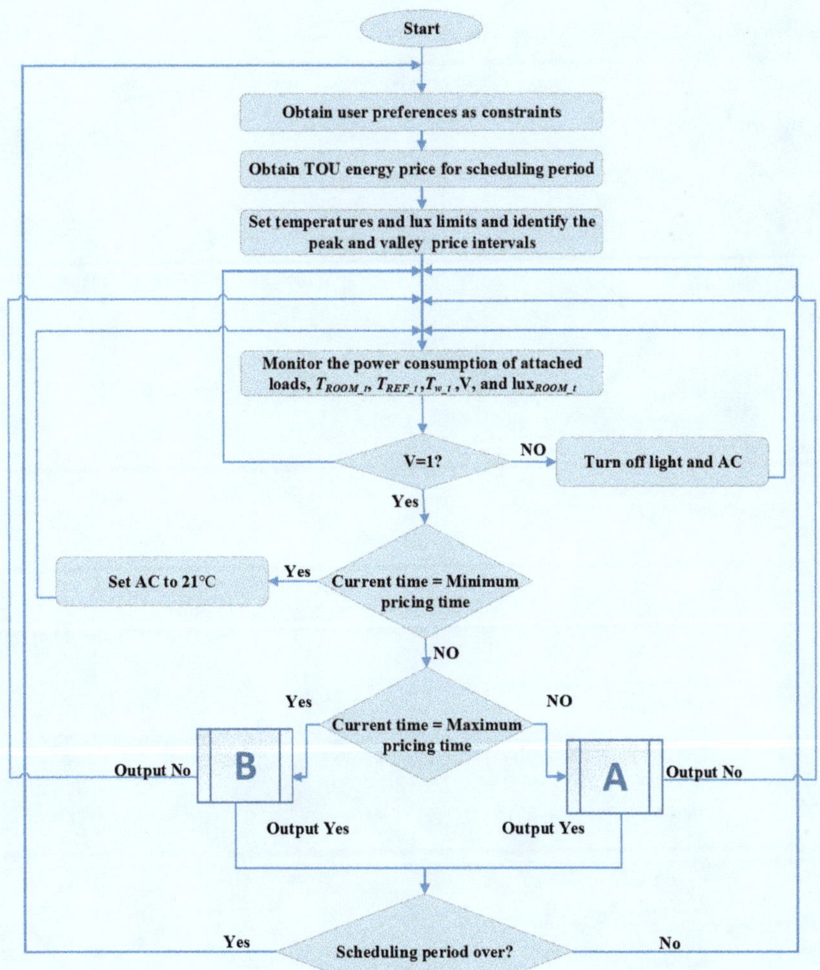

Figure 15. *Cont.*

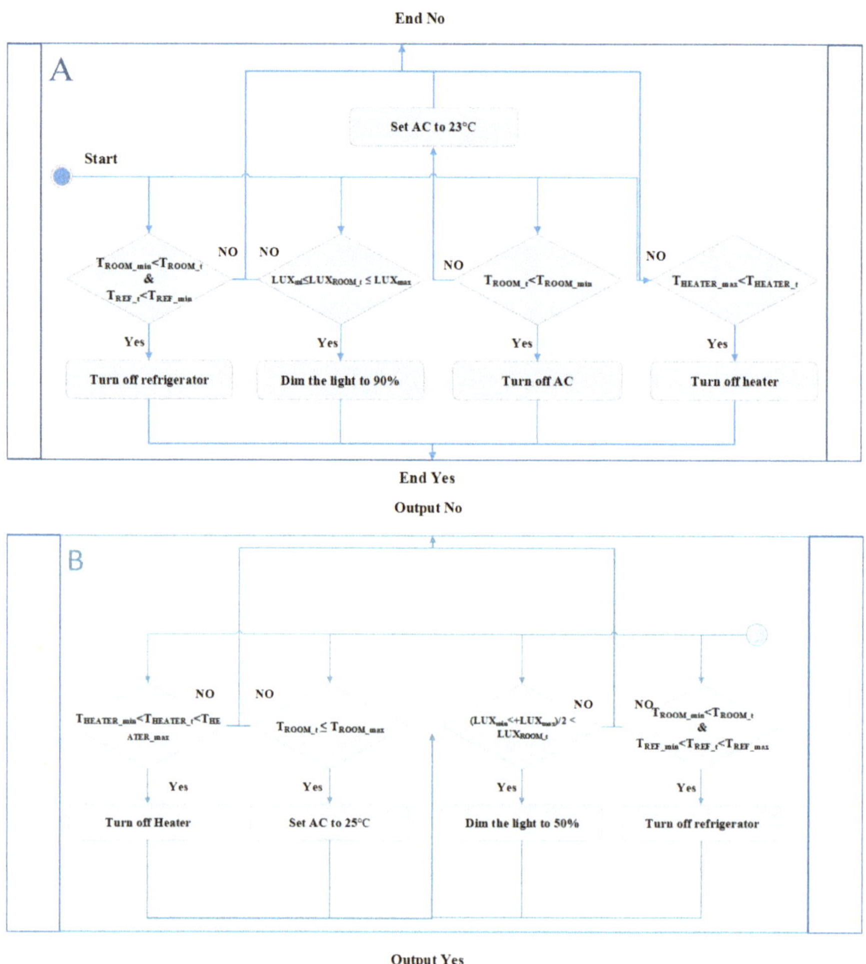

Figure 15. Complete flowchart of the energy management algorithm.

3.4. Experimental Methodology

Two 24-h experiments were performed to demonstrate the ability of the designed circuitries to monitor and control the electrical appliances after setting up the hardware and specifying the inputs and outputs of the energy management algorithm. The economic benefit of the HEM system was also analyzed. Firstly, the power consumptions of the appliances under normal utilization over the duration of a day were monitored and recorded. The temperature and illuminance were also recorded. The HEM system was then applied for another 24 h under exactly the same circumstances of the appliances, and similar recordings as in the previous experiment were obtained.

4. Results and Discussion

The power consumption, output, and daily energy consumption of each appliance under normal operation and when using the HEM system were compared. Figure 16 shows the electricity pricing signal during the use of the HEM system.

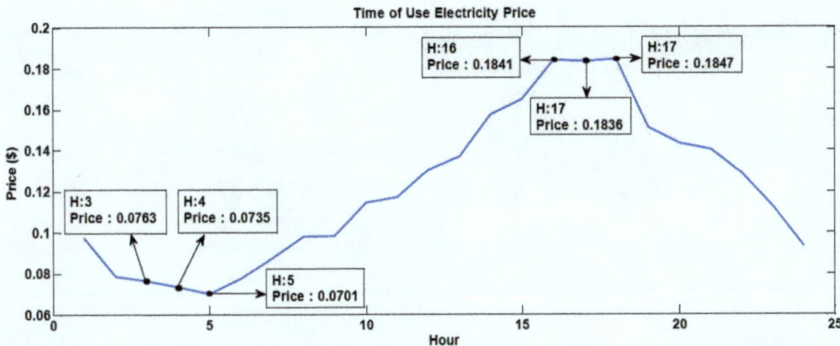

Figure 16. Electricity pricing signals during the use of the home energy management (HEM) system.

The duration of a day during the experiments comprised three main types of hours, namely, the minimum pricing hours (third, fourth, and fifth hours of the day), the maximum pricing hours (16th, 17th, and 18th hours), and the normal pricing hours (all other hours), as can be seen from Figure 16. The HEM algorithm was expected to manage the appliances in different ways during these three types of hours of the day.

4.1. LED Light

The LED light was turned on for 12 h of the day during the experiment, and then turned off for the other 12 h. However, the HEM algorithm could reduce the energy usage during the utilization period. Figure 17 compares the power consumption of the LED light without and with the use of the HEM system, as well as the corresponding illuminance variations.

In Figure 17, the first 12 h correspond to when the LED light was turned on. The light was fully on when the HEM system was not used, while the illuminance was regulated when the system was used. The algorithm turned off the light whenever the room was vacant, as can be observed from the figure. In addition, when the room was occupied, the light was dimmed to 90% of its illuminance capacity during normal electricity pricing hours, but it operated at full capacity during minimum pricing hours. The variation of the illuminance was within the preset preferences in Table 3 during all the scheduling periods, and below the maximum constraint by 10% during non-peak periods, as can be observed from Figure 17b. However, all of the constraints were removed during minimum pricing hours. Figure 18 shows the effects of the HEM algorithm on the daily energy consumption and electricity bill.

The energy and bill savings achieved by the occupancy principle of the HEM algorithm were remarkable, although there were no savings during the minimum pricing period, because the HEM algorithm allowed for the LED light to operate at full capacity at that time. Overall, the algorithm afforded energy and bill savings of 0.133 kWh and \$0.014, respectively, for the day.

4.2. AC Unit

As with the LED light, the operation of the AC unit was based on the room occupancy, as well as the room temperature. Figure 19 shows the power consumptions without and with the use of the HEM system and the corresponding temperature variations. It should be noted that, before the scheduling, the use do not turn off the AC unit even during unoccupied period due to hot climate in UAE. However, the user usually control the AC unit only by just adjusting thermostat set points manually. This is clearly seen in the power consumption pattern and temperature plot represented in the blue line in Figure 19.

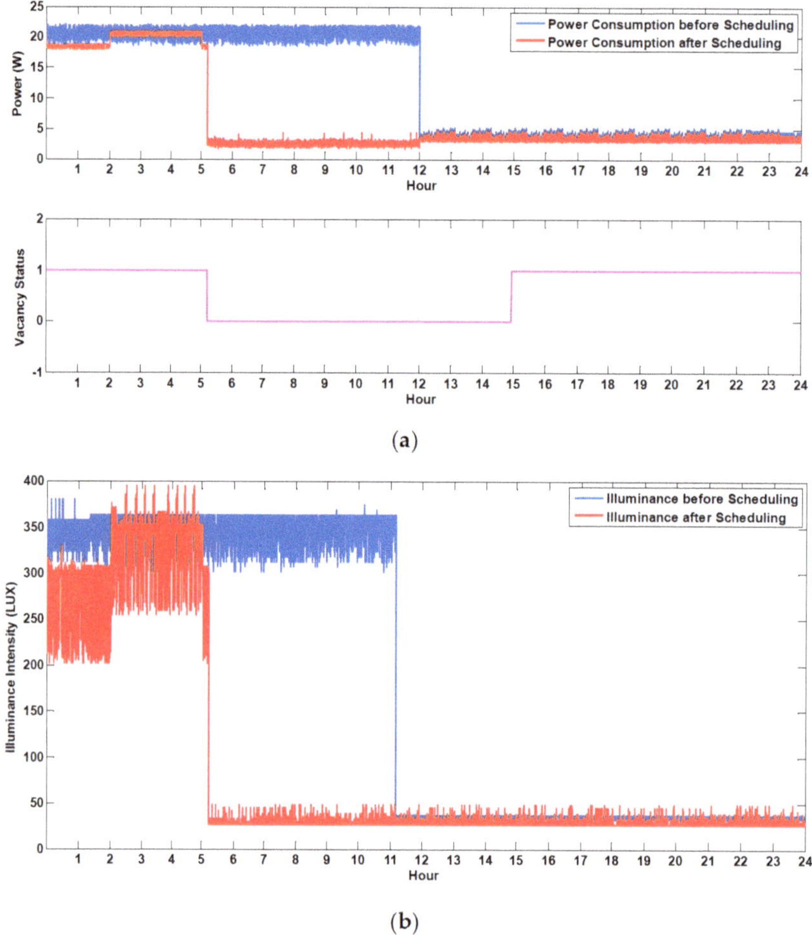

Figure 17. (**a**) Power consumptions of the LED light, room occupancy status and (**b**) illuminance variations without and with the use of the HEM system.

The AC unit was continuously operated until the set temperature was reached, and the compressor was then automatically turned off. Based on the set temperature limits, the unit subsequently alternated between the on and off status to save energy. Under normal energy pricing, the AC was continuously operated, except the consideration of the room occupancy status was manually turned off. However, the HEM algorithm did not allow for the AC to operate until the threshold temperature of the room was reached. Although the temperature threshold varies with the pricing regime (minimum, normal, or maximum pricing), the algorithm was always able to reduce the energy consumption.

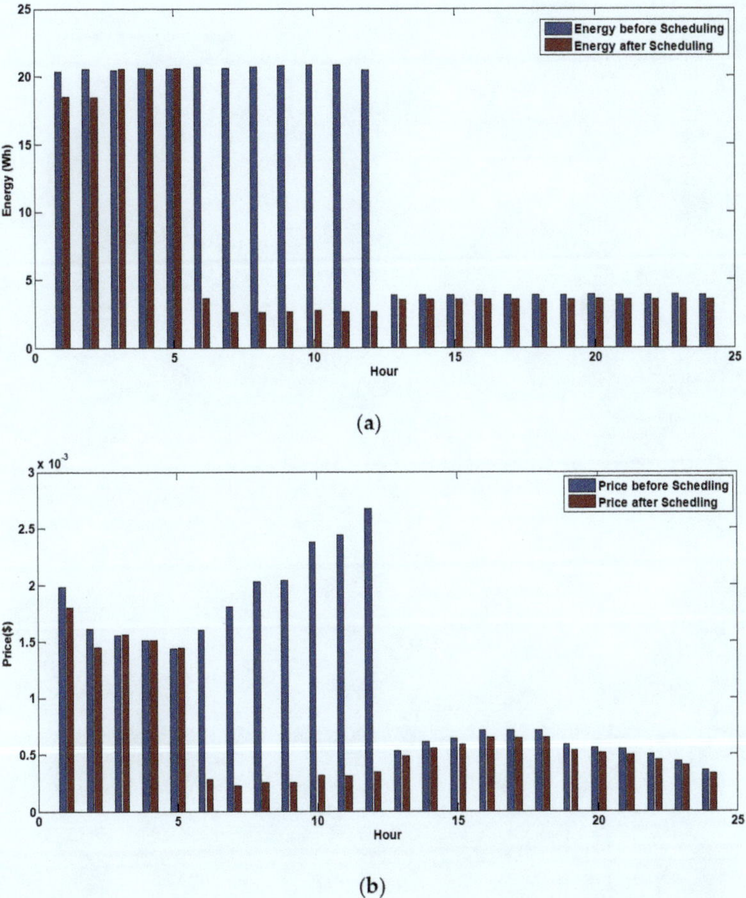

Figure 18. (**a**) Energy consumptions and (**b**) bills incurred by the LED light without and with the use of the HEM system.

During non-peak pricing hours, the AC was allowed to operate, provided that the room was occupied and the room temperature was within the preset limits. In addition, the algorithm controlled the AC to maintain the lowest possible temperature during low pricing hours, so it could be turned off during high pricing hours without affecting the user comfort. However, the AC was allowed to maintain the room temperature at the preset upper temperature during the maximum pricing hours, as can be observed from the third graph of Figure 18a. It was turned off more frequently when the room temperature was further below the preset maximum temperature. This can be observed from the noticeably different cycles of the AC compressor during the maximum pricing hours. Despite all of the applied constraints, the HEM algorithm was able to maintain the room temperature within the preset temperature range.

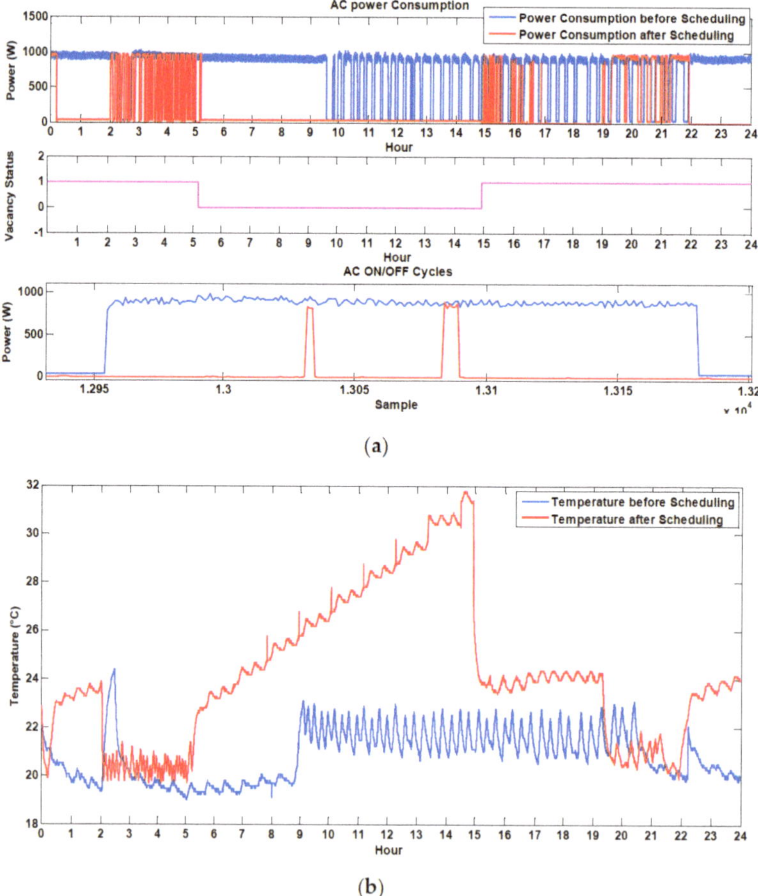

Figure 19. (**a**) Power consumptions of the AC, room occupancy statuses, on/off cycles of the AC and (**b**) room temperature variations without and with the use of the HEM system.

The room temperature during normal pricing hours was only subject to manual operation of the AC. However, this was dependent on the occupancy of the room when using the HEM system, with the temperature being allowed to rise above the maximum comfort level when the room was unoccupied. Moreover, the preset temperatures for the minimum and maximum pricing hours were the same as those for the normal hours. Figure 20 shows the energy consumption and bill trends of the AC unit.

The highest amount of energy was used during the minimum pricing hours, which was because the HEM algorithm maximized the AC use at these times, and reduced it during the maximum pricing hours, as can be observed from Figure 19. Overall, the HEM system significantly reduced the daily energy consumption and bill by 13.9 kWh and $1.64, respectively.

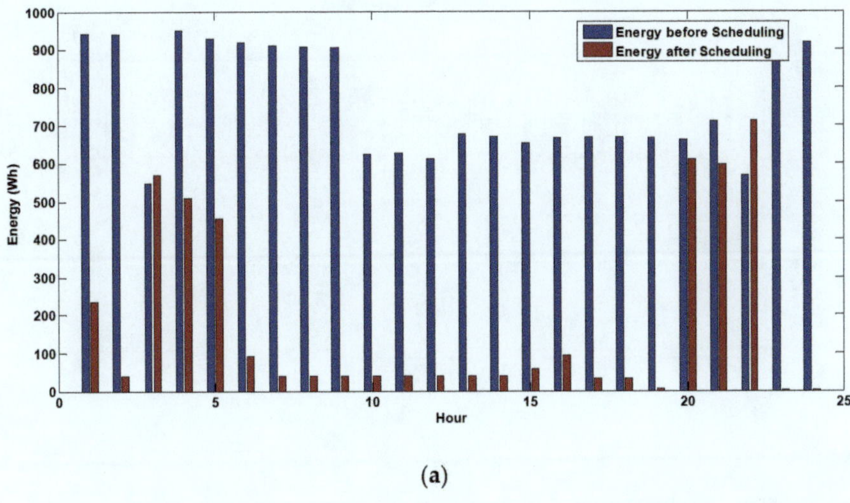

(a)

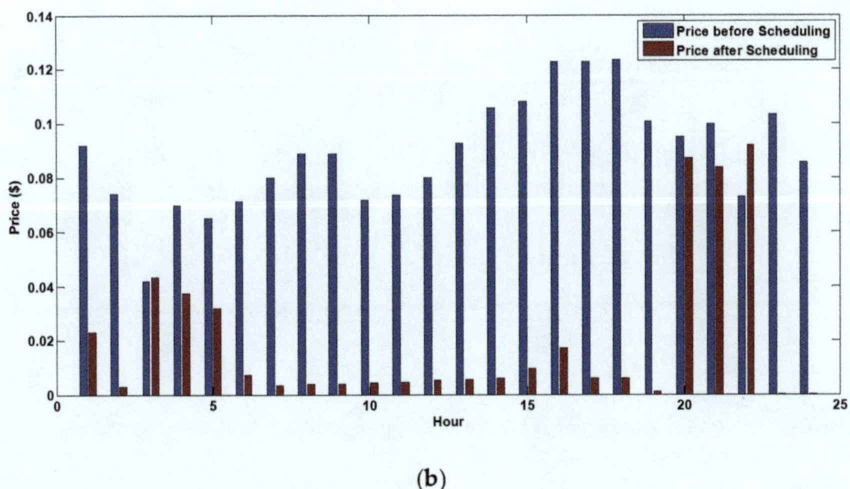

(b)

Figure 20. (**a**) Energy consumptions and (**b**) bills incurred by the AC unit without and with the use of the HEM system.

4.3. Refrigerator

Unlike the two previous home appliances, the use of the refrigerator does not depend on the room occupancy status. Rather, the scheduling of the refrigerator was based on the internal and surrounding temperatures. Figure 21 shows the effect of the HEM system use on the power consumption of the refrigerator and the internal temperature variation of the refrigerator.

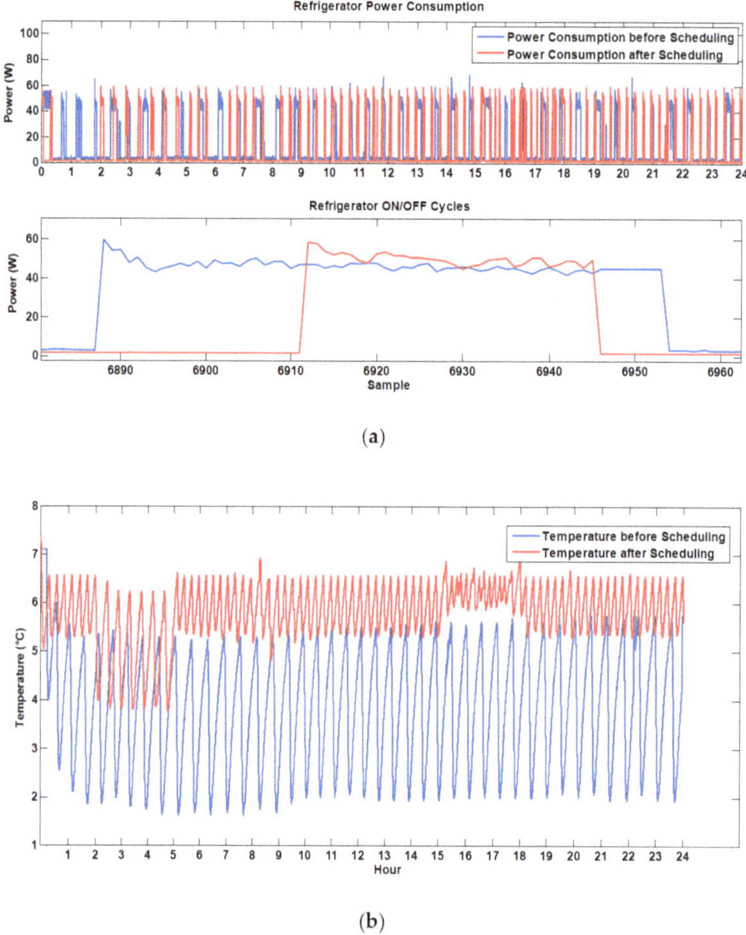

(a)

(b)

Figure 21. (**a**) Power consumptions and (**b**) internal temperatures of the refrigerator without and with the use of the HEM system.

The power consumption of the refrigerator was reduced by decreasing the time during which the refrigerator was switched on. The refrigerator was turned off during normal pricing hours when the lower preset temperature limit was reached, and more often during the maximum pricing time when the internal temperature was within the preset range. This can be clearly observed from the second graph of Figure 20b, where the on cycle while using the HEM system is narrower than that when not using the system during the maximum pricing hours. However, the HEM algorithm effectively managed the internal temperature within the desired range, regardless of the reduced power consumption and on time of the refrigerator. The benefit of the HEM algorithm can be assessed by comparison of the hourly energy consumptions and bills without and with the use of the proposed system, as illustrated in Figure 22.

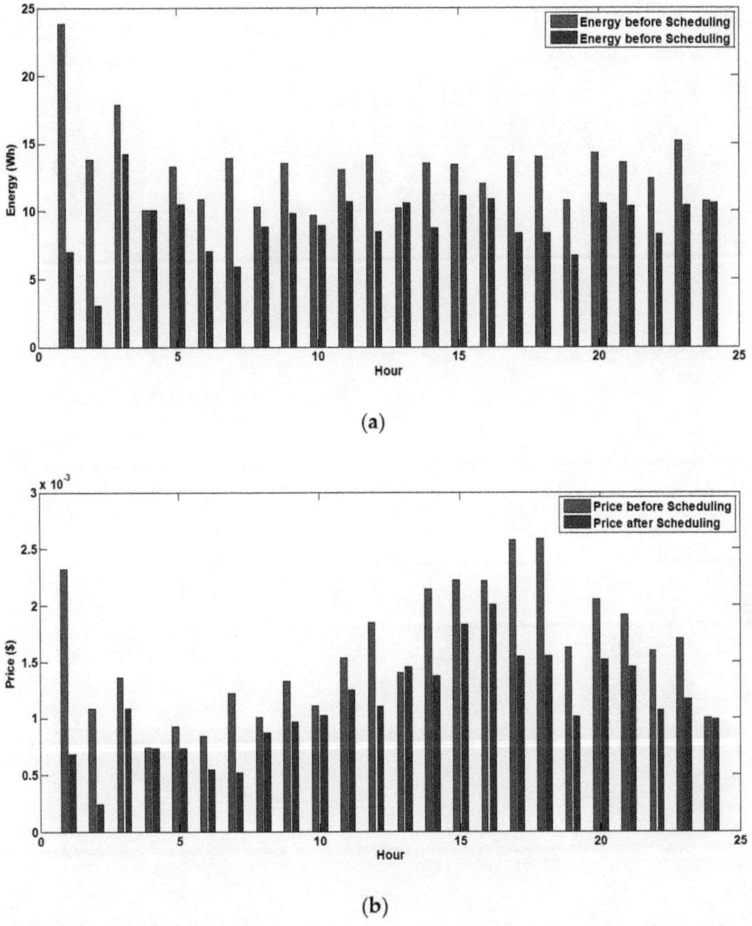

Figure 22. (**a**) Energy consumptions and (**b**) bills incurred by the refrigerator without and with the use of the HEM system.

The system afforded reductions during non-peak pricing hours although the energy consumptions and bills during minimum pricing hours without and with the use of the HEM system are similar. The savings were particularly substantial during maximum pricing hours. Overall, the daily energy and bill savings that were achieved by the system were 0.098 kWh and $0.0116, respectively.

4.4. Water Heater

Similar to the case of the refrigerator, the operation of the water heater is not dependent on the occupancy status of the room. However, greater scheduling flexibility was possible with the water heater, because it could better preserve its internal temperature as compared with the refrigerator when considering that a water heater tank is not frequently opened. Figure 23 shows the power consumptions and temperature variations of the heated water without and with the use of the HEM system.

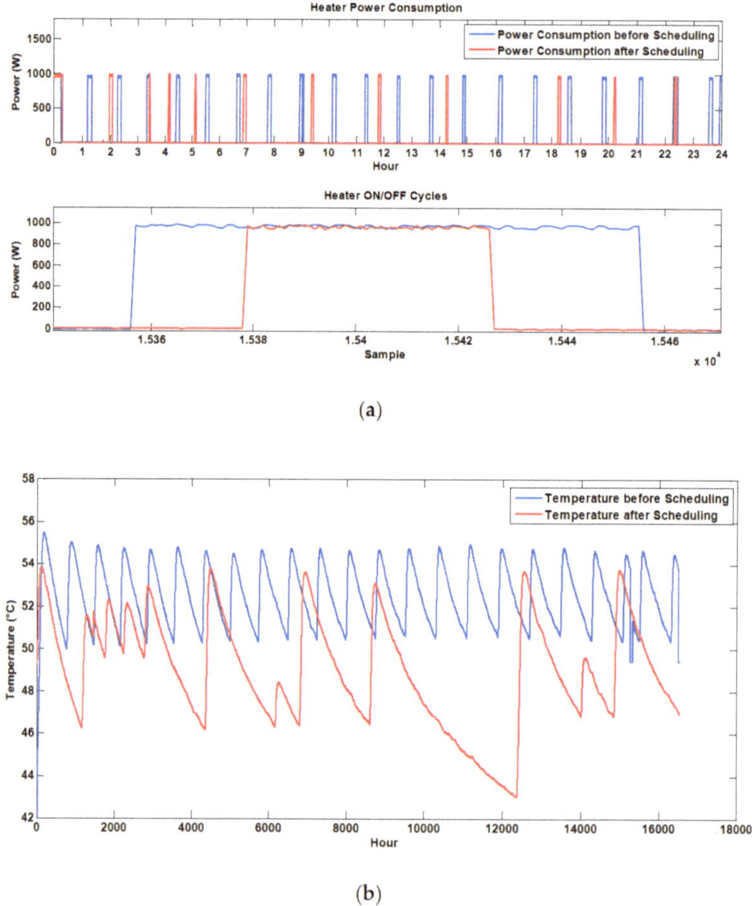

Figure 23. (**a**) Power consumptions and (**b**) heated water temperatures without and with the use of the HEM system.

There was considerable decrease in the on cycles of the water heater when using the HEM system, because the heater could be turned off when the temperature reached the middle of the preset range. This reduced the energy consumption while maintaining the water at a comfortable temperature. Figure 24 shows the water heater energy and bill savings achieved by the HEM system.

Daily energy and bill savings of 1.577 kWh and $0.187, respectively, were achieved by applying the HEM system to the water heater, especially during the maximum pricing hours.

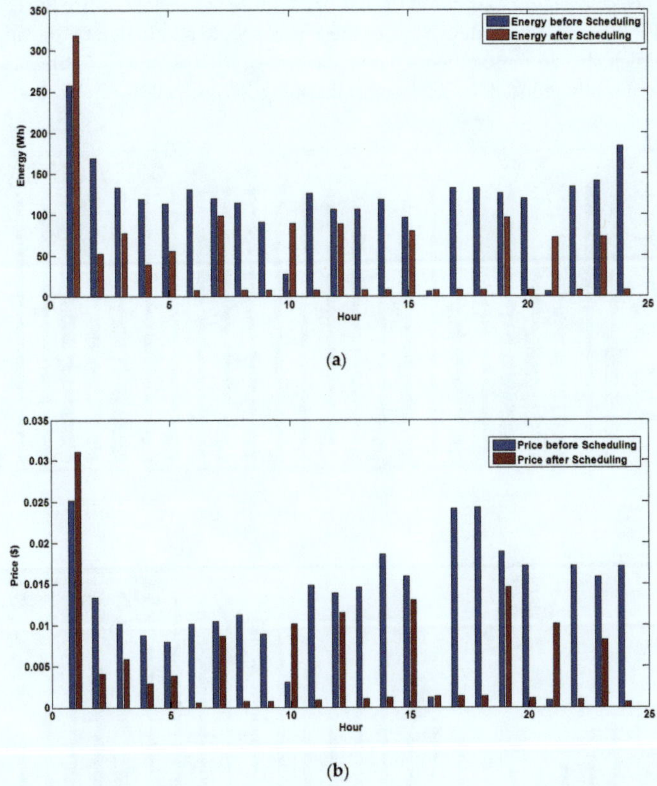

(a)

(b)

Figure 24. (**a**) Energy consumptions and (**b**) bills incurred by the water heater without and with the use of the HEM system.

4.5. Total Energy Consumption Reduction

Figure 25 shows the total power consumption trends for all of the appliances without and with the use of the proposed HEM system.

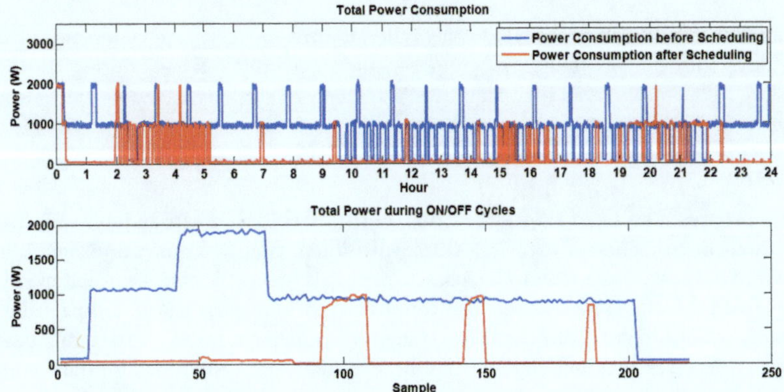

Figure 25. Total power consumptions of all the appliances without and with the use of the HEM system.

The many peaks in Figure 24 without the use of the HEM system are due to the concurrent use of all the appliances throughout the day. Most of these peaks were eliminated by the application of the HEM system, which also decreased the total on cycle time of the appliances. Ultimately, the use of the HEM system significantly reduced the total energy consumption and bill, as can be observed in Figure 26.

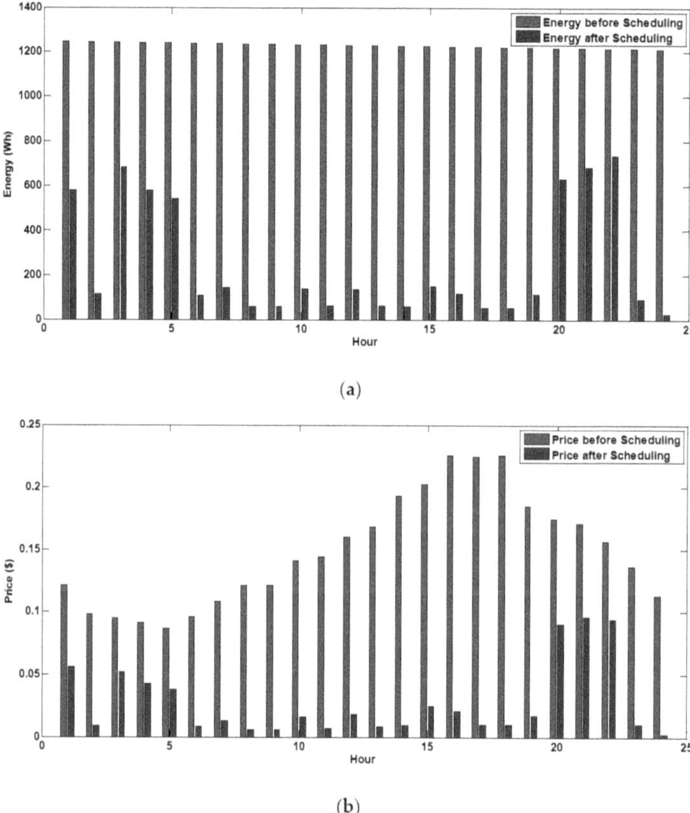

Figure 26. Total (**a**) energy consumptions and (**b**) bills without and with the use of the HEM system.

The application of the proposed HEM system alters the previous constant total energy consumption over the day, with the previous maximum consumption only retained during the minimum pricing hours, as can be observed from Figure 26. The proposed system enabled total daily energy and bill savings of 23.5 kWh and $2.898, respectively.

5. Conclusions

This paper presents a smart HEM system that consists of both hardware and software. The system utilizes various instruments for the wireless transmission of energy, temperature, and illuminance information, which the main controller uses to remotely generate and send commands to the actuating circuitries of the hardware. The commands include turn on/off, temperature control, and illuminance adjustment commands. The system software includes an effective rule-based HEM algorithm that receives real-time inputs from the various system circuitries for the control of the appliances in such a way that reduces their power consumption without affecting the lifestyle of the user. The results of the experimental implementation of the proposed system confirmed its control reliability, with reductions of 23.5 kW and $2.898 being achieved in the daily energy consumption and bill, respectively. In the future work, the hardware design could be improved by adding new supporting circuitries or other functionalities, such as detecting the number of persons in the room. Furthermore, adopting fuzzy or heuristic optimization based approaches could enhance the rule based scheduling algorithm. The proposed HEM system promises to be an asset for households with high energy consumptions.

Author Contributions: Conceptualization, H.S. and E.A.-H.; Methodology, H.S., E.H.; Software, E.A.-H.; Validation, H.S., E.A.-H. and R.S.; Formal analysis, H.S., E.A.-H. and R.S.; Resources, H.S.; Writing—Original draft preparation, E.A.-H., H.S.; Writing—Review and editing, R.S.; Visualization, H.S., E.A.-H. and R.S.; Supervision, H.S.; Project administration, H.S.; Funding acquisition, H.S. and R.S. All authors have read and agreed to the published version of the manuscript.

Funding: The research was funded by the United Arab Emirates University vide fund code 31N259-UPAR (2) 2016.

Conflicts of Interest: The authors declare no conflict of interest.

Appendix A

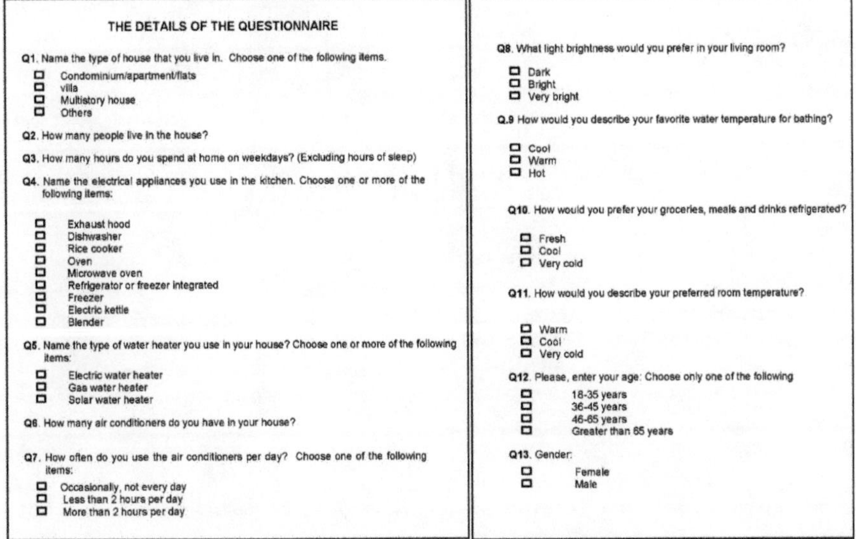

References

1. Yaghmaee, M.H.; Moghaddassian, M.; Leon-Garcia, A. Autonomous two-tier cloud-based demand side management approach with microgrid. *IEEE Trans. Ind. Inform.* **2017**, *13*, 1109–1120. [CrossRef]

2. Chavali, P.; Yang, P.; Nehorai, A. A distributed algorithm of appliance scheduling for home energy management system. *IEEE Trans. Smart Grid* **2014**, *5*, 282–290. [CrossRef]

3. Moon, J.W.; Kim, J. ANN-based thermal control models for residential buildings. *Build. Environ.* **2010**, *45*, 1612–1625. [CrossRef]

4. Cui, B.; Fan, C.; Munk, J.; Mao, N.; Xiao, F.; Dong, J.; Kuruganti, T. A hybrid building thermal modeling approach for predicting temperatures in typical, detached, two-story houses. *Appl. Energy* **2019**, *236*, 101–116. [CrossRef]

5. Zhang, D.; Li, S.; Sun, M.; O'Neill, Z. An optimal and learning-based demand response and home energy management system. *IEEE Trans. Smart Grid* **2016**, *7*, 1790–1801. [CrossRef]

6. Althaher, S.; Mancarella, P.; Mutale, J. Automated demand response from home energy management system under dynamic pricing and power and comfort constraints. *IEEE Trans. Smart Grid* **2015**, *6*, 1874–1883. [CrossRef]

7. Agamah, S.U.; Ekonomou, L. A Heuristic combinatorial optimization algorithm for load-leveling and peak demand reduction using energy storage systems. *Electr. Power Compon. Syst.* **2017**, *45*, 2093–2103. [CrossRef]

8. Adika, C.O.; Wang, L. Autonomous appliance scheduling for household energy management. *IEEE Trans. Smart Grid* **2014**, *5*, 673–682. [CrossRef]

9. Arun, S.L.; Selvan, M.P. Intelligent residential energy management system for dynamic demand response in smart buildings. *IEEE Syst. J.* **2018**, *12*, 1329–1340. [CrossRef]

10. Tsui, K.M.; Chan, S.C. Demand response optimization for smart home scheduling under real-time pricing. *IEEE Trans. Smart Grid* **2012**, *3*, 1812–1821. [CrossRef]

11. Shirazi, E.; Jadid, S. Optimal residential appliance scheduling under dynamic pricing scheme via HEMDAS. *Energy Build.* **2015**, *93*, 40–49. [CrossRef]

12. Anvari-Moghaddam, A.; Monsef, H.; Rahimi-Kian, A. Optimal smart home energy management considering energy saving and a comfortable lifestyle. *IEEE Trans. Smart Grid* **2015**, *6*, 324–332. [CrossRef]

13. Wen, Z.; O'Neill, D.; Maei, H. Optimal demand response using device-based reinforcement learning. *IEEE Trans. Smart Grid* **2015**, *6*, 2312–2324. [CrossRef]

14. Shareef, H.; Ahmed, M.S.; Mohamed, A.; Al Hassan, E. Review on home energy management system considering demand responses, smart technologies, and intelligent controllers. *IEEE Access* **2018**, *6*, 24498–24509. [CrossRef]

15. Li, W.; Yuen, C.; Hassan, N.U.; Tushar, W.; Wen, C.; Wood, K.L.; Hu, K.; Liu, X. Demand response management for residential smart grid: From theory to practice. *IEEE Access* **2015**, *3*, 2431–2440. [CrossRef]

16. Al-Hassan, E.; Shareef, H.; Islam, M.M.; Wahyudie, A.; Abdrabou, A.A. Improved smart power socket for monitoring and controlling electrical home appliances. *IEEE Access* **2018**, *5*, 49292–49305. [CrossRef]

17. Ahmed, M.S.; Mohamed, A.; Khatib, T.; Shareef, H.; Homod, R.Z.; Abd Ali, J. Real time optimal schedule controller for home energy management system using new binary backtracking search algorithm. *Energy Build.* **2017**, *138*, 215–277. [CrossRef]

18. Paterakis, N.G.; Erdinc, O.; Bakirtzis, A.G.; Catalao, J.P.S. Optimal household appliances scheduling under day-ahead pricing and load-shaping demand response strategies. *IEEE Trans. Ind. Inform.* **2015**, *11*, 1509–1519. [CrossRef]

19. Han, J.; Choi, C.; Park, W.; Lee, I.; Kim, S. Smart home energy management system including renewable energy based on ZigBee and PLC. *IEEE Trans. Consum. Electron.* **2014**, *60*, 198–202. [CrossRef]

20. GE Brillion™ Connected Appliances. General Electric. Available online: http://www.geappliances.com/ge/connected-appliances/ (accessed on 20 January 2020).

21. Honda Smart Home US. Honda. Available online: http://www.hondasmarthome.com/tagged/hems (accessed on 20 January 2020).

22. Sumsung Smart Home Support. Sumsung. Available online: https://www.samsung.com/us/support/smart-home/ (accessed on 18 January 2020).

23. Fibaro Smart Home. Fibaro. Available online: https://www.fibaro.com/en/ (accessed on 20 January 2020).

24. Open Access Same-Time Information System. New York Independent System Operator. Available online: http://mis.nyiso.com/public/P-7Alist.htm (accessed on 2 January 2020).

25. DFRobot. Available online: https://wiki.dfrobot.com/CO2_Sensor_SKU_SEN0159 (accessed on 2 January 2020).

26. Martin Almenta, M.; Morrow, D.J.; Best, R.J.; Fox, B.; Foley, A.M. Domestic fridge-freezer load aggregation to support ancillary services. *Renew. Energy* **2016**, *87*, 954–964. [CrossRef]

27. GMP Online Training. Available online: https://www.gmp-compliance.org/gmp-news/ambient-room-temperature-cold-what-is-what (accessed on 25 April 2020).

28. Rea, M. *The IESNA Lighting Handbook: Reference & Application*, 9th ed.; Illuminating Engineering Society of North America: New York, NY, USA, 2000.

29. Nistor, M.; Antunes, C.H. Integrated Management of Energy Resources in Residential Buildings—A Markovian Approach. *IEEE Trans. Smart Grid* **2018**, *9*, 240–251. [CrossRef]

Article

Identification of Major Inefficient Water Consumption Areas Considering Water Consumption, Efficiencies, and Footprints in Australia

Muhammad Atiq Ur Rehman Tariq [1,2,*], **Riley Raimond Damnics** [1], **Zohreh Rajabi** [1], **Muhammad Laiq Ur Rahman Shahid** [3] and **Nitin Muttil** [1,2,*]

[1] College of Engineering and Science, Victoria University, Melbourne, VIC 8001, Australia; riley.damnics@live.vu.edu.au (R.R.D.); zohreh.rajabi@live.vu.edu.au (Z.R.)

[2] Institute for Sustainable Industries & Liveable Cities, Victoria University, P.O. Box 14428, Melbourne, VIC 8001, Australia

[3] Department of Electronic Engineering, University of Engineering and Technology Taxila, Rawalpindi 46000, Pakistan; Laiq.Shahid@uettaxila.edu.pk

* Correspondence: atiq.tariq@yahoo.com (M.A.U.R.T.); Nitin.Muttil@vu.edu.au (N.M.)

Received: 25 July 2020; Accepted: 30 August 2020; Published: 4 September 2020

Abstract: Due to population growth, climatic change, and growing water usage, water scarcity is expected to be a more prevalent issue at the global level. The situation in Australia is even more serious because it is the driest continent and is characterized by larger water footprints in the domestic, agriculture and industrial sectors. Because the largest consumption of freshwater resources is in the agricultural sector (59%), this research undertakes a detailed investigation of the water footprints of agricultural practices in Australia. The analysis of the four highest water footprint crops in Australia revealed that the suitability of various crops is connected to the region and the irrigation efficiencies. A desirable crop in one region may be unsuitable in another. The investigation is further extended to analyze the overall virtual water trade of Australia. Australia's annual virtual water trade balance is adversely biased towards exporting a substantial quantity of water, amounting to 35 km^3, per trade data of 2014. It is evident that there is significant potential to reduce water consumption and footprints, and increase the water usage efficiencies, in all sectors. Based on the investigations conducted, it is recommended that the water footprints at each state level be considered at the strategic level. Further detailed analyses are required to reduce the export of a substantial quantity of virtual water considering local demands, export requirements, and production capabilities of regions.

Keywords: water accounting; resource efficiencies; virtual water trade; Australian trade sustainability; sustainable resource management; governance-engineering nexus

---

1. Introduction

Inefficient and excessive use of water could deplete aquifers, degrade flora and fauna habitats, and cause water supply shortages. Water is the fundamental resource required for sustaining life; however, with the current trends in population growth, climatic disturbances of water distribution, and the mismanagement of water usage, water scarcity is becoming a more prevalent issue not only globally but also within Australia. For example, the total withdrawal of global water resources was 4 trillion m^3 in 2014; however, it was only 1 trillion m^3 in 1934 [1]. A report by the World Water Assessment Programme [2] predicts "By 2025, 1.8 billion people are expected to be living in countries or regions with absolute water scarcity, and two-thirds of the world population could be under water stress conditions". Several examples of water scarcity have occurred over the years, which have highlighted the importance of water sustainability [3]. As water resources become increasingly scarce, the appropriate usage of water needs to be emphasized, particularly in the agricultural sector. This trend is worrying considering the additional factor of global warming, which is further contributing to water scarcity through the increased evaporation of freshwater resources. This phenomenon is generally characterized by less rainfall and drier conditions. This is felt locally in southern Australian cropping zones, which are affected by a severe deficit of water, putting pressure on water supply networks [4,5]. The FAO predicted that the amount of global water withdrawal for agriculture will increase by 11 percent from 2006 to 2050. At present, extraction from aquifers has reached 100% of total groundwater replenishment rates in many parts of the world and agriculture systems are already at risk due to water scarcity [6,7]. As a result, these excessive abstractions of groundwater have led to severe problems with groundwater reserves.

Despite being the driest populated continent on the planet, Australia has the highest rate of internal domestic water use per person [8]. The Australian government initialized a number of project streams including urban water efficiency projects addressing issues such as reduced leakage, sewerage treatment practices, and stormwater capture and recycling [9]. Industrial water efficiency projects such as plant upgrades, processing or product redesign, and implementation and water recycling have also been initiated. Off-farm projects have included dams and water storage, stock and domestic pipelines, and upgraded channel systems [10]. Metering projects comprise flow regulation infrastructure, installing meters and upgrading meters to comply with the Australian Standard. On-farm projects have included drip irrigation systems, replacing open channels with pipes and water-efficient root stock. Various other steps have been taken, such as "New Water" from saline aquifers, water loss management programs, water efficiency audits of steam systems, irrigation network renewal, aquifer recharge with storm water programs, etc. These are effective approaches to address the water crisis by enhancing water availability. However, the smart use of this precious resource is still being ignored at a larger scale. The overall picture of water efficiency still lacks better planning of water resources.

It has been proven that monitoring and controlling the water footprint can reduce water consumption by identifying water-related risks within the supply chain [11]. Water footprint assessments can be used to ensure sustainable water usage and maintain global water security. The concept of the water footprint introduced the idea of measuring water consumption throughout the lifecycle of a product or service, including direct and auxiliary water consumption values throughout the product's lifecycle. This consumption is categorized into three broad classifications: blue, green, or grey water. Blue water is classified as surface or groundwater consumed or lost in the production of an item. Losses from blue water are counted when water can no longer be used in the defined analysis area, e.g., evaporation or pumped away. Green water is water stored in the root-zone of plants and is consumed by plants or evaporated away. Grey water is relevant to the concentration of pollutants released in the production of a good or service. It is the amount of water required to maintain the water quality standard in the pollutant release zone. The sum of these three values provides a metric of the true impact of a product or service on the water network.

In addition to products and services, water footprints of nations can help identify the priority areas for the water security of a country. Hoekstra and Chapagain [12] investigated the water footprints of Morocco and the Netherlands for agricultural production from domestic sources. Yu et al. [13] concluded that water consumption for domestic, industrial, and agriculture sectors must be taken into account in planning water provision and promoting sustainable water consumption in a similar study for England at regional and national level. Feng et al. [14] considered the spatial aspect of the internal and external water footprints for the UK. In recent years, a significant amount of detailed water efficiency, footprints, virtual trade, and accounting research has been carried out for China in the fields of food security, power generation, urban households, and agriculture sectors [15–19].

Unfortunately, Australian water footprints are concerning as the added value of water for industrial usage is lowest in the world, and water footprints for domestic and industrial usages are the fourth highest at the global level [20,21]. Nevertheless, the country exports a large amount of virtual water annually through its trade. Considering the inefficient water usage in almost all sectors in Australia and the capabilities of water footprint assessments to address water security issues, the following research question triggered the current research:

What are the major water consumption sectors and how can the water footprint help in reducing the water accounting-related vulnerabilities in Australia by identifying the most detrimental sector?

The aim of this research is to analyze the substantial water consumption, efficiencies, and footprints in Australia to examine water conservation in different sectors, including trade. The undertaken research identifies possibilities to reduce the water footprints of different uses. A framework is developed to analyze the water security issues of a country/region. The work itself demonstrates an application of combining water footprint and virtual water trade techniques with traditional water accounting methods. The outcomes of this research not only provide guidance to improve water consumption in Australia but are equally beneficial for other countries/regions to review their water consumption practices considering the availability and economic value of water.

The aim of this research is to analyze the substantial water consumption, efficiencies, and footprints in Australia to examine water conservation in different sectors, including trade. The undertaken research identifies possibilities to reduce the water footprints of different uses. A framework is developed to analyze the water security issues of a country/region. The work itself demonstrates an application of combining water footprint and virtual water trade techniques with traditional water accounting methods. The outcomes of this research not only provide guidance to improve water consumption in Australia but are equally beneficial for other countries/regions to review their water consumption practices considering the availability and economic value of water.

1.1. Freshwater Availability in Australia

Australia receives an average of 417 mm of rainfall per year, which generates 3700 km^3 of runoff [22]. Rainfall supports Australia's dryland (non-irrigated) agriculture and a number or domestic water supplies (via rainwater tanks); however, rainfall is not considered a water resource for statutory water management. Only rainfall-runoff into creeks, rivers, and lakes (which accounts for 9% of total water received from rainfall) or recharged groundwater aquifers (2% of total water received from rainfall) is considered a managed resource. About 89% of water received from rainfall is lost through evapotranspiration [23].

Australia's population data (2014) shows that the population has doubled in the last 50 years. As the population has grown, water resources have decreased due to climate change. Renewable internal freshwater resources reduced from 45,802 m^3 in 1962 to 20,971 m^3 per capita as of 2014 [24,25]. Australia has experienced severe droughts in the past, such as the Federation drought (1895–1902), World War II drought (1937–1945), and the Millennium draught (2001–2009) [26]. These droughts mostly impacted Australia's agricultural sector. Large water storages, designed to hold reserves to manage dry years, were also drawn down during these droughts [27].

1.2. Major Water Consumption Sectors

Australia has the fourth largest domestic water footprint globally [20]. It uses more water domestically than could be considered "essential" and may not be utilizing its water resources efficiently in the industrial sector [20,28]. Figure 1 shows that the largest consumer of freshwater resources is the agriculture sector (59%), followed by industry (29%) and then domestic (12%) sectors. Previous research prioritizes the agriculture sector being the largest consumer of Australian water, because an efficient management of water resources in the agriculture sector will have a significant impact on overall water efficiency [29].

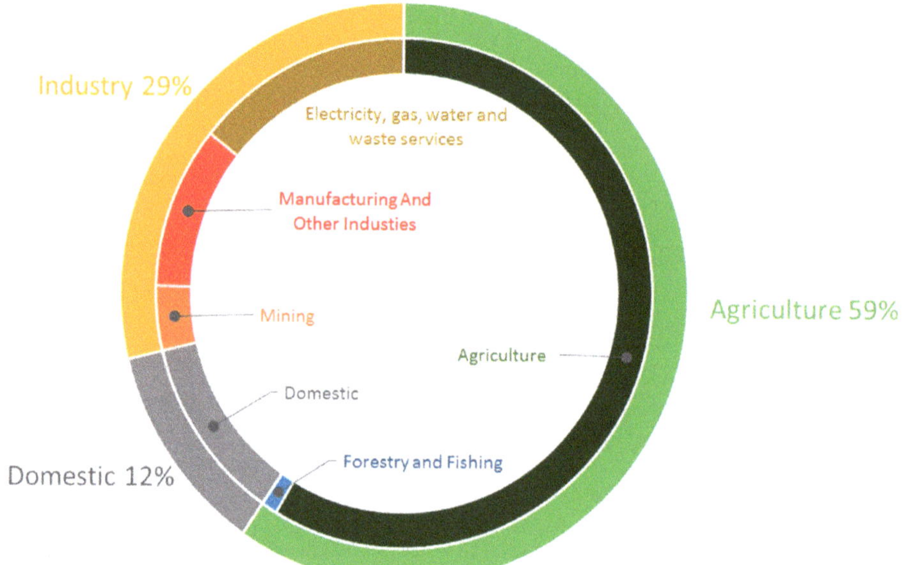

Figure 1. Water consumption share in major sectors 2015–2016 (Data source: Australian Bureau of Statistics, 2017).

2. Water Footprint Accounting

After the analysis of freshwater availability, a detailed analysis of water footprints of major water consuming sectors was carried out. The proposed methodology is explained in Figure 2. A standard water consumption accounting for the domestic and industrial sectors was conducted; however, a detailed and thorough analysis for major crops was performed at the regional level due to high water consumption in the agriculture sector. Australia has one of the highest negative balances of green water, mainly due to its exports [30]. Hoekstra et al. [8] determined the water footprints of Australia at the national level and found that the internal water footprint for domestic water is 341 m^3/Cap/y, and for industrial is 64 m^3/Cap/y internal and 211 m^3/Cap/y external, whereas for agricultural products it is 736 m^3/Cap/y internal and 41 m^3/Cap/y external. Therefore, recalculating the internal and external water footprints is beyond the scope of this research.

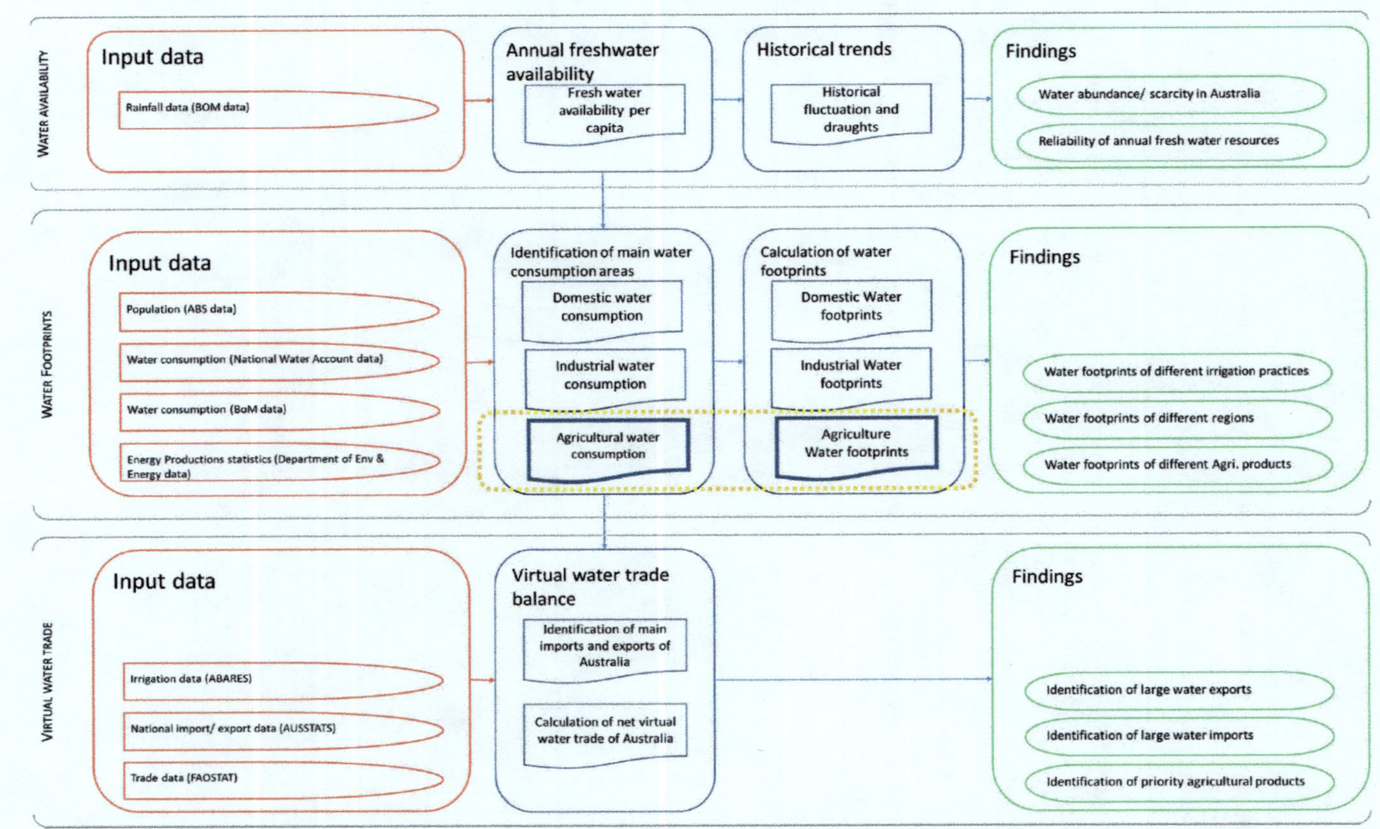

Figure 2. Flowchart of the methodology.

2.1. Domestic Water Consumption Accounting

Population and water consumption data was collected from the National Water Account and the Australian Bureau of Statistics (ABS) [31] to perform the domestic water consumption accounting. Calculation of water use per capita highlighted the states with higher water consumption. The data for end water usage was collected for the states of Northern Territory (NT), Victoria (VIC), Western Australia (WA), Queensland (QLD), and New South Wales (NSW). The states of NT, VIC, and QLD however, only had data available on a city basis; this data was for the largest cities in the state therefore was an accurate representation of water usage trends within the states. Domestic water usage (12%) is a relatively minor portion compared to the agricultural and industrial sectors. However, due to its nature of use and based on direct stakeholders, urban water footprints have the highest priority. Excluding NT and WA, the remaining six states have a mean usage of 74,186 L per capita per year (LCY) (refer Figure 3) which is higher than the global average footprints of 57,000 LCY. WA and NT demonstrate much higher water consumption of 127,929 LCY (72% higher than the middle band average of other states excluding these two) and 158,407 LCY (114% greater than the middle band average of other states excluding these two), respectively. NT and WA have potential to reduce water consumption. Further investigation revealed that the lifestyle of these two states is characterized by larger houses with green areas that need a substantial amount of water for gardening [32]. End water usage data shows that Darwin and West Australia use 70% and 50%, respectively, of domestic water for outdoor use, whereas Brisbane and Melbourne use 5% and 18%, respectively [33].

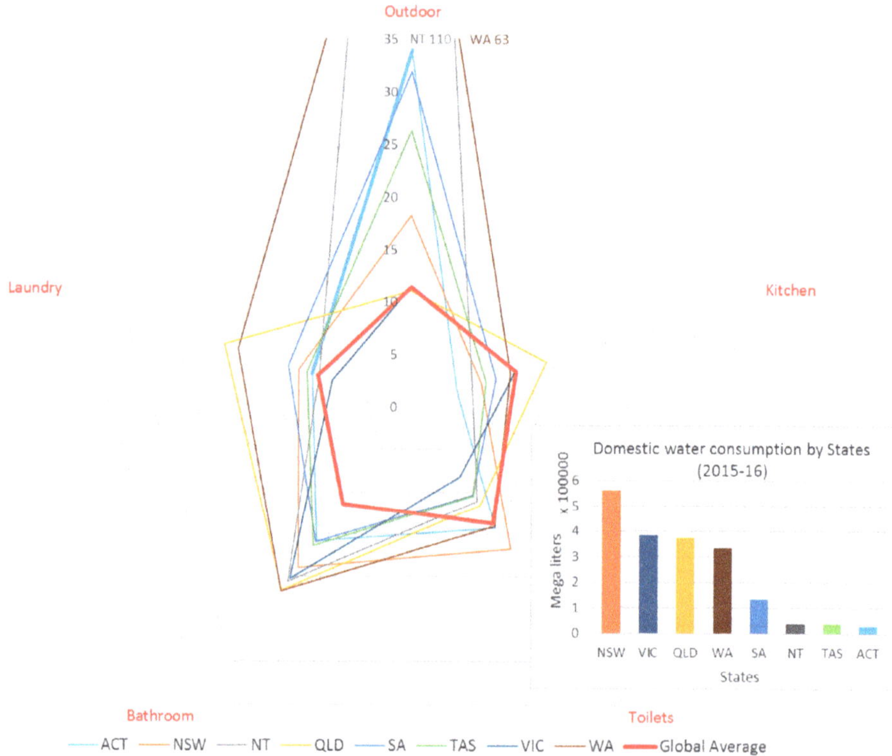

Figure 3. Domestic water consumption by each state; total consumption on right (in frame graph) and consumption patterns in each state on right side (2015–2016) (Data sources: [31,32,34–38]).

2.2. Industrial Water Consumption Accounting

The industrial water consumption accounting was calculated using a combination of the National Water Account and GVA (Gross Value Added) data taken from the Australian Bureau of Statistics [21,28,34,39]. Green and grey water footprints for industrial sector were assumed to be the same [40] and, for the scope of this research, detailed insight of these two was not required. Therefore, blue water consumption was the only value used for the state-wise comparison. Additionally, the population data found for the domestic sector was used in addition to energy production statistics from the Department of Environment and Energy [41]. Data was collected for mining, manufacturing, and utilities industries.

The literature provides details of each industry type (mining, manufacturing, and utilities), however, due to differences in currencies and circumstances, a comparison with global average was not considered to be a useful indicator. Therefore, an interstate comparison was carried out for better understanding of practices within Australia. NSW is the largest consumer in industrial water consumption followed by QLD and then VIC (refer Figure 4). These values represent the extent of industrial activities in each state. When the value added per unit of water used in each state is compared, the results are altogether different. Standardized results are obtained for all of the states except ACT, which had the largest footprints in the industrial sector. Having a relatively small area and being a predominantly government service area, it does not contribute significantly to the economy. From the available data, Tasmania appears to be ahead of the competition but this may be simply due to the type of mineral being extracted. For the industrial sector, it is hard to pinpoint where water efficiency could be increased as limited data is available due to private rights of industries regarding water usage.

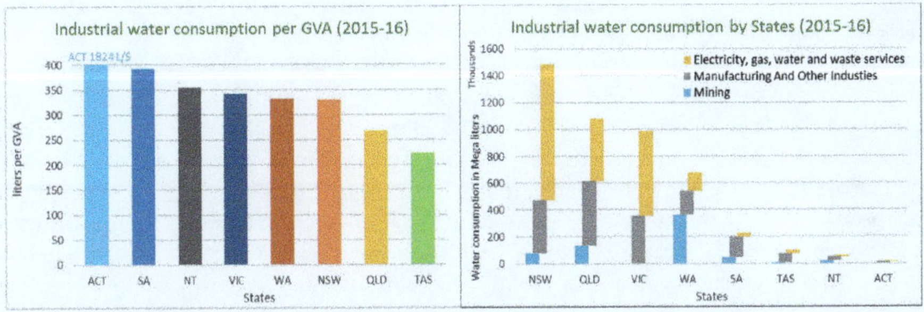

Figure 4. Industrial water consumption (ML) per industry and per Gross Value Added (GVA) excluding agriculture (thousand $) by each state (2015–2016) on left and total consumption by each sector on right (Data source: Australian Bureau of Statistics, 2017).

2.3. Agricultural Footprints

Because many crops are grown in particular areas only, there was not extensive data within the National Water Account database for comparison purposes. In this study, the data was collected from the Water Use on Australian Farms and Agricultural Commodities Australia data by the ABS [31,32,34,42–44]. Where there was insufficient data to compare the states, the two largest Natural Resource Management (NRM) regions within the states were chosen for comparison. Based on the large water footprints involved and the scale of production, four agricultural commodities were selected and their water footprints are compared across the NRM regions:

- Cotton
- Rice
- Sugar cane
- Grapevines

The agricultural footprint includes two distinct components: direct virtual water and water for energy (indirect) [45]. For the sake of simplicity in this study, only direct virtual water component was calculated. It is also worth noting that the green water footprint is only pertinent to the agricultural sector; typically, for the growing of a crop or tree the total water footprint is equivalent to:

$$WF_{proc} = WF_{proc.green} + WF_{proc,blue} + WF_{proc,grey} \tag{1}$$

As mentioned previously, information regarding the blue water ($WF_{proc, Blue}$) footprint in a process was gathered from the ABS, and the grey water ($WF_{proc, grey}$) footprint in a process was considered redundant for the sake of comparison across the states. Thus, only the green water footprint ($WF_{proc, green}$) in a process was required to be calculated. For this purpose, a combination of rainfall data by state from the Bureau of Meteorology (BOM) and the Blaney–Criddle equation for a simple calculation of evapotranspiration yielded the green value:

$$ET_o = p \times (0.457T_{mean} + 8.128) \tag{2}$$

where ET_o is the reference evapotranspiration (mm day^{-1}) (monthly), T_{mean} is the mean daily temperature (°C) given as $T_{mean} = (T_{max} + T_{min})/2$, and p is the mean daily percentage of annual daytime hours. This approximate method is stated as Equation (3):

$$WF_{proc,green} = \frac{(\text{Mean Areal Rainfall (State)} - \text{Evapotranspiration}) \times \text{Crop Area} \times \text{Growing Period}}{\text{Crop Yield}} \tag{3}$$

2.4. Detailed Investigation of Agricultural Practices

The main part of the water usage in the agricultural industry is used for irrigation purposes, representing more than 85% of the total water consumption in the agricultural sector. The remaining 15% is used for other agricultural purposes such as drinking water and cleaning. According to the Australian Bureau of Statistics, NSW was ranked as the state with the largest agricultural water consumption, being just higher than QLD and VIC for the period 2014–2016 (refer to Figure 5). The percentage of land use for agriculture is 58% in Australia. The agricultural activities are driven by the climate, economy, water availability, and soil conditions. The low consumption rate of NT is mainly because 99.5% of the agricultural area is grazing native vegetation [46]. The state experiences a high fluctuation in rainfall, because of which the limited crop production activities use groundwater (82%) [47]. As such, although it may appear that TAS and VIC use more water, they may achieve better yields from the area farmed. This highlights the need to compare each crop individually in addition to comparing the yield for the relevant crop.

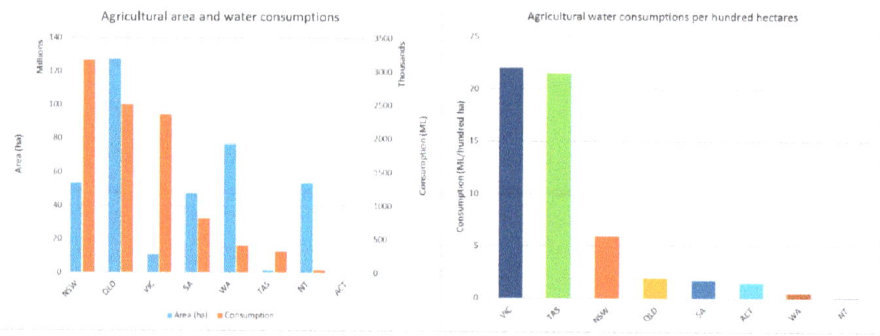

Figure 5. Land use and water consumption for agricultural production.

As mentioned earlier, cotton, rice, sugar cane, and grapevines were selected and their water footprints were compared across the states. It must be noted that the comparison of these commodities is based on water consumption per mass and the value added of water per mass of the commodity.

2.4.1. Comparison of Cotton

Murrumbidgee regions (NSW) and Fitzroy regions (QLD) both consume the most water per tonne of cotton produced, 5.28 ML/tonne and 5.10 ML/tonne and economic value 456 $/ML and 486 $/ML, respectively. These are also above the global water footprint average for the green/blue water footprint of cotton at 3.6 ML/tonne [48–50]. The other two regions, Border Rivers-Gwydir (NSW) and Border Rivers Maranoa-Balonne (QLD) are also above this average having 4.27 ML/tonne and 3.90 ML/tonne, respectively, and economic value of 564 $/ML and 636 $/ML, respectively. Border Rivers-Gwydir and Border Rivers Maranoa-Balonne are in close proximity geographically, whereas Murrumbidgee and Fitzroy are further south and north, respectively (refer to Figure 6). It must be noted that the water efficiency of cotton crops was much lower initially, and improved by approximately 40% from 2003 to 2013 nationwide [51].

2.4.2. Rice Comparison

In the case of rice production, NSW, VIC, and QLD all have a similar water footprint (1.32 ML/tonne (economic value 324 $/ML), 2.3 ML/tonne (economic value 181 $/ML), and 1.82 ML/tonne (economic value 275 $/ML), respectively. With the exception of NSW, the other states are above the global average water footprint of 1.5 ML/tonne as per the Crop Water Footprint Benchmark [48]. NT is a significant over user of water resources, with a water footprint of 3.54 ML/tonne, which is double the next most significant state's water footprint value. The water footprint of the Northern Territory region (NT) (with economic value 127 $/ML) is 126% greater than the average of the other regions. It should also be noted that the other underperformer lies in the Wet Tropics region (QLD); the water footprint of this region lies above the global average and is a 38% above the average of the selected regions excluding the NT.

2.4.3. Sugar Cane Comparison

The calculated water footprints for the Border Rivers-Gwydir (NSW) (0.08 ML/tonne, 327 $/ML), Burdekin (QLD) (0.11 ML/tonne, 248 $/ML), and the Wet Tropics (QLD) (0.08 ML/tonne, 448 $/ML) lie under the global average of 0.197 ML/tonne. The Northern Rivers (NSW) has a significantly small footprint at 0.03 ML/tonne, however, this result appears unlikely and is due to an error in the data (as cautioned by the ABS). On the other hand, the Rangelands (WA) uses 0.17 ML/tonne with an economic value of 68 $/ML, and lies below the global average.

2.4.4. Grape Comparison

A solid average across all the states is observed except the significant outlier of NT (5.1 ML/tonne) which, compared to the remaining states' average (0.42 ML/tonne) is 1214% higher. This average is significantly below the global water footprint at 0.61 ML/tonne. QLD is the next highest value, indicating that there is a correlation between the water footprint and mean temperature for grape production. The substantial water footprint difference is due to the nature of grape production in each of the states; wine production generally has a greater economic value of the water used. NT for example uses 0%, 693 $/ML, TAS uses 100%, 6680 $/ML, and VIC uses 72% 3140 $/ML of their grapes for wine production.

The water consumption figures used for calculation purposes are conservative because water theft is not accounted for in the calculations [52–54].

Detailed analysis of water footprints and economic values of water for major crops across Australia

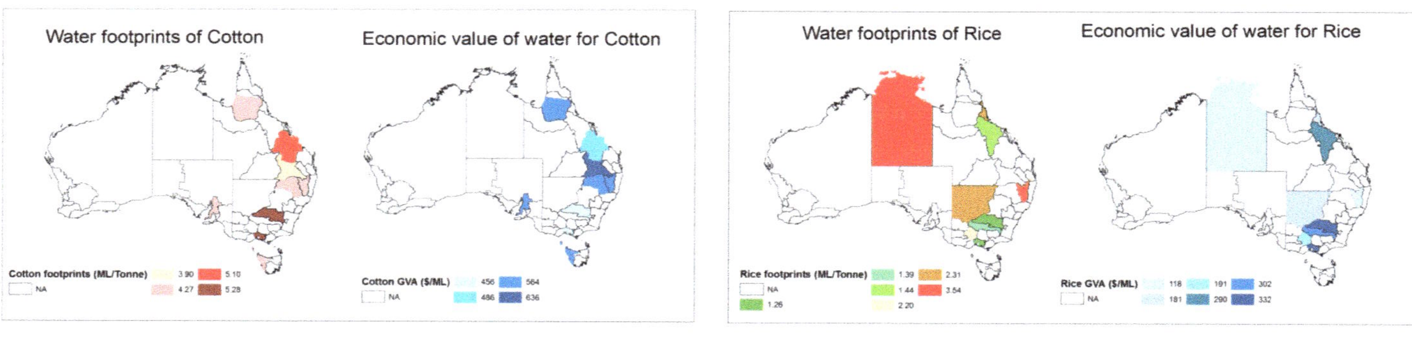

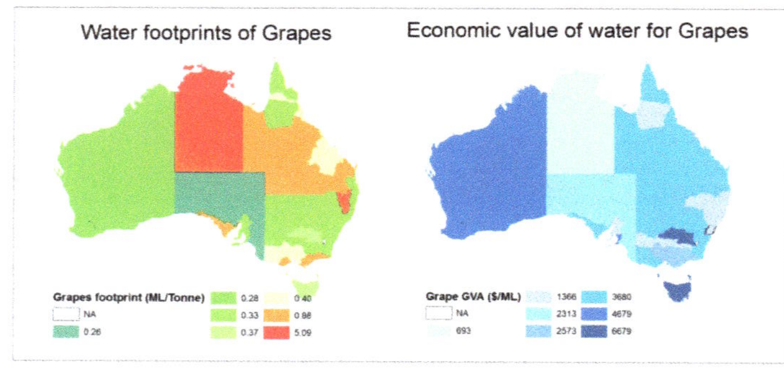

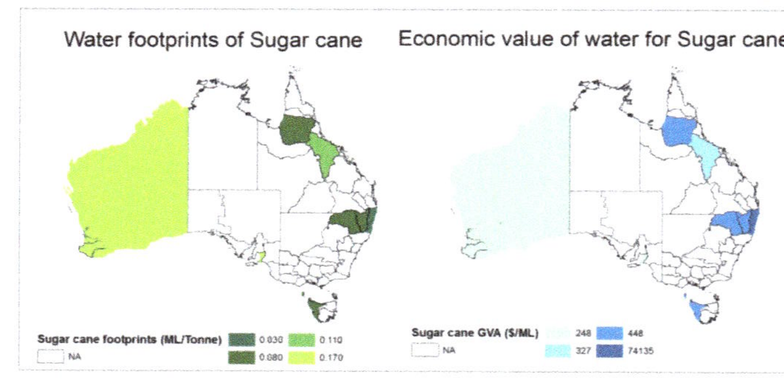

Figure 6. Detailed analysis of water footprints (red shades are above the global average and green are below the global average) and economic value of water for major crops in Australia.

2.5. Irrigation Practices

Because the water footprints of majority of crops were found to be above the global average, it was felt necessary to investigate the irrigation practices in Australia to examine the role of irrigation in higher water footprints. Large-scale irrigation started in Australia in the 1800s. The irrigated area grew steadily during 1920–1960, then increased rapidly number until 1990. The irrigated area decreased from 1996 until 2012 due to drought conditions. The most common method of irrigation is surface irrigation followed by sprinkle irrigation using large mobile machines. The use of drip irrigation has increased in recent years, and overall drip irrigation is the least preferred method. The use of sprinkler irrigation remained similar from 2003 to 2009 [55]. It was found that Australia continues to predominantly use traditional inefficient methods of irrigation.

3. Analysis of Virtual Water Trade

The third area of investigation was the analysis of virtual water trade, because imports and exports can influence the overall water security of a region [48]. The literature was consulted to develop an effective approach to identify the main crops/trade goods contributing to the export of virtual water. Water Embodied in Bilateral Trade (WEBT) and Multi-Regional Input–Output Analysis (MRIO) were used to calculate the embodied water because these methods are traditionally used for energy and carbon footprints as two input–output top-down approaches [56–59]. The bottom-up approaches do not consider the entire industrial supply chain, therefore the results are highly impacted by the recorded data set and the total water footprint can vary by as much as 48% [60,61]. The main exports involving major water content are agricultural products. A basic assessment of virtual water content was performed to estimate the virtual water export. The trade data of the top ten imports and exports were collected from the FAOSTAT website for the year 2014 (latest available data). The mainstream agricultural products were identified, and the calculation of water trade was performed accordingly (refer Equation (4)). The proportions of green, blue, and grey water in each export commodity were calculated.

$$\text{Volume} \times \text{Virtual water content} = \text{Volume of Virtual Water} \qquad (4)$$

Australia's total export of green virtual water was 26.5 km^3, of blue virtual water was 6.4 km^3, and of grey virtual water was 3.6 km^3, thus totaling 36.5 km^3. By comparison, the import of green virtual water was 1.6 km^3, of blue virtual water was 0.184 km^3, and of grey virtual water was 0.078 km^3, for a total water footprint of 1.8 km^3. Thus, in 2014 Australia indirectly exported 35 km^3 of water, which represents a significant indirect loss of freshwater. The total export value data was obtained from the FAO website [62]. For the calculation of usable water value, green water was not counted because green water must be consumed at the location at which the precipitation occurs (refer Equation (5)). Therefore, the usable water value comprised only blue and grey waters [11,59]:

$$\text{Usable Water Value} = \frac{\text{Total Export Value}}{\left(\text{Total Blue Water Export} + \text{Total Grey Water Export}\right)} \qquad (5)$$

According to the Australian Department of Foreign Affairs and Trade, Australia's exports of services and goods totaled $330.3 billion. Iron ore is Australia's largest export commodity [63–66]. Because of Australia's natural resources, coal and natural gas are also significant export resources. About 65% of all agricultural products produced in Australia are exported for a relatively lower economic benefit. Some of the exported products are the most water-intensive crops produced, such as cotton (98% exported on a three-year average). Despite covering 58% of Australian land and accounting for 59% of water consumption, none of the agricultural products are among Australia's top five exports. Furthermore, agricultural products contribute only 2.7% of GDP and account for 2.5% of employment [67,68]. Australia has GDP of $1.561 trillion and per capita GDP of $68,973, and was ranked fifth in the world in 2014 in the latter measure according to data released by the Australia

Bureau of Statistics and Department of Foreign Affairs, Trade, and Industry [65,69,70]. Analysis indicated sheep meat is the most efficient water value agricultural product (14.90 $/ML) as it requires minimal or no blue water usage. The second highest value added agricultural product was found to be wine grapes (5.72 $/ML). The lowest usable water values were obtained for cotton lint (0.48 $/ML), rice (0.53 $/ML), and raw sugar centrifugal (0.69 $/ML), due to the obvious reason of the high consumption of blue water and low export value. The analysis clearly indicates a need to consider the usable water value at the strategic level. Alternative crops, high efficiency crops, crops suiting specific climatic zones, and reduction of the cultivation of low usable water value crops can be considered after detailed analysis.

4. Conclusions and Recommendations

Australian water footprints for each sector (domestic, agricultural, and industrial) are much higher than the average global footprints. Domestic water consumption is higher for bathroom, laundry, and outdoor usage. The outlier is outdoor gardening usage, which is exceptionally higher than the global average. Usage of the states of NT and WA is exceptionally high, at 1000% and 570% of the global average, respectively. To reduce domestic water consumption, it is recommended that these states should reduce gardening water consumption by introducing legislation (water restrictions) to limit its use, educating the populace about water tolerant garden varieties, using recycled water for outdoor usage, and incentivizing the installation of rainwater tanks (which could be used for gardening to reduce the pressure on the freshwater supply). If these states reduced their water usage to the average level of the remaining states, a total of 161,279 ML could be saved annually. Furthermore, it is recommended that more precise data for each domestic component should be compiled to identify the usage subtype where the significant wastage is occurring. Ultimately water mismanagement in the domestic sector not only risks urban water security but retroactively results in financial losses.

The importance of industrial use, which consumes 29% of freshwater resources, cannot be ignored. To identify the scope for improvements in the water efficiency of industries, specific targeting of an industrial process should be investigated. However, this is not possible under the present conditions of private agreements between the government and industries (particularly mining). Therefore, for future studies of the water footprints of mining it would be beneficial if details of water usage, in addition to data on the mass of mineral being extracted per state, is made available. This would allow investigation of which mining practices are operating more efficiently. However, it can be concluded that, among the industry sector, utilities (energy, water supply, and gas, etc.) are the main water consumers, followed by manufacturing. Considering the volatile conditions of freshwater availability in Australia, the government may motivate the industry to carry out water footprint studies for various industries and support those that use water more efficiently by offering effective incentives.

For the agricultural sector, the water footprint and water economic efficiencies can help identify suitable regions for different crops. Australia can undoubtedly increase its agricultural production and water efficiencies if modern irrigation methods are used. However, the selection of the most suitable crops in different regions plays a key role in determining the extent of the water footprint. To account for the economic perspective, economic efficiency was calculated in addition to the water footprints of the agriculture sector. Australia's current approach of crop selection for different regions can be improved if water footprints are considered at the planning stage. Crop cultivation planning must be performed at a central level by considering the availability of green water in different NRM regions, production capabilities of different regions for different crops, local demands, export priorities, climatic impacts, and future perspectives. Because the green water component cannot be allocated to another region without being embedded in a product, agricultural products should absorb all available green water in the region with minimum or no blue water supplement. However, this is not always possible, and it is essential to ensure that only appropriate crop types are planted in different regions to maximize green water utilization. The consumption by agricultural products of less than the available green water will cause resource wastage (if the excess water is not converted into blue water); however, if the product requires more than the available green water then blue water needs to be consumed, which is more valuable due to its capability to be reallocated to other regions. Supplementation of blue water should be to an option to maximize the possible value added of agricultural products. This can be achieved by water pricing and raising the water footprint awareness among the stakeholders, particularly farmers and national planning authorities.

With the exception of grapes, crops are not cultivated in all regions due to various reasons in addition to their suitability to the climate and geography of the area. It can be concluded that the climate is an important factor, however, the type of grape plays an important role in determining the water footprints. Wine production grapes yield a higher economic value of water. Wine producing grapes can be experimented with in NT because this is the only region which does not produce grapes for wine production and has a water footprint that is 12 times higher than the average of the rest of Australia. It is also important to note that the lowest economic value of water (which is in NT, at 693 $/ML) is even higher than the maximum value of water in any other crop and any other region. A further exploration of crop practices will not only identify more efficient water usage possibilities but will provide an opportunity to shift towards more efficient crops in terms of water footprints.

Water usage above the median water consumption rate can be used as an indicator to identify inefficient agricultural practices. Threshold or acceptable water footprint limits for different types of crops must be established by the government to ensure the efficient use of this valuable resource. To achieve further increases in the efficiency of the economic value of water, the concept of an open water market regulated under a focused government strategy can be trialed.

Managing the virtual water trade ensures that trade patterns does not endanger the water balance of the country. The current virtual trade of water is concerning due to the export of higher volumes of water used to generate lower economic values, which is neither economically feasible nor sustainable in the longer term. Although the water security of Australia is extremely volatile and vulnerable to seasonal and annual fluctuations of freshwater availability, the imbalance of virtual water trade requires immediate attention. In the case of another severe drought, such as the previous Millennium drought, the country will not only suffer a water shortage but also face a sharp decline in revenue generation resulting from the negative economic impacts of the hazard. Further detailed analyses are required to reduce the export of a substantial quantity of virtual water considering local demands, export requirements, and production capabilities of Australian resources. A focused vulnerability analysis of the Australian economy, its goods and services trade, and virtual water trade is recommended to comprehend and envisage the impacts of any future drought in Australia.

The current study explored water consumption and efficiencies in three major sectors of water consumption. Further detailed analysis of the major industrial production sectors is already underway. The first-tier industries include the construction, power generation (particularly the renewable sector), and the food processing industries to supplement the missing portion of the agricultural sector. Further investigation of the virtual water trade is being conducted to calculate the internal and external water footprint at a more precise level. This research ultimately aim to develop a comprehensive framework of water usage efficiencies for a country, and Australia in particular.

Author Contributions: Conceptualization, M.A.U.R.T., R.R.D. and N.M. Data curation, R.R.D.; Formal analysis, M.A.U.R.T., R.R.D. and M.L.U.R.S.; Investigation, M.A.U.R.T., R.R.D. and N.M.; Methodology, M.A.U.R.T., R.R.D. and Z.R.; Project administration, M.A.U.R.T. and N.M.; Software, M.A.U.R.T. and Z.R.; Supervision, M.A.U.R.T. and N.M.; Visualization, M.A.U.R.T., Z.R. and M.L.U.R.S.; Writing—original draft, M.A.U.R.T. and R.R.D.; Writing—review & editing, M.A.U.R.T., Z.R., M.L.U.R.S. and N.M. All authors have read and agreed to the published version of the manuscript.

Funding: This research received no external funding.

Conflicts of Interest: The authors declare no conflict of interest.

References

1. Worldometers, Global Water Use. 2019. Available online: https://www.waterscarcitysolutions.org/ (accessed on 11 December 2019).
2. World Water Assessment Programme. *Managing Water Under Uncertainty and Risk*; United Nations Educational Scientific and Cultural Organization: Paris, France, 2012.
3. Mekonnen, M.M.; Hoekstra, A.Y. Sustainability: Four billion people facing severe water scarcity. *Sci. Adv.* **2016**, *2*, e1500323. [CrossRef] [PubMed]
4. Bureau of Meteorology, Recent Rainfall, Drought and Southern Australia's long-Term Rainfall Decline, 2015. Available online: http://www.bom.gov.au/climate/updates/articles/a010-southern-rainfall-decline.shtml#toc1. (accessed on 5 October 2018).
5. Bureau of Meteorology, Indian Ocean influences on Australian Climate, 2018. Available online: http://www.bom.gov.au/climate/iod/ (accessed on 11 October 2018).
6. FAO. *The State of the World's Land and Water Resources for Food and Agriculture (SOLAW)–Managing Systems at Risk*; The Food and Agriculture Organization of the United Nations and Earthscan: New York, NY, USA.
7. FAO AQUASTAT Global Information System of Water and Agriculture: Rome, Italy, 2012. Available online: http://www.fao.org/aquastat/en/ (accessed on 2 September 2020).
8. Hoekstra, A.C.A. Water footprints of nations: Water use by people as a function of their consumption pattern. *Water Resour. Manag.* **2007**, *21*, 35–48. [CrossRef]
9. Department for Environment and Water, Efficiency Measure projects in South Australia, Government of South Australia for Environment and Water. 2017. Available online: https://www.environment.sa.gov.au/topics/river-murray/basin-plan/efficiency-measure-projects (accessed on 1 December 2019).
10. Water services Association of Australia (WSAA). *Water Efficient Australia, Smart Approved WaterMark*; Water Services Association of Australia (WSAA): Canberra, Australia, 2019; Available online: https://www.wsaa.asn.au/publication/water-efficient-australia (accessed on 10 December 2019).
11. Water Foot Print Network, Global Water Footprint Standard, 2011. Available online: https://waterfootprint.org/en/water-footprint/global-water-footprint-standard/ (accessed on 17 December 2019).
12. Hoekstra, A.Y.; Chapagain, A.K. The water footprints of Morocco and the Netherlands: Global water use as a result of domestic consumption of agricultural commodities. *Ecol. Econ.* **2007**, *64*, 143–155. [CrossRef]
13. Yu, Y.; Hubacek, K.; Feng, K.; Guan, D. Assessing regional and global water footprints for the UK. *Ecol. Econ.* **2010**, *69*, 1140–1147. [CrossRef]
14. Feng, K.; Hubacek, K.; Minx, J.; Siu, Y.L.; Chapagain, A.; Yu, Y.; Guan, J. Barrett, Spatially Explicit Analysis of Water Footprints in the UK. *Water* **2010**, *3*, 47–63. [CrossRef]
15. Cai, B.; Liu, B.; Zhang, B. Evolution of Chinese urban household's water footprint. *J. Clean. Prod.* **2019**, *208*, 1–10. [CrossRef]
16. Liu, X.; Shi, L.; Engel, B.A.; Sun, S.; Zhao, X.; Wu, P.; Wang, Y. New challenges of food security in Northwest China: Water footprint and virtual water perspective. *J. Clean. Prod.* **2020**, *245*, 245–118939. [CrossRef]
17. Zhang, S.; Taiebat, M.; Liu, Y.; Qu, S.; Liang, S.; Xu, M. Regional water footprints and interregional virtual water transfers in China. *J. Clean. Prod.* **2019**, *228*, 1401–1412. [CrossRef]
18. Zhai, Y.; Shen, X.; Quan, T.; Ma, X.; Zhang, R.; Ji, C.; Zhang, T.; Hong, J. Impact-oriented water footprint assessment of wheat production in China. *Sci. Total Environ.* **2019**, *689*, 90–98. [CrossRef] [PubMed]
19. Xie, X.; Jiang, X.; Zhang, T.; Huang, Z. Study on impact of electricity production on regional water resource in China by water footprint. *Renew. Energy* **2020**, *152*, 165–178. [CrossRef]
20. Mekonnen, M.M.; Hoekstra, A.Y. *National Water Footprint Accounts: The Green, Blue and Grey Water Footprint of Production and Consumption*; UNESCO-IHE: Delft, The Netherlands, 2011.
21. The Wold Bank, Industry (including Construction), Value Added (Constant 2010 US$), 2018. Available online: https://data.worldbank.org/indicator/NV.IND.TOTL.KD?view=chart (accessed on 18 September 2018).
22. Prosser, I. *Current Water Availability and Use*; Csiro: Melbourne, Australia, 2011.
23. BOM, Water in Australia -Australian Government—Bureau of Meteorology, BOM, 2018–2019. Available online: http://www.bom.gov.au/water/waterinaustralia/ (accessed on 25 December 2019).
24. Index Mundi, Australia—Renewable Internal Freshwater Resources per Capita (cubic meters), 2015. Available online: https://www.indexmundi.com/facts/australia/indicator/ER.H2O.INTR.PC (accessed on 15 December 2019).

25. Hugo, G. Changing Patterns of Population distribution in Australia. *J. Popul. Res. Spec. Ed.* **2002**. Available online: https://search.informit.com.au/documentSummary;dn=395273573294486;res=IELAPA (accessed on 2 September 2020).

26. The Role of Water in Australia's Uncertain Future. The Conversation 2015. Available online: https://theconversation.com/the-role-of-water-in-australias-uncertain-future-45366 (accessed on 24 June 2019).

27. Australian Food History Timeline, 2001—2008 Millennium Drought, 2011. Available online: https://australianfoodtimeline.com.au/millennium-drought/ (accessed on 1 December 2019).

28. The World Bank, Agriculture, Forestry, and Fishing, Value Added (Current US$), 2018. Available online: https://data.worldbank.org/indicator/NV.AGR.TOTL.CD?end=2017&start=2017&view=bar (accessed on 14 September 2018).

29. UNSW-Sydney Connected Waters Initiative, Connected Waters, 10 September 2007. Available online: http://www.connectedwaters.unsw.edu.au/articles/2007/09/national-water-footprints (accessed on 28 November 2019).

30. Fader, M.; Gerten, D.; Thammer, M.; Heinke, J.; Lotze-Campen, H.; Lucht, W.; Cramer, W. Internal and external green-blue agricultural water footprints of nations, and related water and land savings through trade. *Hydrol. Earth Syst. Sci. Discuss.* **2011**, *8*, 483–527. [CrossRef]

31. Australian Bureau of Statistics, Australian Demographic Statistics, Jun 2016, 2017. Available online: http://www.abs.gov.au/AUSSTATS/abs@.nsf/Lookup/3101.0Main+Features1Jun%202016?OpenDocument (accessed on 1 September 2018).

32. Australian Bureau of Statistics, Water Account, Australia, 2015–16, 2017. Available online: http://www.abs.gov.au/AUSSTATS/abs@.nsf/allprimarymainfeatures/49F854E3831E4294CA2580580015E2A6?opendocument. (accessed on 30 August 2018).

33. Hoekstra, A.Y.; Chapagain, A.K.; Aldaya, M.M.; Mekonnen, M.M. *The Water Footprint Assessment Manual: Setting the Global Standard*; Earthscan: London, UK, 2011.

34. Australian Bureau of Statistics, Australian National Accounts: State Accounts, 2015–16, 2017. Available online: http://www.abs.gov.au/AUSSTATS/abs@.nsf/allprimarymainfeatures/8DEE81B03A2C2842CA2581DA000F4790?opendocument (accessed on 5 September 2018).

35. Power and Water Corporation, the Darwin Water Story, 2013. Available online: https://www.powerwater.com.au/__data/assets/pdf_file/0011/1523/Darwin_Water_Story_2013.PDF (accessed on 23 September 2019).

36. Redhead, M. *Melbourne Residential Water End Uses Winter 2010*; summer 2012; Smart Water Fund: Melbourne, Australian, 2013.

37. Griffith University, School of Engineering and Smart Water Research Centre, South East Queensland Residential End Use Study: Baseline Results—Winter 2010, Urban Water Security Research Alliance, City East, 2010. Available online: http://hdl.handle.net/10072/38126http://www.urbanwateralliance.org.au/publications/UWSRA-tr31.pdf (accessed on 25 September 2019).

38. Rickwood, P.; Giurco, D.; Glazebrook, G.; Kazaglis, A.; Thomas, L.; Zeibots, M.; Boydell, S.; White, S. *Integrating Population, Land Use, Transport, Water and Energy-Use Models to Improve the Sustainability of Urban Systems*; University of Technology Sydney: Sydney, Australia, 2007.

39. TeamPoly, Water Price in Australia, 2018. Available online: https://www.teampoly.com.au/2018/06/15/water-prices-in-australia/ (accessed on 2 December 2019).

40. Averink, J. *Global Water Footprint of Industrial Hemp Textile*; University of Twente: Enschede, The Netherlands, 2015.

41. Department of the Environment and Energy. *Australian Energy Statistics: Table O*; Department of the Environment and Energy: Canberra, Australia, 2018.

42. ABARES, Agricultural Commodities, 2019. Available online: http://www.agriculture.gov.au/abares/research-topics/agricultural-commodities (accessed on 3 May 2019).

43. ABARES, Food Consumption and Imports, 2018. Available online: https://public.tableau.com/profile/australian.bureau.of.agricultural.and.resource.economics.and.sci#!/vizhome/FooddemandinAustralia/Indicators-fig3 (accessed on 2 May 2019).

44. PMSEIC Expert Working Group. *Australia and Food Security in a Changing World*; Australian Government: Canberra, Australia, 2010.

45. Chini, M.; Konar, M.; Stillwell, A.S. Direct and indirect urban water footprints of the United States. *Water Resour. Res.* **2016**, *53*, 1.

46. Department of Agriculture Water and the Environment, about My Region-Northern Territory, Department of Agriculture Water and the Environment, 4 February 2020. Available online: https://www.agriculture.gov.au/abares/research-topics/aboutmyregion/nt (accessed on 19 December 2019).

47. Australian Bureau of Statistics, Irrigation on Australian farms, 1301.0—Yearbook Chapter, 2008, 2008. Available online: https://www.abs.gov.au/AUSSTATS/abs@.nsf/Previousproducts/1301.0Feature%20Article16012008?opendocument&tabname=Summary&prodno=1301.0&issue=2008&num=&view= (accessed on 18 February 2020).

48. Frontier Economics. *The Concept of 'Virtual Water'—A Critical Review*; The Frontier Economics Network: Melbourne, Australia, 2008.

49. Mekonnen, M.M.; Hoekstra, A.Y. *Value of Water Research Report Series No. 47*; UNESCO-IHE: Delft, The Netherlands, 2010.

50. Mekonnen, M.M.; Hoekstra, A.Y. *Value of Water Research Report Series No. 48*; UNESCO-IHE: Delft, The Netherlands, 2010.

51. Roth, G.; Harris, G.; Gillies, M.; Montgomery, J. *Water-Use Efficiency and Productivity Trends in Australian Irrigated Cotton: A Review*; Crop and Pasture Science: Collingwood, Australia, 2013; Volume 64, pp. 1033–1048.

52. McNally, Alleged Barwon-Darling Water Thieves to be Prosecuted after ABC Investigation, 2018. Available online: https://www.abc.net.au/news/2018-03-08/nsw-water-theft-barwon-darling-government-prosecuting/9527364 (accessed on 16 May 2019).

53. Ferguson, Cotton Grower Anthony Barlow Pleads Guilty to Illegal Pumping of Murray Darling Water, 2018. Available online: https://www.abc.net.au/news/2018-11-26/cotton-grower-pleads-guilty-to-mdb-illegal-pumping/10553990. (accessed on 12 May 2019).

54. WaterSource, NSW Ombudsman Makes a Move on Murray-Darling Basin Water Thieves, 2018. Available online: https://watersource.awa.asn.au/environment/natural-environment/nsw-ombudsman-makes-a-move-on-murray-darling-basin-water-thieves/ (accessed on 6 May 2019).

55. Irrigation Australia, 2019. Available online: https://www.irrigationaustralia.com.au/about-us/types-of-irrigation/types-of-irrigation (accessed on 10 May 2019).

56. Suh, S.; Lenzen, M.; Treloar, G.J.; Hondo, H.; Horvath, A.; Huppes, G.; Jolliet, O.; Klann, U.; Krewitt, W.; Moriguchi, Y.; et al. *System Boundary Selection in Life-cycle Inventories Using Hybrid Approaches*; Environmental Science & Technology: Bethesda, MD, USA, 2004; pp. 657–664.

57. Weber, L.; Matthews, H.S. Quantifying the Global and Distributional Aspects of American Household Carbon Footprint. *Ecol. Econ.* **2008**, *66*, 379–391. [CrossRef]

58. WFN, Water Footprint: Introduction, 2008. Available online: https://www.waterfootprint.org/en/ (accessed on 21 February 2020).

59. Water Footprint Network, Aims & History, 2018. Available online: http://waterfootprint.org/en/about-us/aims-history/ (accessed on 15 August 2018).

60. Feng, K.; Chapagain, A.; Suh, S.; Pfister, S.; Hubacek, K. *A Comparison of Bottom-Up and Top-Down Approaches to Calculating the Water Footprints of Nations*; Routledge-Taylor and Francis Group: London, UK, 2011; Volume 23, pp. 371–385.

61. Minx, J.C.; Weber, C.L.; Edenhofer, O. *Growth in Emission Transfers International Trade from 1990 to 2008*; National Center for Biotechnology Information: Bethesda, MD, USA, 2011; Volume 108, p. 21.

62. FAO. FAOSTAT, 05 May 2019. Available online: http://www.fao.org/faostat/en/#data/TP (accessed on 5 May 2019).

63. DFAT Australia's Trade Statistic at a Glance, 2018. Available online: https://dfat.gov.au/trade/resources/trade-at-a-glance/Pages/top-goods-services.aspx (accessed on 3 December 2019).

64. Anderson, Fifty Years of Australia's Trade, 12 2014. Available online: https://dfat.gov.au/about-us/publications/Documents/fifty-years-of-Australias-trade.pdf (accessed on 23 March 2019).

65. Ministry of Commerce Country, Ministry of Commerce Country Report Trade Data and Australian Office of Commerce, 2017. Available online: http://www.mofcom.gov.cn/article/tongjiziliao/fuwzn/ckt (accessed on 4 May 2019).

66. Ministry of Commerce Country Report Trade Data and Australian Office of Commerce, Overview of Australian Trade in Goods and Bilateral Trade between China and Australia. 2017. Available online: http://www.mofcom.gov.cn/article/tongjiziliao/fuwzn/ckt (accessed on 4 May 2019).

67. Jackson, T.Z.K. *Snapshot of Australian Agriculture*; ACT 2601; Australian Bureau of Agricultural and Resource Economics and Sciences: Canberra, Australia, 2018.
68. Reserve Bank of Australia, Chart Pack, Regions and Industry, 2019. Available online: https://www.rba.gov.au/chart-pack/regions-industry.html (accessed on 4 May 2019).
69. Sherald, M. Australian's the World's Wealthiest, 2011. Available online: https://www.smh.com.au/lifestyle/australians-the-worlds-wealthiest-20111101-1mt2r.html (accessed on 13 March 2019).
70. Australia Bureau of Statistics. *Balance of Payments and International Investment Position*; Australian Bureau of Statistics: Belconnen, Australia, 2014.

applied sciences

Article

Use of Digital Technologies for Intensifying Knowledge Sharing

Katarína Stachová [1], Zdenko Stacho [1,*] , Dagmar Cagáňová [2] and Augustín Stareček [2]

1 Institut of Civil Society, University of SS. Cyril and Methodius in Trnava, Bučianska 4/A, 917 01 Trnava, Slovakia; katarina.stachova@ucm.sk

2 Institute of Industrial Engineering and Management, Faculty of Materials Science and Technology in Trnava, Slovak University of Technology in Bratislava, J. Bottu 25, 917 01 Trnava, Slovakia; dagmar.caganova@stuba.sk (D.C.); augustin.starecek@stuba.sk (A.S.)

* Correspondence: zdenko.stacho@ucm.sk; Tel.: +421-907-082-448

Received: 4 May 2020; Accepted: 20 June 2020; Published: 22 June 2020

Abstract: The operation of companies in the current environment, introducing the concept of Industry 4.0 and the establishment and expansion of inter-company networks, as well as open and knowledge-based systems in organizations, is a precondition for success in the efficient acquisition, processing, storage, and sharing of key information. The instruments created for this purpose came to the market gradually during the Third Industrial Revolution; however, with the outbreak of the Fourth Industrial Revolution, their development underwent a leap, whereas this stage is associated with the introduction and use of big data in human resources management. The aim of this paper is to analyze the current state of organizations operating with a focus on Slovakia, as well as the importance and knowledge value of the organizations, their way of internal sharing, and the level of knowledge database use for this purpose in the context of the region in which the organizations operate. On one hand, the results show a lack of businesses orientation with regard to this issue, especially in comparison with neighboring countries; however, on the other hand, they also show relatively significant progress in this area, which may increase competitiveness internationally in the near future.

Keywords: digitization; knowledge sharing; human resources management; competitiveness; Industry 4.0

---

1. Introduction

The current business environment determined by globalization and the advent of the Industry 4.0 concept [1–5], where key changes are mainly related to systems, inter-machine relationships, and machine–people relationships [6–10], requires a high degree of focus on the development of disposable human resources and their competences [11–13] in an effort to ensure competitiveness.

Industry 4.0 systems are primarily based on new supporting technologies such as adaptive robotics [14], cybernetic physical systems (CPS) [15], cloud technologies [16], virtualization technologies, etc. [17]. The importance of the integrated enterprise concept in terms of competitiveness and sustainability is increasing in connection with the implementation of the Industry 4.0 concept and the emergence of both inter-company networks and open and knowledge-based systems in organizations [18–21], where the effective acquisition, processing, storage, and sharing of key information is a predictor of success. The creation of such a knowledge database is then the basis for a quick search and obtaining of variables. However, we must be careful when entering the data, and confidentiality must be taken into account before any addition to the database.

At the time of digitization, the potential of big data analysis [22–26] is a topic of expert discussions of the human resources community, because organizations have a wealth of information at their disposal, including employee demographic data, recruitment data, key performance indicator (KPI) metrics, etc. Based on a knowledge database, human resource professionals can make more objective recruitment decisions, reduce adverse effects from employee performance, support employees with a higher likelihood of loyalty to the company, or effectively train employees in line with current needs.

In an effort to meet the expectations of customers and employees, businesses are deploying intelligent features into all internal systems to create the necessary flexibility and capacity. Simple and monotonous processes are automated, while other processes become more complex and interrelated. As a result, the requirements for the competence profile of available human capital are changing. In connection with the advent of Industry 4.0, changes in the requirements for both professional and personal characteristics necessary to perform work tasks can, thus, be expected [27]. Surveys carried out to this effect indicated a priority focus on areas for *personal competencies* (ability to act autonomously on specific incentives), *social/interpersonal competences* (ability to communicate and collaborate with other employees and groups), *competencies necessary for immediate realization of ideas* (ability to translate ideas into practice), *professional competencies* (ability to find and use specific knowledge at work, process understanding, information technology (IT) security understanding) [28], *methodological competencies* (e.g., creativity, problem solving, analytical thinking), and *personality competences* (e.g., flexibility, motivation to learn, ability to work under pressure) [29]. The concept of so-called learning organization and the associated effective knowledge sharing seem to be an ideal approach to acquiring key competences of disposable human resources, while digitization itself and digital systems and applications provide optimal tools that can be used for this purpose.

The solution for remote storage and data availability is provided by the cloud computing system [30]. The role of knowledge workers in the use of IT tools in an effort to remove the boundaries between firmly defined organizational units for the benefit of the overall information flow is becoming increasingly important. In this context, digitization has a significant impact in alternative work regimes, especially when working from home with connections to business systems. There are needed programs (so-called BYOD—bring your own device), which involve creating unified electronic platforms for information sharing for distance employees, which are associated with increased demands for the security of internal data and systems [12]. Security, or more precisely IT risk, was identified by Záležáková [31] as one of the main areas of risk of Industry 4.0, highlighting the risk of security (data) costs, virus protection, protection of sensitive information and business secrets, encryption, firewall server protection, and automatic scanning.

However, data protection is not the only challenge or threat to remote data storage and availability. An important bottleneck in this context is the willingness of individuals to share knowledge and, thus, create an information base for the subsequent use of big data in practice. Organizations are now realizing the fact that knowledge sharing is important because it provides a link between individuals and organizations [32]. It is the transfer of the knowledge that individuals have that transform to the organizational level into economic and competitive value for organizations [33]. Organizations recognize the need to focus on motivating individuals to share knowledge, as these are confidential and inextricably linked to human egos and commitment and do not flow easily through the organization [34,35]. However, organizations often make the mistake of setting an environment opposite to the goal of information sharing, where employees are rewarded for owning the information. This has a demotivating effect on an individual sharing data, as being the owner of knowledge has the consequence of gaining a reward, a shift in the organizational structure, job security, etc. [34]. If individuals realize that their strength in the organization derives from the knowledge they own, it is likely that they will accumulate knowledge instead of sharing knowledge [34,36,37]. According to Brown and Woodland [38], positive results in the context of willingness to share knowledge were reported in several surveys, pointing to the need for awareness of mutuality and reciprocity. Reciprocity or knowledge transfer can facilitate knowledge sharing if individuals see that their added value depends on the extent to which they share their own knowledge with others [33,39]. Reciprocity as a knowledge-sharing motivator means that individuals must be able to predict that knowledge-sharing will prove useful [40], although they are not exactly sure what the outcome will be [41]. The correlation between reciprocity and sharing of knowledge boundaries implies that receiving knowledge from others stimulates the flow of knowledge [40]. The relationship between knowledge sharing and incentives was further supported by studies [36,42], which found that there were significant changes in the stimulus system that encouraged individuals to share their knowledge, especially through technology. The forthcoming Fourth Industrial Revolution provides comprehensive systems and tools aimed at promoting information sharing at the individual, team, and organization levels, as well as at a cluster level [12,16,17,20,22,30]. For this reason, this paper discusses the current state of focus of organizations operating in Slovakia with regard to the importance and value of knowledge in the organization, their way of internal sharing, and the level of use of knowledge databases for this purpose.

The increase in disparities in individual regions of Slovakia represents an important barrier in the effort to build a competitive economy. Considering the increase in disparities found in other surveys in the analysis of the impact of employee involvement in innovation processes [11], awareness of the smart city concept [43], or employee motivation in creative behavior [44], in the context of individual regions, the authors of this paper found out whether similar results were published in other researches.

Other authors focusing on the issues of organizations operating in Slovakia in terms of the analysis of the principle of individualization in human resource management [12], employee motivation [37], human capital building and development [45], or the use of effective strategic tools for human resource development [46,47] confirmed increase in disparities. Based on the fact above, we analyzed individual research questions in the context of relation to the region in which the organizations operate.

2. Literature Review

Veber [48] defined knowledge as dynamic systems including interactions among experience, skills, facts, relationships, values, thinking processes, and importance. These systems are created by information together with experience, skills, intuition, personal imaginations, and mental models. Knowledge has two dimensions, explicit and tacit. According to Mládeková [49], explicit knowledge is expressed by formal and systematic language, via data. It is possible to verbalize, to write, or to draw it, while it can also be stored and transferred. The explicit dimension of knowledge is information. Tacit knowledge is created via the interaction of explicit knowledge with experience, skill, intuition, imaginations, mental models, etc. It is connected to activities, procedures, routines, ideas, values, or emotions of a certain person. Therefore, it is very difficult to express and share it. Its character is highly personal. The employee who is its proprietor does not have to know about it.

In essence, knowledge continuity management is based on communication. It is important for employees to know what it means to the organization when they know what the others need to know and what information should be shared; then, as a result of that knowledge, they transfer knowledge to others. While the idea itself is simple, setting up effective knowledge continuity management is a complex matter that requires technical, organizational, and managerial steps, as well as strong support from the top management. Management must understand the effort dedicated to knowledge transfer as an integral and common part of an organization's operations [50].

The employees who have the knowledge and experience should be considered as the experts in the organization, and company management should be aware of these employees. The organization should encourage its employees to transfer knowledge and experience and try to eliminate the unwillingness to share knowledge for fear of being replaced [51].

There is often knowledge loss in organizations; for example, when workers leave, there is a brain drain, which is an adverse effect. The preservation of knowledge can be achieved using strategic knowledge management and ensuring knowledge continuity. An organization's knowledge strategy determines whether the organization operates more with tacit or explicit knowledge. In predominant work with explicit knowledge, a codification strategy is used according to References [52,53]. At present, in the era of digitization intensification, organizations have a wide range of software or applications available that facilitate the knowledge-sharing process. Among examples of technical support, it is possible to include knowledge systems, knowledge databases, e-learning applications, and electronic data interchange (EDI) support systems [54], which allow archiving and increasing the knowledge of employees provided.

Within the predominant work with tacit knowledge, a personalization strategy, creativity of employees, individual approach to the product and the customer, support for knowledge sharing, and databases have a supporting role. Hansen [52] further added that organizations applying codification strategies achieve savings in labor and communication costs through the re-use of knowledge. It should be added that References [52,53] agree that effective organizations must focus on one of these strategies and use the other in a supporting role. It is not possible to use both approaches at the same level or to reject one of these approaches completely. While ensuring the continuity of knowledge, the personalization strategy prevails.

Based on the above-mentioned findings, we focused our research attention on identifying the interest of companies operating in Slovakia in knowledge sharing, as we consider this interest to be a basic assumption for effective knowledge continuity management. On the other hand, we focused on the level of knowledge database use in order to identify the readiness of enterprises to implement either personalization or codification knowledge strategies.

3. Materials and Methods

We used several research and statistical methods to achieve the goal of this paper. We primarily used the analysis of current scientific research publications and confronted individual expert opinions on the issue with their own conclusions.

In order to obtain and process research data, we mainly used questionnaire surveys conducted in 2014–2018. The respondents of the survey were employees primarily responsible for human resources management and development in companies operating in Slovakia. A questionnaire with 90 questions related to human resources management was used as a tool to examine the current state of implementation of digital technologies in order to increase knowledge sharing in the enterprise. However, for the purpose of this paper, we only processed answers to questions related to knowledge sharing in the company within the context of new IT tools used. The number of respondents oscillated around 750 each year, with the rate of correctly completed questionnaires ranging from 65% to 75%. Before the actual phase of data acquisition, we set two stratification criteria based on which we subsequently specified a basic set of potential respondents. The first criterion was the region of the enterprise operation within Slovakia under the Nomenclature of Territorial Units for Statistics (NUTS) system. Specifically, Slovakia was divided into the NUTS 2 category (Bratislava region, western Slovakia, central Slovakia, eastern Slovakia), whereas the subsequent structural composition of the research sample was based on data from the Statistical Office (ŠÚ) of the Slovak Republic (SR). The second stratification criterion was the number of employees in the enterprises under investigation. The lower limit was set at 50 employees. The main purpose of establishing this stratification criterion was to primarily address to medium-sized enterprises, where we assumed, on the one hand, the existence of a formalized department and agenda for the area of human resources management and, on the other hand, given the number of employees, we predicted an increase in problems in the area of effective knowledge sharing for all employees. Based on data from the Statistical Office (ŠÚ) of SR for the period under review, it can be stated that the number of enterprises with 50 or more employees in individual regions oscillated around the same value. The specific regional structure of enterprises with more than 50 employees in the observed years is shown in Table 1.

Table 1. Regional structure of enterprises with over 50 employees. NUTS—Nomenclature of Territorial Units for Statistics.

Region (NUTS II)	Bratislava Region	Western Slovakia	Central Slovakia	Eastern Slovakia
Regions	BA	TT, TN, NR	BB, ZA	KE, PO
Number of enterprises in 2013	1074	895	639	603
Number of enterprises in 2014	1098	904	644	612
Number of enterprises in 2015	1105	916	651	613
Number of enterprises in 2016	1114	923	649	621
Number of enterprises in 2017	1123	926	654	623

Source: own elaboration based on the Statistical Office (ŠÚ) of the Slovak Republic (SR) [55]. Note: BA—Bratislava Region, TT—Trnava region, TN—Trenčín Region, NR—Nitra Region, BB—Banská Bystrica Region, ZA—Žilina Region, KE—Košice Region, PO—Prešov Region.

The determination of the optimal research sample size from the population set out in Table 1 was performed at the 95% confidence level and confidence interval H = ±0.10. The size structure of a sufficient research sample, as well as the real size of the research sample in the analyzed years for individual regions of Slovakia, is shown in the Table 2.

Table 2. Determination of research sample for individual Slovak regions.

Region (NUTS II)	Bratislava Region	West Slovakia	Central Slovakia	East Slovakia
Regions	BA	TT, TN, NR	BB, ZA	KE, PO
Sufficient size of research sample	88	87	84	83
Real size of research sample 2013–2017 (n)	151	132	126	123

Source: own elaboration.

Companies from all sectors of economy were represented in the research each year (Table 3), whereas the research results did not show any significant differences at cross-sector comparison. For this reason, the research results were evaluated cumulatively, i.e., regardless of the sectors that companies operate in.

Table 3. Percentage share of companies operating in individual sectors.

Sector	Share of Companies as a %				
	2013	2014	2015	2016	2017
Manufacturing	40.2	37.9	40.1	37.6	39.1
Agriculture, forestry, and fishery	8.2	9.0	7.3	10.1	8.0
Power industry and water management	4.4	3.4	4.1	2.8	4.2
Services	34.0	36.7	37.2	38.1	38.3
Other	13.2	14.0	11.3	11.4	10.4

Source: own research.

In order to evaluate the obtained data, we firstly used the methods of descriptive statistics, where we found the percentage of individual analyzed indicators. Subsequently, we statistically processed and evaluated the observed values through the analysis of basic indices (changes in values from the first year) and chain indices (changes in values monitored from the previous year) in order to identify the development trends of individual indicators over time. Finally, we verified the established hypotheses by means of a chi-square comparative analysis.

In accordance with the objectives of the paper, we formulated two research questions, on the basis of which we set up two hypotheses.

- RQ1: To what extent is knowledge shared in organizations operating in Slovakia?
- RQ2: To what extent are knowledge databases used in the conditions of Slovak organizations?
- H1: There is a statistically significant difference in the level of knowledge sharing in Slovak enterprises based on the geographical location of the organization.

- H2: There is a significant difference in the use of knowledge databases based on the geographical operation of the organization.

4. Results

In order to answer the research questions, we focused primarily on identifying the level of knowledge sharing in companies operating in Slovakia in the period 2013 to 2017, and we found that, in most companies, knowledge is shared, but more than half of the respondents stated that they only shared knowledge necessary to work (Table 4).

Table 4. Level of knowledge sharing in enterprises.

Level of Knowledge Sharing in Enterprises	Share of Enterprises as a %				
	2013	2014	2015	2016	2017
Completely shared	7.2	14.4	14.1	15.9	17.7
Almost completely shared	31.8	33.3	33.1	31.1	33.0
Only the knowledge that is essential to work is shared	59.8	51.6	50.4	50.5	49.8
Not shared, knowledge is a means of securing power	0.2	1.4	1.4	1.7	2.4
Not shared, knowledge is a means of securing a monopoly from fear of losing a job	1.0	0.3	1.0	0.8	1.1

Source: own elaboration.

The above findings are positive, on one hand, mainly because most respondents gave a positive response; however, on the other hand, the fact that about half of the respondents were positive only in terms of sharing knowledge needed to work is very disturbing. This result indicates a lack of readiness in companies operating within Slovakia to use disposable human resources and their potential to the maximum extent possible, which ultimately reduces their overall competitiveness vis-à-vis companies that use this potential more effectively.

We also focused on the analysis dealing with the basic index of changes in the monitored period in the area of knowledge sharing in companies (Table 5).

Table 5. Basic index of changes—rate of knowledge sharing in enterprises.

Basic Index of Changes—Rate of Knowledge Sharing in Enterprises	bi14/13	bi15/13	bi16/13	bi17/13
Completely shared	2.000	1.958	2.208	2.458
Almost completely shared	1.047	1.041	0.984	1.038
Only the knowledge that is essential to work is shared	0.868	0.842	0.845	0.833
Not shared, knowledge is a means of securing power	7.000	7.000	8.500	12.000
Not shared, knowledge is a means of securing a monopoly from fear of losing a job	0.300	0.000	0.800	1.100

Source: own elaboration.

Based on the above results, we observed two trends. On one hand, there is a positive trend in the increasing number of enterprises in which knowledge among employees is fully shared at the expense of a decreasing number of enterprises in which only the knowledge needed to work is shared. On the other hand, there was a negative trend, especially in the area of information use retention with the intention to keep working in a decision-making position.

Most survey respondents in the question aimed at identifying assessment rates and forms of knowledge remuneration stated that the amount and quality of information sharing is not monitored or remunerated in any way (Table 6).

Table 6. Form of evaluation and remuneration of knowledge.

In What Form Is Knowledge Evaluated and Remunerated?	Share of Enterprises as a %				
	2013	2014	2015	2016	2017
Knowledge sharing is not tracked or remunerated	62.6	63.7	61.2	62.4	58.7
The amount and quality of shared knowledge for which the employee is remunerated in cash is closely monitored	28.2	25.0	25.6	27.4	28.8
The amount and quality of shared knowledge for which the employee is remunerated in non-cash is closely monitored	8.1	9.1	10.9	8.2	10.6
Otherwise	1.1	2.2	2.3	2.0	1.9

Source: own elaboration.

The above results show a very low level of interest and initiatives of companies operating in Slovakia in the area of employee motivation to share knowledge, as only about 40% of respondents paid attention to some extent to this issue in the monitored period.

We also placed emphasis on the analysis dealing with the basic index of changes in the period under review in the field of assessment rate and forms of knowledge remuneration (Table 7).

Table 7. Basic index of changes—form of knowledge evaluation and remuneration.

Basic Index of Changes—Form of Knowledge Evaluation and Remuneration	bi14/13	bi15/13	bi16/13	bi17/13
Knowledge sharing is not remunerated	1.018	0.977	0.997	0.938
The amount and quality of shared knowledge for which the employee is remunerated in cash is closely monitored	0.887	0.908	0.972	1.021
The amount and quality of shared knowledge for which the employee is remunerated in non-cash is closely monitored	1.123	1.346	1.012	1.309
Otherwise	2.000	2.091	1.181	1.727

Source: own elaboration.

We observed only a minimal positive increase in all the monitored attributes and, therefore, we observed that their state remained almost unchanged during the survey period.

In order to identify the degree of IT technology use in knowledge sharing, we focused our attention in the survey on the analysis dealing with the basic index of changes in the monitored period in the field of the use of knowledge databases, with respect to information systems for collecting and sharing knowledge in the enterprise (Table 8). As knowledge databases represent only a tool to support knowledge continuity management, whereas the level of their use depends on specific parameters such as the amount and form of information shared or the knowledge sharing strategy, we simplified the respondents' responses to yes/no.

Table 8. Basic index of changes—use of knowledge databases (information systems for collecting and sharing knowledge in an organization.

Use of Knowledge Databases (Information Systems for Collecting and Sharing Knowledge in an Organization)				
2013	2014	2015	2016	2017
22.1	22.7	24.5	25.9	31.3

Basic index of changes—use of knowledge databases (information systems for collecting and sharing knowledge in an organization			
bi14/13	bi15/13	bi16/13	bi17/13
1.026	1.111	1.179	1.427

Source: own elaboration.

We observed a positive continuous increase in the monitored attribute and, therefore, we can state that its condition improved during the period of the survey; we expect a similar trend in the upcoming period.

In addition to the overall current status of the analyzed attributes or their development in the period under review in companies operating in Slovakia, we investigated whether there is a relationship between knowledge sharing and the region of business activity, where disparities in various other areas of business management were relatively significant [56,57].

- H1: Based on the chi-square test (chi = 5.826, degrees of freedom (df) = 3, $p = 0.12$) (dichotomic variable = knowledge is not shared in the enterprise; chi categorical variable = regions), we can conclude that there is no statistically significant relationship between the variables. Based on this result, we rejected hypothesis H1 (existence of a statistically significant difference in the level of knowledge sharing in companies operating in Slovakia based on the geographical seat of the organization.

Accordingly, the statistical evaluation of dependencies under the H1 hypothesis showed that the rate of knowledge sharing is not related to the regional activity of the company. This is a positive fact, since it is possible to find a consistent approach across the country for businesses in this area. It can, therefore, be assumed that, in the case of the intensification of knowledge sharing, both within and between business entities, it can be assumed that such intensification will be equally reflected regardless of the region of activity.

- H2: Based on the chi-square test (chi = 10.675, df = 3, $p = 0.014$) (dichotomic variable = the enterprise does/does not shared knowledge databases; chi categorical variable = regions), we can state that there is a significant relationship. Based on this result, we confirmed hypothesis H2 (existence of a significant difference in the use of knowledge databases based on the geographical activity of the organization).

Accordingly, the statistical evaluation of dependencies under the H2 hypothesis showed that the rate of use of shared knowledge databases is related to the regional activity of the company. When comparing the Bratislava region with the East Slovak region, this difference was as high as 25%. These results indicate a relatively significant difference in the approach of enterprises operating in Slovakia to the implementation of new technologies in order to increase knowledge sharing across the enterprise. As a result of such an approach, disparities between businesses operating in individual regions may increase in the future, thus again affecting the level of development and economic productivity of these regions. As a result of such an approach, in the future, there may be an increase in disparities between companies operating in individual regions, which will again negatively affect the level of development and economic productivity of individual regions of Slovakia, as less developed regions (Eastern Slovakia) progress more slowly than those more developed (Bratislava region).

5. Discussion

The Fourth Industrial Revolution is characterized by the onset of radical changes associated with the advent of new technologies that enable the dissemination and sharing of innovation much faster than ever before [58,59]. At a time when the need for efficient data sharing is on the rise, the goal is to achieve a mostly potential positive setting on Tidd's five-tier scale of stages of building an innovative enterprise with high employee involvement. Reaching the top, this scale will ensure that every employee is fully involved in experimenting and improving things, in sharing information, and in creating an active learning organization [60]. Thus, the innovations that currently strongly determine the competitiveness of businesses are stimulated by free communication, decentralization, and trust between different hierarchical levels and, thus, the ability to communicate ideas and share knowledge across all business structures. Unfortunately, a survey based on an interview with the executives of organizations operating in Slovakia revealed that the term knowledge sharing is understood by most employees to only transmit explicit information, while this topic is not addressed in depth from the perspective of transferring practical or tacit knowledge [61]. The survey showed a low level of interest and initiatives of companies operating in Slovakia in the field of motivating employees to share knowledge (only about 40% of participating respondents), which, in the context of the Bencsik survey [61], is negative for companies, as it can be assumed that if companies do not focus on supporting and motivating employees to share knowledge, they also do not focus on monitoring the content of shared knowledge.

The survey of this paper showed that, in 2013–2017, a positive trend of full information sharing began in enterprises, whereby the percentage of positive respondents increased from 7% to almost 18% (Table 3), which suggests a positive trend in the future. This may be reflected in the future in increasing the country's competitiveness, as the degree of knowledge sharing is also one of the factors influencing the ability to succeed. This trend is also confirmed by the positive progress of Slovakia in the Global Competitiveness Index, which analyzes the pillars of the institution, infrastructure, macroeconomic environment, health and basic education, higher education and training, commodity market efficiency, labor market efficiency, financial market maturity, technological readiness, market size, business sophistication, and innovation [62]. Based on the World Economic Forum 2017, the Slovak Republic ranked 65 out of 138 countries evaluated (Table 9) [63].

Table 9. Global Competitiveness Index [63].

Global Competitiveness Index	2012–2013	2015–2016	2016–2017
Slovakia ranked	71/144	67/140	65/138

Since the aim of the Global Competitiveness Index is to highlight what is important for the long-term growth of a country's competitiveness, the World Economic Forum reviewed the content of analyzed pillars in 2018 (institutions, infrastructure, information and communication technology (ICT) adoption, macroeconomic stability, health, skills, commodity market efficiency, labor market efficiency, financial market maturity, market size, business dynamics, ability to innovate), where Slovakia ranked 42nd in 2019 [64]. A significant shortage compared to the other analyzed countries in Slovakia is, for example, in the pillar of skills, namely, in the possibility of getting a qualified employee on the labor market (up to 127th place) [64]. This is also confirmed by the KPMG study, where Slovakia received the lowest human development index of only 3.88 out of 10, a significantly below average score (all other V4 countries gained 6.6 points or more). This implies that companies often have to settle for less qualified employees who need additional internal or external training [65]. In the context of the above, there is a need to store and share practical and tacit knowledge, which is an important variable, especially in the training of new employees during adaptation. However, the presented survey pointed out shortcomings in the current state of knowledge sharing in the analyzed organizations. On the other hand, when analyzing the level of digital skills, Slovakia is ranked 48th [64]. It follows that companies operating in Slovakia have the potential to use the incoming opportunities of the Industrial Revolution 4.0 in order to acquire and preserve skills (knowledge and information) in the company to be able to use them in the development of internal and newly acquired external employees, whereas they are currently unable to rely on the recruitment of sufficiently qualified staff from external sources. The survey presented herein pointed to significant differences in the use of knowledge databases and the geographical operation of the organization, thus confirming the existence of disparities in Slovakia, as Slovakia is one of the Organization for Economic Co-operation and Development (OECD) countries with the largest regional differences in income and unemployment. Significant diversity within the regions of Slovakia also exists in economic growth and gross domestic product (GDP) per capita, as well as in the educational level [66]. It follows from the above that it is desirable to focus on reducing the disparities of Slovakia's regions to support the use of knowledge databases in companies operating outside the Bratislava region, as this is one of the important variables in increasing the company's competitiveness, and a high percentage of employees have digital skills [64].

Several surveys conducted in Slovakia showed that businesses most often use information technologies such as the internet environment, e-mail, databases, and intranet to disseminate knowledge [67]. The above-mentioned results align with our research, which also showed an increase in the use of knowledge databases or, more precisely, information systems for collecting and sharing knowledge in the company in 2013–2017 from 22% to more than 30% (Table 7). For the effective functioning of knowledge management, it is important to harmonize its elements into one unit. Therefore, it is necessary to create a reliable common technology infrastructure that will allow easy capture of knowledge validation, associated with the creation of a knowledge base and the subsequent effective sharing and extraction of knowledge [68]. For this purpose, companies are using or will have to start using big data processing tools. It is the use of big data processing tools that enhances the measurability of soft indicators needed for the analysis of human resources management and development. There are currently several big data analysis concepts on the market, such as Data Intelligence, Python, and others. These are predictive analytics tools that companies use to shape employee experience in relation to key business performance indicators such as customer satisfaction (net promoter score evaluated by customers), employee satisfaction (net promoter score evaluated by employees), financial performance indicators regarding share on the market [69], the probability of employees leaving, and others.

The new reality of competition will fundamentally require new ways of thinking about procedures in human resources management, changes in human resources, and changes in the mindset of people management specialists [70]. Based on expert estimates, four significant changes can be expected in the area of human resources [71]. These relate to the reification of human resources, which implies that tasks in the area of human resources management will be increasingly transferred to intelligent smart things, such as intelligent industrial manufacturing tools that can perform time and motion studies and autonomously collect and distribute data associated with normal hours and breaks, but which can also autonomously introduce employees to their functions and train them in proper use [44]. Furthermore, it is a human resources approach that uses sensors in intelligent matters to identify various data relevant to human resource management, eliminating manual labor-intensive, expensive, and error-prone data recording used so far. Changes will also occur in the human resources management datafication, given that human resources management databases will grow exponentially and the data will describe numerous human resources issues in a comprehensive and detailed way [71]. The fourth change that technological developments in human resources management will bring is the technical integration of human resources management, which will coordinate human resources management measures with operational business requirements, which will be implemented automatically [44]. The vast majority of experts expect the direct technical integration of human resources (HR) management, i.e., the direct interaction of HR software with sensors and actuators in intelligent things used by employees, e.g., to directly provide adequate training for the situation, when the demand for training occurs [71].

6. Conclusions

The human resources department, which was previously based on soft information and tools, is gaining a whole new perspective with the advent of the Fourth Industrial Revolution. Currently, these departments, based on available and emerging new technologies, will have a wealth of information including employee demographics, recruitment and KPI performance data, the level of knowledge and skills at work, the quality and quantity of shared information, employee loyalty, and so on. Based on these data, human resources experts can make decisions based not only on soft, but also on hard data. With the advent of the Fourth Industrial Revolution, the human resources management area gained new tools for collecting and analyzing information, enabling data-based decision-making in all dimensions of the human resources area. The basic variable in the use of new opportunities is the involvement of employees dedicated to human resources management in organizations, as well as technical support and the competence of employees to work with new tools, to select and share appropriate information through appropriate communication channels, and to use them to streamline employees. The survey presented herein focused on individual manifestations of information digitization, to which the company responds via technological innovations toward customers and employees, such as digitization of analog and biometric data, digital interaction platforms, networking, big data analytics, rapid analytics, predictive analytics, and the use of social networks when searching and selecting employees. In this survey, the current and expected future state was analyzed, which should predict the level of awareness of existence and, accordingly, the need to acquire the competence of employees working in human resources management departments to work with new tools that can be used for both effective storage and sharing of necessary knowledge. The 2013–2017 survey revealed an insufficient focus of organizations on both knowledge sharing and the use of knowledge databases in organizations operating in Slovakia. For this reason, we feel that there is a need to focus on a more specific analysis, which was, therefore, launched in January 2020 and data collection is currently underway. The results of this survey will be published, where we will try to define the development trend in companies.

Author Contributions: Conceptualization, K.S. and Z.S.; methodology, A.S.; software, A.S.; validation, A.S.; formal analysis, Z.S.; investigation, Z.S.; resources, D.C.; data curation, K.S., A.S., and Z.S.; writing—original draft preparation, K.S., A.S., and Z.S.; writing—review and editing, D.C. and Z.S.; visualization, Z.S.; supervision, D.C.; project administration, K.S. and D.C.; funding acquisition, D.C. All authors read and agreed to the published version of the manuscript.

Funding: This research was funded by the VEGA 2/0077/19 project titled "Work competencies in the context of Industry 4.0".

Acknowledgments: This research was supported by project VEGA 1/0412/19 Systems of Human Resources Management in the 4.0 Industry Era, with additional support of the project APVV-17-0656 "Transformation of Paradigm in Management of Organizations in the Context of Industry 4.0", and with support of the VEGA 2/0077/19 project titled "Work competencies in the context of Industry 4.0".

Conflicts of Interest: The authors declare no conflicts of interest.

References

1. Xu, Y.; Wang, Y.; Tao, X.; Lizbetinova, L. Evidence of Chinese income dynamics and its effects on income scaling law. *Phys. A Stat. Mech. Appl.* **2017**, *487*, 143–152. [CrossRef]

2. Gazova, A.; Papulova, Z.; Papula, J. The application of concepts and methods based on process approach to increase business process efficiency. *Procedia Econ. Financ.* **2016**, *39*, 197–205. [CrossRef]

3. Kucharčíková, A.; Tokarčíková, E.; Blašková, M. Human Capital Management—Aspect of the Human Capital Efficiency in University Education. *Procedia Soc. Behav. Sci.* **2015**, *177*, 48–60. Available online: http://www.sciencedirect.com/science/article/pii/S1877042815016869# (accessed on 25 November 2019).

4. Babelova, Z.G.; Starecek, A.; Caganova, D.; Fero, M.; Cambal, M. Perceived Serviceability of Outplacement Programs as a Part of Sustainable Human Resource Management. *Sustainability* **2019**, *11*. [CrossRef]

5. Lorincová, S. The Improvement of the Effectiveness in the Recruitment Process in the Slovak Public Administration. *Procedia Econ. Financ.* **2016**, *34*, 382–389. [CrossRef]

6. Chen, Y.R.R.; Schulz, P.J. The effect of information communication technology interventions on reducing social isolation in the elderly: A systematic review. *J. Med. Internet Res.* **2016**, *18*. [CrossRef] [PubMed]

7. Ustundan, A.; Cevikcan, E. *Industry 4.0: Managing The Digital Transformation*; Springer International Publishing: Cham, Switzerland, 2018. [CrossRef]

8. Hudakova, M.; Urbancova, H.; Vnouckova, L. Key Criteria and Competences Defining the Sustainability of Start-Up Teams and Projects in the Incubation and Acceleration Phase. *Sustainability* **2019**, *11*. [CrossRef]

9. Taylor, M.P.; Boxall, P.; Chen, J.J.; Xu, X.; Liew, A.; Adeniji, A. Operator 4.0 or Maker 1.0? Exploring the implications of Industrie 4.0 for innovation, safety and quality of work in small economies and enterprises. *Comput. Ind. Eng.* **2018**, *139*, 105486. [CrossRef]

10. Hitka, M.; Lorincova, S.; Bartakova, G.P.; Lizbetinova, L.; Starchon, P.; Li, C.; Zaborova, E.; Markova, T.; Schmidtova, J.; Mura, L. Strategic Tool of Human Resource Management for Operation of SMEs in the Wood-processing Industry. *BioResources* **2018**, *13*, 2759–2774. [CrossRef]

11. Stacho, Z.; Stachova, K.; Caganova, D. Participation of all Employee Categories in Innovation Processes in Slovak Organisations. *Mob. Netw. Appl.* **2020**, 1–7. [CrossRef]

12. Blstakova, J.; Joniakova, Z.; Skorkova, Z.; Némethova, I.; Bednar, R. Causes and Implications of the Applications of the Individualisation Principle in Human Resources Management. *AD ALTA J. Interdiscip. Res.* **2019**, *9*, 323–327.

13. Pavlendova, G.; Sujanova, J.; Caganova, D.; Novakova, R. Smart Cities Concept for Historical Buildings. *Mob. Netw. Appl.* **2020**, *25*. [CrossRef]
14. Wittenberg, C. Cause the Trend Industry 4.0 in the Automated Industry to New Requirements on User Interfaces. In *International Conference on Human-Computer Interaction*; Kurosu, M., Ed.; Springer: Cham, Switzerland, 2015. [CrossRef]
15. Bagheri, B.; Yang, S.; Kao, H.A.; Lee, J. Cyber-physical systems architecture for self-aware machines in Industry 4.0 environment. *IFAC-Pap Online* **2015**, *48*, 1622–1627. [CrossRef]
16. Thames, J.L.; Schaefer, D. Cybersecurity for Industry 4.0 and advanced manufacturing enivronments with esemble intelligence. In *Cybersecurity for Industry 4.0*; Thames, J.L., Schaefer, D., Eds.; Springer: Berlin, Germany, 2017; pp. 1–33. [CrossRef]
17. Paelke, V. Augmented reality in the smart factory supporting workers in an Industry 40. Environment. In Proceedings of the 2014 IEEE Emerging Technology and Factory Automation (ETFA), Barcelona, Spain, 16–19 September 2014; pp. 1–4. [CrossRef]
18. Wilkesmann, M.; Wilkesmann, U. Industry 4.0—organizing routines or innovations? *Vine J. Inf. Knowl. Manag. Syst.* **2018**, *48*, 238–254. [CrossRef]
19. Sivathanu, B.; Pillai, R. Smart HR 4.0—How industry 4.0 is disrupting HR. *Hum. Resour. Manag. Int. Dig.* **2018**, *26*, 7–11. [CrossRef]
20. Martinez, M.G.; Zouaghi, F.; Garcia, M.S. Capturing value from alliance portfolio diversity: The mediating role of R&D human capital in high and low tech industries. *Technovation* **2017**, *59*, 55–67. [CrossRef]
21. Lizbetinova, L.; Lorincova, S.; Caha, Z. The Application of the Organizational Culture Assessment Instrument (OCAI) to Logistics Enterprises. *Nase More* **2016**, *63*, 170–176. [CrossRef]
22. Bencsik, A.; Juhász, T.; Mura, L.; Csanádi, A. Formal and Informal Knowledge Sharing in Organisations from Slovakia and Hungary. *Entrep. Bus. Econ. Rev.* **2019**, *7*, 25–42. [CrossRef]
23. Remisova, A.; Lasakova, A.; Kirchmayer, Z. Influence of Formal Ethics Program Components on Managerial Ethical Behavior. *J. Bus. Ethics* **2019**, *160*, 151–166. [CrossRef]
24. Jenco, M.; Droppa, M.; Lysa, L.; Križo, P. Assessing the Quality of Employees in Terms of their Resistance. *Calitatea* **2018**, *19*, 48–53.
25. Jankelová, N.; Joniaková, Z.; Blštáková, J.; Némethová, I. Readiness of human resource departments of agricultural enterprises for implementation of the new roles of human resource professionals. *Agric. Econ.* **2017**, *63*, 461–470. [CrossRef]
26. Spisakova, E.D.; Mura, L.; Gontkovicova, B.; Hajduova, Z. R&D in the Context of Europe 2020 in Selected Countries. *Econ. Comput. Econ. Cybern. Stud. Res.* **2017**, *51*, 243–261.
27. Grzybowska, K.; Łupicka, A. Key competencies for Industry 4.0. *Econ. Manag. Innov.* **2017**, *1*, 250–253.
28. Agolla, J.E. Human Capital in the Smart Manufacturing and Industry 4.0 Revolution. *Digit. Transform. Smart Manuf.* **2018**, 41–58. [CrossRef]
29. Hecklau, F.; Galeitzke, M.; Flachs, S.; Kohl, H. Holistic approach for human resource management in Industry 4.0. *Procedia CIRP* **2016**, *54*, 1–6. [CrossRef]
30. Gilchrist, A. *Industry 4.0: The Industrial Internet of Things*; Apress: Berkeley, CA, USA, 2016. [CrossRef]
31. Zalezakova, E. Nástup Industry 4.0. Available online: http://files.sam-km.sk/200000375-6837469375/%20Zalezakova_Nastup_industry_4.0.pdf (accessed on 12 December 2018).
32. Ipe, M. Knowledge Sharing in Organizations: A Conceptual Framework. *Hum. Resour. Dev. Rev.* **2003**, *2*, 337–359. [CrossRef]
33. Hendriks, P. Why share knowledge? The influence of ICT on the motivation for knowledgesharing. *Knowl. Process Manag.* **1999**, *6*, 91–100. [CrossRef]
34. Davenport, T.H.; De Long, D.W.; Beers, M.C. Successful Knowledge Management Projects. *Sloan Manag. Rev.* **1998**, *39*, 43–57.
35. Hitka, M.; Závadská, Z.; Jelačić, D.; Balážová, Ž. Qualitative indicators of company employee satisfaction and their development in a particular period of time. *Drv. Ind. Znan. Časopis Za Pitanja Drv. Tehnol.* **2015**, *66*, 235–239. [CrossRef]
36. Gupta, A.K.; Govindarajan, V. Knowledge management's social dimension: Lessons from Nucor Steel. *Mit Sloan Manag. Rev.* **2000**, *42*, 71.

37. Lizbetinova, L.; Hitka, M.; Li, C.; Caha, Z. Motivation of employees of transport and logistics companies in the Czech Republic and in a Selected Region of the PRC. In *MATEC Web of Conferences*; EDP Sciences: Les Ulis, France, 2017. [CrossRef]

38. Brown, R.B.; Woodland, M.J. Managing knowledge wisely: A case study in organiza-tional behavior. *J. Appl. Manag. Stud.* **1999**, *6*, 175–198.

39. Weiss, L. Collection and connection: The anatomy of knowledge sharing in professional ser-vice. *Organ. Dev. J.* **1999**, *17*, 61–72. [CrossRef]

40. Schultz, M. The uncertain relevance of newness: Organizational learning and knowledgeflows. *Acad. Manag. J.* **2001**, *44*, 661–681. [CrossRef]

41. Nahapiet, J.; Ghoshal, S. Social capital, intellectual capital and the organizational advan-tage. *Acad. Manag. Rev.* **1998**, *23*, 242–266. [CrossRef]

42. Quinn, J.B.; Anderson, P.; Finkelstein, S. Leveraging intellect. *Acad. Manag. Exec.* **1996**, *10*, 7–27. [CrossRef]

43. Caganova, D.; Starecek, A.; Hornakova, N.; Hlasnikova, P. The Analysis of the Slovak Citizens' Awareness about the Smart City Concept. *Mob. Netw. Appl.* **2019**, *24*, 2050–2058. [CrossRef]

44. Stachova, K.; Stacho, Z.; Blstakova, J.; Hlatká, M.; Kapustina, L.M. Motivation of Employees for Creativity as a Form of Support to Manage Innovation Processes in Transportation-Logistics Companies. *Naše More* **2018**, *65*, 180–186. [CrossRef]

45. Kucharčíková, A.; Mičiak, M. Human capital management in transport enterprise. In Proceedings of the MATEC Web of Conferences: LOGI 2017 18th International Scientific Conference, České Budějovice, Czech Republic, 19 October 2017. [CrossRef]

46. Lorincová, S.; Hitka, M.; Štarchoň, P.; Stachová, K. Strategic instrument for sustainability of human resource management in small and medium-sized enterprises using management data. *Sustainability* **2018**, *10*, 3687. [CrossRef]

47. Hitka, M.; Lorincová, S.; Ližbetinová, L.; Bartáková, G.P.; Merková, M. Cluster analysis used as the strategic advantage of human resource management in small and medium-sized enterprises in the wood-processing industry. *BioResources* **2017**, *12*, 7884–7897. [CrossRef]

48. Veber, J.; Kol, A. *Management, Základy, Prosperita, Globalizace*; Management Press: Prague, Czech Republic, 2000.

49. Mládková, L. *Manegement Znalostních Pracovníků*; C.H. Beck: Prague, Czech Republic, 2008.

50. Beazley, H.; Boenisch, J.; Harden, D. *Continuity Management: Preserving Corporate Knowledge and Productivity When Employees Leave*; Wiley: London, UK, 2002.

51. Urbancová, H. *Kontinuita Znalostí: Jak Uchovat Znalosti Klíčových Pracovníků v Organizaci 3. Rozšířené Vydání*; Adart spol. s ro: Prague, Czech Republic, 2013.

52. Hansen, M.T.; Nohria, N.; Tierney, T. What´s your Strategy for managing knowledge? *Harv. Bus. Rev.* **1999**, *1*, 106–116.

53. Shih, H.; Chiang, Y. Strategy alignment between HRM, KM and corporate development. *Int. J. Manpow.* **2005**, *26*, 5–19. [CrossRef]

54. Sangjae, L.; Kun Chang, L. The relationship among formal EDI controls, knowledge of EDI controls, and EDI performance. *Inf. Technol. Manag.* **2010**, *11*, 43.

55. Šú, S.R. Štatistický úrad Slovenskej Republiky. Available online: https://slovak.statistics.sk (accessed on 5 February 2019).

56. Korec, P.; Polonyová, E. Zaostávajúce regióny Slovenska-pokus o identifikáciu a poukázanie na príčiny. *Acta Geogr. Univ. Comen.* **2011**, *55*, 165–190. Available online: http://www.actageographica.sk/stiahnutie/55_2_03_Korec_Polonyova.pdf (accessed on 10 October 2018).

57. Stacho, Z.; Stachová, K.; Raišienė, A.G. Change in approach to employee development in organizations on a regional scale. *J. Int. Stud.* **2019**, *12*, 299–308. [CrossRef]

58. Kohnová, L.; Salajová, N.; Šlenker, M. Changes in the necessary skills and knowledge of employees under the influence of technological changes in the industrial revolutions. *J. Cult.* **2019**, *9*, 26–33.

59. Urbancova, H.; Hudáková, M. Benefits of employer brand and the supporting trends. *Econ. Sociol.* **2017**, *10*, 41–50. [CrossRef]

60. Tidd, J.; Bessant, J.; Pavitt, K. *Řízení Inovací*; ComputerPress: Brno, Czech Republic, 2007; pp. 250–312.

61. Bencsik, A.; Machová, R. Znalostný Manažment a Etika z Pohľadu Podnikania. Available online: https://www.tvp.zcu.cz/cd/2012/PDF_sbornik/014.pdf (accessed on 12 March 2020).

62. Fifek, E.; Krajčík, D.; Steinhauser, D.; Zábojník, S. *Hodnotenie Konkurencieschopnosti Ekonomiky v Medzinárodnom Porovnaní*; Ekonóm: Bratislava, Slovak Republic, 2015; pp. 64–153.
63. Chmelová, M. Súčasné postavenie vybraných krajín vrámci ukazovateľa konkurenciechopnosti avýzvy do budúcnosti. *Finančné Trhy* **2017**, *14*, 1–17. Available online: http://www.derivat.sk/files/2017%20financne%20trhy/FT_3_2017_Chmelova.pdf (accessed on 17 October 2019).
64. Schwab, K. The Global Competitiveness Report 2019 WEF. Available online: http://www3.weforum.org/docs/WEF_TheGlobalCompetitivenessReport2019.pdf (accessed on 10 February 2020).
65. KPMG. Growth Promise Indicators: 2018 Report. Available online: https://home.kpmg/content/dam/kpmg/uk/pdf/2018/01/growth-promise-indicators-report-2018.pdf. (accessed on 1 May 2019).
66. EK Odborníci Diskutovali, ako Znižovať Zaostávanie Regiónov. Zastúpenie na Slovensku. Available online: https://ec.europa.eu/slovakia/news/seminar_27042017_sk (accessed on 15 April 2020).
67. Štofková, K. Možnosti implementácie znalostného manažmentu vorganizáciách. *Podn. Ekon. Amanažment* **2013**, *3*, 9–14.
68. Collinson, C.; Parcel, L.G. *Knowledge Management*; Computer Press: Brno, Czech Republic, 2005; pp. 15–135.
69. Jeleníková, S. Nové trendy v HR. *Hr Live Mag.* **2017**, *2*, 14–18.
70. Dvořáková, Z.; Kol, A. (Eds.) *Řízení Lidských Zdrojů*; C.H. Beck: Prague, Czech Republic, 2012; pp. 248–486.
71. Strohmeier, S.; Piazza, F.; Majstorovic, D.; Schreiner, J. Smart HRM—A Delphi Study on the Future of Digital Human Resource Management (HRM 4.0). MIS—Saarland University. Available online: https://www.uni-saarland.de/fileadmin/user_upload/Professoren/fr13_ProfStrohmeier/Aktuelles/Final_report_Smart_HRM_EN.pdf (accessed on 3 April 2017).

MDPI

St. Alban-Anlage 66

4052 Basel

Switzerland

Tel. +41 61 683 77 34

Fax +41 61 302 89 18

www.mdpi.com

Applied Sciences Editorial Office

E-mail: applsci@mdpi.com

www.mdpi.com/journal/applsci